Handbook of Experimental Pharmacology

Volume 133

Springer
Berlin
Heidelberg
New York
Barcelona
Hong Kong
London
Milan
Paris
Singapore
Tokyo

Vaccines

Contributors

T. Bergström, M. Corbel, P. Corran, J.J. Donnelly, L. Hellman, I. Hellström, K.E. Hellström, S.L. Hoffman, J. Holmgren, R. Janssen, D.C. Kaslow, S.H.E. Kaufmann, S. Kumar, C. Laferrière, P.-H. Lambert, M.A. Liu, P. Minor, E. Norrby, M. Pizza, J.T. Poolman, R. Rappuoli, J. Reimann, E.G. Sandström, G.C. Schild, F. Schödel, A.M. Schultz, C.-A. Siegrist, A.J. Slavin, V.C. Stevens, A.-M. Svennerholm, B. Svennerholm, S.B. Svenson, J. Thole, J.B. Ulmer, A. Vahlne, H.L. Weiner, D. Young

Editors

P. Perlmann and H. Wigzell

Springer

Professor
PETER PERLMANN
The Wenner-Gren Institute
Department of Immunology
Stockholm University
The Arrhenius Laboratories F5
S-106 91 Stockholm
SWEDEN

Professor
HANS WIGZELL
MTC, Karolinska Institute
S-171 77 Stockholm
SWEDEN

With 16 Figures and 20 Tables

ISBN 978-3-642-64196-1 Springer-Verlag Berlin Heidelberg New York

Library of Congress Cataloging-in-Publication Data

Vaccines / contributors, T. Bergström . . . [et al.]; editors, P.
Perlmann and H. Wigzell.
p. cm. – (Handbook of experimental pharmacology; v. 133)
Includes bibliographical references and index.
ISBN978-3-642-64196-1(hardcover: alk. paper)
1. Vaccines. I. Bergström, T. (Tomas) II. Perlmann, Peter.
III. Wigzell, Hans, 1938– . IV. Series.
[DNLM: 1. Vaccines–therapeutic use. 2. Vaccines–pharmacology.
3. Immune System–physiopathology. W1HA51L v. 133 1999]
QP905.H3 vol. 133
[RM281]
651′.1 s–dc21
[651′.372]
DNLM/DLC
for Library of Congress 98–42189
CIP

ISBN-13: 978-3-642-64196-1
DOI:10.1007/978-3-642-59955-2
e-ISBN-13:978-3-642-59955-2

Softcover reprint of the hardcover 1st edition 1999

Cover design: *design & production* GmbH, Heidelberg

Typesetting: Best-set Typesetter Ltd., Hong Kong

Production Editor: Angélique Gcouta

SPIN: 10537693 27/3020 – 5 4 3 2 1 0 – Printed on acid-free paper

Preface

The immune system is the only organ system in the body besides the central nervous systems endowed with memory. Both types of memories are specific and long-lasting, sometimes life long. This memory capacity of the immune system provides the basis for the most cost-efficient of all medical interventions, successful vaccinations against many common infectious diseases. Such a success requires the isolation of the infectious agent or toxic substance, methods to grow and/or purify the relevant antigen and change it into something innocuous whilst maintaining its immunogenicity. Whereas the early vaccines could only use the enhanced resistance against infectious disease as a measure of vaccine efficacy, most modern vaccines rely upon standardized laboratory tests accepted to parallel the *in vivo* protective capacity to confirm the quality and potency of the respective vaccine.

We are presently experiencing an explosion in the development of new and/or improved vaccines. This is largely due to a parallel rapid expansion in our knowledge of the immune system and of the detailed molecular structure and function of microorganisms. Using this knowledge it is now possible to compose vaccines of new types where only certain molecules (or parts of molecules) derived from a pathogen are included, excluding other potentially harmful ones. Whereas earlier attenuated live vaccines were created by empirical means such vaccines can now be created by molecularly defined interventions in the genome of the microorganism. In fact, recent developments in molecular biology are now also opening up ways to induce immune protection by transfecting the human host with relevant genes from the pathogens, s.c. DNA vaccines.

Our understanding of the way immunological memory is being generated at the cellular level has allowed for the construction of combination vaccines where i.a. carbohydrates linked to foreign proteins can be converted into efficient inducers of long-lasting specific immune memory. The detailed knowledge of the genetic build up and function of many organisms has also been helpful in understanding the impressive capacity of many microorganisms to display a high degree of variability with regard to display of antigens. It is in fact still quite open whether certain infectious diseases will ever be able to be controlled by vaccinations. Most common, lethal infections are predominantly present in the poorer parts of the human population where expensive drugs can not be afforded making research into vaccines against these diseases an urgent research priority.

In the present book a background into our basic understanding how the immune system is functioning during infection and immunization is offered. It is then followed by a series of articles dealing with most of the infectious diseases in man against which vaccines exist or are being developed. All chapters are written by well-known experts in their fields. The present situation of vaccines developed to prevent pregnancy, alleviate autoimmune diseases or eliminate cancer cells is also summarized. We hope that you will find the reading of this book informative as well as stimulatory.

PETER PERLMANN
HANS WIGZELL

List of Contributors

Bergström, T., Department of Clinical Virology, University of Göteborg, S-413 46 Göteborg, Sweden

Corbel, M., NIBSC, Blanche Lane, South Mimms, Potters Bar, Herts EN6 3QG, United Kingdom

Corran, P., NIBSC, Blanche Lane, South Mimms, Potters Bar, Herts EN6 3QG, United Kingdom

Donnelly, J.J., Department of Virus and Cell Biology, Merck Research Laboratories, West Point, PA 19486, USA

Hellman, L., Department of Medical Immunology and Microbiology, BMC, Box 582, S-751 23 Uppsala, Sweden

Hellström, I., Pacific NW Research Foundations, 720 Broadway, Seattle, WA 98121, USA

Hellström, K.E., Pacific NW Research Foundations, 720 Broadway, Seattle, WA 98121, USA

Hoffman, S.L., Naval Medical Research Institute, Malaria Program, 12300 Washington Ave., Rockville, MD 20852, USA

Holmgren, J., Department of Medical Microbiology and Immunology, Göteborg University, Gudlhedsgatan 10, S-413 46 Göteborg, Sweden

Janssen, R., Department of Infectious Diseases and Microbiology, Imperial College School of Medicine, Norfolk Place, London W2 1PG, United Kingdom

Kaslow, D.C., Laboratory of Parasitic Diseases, National Institutes of Health, Bethesda, MD 20852, USA

Kaufmann, S.H.E., Max-Planck-Institute for Infection Biology, Monbijoustr. 2, D-10117 Berlin, Germany

Kumar, S., Naval Medical Research Institute, Malaria Program, 12300 Washington Ave., Rockville, MD 20852, USA

LAFERRIÈRE, C., Pre-Clinical Bacteriology, Research & Development, Smith Kline Beecham Biologicals, 89, rue de l'Institut, B-1330 Rixensart, Belgium

LAMBERT, P.-H., WHO Global Programme for Vaccines and Immunization, 1 rue Michel Servet, CH-1211 Geneva, Switzerland

LIU, M.A., Chiron Corporation, 4560 Horton St., MS 4.657, Emeryville, CA 94608-2916, USA

MINOR, P., NIBSC, Blanche Lane, South Mimms, Potters Bar, Herts EN6 3QG, United Kingdom

NORRBY, E., Secretary General, The Royal Swedish Academy of Sciences, PO Box 50005, S-104 05 Stockholm, Sweden

PIZZA, M., IRIS, Chiron S.p.A., Via Fiorentina 1, I-53100 Siena, Italy

POOLMAN, J.T., Pre-Clinical Bacteriology, Research & Development, Smith Kline Beecham Biologicals, 89, rue de l'Institut, B-1330 Rixensart, Belgium

RAPPUOLI, R., IRIS, Chiron S.p.A., Via Fiorentina 1, I-53100 Siena, Italy

REIMANN, J., Institute of Medical Microbiology and Immunology, University of Ulm, Albert-Einstein-Allee 11, D-89070 Ulm, Germany

SANDSTRÖM, E.G., Department of Dermatovenereology, Söder Hospital, S-118 83 Stockholm, Sweden

SCHILD, G.C., NIBSC, Blanche Lane, South Mimms, Potters Bar, Herts EN6 3QG, United Kingdom

SCHÖDEL, F., INSERM U 80, EVAX Technologies GmbH, Fraunhoferstr. 10, D-82152 Martinsried, Germany

SCHULTZ, A.M., Vaccine and Prevention Research Program, Division of AIDS, NIAID, 6003 Executive Blvd., Rockville, MD 20892, USA

SIEGRIST, C.-A., Centre for Neonatal Vaccinology, C.M.U., 1 rue Michel Servet, CH-1211 Geneva, Switzerland

SLAVIN, A.J., Center for Neurologic Diseases, Brigham & Women's Hospital, Harvard Medical School, 77 Ave. Louis Pasteur, HIM 730, Boston, MA 02115, USA

STEVENS, V.C., Department of Obstetrics & Gynecology, College of Medicine, The Ohio State University, Fifth Floor, Means Hall, Columbus, OH 43210-1228, USA

SVENNERHOLM, A.-M., Department of Medical Microbiology and Immunology, Göteborg University, Guldhedsgatan 10, S-413 46 Göteborg, Sweden

SVENNERHOLM, B., Department of Clinical Virology, Göteborg University, S-413 46 Göteborg, Sweden

SVENSON, S.B., Swedish Institute for Infectious Disease Control, S-10521 Stockholm, Sweden

THOLE, J., Division of Immunological and Infectious Diseases, TNO Prevention and Health, Zernikedreef 9, 2301 CE Leiden, The Netherlands

ULMER, J.B., Chiron Corporation, 4560 Horton Street, MS 4.355, Emeryville, CA 94608-2916, USA

VAHLNE, A., Division of Clinical Virology, Department of Immunology, Microbiology, Pathology and Infectious Diseases, Karolinska Institutet, S-141 86 Stockholm, Sweden

WEINER, H.L., Center for Neurologic Diseases, Brigham & Women's Hospital, Harvard Medical School, 77 Ave. Louis Pasteur, HIM 730, Boston, MA 02115, USA

YOUNG, D., Department Infectious Diseases and Microbiology, Imperial College School of Medicine, Norfolk Place, London W2 1PG, United Kingdom

Contents

CHAPTER 2

Immunology of Infection

CHAPTER 3

DNA Vaccines: Immunogenicity and Preclinical Efficacy

CHAPTER 4

Antigen-Presentation Systems, Immunomodulators, and Immune Responses to Vaccines

CHATER 9

Outer Membrane Protein Vaccines

CHAPTER 10

Carbohydrate-Based Bacterial Vaccines

CHAPTER 19

Vaccines Against Allergies

CHAPTER 1

Vaccines: Past, Present and Future

G. SCHILD, M. CORBEL, P. CORRAN, and P. MINOR

A. Bacterial Vaccines: Past, Present and Future

I. Past

The earliest bacterial vaccines were live strains derived from virulent cultures isolated from cases of natural infection. These were attenuated by empirical methods based on sub-culture under adverse conditions. Sometimes this involved exposure to unusually high temperatures or culture on media that contained substances inhibitory to wild-type strains. Occasionally attempts were made to select strains by passage in unnatural host species, but generally this approach was less successful than for viral attenuation. Examples of vaccines produced by these processes include the fowl cholera (*Pasteurella multocida*) and anthrax vaccines developed by PASTEUR (1880, and PASTEUR et al. 1881) and the bacille Calmette-Guérin (BCG) tuberculosis vaccine developed by CALMETTE et al. (1928). The Pasteur anthrax vaccine strains were selected by growth at high temperature (42°C) and were notoriously unstable and difficult to administer consistently.

The reasons for this did not become apparent until quite recently when the genetic basis of the pathogenicity of *Bacillus anthracis* was established. It now seems probable that the Pasteur anthrax vaccine consisted of organisms from which the plasmid pXO 1 encoding the toxin complex had been deleted. Such a strain would still synthesize the other major virulence determinant, the poly-D-glutamic acid capsule encoded by the pXO2 plasmid, but would not stimulate a protective response as this material is not antigenic. It is assumed that the protective activity resulted from the presence of small numbers of virulent organisms, hence the difficulty of achieving both safe and effective batches. This problem was not resolved until the development of the Sterne strain of *B. anthracis* from which the pXO2 capsule plasmid has been deleted but in the toxin-encoding pXO1 plasmid is retained. This strain was also selected by empirical methods, but the use of effective and reliable in vivo assays assured attenuation and efficacy.

The BCG strain was selected by repeated sub-culture in the presence of ox bile of a mycobacterial strain isolated from bovine milk (CALMETTE et al. 1921). The basis of attenuation of this strain is still not understood, nor is the basis of its immunizing activity. Furthermore, numerous variants have arisen during the course of repeated serial sub-culture, and there is evidence that these differ in protective activity. This has resulted in problems in assuring consistency between vaccines produced by different manufacturers and even by the same manufacturer over an extended time interval. The problem is exacerbated by the absence, even today, of a potency assay which predicts immunizing activity in humans.

Many other live attenuated bacterial vaccines were developed against a variety of diseases including cholera, dysentery, paratyphoid, plague, typhoid and tularaemia. Most proved inconsistent in safety, efficacy or both. In general, because of these problems the development of bacterial vaccines tended to move away from live attenuated strains. Two other main approaches were followed, the use of killed virulent bacteria and the production of detoxified toxins (anatoxins or toxoids).

The development of killed vaccines required methods that would ensure loss of viability but not destroy protective antigens. Usually mild heat treatment was used, but in some cases this needed to be supplemented with chemical treatment. The inactivation procedures were all developed empirically and still tend to vary in detail between manufacturers even for similar types of product. Nevertheless, this approach was successful in producing effective vaccines against a number of diseases. Currently available bacterial vaccines prepared from killed organisms include cholera, pertussis, plague and typhoid whole-cell vaccines. Attempts were also made to apply this approach to many other organisms including gonococci, meningococci, pneumococci and mycobacteria but without success. The reasons for these failures were not understood at the time because of lack of basic knowledge of bacterial structure and the immunology of infection. They included inability to stimulate protection and also unacceptable toxicity in the case of vaccines prepared from gram negative cocci. The lack of efficacy resulted from the wide diversity of antigenic types, the modulation of expression of antigens during growth in vivo and the need in the case of certain organisms such as mycobacteria, to stimulate cell-mediated immunity to relevant antigens. None of these factors was identified until many years later, and this lack of knowledge impeded the development of vaccines against many infectious agents.

Even for bacteria that were amenable to the killed cell approach, variations in expression of protective factors by different strains and by the same strains grown under different cultural conditions caused inconsistencies between manufacturers and even between different batches produced by the same manufacturer. This emerged as a particular problem with killed pertussis and typhoid vaccines. In the case of the former this was not resolved until an effective potency test was developed. In the case of typhoid vaccines no satisfactory potency test predictive of performance in humans has yet been developed and the production of whole-cell vaccines followed a very uneven course. After the initial successes reported by Wright and colleagues, subsequent studies indicated poor efficacy. Eventually it was established that the inactivation procedure should preserve the Vi antigen intact (and probably coincidentally many protein antigens). This was achieved by inactivation with cold ethanol or acetone. Interestingly, although acetone-inactivated preparations showed the highest efficacy in field trials (YUGOSLAVIA TYPHOID COMMISSION 1964), this type of preparation was not developed commercially, most manufacturers preferring to retain heat-phenolized preparations. However, these are now being superseded by purified sub-unit or live attenuated strains (KLUGMAN et al. 1987; LEVINE et al. 1990).

Because of their ill-defined composition killed whole-cell vaccines can cause difficulties. Nevertheless, some are still in use and provide protection against important infectious diseases. Possibly the best example is killed whole-cell pertussis vaccine. The results of recent clinical trials in Germany, Senegal and Sweden have confirmed that such vaccines, when subject to effective quality control, are capable of achieving levels of efficacy which match or exceed those of even the most effective sub-unit preparations (OLIN et al. 1997).

Somewhat better defined are the bacterial toxoid vaccines. These represented the first successful attempt to develop vaccines targetted against defined bacterial products and in particular against the factors primarily responsible for inducing the disease process. Once it was recognized that the pathological effects of certain infections could be reproduced by administration of cell-free extracts of the organisms or their culture fluids, attempts were made to render these non-toxic while retaining immunizing capacity. This did not prove entirely straightforward and the selection of reliable inactivating agents was difficult. Neutralization with antitoxin was used to produce a toxin-antitoxin complex for immunization. This approach was later combined with early toxoided preparations to protect against reversion. Although preparations of this type remained in production for many years, they required very careful preparation with precise control of the proportions of toxin and antitoxin used.

Many problems resulted from dissociation or incomplete neutralization of the toxin and the use of chemical inactivating agents was favoured. Early attempts to produce a diphtheria vaccine used iodine to inactivate the toxin. It proved difficult to achieve consistent results with this and frequent problems were encountered with inadequate detoxification or reversion to toxicity. Conversely, excessive treatment led to a degraded and non-protective product. The introduction of formaldehyde for detoxification represented a major advance and made possible the large-scale production of effective vaccines against diphtheria, tetanus and other toxigenic clostridial diseases (RAMON 1923). However, the interaction of formaldehyde with proteins is complex, and careful control of the process is essential. Nevertheless, formol toxoids are still favoured for the production of diphtheria and clostridial vaccines, including tetanus. Formaldehyde has also been adopted for the detoxification of pertussis toxin used in some current sub-unit vaccines.

Toxoids of this type are fairly easy and relatively inexpensive to manufacture. As diphtheria and tetanus vaccines produced by this means have proved safe and very effective, there has been considerable reluctance on the part of manufacturers to apply more sophisticated approaches, such as genetic detoxification, to these vaccines even though this has been technically feasible for some considerable time.

II. Present

The types of bacterial vaccine currently in use, together with some examples are listed in Table 1. Of these the live attenuated, killed whole-cell and toxoid categories were developed many years ago and include preparations with a well-established track record. BCG vaccine has had a rather controversial history in relation to efficacy against post-primary tuberculosis but persists as no more effective alternative has yet been introduced.

Table 1. Currently available bacterial vaccines

Vaccine type	Reactogenicity	Efficacy
Killed whole cell		
Cholera	Moderate-high	Low (<50%)
Pertussis	Moderate-high	Up to 95%
Plague	High	Uncertain, short term
Q fever	Moderate	Approx. 80%
Typhoid	High	Up to 75%
Live attenuated		
BCG (tuberculosis)	Low	Variable (0%–75%)
Cholera CVD103HgR	Low	Still in trials
Tularaemia LVS	Low	Uncertain
Typhoid Ty21a	Low	Variable (20%–70%)
Purified sub-unit		
Anthrax	Moderate-high	Uncertain
Acellular pertusssis	Low	70%–85%
Meningococcal group B OMV	Low	Approx. 55%[c]
Toxoid		
Botulism (1–5 component)	Moderate	Probably high
Diphtheria	Low-moderate	>90%
Tetanus	Low-moderate	>95%
Polysaccharide		
Meningococcal A, C, AC, $ACW_{135}Y$	Low	Up to 90%, short term[a]
Pneumococcal 23-component	Low	56%–67%[a]
Typhoid Vi	Low	Up to 70%
Polysaccharide-protein conjugate		
Haemophilus influenzae b (various compositions in use)	Low	>95%
Meningococcal C	Low	In trials[b]
Meningococcal AC	Low	In trials[b]
Pneumococcal 7-valent	Low	In trials[b]

[a] Not suitable for use in recipients aged under 2 years.
[b] Preliminary results predict high efficacy based on immunogenicity data.
[c] In clinical trials.

More recently the emphasis has been on the development of vaccines of more defined composition and properties. This has included the elaboration of vaccines based on purified sub-units and production of attenuated strains containing predetermined genetic modifications. An approach has also been made to improvement of the immunogenicity of natural antigens by chemical modification. This has led to the development of the polysaccharide-protein conjugate vaccines.

1. Purified Sub-unit Vaccines

a) Polysaccharides

These are prepared from the capsular polysaccharides of encapsulated bacterial strains. While many pathogenic bacterial species synthesize capsular polysaccharides, these have been exploited for vaccine production in only a few instances. These include the A, C, W and Y serogroup antigens of *Neisseria meningitidis*, the capsular antigens of a variety of *Streptococcus pneumoniae* serotypes, the capsular antigen of *Haemophilus influenzae* type b (Hib) and the Vi antigen of *Salmonella typhi*. These polysaccharides can be produced to a high level of purity and are essentially free of toxic or reactogenic components. However, to possess immunogenicity they must be of a minimum molecular size as well as retaining their original repeating sub-unit composition. Such vaccines have been used successfully in certain situations but have certain limitations. In particular, they behave as T cell independent antigens and are inefficient in stimulating isotype and sub-class switching, tend to evoke antibodies of low avidity and do not stimulate a memory response. Of even greater importance is their failure to evoke responses in the very young (AUSTRIAN 1981). This has proved a particular disadvantage in the case of Hib, meningococcal and pneumococcal polysaccharides where much of the burden of disease caused by the corresponding organisms falls on infants. This problem has stimulated the search for ways of making these antigens more effective. This has been achieved by conjugating them to proteins to produce semi-synthetic glycoconjugates (SCHNEERSON et al. 1989).

b) Protein-Polysaccharide Conjugate Vaccines

These are produced from purified bacterial polysaccharides or oligosaccharide sub-units by chemical bonding to a protein carrier which acts as a source of T_H epitopes (MOREAU 1996). Either of two main conjugation strategies may be adopted: (a) direct conjugation of activated polysaccharide to an activated protein, (b) indirect conjugation of activated polysaccharide to a linker molecule followed by conjugation to an activated protein which itself may or may not be attached to a linker molecule. Option (a) is to be preferred when possible as the use of linker molecules may induce a substantial response to protein-linker neoepitopes at the expense of the saccharide. With successful construction polysaccharide-protein conjugates can induce excellent T cell dependent responses to carbohydrate antigens. They have been outstandingly successful in the case of Hib conjugates SCHNEERSON et al. (1980) and pre-clinical animal studies and clinical trial data suggest that similar achievements are possible with meningococcal group C and multivalent pneumococcal conjugates. Many other conjugates have been prepared and demonstrated experimentally to stimulate antibodies against a variety of bacteria including *Bacteroides*, *Escherichia coli*, *Klebsiella*, *Pseudomonas*, *Salmonella* and *Streptococcus* group B. Vaccines can also be prepared against surface polysaccharides of non-encapsulated bacteria, for example, *Brucella*, *Shigella*, by conjugating detoxified lipopolysaccharide to a carrier protein (TAYLOR et al. 1993).

The number of carrier proteins used to date has been limited. Manufacturers have favoured established vaccine components such as diphtheria or tetanus toxoids or the non-toxic diphtheria mutant protein CRM 197 as carriers in the belief that this would more easily satisfy regulatory requirements. However, this is already beginning to present problems of potential overload of these components in complex combination vaccines. Increased reactogenicity and epitope suppression or enhancement are possible complications of excessive exposure to individual antigens. This has led to the search for additional carrier proteins (ROBBINS et al. 1996).

2. Purified Protein Sub-units

These target individual protective antigens. This is relatively straightforward in the case of toxigenic infections such as diphtheria or tetanus but the identification of candidate antigens has proved more elusive for infections where the mechanism of pathogenesis is more complex, for example, *N. meningitidis* serogroup B. For *Bordetella pertussis* various combinations have been used in formulating acellular vaccines. All include detoxified pertussis toxin in some form but with or without the addition of other components including filamentous haemagglutinin, pertactin and fimbriae (fim 2 and 3). Clinical trial data suggest that efficacy is improved with increasing number of components, but that the method of detoxification can also influence the results (OLIN et al. 1997).

Protein components for vaccines can be produced by purification from the natural strain or by rDNA technology in a heterologous host. The latter system can circumvent some purification problems but rDNA proteins may show post-translational modifications distinct from those seen in the natural product. Some complex proteins may also be difficult to express and assemble in heterologous systems, an example being pertussis toxin. The genetically detoxified protein must be expressed in a *B. pertussis* strain from which the wild-type gene has been deleted (PIZZA et al. 1989).

The rDNA approach may be the only practicable way of producing antigens from organisms that are not amenable to in vitro culture, for example, *Treponema pallidum* and *Mycobacterium leprae*.

An increasing number of vaccines are being developed from recombinant proteins, for example, new anthrax, plague, *Helicobacter* and Lyme disease vaccines, but with the exceptions of isolated examples of meningococcal group B and acellular pertussis vaccines none is in routine production.

3. Genetically Modified Strains

Current policy is to develop live attenuated strains containing defined mutations. An early attempt to do this produced the Ty 21a live typhoid vaccine. This strain was thought to contain a mutation in the galactose epimerase gene as the attenuating modification. However, this has been demonstrated not to be the case. The Ty 21a strain was produced by mutagenesis with nitrosoguanidine and contains numerous mutations, many of which may contribute to attenuation.

It is now the practice to introduce specific mutations in identified genes by site-directed mutagenesis. These usually involve key metabolic or "housekeeping "genes such as those responsible for the shikimic acid pathway (Aro mutants) or pho or cya mutants (HOISETH and STOCKER 1981). Another approach is to produce deletions or modifications to genes encoding specific virulence determinants, for example, the CVD103HgR mutant of *Vibrio cholerae*. In this the cholera toxin, zonula occludens toxin and haemolysin genes have been deleted or rendered inoperative. This strain and aro mutants of *Salmonella typhi* have reached marketing authorization application stage in some countries. Many other genetically modified strains are currently undergoing preclinical evaluation.

The application of such organisms as delivery systems for a variety of protective antigens has attracted considerable interest. Modified *Salmonella* and BCG strains have been specifically targetted although naturally nonpathogenic bacteria such as *Streptococcus lactis* have also been studied (CHATFIELD et al. 1993). Although some success has been obtained with these systems in experimental studies, none has so far achieved routine application. Similarly, many other non-living delivery systems and new adjuvants are currently under investigation (POWELL and NEWMAN 1995). These are more appropriately considered in terms of future developments.

III. Future

Approaches to vaccine design that are under examination include: synthetic peptides, anti-idiotypes, nucleic acids, live recombinants, multivalent combinations, slow-release/single-dose presentations, and oral/mucosal delivery systems.

1. Synthetic Peptides

Linear peptides reproducing part of the primary sequence of bacterial antigens have met with little success in achieving protection against bacterial diseases. They tend to stimulate responses which are too narrow in terms of HLA restriction. Circular and complex peptides have shown rather more promising results in experimental studies, for example, in stimulating responses to the meningococcal group B por A variable regions. They are also more likely to stimulate adequate T and B lymphocyte responses (FITZMAURICE et al. 1996). However, potentially the most useful is the mimeotope approach. This uses non-linear arrays of peptide sequence to simulate the three-dimensional structure and charge distribution of epitopes which may be of protein or non-protein composition. Modification of the structure allows "fine tuning" to reduce or increase cross reactivity and avidity. This approach could be extremely useful for simulating non-protein antigens that may be impossible to purify or to produce by recombinant methods.

2. Anti-idiotypes

These are natural mimeotopes that use monoclonal antibodies to the combining site of antibody to the native antigen. Again this approach is potentially most useful for non-protein antigens such as polysaccharides. There are many problems associated with the use of such materials as vaccines, such as assuring freedom from extraneous agents, and none intended for human use has so far advanced beyond the experimental stage. The refinement of recombinant antibody production methods may overcome these problems.

3. Nucleic Acids

Vaccines may be prepared by construction of plasmids containing the cloned gene for an antigen together with the sequences necessary for transcription and expression within eukaryotic cells. These, when suitably presented, may trigger protective immune responses against a variety of antigens. This approach has given encouraging results with a number of bacterial pathogens. However, there are many regulatory and safety issues to be addressed with this type of preparation. Nucleic acid vaccines will probably prove most useful for those infections which are not easily amenable to prevention by other methods. An example is tuberculosis, against which DNA vaccines are hitherto the only preparations to have stimulated protection approaching that given by BCG (HUYGEN et al. 1996). They may also provide a means of constructing compatible multivalent vaccines which can be delivered by the mucosal route.

4. Live Recombinants

These may consist of individual pathogens attenuated by defined mutations, for example, AroAC$^-$ mutants of *S. typhi* or more complex constructs expressing protective antigens for multiple pathogens. Thus in the future it may be possible to design a "universal" enteric vaccine which expresses the antigens of *Campylobacter*, *Entamoeba*, *Giardia*, *S. typhi*, *Shigella*, and *V. cholerae* in a single preparation. Hitherto the development of such constructs has been impeded by interference between some of the components, for example, the long O chains of *Salmonella* have masked the shorter O chains of *V. cholerae*.

However, these problems may be overcome by targetting different antigens or by using a different presenting system.

5. Multivalent Combinations

The proliferation of new vaccines and the difficulty of integrating them into acceptable immunization schedules has stimulated the development of complex combinations, for example, a "universal paediatric vaccine" containing acellular pertussis, diphtheria, tetanus, Hib, hepatitis B, and inactivated poliomyelitis in a single formulation. This has not proved straightforward as some incompatibilities have emerged, such as interference between acellular pertussis and Hib components. The future trend may be to target syndromes, for example, "meningitis vaccines" containing Hib, meningococcal and pneumococcal antigens, "enteric vaccines" and "sexually transmitted disease vaccines" each covering the major diseases in these groups. It is also likely that combinations will be formulated with delivery systems appropriate to the route of entry of the pathogen.

6. Slow-Release/Single-Dose Vaccines

Related to the problems addressed by multivalent vaccines is the concept of single-dose preparations. For much of the world population access to medical facilities is limited, and the implementation of complex multi-dose immunization schedules impracticable. This has directed efforts towards vaccines which will produce lasting immunity after a single dose. These have been based on systems which either release antigens slowly over a prolonged period, or which produce pulsed release to simulate multiple doses. In practice, sustained release has been more easily achieved. Current developments favour biodegradable microspheres (GANDER and MERKLE 1998), but future preparations are likely to use more sophisticated vehicles, particularly those that achieve high-grade responses when given by the oral route. Single dose preparations will also be designed to provide simultaneous immunization against multiple pathogens.

7. Oral/Mucosal Delivery Systems

The ideal vaccine would stimulate lasting immunity against a variety of pathogens without adverse side effects when given as a single dose by the oral route. Such an objective has yet to be achieved but may not be totally unattainable. Investigations are proceeding along the lines of live vectors and non-living delivery systems. The ideal will incorporate multiple antigens and single-dose schedules. Systems employing antigens encapsulated in microparticles or nanopartcles have shown promising results when delivered by oral, respiratory or parenteral routes (GANDER and MERKLE 1998). Live delivery systems, although offering some advantages, also have limitations and these may prove crucial.

The cloning and expression of genes encoding bacterial antigens in plants offers the prospect of delivering vaccines in foods. There are many unresolved issues over dosage and regulation of such products, and their greatest future advantage may lie in the bulk production of vaccine components without the need for elaborate and expensive fermentation facilities.

B. Viral Vaccines

I. Past

Vaccines are among the most effective medical intervention against viral diseases. Partly this is because specific anti-viral antibiotics are rare as a result of the intimate relationship between the cell and the virus, and partly because it is clear infection with a particular pathogen often gives long-lived protection if the patient survives. Small pox is usually regarded as the first human disease for which vaccination was developed as a deliberate intervention in a healthy subject to prevent subsequent infection. Jenner not only vaccinated James Phipps with cowpox virus in 1796, he also challenged him with live smallpox afterwards. While this was somewhat dubious ethically, it is generally considered to have set the scene for the eventual eradication of smallpox from the world. Much of the vaccine used was primitive by modern standard, much of it grown on the scarified flanks of cattle with poor control of the seed virus or contamination. Adverse reactions were common; fever and swollen lymph glands were inevitable in first time vaccinees and disseminated vaccinia and encephalitis occurred at frequencies which would be entirely unacceptable nowadays. Nonetheless the use of the vaccine eradicated the disease from the globe. It is probable that the selection of more appropriate strains and better production methods could have limited the adverse reactions seen. Current attempts to use the smallpox vaccine as a vector for other antigens are handicapped by its undesirable side effects, but it can be argued that it was the most successful vaccine so far, in that it no longer needs to be used.

While Jenner used an existing virus of animals to vaccinate humans, Pasteur is credited with the first scientific development of a vaccine. The material which was developed as a post-exposure treatment consisted of the central nervous system of rabbits infected with human rabies, a procedure thought to attenuate the virulence of the virus for humans. Similar types of killed vaccine, usually phenol-treated extracts, are still in use in many parts of the world, despite their known hazards, which include the induction of immune encephalitis because of the contamination with the donor animal CNS. The potency is also questionable in many cases. Slightly better versions included rabies virus grown in duck eggs, which while not causing encephalitis did not induce much protection either as the virus titre was so low. Many of the problems associated with the early types of vaccine were to do with poorly defined production conditions, and this is the main area of improvement. Certainly the mechanism by which the vaccines against viral diseases in current use confer protection rather than cause disease is still very poorly understood.

II. Present

In the context of viral vaccines the present may be taken to have started in 1937 when Theiler developed the 17D strain of yellow fever vaccine (Barrett 1997). The strain was developed from a wild-type isolate, Asibi, by repeated passage in mouse embryo, chick embryo and minced chick embryo devoid of nervous tissue. After 176 passages the virus was attenuated and is among the safest and most effective vaccines in use today, despite the fact that it has been passaged to different degrees in different laboratories, so that different vaccines are by no means identical. The vaccine is still grown in embryonated hens-eggs, which also form the production substrate for the second vaccine of the modern era, that against influenza. Influenza vaccines, however, are killed. In the early days all influenza vaccines consisted of whole inactivated virus particles. These preparations were associated with adverse reactions, such as mild fever, and it was found that splitting the virus with detergent and solvent reduced the reactions seen. In recent times the reaction rate has been reduced still further by the use of subunit vaccines which consist of just the two outer glycoproteins the haemagglutinin and neuraminidase. While an immune response to these two proteins confers protection, they are not as immunogenic as the whole virus preparations. A reduced immunogenicity of subunit vaccines seems to be a general feature of vaccines against viral diseases.

Polio vaccines were developed in the 1950s by Salk and Sabin (Minor 1997a,b). Salk devised a method for very careful treatment of a polio virus preparation with dilute formalin such that the infectivity was destroyed while the antigenic structure of the virus, which is somewhat fragile, was retained. There were immediate difficulties with the inactivation process, leading to the Cutter incident, in which a number of recipients contracted poliomyelitis from the vaccine. However the production process was modified very rapidly, and current vaccines are based on much the same process. They have proved both safe and effective in the course of over forty years" use. The live attenuated strains developed by Sabin were the product of a painstakingly careful consideration of the pathogenesis of poliomyelitis.

Polio virus normally infects the human gut without ill effects or detectable symptoms. Occasionally there is a viraemic phase when sites away from the gut become infected, and even more rarely the central nervous system, specifically the motor neurons of the spinal cord become infected and destroyed, leading to paralytic poliomyelitis. It was reasoned by Sabin that the viruses growing in the different locations are different in character, and that the type of virus required as a vaccine is one which is able to grow in the human gut but not the CNS. The strains in current use, and which are likely to eliminate the wild-type polio virus from the world in the near future have these properties. While they adapt as they replicate in the recipient and very rarely are able to change sufficiently to cause paralytic disease, their great effectiveness indicates the important of an understanding of pathogenesis in vaccine development if serious vaccine associated disease is to be avoided.

Measles is a common childhood infection which in developed countries is regarded as unpleasant but largely trivial. In developing countries, however, it is a major cause of death in childhood, with an estimated 1.4 million deaths occurring annually. The vaccines used generally derive from the Edmonston strain isolated by Enders in the United States in the 1950s. They are live attenuated vaccine strains given by intramuscular injection, and if correctly used, are highly effective. However, they do not work in the presence of maternal antibody, so that the recommended age of first vaccination is 12–15 months in developed countries and 9 months in developing areas. The vaccine strains are stability attenuated, and while they may cause a low-grade fever some 7–10 days post-immunization, the incidence of serious reactions is low. Understanding of their mode of action is poor, and this handicaps efforts to produce vaccines which can be safely and effectively given at an earlier age, as is discussed below.

Mumps is a childhood infection which is also widely regarded as trivial, although in the absence of vaccination strategies it is the most common cause of hospitalization due to meningitis. The vaccines against mumps are in general effective but virologically poorly defined (MINOR 1997a,b). The Urabe strain was implicated in aseptic meningitis in recipients but is fairly well characterized. Two other commonly used live vaccine strains are Rubini, which by its passage history should be derived from a wild-type isolate from a Swiss Italian child but is in fact almost certainly derived from the Enders strain, a commonly used laboratory virus, and the Jeryl Lynn strain, which is a mixture of two distinct strains, both probably originating from America.

One of the few viral vaccines in current use which is a product of a modern molecular virology is the hepatitis B vaccine. Early vaccines were derived from the plasma of chronic carriers of hepatitis B, who produce an excess of the viral surface antigen, which can be purified in the form of immunogenic particles, and after appropriate treatments to remove any infectious virus is extremely effective. Plasma-derived vaccines of this type are in use in certain countries at present, but other products are manufactured by expression of the gene encoding the surface antigen in yeast. This results in the formation of similar highly immunogenic particles without any conceivable risk of infection. Production depended on the identification of the surface antigen as the protective immunogen, the identification of the viral gene encoding it and its subsequent expression in a suitable host cell. Bacterial expression was not satisfactory. Hepatitis B vaccines therefore afford one of the few examples of a viral vaccine produced by recombinant technology as well as one of the few where a subunit vaccine is entirely effective. Shortly after the licensure of the hepatitis B vaccine a vaccine against hepatitis A was developed, hepatitis A is caused by a virus similar to polio virus, and the vaccine is a formalin-inactivated preparation similar in nature to inactivated polio vaccine. The well-tested empirical methods of development of vaccines against viral diseases are therefore still effective.

III. Future

In the near future a number of classical type viral vaccines are likely to be used, including a rotavirus vaccine consisting of a simian rotavirus strain and other strains derived from it. This is essentially the same approach as that used by Jenner in 1796, of taking a non-human virus and using it in human recipients. While the classical approach has been very successful in a large number of important instances, there have been a number of notable insuperable problems.

No vaccine has been developed against the common cold despite its economic importance and an obvious public demand. This is attributed to the large number of viruses which can cause the disease, including approximately 100 antigenically distinguishable strains of rhinovirus, as well as adenoviruses and coronaviruses. Similarly, while there is good evidence that influenza vaccines afford protection against disease if they match the circulating epidemic strains, the continual drift in the wild-type virus means that maintaining the match requires a large amount of effort. The difficulty of developing vaccines against a polymorphous collection of viruses is also regarded as an important consideration in the development of vaccines against HIV.

In some instances the understanding of the immune response or the pathogenesis of the disease is so limited that vaccine development could conceivably be dangerous. In the 1960s formalin-inactivated, alum-adsorbed vaccines were developed against measles and respiratory syncytial virus, both of which are in the paramyxovirus family. When vaccinees were subsequently exposed to wild-type virus, they experienced serious atypical disease (Fulginiti et al. 1967; Kakak et al. 1993); in the case of respiratory syncytial virus several children died. The effects were long lasting and suggested that some inappropriate cellular immune response had been generated. It is still not clear how this had been effected, but this has severely restricted efforts to develop new measles vaccines which could be given in the presence of maternal antibody, or any kind of vaccine against respiratory syncytial virus, although live attenuated strains are currently receiving close attention.

A further example in which understanding of the protective response is lacking is afforded by herpes virus, where recurrences of herpetic lesions can be a regular and painful consequence of primary infection. Clearly between recurrences the infection is somehow controlled, presumably by an immune response. This has led to the trial of various types of vaccine to stimulate immunity. In view of the amount of immunogen which must be generated during a recurrence it is not clear how effective this approach is likely to be, and some trials have been clear failures. However, improving the immune response may be what is required.

Thus in general a better understanding of immunity and virulence and pathogenesis is likely to be required to generate better viral vaccines. Specific future developments could include improved adjuvants to generate the appropriate protective immune response and, once this has been identified, various presentation systems including DNA vaccines and the development of therapeutic vaccines, especially vaccines against papilloma viruses, the major cause of cancer of the cervix.

C. Parasite Vaccines

I. Past

Parasite diseases have traditionally been controlled, if at all, by vector control and by drug prophylaxis and treatment. Vaccination has not been an option for virtually any parasitic disease. The single exception is cutaneous leishmaniasis, where the practice of deliberate infection of a sacrificial site – such as the buttocks – with material derived from an active sore has been widespread in the Middle East for centuries (HANDMAN 1986). During this century the use of live cultured promastigote preparations has been introduced, and large numbers of vaccinations have been carried out in the former Soviet Union and Israel (HANDMAN 1986) and recently Iran (ENGERS et al. 1996). However, the difficulties involved in standardizing and controlling what is in effect a naturally virulent strain have been considerable, and attempts to produce attenuated varieties by a range of techniques have tended to result in strains which sensitize but do not protect.

II. Present

Although no vaccines for prevention of parasitic disease in man have yet reached the stage of routine use, significant successes have been achieved in the veterinary field which give grounds for optimism. Thus effective vaccines have been developed against the protozoan parasites *Babesia bovis*, *B. bigemina* and *Theileria parva*, the cestode *Taenia ovis* and the nematodes *Dictyocaulus filaria* and *D. viviparus* (TELLAM et al. 1997).

Prospects for the development of parasite vaccines have been transformed over the last 15 years by the development of recombinant DNA techniques and, even more recently, by the great strides made in the understanding of the immune system. With the exception of leishmaniasis, immunization with attenuated or killed whole parasites is not likely to be a practical option, and attention has concentrated almost entirely on the identification of suitable candidate parasite components and antigenic determinants, informed by a rapidly increasing understanding of the immune response during infection.

1. Killed and Attenuated Vaccines

The only important parasitic disease for which large-scale culture outside a live host is practical is for *Leishmania* spp. A number of trials are proceeding using killed organisms either as prophylactic vaccines or as immunotherapeutic treatment (as an alternative or adjunct to chemotherapy) for a number of different species of *leishmania*, both old-world and new (ENGERS et al. 1996). Most of these trials combine the killed organisms with a low dose of BCG as adjuvant. Phase III trials for single-dose regimes against cutaneous leishmaniasis in Iran and Pakistan and visceral leishmaniasis in Sudan were due for completion in 1997. In addition, there are attempts to produce stable attenuated strains by removing essential virulence genes (ENGERS et al. 1996).

2. Peptide Vaccines

Because of the difficulty of obtaining parasite components for immunization, there have been several attempts to construct synthetic vaccines combining epitopes identified as being possibly protective. Peptide constructs based on the immunodominant repeating sequence of the *P. falciparum* circumsporozooite coat protein conferred slight protection to challenge in volunteers but showed no efficacy in field condition. However, a vaccine based on empirically identified blood-stage epitopes, Spf66, has been extensively clinically tested (AMADOR et al. 1996). Early trials were compromised by poor design, but four recent double-blind trials have given conflicting results, two (in Colombia and Tanzania) showing about 30% protection against clinical malaria and two others (in Thailand and the Gambia) showing none (FACER and TANNER 1997). A further trial amongst those aged under 5 years is in progress in Tanzania.

3. Vaccines Based on Expressed Parasite Proteins

Almost all current efforts to develop vaccines depend on the expression of cloned parasite genes in a suitable system. Of the many under development, one of the most promising is the *P. falciparum* pre-erythrocytic stage vaccine RTS,S, in which the circumsporozooite protein is expressed fused to the HbSAg protein of hepatitis B virus. This construct showed exceptional protection to challenge in immunized volunteers (six of seven protected), but only when formulated with a complex adjuvant containing monophosphoryl lipid A and the saponin, QS-21, in an oil-in-water emulsion (STOUTE et al. 1997). This combination is currently entering clinical trial in the Gambia. There are large numbers of possible candidate blood-stage malaria antigens under development, including the invasion-associated protein MSP-1 (HOLDER 1994).

One interesting approach is the development of transmission-blocking vaccines. A trial is underway of a vaccine based on *P. falciparum* Pfs25, a protein expressed only in the mosquito-specific developmental stages and thus not subject to immune selection in the mammalian host. Antibodies present in the blood meal may interfere with the development of the parasite in the mosquito gut, thus preventing onward transmission (KASLOW 1997). While such vaccines may not help the recipients directly, they may greatly reduce transmission to other family members. Almost all the possible routes of expression and delivery of recombinant proteins appear to be under investigation for candidate malaria antigens. For other parasite diseases the proteins under investigation include juvenile components, which may be of importance in blocking infection, and also conserved parasite antigens such as glutathione-*S*-transferase, triosephosphate isomerase and actin, immunity to which may significantly limit parasite burden.

4. Live Carriers

Several different approaches to the construction of vaccinia recombinants containing parasite are under investigation. The most spectacular of these is the malaria candidate vaccine NYVAC-7, in which no less than seven genes coding for pre-erythrocytic, blood-stage and mosquito-specific antigens have been incorporated (TINE 1996). NYVAC-7 is at present in phase I and II trials. There is also interest in *Salmonella* as a carrier for protective antigens from *Leishmania* and *Schistosoma* and in the use of transformed BCG as a vehicle rather than simply as an adjuvant for protective antigens (ENGERS et al. 1996; WAINE and MCMANUS 1997).

5. Other Approaches

Since the chief factor underlying development of cerebral malaria seems to be the abnormal production of TNF, stimulated by endotoxin-like *Plasmodium* products, one proposal is to protect against this, the major cause of morbidity in falciparum malaria, by immunization against toxins (PLAYFAIR 1996). A second interesting approach, for which there is already precedent in veterinary vaccines, is to block transmission by vaccinating against the arthropod vector, rather than the parasite, and thus interfere either with ingestion or digestion.

III. Future

Clearly any prediction for the future is made against a much more tentative background than for groups of organisms for which there is an extensive background of successful vaccination. There are uncertainties about such basic issues of what degree of protection is reasonable to expect – high or limited – what kind of protection should be looked for, what effect this would have on the epidemiology (GREENWOOD 1996; RILEY 1997; CHAN et al. 1997), and whether the product is likely to be affordable in the context of the resources available to health care. Mundane issues such as safety, stability and reproducibility have hardly been touched on. Any likely vaccine must probably be expected to take its place as part of a package of disease-management measures rather than as the main solution. Two areas distinct from those under current development stand out:

1. DNA Vaccines

The experimental use of DNA to induce immune responses shows great promise, but raises technical as well as regulatory issues. Recent results suggest that the combination of DNA priming followed by boosting with the same immunogen delivered by an alternative means can result in immunity which cannot be obtained with either alone. Clearly there is much to be learnt about the way in which DNA vaccines are best used.

2. Adjuvants

This is a complex area, and a recent compendium of adjuvants runs to over 80 pages (VOGEL and POWELL 1995). Nevertheless it is clear, from the example of RTS,S above, that adjuants can be critical in directing the immune response in a productive direction. The use of adjuvants is still highly empirical, but clearly this is an area in which progress is very rapid.

Finally, one of the lessons of the past 10 years is that, just as synthesizing a peptide epitope does not deliver a vaccine automatically, nor does cloning a protein, a better understanding of the interaction of parasites with the immune system is not only an important but an essential part of developing a parasite vaccine. Fortunately, the pace of increase in knowledge of immunity is breathtakingly fast at present.

References

Amador A, Aponte JJ, Patarroyo ME (1996) Development and field-testing of the synthetic Spf66 malaria vaccine. In: Hoffman SL (ed) Malaria vaccine development. ASM, Washington, pp 229–248

Austrian R (1981) Some observations on the pneumococcus and on the current status of pneumococcal disease and its prevention. Rev Infect Dis Suppl:SI–SI7

Barrett ADT (1997) Yellow fever vaccines. Biologicals 25:17–25

Calmette A, Bocquet A, Nègre L (1921) Contribution à l'étude du bacille tuberculeux bilié. Ann Inst Pasteur 9:561–570

Calmette A, Guérin C, Nègre L, Bocquet A (1928) Prémunition des nouveau – nés contre la tuberulose par le vaccin BCG (1921–1926). Ann Inst Pasteur 2:89–120

Chan M-S, Woolhouse MEJ, Bundy DAP (1997) Human schistosomiasis: potential long-term consequences of vaccination programmes Vaccine 15:1545–1550

Chatfield S, Roberts M, Londono P, Cropley I, Douce G, Dougan G (1993) The development of oral vaccines based on live attenuated *Salmonella* strains – FEMS. Immunol Med Microb 7:108

Corbel MJ (1994) Control testing of combined vaccines: a consideration of potential problems and approaches. Biologicals 22:353–360

Engers HD, Bergquist R, Modabber F (1996) Progress on vaccines against parasites. Dev Biol Stand 87:73–84

Facer CA, Tanner M (1997) Clinical trials of malaria vaccines. Adv Parasitol 39:2–68

Fitzmaurice CJ, Brown LE, McInerney TL, Jackson DC (1996) The assembly and immunological properties of non-linear synthetic immunogens containing T cell and B cell determinants. Vaccine 14:553–560
Fulginiti VA, Eller JJ, Downie AW, Kempe CH (1967) Altered reactivity to measles virus. Atypical measles in children previously immunised with inactivated measles virus vaccines. JAMA 202:101–106
Gander B, Merkle HP, Corradin G (eds) (1998) Antigen delivery systems. Immunological and technological issues. Harwood Academic Publishers, Switzerland
Greenwood B (1996) What can be expected of malaria vaccines? In: Hoffman SL (ed) Malaria vaccine development. ASM, Washington, pp 277–301
Handman E (1986) Leishmaniasis: Antigens and host-parasite interactions. In: Pearson TW (ed) Parasite antigens: Towards new strategies for vaccines. Dekker, New York, pp 5–48
Hoiseth SK, Stocker BAD (1981) Aromatic-dependent *Salmonella typhimurium* are non-virulent and effective as live vaccines. Nature 291:238–239
Holder AA (1994) Proteins on the surface of the malaria parasite and cell invasion. Parasitol 108:S5–S18
Huygen K, Content J, Denis O, Montgomery DL et al (1996) Immunogenicity and protective efficacy of tuberculosis DNA vaccine. Nat Med 2:893–898
Ivanoff B, Levine MM, Lambert P-H (1994) Vaccination against typhoid fever; present status. Bull WHO 72:954–971
Kakak TJ, Soike K, Brideau RJ, Zayd RM, Cole SL, Zhang J-Y, Roberts ED, Wells PA, Wathen MW (1993) A human respiratory syncytial virus (RSV) primate model of enhanced pulmonary pathology induced with a formalin inactivated RSV vaccine but not a recombinant FG subunit vaccine. J Infect Dis 167:553–561
Kaslow DC (1997) Transmission-blocking vaccines: uses and ccurrent status of development. Int J Parasitol 27:183–189
Klugman KP, Gilbertson IT, Koornhof HJ (1987) Protective activity of Vi-capsular polysaccharide vaccine against typhoid fever. Lancet II:1165–1169
Levine MM, Ferreccio C, Cryz S, Ortiz E (1990) Comparison of enteric-coated capsules and liquid formulation of Ty21a typhoid vaccine in a randomized controlled field trial. Lancet 336:891–894
Medical Research Council (1957) The prevention of whooping cough by vaccination. BMJ 1:1463–1471
Minor PD (1997a) Poliovirus. In: Nathanson N, Ahmed R, Gonzalez-Scararo F, Griffin DE, Holmes KV, Murphy FA, Robinson HL (eds) Viral pathogenesis. Lipincott-Raven, Philadelphia, New York, pp 555–574
Minor PD (1997b) Laboratory tests of mumps vaccines. Biologicals 25:35–40
Moreau M (1996) Conjugation technologies. In: Plotkin S, Fantini B (eds) Vaccinia, vaccination, and vaccinology; Jenner, Pasteur and their successors. Elsevier, Paris, pp 145–149
Olin P, Gustafson L, Rasmussen F, Hallander H, Heijbel H, Gottfarb P (1997) Efficacy trial of acellular pertussis vaccines. Technical Report II. Swedish Institute for Infectious Disease Control, Stockholm
Pasteur L (1880) De l'attenuation du virus du choléra des poules. C R Acad Sci Paris 91:673–680
Pasteur L, Chamberland C, Roux E (1881) Le vaccin de charbon. C R Acad Sci Paris 92:666–668
Pizza M, Covacci A, Bartoloni A, Perugini M, Nencioni L, De Magistris MT, Villa L, Nucci D, Manetti R, Bugnoli M, Giovannoni F, Olivieri R, Barbieri JT, Sato H, Rappuoli R (1989) Mutants of pertussis toxin suitable for vaccine development. Science 246:497–500
Playfair JHL (1996) An antitoxic vaccine for malaria? In: Hoffman SL (ed) Malaria vaccine development. ASM, Washington, pp 167–179
Powell MF, Newman MJ (eds) (1995) Vaccine design. The subunit and adjuvant approach. Pharm Biotechnol vol 6

Ramon G (1923) Sur le pouvoir floculant et sur les propriétés immunisantes d'une toxine diphtherique rendue anatoxique (anatoxine). C R Acad Sci Paris 177:1338–1340

Riley E (1997) Malaria vaccines: Current status and future prospects. J Pharm Pharmacol 49:21–27

Robbins JB, Schneerson R, Szu SC, Pozsgay V (1996) Polysaccharide – protein conjugate vaccines. In: Plotkin S, Fantini B (eds) Vaccinia, vaccination and vaccinology: Jenner, Pasteur and their successors. Elsevier, Paris, pp 135–143

Schneerson R, Barrera O, Sutton A, Robbins JB (1980) Preparation, characterization, and immunogenicity of *Haemophilus influenzae* type b polysaccharide protein conjugates. J Exp Med 152:361–376

Stoute JA, Slaoui M, Heppner G, Momin P et al (1997) A preliminary evaluation of a recombinant circumsporozooite protein vaccine against plasmodium falciparum malaria. N Engl J Med 336:86–91

Taylor DN, Trofa AC, Sadoff J et al (1993) Synthesis, characterization and clinical evaluation of conjugate vaccines composed of the O-specific polysaccharides of *Shigella dysenteriae* type 1, *Shigella flexneri* type 2a, and *Shigella sonnei* (*Plesiomonas shigelloides*) bound to bacterial toxoids. Infect Immun 61:3678–3687

Tellam R, Wright I, Johnson KS (1997) Categories of vaccines according to their antigenic target. III. In: Pastoret PP, Blancou J, Vannier P, Verschueren C (eds) Veterinary vaccinology, Chap 14. Elsevier, Paris, pp 470–489

Tine JA (1996) NYVAC-Pf7: a poxvirus-vectored, multiantigen, multistage vaccine candidate for Plasmodium falciparum malaria. Infect Immun 64:3833–3844

Vogel FR, Powell MF (1995) A compendium of vaccine adjuvants and excipients. In: Powell MF, Newman MJ (eds) Vaccine design: The subunit and adjuvant approach. Plenum Press, New York, pp 141–228

Waine GJ, McManus DP (1997) Schistosomiasis vaccine development – the current picture. Bioessays 19:435–443

Yugoslavia Typhoid Commission (1964) A controlled field trial of the effectiveness of acetone-tried and inactivated and heat-phenol inactivated typhoid vaccines in Yugoslavia. Bull WHO 30:623–630

CHAPTER 2
Immunology of Infection

S.H.E. KAUFMANN and J. REIMANN

A. Introduction

Vaccinology can be viewed as an application of immunology. The two disciplines arose from a common root in the late nineteenth century, namely investigations of the host response to infectious agents. Emil von Behring, in studying the humoral immune response to the toxin-producing pathogens *Clostridium tetani* and *Corynebacterium diphtheriae*, realized the implication of his observation that protection can be transferred with sera from immunized animals (VON BEHRING 1915). In developing the therapeutic strategy of passive vaccination he relied on antibody quantifications developed by Paul EHRLICH (1904). Soon, however, vaccinology and immunology diverged. The vaccines in current use were often developed empirically and virtually independently of the progress made in basic immunology. This approach has now reached its limits. Novel vaccine generations are required for infectious diseases that still cause enormous health problems, and for which vaccines are not yet available. These vaccines can be developed only in the realm of modern immunology. Recent advances in immunology have provided insights into the specific, anti-infective protection and have thereby supplied the tools for selective stimulation of appropriate immune responses (KAUFMANN 1996a).

Infectious disease is one possible outcome of the replication of a pathogen in the mammalian host. By definition, infectious diseases harm the host and may even cause death. The immune system aims at eradicating the pathogen and thus represents an equally harmful menace to infectious agents. Some infectious diseases have dramatically reduced the human population. For example, one-quarter of the population in Western Europe was killed in the Middle Ages by the plague, better known in those days as the Black Death. However, obviously, none has succeeded in eradicating us. Conversely, we have also not yet accomplished the complete elimination of any infectious agent responsible for a medically important disease (with the exception of the smallpox virus). Pathogens and the immune system of their host are best viewed as competing and interacting partners in a coevolutionary process. Protecting the host from infection is not an easy task for the immune system, which has had to develop some type of specialization. The host response encompasses an innate, nonspecific and an adaptive (or acquired), specific immune response. It has been known for a long time that the innate immune system not only acts prior to and independently of adaptive immunity, but that it is also often amplified by the latter. More recently, experimental evidence has been provided that the innate response regulates adaptive immunity (Fearon and Locksley 1996).

The specific immune system is the prime target for vaccines because it is antigen-specific and possesses memory. Specificity promotes focus on a single infectious agent, thus avoiding the damage caused by an exaggerated immune response. Memory allows the more efficient control of the same agent upon second encounter. Increasing knowledge about the instructive role of the innate on the adaptive immune response has revealed that the success of a vaccine is decided at an early stage in the nonspecific host antigen interaction.

The range of agents that can infect the mammalian host is extremely broad, encompassing prions (which are not considered here further), viruses, bacteria, fungi, protozoa, and helminths. Each species of infectious agent has developed its unique strategy for invasion of and survival in the host. Despite the great diversity of strategies used by pathogens to invade the host, some common features can be delineated, with the risk of oversimplification. First, pathogens exploit for their purposes either the intra- or the extracellular milieu in the host. Bacteria, fungi, and protozoa utilize either habitat (Kaufmann 1993; Gotschlich 1993; Locksley 1996). Virus replication takes place within the host cell, and hence viruses are strictly intracellular pathogens. While several viruses remain virtually hidden within host cells during most of their life cycle, large numbers of some viruses are detectable in the extracellular space for extended time periods (Zinkernagel 1993). Helminths are too large to live within cells and are hence extracellular pathogens (perhaps with the exception of *Trichinella*) (Nutman 1996). A second distinguishing feature is related to the cytopathogenicity or toxicity of the infectious agent. Some pathogens are highly harmful to host cells and typically cause acute diseases. Others are noncytopathic or nontoxic and may establish persistent infection that eventually results in disease (Aichele et al. 1996; Zinkernagel 1996). In these cases pathology is often caused indirectly by the immune response.

B. The Adaptive (Acquired) Immune Response

The adaptive, or acquired, immune response is the specific attack of the host against a pathogen which ideally results in the sterile eradication of this agent. Specific immune reactions may also account for pathological sequelae that arise during infection or sometimes even long after sterile eradication of the pathogen. The latter corollae may unfold as autoimmune disease (Aichele et al. 1996; Kaufmann and Schoel 1994). Because the adaptive immune response represents the target of vaccination, it is the focus of this chapter. To counteract the different types of survival strategies of pathogens the adaptive immune response has developed a division of tasks. The two major lymphocyte populations, B cells and T cells, are the mediators of adaptive immunity.

Antibodies produced by B lymphocytes interact with antigens directly. Antibody-producing B cells recognize native, extracellular antigens and respond to exogenous pathogens and their toxins. They counteract extracellular helminths, bacteria, fungi and protozoa, and neutralize their toxic products. Virus-specific antibodies capture and neutralize free virions during viremia before entry into host cells. In contrast, T lymphocytes recognize fragments of foreign antigens that are presented on the surface of the infected host cell. They recognize host cells infected with various intracellular microbes, i.e., viruses, bacteria, fungi, and protozoa. Following specific recognition T cells can selectively eliminate infected cells from various organ sites or activate effector mechanisms that destroy pathogens. T lymphocytes recognize antigen by virtue of the T cell receptor for antigen (TCR) that is associated with the multimeric CD3 complex (WEGENER et al. 1992; CHAN et al. 1994). Components of the CD3 complex transduce antigen-specific activation signals into T cells following ligand recognition by the TCR. The TCR/CD3 molecular complex therefore represents the characteristic and unique marker of T lymphocytes.

T cells are further divided into subpopulations according to the type of TCR and the profile of surface markers they express. The $CD4^+$ and $CD8^+$ T cell subsets expressing the $\alpha\beta$ TCR are the most important subsets (JANEWAY 1992). These are referred to below as "conventional T cells" (Fig. 1). Because production of most immunoglobulin (Ig) isotypes requires T cell "help," $CD4^+$ T cells are involved in most antibody-mediated immune responses. In contrast, $CD8^+$ T cells generally lyse target cells and hence are termed cytotoxic T lymphocytes (CTL). Several additional T cell populations have been identified that specifically recognize antigens from infectious agents either through an alternative antigen receptor or in the context of an alternative restriction element. These T cells are termed "unconventional T cells" here (Fig. 1). Their contribution to the anti-infectious immune response has either been clearly demonstrated or appears likely.

C. The Cellular Immune System: T Cells

I. MHC-Restricted Presentation of Epitopes to T Cells

MHC products (e.g., the human leukocyte antigens, HLA, in man) present antigenic peptides to T lymphocytes (GERMAIN and MARGULIES 1993; CRESSWELL 1994; YORK and ROCK 1996). Heterodimeric MHC class II (MHC-II) molecules are composed of polymorphic α- and β-chains. MHC class I (MHC-I) molecules are also heterodimeric, composed of a polymorphic α-(heavy) chain noncovalently bound to the invariant β_2-microglobulin (β_2-m) light chain. The conventional MHC-I molecules are all encoded by three loci within the MHC and are designated MHC-Ia molecules. All MHC-Ia molecules are polymorphic, but products from different loci display a different degree of variability. The polymorphism of these integral membrane proteins is clustered in the peptide-binding domains. There are more than 30 loci within

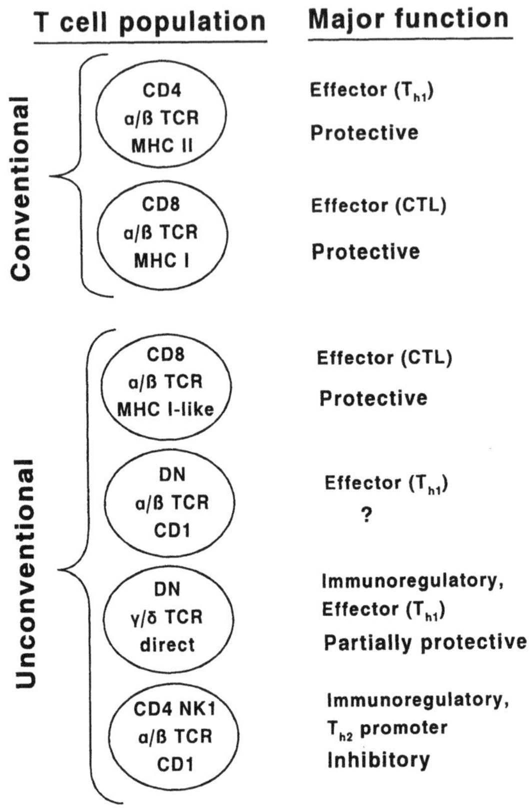

Fig. 1. Major T cell populations and functions in the anti-infective immune response

the MHC and outside the MHC that encode MHC-I-like molecules, designated MHC-Ib molecules (BECKMAN and BRENNER 1995; BLUMBERG et al. 1995; PORCELLI 1995; LENZ and BEVAN 1996). MHC-Ib molecules are of low polymorphism or are nonpolymorphic, i.e., they are shared by different individuals. The CD1 glycoproteins encoded outside of the MHC have sequence homology to MHC-I molecules, show a restricted tissue distribution, and are associated with β_2-m (BLUMBERG et al. 1995; PORCELLI 1995; MELIAN et al. 1996). Some MHC-Ib molecules have been shown to present antigenic epitopes to T cells. The MHC-Ia, MHC-Ib and CD1 molecules may play important roles in presenting antigenic epitopes to distinct subsets of the T cell compartment of the adaptive immune system (Fig. 2).

II. Conventional MHC-I or MHC-II Restricted TCR$\alpha\beta$ T Cells

The vast majority of T cells (>90%) in peripheral blood and lymphoid organs of man and the most widely used experimental animal, the mouse, express a TCR composed of an α- and a β-chain (HEDRICK and EIDELMAN 1993). These TCR$\alpha\beta$ T cells coexpress either the CD4, or the CD8 coreceptor molecule in

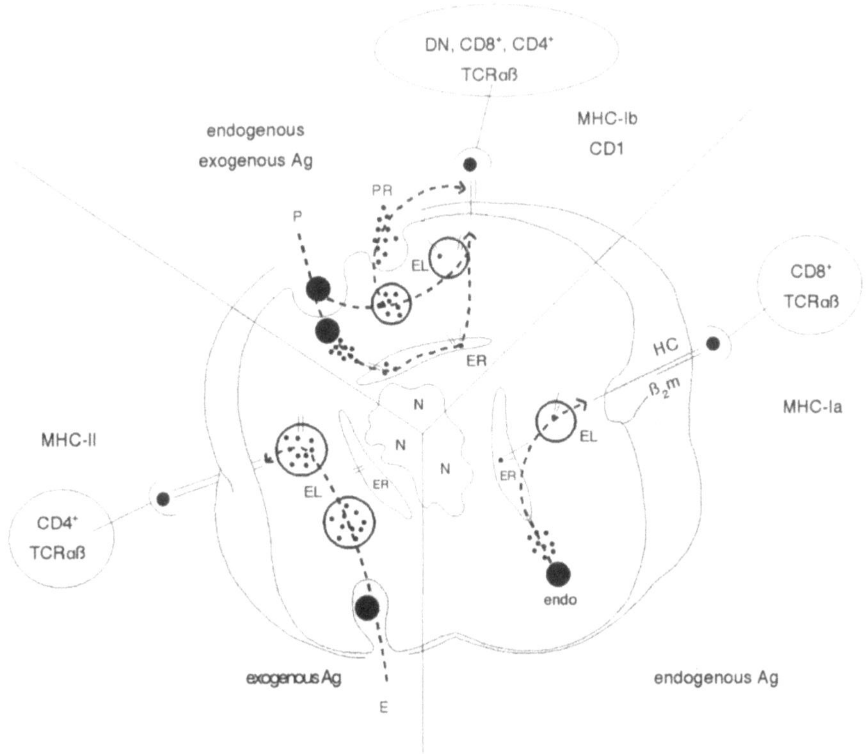

Fig. 2. Antigen processing and epitope presentation by cells. Abbreviations: MHC, major histocompatibility complex; TCR, T cell receptor for antigen; AG, antigen; endo, endogeneous antigen; DN, double negative; ER, endoplasmic reticulum; EL, endolysosome; N, nucleus; HC, MHC class I heavy chain; β2m, β2 microglobulin; P, phagocytosis; PR, peptide regurgitation

a mutually exclusive way. The $CD4^+$ and $CD8^+$ $TCR\alpha\beta^+$ T cells are the most abundant T lymphocytes. In the peripheral "conventional" $TCR\alpha\beta$ T cell population approximately 60% express the CD4 molecule and ca. 40% express the CD8 molecule. The $\alpha\beta$ T lymphocytes recognize antigenic peptides in the context of polymorphic MHC products. CD4 T cells are restricted by MHC-II molecules; CD8 T cells are restricted by MHC-I gene products.

The $CD8^+$ T cells recognize peptides that are presented by MHC-I surface glycoproteins. The antigenic peptides presented by MHC-I molecules are generally derived from endogenous antigens present in the cytosol of the presenting cell (Fig. 2). $CD8^+$ T cells are often cytolytic for the cells that they specifically recognize, lysing virus-infected cells and thus terminating their replication. The control of virus infection is therefore thought to be the central role of $CD8^+$ CTL (Zinkernagel 1993).

$CD4^+$ T lymphocytes are often designated T "helper" (Th) cells because they produce various cytokines that support clonal expansion, differentiation, and activation of cells of the immune system. CD4$^+$ T cells recognize peptides presented by heterodimeric surface glycoproteins encoded by MHC-II loci. Peptides presented by MHC-II molecules are generally derived from exogenous antigens that are engulfed by the presenting cell. Small antigenic complexes are taken up by endocytosis or pinocytosis and are partially digested in endosomes to generate immunogenic peptides; large antigenic complexes (such as microbes) are taken up by phagocytosis and are partially digested in phagolysosomes (BENHAM et al. 1995; GERMAIN et al. 1996). As shown in Fig. 3, distinct functional subsets (i.e., Th1 and Th2 cells) can be distinguished within the CD4$^+$ Th cell population (ABBAS et al. 1996). Th1 cells produce cytokines that primarily activate macrophages and CTL. Because intracellular bacteria, fungi, and protozoa typically reside in the phagosome of macrophages, activation of antimicrobial activities in infected macrophages represents the domain of Th1 cells. Th2 cells produce cytokines that directly activate eosinophils and induce IgE antibodies that stimulate basophils, eosinophils, and mast cells. These are important effector mechanisms for the control of helminth infections. Hence Th2 lymphocytes are responsible for this kind of pathogens. Th1 and Th2 cytokines affect antibody production by B cells. Th2 cytokines stimulate B cell maturation and induce Ig switch to neutralizing Ig isotypes, whereas

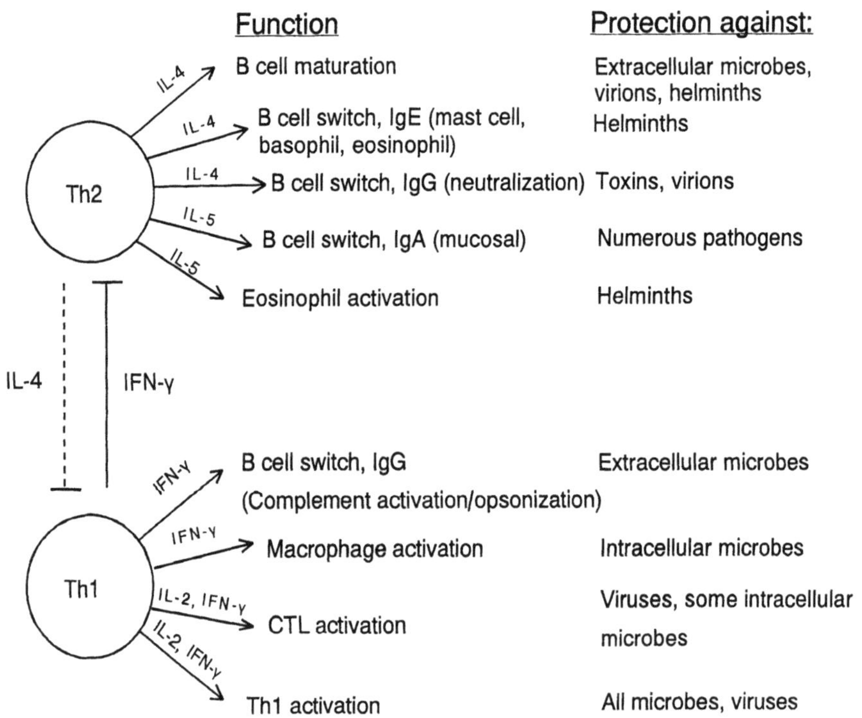

Fig. 3. Effector functions of anti-infective immunity controlled by Th1 or Th2 cells and their respective cytokines

Th1 cytokines stimulate Ig isotypes that promote microbial phagocytosis (RAJEWSKY 1996).

Pathogens residing in the phagosome or endosome are termed "endosomal pathogens," and infectious agents living in the cytosol are conveniently referred to as "cytosolic pathogens" (KAUFMANN 1993). The subcellular localization of an infectious agent markedly influences the type of T cell that is activated, an issue which is discussed in more detail below because of its great importance for rational vaccine design. Antigenic peptides presented by individual sets of MHC-Ia or MHC-II molecules of different individuals differ considerably. This observation led to the designation of the "MHC-restricted T cell response." Individual amino acid residues have recently been identified within an antigenic peptide that bind to a given MHC haplotype (anchor motifs) (RAMMENSEE et al. 1993, 1995). The high degree of MHC polymorphism and the extensive heterogeneity of antigenic peptides requires that a peptide vaccine covers a large repertoire of different peptides that are recognized by different individuals within the vaccinee population. This imposes severe restrictions to any single peptide vaccine.

III. Conventional MHC-Ia Restricted $CD8^+$ TCR$\alpha\beta$ T Cells

Most $CD8^+$ CTL effectors express the TCR$\alpha\beta$ antigen receptor and recognize peptides in the context of MHC-Ia glycoproteins expressed on most cells throughout the body. By binding to the constant $\alpha 3$ region of the MHC-Ia glycoprotein the CD8 molecule guides the CTL to the antigen-presenting stimulator or target cell. $CD8^+$ CTL are the predominant lymphoid cell subset in inflammatory reactions triggered by many intracellular pathogens, in particular viruses. $CD8^+$ effector CTL can eliminate infected cells and thereby limit the further spread of the pathogen. This cytotoxicity is mediated by pore-forming proteins (perforins) and enzymes (granzymes) that are stored in granules of activated CTL (PODACK and KUPFER 1991; CARTER and DUTTON 1995; LIU et al. 1996). Perforins form pores in target cells in a similar way as polymerized C9 complement components. This leads to the death of target cells by necrosis. The granzymes enter targets through perforin pores and induce rapid apoptosis in these cells. In addition, CTL express on their surface the Fas ligand (FasL). Binding of FasL to its receptor on the target cell, the Fas (or CD95, APO-1) determinant can also induce apoptotic death of the target cell. Specific recognition of targets by CTL requires a close contact between effector and infected cell. This optimizes the chances of a specific and direct attack of the infected cell by the CTL and suggests a major role of this T cell subset in virus clearance. CTL furthermore optimize their chance of recognizing MHC-I bound peptides by secreting interferon (IFN) γ, a potent inducer of both class I and class II MHC glycoprotein expression.

$CD8^+$ CTL responses are often directed against a small number of immunodominant epitopes of a large and complex pathogen. After the specific deletion of immunodominant epitopes (from the pathogen) or of the presenting

MHC-I molecule (from the host) multiple CTL responses against "cryptic" or "subdominant" epitopes of the same antigen or another antigen of the same pathogen often arise (SERCARZ et al. 1993). The epitope-specific immunoregulatory events underlying this additional "restriction" of specific T cell activation in vivo are ill defined. The implication of this phenomenon is that it is difficult for a pathogen successfully to develop escape variants by mutating a single epitope. Nevertheless, the suppression of CTL responses in the infected host is a frequent event which can lead to many asymptomatic carriers. Novel, carefully designed vaccination strategies may offer a chance to prime responses against a large repertoire of alternative epitopes of the pathogen that are usually not stimulated during natural infection.

Experimental models have demonstrated that $CD8^+$ CTL responses can control many infections involving intracellular pathogens, in particular viruses, but they can also mediate severe immunopathology (ZINKERNAGEL 1996). This may be a direct result of the acute destruction of large numbers of infected cells by CTL or may involve more complex immune dysregulations.

$CD8^+$ CTL and $CD4^+$ Th cells frequently interact in the course of an immune response. The relative dependence of $CD8^+$ CTL responses on $CD4^+$ T cell "help" in different cell-mediated immune responses varies with both the pathogen and the organ infected by the pathogen.

IV. Unconventional MHC-Ib Restricted $CD8^+$ TCR$\alpha\beta$ T Cells

In the mouse a subgroup of unconventional $CD8^+$ TCR$\alpha\beta$ T cells responds to bacterial peptides presented by MHC-I-like molecules (HARTY et al. 1996). In experimental *Listeria monocytogenes* infection of mice these cells are protective. This has been shown by adoptively transferring protection against listeriosis with this primed T cell subset into syngeneic, naive hosts. The relevant peptides contain the *N*-formyl-methionine (*N*-f-Met) N-terminal residue that represents a characteristic leader sequence in bacteria. The sequence is virtually absent in mammalian proteins with the exception of a few mitochondrial polypeptides. These *N*-f-met containing peptides are hydrophobic. Although their processing partially resembles that of exogenous protein antigens partially degraded in an endosomal compartment for MHC class II restricted peptide presentation, the *N*-f-Met containing peptides are presented by MHC-Ib molecules (Fig. 2). Cognates of the murine MHC-Ib molecule H-2M3 responsible for *N*-f-Met peptide presentation have not been identified in humans. It therefore remains unclear whether these immune mechanisms play a role in the anti-infectious immune response of man.

V. Unconventional Double-Negative (DN, $CD4^-$ $CD8^-$), CD1-Restricted TCR$\alpha\beta$ T Cells

Double-negative (DN) and some $CD8^+$ TCR$\alpha\beta$ T cells form a group of unconventional T cells that have been identified in the human system. These T cells recognize mycobacterial lipoglycans and lipids presented by CD1 molecules, demonstrating that T cells can specifically recognize nonpeptide ligands (BLUMBERG et al. 1995; BRANCH MOODY et al. 1997). The DN TCR$\alpha\beta$ T cells represent a minor population (<2%) of all TCR$\alpha\beta^+$ T cells in the peripheral blood of humans. Although the CD1 molecule is expressed on the cell surface in association with β_2-m, processing of glycolipids or lipids for CD1-restricted epitope presentation resembles processing of "exogenous protein antigens" for MHC class II restricted peptide presentation rather than "endogenous processing" for MHC-Ia restricted peptide presentation (Fig. 2). DN TCR$\alpha\beta$ T cells with specificity for lipoglycans and lipids have not been identified in mice, possibly because the responsible CD1 molecules (a, b and c) are absent in this species. However, circumstantial evidence for presentation of non-proteinaceous ligand by MHC-I-like molecules has been obtained also in the mouse system (NATARAJ et al. 1996).

VI. Unconventional CD1-Restricted CD4$^+$ or DN NK1$^+$ TCR$\alpha\beta$ T Cells

An unconventional TCR$\alpha\beta$ T cell population has been identified in mice that coexpresses a marker (NK1) characteristic for natural killer (NK) cells (BENDELAC 1995; MACDONALD 1995). Although most of these NK1$^+$ TCR$\alpha\beta$ T cells are of the CD4$^+$ phenotype, some are DN (CD4$^-$ CD8$^-$). These T cells express a restricted TCR repertoire with an invariant Vα chain and a biased usage of certain Vβ chains, suggesting specificity for a limited number of ligands. These NK T cells are specific for the only CD1 isotype which exists in mice, i.e., CD1d. It remains to be established whether they recognize CD1 proper or a microbial ligand presented by CD1. In humans $\alpha\beta$ T cells expressing the homologous invariant Vα chain have been identified, providing indirect evidence that similar cells are operative in man.

VII. TCR$\gamma\delta$ T Cells

Amongt the unconventional T cells TCR$\gamma\delta$ T cells are the most prominent subpopulation (KAUFMANN 1996b). TCR$\gamma\delta$ T cells use an alternative heterodimeric TCR composed of a γ- and a δ-chain. Generally TCR$\gamma\delta$ cells lack CD4 and CD8, i.e., are DN. In mouse and man they represent fewer than 10% of the T cells in secondary lymphoid tissues but are much more prevalent (20%–30%) in pigs, ruminants, and chickens. Large numbers of TCR$\gamma\delta$ T cells are found in the mucosa and skin. Thus in mice one-third of all intraepithelial lymphocytes in the mucosal epithelia are TCR$\gamma\delta$ T cells. The precise role of TCR$\gamma\delta$ T cells in infections is not yet understood. Especially in chronic immune responses these T lymphocytes are known to accumulate at the site of reaction and may play a role in regulating the inflammatory process.

The majority of the peripheral TCR$\gamma\delta$ T cells in man are characterized by the Vγ9δ2 chain combination (Vγ2δ2 in another nomenclature). These Vγ9δ2 T cells respond to nonproteinaceous microbial ligands which contain phosphate coupled to a backbone of alkyl derivatives or perhaps carbohydrates or nucleotides (KAUFMANN 1996b). Recognition of these $\gamma\delta$ T cells is independent of the clonal specificity of this TCR$\gamma\delta$, thus explaining the capacity of these ligands to stimulate all Vγ9δ2 T cells. Current evidence suggests that presentation of the phospholigands on the cell surface is independent of accessory molecules. It remains to be established whether the phospholigands are derived from microbes residing inside host cells and thus undergo intracellular processing, or whether they are attached to the cell surface from the extracellular milieu. Although the murine $\gamma\delta$ T cells do not respond to nonproteinaceous phospholigands, experiments using gene deletion mutant mice deficient for $\gamma\delta$ T cells have revealed a protective role of $\gamma\delta$ T cells in different infection models (KAUFMANN 1996b). In the protective immune responses against intracellular bacteria and protozoa $\gamma\delta$ T cells participate at an early stage, and they further contribute to protection against some viral infections at late stages.

Although the participation of unconventional T lymphocytes in the anti-infective immune response appears likely, the importance of their contribution remains to be established. These T cells represent minor fractions of the total T lymphocyte population, but their broader antigen reactivity could compensate at least partially for this. Moreover, antigen recognition by these cells is not controlled by polymorphic molecules, allowing stimulation by few defined ligands in a virtually MHC unrestricted fashion. Focus of unconventional T cells on nonproteinaceous ligands suggests that such components should be considered for inclusion in subunit vaccines.

D. Cytokines

Cytokines are soluble mediators that transduce signals between immature and mature immune cells and other host cells. Cytokines are produced by many different cell types, including lymphoid and myeloid cells of the innate or the adaptive immune system. Cytokines express pleiotropic functions, for example, they act on multiple host cells and induce many different biological functions. They regulate differentiation of lymphomyeloid cells from progenitor cells, and they determine the type of specific effector functions activated during anti-infective immune responses (Fig. 4). The latter include:

- Proinflammatory activities that promote inflammatory responses during innate and specific immune reactions
- Activities that alleviate inflammatory reactions
- Mobilization of effector functions of specific T cell subsets
- Regulatory signals which determine the outcome of the adaptive immune response

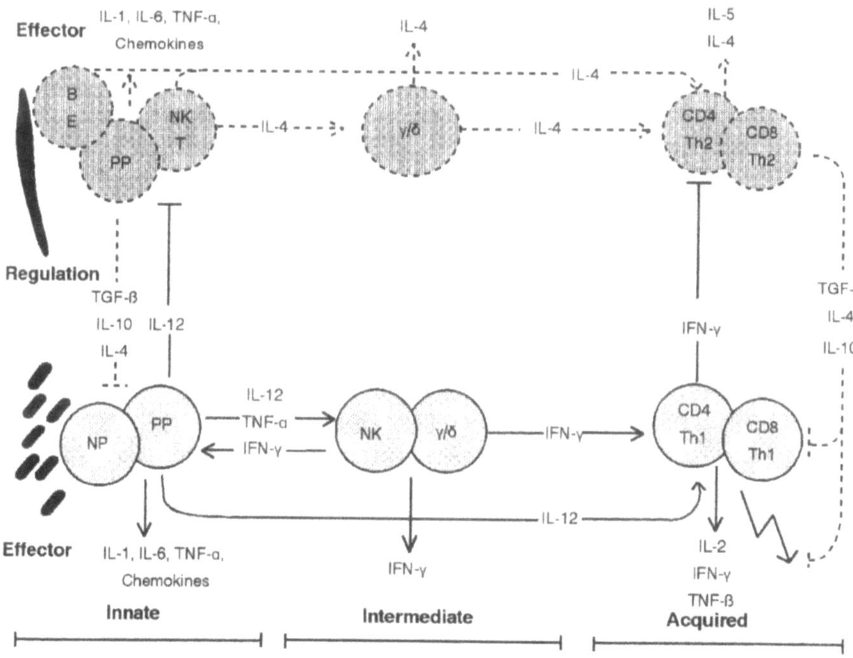

Fig. 4. Simplified scheme illustrating the development of Th1 and Th2 dominated immune response to infectious agents. *B*, Basophil (and mast cell); *E*, eosinophil; *PP*, professional phagocyte; *NP*, nonprofessional phagocyte (e.g., endothelial or epithelial cell)

The proinflammatory cytokines include tumor necrosis factor (TNF), interleukin (IL) 1, IL-6, and the large group of chemokines (Dinarello 1992; Akira et al. 1993; Tracey and Cerami 1993). These cytokines are produced rapidly after infection and induce inflammatory responses by attracting leukocytes to the site of pathogen growth. Although inflammation precedes the adaptive immune response, proinflammatory cytokines are also produced at later stages and frequently cooperate with cytokines of the adaptive response in the activation of effector functions. Finally, some proinflammatory cytokines modulate the development of adaptive immunity. For example, TNF participates in early monocyte influx to sites of microbial replication, activates NK cells which are potent IFN-γ producers, and synergizes with IFN-γ in macrophage activation. The anti-inflammatory cytokines, such as transforming growth factor (TGF) β or IL-10 are important counterregulators of inflammation (Moore et al. 1993; Wahl 1994). They promote termination of the anti-infective immune response and in this way prevent exaggerated tissue damage.

By producing different cytokines Th cells stimulate adequate effector mechanisms for the control of different infectious agents. The major Th cytokines are IL-2, IFN-γ, IL-4, and IL-5. Although exceptions exist, there prevails a trend towards gradual polarization of an ongoing immune response into one of two alternative effector cell populations that mediate different immune reactivities. These responses are often designated as Th1-type and Th2-type responses induced by the predominant production of either IL-2 and IFN-γ, or IL-4 and IL-10 (Fig. 3). At present Th1 and Th2 cells are characterized operationally on the basis of the cytokines that they produce (Mosmann and Sad 1996). Attempts to identify stable surface markers unique for either Th type have thus far failed. Although evidence has been presented for preferential expression of CD30 on human Th2 cells, current evidence suggests that CD30 does not qualify as an exclusive Th2 marker.

Macrophage activation represents the essential step in the control of intracellular microbes, be they bacteria, fungi, or protozoa. IFN-γ is the major macrophage-activating cytokine and synergizes in this task with other cytokines, in particular TNF. IL-2 is central to the maturational activation of CTL with further support by IFN-γ. CTL are crucial for antiviral defence and also contribute to control of some intracellular microbes. Therefore control of intracellular pathogens is considered a domain of Th1 cells. B cells are the prime targets of IL-4 (Kopf et al. 1993). This cytokine is not only responsible for B cell maturation but also promotes B cell switch to IgE isotype secretion. IL-5 and IL-4 activate eosinophils and mast cells, respectively (Takatsu et al. 1994). Cross-linking of Fcε receptors on mast cells, basophils, and eosinophils by IgE causes their degranulation and leads to the release of toxic effector molecules (Wardlaw et al. 1995). In this way IgE-coated helminths are attacked by eosinophils, an attack that is supported by basophils and mast cells. Th2 cells therefore play a prominent role in antihelminthic immune responses. Moreover, IL-5 induces B cell switching to production of IgA, which is the main mucosal antibody isotype, and IL-4 promotes the B cell switch to IgG subclasses (e.g., IgG1 in the mouse) that neither bind complement nor opsonize but are capable of neutralizing foreign molecules in the circulation. Hence Th2 controlled immune responses participate in the neutralization, inactivation, and elimination of microbial products in the circulation. IFN-γ also affects antibody responses in the mouse. First, it promotes secretion of Ig isotypes with complement fixing and opsonizing activity, for example, IgG2a and IgG3 and thus supports microbial elimination by professional phagocytes. Second, IFN-γ induces synthesis of the secretory component in epithelial cells and hence contributes to transport of IgA antibodies through epithelial layers to the mucosal surface. Thus secretion of antibodies is under the control of both Th1 and Th2 cytokines.

In conclusion, depending on their cytokine secretion pattern Th cells can be distinguished into two polar forms which are focused on different infectious agents (Figs. 3, 4). The Th1 cells are responsible primarily for microbes which are controlled by phagocytosis with an emphasis on intracellular pathogens. In contrast, the Th2 cells mobilize anti-infective mechanisms which are independent of phagocytosis.

E. The Early Decision on the Type of Effector Functions that Prevail in an Immune Response

How is the differentiation of Th1 and Th2 cells regulated? We have recently witnessed a number of findings which suggest that this decision is made at the outset of infection through cytokines produced by cells of the innate immune system (ABBAS et al. 1996) (Fig. 4). Early production of IL-12 favors development of Th1 cells, whereas early production of IL-4 supports Th2 cell development. Particular microbial components seem to guide the differentiation of the most appropriate T cell response. Viral, bacterial, and some protozoal components stimulate IL-12 production in macrophages and result in an anti-infective response. In contrast, soluble proteins, particularly those derived from helminths, favor early IL-4 secretion. Although the cellular sources of this early IL-4 response are still ill defined, several candidates are known, including $CD4^+$ $NK1^+$ T cells, mast cells, basophils, and eosinophils. Although the two pathways are still interconvertible during the initial phase of Th cell development, the polarization gradually becomes more stable. This results from the negative control of IFN-γ on Th2 T cell development and of IL-4 on Th1 T cell development.

Many factors of very different nature have been identified that influence the differentiation of mature effector functions in primed $CD4^+$ T cells. The genetic background clearly has a major effect on the Th1/Th2 pathways of T cell differentiation, although none of the involved genes has yet been identified (GORHAM et al. 1996). Both the route of antigen administration and the type and dose of the antigen injected exert a major effect on the preferential or exclusive activation of the Th1 or Th2 pathway (BRETSCHER et al. 1992). Cell interaction or adhesion molecules seem to play an instructive role for the induction of Th1 and Th2 $CD4^+$ T cell effector populations. The B7-1 (CD80) and B7-2 (CD86) costimulator molecules may differentially activate the Th1/Th2 pathways (KUCHROO et al. 1995; OZAWA et al. 1996). Ligation of the CD40 receptor on antigen-presenting cells (APC), such as dendritic cells, by soluble or cell-associated (TNF-like) CD40L is one of the most potent triggers known for IL-12 release (STUBER et al. 1996; FOY et al. 1996).

Although the preferential priming of a protective immune response can be achieved in many instances, some infectious agents have developed ways to outwit the immune response by priming a nonprotective type of response. The most notable examples of this are *Leishmania major* and *Mycobacterium leprae*. In experimental leishmaniasis of mice both susceptible and resistant strains exist. Resistance accompanies the development of a potent Th1 response whereas susceptibility is paralleled by predominance of Th2 cells. Susceptible mice suffer from reduced IL-12 and elevated IL-4 production at the outset of infection. This immune dysregulation results in default Th2 development with lethal outcome. In contrast, early IL-12 produced in resistant mice causes Th1 cell differentiation and control of disease. Although susceptibility to infectious agents of man is often multifactorial, similar mechanisms may also affect the course of certain infectious diseases, such as leprosy. Obviously any efficacious vaccine must stimulate the type of immune response which is most appropriate for control of the given infectious agent and must counteract any mechanism which might increase susceptibility.

F. The Humoral Immune System: B Cells

B cells produce antibodies but they are also key APC for $CD4^+$ T cells. B cells express surface Ig molecules, and their receptor for antigen. After the initial contact with antigen surface Ig is usually cross-linked as a consequence of multipoint binding of complex antigens (RAJEWSKY 1996). This generates an activation and differentiation signal for the B cells to expand clonally, to generate a memory population, or to differentiate into antibody-secreting plasma cells. At the same time, the Ig-bound antigen is internalized and degraded in an endolysosomal compartment, and peptides derived from these proteins are presented in the context of MHC class II glycoproteins on the cell surface. $CD4^+$ T cells specific for the peptide can then be recruited into the response to provide help for further B cell development.

In addition to membrane-bound Ig that capture antigen efficiently and specifically for presentation to $CD4^+$ T cells, soluble (secreted) Ig are major effector molecules. The type of effector function mediated by an antibody depends on its isotype. Antibody class switching, i.e., isotype switching, is controlled by cytokines. Several cytokines that control isotype switching have been identified in the murine system: IL-4 induces IgG1 and IgE, IL-5 and TGF-β induce IgA, and IFN-γ induces IgG2a and IgG3 expression (STAVNEZER 1996). Antibodies of the IgM, IgG2a, and IgG3 isotypes activate complement. They direct cytolytic components of the complement system to pathogens and thereby lyse either the pathogen or the infected cell (LISZEWSKI et al. 1996). Moreover, complement activation generates the complement components C4a, C5a, and C3a that are proinflammatory; these so-called anaphylatoxins attract leukocytes to the site of their production and hence contribute to the anti-infective defense. Antibody-coated cells can be directly destroyed by cytolytic effector cells mediating antibody-dependent cellular cytotoxicity (ADCC). Large granular lymphocytes that express receptors for the constant region for IgG (FcγR) lyse IgG-coated host cells. Furthermore, IgG coating facilitates uptake of FcγR-bearing microbes by macrophages which otherwise evade phagocytosis. Similarly, opsonization of pathogens by the complement component C3b mediates their engulfment by macrophages expressing C3b receptors. Eosinophils, basophils, and mast cells destroy cells or micro-organisms that have bound IgE antibodies through FcεR; for example, they attack IgE-coated helminths. Nonopsonizing antibodies, such as IgG1, contribute to the antimicrobial host responses by neutralizing the infectivity of pathogens or the toxicity of microbial products.

G. T and B Cell Memory

One of the hallmarks of the specific immune system is memory, i.e., the ability of the system to store specific information about previous encounters of the same "foreign" antigen. Both T and B cells can develop memory cells (RAJEWSKY 1996; MACKAY 1993; AHMED and GRAY 1996; ZINKERNAGEL et al. 1996). Memory involves (a) the generation of increased numbers of clonal progeny of antigen-stimulated T or B cells; (b) the development of a surface marker profile that is different in naive and memory cells; (c) the expression of an antigen receptor specificity that is similar to that of the naive precursor cell but may be of higher avidity; and (d) the development of new and often less demanding requirements for activation and differentiation stimuli. It is controversial whether the persistence of antigen is a necessary prerequisite for the establishment of a long-lived B and/or T cell memory.

Following the secondary challenge, several days are needed to specifically "restimulate" in vivo the pool of memory T cells to fully functional effectors. This implies that secondary infections are not prevented by T cell mediated immunity. Rather, memory T cells clear a secondary infection more efficiently, but cannot prevent it. This is unlike the operation of the B cell memory response, in which an existing level of circulating antibodies can prevent the infection of the host cell by the pathogen.

H. Mucosal Immunity

Lymphoid cells are abundantly present in mucosa-associated lymphoid tissues (MALT). In terms of total numbers T and B lymphocytes in gut-associated lymphoid tissues (GALT) represent the largest, organ-specific lymphoid cell population as well as a major proportion of the total peripheral lymphocyte population in both man and mouse. The GALT consists of organized lymphoid aggregates such as Peyer's patches, appendix, mesenteric lymph nodes, and solitary lymphoid nodules. In addition, the nonorganized (diffuse) lymphoid elements in the epithelium and in the lamina propria are part of GALT (CROITORU and BIENENSTOCK 1994). Marked heterogeneity of lymphoid cell subsets is evident within individual GALT compartments (CROITORU and BIENENSTOCK 1994; GUY-GRAND and VASSALLI 1993; FANTRY and JAMES 1994; JAMES and SEITZ 1994; POUSSIER and JULIUS 1994). Furthermore, evidence of marked regional specialization of the GALT compartments is emerging (CAMERINI et al. 1993; BEAGLEY et al. 1995). The mucosal surfaces of the gastrointestinal, respiratory, and genitourinary tracts are the sites of entry of many species of virus, bacteria, fungi and protozoa. These include: (a) food-borne infections which either are contained locally (causing different forms of gastroenteritis), or spread systemically (such as typhoid fever); (b) air-borne infections (a notable example of which is tuberculosis); (c) urogenital infections; and (d) genital infections (sexually transmitted diseases).

The specific mucosal immune response involves specialized, local lymphoid tissues that contain B and T cells. The regulation of the cellular (T) immune responses in mucous membranes is complex and not yet adequately understood. In particular the question of which conditions favor the establishment of oral tolerance *versus* the induction of a mucosal T cell response is unresolved. The major humoral immune response in the mucosa is the induction of secretory IgA. This Ig class is actively secreted by mucosal epithelia and protects the mucosal surfaces. Because IgA is resistant to proteolytic degradation, it seems well suited as a first line of defense.

Ideally, mucosal vaccination provides the possibility to prevent colonization at mucosal tissue sites prior to stable infection. Such early intervention would prevent many harmful sequelae of infection. Evidence exists that the different parts of the MALT in the gut, urogenital tract, and lung are interconnected. Accordingly, stimulation of the gut-associated immune system by oral vaccination may also activate mucosal immune responses in urogenital and bronchoalveolar tissues. Hence the design of mucosal vaccines is a promising field of research.

I. Unsuccessful Induction of Specific Immune Responses

T cells are central for the induction and regulation of specific immune responses. Recently many studies have focused on the mechanisms that prevent efficient T cell priming during natural infections or immunization with a defined antigen. Host-derived and pathogen-derived factors have been identified that block T cell stimulation at an early stage of the specific response. The host-derived mechanisms include specific immune tolerance, immune deviation, and the genetically determined nonresponder status. Contact with antigen may result in specific tolerance instead of a cellular immune response, especially in the neonatal period. If the initial contact of the individual takes place at "immune-privileged" sites of the organism (e.g., the anterior chamber of the eye or testicular tissues), stable and specific tolerance is induced consistently. The pattern of the expressed MHC alleles may not allow the individual to present epitopes from a given parasite to the T cell system. These examples can be considered states of unresponsiveness which are inherent in the adaptive immune system.

Many pathogens have developed means to block or circumvent the immune system of the host or to prime inefficient immune responses that are not protective. The pathogen-induced mechanisms are complex and differ widely between pathogens even from the same genus. In various groups of viruses mechanisms have been identified that block the biogenesis or surface expression of MHC-I molecules. Antigen processing in the cytosol, the transport of peptides from the cytosol into the lumen of the endoplasmic reticulum, the assembly of trimeric MHC-I molecules in the lumen of the endoplasmic reticulum, their efficient transport to the cell surface, and their stable surface display can be inhibited in infected cells by viral proteins (Früh et al. 1997). Pathogens have evolved products which neutralize cytokines or mimic the action of cytokines that prime nonprotective immune responses. Direct infection of immune effector cells or of professional APC can eliminate functional immunity, as dramatically demonstrated by HIV infections (Fauci 1996).

J. Implications for Rational Vaccine Design

"New" diseases (e.g., AIDS), "old" diseases (e.g., tuberculosis), and the identification of the infectious nature of widespread diseases considered nontransmissible until recently (e.g., gastritis and ulcers associated with *Helicobacter pylori*) stress the need for a new generation of effective vaccines (Kaufmann 1996a). The dramatic advances in biotechnology made possible the routine production of even complex antigens in recombinant form on a large scale. This has laid the basis for new generations of subunit vaccines. At the same time, live vaccines may be required for satisfactory immunization against certain infectious diseases. In these instances the possibility of recombinant live carriers expressing defined antigens will be an interesting option. Adjuvants will be of critical importance in augmenting or modulating the immunogenicity of recombinant antigens of very different chemical nature and structure available through biotechnological techniques (Morein et al. 1996). The concept of an "adjuvant" will achieve a rational basis. The choice of an agent or cytokine that enhances priming of desirable immune effector functions and suppresses development of undesirable effector functions will become more rational. Insights into the molecular basis of specific immune reactions, in particular T cell-mediated responses, are expanding continuously. These two developments have allowed first attempts to design vaccines that preferentially stimulate the particular type of immune response required for protection.

The concept of vaccination has been furthermore enlarged. "Vaccination" is currently considered both as a prophylactic and as a therapeutic intervention. Designs of vaccines or immunotherapies have been of interest not only in the field of infectious disease but also in those of cancer research and autoimmune diseases (Pardoll 1995; Boon and van der Bruggen 1996; Hellström et al. 1996; Cohen and Mor 1996). While vaccines against tumors and infectious diseases are aimed primarily at the stimulation or potentiation of a protective immune response capable of eliminating the threatening agent, vaccines against autoimmune diseases are directed mostly at interfering with an exaggerated immune response. Hence the spectrum of potential vaccine applications has been considerably enlarged.

The currently available vaccines are most efficient in the antibody-mediated elimination of extracellular pathogens (either bacteria or viruses) or in the neutralization of toxins. In contrast, the available vaccines against intracellular pathogens that rely on T cell mediated immunity are strikingly less efficient. To design vaccines that efficiently and preferentially stimulate selected compartments of the T cell system, particular attention must be given to the preparation of the antigen and its mode of delivery. These two aspects of vaccine development are the most essential factors because they critically affect the type and magnitude of the immune response(s) elicited. Today's major challenge in vaccine development is the design of recombinant antigen constructs and ways to deliver them to induce stable, protective responses and to avoid harmful responses. To achieve this goal the following issues on candidate vaccine antigen formulations must be addressed in experimental animal models:

1. What is the biodistribution of the vaccine following different routes of delivery? This will lead to information on the type of the APC selected by the vaccination technique.
2. Within the relevant APC, what type of processing is the antigen subjected to? This will elucidate the type and repertoire of epitopes presented in the context of conventional MHC molecules or alternative MHC-like molecules on the surface of APC.
3. What repertoire of costimulatory activities (cytokines and adhesion molecules) are expressed by the activated, antigen-processing and antigen-presenting cell that can facilitate and modulate specific priming of particular lymphocyte subsets?
4. What kind of immunoregulatory lymphoid cells are recruited into the priming of a specific response during the in situ presentation of the antigen by the activated APC? Such regulatory cells have been shown to have a decisive effect on the type of the ensuing immune response. This is particularly relevant to the question as to which type of $CD4^+$ T cell (Th1 versus Th2) predominates in response to vaccination.
5. What is the spectrum of humoral and cellular immune effector functions (antibodies and their isotypes, T cells and their functional phenotypes) specifically activated in response to the candidate vaccine? Generation of the desired immune reactivities to several epitopes should not be limited by the variation in immune responsiveness within the population due to MHC polymorphism.
6. Is it possible to establish a stable and specific B and T cell memory with the vaccine? This may require the persistence of antigen on APC in lymphoid tissues to continuously recruit and to restimulate memory cells and to continuously develop antibody-secreting cells to ensure the continuous presence of antibody.

Vaccination remains the best answer to most infectious diseases. The hallmark of the efficacy of an acceptable vaccine is its ability to confer protective immunity to the majority of vaccinees. Advances in understanding the cell biology of immune responses are just beginning to have a major effect on vaccine development. Vaccinology of the next decade will provide techniques to specifically and efficiently prime immune responses against recombinant antigens that provide significant protection against natural challenge infections and avoid many facets of immunopathology.

Acknowledgements. S.H.E.K. acknowledges the financial support from the Sonderforschungsbereich 322 "Lympho-Hemopoiesis," the German Science Foundation, project Ka 573/3-2; the Interdisciplinary Centre for Clinical Research; the BMBF Joint Project "Mycobacterial Infections"; and the German Leprosy Relief Association. J.R. acknowledges support from the German Federal Ministry for Research and Technology (contract 01 K 9482), the German-Israeli Foundation (contract I 228 019.02/92), the German Science Foundation (Re549/4-2 and Re549/6-1), and the Fritz-Thyssen Foundation (contract K 32 476). The authors wish to thank R. Mahmoudi for excellent secretarial help.

References

Abbas AK, Murphy KM, Sher A (1996) Functional diversity of helper T lymphocytes. Nature 383:787–793

Ahmed R, Gray D (1996) Immunological memory and protective immunity: understanding their relation. Science 272:54–60

Aichele P, Bachmann MF, Hengartner H, Zinkernagel RM (1996) Immunopathology or organ-specific autoimmunity as a consequence of virus infection. Immunol Rev 152:21–45

Akira S, Taga T, Kishimoto T (1993) Interleukin-6 in biology and medicine. Adv Immunol 54:1–78

Beagley KW, Fujihashi K, Lagoo AS, Lagoo-Deenadaylan S, Black CA, Murray AM, Sharmanov AT, Yamamoto M, McGhee JR, Elson CO, Kiyono H (1995) Differences in intraepithelial lymphocyte T cell subsets isolated from murine small versus large intestine. J Immunol 154:5611–5619

Beckman EM, Brenner MB (1995) MHC class I-like, class II-like and CD1 molecules: distinct roles in immunity. Immunol Today 16:349–352

Bendelac A (1995) Mouse $NK1^+$ T cells. Curr Opin Immunol 7:367–374

Benham A, Tulp A, Neefjes J (1995) Synthesis and assembly of MHC-peptide complexes. Immunol Today 16:359–362

Blumberg RS, Gerdes D, Chott A, Porcelli SA, Balk SP (1995) Structure and function of the CD1 family of MHC-like cell surface proteins. Immunol Rev 147:5–29

Boon T, Van der Bruggen P (1996) Human tumor antigens recognized by T lymphocytes. J Exp Med 183:725–729

Branch Moody D, Sugita M, Peters PJ, Brenner MB, Porcelli SA (1996) The CD1 restricted T cell response to mycobacteria. Res Immunol 147:550–559

Bretscher PA, Wei G, Menon JN, Bielefeldt-Ohmann H, Bielefeldt Ohmann H (1992) Establishment of stable, cell-mediated immunity that makes "susceptible" mice resistant to Leishmania major. Science 257:539–542

Camerini V, Panwala C, Kronenberg M (1993) Regional specialization of the mucosal immune system. Intraepithelial lymphocytes of the large intestine have a different phenotype and function than those of the small intestine. J Immunol 151:1765–1776

Carter LL, Dutton RW (1995) Relative perforin- and Fas-mediated lysis in T1 and T2 CD8 effector populations. J Immunol 155:1028–1031

Chan AC, Desai DM, Weiss A (1994) The role of protein tyrosine kinases and protein tyrosine phosphatases in T cell antigen receptor signal transduction. Annu Rev Immunol 12:555–592

Cohen IR, Mor F (1996) Vaccination therapy against autoimmune diseases. In: Kaufmann SHE (ed) Concepts in vaccine development. De Gruyter, Berlin, pp 559–572

Cresswell P (1994) Assembly, transport, and function of MHC class II molecules. Annu Rev Immunol 12:259–294
Croitoru K, Bienenstock J (1994) Characteristics and functions of mucosa-associated lymphoid tissue. In: Ogra PL, Lamm ME, McGhee JR, Mestecky J, Strober W, Bienenstock J (eds) Handbook of mucosal immunology. Academic, San Diego, pp 141–150
Dinarello CA (1992) Role of interleukin-1 in infectious diseases. Immunol Rev 127:119–146
Ehrlich P (ed) (1904) Gesammelte Arbeiten zur Immunitätsforschung. Hirschwald, Berlin
Fantry GT, James SPJ (1994) Cell-mediated immunity and mucosal immunity. Curr Opin Gastroenterol 10:365–373
Fauci AS (1996) Host factors and the pathogenesis of HIV-induced disease. Nature 384:529–534
Fearon DT, Locksley RM (1996) The instructive role of innate immunity in the acquired immune response. Science 272:50–54
Foy TM, Aruffo A, Bajorath J, Buhlmann JE, Noelle RJ (1996) Immune regulation by CD40 and its ligand gp39. Annu Rev Immunol 14:591–617
Früh K, Ahn K, Peterson PA (1997) Inhibition of MHC class I antigen presentation by viral proteins. J Exp Med 75:18–27
Germain RN, Margulies DH (1993) The biochemistry and cell biology of antigen processing and presentation. Annu Rev Immunol 11:403–450
Germain RN, Castellino F, Han R, Reis e Sousa C, Romagnoli P, Sadegh-Nasseri S, Zhong T-M (1996) Processing and presentation of endocytically acquired protein antigens by MHC class II and class I molecules. Immunol Rev 151:5–30
Gorham JD, Guler ML, Steen RG, Mackey AJ, Daly MJ, Frederick K, Dietrich WF, Murphy KM (1996) Genetic mapping of a murine locus controlling development of T helper 1/T helper 2 type responses. Proc Natl Acad Sci USA 93:12467–12472
Gotschlich EC (1993) Immunity to extracellular bacteria. In: Paul WE (ed) Fundamental immunology. Raven, New York, pp 1287–1308
Guy-Grand D, Vassalli P (1993) Gut intraepithelial T lymphocytes. Curr Opin Immunol 5:247–252
Harty JT, Lenz LL, Bevan MJ (1996) Primary and secondary immune responses to Listeria monocytogenes. Curr Opin Immunol 9:526–530
Hedrick SM, Eidelman FJ (1993) T lymphocyte antigen receptors. In: Paul WE (ed) Fundamental immunology. Raven, New York, pp 383–420
Hellström KE, Hellström I, Chen L (1996) Can costimulated tumor immunity be therapeutically efficacious? Immunol Rev 145:123–145
James SP, Zeitz M (1994) Human gastrointestinal mucosal T cells. In: Ogra PL, Lamm ME, McGhee JR, Mestecky J, Strober W, Bienenstock J (eds) Handbook of mucosal immunology. Academic, San Diego, pp 275–285
Janeway CA Jr (1992) The T cell receptor as a multicomponent signalling machine: CD4/CD8 coreceptors and CD45 in T cell activation. Annu Rev Immunol 10:645–674
Kaufmann SHE (1993) Immunity to intracellular bacteria. In: Paul, WE (ed) Fundamental immunology. Raven, New York, pp 1251–1286
Kaufmann SHE (ed) (1996a) Concepts in vaccine development. De Gruyter, Berlin
Kaufmann SHE (1996b) $\gamma\delta$ and other unconventional T lymphocytes: what do they see and what do they do? Proc Natl Acad Sci USA 93:2272–2279
Kaufmann SHE, Schoel B (1994) Heat shock proteins as antigens in immunity against infection and self. In: Morimoto RI, Tissières A, Georgopoulos C (eds) The biology of heat shock proteins and molecular chaperones, vol II. Cold Spring Harbor Laboratory Press, Cold Spring Harbor, pp 495–531
Kopf M, Le Gros G, Bachmann MF, Lamers MC, Blüthmann H, Köhler G (1993) Disruption of the murine IL-4 gene blocks Th2 cytokine responses. Nature 362:245–248

Kuchroo VK, Das MP, Brown JA, Ranger AM, Zamvil SS, Sobel RA, Weiner HL, Nabavi N, Glimcher LH (1995) B7-1 and B7-2 costimulatory molecules activate differentially the Th1/Th2 developmental pathways: application to autoimmune disease therapy. Cell 80:707–718

Lenz LL, Bevan MJ (1996) H2-M3 restricted presentation of *Listeria monocytogenes* antigens. Immunol Rev 151:107–121

Liszewski MK, Farries TC, Lublin DM, Rooney IA, Atkinson JP (1996) Control of the complement system. Adv Immunol 61:201–283

Liu C-C, Young LHY, Young JDE (1996) Mechanisms of disease. Lymphocyte-mediated cytolysis and disease. N Engl J Med 335:1651–1659

Locksley RM (1996) Immune responses to protozoa. In: Rich RR, Fleisher TA, Schwartz BD, Shearer WT, Strober W (eds) Clinical immunology: principles and practice. Mosby Year Book, St Louis, pp 550–560

MacDonald HR (1995) NK1.1^{+} T cell receptor-$\alpha\beta^{+}$ cells: new clues to their origin, specificity, and function. J Exp Med 182:633–638

Mackay CR (1993) Immunological memory. Adv Immunol 53:217–265

Melian A, Beckman EM, Porcelli SA, Brenner MB (1996) Antigen presentation by CD1 and MHC-encoded class I-like molecules. Curr Opin Immunol 8:82–88

Moore KW, O'Garra A, De Waal Malefyt R, Vieira P, Mosmann TR (1993) Interleukin-10. Annu Rev Immunol 11:165–190

Morein B, Lövgren K, Cox JH (1996) Modern adjuvants, functional aspects. In: Kaufmann SHE (ed) Concepts in vaccine development. De Gruyter, Berlin, pp 243–264

Mosmann TR, Sad S (1996) The expanding universe of T-cell subsets: Th1, Th2 and more. Immunol Today 17:138–146

Nataraj C, Brown ML, Poston RM, Shawar SM, Rich RR, Lindahl KF, Kurlander RJ (1996) H2-M3wt-restricted, Listeria monocytogenes-specific CD8 T cells recognize a novel, hydrophobic, protease-resistant, periodate-sensitive antigen. Int Immunol 8:367–378

Nutman TB (1996) Immune responses to helminth infection. In: Rich RR, Fleisher TA, Schwartz BD, Shearer WT, Strober W (eds) Clinical immunology: principles and practice. Mosby Year Book, St Louis, pp 561–570

Ozawa H, Aiba S, Nakagawa, Tagami H (1996) Interferon-γ and interleukin-10 inhibit antigen presentation by Langerhans cells for T helper type 1 cells by suppressing their CD80 (B7-1) expression. Eur J Immunol 26:648–652

Pardoll DM (1995) Paracrine cytokine adjuvants in cancer immunotherapy. Annu Rev Immunol 13:399–416

Podack ER, Kupfer A (1991) T-cell effector functions: mechanisms for delivery of cytotoxicity and help. Annu Rev Cell Biol 7:479–504

Porcelli SA (1995) The CD1 family: a third lineage of antigen-presenting molecules. Adv Immunol 59:1–98

Poussier P, Julius MH (1994) Thymus-independent T cell development and selection in the intestinal epithelium. Annu Rev Immunol 12:521–553

Rajewsky K (1996) Clonal selection and learning in the antibody system. Nature 381:751–758

Rammensee HG, Falk K, Rötzschke O (1993) Peptides naturally presented by MHC class I molecules. Annu Rev Immunol 11:213–244

Rammensee HG, Friede T, Stevanoviic S (1995) MHC ligands and peptide motifs: first listing. Immunogenetics 41:178–228

Sercarz EE, Lehmann PV, Ametani A, Benichou G, Miller A, Moudgil K (1993) Dominance and crypticity of T cell antigenic determinants. Annu Rev Immunol 11:729–766

Stavnezer J (1996) Antibody class switching. Adv Immunol 61:79–146

Stuber E, Strober W, Neurath M (1996) Blocking the CD40L-CD40 interaction in vivo specifically prevents the priming of T helper 1 cells through the inhibition of interleukin 12 secretion. J Exp Med 183:693–698

Takatsu K, Takaki S, Hitoshi Y (1994) Interleukin 5 and its receptor system: implications in the immune system and inflammation. Adv Immunol 57:145–190
Tracey KJ, Cerami A (1993) Tumor necrosis factor, other cytokines and disease. Annu Rev Cell Biol 9:317–343
von Behring E (1915) Gesammelte Abhandlungen. Marcus and Webers, Bonn
Wahl SM (1994) Transforming growth factor β: the good, the bad, and the ugly. J Exp Med 180:1587–1590
Wardlaw AJ, Moqbel R, Kay AB (1995) Eosinophils: biology and role in disease. Adv Immunol 60:151–266
Wegener A-MK, Letourneur F, Hoeveler A, Brocker T, Luton F, Malissen B (1992) The T cell receptor/CD3 complex is composed of at least two autonomous transduction modules. Cell 68:83–95
York IA, Rock KL (1996) Antigen processing and presentation by the class I major histocompatibility complex. Annu Rev Immunol 14:369–396
Zinkernagel RM (1993) Immunity to viruses. In: Paul WE (ed) Fundamental immunology. Raven, New York, pp 1211–1250
Zinkernagel RM (1996) Immunology taught by viruses. Science 271:173–178
Zinkernagel RM, Bachmann MF, Kundig TM, Oehen S, Pirchet H, Hengartner H (1996) On immunological memory. Annu Rev Immunol 14:333–367

CHAPTER 3

DNA Vaccines: Immunogenicity and Preclinical Efficacy

J.B. ULMER, J.J. DONNELLY, and M.A. LIU

A. Overview

DNA vaccination is a novel vaccine technology that is being developed for a variety of clinical applications, as well as being utilized as a tool for exploring basic questions related to antigen processing and presentation. DNA vaccines differ from traditional vaccines in that, rather than consisting of the antigen itself (in the form of an inactivated or attenuated pathogen) or of a purified or recombinant component of the pathogen (e.g., bacterial polysaccharide or viral protein), they are bacterial plasmids composed of a strong promoter such as a viral promoter which is active in mammalian cells, a gene encoding the protein antigen of interest, and a transcriptional terminator and polyadenylation sequence. The broader term of nucleic acid vaccines refers to both DNA and RNA vaccines; however, due to the relative instability of RNA compared to DNA, RNA vaccines are less developed, but could have certain advantages over DNA such as not requiring delivery of the polynucleotide to the nucleus.

Several reviews on DNA vaccines have been published, including a recent comprehensive review of the literature (DONNELLY et al. 1997b). As perhaps a more instructive approach for the purposes of this publication, the present review describes the immunogenicity and preclinical efficacy of DNA vaccines in various animal models to illustrate the characteristics of these vaccines from a pharmacological standpoint. Indeed, because DNA could be considered a chemical entity, in contrast to traditional vaccines which are clearly biological entities, this may facilitate its development as a vaccine approach. The generic nature of DNA plasmids facilitates their construction and purification; hence one can readily make constructs encoding a variety of proteins from different pathogens. Thus for disease models in which different types, kinetics, or localization of immunity may determine the efficacy of a vaccine, DNA vaccines provide a common technological approach to address these issues.

B. Background

The initial observation that plasmids encoding reporter genes formulated in saline could be simply injected into muscle with resultant expression of the encoded proteins (WOLFF et al. 1990) was surprising, given the manipulations and formulations that are required for transfection of cells in vitro. Muscle appeared to be the tissue in which the highest levels of protein are expressed, presumably due to a higher level of transfection than in other tissues, but even in muscle the transfection rate was quite low (approx. 1% of myocytes). Nevertheless, these observations suggested that this approach could be used to generate an immune response against the protein synthesized in situ. Indeed, a subsequent paper reported that antibody responses could be generated utilizing a biolistic device to deliver plasmid DNA encoding human growth hormone coated onto gold beads into the skin (TANG et al. 1992).

In addition to generating protein antigens leading to production of antibody responses, expression of antigens within a cell could lead to the processing and presentation of the antigen in association with MHC class I molecules and result in the induction of CTL. Hence the full potential of this technology is to induce both humoral and cellular immune responses. In fact, the first published demonstration of the efficacy of DNA vaccines to protect against challenge with a pathogen was based on the induction of MHC class I restricted CTL (ULMER et al. 1993). Immunization of mice with a plasmid encoding the nucleoprotein of influenza A virus (NP DNA) resulted in the generation of CTL specific for the peptide epitope restricted by the $H2K^d$ MHC class I haplotype of the mice. Moreover, the mice were protected from both morbidity and mortality upon subsequent challenge with influenza virus. The protection observed was cross-strain in that the strain from which the NP was cloned was an H1N1 strain arising in 1934 (A/PR/08/34) while the challenge strain of virus was a different subtype, H3N2, and arose 34 years later in 1968 (A/HK/68). While substantial antibody responses against the NP were also generated, the protection was demonstrated to be due to cellular immunity and not to the antibodies (MONTGOMERY et al. 1993; ULMER et al. 1993; ULMER et al. 1998).

However, muscle cells, which appear to most readily internalize plasmid and express the encoded protein, lack key cell surface proteins present on professional antigen-presenting cells (APCs) and which are necessary for the induction of these CTLs. Recent studies have shown that while i.m. immunization of plasmid results in the generation of stronger cellular immune responses (CTL and/or T helper responses) than does immunization by other routes (i.v., i.d., s.c, or i.p.) (DONNELLY et al. 1994), professional APCs rather than muscle cells are responsible for the direct presentation of antigen for the priming of CTL (CORR et al. 1996; FU et al. 1997; IWASAKI et al. 1997). While it is not clear whether professional APCs are directly transfected after i.m. injection of DNA vaccines, this is not a requirement since expression of antigen in muscle cells alone is sufficient to induce CTL (ULMER et al. 1996), and this transfer of antigen from transfected muscle cells to APCs can result in induction of CTL (DOE et al. 1996; FU et al. 1997; HARPIN et al. 1997).

DNA vaccines encoding the influenza virus surface glycoprotein haemagglutinin (HA) have also been shown to be effective for the generation of functional neutralizing antibodies which could likewise provide protection from challenge with the homologous strain of virus (FYNAN et al. 1993;

MONTGOMERY et al. 1993; ULMER et al. 1994; JUSTEWICZ and WEBSTER 1996; DECK et al. 1997). Thus the potential of the technology for the generation of both humoral and cellular protective immune responses has been well documented in the influenza model. Given the relative simplicity of inserting different genes into a plasmid and of altering the coding region to direct the encoded protein to various cellular locations, DNA vaccines provide a tool both to explore issues of antigen processing and presentation by easily manipulating the antigen and to address different models of immunity relevant to different disease processes.

C. Effectiveness of DNA Vaccines in Animal Models

I. Miscellaneous

An early issue to address was whether the observations regarding the uptake of plasmid DNA and the resultant expression of protein and immune responses were phenomena restricted to rodent species. A number of publications soon documented the ability of plasmids to result in the expression of reporter genes and the induction of antibody responses in a variety of species including ferrets, rats, rabbits, guinea pigs, chickens, cows, fish, and nonhuman primates. Protection against viral challenge was demonstrated to occur in chickens and ferrets immunized with DNA encoding the HA of influenza A virus (DONNELLY et al. 1995a; FYNAN et al. 1993; ROBINSON et al. 1993), rabbits immunized with papillomavirus DNA vaccines (DONNELLY et al. 1996; SUNDARAM et al. 1997), and guinea pigs immunized with HSV DNA vaccines (BOURNE et al. 1996; MANICKAN et al. 1995; MCCLEMENTS et al. 1996). Furthermore, antibodies and protection were elicited in cows following immunization with DNA encoding a surface glycoprotein D of bovine herpesvirus-1 (COX et al. 1993), thereby demonstrating that DNA vaccines are efficacious in animals of much greater mass than the standard laboratory animals. Subsequently immunogenicity of DNA vaccines was also demonstrated in pigs, horses, and dogs (unpublished observations). Because such vaccines may have utility for veterinary as well as human applications, these observations have added significance in demonstrating the breadth of species in which the technology has proven to be effective.

II. Ferrets

In order to gain a sense of the immunogenicity of DNA vaccines compared to standard vaccines, such as inactivated whole virus vaccines and subunit vaccines, studies were performed in nonhuman primates and in ferrets to evaluate appropriate immunogenicity and, in the case of ferrets, to carry out challenge studies with relevant strains of virus. Because ferrets are used as the standard model for the assessment of antibody responses against influenza vaccines designed for human use, the antibody responses observed following immunization with DNA vaccines and protection (as measured by virus titer in nasal washings) following subsequent viral challenge could be compared with those obtained after immunization with commercially obtained, licensed inactivated virus vaccines. The DNA vaccines compared quite favorably in that, even following challenge with naturally drifted human strains of influenza that were different from the strain from which the vaccine was derived, the nasal viral washings were less (i.e., the protection was greater) than when the commercial vaccine was utilized (DONNELLY et al. 1995a, 1997a).

III. Nonhuman Primates

Further comparisons were performed between DNA vaccines and licensed influenza vaccines in primates (DONNELLY et al. 1995a; LIU et al. 1996) to assess their relative immunogenicity and to determine whether cocktails of different DNA plasmids could be given without resulting in interference with the immune response directed against a particular antigen. African green monkeys were immunized with a mixture of plasmids encoding various influenza antigens (HAs from the three strains present in the 1989–1993 or 1994–1995 licensed vaccines, plus NP and matrix from both an A and in some cases a B strain as well) for a final mixture of five to seven separate DNA constructs. Control monkeys were immunized with the full human dose of commercially obtained trivalent inactvated vaccine (that is, two A strains and one B strain of influenza virus). Following a priming set of two doses for each vaccine, as is recommended for humans suspected of having no prior infection with influenza virus, the hemagglutination-inhibiting (HI) antibody titer, which is the universally utilized surrogate measure of neutralizing antibody, was determined.

The HI antibody titers obtained with the DNA vaccines were found to be as high as or higher than those seen with the corresponding licensed vaccines. Unlike in mice, in which antibody titers even following a single injection of DNA vaccine or inactivated vaccine persisted for essentially the lifetime of the mouse, in the primates the titers resulting from the DNA vaccines, as well as both types of licensed vaccines (inactivated virus and partially purified inactivated virus), declined over a matter of weeks such that by 26 weeks the titers were near background levels. However, boosting after 1 year resulted in a large increase in antibody responses for animals receiving either commercial or DNA vaccine, but the responses in the DNA-immunized group were generally higher. Vaccination of African green monkeys with DNA vaccines encoding HSV gD and gB resulted in enzyme-linked immunosorbent assay and neutralizing antibodies with as little as 10 μg per construct given twice (LIU et al. 1997). These responses were long-lived (at least 24 weeks) and showed a strong boosting effect at 24 weeks.

DNA vaccines encoding the circumsporozoite (CSP) protein of *Plasmodium yoelli* (PyCSP) or *P. falciparum* (PfCSP) were used to immunize *Aotus* monkeys. This species has been the preclinical model for the study of vaccines for *P. falciparum*, the parasite responsible for malaria (COLLINS 1996). Interestingly, the PyCSP construct which had been as immunogenic in mice as the best peptide/adjuvant combination previously tested (SEDEGAH et al. 1994) did not result in development of antibodies after i.m. injection of the monkeys. In contrast, immunization by the intradermal route (i.d.) induced antibodies in a dose-dependent manner (GRAMZINSKI et al. 1997). Similarly, with the PfCSP DNA vaccine, which had generated antibodies in mice by either i.m. or i.d. injection, the monkeys responded only to the i.d. dose. These observations underscore the potential differences in responses which may be seen between species as noted above.

HIV DNA vaccines have been shown to induce both humoral and cellular immune responses in nonhuman primates. These include antibody responses against HIV or simian immunodeficiency virus (SIV) envelope in sera from rhesus or cynomologous monkeys immunized with DNA encoding the envelope protein from HIV or SIV (which were capable of neutralizing homologous virus in vitro) (WANG et al. 1995a; LU et al. 1996; SHIVER et al. 1996). With respect to cellular responses, both helper and cytotoxic T lymphocyte responses have been demonstrated with DNA constructs despite the challenge haplotype diversity presents. For example, CTL were generated following in vitro restimulation of peripheral blood lymphocytes from rhesus monkeys injected with HIV env DNA. These CTL were of the $CD8^+$ phenotype and levels increased following boosting and persisted for months (WANG et al. 1995a; LIU et al. 1996; LU et al. 1996; SHIVER et al. 1996). MHC class II restricted $CD4^+$ helper T cells were also generated following immunization of rhesus monkeys with env DNA. These $CD4^+$ T cells from peripheral blood lymphocytes secreted interferon-γ and tumor necrosis factor-α in response to antigen-specific restimulation in vitro (LEKUTIS et al. 1997). Lymphoproliferative and/or CTL responses were also observed in cynomologous monkeys inoculated with HIV env DNA (WANG et al. 1995a) and rhesus monkeys inoculated with various SIV env constructs (LU et al. 1996).

Utilizing a chimeric virus system, SHIV, wherein the envelope is composed of HIV envelope, but the internal portion of the virus is derived from SIV, permits an evaluation of DNA vaccine efficacy in primates. Immunization with HIV env DNA (multiple immunizations were carried out with various constructs as efforts were made to improve the expression levels of HIV env) including a final boost with a combination of env DNA and oligomeric env protein, resulted in the generation of both neutralizing antibodies and CTL in two experimental monkeys. When subsequently challenged with 24 $TCID_{50}$ of SHIV (>99% infection), the monkeys were protected from infection, as demonstrated by lack of detectable virus in serum samples for at least 1 year, lack of generation of antibodies against the SIV core, and lack of detectable SIV genes by PCR (LETVIN et al. 1997).

Studies in primates have now been extended to include chimpanzees, in which both newborn and adult animals were vaccinated with DNA encoding hepatitis B surface antigen (HBsAg) or HIV proteins. Utilizing a device called a Biojector, which delivers fluids i.m. as a jet of high-pressure liquid rather than by the use of a needle, to immunize with a construct encoding the HBsAg, two chimpanzees were shown to develop high titers of antibody specific for HBsAg following one or two immunizations (DAVIS et al. 1996). These titers compared favorably with those induced by vaccination with various other forms of HBsAg, including licensed recombinant protein vaccine. In a different study newborn chimpanzees were immunized on the day of birth and then boosted after 6 and 24 weeks by i.m. injection of a DNA construct encoding the HBsAg S and preS2 sequences (PRINCE et al. 1997). The chimpanzees developed antibody responses to the surface protein after vaccination and anamnestic responses after virus challenge. The animals did not produce detectable HBsAg or antibodies to the core protein and thus were shown to be protected from challenge with hepatitis B virus. Chimpanzees have also been used to test the immunogenicity and efficacy of HIV DNA vaccines. Animals immunized with DNA encoding various HIV proteins (gag/pol/env) were shown to have low levels of specific antibodies (some of which were neutralizing in vitro) and to be protected from subsequent challenge with a laboratory isolate of HIV-1 (SF-2) (BOYER et al. 1997).

D. Mucosal Immunization

Most pathogens enter the host via a mucosal surface. Therefore the ability to induce immune responses at mucosal surfaces is of intense interest to vaccinologists. The ability of DNA vaccines to provide mucosal immunity, by either parenteral inoculation or mucosal administration of DNA, is beginning to be assessed. Parenteral immunization with DNA vaccines has been shown to protect against influenza challenge in ferrets (DONNELLY et al. 1995a; WEBSTER et al. 1994) and chickens (FYNAN et al. 1993), as well as against infectious bovine rhinotracheitis caused by bovine herpesviruses (BABIUK et al. 1995; COX et al. 1993), all of which are mucosal challenge systems (although in the case of avian influenza systemic spread of the virus also occurs). Of particular note is that immunization of ferrets with DNA encoding the HA of A/PR/8/34 is able in some instances to completely prevent virus shedding after challenge with the homologous strain of influenza virus (DONNELLY et al. 1995b).

A unique feature of DNA vaccines against influenza is their ability in both ferrets (DONNELLY et al. 1995a) and chickens (KODIHALLI et al. 1997) to significantly reduce virus shedding upon challenge with antigenic drift variants that are related to the immunizing strain but not antigenically cross-reactive as determined by hemagglutination inhibition. However, at least in the case of influenza virus, the ability of parenteral immunization to provide protection from mucosal challenge has been well documented and indeed forms the basis of the currently licensed vaccine. A more rigorous test of mucosal immunity is provided by genital infections with herpes simplex virus and human immunodeficiency virus, and by enteric pathogens such as *Vibrio* and *Shigella* spp., against which parenteral immunization is generally only marginally effective.

A challenge model of herpesvirus in which the virus is applied vaginally in guinea pigs is a well-developed model for human genital herpes infection and provides the opportunity to assess mucosal lesions caused by the infection in addition to measuring other parameters such as hind limb paralysis and urinary retention, and determining the presence of latent virus in neural ganglia of the spinal cord. Immunization with DNA encoding HSV-2 envelope glycoprotein, generated levels of serum antibodies similar to those resulting from vaginal infection with HSV-2 (BOURNE et al. 1996). Guinea pigs immunized with gD and gB DNA, either separately or in combination, had significantly reduced genital mucosal lesions following vaginal challenge with human HSV-2 (BOURNE et al. 1996; MCCLEMENTS et al. 1996, 1997). Vaginal secretions obtained prior to viral challenge contained gD- and gB-specific antibodies of the IgG subclass and thus may have resulted from transudation of serum antibodies (MCCLEMENTS et al. 1997). While the protection seen in the guinea pig vaginal challenge model is presumed to be due to antibody, it is possible that cellular immune responses play a role; lymphocytes from HSV-2 gD or gB DNA-immunzed mice or guinea pigs are capable of antigen-specific proliferation, indicating some cellular responses. Furthermore, gB-specific CTL responses have been demonstrated in gB DNA-immunzed C57BL/6 mice (Armstrong, McClements, and Liu, unpublished observations). Rouse and colleagues have shown that a DNA vaccine encoding HSV-1 gB induces $CD4^+$ cytotoxic T lymphocytes which are protective in the murine zosteriform model of HSV-1 infection (MANICKAN et al. 1995). In the same infection model they also demonstrated CTL responses and protection with a construct encoding the immediate-early transactivator ICP 27.

Efforts have been made to directly immunize with DNA via mucosal routes, on the hypothesis that this route of vaccination results in the more effective generation of mucosal immunity. In addition, it is possible that oral delivery is more practical for the widespread usage of vaccines since they may be more readily administered than injectable vaccines and may have lower cost (although this would need to be demonstrated since particular formulations might increase the cost substantially compared to "naked" DNA in saline). DNA vaccines have been administerered mucosally by direct application of saline-formulated plasmids intranasally or intravaginally. Administration by these routes has resulted in generation of local antibodies against HIV antigens (ASAKURA et al. 1996; HINKULA et al. 1997; Y. WANG et al. 1997) and HSV antigens (KUKLIN et al. 1997), while i.n. inoculation with influenza HA DNA resulted in protection from challenge with virus (FYNAN et al. 1993). Addition of a mucosal adjuvant, such as cholera toxin (BAN et al. 1997; KUKLIN et al. 1997), incorporation of DNA plasmids into liposomes through the use of cationic lipids (KLAVINSKIS et al. 1997) or encapsulation in poly(DL-lactide coglycolide) microparticles (JONES et al. 1997) have all been successfully utilized as modalities to facilitate delivery of DNA vaccines mucosally. These prototype systems are yet to be tested against mucosal challenges with pathogens of the gastrointestinal and reproductive tracts.

E. Delivery Systems and Adjuvants

I. DNA Delivery

Another approach has been to deliver DNA encoding antigens via live vectors, in which, rather than the heterologous gene being incorporated into the genome of a virus or bacterium, the plasmid is in the bacterium, but then is released into a cell following infection of the cell. An example of this approach is to use an attenuated bacterium, such as an auxotrophic mutant of *Shigella* containing a mammalian DNA expression plasmid, to deliver that plasmid into cells in vivo (SIZEMORE et al. 1995). The bacterial construct was a strain which is unable to replicate in the absence of diaminopimelic acid (DAP) into which a plasmid encoding a reporter protein β-galactosidase was introduced. Such a construct is able to invade cells, but since DAP is a critical component of the cell wall, and because mammalian tissue does not contain DAP, such mutants lyse upon cell division, releasing the plasmid DNA into the cytoplasm of the infected cell (NAKAYAMA et al. 1988). Thus the constructs should be safe from the perspective of not causing disease due to *Shigella*. Following two administrations of this recombinant vector (given with or without concomitant administration of DAP), antibody responses against the antigen were seen in the sera (although mucosal antibody was not measured), as was antigen-specific proliferation of splenocytes (SIZEMORE et al. 1995).

II. DNA as Adjuvant

A characteristic of DNA vaccines that is interesting from a pharmacological aspect is that the DNA plasmid itself may not simply be an inert vector delivering the polynucleotide sequence encoding the antigen of interest. Viral and bacterial vectors have an inherent potential to induce immune responses against themselves (often directed against the surface proteins), thus raising a concern regarding the ability to reuse a particular vector. Conversely, expression of certain viral proteins which have specific immune activities including the induction of cytokines, or the alteration in expression of certain proteins (e.g., the down-regulation of the expression of MHC class I molecules), can limit the induction of immune responses against the coexpressed antigen of interest. The plasmid backbone of the DNA vaccines is thought to be less inherently immunogenic; however, evidence supports an immune effect of the DNA itself on the response to the encoded antigen. While it has been known for some years that bacterial DNA is mitogenic for lymphocytes, and that these stimulatory effects may be related to the sequence and hence methylation pattern of bacterial DNA (YAMAMOTO et al. 1992; MESSINA et al. 1993; KRIEG et al. 1995), it is only recently that evidence for a role for these effects on the immunogenicity of DNA vaccines has emerged.

In general, DNA vaccines administered i.m. result in the induction of Th1-type responses as indicated by antigen-specific stimulated secretion of IL-2 and interferon-γ by splenocytes from animals immunized with various DNA vaccines (for review see DONNELLY et al. 1997b). Indirect indication of this is seen by the production of predominantly IgG2a subclass of antibodies following immunization with various DNA constructs. In contrast, repeated immunization with DNA coated onto gold beads into the skin by a biolistics device gradually raises a Th2-type of helper response (FELTQUATE et al. 1997). The bias in the helper response to Th1-type may be due to a number of factors related to antigen expression rather than to any direct effects of the plasmid itself. Yet, it is interesting that coadministration of a plasmid not encoding any protein with a recombinant protein alters the isotype profile of the ensuing antibody response (DONNELLY et al. 1997b). This would indicate that even noncoding plasmid when coinjected with a protein antigen has an effect on the immune response, thus implicating the DNA itself as having an immunological effect. Furthermore, coadministration of noncoding plasmid with a DNA vaccine encoding influenza HA was shown to result in a significant increase in antibody titers in African green monkeys (DONNELLY et al. 1997b). It should be noted that for certain challenge protection studies, control animals were immunized with noncoding plasmid to demonstrate that protection by DNA vaccines does indeed require that the plasmid encode a relevant protein antigen rather than the protection simply being due to nonspecific immune responses (ULMER et al. 1993).

The mechanism for this immunoenhancing effect of DNA may be specific sequences in the plasmid and the consequent methylation pattern. The particular motif (purine-purine-C-G-pyrimidine-pyrimidine) is stimulatory for NK cells and lymphocytes in vitro, inducing them to secrete cytokines (KLINMAN et al. 1996), and CpG oligonucleotides coinjected with antigens can have pronounced effects in vivo (Caulfield et al., unpublished observations). Whether such sequences have similar effects in the context of a double-stranded plasmid remains to be fully elucidated. A specific motif (AACGTT), when inserted in a DNA vaccine encoding β-galactosidase, has been shown to enhance immune responses against the encoded protein (SATO et al. 1996). However, similar vectors encoding other antigens have not demonstrated enhancing qualities (unpublished observations). Therefore potential mitigating factors, such as position and context of CpG motifs, inherent antigenicity of encoded protein, and presence of interfering or competing nucleotide sequences, may play a role in whether CpG motifs have a measurable effect on DNA vaccine potency.

F. Summary

In summary, the general applicability of DNA vaccines to induce protective immune responses has been well established in various animal models, and this has led to early phase I human clinical trials. Current preclinical studies have included investigation of alternative ways to administer DNA vaccines in order to induce specific immune responses (e.g., mucosal) and to modulate or enhance immune responses (e.g., using CpG motifs or coadministration of adjuvants or cytokines). In addition, DNA vaccines have proven to be a useful tool for exploring mechanisms of antigen processing and presentation. Because constructs can readily be made with different targeting motifs incorporated into the protein (e.g., signal sequences for secretion, proteasome targeting for MHC class I presentation, lysosome targeting for MHC class II presentation) or the protein can be altered by the deletion of certain regions such as transmembrane domains, the effects of such changes upon antibody or cellular responses can be assessed in a fairly systematic fashion for a given antigen. Thus the technology holds promise both as a mode of vaccination and as a means of understanding basic issues in immunology.

References

Asakura Y, Hamajima K, Fukushima J, Mohri H, Okubo T, Okuda K (1996) Induction of HIV-1 Nef-specific cytotoxic T lymphocytes by Nef-expressing DNA vaccine. Am J Hematol 53:116–117

Babiuk LA, Lewis PJ, Cox G, van Drunen Littel-van den Hurk S, Baca-Estrada M, Tikoo SK (1995) DNA immunization with bovine herpesvirus-1 genes. Ann NY Acad Sci 772:47–63

Ban EM, van Ginkel FW, Simecka JW, Kiyono H, Robinson MHL, McGhee JR (1997) Mucosal immunization with DNA encoding infleunza hemagglutinin. Vaccine 15:811–813

Bourne N, Stanberry LR, Bernstein DI, Lew D (1996) DNA immunization against experimental genital herpes simplex virus infection. J Infect Dis 173:800–807

Boyer J, Ugen K, Wang B, Agadjanyan M, Gilbert L, Bagarazzi M, Chattergoon M, Frost P, Javadian A, Williams W, Refaeli Y, Ciccarelli R, McCallus D, Coney L, Weiner D (1997) Protection of chimpanzees from high-dose heterologous HIV-1 challenge by DNA vaccination. Nat Med 3:526–532

Collins WE, Galland GG, Sullivan JS, Morris CL, Richardson BB, Roberts JM (1996) The Santa Lucia strain of Plasmodium falciparum as a model for vaccine studies. I. Development in Aotus lemurinus griseimembra monkeys. Am J Trop Med Hyg 54:372–379

Corr M, Lee DJ, Carson DA, Tighe H (1996) Gene vaccination with naked plasmid DNA: mechanism of CTL priming. J Exp Med 184:1555–1560

Cox G, Zamb TJ, Babiuk LA (1993) Bovine herpesvirus-1 – immune-responses in mice and cattle injected with plasmid DNA. J Virol 67:5664–5667

Davis HL, McCluskie MJ, Gerin JL, Purcell RH (1996) DNA vaccine for hepatitis B: evidence for immunogenicity in chimpanzees and comparison with other vaccines. Proc Natl Acad Sci USA 93:7213–7218

Deck RR, DeWitt CM, Donnelly JJ, Liu MA, Ulmer JB (1997) Characterization of humoral immune responses induced by an influenza hemagglutinin DNA vaccine. Vaccine 15:71–78

Doe B, Selby M, Baenziger J, Walker CM (1996) Induction of cytotoxic T lymphocytes by intramuscular immunization with plasmid DNA is facilitated by bome marrow-derived cells. Proc Natl Acad Sci USA 93:8578–8583
Donnelly JJ, Ulmer JB, Liu MA (1994) Immunization with DNA. J Immunol Methods 176:145–152
Donnelly JJ, Friedman A, Martinez D, Montgomery DL, Shiver JW, Motzel SL, Ulmer JB, Liu MA (1995a) Preclinical efficacy of a prototype DNA vaccine – enhanced protection against antigenic drift in influenza-virus. Nature Med 1:583–587
Donnelly JJ, Ulmer JB, Liu MA (1995b) Protective efficacy of intramuscular immunization with naked DNA. Ann NY Acad Sci 772:40–46
Donnelly JJ, Martinez D, Jansen KU, Ellis RW, Montgomery DL, Liu MA (1996) Protection against papillomavirus with a polynucleotide vaccine. J Infect Dis 173:314–320
Donnelly JJ, Friedman A, Ulmer JB, Liu MA (1997a) Further protection against antigenic drift of influenza virus in a ferret model. Vaccine 15:865–873
Donnelly JJ, Ulmer JB, Shiver JW, Liu MA (1997b) DNA vaccines. Annu Rev Immunol 15:617–648
Feltquate DM, Heaney S, Webster RG, Robinson HL (1997) Different T helper cell types and antibody isotypes generated by saline andgene gun DNA immunization. J Immunol 158:2278–2284
Fu TM, Ulmer JB, Caulfield MJ, Deck RR, Friedman A, Wang S, Liu X, Donnelly JJ, Liu MA (1997) Transfer of antigen for priming CTL responses: requirement for bone marrow-derived antigen presenting cells for infection and DNA vaccines. Mol Med 3:362–371
Fynan EF, Webster RG, Fuller DH, Haynes JR, Santoro JC, Robinson HL (1993) DNA vaccines – protective immunizations by parenteral, mucosal, and gene-gun inoculations. Proc Natl Acad Sci USA 90:11478–11482
Gramzinski RA, Maris DC, Doolan D, Charoenvit Y, Obaldia N, Rossan R, Sedegah M, Wang R, Hobart P, Margalith M, Hoffman S (1997) Malaria DNA vaccination in Aotus monkeys. Vaccine 15:913–915
Harpin S, Talbot B, Mbikay M, Elazhary Y (1997) Immune response to vaccination with DNA encoding the bovine viral diarrhea virus major glycoprotein gp53 (E2). FEMS Microbiol Lett 146:229–234
Hinkula J, Lundholm P, Wahren B (1997) Nucleic acid vaccination with HIV regulatory genes: a combination of HIV-1 genes in separate plasmids induces strong immune responses. Vaccine 15:874–878
Iwasaki A, Stiernholm BJ, Chan AK, Berinstein NL, Barber BH (1997) Enhanced cytotoxic T-lymphocyte responses mediated by plasmid DNA immunogens encoding co-stimulatory molecules and cytokines. J Immunol 158:4591–4601
Jones DH, Corris S, McDonald S, Clegg JCS, Farrar GH (1997) Poly (DL-lactide-co-glycolide)-encapsulated plasmid DNA elicits systemic and mucosal antibody responses to encoded protein after oral administration. Vaccine 18:814–817
Justewicz DM, Webster RG (1996) Long-term maintenance of B cell immunity to influenza virus hemagglutinin in mice following DNA-based immunization. Virology 224:10–17
Klavinskis LS, Gao L, Barnfield C, Lehner T, Parker S (1997) Mucosal immunization with DNA-liposome complexes. Vaccine 15:818–820
Klinman DM, Yi AK, Beaucage SL, Conover J, Krieg AM (1996) CpG motifs present in bacteria DNA rapidly induce lymphocytes to secrete interleukin 6, interleukin 12, and interferon gamma. Proc Natl Acad Sci USA 93:2879–2883
Kodihalli S, Haynes J, Robinson H, Webster R (1997) Cross-protection among lethal H5N2 l-Influenza viruses induced by DNA vaccine to the hemagglutinin. J Virol 71:4888–4897
Krieg AM, Yi AK, Matson S, Waldschmidt TJ, Bishop GA, Teasdale R, Koretzky GA, Klinman DM (1995) CpG motifs in bacterial DNA trigger direct B-cell activation. Nature 374:546–549

Kuklin N, Daheshia M, Karem K, Manickan E, Rouse B (1997) Induction of mucosal immunity against herpes simplex virus by plasmid DNA immunization. J Virol 71:3138–3145
Lekutis C, Shiver JW, Liu MA, Letvin NL (1997) HIV-1 env DNA vaccine administered to rhesus monkeys elicits MHC class II-restricted CD4+ T helper cells that secrete IFN-gamma and TNF-alpha. J Immunol 158:4471–4477
Letvin NL, Montefiori DC, Yasutomi Y, Perry HC, Davies M-E, Lekutis C, Alroy M, Freed DC, Lord CI, Handt LK, Liu MA, Shiver JW (1997) Potent, protective anti-HIV immune responses generated by bimodal HIV envelope DNA plus protein vaccination. Proc Natl Acad Sci USA 94:9378–9383
Liu MA, Yasutomi Y, Davies ME, Perry HC, Freed DC, Letvin NL, Shiver JW (1996) Vaccination of mice and nonhuman primates using HIV-gene-containing DNA. Antibiot Chemother 48:100–104
Liu MA, McClements W, Ulmer JB, Donnelly JJ (1997) Immunization of non-human primates with DNA vaccines. Vaccine 15:909–912
Lu S, Arthos J, Montefiori DC, Yasutomi Y, Manson K, Mustafa F, Johnson E, Santoro JC, Wissink J, Mullins JI, Haynes JR, Letvin NL, Wyand M, Robinson HL (1996) Simian immunodeficiency virus DNA vaccine trial in macaques. J Virol 70:3978–3991
Manickan E, Rouse R, Yu ZY, Wire WS, Rouse BT (1995) Genetic immunization against herpes-simplex-virus protection is mediated by CD4(+) T-lymphocytes. J Immunol 155:259–265
McClements WL, Armstrong ME, Keys RD, Liu MA (1996) Immunization with DNA vaccines encoding glycoprotein D or glycoprotein B, alone or in combination, induces protective immunity in animal models of herpes simplex virus-2 disease. Proc Natl Acad Sci USA 93:11414–11420
McClements WL, Armstrong ME, Keys RD, Liu MA (1997) The prophylactic effect of immunization with DNA encoding herpes simplex virus glycoproteins on HSV-induced disease in guinea pigs. Vaccine 15:857–860
Messina JP, Gilkeson GS, Pisetsky DS (1993) The influence of DNA structure on the in vitro stimulation of murine lymphocytes by natural and synthetic polynucleotide antigens. Cell Immunol 147:148–157
Montgomery DL, Shiver JW, Leander KR, Perry HC, Friedman A, Martinez D, Ulmer JB, Donnelly JJ, Liu MA (1993) Heterologous and homologous protection against influenza A by DNA vaccination: optimization of DNA vectors. DNA Cell Biol 12:777–783
Nakayama K, Kelly SM, Curtiss RI (1988) Construction of and ASD^+ expression-cloning vector: stable maintenance and high level expression of cloned genes in a Salmonella vaccine strain. Biotechnology 6:693–697
Prince A, Whalen RG, Brotman B (1997) Successful nucleic acid based immunization of newborn chimpanzees against HBV. Vaccine 15:916–919
Robinson HL, Hunt LA, Webster RG (1993) Protection against a lethal influenza virus challenge by immunization with a haemagglutinin-expressing plasmid DNA. Vaccine 11:957–960
Sato Y, Roman M, Tighe H, Lee D, Corr M, Nguyen MD, Silverman GJ, Lotz M, Carson DA, Raz E (1996) Immunostimulatory DNA sequences necessary for effective intradermal gene immunization. Science 273:352–354
Sedegah M, Hedstrom R, Hobart P, Hoffman SL (1994) Protection against malaria by immunization with plasmid DNA encoding circumsporozoite protein. Proc Natl Acad Sci USA 91:9866–9870
Shiver J, Ulmer J, Donnelly J, Liu M (1996) Humoral and cellular immunities elicited by DNA vaccines: application to the human immunodeficiency virus and influenza. Adv Drug Deliv Rev 21:19–31
Sizemore DR, Branstrom AA, Sadoff JC (1995) Attenuated Shigella as a DNA delivery vehicle for DNA-mediated immunization. Science 270:299–302

Sundaram P, Tigelaar RE, Brandsma JL (1997) Intracutaneous vaccination of rabbits with the cottontail rabbit papillomavirus (CRPV) L1 gene protects against virus challenge. Vaccine 15:664–671

Tang DC, De Vit M, Johnston SA (1992) Genetic immunization is a simple method for eliciting an immune response. Nature 356:152–154

Ulmer JB, Donnelly JJ, Parker SE, Rhodes GH, Felgner PL, Dwarki VJ, Gromkowski SH, Deck RR, De Witt CM, Friedman A, Hawe L, Leander K, Martinez D, Perry HC, Shiver JW, Montgomery DL, Liu MA (1993) Heterologous protection against influenza by injection of DNA encoding a viral protein. Science 259:1745–1749

Ulmer JB, Deck RR, DeWitt CM, Friedman A, Donnelly JJ, Liu MA (1994) Protective immunity by intramuscular injection of low-doses of influenza-virus DNA vaccines. Vaccine 12:1541–1544

Ulmer J, Deck RR, DeWitt CM, Donnelly JJ, Liu M (1996) Generation of MHC class I-restricted cytotoxic T lymphocytes by expression of a viral protein in muscle cells: antigen presentation by non-muscle cells. Immunology 89:59–67

Ulmer JB, Fu T-M, Deck RR, Friedman A, Guan L, DeWitt CM, Liu X, Wang S, Liu MA, Donnelly JJ (1998) Protective $CD4^+$ and $CD8^+$ T cells against influenza induced by vaccination with nucleoprotein DNA. J Virol 72:5648–5653

Wang B, Boyer J, Srikantan V, Ugen K, Gilbert L, Phan C, Dang K, Merva M, Agadjanyan MG, Newman M, Carrano R, Mccallus D, Coney L, Williams WV, Weiner DB (1995a) Induction of humoral and cellular immune-responses to the human immunodeficiency type-1 virus in nonhuman-primates by in-vivo dna inoculation. Virology 211:102–112

Wang B, Boyer JD, Ugen KE, Srikantan V, Ayyaroo V, Agadjanyan MG, Williams WV, Newman M, Coney L, Carrano R et al (1995b) Nucleic acid-based immunization against HIV-I: induction of protective in vivo immune responses. AIDS 9:S159–S170

Wang B, Dang K, Agadjanyan MG, Srikantan V, Li F, Ugen KE, Boyer J, Merva M, Williams WV, Weiner DB (1997) Mucosal immunization with a DNA vaccine induces immune responses against HIV-1 at a mucosal site. Vaccine 15:821–816

Webster RG, Fynan EF, Santoro JC, Robinson H (1994) Protection of ferrets against influenza challenge with a DNA vaccine to the hemagglutinin. Vaccine 12:1495–1498

Wolff JA, Malone RW, Wiliams P, Chong W, Acsadi G, Jani A, Felgner PL (1990) Direct gene transfer into mouse muscle in vivo. Science 247:1465–1468

Yamamoto S, Yamamoto T, Kataoka T, Kuramoto E, Yano O, Tokunaga T (1992) Unique palindromic sequences in synthetic oligonucleotides are required to induce TNF and augment TNF-mediated natural killer activity. J Immunol 148: 4072–4076

CHAPTER 4

Antigen-Presentation Systems, Immunomodulators, and Immune Responses to Vaccines

C.-A. Siegrist and P.-H. Lambert

A. From Empirical Approaches to Rationally Selected Antigen-Presentation Systems and Immunomodulators for Vaccine-Mediated Protection Against Infectious Diseases

Historically vaccines have been developed and widely introduced into clinical use on the empirical assumption that postexposure immunity can be induced in healthy subjects by exposure to live micro-organisms with a reduced virulence profile obtained either by use of the corresponding pathogen of another animal species (e.g., smallpox, rotavirus vaccines) or by in vitro attenuation [e.g., poliomyelitis, measles, mumps, rubella or bacille Calmette-Guérin (BCG) vaccines]. These empirical strategies have proved useful for protection against a number of infectious diseases but are not applicable in a number of circumstances: (a) when attenuated micro-organisms with appropriate immunogenicity/safety profile cannot be generated (e.g., pertussis, hepatitis B), (b) when exposure to an entire micro-organism does not result in anti-toxic protective immunity (e.g., tetanus), (c) when antigenic variation precludes the use of a specific strain as a vaccine source (e.g., pneumococci), (d) when concern exists related to the use of particular live micro-organisms in populations with variable levels of immunocompetence.

Apart from a few whole cell vaccines which contain killed viruses (e.g., influenza) or bacteria (e.g., cholera, pertussis), subunit vaccines containing one or several antigens identified as capable of eliciting protective immune responses represent an obvious alternative to the use of live attenuated micro-organisms. Subunit vaccines based on the use of purified proteins or carbohydrates or of antigens generated by recombinant and biosynthetic technologies offer several advantages, such as increased purity and safety. However, their immunogenicity is frequently lower than that of antigens presented by the corresponding live organisms. This is likely to reflect the absence of additional signals provided by live replicating organisms, which may nonspecifically activate "innate" immune mechanisms at the onset of specific immune responses. Thus protein vaccines (e.g., tetanus/diphtheria toxoids) have been traditionally used in combination with additional substances, called "adjuvants" (from the Latin word *adjuvare* = to help), expected to increase their immunogenicity.

A better understanding of how the immune system takes up, processes, presents, and recognizes antigens and the progressive identification of factors which specifically modulate the induction of the respective arms of the immune system is now leading to the identification of hundreds of molecules with a potential adjuvant effect. At this time, vaccine design has changed from an empirical to a rational approach, and the selection of adjuvant is now becoming an integral part of vaccine development.

The availability of a variety of antigen-presentation systems and the possibility to include into vaccine formulations immunomodulatory substances capable of specifically enhancing selected immune responses are opening new perspectives in vaccine design. Novel approaches should be of particular importance (a) to induce or enhance vaccine responses to weak immunogens, (b) to specifically enhance a given arm of the immune response to some immunogens [e.g., T helper (TH)-1 or TH-2 cells, or cytotoxic T lymphocytes (CTL)], (c) to target immune responses to specific compartments of the immune system (e.g., mucosal), and (d) to enhance immune responses in individuals at an early stage of immunological maturation or with variable levels of immunocompetence.

Thus the whole concept of vaccine adjuvanticity should now be broadened and include a variety of factors (Table 1) which may ensure and enhance the protective capacity of newly designed vaccines. One should distinguish between "antigen-presentation systems," for example, the delivery systems

Table 1. Classification of "vaccine adjuvants" into antigen-presentation systems and immunomodulatory substances

A. Antigen-presentation systems
  1. Live vectors
     Bacterial vectors
     Viral vectors
  2. DNA vaccines
  3. Particulate substances
     Aluminium/calcium salts
     Water-in-oil emulsions
     Oil-in-water emulsions
     Liposomes and virosomes
     Proteosomes
     Biodegradable polymers (microspheres)

B. Immunomodulators
  1. Lipid A derivatives
  2. Saponins derivatives
  3. Non ionic block copolymers
  4. Muramyl dipeptide (MDP) derivatives
  5. Cytokines and interferons

C. Combination of Antigen Presentation Systems and Immunomodulators
  1. ISCOMS
  2. W/O emulsions with built-in immunomodulators
  3. O/W emulsions with built-in immunomodulators
  4. Liposomes with built-in immunomodulators

into which the antigen is incorporated, and which determines its uptake, processing, and presentation by antigen-presenting cells (APC) to B and T lymphocytes, and "immunomodulators," directly activating APC or lymphocytes to specifically enhance or modulate immune responses. Selected antigen-presentation systems and immunomodulators can and have been combined to simultaneously act at both levels of antigen presentation and lymphocyte recognition/differentiation.

The aim of this review is to analyze the way in which antigen-presentation systems and immunomodulators can affect vaccine responses and be used to selectively generate appropriate effector mechanisms. For space limitation this review is essentially limited to parenteral vaccines and does not cover vaccine antigen-presentation systems/immunostimulants suitable for mucosal immunization.

B. What Are the Essential Parameters of Vaccine-Induced Effector Mechanisms?

The protective mechanisms induced by the most extensively used vaccines are yet to be fully deciphered. However, several important immunological parameters of vaccine efficacy have nevertheless been identified over the past 40 years.

I. Antibody-Mediated Vaccine Responses

Antibody-mediated protection plays a crucial role in defense against extracellular bacteria by the induction (a) of antitoxin antibodies capable of blocking the active site(s) of a toxin (diphtheria, tetanus) prior to binding to its receptor, (b) of neutralizing antibodies which prevent adhesion of the microorganisms to its cellular/tissular targets (hepatitis A and B, poliomyelitis), or (c) of antibodies accelerating bacterial clearance by bacterial opsonization, complement fixation and/or antibody-mediated cellular cytotoxicity (*H. influenzae*, *S. Pneumoniae*, *N. Meningitidis*).

The protective effects of vaccine-triggered antibodies are correlated with several parameters (Table 2). Vaccine efficacy depends on quantitative parameters such as the magnitude of antibody responses. The existence of an antibody threshold associated with protection has been clearly defined for prevention against diphtheria, tetanus, and hepatitis B. Determination of antibody levels is widely used to assess vaccine immunogenicity, during the initial clinical phases of vaccine development, and later under routine conditions to identify subjects qualified as poor responders who should benefit from additional preventive measures (e.g., health care workers immunized with hepatitis B vaccine). Due to the relative technical ease with which vaccine antibodies can be measured, determination of antibody levels is often the sole parameter of immunogenicity measured, even in paradoxical situations in

Table 2. Parameters of vaccine – induced effector mechanisms which may be modulated by selection of antigen-presentation systems/immunomodulators

Effector mechanisms	Parameters	Consequences
Antibodies	Magnitude (antibody titers)	Protective threshold
	Kinetics of induction	Delay until protection
	Persistence (+ memory B cells?)	Duration of protection
	Functional activity	Correlates of protection
	Specificity/avidity	Correlates of protection (?)
	Isotypes	Half-life
		Complement fixation
	Localization	Systemic/mucosae
CD4 T cells	Magnitude	Cytokine production
	Kinetics of induction (?)	Delay until protection
	Induction of memory cells	Duration of protection
	TH1 cytokine production	Clearance of intracellular organisms
	TH2 cytokine production	B cell help
CD8 T cells	Magnitude (?)	Clearance of infected cells
	Kinetics of induction (?)	Delay until protection
	Induction of memory cells	Duration of protection

which they are known not to be correlated at all with protection (pertussis). The kinetics of antibody induction affects the delay required between immunization and vaccine-mediated protection, a factor which may be critical to protect infants and travelers and in epidemic situations. Antibody persistence over time in absence of reexposure to antigen is often critical to ensure a long-lasting protective effect (e.g., diphtheria, tetanus) and to determine the frequency at which booster doses should be administered.

Although less often assessed, qualitative parameters of antibody responses are essential parameters of antibody-mediated protection against disease. Antigen specificity is obviously critical for vaccine-induced protection. When available, functional assays (opsonization, neutralization, passive protection in animal models) represent the most valuable immunological indicators of protection. Variable degrees of avidity of antibodies for their antigenic target have been described following the use of various *Haemophilus influenzae* b conjugate vaccines, possibly leading to different levels of bactericidal activity. The distribution of immunoglobulin (Ig) isotypes determines their half-life, which may range from 7 days (IgG3) to 28 days (IgG1) and their complement binding capacity. Human antibodies belonging to the IgG1, IgG2, IgG3, and IgM isotypes are capable of initiating the classic complement pathway (IgG3 > IgG1 > IgG2 > IgM), whereas IgG4 and IgE are inactive in this regards. Lastly, the localization of antibody production, either in the bone marrow, in lymph nodes, or at a given mucosal surface may clearly differentially affect antibody-mediated protection against diseases for which microorganisms should be counteracted at their mucosal site of entrance.

II. Vaccine-Specific T Cell Responses

In addition to the requirement of T cell help for most antibody responses, antigen-specific effector T cells are required for full protection against most viral pathogens (measles, mumps, rubella, varicella), as well as bacterial and parasitic agents capable of surviving within the cellular environment (*M. tuberculosis*, *Leishmania*).

1. CD4 T Cell Vaccine Responses

The effect of vaccine-triggered CD4 T cell responses can be analyzed (Table 2) quantitatively, principally by determination of the intensity of antigen specific proliferation of peripheral blood mononuclear cells, kinetics of induction and persistence over time (frequency of memory T cells) in the absence of in vivo reexposure to antigen. The relative importance of these factors for vaccine-mediated protection has been studied only little in man, mostly for logistical reasons. It can be inferred, however, that vaccine efficacy, delay prior to protection induction, and duration of protection often reflect variations in these parameters. CD4 T cell responses can also be analyzed qualitatively in terms of their cytokine production following in vitro restimulation with specific antigen. These analyses have led to the identification of interleukin (IL) 2, interferon (IFN) γ), and tumor necrosis factor (TNF) α producing T helper (TH)-1 cells and of IL-4, IL-5, IL-6, and IL-10 producing TH2 cells (Cherwinski et al. 1987; Mosmann and Moore 1991; Romagnani 1991). The respective roles of TH1 cells, also assessed by the induction of delayed-type hypersensitivity reactions, in the clearance of intracellular organisms, and of TH2 cells in providing B cell help for effective induction of circulating and secretory antibodies have been well characterized. Some of the factors which affect the preferential differentiation of CD4 responses along either one of these two pathways have been identified and have opened new possibilities for a more rational design of vaccine formulations.

2. Characteristics of Vaccine-Specific CD8 Cytotoxic Responses

Few parameters assessing the quality of vaccine-specific CD8 CTL responses have been identified (Table 2). These responses are rarely measured in man and are usually described in terms of the presence or absence of CTLs, given the technical difficulties associated with the precise determination of the frequency of CTL precursors in clinical studies. The kinetics of CTL induction and the factors controlling the induction of CD8 memory responses are largely unknown, although they can be postulated as important for vaccine-mediated protection. Since CTL induction requires major histocompatibility (MHC) class I restricted antigen presentation and help from CD4 TH1 cells, antigen-presentation systems and vaccine formulations aiming at the induction of CTLs are often evaluated for their capacity to introduce antigen into the cytosolic presentation pathway and to induce TH1 differentiation of vaccine CD4 responses.

Thus the effect of various antigen-presentation systems and immunomodulators on vaccine-driven antibody responses can be analyzed quantitatively in terms of their respective effects on antibody titers, kinetics, and persistence, and qualitatively in terms of functional efficacy, specificity, avidity, isotypes, and localization. The effect of vaccine formulations on vaccine-driven T cell responses is best analyzed in terms of modulation of the TH1/TH2 differentiation of vaccine CD4 responses, reflected by their antigen-specific cytokine production, and of induction of CD8 cytotoxic T cells.

C. What Fundamental Immunological Mechanisms Can Be Modulated by Antigen-Delivery Systems and Immunomodulators?

The current understanding of the mode of action by which "adjuvants" mediate their effects on vaccine responses is yet limited. However, in spite of the technical difficulties of in vivo studies and of the potential limitations of the relevance of in vitro analyses, there is a growing interest for deciphering the mechanisms by which adjuvants work. It is obvious that progress in the identification of fundamental mechanisms of antigen processing, presentation, and recognition by B and T lymphocytes is rapidly improving our understanding of adjuvanticity (Table 3).

Antigen-delivery systems and immunomodulators can be expected (a) to affect the presentation of vaccine antigens to B and T lymphocytes and (b) to affect lymphocyte responses by modification of the microenvironment in which they take place.

I. Modulation of Antigen Presentation to T/B Cells

1. Antigen Conformation

The capacity of an antigen-delivery system to preserve the conformational integrity of a specific antigen is an essential parameter. Since antigenic sites recognized by neutralizing antibodies are often conformational rather than linear structures, the integrity of conformational epitopes for B cell recognition is an important requirement of vaccine-mediated protection. This integrity can be easily lost during in vitro vaccine preparation depending upon the stability of the native antigen and the nature of its physical interactions with the antigen-delivery system. The biochemical features of a protein, including its hydrophilic/amphipathic/lipophilic nature, directly affect the type of antigen-delivery systems which can be used. For example, the adsorption of basic proteins to aluminum hydroxide is low, and it is easier to incorporate

Table 3. Mode of action of antigen-presentation systems/vaccine formulations

Basic mechanisms	Parameters	Influenced by
Antigen presentation	Conformational integrity	Antigen presentation systems (all)
Antigen persistence	Short-term depot	Mineral compounds
		Oil emulsions
		Liposomes
		Microspheres
		Block copolymers
	Long-term depot	Large (>10 μm) microspheres
	In vivo replication	Live vectors
	Prolonged expression	DNA vaccines
Antigen targeting of APC	Receptor binding	CT, LT, others
	Antigen uptake/ processing	Particulate substances
		Saponin-containing systems
APC activation	MHC expression	Block copolymers
	Cytokine production	Lipid A derivatives
		Saponins (?)
		MDP derivatives
		ISCOMS
	Cytosolic processing	Live vectors
		DNA vaccines
		Liposomes

hydrophilic antigens into water-in-oil rather than into oil-in-water emulsions. As a direct consequence, a universal antigen-presentation system suitable for use with any given immunogen is unlikely to be developed.

The availability of antigen in native conformation is also an advantage for its long-term depot as antigen-antibody complexes on follicular dendritic cells, which appears essential for both affinity maturation and persistence of antibody responses. Thus the capacity of an antigen-delivery system to optimally present a protein into its native conformation directly affects the quality (affinity, biological function) and possibly the duration of antibody responses.

2. Antigen Persistence

The persistence of antigen exposure can be enhanced by mineral compounds, oil-based formulations, and liposomes which cause a short-term depot of antigen at the site of injection. Large (>10 mm) biodegradable polymer microspheres can extend this depot period, probably up to several weeks or months and even control the release of antigen over time so as to mimic the administration of a second vaccine dose. Depot formation extends antigen availability for uptake by APC, typically dendritic and Langerhans cells, into which antigen is taken up by either fluid-phase pinocytosis or receptor-mediated endocytosis.

The duration of antigen presentation to the immune system can also be increased by the use of live vectors with prolonged replication patterns such as vaccinia or mycobacteria. Prolonged in vivo antigen expression is also considered as an essential feature of the immunogenicity of DNA vaccines.

3. Targeting of Professional Antigen-Presenting Cells

In contrast to B lymphocytes, T lymphocyte recognition of antigen is based on binding of small peptides issued from the processing of native proteins within the endosomal or cytosolic cellular compartments, followed by MHC class II or class I presentation to CD4 or CD8 T cells, respectively. Antigen uptake and processing within APC and presentation in the appropriate molecular context is thus a requirement for adequate T cell recognition.

The uptake of antigen can be enhanced by the formation of aggregates which is the main mechanism by which the particulate antigen-presentation system enhances immune responses. Targeting of dendritic cells can also be increased by the presence of sugar moieties or other molecules which can be bound by receptors at the surface of APC. Binding of GM-1 ganglioside by cholera toxin (CT) and *E. coli* labile toxin (LT) (RIBI et al. 1988) or of the DEC-205 receptor at the surface of dendritic cells by carbohydrate-glycoprotein ligands were shown to lead to enhanced endocytosis and presentation to T cells (JIANG et al. 1995). This DEC-205 mediated endocytosis could represent one of the mechanisms at the basis of the efficacy of Saponin-containing delivery systems such as ISCOMS.

II. Modulation of CD4 T Cell Responses

Activation of APC by various external signals such as granulocyte-macrophage colony-stimulating factor (GM-CSF) results in production of inflammatory cytokines such as IFN-γ, TNF-α, and IL-1β which in turn attract CD4 T cells to the site of antigen presentation. Immunomodulatory substances can globally enhance APC activation, resulting in an increase in MHC class II expression or in costimulatory signals required for optimal T cell activation. Increase in MHC class II expression on APC has been reported upon use of substances capable of APC activation. In contrast, little is yet known about specific enhancement of costimulatory molecules such as CD40, B7-1, and B7-2 or CD28 by immunomodulators.

Immunomodulatory substances can also be used to affect the differentiation of CD4 T cells preferentially along the TH1 or the TH2 pathway. As an example, bacterial endotoxins and their derivatives such as monophosphoryl lipid A induce preferential TH1 responses (JOHNSON and TOMAI 1990) whereas aluminum salts have a strong TH2 driving effect (MANCINO and OVARY 1980). It should be remembered that the differentiation into either TH1 or TH2 vaccine CD4 T cells directly affects both antibody switching and the resulting isotype of vaccine-induced antibodies and CTL induction.

III. Induction of CD8 T Cell Responses

Induction of CD8 T cell responses usually requires introduction of antigen into the cytosolic processing pathway for MHC class I presentation. This is best achieved by in vivo antigen expression following use of live viruses or DNA vaccines. It can also occur following interaction of the antigen-delivery system with cell membranes, either by fusion to the external cell membrane or by escape from the endosomal into the cytosolic compartment. Alternatively, short peptides can bind directly to MHC molecules at the surface of APC if provided externally, for example, by a water-in-oil emulsion (SCALZO et al. 1995).

D. What Is the Effect of Specific Delivery Systems and Immunomodulators on Responses to Vaccine Antigens?

In order to increase the immunogenicity of peptide/protein antigens, immunogens can be included into various antigen-presentation systems and/or administered together with immunomodulatory molecules.

I. Effect of Antigen-Delivery Systems

1. Live Bacterial and Viral Vectors

Given the unique immunogenic properties of live vaccines, especially in terms of the duration of the elicited protective immune responses, numerous microorganisms have been attenuated in vitro in an attempt to develop safe vectors into which heterologous antigens could be efficiently expressed.

a) Bacterial Vectors

Recent developments have made it possible to construct nonreverting, attenuated live bacterial vaccines which have been proven safe and immunogenic in human volunteers (review by LINDBERG and PILLAI 1996). These attenuated strains can often be further manipulated to express heterologous antigens and thus function as live vaccine vectors. The main interest of such approaches is to make use of the properties of these bacteria to target specific APC and to efficiently trigger both the innate and specific immune system. Their inherent limitation is the frequent interference of preexisting antivector immunity.

Mycobacterial vectors such as the BCG strains have been considered in view of their relative safety profile, of their capacity to persist in tissues for weeks or months, and to elicit strong cell-mediated immune responses. Animal studies including nonhuman primate species have indicated the induction of potent CD4 and CD8 responses to foreign antigens, which could also be elicited when given by the mucosal route (review by CRYZ 1996). Preexisting immunity, which can be expected in most individuals following early exposure to nontuberculosis environmental mycobacteria, however, has been found significantly to decrease T cell proliferative responses to the expressed foreign antigen. Clinical studies are expected to be initiated soon to evaluate the feasibility of this approach in humans, whereas efforts are being made to further attenuate the BCG vector.

A variety of *S. typhi* attenuated strains constructed by recombinant DNA technology have been found to be nonpathogenic in humans following mucosal immunization and to induce strong local secretory, humoral, and cell-mediated immune response to the vector strain. Induction of antibodies against epitopes of antigen(s) corresponding to the inserted foreign gene(s) has been observed in several animal species (Barry et al. 1996), but this approach appears less effective in clinical studies assessing *S. typhi* recombinant strains encoding *P. falciparum* circumsporozoite protein or hepatitis B antigen (Nardelli-Haefliger et al. 1996).

Avirulent bacterial species such as those which constitute the normal flora (*Lactobacillus*, *Streptococcus*, *Enterococcus*, *Staphylococcus*) offer obvious safety advantages. They have been recently shown in animal models to be capable of competing with their normal flora counterparts and occasionally to induce antibody responses to foreign antigens (Fischetti et al. 1996). Further studies are ongoing to better delineate their immunogenic potential.

b) Viral Vectors

Several groups of viruses have potential as live recombinant vectors, although they must often face limitations in the size of foreign genes which they may accommodate (influenza, polio, adenovirus). They may also be relatively unstable, present a risk of in vivo recombination (adenovirus, polio, influenza), and be limited by the induction of anti-vector immunity and by the risk of establishing latent infection (herpes) (Minor 1996).

In view of their large capacity for foreign DNA, efficient expression, and authentic processing of incorporated foreign proteins, pox viruses, either derived from the vaccinia strain and genetically manipulated for enhanced safety (NYVAC, MVA strains) or derived from avipox viruses (e.g., canarypox), have an interesting potential for use as live recombinant viral vectors (Pincus et al. 1995; Plotkin et al. 1995; Paoletti 1996). They have been found to induce antibody, CD4, and CTL responses to encoded foreign antigens in numerous animal studies. A vaccinia virus recombinant expressing the rabies virus glycoprotein (RG) has been shown to vaccinate the red fox in Europe and raccoons in the United States. An attenuated recombinant vaccinia virus expressing multiple malaria antigens (NYVAC-7) is under testing in humans. The canarypox (ALVAC)-RG, which is restricted for replication to avian species, confers protection against rabies in dogs (Taylor et al. 1994) and induces protective antibody levels and T cell proliferative responses in humans (Fries et al. 1996). ALVAC canarypox recombinants expressing measles hemagglutinin and fusion proteins induce only modest responses in seropositive adults and are still under evaluation. Human trials with ALVAC-gp160 from HIV-1 proved immunogenic only in a subset of volunteers (EGAN et al. 1995), but weak primary responses are significantly enhanced by administration of a protein boost (PIALOUX et al. 1995; FLEURY et al. 1996), thus indicating successful T cell priming. HIV-1 gp160 specific CTLs were demonstrated in 39% of 18 subjects and reported by the authors as the best human CTL responses ever observed with any tested HIV vaccine (FLEURY et al. 1996). Interestingly, vector-induced immunity does not seem to be a major limiting factor with canarypox, thus allowing immunization to be repeated with the same vector, in contrast to many live vectors, including vaccinia-derived strains.

Thus live viral vectors can be developed so as to be safe and immunogenic in humans. Their modes of action rely essentially on their capacity to target professional APC efficiently and to induce both MHC class II and class I antigen presentation, leading to priming of both CD4 and CD8 T cells. Their use as antigen-delivery systems can also be optimized by incorporation of genes encoding specific cytokines, and this further broadens their potential for vaccine design.

2. DNA Vaccines

The unexpected observation that direct intramuscular or transdermal injection of a DNA plasmid can lead to in vivo expression of antigen by transfected cells (WOLFF et al. 1990), to antigen uptake by APC, and to the efficient induction of antigen-specific immune responses (ULMER et al. 1993) led to a novel generic approach of immunization.

The bases and characteristics of DNA immunization are detailed in Chap. 3 of this volume. As antigen-presentation systems, the main interest of DNA vaccines is that they induce direct in vivo synthesis of the antigen under its native form, followed by authentic antigen processing and prolonged presentation by surface MHC class I molecules to $CD8^+$ T cells. The rather low level of MHC class I expression at the muscle cell surface may contribute to enabling muscle cells to avoid rapid immune detection and destruction, thus increasing the time of efficient antigen synthesis and uptake by APC. In animal models in which they have been tested so far, DNA vaccines thus appear as the most potent inducers of CD8 cytotoxic T cells. This potential is likely to be enhanced by the "adjuvant" property of DNA itself, which induces strong differentiation of CD4 T cells towards the TH1-like phenotype. This was recently reported as due to the presence of specific immunomodulatory motives within bacterial DNA, capable of rapidly enhancing transcription of IL-6, IFN-α, IFN-γ, and IL-12 in APC (SATO et al. 1996) and thus to activate natural killer and TH1 cell responses.

Another potential interest of DNA vaccines as antigen-presentation systems is the prolonged period during which vaccine responses can be demonstrated, sometimes after a single DNA immunization. However, weaker and shorter antibody responses recently demonstrated in nonhuman primates suggest that this feature reflects peculiarities of rodent DNA immunization. The first clinical trials currently ongoing with DNA vaccines will hopefully soon yield initial information on both the safety and the immunogenicity of this novel antigen-presentation system.

II. Effect of Particulate Substances on Vaccine Responses

1. Aluminum and Calcium Salts

Aluminum salts have been used for decades in both veterinary and human applications, including in routine childhood vaccines since 1930. Aluminum hydroxide has been described as more potent than aluminum phosphate (review by GUPTA and SIBER 1994). Both form insoluble gel precipitates to which proteins bind with variable efficacy depending on the strength of the electrostatic interactions (HEM and WHITE 1995). The adjuvanticity of aluminum salts thus essentially depends on the biochemical properties of the antigen and its resulting adsorption capacity. As examples, use of aluminum hydroxide is not found to increase antibody responses to influenza hemagglutinin (DAVENPORT et al. 1968) or to *Haemophilus influenzae* type b (HIB)-tetanus (TT) conjugate vaccine (CLAESSON et al. 1988) although aluminum hydroxide is well known to enhance the immunogenicity of tetanus toxoid. Significantly enhanced specific immune responses are associated with pretreatment with phosphate anions to circumvent the effect of electrostatic repulsive forces which reduce the ability of aluminum hydroxide to adsorb basic proteins (RINELLA et al. 1996) and with adsorption of a specific monoclonal anti-HA antibody to the aluminum preparation prior to addition of influenza virus hemagglutinin (SKEA and BARBER 1993). Whether similar strategies will prove useful for other vaccine immunogens with poor adsorption capacity to aluminum salts is open for study. An apparent need for excess free aluminum adjuvant has also been shown necessary for reaching optimal immunogenicity (GUPTA et al. 1995).

Aluminum compounds form a depot at the site of injection, from which antigen is slowly released. Antigen uptake by APC is enhanced by the particulate nature of the formulation, which also activates macrophages and complement (GUPTA et al. 1995).

The use of aluminum salts results in the development of faster, higher, and longer lasting primary antibody responses than soluble vaccines (APRILE and WARDLAW 1966). In contrast, their usefulness for the induction of booster or secondary responses has not been established (GUPTA et al. 1995).

Aluminum salts have been demonstrated as influencing the differentiation of T lymphocyte responses towards the TH2 phenotype both in mice and humans (AUDIBERT and LISE 1993), which explains their enhancing effects on antibody responses, the generation of IgE-mediated allergic reactions, and their inability to elicit cell-mediated TH1 and CTL responses.

The safety of aluminum salts has been well established both in adults and infants, even if severe local reactions occasionally occur (GUPTA et al. 1993).

Their toxicity has been ascribed to their capacity to increase vascular permeability and to a toxic effects on macrophages (GOTO et al. 1993), resulting in a histiocytic foreign body reaction (GARCIA-PATOS et al. 1995). Hypersensitivity reactions following repeated use of aluminum hydroxide result in sometimes painful and persistent subcutaneous nodules. Calcium salts elicit hemolytic effects (GOTO et al. 1993) and, although effective, have not proven less reactogenic than aluminum hydroxide when combined with diphtheria and tetanus toxoid vaccines (AGGERBECK and HERON 1995; AGGERBECK et al. 1995).

2. Water and Oil Emulsions

Water and oil emulsions have been used extensively in animal vaccines, either initially as the classical Freund's complete adjuvant (FCA) containing paraffin oil mixed with killed mycobacteria (FREUND et al. 1937) or as the water-in-oil emulsion deprived of mycobacteria (Freund's incomplete adjuvant, FIA) (EDELMAN 1980; HILLEMAN 1966). FIA has been used in humans to enhance the immunogenicity of influenza and TT vaccines (EDELMAN 1980; HILLEMAN 1966) but discontinued because of granuloma/cyst formation at site of injection and fear of carcinogenicity following studies in mice.

More recently, various mineral oils incorporated in numerous types of water-in-oil (W/O) and oil-in-water (O/W) emulsions stabilized by emulsifiers have been developed and evaluated for their relative adjuvanticity/reactogenicity profile (review by GUPTA et al. 1993).

W/O emulsions have little specific immunogenic properties but can nevertheless enhance antibody responses to hydrophilic antigens by constituting a short-term depot from which antigen is slowly released (HERBERT 1966). One such W/O emulsion (Specol) has shown a positive adjuvanticity/reactogenicity ratio when combined with weak immunogens prior to immunization of mice (LEENAARS et al. 1995) and rabbits (LEENAARS et al. 1994). W/O emulsions with external antigenic peptides have been described as capable of CTL induction, in studies including in a few human subjects (ACHOUR et al. 1995). However, they fail to confer protection in mice against blood stages of *Plasmodium yoelii*, and this may be related to their incapacity to induce parasite specific IgG2a antibodies (TEN HAGEN et al. 1993).

O/W emulsions are considered to enhance antigen presentation of amphipathic immunogens, possibly by the targeting of APC and increased antigen uptake. Most of the emulsions currently in preclinical or clinical trials are complex formulations incorporating immunomodulatory substances (see section D.IV).

3. Liposomes and Virosomes

Liposomes microspheres consist of cholesterol and phospholipid bilayers alternating with aqueous spaces allowing surface-linked or encapsulation of either water-soluble or lipid-soluble molecules (GREGORIADIS and PANAGIOTIDI 1989). Their adjuvanticity has been attributed to slow release of antigen at the injection site and to enhanced uptake by APC (FORTIN and THERIEN 1993), which enhances both humoral and cell-mediated immune responses (LOWELL 1990; PIETROBON 1995; ALVING et al. 1995). Surface-linked antigens have been found to remain preferentially in the endosomal compartment and to stimulate CD4 T cells, whereas encapsulated antigen more readily escapes into the cytosol and the MHC class I pathway for presentation to CD8 T cells (FORTIN et al. 1996). Liposomes have also been shown suitable for mucosal immunization (DE HAAN et al. 1995a,b; VADOLAS et al. 1995; ARAMAKI et al. 1994). Their use is limited by a difficult entrapment of antigen, but their efficacy can be enhanced by incorporation of immunomodulators (see following section).

A specific type of liposomes referred to as virosomes has been generated by insertion of virus fusion proteins (essentially influenza virus glycoproteins) into the liposomal bilayers (GLUCK et al. 1992). These fusion proteins enhance cell binding and endocytic uptake. Delivery of the virosome content into the cytosol could potentially be enhanced by fusion of the viral protein to the endosomal membrane, but CTL induction has not been reported, possibly following preferential induction of TH2 rather than TH1 responses. The immunogenicity and safety of virosomes containing inactivated hepatitis A vaccine have already been demonstrated in clinical studies (GLUCK et al. 1994; LOUTAN et al. 1994; POOVORAWAN et al. 1995). However, when used as carriers for multiple antigens as a new type of combined vaccines, antigenic competition limits their immunogenicity (MENGIARDI et al. 1995).

4. Proteosomes

Proteosomes are preparations of outer membranes of meningococci which interact to form structures into which amphipathic antigens are incorporated (LOWELL 1990). As carrier particles they enhance targeting to APC and antigen presentation to CD4 T cells. They do not enhance MHC class I restricted presentation and subsequent CTL induction but have been reported useful for mucosal immunization (ORR et al. 1993), especially after addition of cholera toxin B subunit (ORR et al. 1994), which suggests that mucosal delivery of proteosome vaccines enhances both mucosal and systemic immunity.

5. Micro- and Nanospheres of Biodegradable Polymers

Microspheres of biodegradable polymers (polylactide-coglycolide, polyphosphazenes) can be used for delivery of vaccine antigen (LANGER 1990; ELDRIDGE et al. 1991; AGUADO and LAMBERT 1992). Their mode of action depends on their size: particles smaller than 10 μm are efficient in targeting antigens to APC, including on mucosal surfaces, whereas large microspheres can be used for long-term depot of antigen. Polymers can be selected with various rates of in vivo degradation so as to achieve either progressive or pulsate antigen release. If technical difficulties still inherent to their manufacturing process can be solved, a single injection of these formulations could avoid the requirement for booster injections of alum-adsorbed proteins (ALONSO et al. 1994).

III. Effect of Immunomodulators that Can Be Incorporated into Antigen-Presentation Systems

Several substances with immunoregulatory properties have been identified and efforts to generate safer novel derivatives of the original molecules have often been successful (Table 4).

Table 4. Influence of antigen-presentation systems on vaccine-induced effector mechanisms

Category of formulation	Influence on vaccine responses	Demonstrated in humans
1. Live vectors		
Bacterial vectors		
BCG derived	Strong enhancement of CD4 TH I responses	
	Induction of CTL responses	
S. typhi	Enhancement of antibody responses	
	Enhancement of CD4 responses	
	Induction of CTL responses	
Viral vectors		
Pox derived	Enhancement of antibody responses (weak)	Yes
	Enhancement of CD4 responses	Yes
	Induction of CTL responses	Yes
2. DNA vaccines	Enhancement of antibody responses	
	Enhancement of CD4 responses	
	Induction of CTL responses	
3. Particulate substances		
Aluminium salts	Enhancement of primary antibody responses	Yes
	Induction of strong TH2 CD4 responses	Yes
W/O emulsions	Enhancement of antibody responses	
O/W emulsions	Enhancement of antibody/CD4 responses (weak)	
Liposomes	Enhancement of antibody responses	Yes
	Enhancement of TH1/TH2 responses (?)	
	Induction of CTL responses	Yes
	Induction of mucosal responses	
Virosomes	Enhancement of antibody responses	Yes
Proteosomes	Enhancement of antibody responses	
	Induction of mucosal. responses	
	Enhancement of TH2 responses (?)	
	Induction of mucosal. responses	
Polymer microspheres	Enhancement of antibody responses	
	Induction of mucosal responses	

1. Lipid A and Derivatives

Lipopolysaccharides of gram-negative bacteria, in particular the lipid A region of lipopolysaccharide, exert a strong adjuvant effect on antibody responses (JOHNSON et al. 1956). Their toxicity, which precludes their use in human vaccines, led to identification of less reactogenic lipid A derivatives, among which monophosphoryl lipid A (MPL) appears the most promising (RIBI et al. 1985; RUDBACH et al. 1990).

The mode of action of MPL is primarily by activation of APC to produce increased levels of cytokines such as IFN-γ, TNF-α, and IL-1β (PERERA et al. 1993), leading to proliferation of CD4 T cells expressing IFN-γ and IL-2 but not IL-4, IL-5, or IL-10, i.e., of a TH1 rather than a TH2 phenotype (MATTERN et al. 1994). MPL has been used in numerous formulations based on liposome or emulsions, alone or with additional immunomodulatory substances (see Sect. D.IV).

2. Saponins (Quil A, QS21)

Plant extracts containing saponin, initially isolated from the bark of the *Quillaja saponaria* tree, have been found to exhibit adjuvant effects but are associated with an unacceptably high rate of side effects (DALSGAARD 1978). This led to purification of a less toxic Quil A moiety which was further purified into several different saponins. Among these, QS21 revealed a potent adjuvant with reduced toxicity (KENSIL et al. 1991, 1995), whose precise mode of action has not yet been described.

Enhanced antibody responses of a preferential IgG2a isotype together with strong TH1 and CTL responses to subunit antigens have been elicited in various animal species (WU et al. 1992; KENSIL et al. 1995; KENSIL 1996). Protection of dogs from Lyme disease has been achieved with a QS21-containing vaccine (COUGHLIN et al. 1995). In baboons the use of QS21 elicits earlier seroconversion and higher neutralizing antibody titers to HIV gp120 (POWELL et al. 1994), whereas simian immunodeficiency virus gag- and env-specific CTLs have been elicited in rhesus macaques (NEWMAN et al. 1994). The reactogenicity of QS21 was found acceptable in initial human studies performed in melanoma patients and is now under evaluation in various vaccine studies.

3. Nonionic Block Copolymers

Incorporation of nonionic block copolymers in antigen-presentation systems such as W/O or O/W emulsions enhances antibody responses to a variety of viral, parasite, or bacterial antigens (HUNTER et al. 1991; BREY 1995a,b). Their primary mode of action is to enhance antigen presentation by adherence to lipids, retention of protein antigen to surfaces, activation of complement, and increased macrophage MHC class II expression (HUNTER et al. 1994). They do appear to support induction of CD4 TH1 and TH2 cells but not that of CD8 cytotoxic lymphocytes. The reactogenicity of initially developed products such as TiterMax (CytRx) is considered too high for use in humans, but numerous other compounds are currently being developed.

4. MDP and Derivatives

Attempts to purify components of *Mycobacteria* that would be deprived of the toxicity of FCA led to the identification of the minimal structure needed for adjuvanticity: *N*-acetyl muramyl-L-alanyl-D-isoglutamine (MDP) (ELLOUZ et al. 1974). MDP proved still too pyrogenic for human use, and additional derivatives have been developed (BAHR et al. 1995). The lack of toxicity of murabutide unfortunately is correlated to its weak adjuvant effect (GUPTA et al. 1993). Romurtide is a stearoyl derivative of MDP which is currently under evaluation.

In vitro studies with muramyl peptides were shown to activate murine macrophages and to enhance their level of MHC class II expression (KHAIDUKOV et al. 1995), to induce cell adhesion molecules on human leukocytes (ASANO et al. 1995; DARCISSAC et al. 1996) and to stimulate the production of IL-1α and IL-1β, IL-6, IL-8, TNF-α and IL-1 receptor antagonist by human monocytes (SUZUKI et al. 1994; ASANO et al. 1994). Their use in W/O emulsions favors TH2 responses, whereas incorporation into O/W emulsions is reported to enhance TH1 responses (DOZMOROV et al. 1994). Other MDP derivatives are included in complex vaccine formulations: threonyl-MDP is included in the preparation of SAF-1 and muramyl tripeptide in the O/W MF-59 emulsion (see below).

5. Cytokines and Interferons

Incorporation of cytokines in antigen-presentation systems or simultaneous administration with vaccines has been envisaged and performed in a few animal models (HEATH and PLAYFAIR 1992). They can be selected for their specific action on dendritic cells (GM-CSF), on T and B cell maturation (IL-1), or on differentiation of TH1 (IL-2, IFN-γ, IL-12) or TH2 (IL-4) CD4 T cells. Coadministration of IFN-γ induces earlier and stronger antibody responses to hepatitis B in hemodialysis patients (QUIROGAJA et al. 1990). The successful use of IL-12 in a number of experimental models is of particular interest in view of the importance and difficulties of raising strong TH1 responses with subunit vaccines (LAMONT and ADORINI 1996; SKEEN et al. 1996; O'TOOLE et al. 1996; BLISS et al. 1996; METZGER et al. 1996). IL-12 is being studied in non-human primates as an adjuvant for use with killed *Leishmania* vaccine.

The use of GM-CSF to increase dendritic cell recruitment and activation and thus enhance vaccine responses appears sufficiently promising in animal models (DISIS et al. 1996) to have led to initial pilot studies in humans. Administration to hemodialysis patients previously nonresponsive to at least three high doses of hepatitis B vaccine resulted in seroconversion of 50% of patients (HESS et al. 1996). In another study protective antibody titers to hepatitis surface antigen were observed after a single vaccine dose in 11 of 81 subjects receiving GM-CSF, compared to none of the controls (Tarr et al. 1996). Injections were reported as well having been tolerated. Adjuvant application of IFNs is also currently under evaluation (El Kassas and Kirkwood 1996).

Concerns related to the many potential limitations of cytokine use (species specificity, stability, toxicity, and cost) have found elegant solution by the use of cytokine-encoding DNA plasmids delivering minute quantities of cytokine within the microenvironment at the onset of a vaccine response (Kim et al. 1997; Geissler et al. 1997).

IV. Effect of Vaccine Formulations that Combine Antigen-Delivery Systems and Immunomodulatory Substances

Following the example of FCA, which contains paraffin oil mixed with killed mycobacteria (Freund et al. 1937), the most effective way to modulate immune responses to vaccine antigens specifically is to act at both levels of APC and T/B lymphocytes by specifically incorporating one or several immunomodulator(s) into an antigen-delivery system (Table 5).

1. Immune-Stimulating Complexes

Immune stimulating complexes (ISCOMS) are antigen-presenting systems of a particulate nature which results from the interaction of cholesterol and phospholipids. They form cagelike structures into which amphipathic antigens can be multimerized and include a saponin derivative (Quil-A) as a built-in immunomodulator, which is responsible for their immunoenhancing activity (Morein et al. 1984; Rimmelzwaan and Osterhaus 1995; Barr and Mitchell 1996). They have been shown in mice to lead to specific targeting of macrophages (Claassen et al. 1995) and to enhanced uptake and processing by several types of APC (Villacres-Eriksson 1995). This could result from association of the saponin carbohydrates with the DEC-205 receptor capable of binding carbohydrate-glycoprotein ligands and leading to enhanced endocytosis (Jiang et al. 1995). ISCOMS have been shown to induce APC activation (reflected by enhanced production of inflammatory cytokines) (Behboudi et al. 1996). Degradation of antigen into the endosomal pathway results in MHC class II restricted antigen presentation to specific CD4 T cells, and strong TH1/TH2 proliferation and cytokine production have been induced by ISCOMS (Villacres-Eriksson 1995). Induction of CD8 cytotoxic T cells has also been reported (Takahashi et al. 1990; Heeg et al. 1991; Mowat et al. 1991). This seems to follow ISCOM fusion with endosomal membranes and antigen delivery to the cytosol.

The interesting properties of ISCOMS have been confirmed with viral and nonviral antigens in numerous immunization models including use of mucosal routes, albeit with less consistent results (Barr and Mitchell 1996).

Table 5. Influence of immunomodulators on vaccine-induced effector mechanisms

Category of formulation	Influence on vaccine responses	Demonstrated in humans
Immunomodulators		
1. Lipid A derivatives (MPL)	Enhancement of antibody responses	Yes
	Induction of strong THI responses	
	Induction of CTL responses	
2. Saponin derivatives (QS21)	Enhancement of antibody responses	Yes
	Induction of strong THI responses	
	Induction of CTL responses	
3. Nonionic block copolymers	Enhancement of antibody responses	
	Induction of TH1/TH2 responses	
4. MDP derivatives	Enhancement of antibody responses	
	Strong induction of TH1 (lipophilic MDP) responses	
	Strong induction of TH2 (hydrophilic MIP) responses	
5. Cytokines	Strong induction of TH1 (IL-12, IL-2, IFN-γ) responses	
	Strong induction of TH2 (IL-4) responses	
	Attraction, migration, activation of dendritic cells (GM–CSF)	
Antigen presentation systems with built-in immunomodulators		
1. ISCOMS	Enhancement of antibody responses	
	Strong enhancement of THI/TH2 responses	
	Strong induction of CTL responses	
	Induction of mucosal responses	
	Escape from inhibition by maternal antibodies	
2. DETOX	Enhancement of antibody responses	Yes
3. SAF	Enhancement of antibody responses	
	Enhancement of CD4 responses	
4. MF-59	Enhancement of antibody responses	Yes
	Enhancement of CD4 responses	Yes
	Induction of CTL responses	Yes
5. SB-AS2	Enhancement of antibody responses	Yes
	Enhancement of CD4 TH1 responses	Yes

Protection against challenge has been demonstrated in numerous species, including primates (Rimmelzwaan and Osterhaus 1995; Barr and Mitchell 1996). Their human use awaits demonstration of the safety of Quil-A, which, interestingly, appears devoid of significant toxicity when incorporated into ISCOMS (Ronnberg et al. 1995).

2. W/O Emulsions with Built-In Immunomodulators

A modification of FCA using squalene oil and nor-MDP has been used successfully in a phase I clinical trial of a birth control vaccine (Jones et al. 1988).

Other formulations have been found successful in animal models but with unacceptable reactogenicity for human use (TiterMax, CytRx). In spite of the limitation of W/O emulsion which form only short-term antigen depot and do not readily support the generation of CTL responses, W/O based emulsions with built-in immunomodulators would nevertheless be suitable for use with hydrophilic immunogens. Several formulations are currently in either advanced preclinical or clinical trials.

3. O/W Emulsions with Built-In Immunomodulators

In contrast to hydrophilic antigens, amphipathic immunogens are best incorporated into O/W emulsions. Several formulations with interesting properties have been developed, some of which have reached the stage of clinical trials.

Squalene-based O/W emulsions containing MPL and mycobacterial cell wall skeleton (DETOX, Ribi) have reached the phase of clinical trials in the fields of malaria (HOFFMAN et al. 1994) and cancer (SCHULTZ et al. 1995; MILES et al. 1996) vaccines. Tolerance was judged acceptable, and antibody levels – but not delayed-type hypersensitivity responses – were higher than with alum.

Another formulation has been developed containing nonionic block copolymer and threonyl MDP, the syntex adjuvant formulation (SAF) (ALLISON and BYARS 1986). Use in rhesus monkeys in the evaluation of a prototype AIDS vaccine demonstrated enhanced titers of neutralizing antibodies capable of complement mediated lysis but not of antibody-dependent cell-mediated cytolysis, and increased blastogenic responses but lack of CTL induction (OHKAWA et al. 1994). Results of use of SAF with HIV-1 antigen in humans have not been published.

A squalene oil emulsified with Tween 80 and Span 85 (MF-59, Chiron) has been shown to be a safe and potent stimulator of animal and human cellular and humoral responses to subunit antigens such as herpes simplex virus, HIV, hepatitis B and influenza (OTT et al. 1995; LANGENBERG et al. 1995; TRAQUINA et al. 1996). Addition of the MTP-TE MDP derivative increased reactogenicity, which was felt to outweigh the small increase in immunogenicity that it conferred (GRAHAM et al. 1996; KEEFER et al. 1996). Evaluation of other formulations adding various additional immunomodulators to MF-59 have been reported in mice (VALENSI et al. 1994).

Most recently an experimental vaccine containing a recombinant circumsporozoite protein (RTS,S) was reported to protect 6/7 subjects against *P. falciparum* malaria (STOUTE et al. 1997) when administered in the SB-AS2 (SmithKline Beecham Biologicals) formulation, an O/W emulsion containing both MPL and QS21. In contrast, two RTS,S formulations containing either alum and deacylated MPL (SB-AS4) or RTS,S in a O/W emulsion without additional immunostimulants (SB-AS3) failed to induce protection. The most common adverse reaction was pain at the site of injection, which increased with the number of vaccine doses and was controlled by decrease in the volume of vaccine. These promising results validate the efforts of designing rational novel vaccine formulations against diseases which have so far eluded prevention by vaccination.

4. Formulations Based on Liposomes with Built-In Immunomodulators

MPL has proven its interesting properties when incorporated into liposomes, including the induction of CTLs to soluble protein antigens entrapped in liposomes (ZHOU and HUANG 1993; ALVING 1993; ULRICH and MYERS 1995). Liposomal formulation incorporating MDP (LINUMA et al. 1995), Quil A (LIPFORD et al. 1994) or a mixture of MPL and QS21 (PRIEELS et al. 1994) offer interesting properties for use with amphipathic immunogens, and several formulations are currently under trial.

E. How Can Vaccines Be Designed for Selected Target Populations with Variable Levels of Immunocompetence?

Studies on the immunogenicity, safety, and protective efficacy of antigen-presentation systems and immunomodulators have been largely performed in either animal species or humans subjects with a fully mature and functional immune system. However, vaccination targets are most often infants and elderly with some degree of physiological immunodeficiency, who may fail to raise protective immune responses to conventional vaccine formulations and require particular attention regarding the choice of antigen-delivery system or immunomodulators in order to induce appropriate responses. Special safety considerations should also affect the selection of live vaccines to be used in populations or in individuals within a context of varying levels of immunocompetence.

I. Inducing Efficient Vaccine Responses in Early Life

1. Characteristics of Immune Responses in Early Life

Neonatal or early infant immunizations are required to protect from diseases caused by pathogens to which they are exposed in early life, and which result in an enhanced risk of severe or persistent infections due to physiological deficiencies of both their innate and specific immune system (review by HAENEY 1994). Antibody responses raised by newborns and infants to infectious agents or vaccines are often slower, weaker, of shorter duration, and of lower avidity than those elicited later in life. Importantly, antibody responses to carbohydrate determinants, for example, bacterial capsules, cannot be elicited before 18–24 months of age. The impaired function of neonatal APC and natural killer cells is considered to directly affect the development of specific neonatal T cell responses, decreasing the capacity to generate the strong cellular responses required for clearance of intracellular micro-organisms. In vitro studies of neonatal human T cells have indicated a low frequency of T cell precursors, increased requirements for costimulation, poor B cell help capacity in terms of both cytokine production and CD40L expression by activated T cells, and low cytokine production (mainly IL-2 and IFN-γ). Recently, neonatal T cells were shown to follow a preferential development into IL-4 and IL-5 producing effector cells (DEMEURE et al. 1995; YANG et al. 1995) and to respond to specific cytokines such as IL-12 not only by IFN-γ but also by IL-4 production (SHU et al. 1994).

Similar observations recently generated in murine models of neonatal immunization with conventional vaccines allow a progressive deciphering of fundamental mechanisms at the basis of immune maturation in early life. Neonatal and early life responses to a selection of vaccine antigens and presentation systems (tetanus and measles virus peptides, tetanus toxoid, live viral attenuated measles virus, canarypox recombinant measles vector, and BCG) have been shown to differ qualitatively from adult responses by a decreased IgG2a/IgG1 ratio of vaccine-specific antibodies, the secretion of significantly higher IL-5 and lower IFN-γ levels by vaccine-specific T cells, and an impaired induction of cytotoxic T cell precursors (BARRIOS et al. 1996a). This is interpreted as reflecting a preferential neonatal polarization towards TH2 versus TH1 responses induced upon early exposure to vaccines. This differentiation bias is believed to result from defective neonatal APC-T cell interactions, whose optimization could thus represent the main target for novel vaccine formulations.

2. Selecting Antigen-Presentation Systems/Immunomodulators for Use in Early Life

Optimization of early life vaccines implies (a) the induction of vaccine responses at a very young age, so as to confer protection before age of exposure to the micro-organisms, (b) the generation of protective antibody responses, including to polysaccharide determinants of encapsulated bacteria, (c) the induction of strong TH1 and CTL responses against intracellular micro-organisms, and (d) the induction of infant vaccine responses in spite of the persistence of antibodies of maternal origin. In addition, some vaccine safety issues are specifically associated with this early-age relative immunodeficiency, and this is not without relevance for the level of parental acceptance of the proposed vaccinations.

With a few notable exceptions (hepatitis B), subunit vaccines are poorly immunogenic until several weeks after birth, and their use in young infants requires the administration of multiple vaccine doses. In order to protect from agents such as respiratory syncytial virus, which is responsible for early (<3–4 months) disease in a majority of infants, novel vaccine formulations should be immunogenic rapidly after birth. Parameters for early neonatal immunogenicity are not yet well defined. With a few exceptions (polio, BCG), live attenuated viral vaccines cannot be used easily in very early life due to excessive reactogenicity, insufficient infant immunogenicity, or persistence of antibodies of maternal origin. Whether novel live vectors will meet all these challenges cannot yet be determined.

a) Inducing Protective Antibody Responses in Infants and Neonates

The induction of protective infant antibody responses to polysaccharide determinants has been achieved by their conjugation to carrier proteins. Given the number of bacterial strains and serotypes (*H. influenzae*, *S. pneumoniae*, *N. meningitidis*, *Streptococcus* B) for which conjugate vaccines represent a potentially useful vaccine strategy, novel carrier molecules are needed to avoid antigenic competition and epitope suppression.

The TH2-driving capacity of alum, the sole adjuvant currently included in infant vaccines, together with a preferential TH2 differentiation of neonatal T cells, probably accounts for its enhancement of infant vaccine antibody responses. However, unless the development of combined vaccines continues to progress significantly, the number of vaccine doses required in the first 2 years of life could soon reach the limit above which alum-triggered reactogenicity becomes an issue.

Identification of immunomodulatory substances which would accelerate vaccine responses in early life would represent an important progress. Preliminary results obtained in murine models of neonatal immunization suggest that the slow kinetics of antibody production observed in early life is not easily modified: although use of specific adjuvants have been shown to induce adultlike antibody titers upon neonatal immunization, primary responses remain significantly weaker and slower than those observed in adult animals (Barrios et al. 1996b). Unfortunately, the kinetics of immune responses triggered by DNA vaccines appear even slower than conventional vaccines.

b) Inducing Strong TH1 and CTL Responses in Infants and Neonates

Induction of early vaccine TH1 and CTL responses cannot be achieved by the use of alum-adjuvanted subunit vaccines. Few studies assessing the capacity of various adjuvants to enhance neonatal responses to vaccine antigens have yet been reported in animal models. Induction of TH1-like vaccine responses can be induced with FCA (Forsthuber et al. 1996), and thus APC/T cells interactions can be optimized in spite of neonatal immaturity. Since the unacceptable reactogenicity of FCA precludes its use in humans, other vaccine formulations should be evaluated. Administration of a tetanus immunodominant peptide emulsified in a W/O emulsion containing a block copolymer (TiterMax) successfully induced a similar TH1-TH2 pattern of antibodies and cytokine production in either neonatal, juvenile, or adult mice (Barrios et al. 1996b). However, in spite of the lack of toxicity in adult animals, the use of this adjuvant is limited by unsuspected excessive neonatal toxicity. Other vaccine formulations are thus currently evaluated in animal models for their capacity to induce TH1 as well as CTL responses in the neonatal period. It can be anticipated from preliminary results that the correct balance between adjuvanticity and reactogenicity might be even more difficult to achieve in the neonatal period than later in life.

Experiments performed in mice have shown that replication-deficient vectors such as canarypox vaccines, which offer the desired safety profile for use in early life, are not capable of inducing the strong TH1 and CTL responses observed in adult animals (BARRIOS et al. 1996a), in contrast to vaccine strains with a more prolonged replication pattern (SARZOTTI et al. 1996 and unpublished observations). This could be in favor of vectors with prolonged replication patterns such as those derived from mycobacteria. However, although the neonatal use of BCG has been demonstrated to be generally safe, the use of mycobacterial vectors for routine infant immunization will definitely require the generation of further attenuated vectors.

Evaluation of the neonatal and early life immunogenicity of DNA vaccines encoding a number of viral and bacterial antigens has recently demonstrated their unique capacity to induce murine adultlike TH1 and CTL responses at a time of still significant immune immaturity (MARTINEZ et al. 1997; SIEGRIST and LAMBERT 1997). Thus DNA immunization could represent an interesting strategy for the prevention of infections with intracellular agents responsible for severe/persistent infections upon early exposure, which require the induction of TH1 and CTL responses, and for which no current efficient and safe conventional vaccine exists.

c) Inducing Vaccine Responses in Presence of Maternal Antibodies

Infant responses to vaccines, in particular to live vaccines, can be limited by the persistence of antibodies of maternal origin during the first year of life. This is believed to reflect first a rapid neutralization of attenuated viruses or the formation of rapidly eliminated antigen-antibody complexes, which decreases the vaccine load available for uptake and presentation by APC and, secondly, the masking of epitopes by maternal antibodies.

Among novel antigen-delivery systems, ISCOM subunit vaccine formulations have been reported to overcome the interference of maternal antibodies in monkeys and in juvenile foals (NORDENGRAHN et al. 1996). This could represent a significant advantage which warrants their further evaluation in neonatal models of immunization. In contrast, antibody responses to live recombinant canarypox vectors and DNA vaccines expressing the measles virus hemagglutinin are both inhibited by the presence of high titers of maternal antibodies (SIEGRIST and LAMBERT 1997). Other antigen-delivery systems could be required to meet the challenge of circumventing the inhibitory effects of maternal antibodies. Whether mucosal immunization is less susceptible to maternal antibody mediated inhibition of vaccine responses is currently being studied.

II. Inducing Efficient Vaccine Responses in the Elderly

Immunosenescence is considered to contribute significantly to morbidity and mortality in the elderly, although little is known of the biochemical, cellular, and molecular bases affecting the immune system of aged humans (reviewed by HAENEY 1994; PAWELEC et al. 1995).

In contrast to neonates, the effect of aging on the immune response is more variable, but the changes in cell-mediated immunity have the greatest clinical impact. The elderly are more susceptible to infections with various pathogenic organisms that are young adults or middle-aged individuals. This heightened susceptibility of the elderly is most likely a reflection of the age-associated decline in the competence of the immune system. It appears that the population of T cells is altered with advanced age, and possibly the efficiency of the monocyte/macrophage cells to destroy microbial invaders declines. In mice there is a relatively weak ability of aged animals to generate cytokines associated with the TH1 subset of $CD4^+$ T cells, in particular, IL-2 and IFN-γ (ALBRIGHT and ALBRIGHT 1994). This decline in T cell function certainly alters the balance between the immune system and intracellular organisms and leads to reactivation of previously controlled infections.

Impaired function of immunoregulatory cells also contributes to the poor and unsustained antibody responses of the elderly to primary and booster immunizations (HAENEY 1994). A shift to TH0 cytokine production by $CD4^+$ cells was also shown in human studies on healthy centenarians. The prevalence of T cell clones producing IFN-γ decreased from 71% in young subjects to 39% in centenarians, whereas the number of clones showing the opposite TH2 pattern is similar in young and aged donors (PAGANELLI et al. 1996). However, morphologically and functionally intact dendritic cells (DC) can be derived from the peripheral blood of aged individuals. These cell populations have been shown to express HLA class I and class II, CD23, CD32, CD40, CD44, and CD54 but not CD3 or CD19. When tested for their antigen-presenting capacity, dendritic cells from young and old individuals prove equally effective in inducing the proliferation of tetanus toxoid-specific T cell clones after antigenic stimulation (STEGER et al. 1996).

In spite of evidence for the clinical efficacy of influenza vaccine in subjects aged over 65 years (NICHOL et al. 1994), vaccine antibody responses to influenza have been variably reported as either satisfactory or significantly lower than those elicited in younger subjects. These contradictory observations might be explained by several confounding factors, including variations in age at immunization, preexisting medical conditions, strains of influenza virus, and types of vaccine (TREANOR et al. 1994, 1996; MORGAN and KING 1996; VAN HOECKE et al. 1996; MCELHANEY et al. 1993; GLATHE et al. 1993). Lower antibody responses of elderly subjects to tetanus toxoid have also been reported and shown to be associated with decreased numbers of specific antibody secreting B cells and with reduced production of antibodies by

TT-specific B cells (BURNS et al. 1993). In contrast, no association was found between aging and postvaccination levels of antibodies to capsular polysaccharides of *Streptococcus pneumoniae* (MUSHER et al. 1996; SANKILAMPI et al. 1996).

In spite of the lack of understanding of the mechanisms at the basis of immunosenescence, a few studies showing the enhancement of elderly vaccine responses by use of specific vaccine formulations have already been reported. For example, liposomes are reported to enhance influenza specific CTLs (POWERS et al. 1995), and a virosome influenza vaccine has been shown effective in elderly people (GLUCK et al. 1994). Numerous studies with novel antigen-delivery systems/immunomodulatory substances are currently planned or in clinical trials in the aged population, and results can be expected within the next few years.

III. Inducing Efficient Vaccine Responses in Immunodeficient Patients

Vaccination in the context of rare congenital immunodeficiencies represents an extreme situation which does not affect global immunization strategies. In contrast, the increasing number of patients with or at risk of acquired immunodeficiencies cannot be ignored, and the safe induction of effective immune responses in presence of variable levels of immunocompetence is a real challenge. Given the risks inherent to the use of live vaccines in patients with varying levels of immunocompetence, studies with other antigen-delivery systems such as replication-deficient vectors or DNA vaccines as well as of specific immunomodulators are of increasing importance.

1. Enhancing Vaccine Responses in Immunosuppressed Patients

Few studies comparing various antigen-presenting systems/immunomodulators in immunosuppressed patients have yet been performed, essentially for logistic reasons. Enhancing the immunogenicity of hepatitis B vaccine in hemodialyzed, cancer, or transplanted patients has been achieved by a simple increase in the antigen content and administration of additional vaccine doses, which indicates their potential for improved vaccine responses. The markedly enhanced immunogenicity of recombinant hepatitis B vaccines given with MF-59 to nonhuman primates also suggests that it could be a desirable immunomodulatory substance for use with hepatitis B vaccines in such patients (TRAQUINA et al. 1996).

2. Enhancing Vaccine Responses in HIV-1 Infected Patients

The occurrence of the HIV pandemic has raised questions about the safety and efficacy of routine childhood immunizations (BCG, oral polio vaccine, diphtheria-tetanus-pertussis vaccine, measles, hepatitis B virus) in this context. Fortunately, the large majority of infants born to HIV-infected mothers are either uninfected or asymptomatic during the first year of life, and although their immune responses to most vaccines are weaker than those of uninfected children, experience has accumulated that they can be safely immunized, including with live attenuated measles vaccine. However, disease progression is clearly associated first with impaired immune responses to vaccines and subsequently with an increased risk of undesirable side effects, which underlines the importance of early immunization. The vaccine-mediated prevention of infectious diseases in adult HIV-1 infected patients follows the same rule: T cell dependent vaccine responses become more difficult to induce as disease progresses towards a decrease in CD4 T cells below $200/mm^3$ (BORLEFFS and SCHNEIDER 1994). Whether the use of vaccine formulations able to enhance APC/T cell interactions in normal individuals can extend the period of time during which immunization remains effective in spite of progression towards AIDS is open for study. Since administration of specific immunomodulators to HIV-1 infected patients is currently receiving considerable attention, studies performed within the next few years can be expected to provide information on their usefulness for enhancement of vaccine responses. Although an increase in HIV-1 replication has sometimes been observed after immunization, this appears a transient phenomenon whose potential consequences are not considered to question the benefit of vaccine-mediated prevention of infectious diseases in this at risk population.

F. Conclusions

Numerous antigen-delivery systems/formulations are now available to induce or enhance vaccine responses, and many more will certainly be identified. The determination of the optimal vaccine formulation can be expected to depend on the nature of vaccine-induced effector mechanisms which are required for protection against a given micro-organism and on the biochemical nature of the immunogen itself.

A better understanding at a molecular level of the basic mechanisms involved in the cellular interactions which take place at the onset of the immune response is certainly critical for the definition of optimal approaches to generate protective responses by vaccination. This is also likely to open the way towards vaccines which may circumvent limitations due to immunological immaturity, immunosenescence, and varying degrees of immunodeficiency.

Enhancing the magnitude of vaccine antibody responses has already been achieved in several situations. In contrast, none of these vaccine formulations has yet been found capable of extending the period of antibody production sufficiently to decrease the requirement for booster doses of subunit vaccines. The induction of strong specific human TH1 responses has been reported with fewer vaccine formulations, particularly live attenuated vaccines and antigen-delivery systems with built-in immunomodulators such as MPL, QS21, MDP derivatives or nonionic block copolymers. The panel of candidates able to induce vaccine-specific CTL responses is probably more limited since it has so far been reported only in human following use of live vectors or of formulations, including MPL or QS21. If the strong capacity for TH1 and CTL priming of DNA vaccines demonstrated in animal models is observed in humans, this antigen-delivery system could represent a significant advantage. Further studies should tell us whether the same or other antigen-delivery systems/ immunomodulators will be capable to increase immune responses in neonates, elderly, and patients with variable levels of immunocompetence.

Although only few of the recently identified delivery systems and new adjuvants have reached the stage of human use, it is now reasonable to expect that the remarkable diversification of tools which can potentially enhance or appropriately polarize immune responses will significantly contribute to accelerate the development of vaccines against some of the most challenging scourge of humanity, such as tuberculosis, malaria, and AIDS.

Acknowledgements. C.A. Siegrist is supported by the Swiss National Science Foundation (SCORE A). The work from the laboratory mentionned in this review was supported by grants from the Swiss National Science Foundation, from WHO Global Programme for Vaccines and Immunization, from the Sandoz Foundation and from Pasteur-Mérieux Serums et Vaccins.

References

Achour A, Moukrim Z, Picard O, Bizzini B, Burny A, Zagury D (1995) HIV-1 soluble antigens induced CD8+ cytotoxic T-cell responses in an immunized individual. Cell Mol Biol Noisy le grand 41:395–400

Aggerbeck H, Heron I (1995) Adjuvanticity of aluminium hydroxide and calcium phosphate in diphtheria-tetanus vaccines I. Vaccine 13:1360–1365

Aggerbeck H, Fenger C, Heron I (1995) Booster vaccination against diphtheria and tetanus in man. Comparison of calcium phosphate and aluminium hydroxide as adjuvants II. Vaccine 13:1366–1374

Aguado MT, Lambert PH (1992) Controlled-release vaccines-biodegradable polylactide/polyglycolide (PL/PG) microspheres as antigen vehicles. Immunobiology 184:113–125

Albright JW, Albright JF (1994) Ageing alters the competence of the immune system to control parasitic infection. Immunol Lett 40:279–285

Allison AC, Byars NE (1986) An adjuvant formulation that selectively elicits the formation of antibodies of protective isotypes and of cell-mediated immunity. J Immunol Methods 95:157–168

Alonso MJ, Gupta RK, Min C, Siber GR, Langer R (1994) Biodegradable microspheres as controlled-release tetanus toxoid delivery systems. Vaccine 12:299–306

Alving CR (1993) Lipopolysaccharide, lipid A, and liposomes containing lipid A as immunologic adjuvants. Immunobiology 187:430–446

Alving CR, Koulchin V, Glenn GM, Rao M (1995) Liposomes as carriers of peptide antigens: induction of antibodies and cytotoxic T lymphocytes to conjugated and unconjugated peptides. Immunol Rev 145:5–31

Aprile MA, Wardlaw AC (1966) Aluminium compounds as adjuvants for vaccines and toxoids in man: a review. Can J Public Health 57:343–360

Aramaki Y, Fujii Y, Yachi K, Kikuchi H, Tsuchiya S (1994) Activation of systemic and mucosal immune response following nasal administration of liposomes. Vaccine 12:1241–1245

Asano T, McWatters A, An T, Matsushima K, Kleinerman ES (1994) Liposomal muramyl tripeptide up-regulates interleukin-1 alpha, interleukin-1 beta, tumor necrosis factor-alpha, interleukin-6 and interleukin-8 gene expression in human monocytes. J Pharmacol Exp Ther 268:1032–1039

Asano T, McIntyre BW, Bednarczyk JL, Wygant JN, Kleinerman ES (1995) Liposomal muramyl tripeptide upregulates adhesion molecules on the surface of human monocytes. Oncol Res 7:253–257

Audibert FM, Lise LD (1993) Adjuvants: current status, clinical perspectives and future prospects. Immunol Today 14:281–284

Bahr GM, Darcissac E, Bevec D, Dukor P, Chedid L (1995) Immunopharmacological activities and clinical development of muramyl peptides with particular emphasis on murabutide. Int J Immunopharmacol 17:117–131

Barr IG, Mitchell GF (1996) ISCOMs (immunostimulating complexes): the first decade. Immunol Cell Biol 74:8–25

Barrios C, Brandt C, Berney M, Lambert PH, Siegrist CA (1996a) Partial correction of the TH2/TH1 imbalance in neonatal murine responses to vaccine antigens through selective adjuvant effects. Eur J Immunol 26:2666–2670

Barrios C, Brawand P, Berney M, Brandt C, Lambert PH, Siegrist CA (1996b) Neonatal and early life immune responses to various forms of vaccine antigens qualitatively differ from adult responses: predominance of a Th2-biased pattern which persists after adult boosting. Eur J Immunol 26:1489–1496

Barry EM, Gomez Duarte O, Chatfield S, Rappuoli R, Pizza M, Losonsky G, Galen J, Levine MM (1996) Expression and immunogenicity of pertussis toxin S1 subunit-tetanus toxin fragment C fusions in Salmonella typhi vaccine strain CVD 908. Infect Immun 64:4172–4181

Behboudi S, Morein B, Villacres Eriksson M (1996) In vitro activation of antigen-presenting cells (APC) by defined composition of Quillaja saponaria Molina triterpenoids. Clin Exp Immunol 105:26–30

Bliss J, Maylor R, Stokes K, Murray KS, Ketchum MA, Wolf SF (1996) Interleukin-12 as vaccine adjuvant. Characteristics of primary, recall, and long-term responses. Ann NY Acad Sci 795:26–35

Borleffs JC, Schneider MM (1994) Routine vaccination in HIV-infected patients: a review. Nether J Med 45:244–249

Brey RN (1995a) Development of vaccines based on formulations containing nonionic block copolymers. Pharm Biotechnol 6:297–311

Brey RN (1995b) Copolymer adjuvants. Adv Exp Med Biol 383:23–34

Burns EA, Lum LG, L'Hommedieu G, Goodwin JS (1993) Specific humoral immunity in the elderly: in vivo and in vitro response to vaccination. J Gerontol 48:B231–B236

Cherwinski HM, Schumacher JH, Brown KD, Mosmann TR (1987) Two types of mouse helper T cell clone. III. Further differences in lymphokine synthesis between Th1 and Th2 clones revealed by RNA hybridization, functionally monospecific bioassays, and monoclonal antibodies. J Exp Med 166:1229–1244

Claassen IH, Osterhaus AD, Claassen E (1995) Antigen detection in vivo after immunization with different presentation forms of rabies virus antigen: involvement of marginalmetallophilic macrophages in the uptake of immune stimulating complexes. Eur J Immunol 25(5):1446–1452

Claesson BA, Trollfors B, Lagergard T, Taranger J, Bryla D, Otterman G, Cramton T, Yang Y, Reimer CB, Robbins JB et al (1988) Clinical and immunologic responses to the capsular polysaccharide of Haemophilus influenzae type b alone or conjugated to tetanus toxoid in 18- to 23-month-old children. J Pediatr 112:695–702

Coughlin RT, Fish D, Mather TN, Ma J, Pavia C, Bulger P (1995) Protection of dogs from Lyme disease with a vaccine containing outer surface protein (Osp) A, OspB, and the saponin adjuvant QS21. J Infect Dis 171:1049–1052

Cryz SJ (1996) Live attenuated bacterial vaccine vector systems. Vaccine 14(7):676–680
Dalsgaard K (1978) A study of the isolation and characcterization of the saponin quil A. Evaluation of its adjuvant activity, with a special reference to the apllication in the vaccination of cattel against foot-and-mouth disease. Acta Vet Scand 69:1–40
Darcissac EC, Bahr GM, Parant MA, Chedid LA, Riveau GJ (1996) Selective induction of CD11a,b,c/CD18 and CD54 expression at the cell surface of human leukocytes by muramyl peptides. Cell Immunol 169:294–301
Davenport FM, Hennessy AV, Askin FB (1968) Lack of adjuvant effect of A1PO4 on purified influenza virus hemagglutinins in man. J Immunol 100:1139–1140
de Haan A, Geerligs HJ, Huchshorn JP, van Scharrenburg GJ, Palache AM, Wilschut J (1995a) Mucosal immunoadjuvant activity of liposomes: induction of systemic IgG and secretory IgA responses in mice by intranasal immunization with an influenza subunit vaccine and coadministered liposomes. Vaccine 13:155–162
de Haan A, Tomee JF, Huchshorn JP, Wilschut J (1995b) Liposomes as an immunoadjuvant system for stimulation of mucosal and systemic antibody responses against inactivated measles virus administered intranasally to mice. Vaccine 13:1320–1324
Demeure CE, Yang LP, Byun DG, Ishihara H, Vezzio N, Delespesse G (1995) Human naive CD4 T cells produce interleukin-4 at priming and acquire a Th2 phenotype upon repetitive stimulations in neutral conditions. Eur J Immunol 25:2722–2725
Disis ML, Bernhard H, Shiota FM, Hand SL, Gralow JR, Huseby ES, Gillis S, Cheever MA (1996) Granulocyte-macrophage colony-stimulating factor: an effective adjuvant for protein and peptide-based vaccines. Blood 88:202–210
Dozmorov IM, Kuzin II, Lutsan NI, Lutsenko GV, Prokhorova AL, Sapozhnikov AM, Andronova TM, Ivanov VT (1994) Comparative study of immunomodulatory properties of muramyl peptides on immune system cells of young and old mice. Immunopharmacol Immunotoxicol 16:149–163
Edelman R (1980) Vaccine adjuvants. Rev Infect Dis 2:370–383
Egan MA, Pavlat WA, Tartaglia J, Paoletti E, Weinhold KJ, Clements ML, Siliciano RF (1995) Induction of human immunodeficiency virus type 1 (HIV-1)-specific cytolytic T lymphocyte responses in seronegative adults by a nonreplicating, host-range-restricted canarypox vector (ALVAC) carrying the HIV-1MN env gene. J Infect Dis 171:1623–1627
Eldridge JH, Staas JK, Meulbroek JA, McGhee JR, Tice TR, Gilley RM (1991) Biodegradable microspheres as a vaccine delivery system. Mol Immunol 28:287–294
El Kassas H, Kirkwood JM (1996) Adjuvant application of interferons. Semin Oncol 23:737–743
Ellouz F, Adam A, Ciorbaru R, Lederer E (1974) Minimal structural requirements for adjuvant activity of bacterial peptidoglycan derivatives. Biochem Biophys Res Commun 59:1317–1325
Fischetti VA, Medaglini D, Pozzi G (1996) Gram-positive commensal bacteria for mucosal vaccine delivery. Curr Opin Biotechnol 7:659–666
Fleury B, Janvier G, Pialoux G, Buseyne F, Robertson MN, Tartaglia J, Paoletti E, Kieny MP, Excler JL, Riviere Y (1996) Memory cytotoxic T lymphocyte responses in human immunodeficiency virus type 1 (HIV-1)-negative volunteers immunized with a recombinant canarypox expressing gp 160 of HIV-1 and boosted with a recombinant gp160. J Infect Dis 174:734–738
Forsthuber T, Yip HC, Lehmann PV (1996) Induction of TH1 and TH2 immunity in neonatal mice. Science 271:1728–1730
Fortin A, Therien HM (1993) Mechanism of liposome adjuvanticity: an in vivo approach. Immunobiology 188:316–322
Fortin A, Shahum E, Krzystyniak K, Therien HM (1996) Differential activation of cell-mediated immune functions by encapsulated and surface-linked liposomal antigens. Cell Immunol 169:208–217

Freund J, Casals J, Hosmer EP (1937) Sensitization and antibody formation after injection of tubercle bacilli and paraffin oil. Proc Soc Exp Biol Med 37:509–513

Fries LF, Tartaglia J, Taylor J, Kauffman EK, Meignier B, Paoletti E, Plotkin S (1996) Human safety and immunogenicity of a canarypox-rabies glycoprotein recombinant vaccine: an alternative poxvirus vector system. Vaccine 14:428–434

Garcia Patos V, Pujol RM, Alomar A, Cistero A, Curell R, Fernandez Figueras MT, de Moragas JM (1995) Persistent subcutaneous nodules in patients hyposensitized with aluminum-containing allergen extracts. Arch Dermatol 131:1421–1424

Geissler M, Gesien A, Tokushige K, Wands JR (1997) Enhancement of cellular and humoral immune responses to hepatitis C virus core protein using DNA-based vaccines augmented with cytokine-expressing plasmids. J Immunol 158:1231–1237

Glathe H, Bigl S, Grosche A (1993) Comparison of humoral immune responses to trivalent influenza split vaccine in young, middle-aged and elderly people. Vaccine 11:702–705

Gluck R, Mischler R, Brantschen S, Just M, Althaus B, Cryz SJJ (1992) Immunopotentiating reconstituted influenza virus virosome vaccine delivery system for immunization against hepatitis A. J Clin Invest 90:2491–2495

Gluck R, Mischler R, Finkel B, Que JU, Scarpa B, Cryz SJJ (1994) Immunogenicity of new virosome influenza vaccine in elderly people. Lancet 344:160–163

Goto N, Kato H, Maeyama J, Eto K, Yoshihara S (1993) Studies on the toxicities of aluminium hydroxide and calcium phosphate as immunological adjuvants for vaccines. Vaccine 11:914–918

Graham BS, Keefer MC, McElrath MJ, Gorse GJ, Schwartz DH, Weinhold K, Matthews TJ, Esterlitz JR, Sinangil F, Fast PE (1996) Safety and immunogenicity of a candidate HIV-1 vaccine in healthy adults: recombinant glycoprotein (rgp) 120. A randomized, double-blind trial. NIAID AIDS Vaccine Evaluation Group. Ann Intern Med 125:270–279

Gregoriadis G, Panagiotidi C (1989) Immunoadjuvant action of liposomes: comparison with other adjuvants. Immunol Lett 20:237–240

Gupta RK, Siber GR (1994) Comparison of adjuvant activities of aluminium phosphate, calcium phosphate and stearyl tyrosine for tetanus toxoid. Biologicals 22:53–63

Gupta RK, Relyveld EH, Lindblad EB, Bizzini B, Ben Efraim S, Gupta CK (1993) Adjuvants – a balance between toxicity and adjuvanticity. Vaccine 11:293–306

Gupta RK, Rost BE, Relyveld E, Siber GR (1995) Adjuvant properties of aluminium and calcium compounds. In: Vaccine design: the subunit and adjuvant approach. Plenum, New York, pp 229–248

Haeney M (1994) Infection determinants at extremes of age. J Antimicrob Chemother 34 [Suppl A]:1–9

Heath AW, Playfair JH (1992) Cytokines as immunological adjuvants. Vaccine 10:427–434

Heeg K, Kuon W, Wagner H (1991) Vaccination of class I major histocompatibility complex (MHC)-restricted murine CD8+ cytotoxic T lymphocytes towards soluble antigens: immunostimulating-ovalbumin complexes enter the class I MHC-restricted antigen pathway and allow sensitization against the immunodominant peptide. Eur J Immunol 21:1523–1527

Hem SL, White JL (1995) Structure and properties of aluminum-containing adjuvants. Pharm Biotechnol 6:249–276

Herbert WJ (1966) Antigenicity of soluble protein in the presence of high levels of antibody: a possible mode of action of the antigen adjuvants. Nature 210:747–748

Hess G, Kreiter F, Kosters W, Deusch K (1996) The effect of granulocyte-macrophage colony-stimulating factor (GM-CSF) on hepatitis B vaccination in haemodialysis patients. J Viral Hepat 3:149–153

Hilleman MR (1966) Critical appraisal of emulsified oil adjuvants applied to viral vaccines. Prog Med Virol 8:131–182

Hoffman SL, Edelman R, Bryan JP, Schneider I, Davis J, Sedegah M, Gordon D, Church P, Gross M, Silverman C et al (1994) Safety, immunogenicity, and efficacy of a malaria sporozoite vaccine administered with monophosphoryl lipid A, cell wall skeleton of mycobacteria, and squalane as adjuvant. Am J Trop Med Hyg 51:603–612
Hunter R, Olsen M, Buynitzky S (1991) Adjuvant activity of non-ionic block copolymers. IV. Effect of molecular weight and formulation on titre and isotype of antibody. Vaccine 9:250–256
Hunter RL, McNicholl J, Lal AA (1994) Mechanisms of action of nonionic block copolymer adjuvants. AIDS Res Hum Retroviruses 10 [Suppl 2]:S95–S98
Jiang W, Swiggard WJ, Heufler C, Peng M, Mirza A, Steinman RM, Nussenzweig MC (1995) The receptor DEC-205 expressed by dendritic cells and thymic epithelial cells is involved in antigen processing. Nature 375:151–155
Johnson AG, Tomai MA (1990) A study of the cellular and molecular mediators of the adjuvant action of a nontoxic monophosphoryl lipid A. Adv Exp Med Biol 256:567–579
Johnson AG, Gaines S, Landy M (1956) Studies on the O antigen of Salmonella typhosa V. Enhancement of antibody response toproteinantigens by the purified lipopolysaccharide. J Exp Med 103:225–246
Jones WR, Bradley J, Judd SJ, Denholm EH, Ing RM, Mueller UW, Powell J, Griffin PD, Stevens VC (1988) Phase I clinical trial of a World Health Organisation birth control vaccine. Lancet 1:1295–1298
Keefer MC, Graham BS, McElrath MJ, Matthews TJ, Stablein DM, Corey L, Wright PF, Lawrence D, Fast PE, Weinhold K, Hsieh RH, Chernoff D, Dekker C, Dolin R (1996) Safety and immunogenicity of Env 2-3, a human immunodeficiency virus type 1 candidate vaccine, in combination with a novel adjuvant, MTP-PE/MF59. NIAID AIDS Vaccine Evaluation Group. AIDS Res Hum Retroviruses 12:683–693
Kensil CR, Patel U, Lennick M, Marciani D (1991) Separation and characterization of saponins with adjuvant activity from Quillaja saponaria Molina cortex. J Immunol 146:431–437
Kensil CR, Wu JY, Soltysik S (1995) Structural and immunological characterization of the vaccine adjuvant QS-21. Pharm Biotechnol 6:525–541
Kensil CR (1996) Saponins as vaccine adjuvants. Crit Rev Ther Drug Carrier Syst 13: 1–55
Khaidukov SV, Komaleva RL, Nesmeyanov VA (1995) N-acetylglucosamine-containing muramyl peptides directly affect macrophages. Int J Immunopharmacol 17:903–911
Kim JJ, Ayyavoo V, Bagarazzi ML, Chattergoon MA, Dang K, Wang B, Boyer JD, Weiner DB (1997) In vivo engineering of a cellular immune response by coadministration of IL-12 expression vector with a DNA immunogen. J Immunol 158:816–826
Lamont AG, Adorini L (1996) IL-12: a key cytokine in immune regulation. Immunol Today 17:214–217
Langenberg AG, Burke RL, Adair SF, Sekulovich R, Tigges M, Dekker CL, Corey L (1995) A recombinant glycoprotein vaccine for herpes simplex virus type 2: safety and immunogenicity. Ann Intern Med 122:889–898
Langer R (1990) New methods of drug delivery. Science 249:1527–1533
Leenaars PP, Hendriksen CF, Angulo AF, Koedam MA, Claassen E (1994) Evaluation of several adjuvants as alternatives to the use of Freund's adjuvant in rabbits. Vet Immunol Immunopathol 40:225–241
Leenaars PP, Hendriksen CF, Koedam MA, Claassen I, Claassen E (1995) Comparison of adjuvants for immune potentiating properties and side effects in mice. Vet Immunol Immunopathol 48:123–138
Lindberg AA, Pillai S (1996) Recent trends in the developments of bacterial vaccines. Dev Biol Stand 87:59–71

Linuma H, Nerome K, Yoshioka Y,Okinaga K (1995) Characteristics of cytotoxic T lymphocytes directed to influenza virus haemagglutinin elicited by immunisation with muramyl dipeptide-influenza liposome vaccine. Scand J Immunol 41:1–10
Lipford GB, Wagner H, Heeg K (1994) Vaccination with immunodominant peptides encapsulated in Quil A-containing liposomes induces peptide-specific primary CD8+ cytotoxic T cells. Vaccine 12:73–80
Loutan L, Bovier P, Althaus B, Gluck R (1994) Inactivated virosome hepatitis A vaccine. Lancet 343:322–324
Lowell GH (1990) Proteosomes, hydrophobic anchors, iscoms and liposomes for improved presentation of peptide and protein vaccines. In: Powell MF, Newman MJ (eds) New generation vaccines. Dekker, New York, pp 141–1660
Mancino D, Ovary Z (1980) Adjuvant effects of amorphous silica and of aluminium hydroxide on IgE and IgG1 antibody production in different inbred mouse strains. Int Arch Allergy Appl Immunol 61:253–258
Martinez X, Brandt C, Saddallah F, Tougne C, Barrios C, Wild F, Dougan G, Lambert PH, Siegrist CA (1997) DNA immunization circumvents deficient induction of TH1 and CTL responses in neonates and during early life. Proc Natl Acad Sci USA 94:8726–8731
Mattern T, Thanhauser A, Reiling N, Toellner KM, Duchrow M, Kusumoto S, Rietschel ET, Ernst M, Brade H, Flad HD et al (1994) Endotoxin and lipid A stimulate proliferation of human T cells in the presence of autologous monocytes. J Immunol 153:2996–3004
McElhaney JE, Meneilly GS, Lechelt KE, Beattie BL, Bleackley RC (1993) Antibody response to whole-virus and split-virus influenza vaccines in successful ageing. Vaccine 11:1055–1060
Mengiardi B, Berger R, Just M, Gluck R (1995) Virosomes as carriers for combined vaccines. Vaccine 13:1306–1315
Metzger DW, Buchanan JM, Collins JT, Lester TL, Murray KS, Van Cleave VH, Vogel LA, Dunnick WA (1996) Enhancement of humoral immunity by interleukin-12. Ann NY Acad Sci 795:100–115
Miles DW, Towlson KE, Graham R, Reddish M, Longenecker BM, Taylor Papadimitriou J, Rubens RD (1996) A randomised phase II study of sialyl-Tn and DETOX-B adjuvant with or without cyclophosphamide pretreatment for the active specific immunotherapy of breast cancer. Br J Cancer 74:1292–1296
Minor PD (1996) Recent developments in the design of viral vaccines. Dev Biol Stand 87:51–57
Morein B, Sundquist B, Höglund S, Dalsgaard K, Osterhaus A (1984) Iscom, a novel structure for antigenic presentation of membrane proteins from envelopped viruses. Nature 308:457–460
Morgan R, King D (1996) Influenza vaccination in the elderly. Postgrad Med J 72:339–342
Mosmann TR, Moore KW (1991) The role of IL-10 in crossregulation of TH1 and TH2 responses. Immunol Today 12:A49–A53
Mowat AM, Donachie AM, Reid G, Jarrett O (1991) Immune-stimulating complexes containing Quil A and protein antigen prime class I MHC-restricted T lymphocytes in vivo and are immunogenic by the oral route. Immunology 72:317–322
Musher DM, Groover JE, Graviss EA, Baughn RE (1996) The lack of association between aging and postvaccination levels of IgG antibody to capsular polysaccharides of Streptococcus pneumoniae. Clin Infect Dis 22:165–167
Nardelli-Haefliger D, Kraehenbuhl JP, Curtiss R, Schodel F, Potts A, Kelly S, De Grandi P (1996) Oral and rectal immunization of adult female volunteers with a recombinant attenuated Salmonella typhi vaccine strain. Infect Immun 64:5219–5224
Newman MJ, Munroe KJ, Anderson CA, Murphy CI, Panicali DL, Seals JR, Wu JY, Wyand MS, Kensil CR (1994) Induction of antigen-specific killer T lymphocyte

responses using subunit SIVmac251 gag and env vaccines containing QS-21 saponin adjuvant. AIDS Res Hum Retroviruses 10:853–861
Nichol KL, Margolis KL, Wuorenma J, Von Sternberg T (1994) The efficacy and cost effectiveness of vaccination against influenza among elderly persons living in the community. N Engl J Med 331:778–784
Nordengrahn A, Rusvai M, Merza M, Ekstrom J, Morein B, Belak S (1996) Equine herpesvirus type 2 (EHV-2) as a predisposing factor for Rhodococcus equi pneumonia in foals: prevention of the bifactorial disease with EHV-2 immunostimulating complexes. Vet Microbiol 51:55–68
Ohkawa S, Wilson LA, Larosa G, Javaherian K, Martin LN, Murphey Corb M (1994) Immune responses induced by prototype vaccines for AIDS in rhesus monkeys. AIDS Res Hum Retroviruses 10:27–38
Orr N, Robin G, Cohen D, Arnon R, Lowell GH (1993) Immunogenicity and efficacy of oral or intranasal Shigella flexneri 2a and Shigella sonnei proteosome-lipopolysaccharide vaccines in animal models. Infect Immun 61:2390–2395
Orr N, Arnon R, Rubin G, Cohen D, Bercovier H, Lowell GH (1994) Enhancement of anti-Shigella lipopolysaccharide (LPS) response by addition of the cholera toxin B subunit to oral and intranasal proteosome-Shigella flexneri 2a LPS vaccines. Infect Immun 62:5198–5200
O'Toole M, Wooters J, Brown E, Swiniarski H, Cull G, Leger L, Herrmann S (1996) Interleukin-12 as adjuvant in peptide vaccines. Ann NY Acad Sci 795:379–381
Ott G, Barchfeld GL, Chernoff D, Radhakrishnan R, van Hoogevest P, Van Nest G (1995) MF59. Design and evaluation of a safe and potent adjuvant for human vaccines. Pharm Biotechnol 6:277–296
Paganelli R, Scala E, Rosso R, Cossarizza A, Bertollo L, Barbieri D, Fabrizi A, Lusi EA, Fagiolo U, Franceschi C (1996) A shift to Th0 cytokine production by CD4+ cells in human longevity: studies on two healthy centenarians. Eur J Immunol 26:2030–2034
Paoletti E (1996) Applications of pox virus vectors to vaccination: an update. Proc Natl Acad Sci USA 93:11349–11353
Pawelec G, Adibzadeh M, Pohla H, Schaudt K (1995) Immunosenescence: ageing of the immune system. Immunol Today 16:420–422
Perera PY, Manthey CL, Stutz PL, Hildebrandt J, Vogel SN (1993) Induction of early gene expression in murine macrophages by synthetic lipid A analogs with differing endotoxic potentials. Infect Immun 61:2015–2023
Pialoux G, Excler JL, Riviere Y, Gonzalez Canali G, Feuillie V, Coulaud P, Gluckman JC, Matthews TJ, Meignier B, Kieny MP et al (1995) A prime-boost approach to HIV preventive vaccine using a recombinant canarypox virus expressing glycoprotein 160 (MN) followed by a recombinant glycoprotein 160 (MN/LAI). The AGIS Group, and l'Agence Nationale de Recherche sur le SIDA. AIDS Res Hum Retroviruses 11:373–381
Pietrobon PJF (1995) Liposome design and vaccine development. In: Vaccine design: the subunit and adjuvant approach. Plenum, New York, pp 347–361
Pincus S, Tartaglia J, Paoletti E (1995) Poxvirus-based vectors as vaccine candidates. Biologicals 23:159–164
Plotkin SA, Cadoz M, Meignier B, Meric C, Leroy O, Excler JL, Tartaglia J, Paoletti E, Gonczol E, Chappuis G (1995) The safety and use of canarypox vectored vaccines. Dev Biol Stand 84:165–170
Poovorawan Y, Theamboonlers A, Chumdermpadetsuk S, Gluck R, Cryz SJJ (1995) Safety, immunogenicity, and kinetics of the immune response to a single dose of virosome-formulated hepatitis A vaccine in Thais. Vaccine 13:891–893
Powell MF, Cleland JL, Eastman DJ, Lim A, Murthy K, Newman MJ, Nunberg JH, Weissburg RP, Vennari JC, Wrin T et al (1994) Immunogenicity and HIV-1 virus neutralization of MN recombinant glycoprotein 120/HIV-1 QS21 vaccine in baboons. AIDS Res Hum Retroviruses 10 [Suppl 2]:S105–S108

Powers DC, Manning MC, Hanscome PJ, Pietrobon PJ (1995) Cytotoxic T lymphocyte responses to a liposome-adjuvanted influenza A virus vaccine in the elderly. J Infect Dis 172:1103–1107
Prieels JP, Garcon-Johnson NM-JC, Slaoui M, Pala P (1994) Patent. Vaccine composition containing adjuvants. PCT/EP93/01524
Quiroga JA, Castillo I, Porres JC, Casado S, Saez F, Gracia Martinez M, Gomez M, Inglada L, Sanchez Sicilia L, Mora A et al (1990) Recombinant gamma-interferon as adjuvant to hepatitis B vaccine in hemodialysis patients. Hepatology 12:661–663
Ribi E, Cantrell J, Takayama K (1985) A new immunomodulator with potential clinical applications: monophosphoryl lipid A, a detoxified endotoxin. Clin Immun Newslett 6:33–36
Ribi HO, Ludwig DS, Mercer KL, Schoolnik GK, Kornberg RD (1988) Three-dimensional structure of cholera toxin penetrating a lipid membrane. Science 239:1272–1276
Rimmelzwaan GF, Osterhaus AD (1995) A novel generation of viral vaccines based on the ISCOM matrix. Pharm Biotechnol 6:543–558
Rinella JVJ, White JL, Hem SL (1996) Treatment of aluminium hydroxide adjuvant to optimize the adsorption of basic proteins. Vaccine 14:298–300
Romagnani S (1991) Human TH1 and TH2 subsets: doubt no more. Immunol Today 12:256–257
Ronnberg B, Fekadu M, Morein B (1995) Adjuvant activity of non-toxic Quillaja saponaria Molina components for use in ISCOM matrix. Vaccine 13:1375–1382
Rudbach JA, Cantrell JL, Ulrich JT, Mitchell MS (1990) Immunotherapy with bacterialendotoxin. In: Endotoxins. Plenum, New York, pp 665–676
Sankilampi U, Honkanen PO, Bloigu A, Herva E, Leinonen M (1996) Antibody response to pneumococcal capsular polysaccharide vaccine in the elderly. J Infect Dis 173:387–393
Sarzotti M, Robbins DS, Hoffman PM (1996) Induction of protective CTL responses in newborn mice by a murine retrovirus. Science 271:1726–1728
Sato Y, Roman M, Tighe H, Lee D, Corr M, Nguyen MD, Silverman GJ, Lotz M, Carson DA, Raz E (1996) Immunostimulatory DNA sequences necessary for effective intradermal gene immunization. Science 273:352–354
Scalzo AA, Elliott SL, Cox J, Gardner J, Moss DJ, Suhrbier A (1995) Induction of protective cytotoxic T cells to murine cytomegalovirus by using a nonapeptide and a human-compatible adjuvant (Montanide ISA 720). J Virol 69:1306–1309
Schultz N, Oratz R, Chen D, Zeleniuch Jacquotte A, Abeles G, Bystryn JC (1995) Effect of DETOX as an adjuvant for melanoma vaccine. Vaccine 13:503–508
Shu U, Demeure CE, Byun DG, Podlaski F, Stern AS, Delespesse G (1994) Interleukin 12 exerts a differential effect on the maturation of neonatal and adult human CD45R0- CD4 T cells. J Clin Invest 94:1352–1358
Siegrist CA, Lambert PH (1997) Immunization with DNA vaccines in early life: advantages and limitations as compared to conventional vaccines. Springer Semin Immunopathol 19(2):233–243
Skea DL, Barber BH (1993) Adhesion-mediated enhancement of the adjuvant activity of alum. Vaccine 11:1018–1026
Skeen MJ, Miller MA, Ziegler HK (1996) Interleukin-12 as an adjuvant in the generation of protective immunity to an intracellular pathogen. Ann NY Acad Sci 795:416–419
Steger MM, Maczek C, Grubeck Loebenstein B (1996) Morphologically and functionally intact dendritic cells can be derived from the peripheral blood of aged individuals. Clin Exp Immunol 105:544–550
Stoute JA, Slaoui M, Heppner DG, Momin P, Kester KE, Desmons P, Wellde BT, Garcon N, Krzych U, Marchand M, Ballou WR, Cohen JD (1997) A preliminary evaluation of a recombinant circumsporozoite protein vaccine against Plasmodium falciparum malaria. RTS,S Malaria Vaccine Evaluation Group. N Engl J Med 336:86–91

Suzuki K, Torii K, Hida S, Hayashi H, Hiyama Y, Oomoto Y, Takii T, Chiba T, Onozaki K (1994) Differences in interleukin 1 (IL-1), IL-6, tumor necrosis factor and IL-1 receptor antagonist production by human monocytes stimulated with muramyl dipeptide (MDP) and its stearoyl derivative, romurtide. Immunopharmacology 28:31–38
Takahashi H, Takeshita T, Morein B, Putney S, Germain RN, Berzofsky JA (1990) Induction of CD8+ cytotoxic T cells by immunization with purified HIV-1 envelope protein in ISCOMs. Nature 344:873–875
Tarr PE, Lin R, Mueller EA, Kovarik JM, Guillaume M, Jones TC (1996) Evaluation of tolerability and antibody response after recombinant human granulocyyte-macrophage colony-stimulating factor (rhGM-CSF) and a single dose of recombinant hepatitis B vaccine. Vaccine 14(13):1199–1204
Taylor J, Tartaglia J, Riviere M, Duret C, Languet B, Chappuis G, Paoletti E (1994) Applications of canarypox (ALVAC) vectors in human and veterinary vaccination. Dev Biol Stand 82:131–135
Ten Hagen TL, Sulzer AJ, Kidd MR, Lal AA, Hunter RL (1993) Role of adjuvants in the modulation of antibody isotype, specificity, and induction of protection by whole blood-stage Plasmodium yoelii vaccines. J Immunol 151:7077–7085
Traquina P, Morandi M, Contorni M, Van Nest G (1996) MF59 adjuvant enhances the antibody response to recombinant hepatitis B surface antigen vaccine in primates. J Infect Dis 174:1168–1175
Treanor J, Dumyati G, O'Brien D, Riley MA, Riley G, Erb S, Betts R (1994) Evaluation of cold-adapted, reassortant influenza B virus vaccines in elderly and chronically ill adults. J Infect Dis 169(2):402–7
Treanor JJ, Betts RF, Smith GE, Anderson EL, Hackett CS, Wilkinson BE, Belshe RB, Powers DC (1996) Evaluation of a recombinant hemagglutinin expressed in insect cells as an influenza vaccine in young and elderly adults. J Infect Dis 173:1467–1470
Ulmer JB, Donnelly JJ, Parker SE, Rhodes GH, Felgner PL, Dwarki VJ, Gromkowski SH, Deck RR, DeWitt CM, Friedman A et al (1993) Heterologous protection against influenza by injection of DNA encoding a viral protein. Science 259:1745–1749
Ulrich JT, Myers KR (1995) Monophosphoryl lipid A as an adjuvant. Past experiences and new directions. Pharm Biotechnol 6:495–524
Vadolas J, Davies JK, Wright PJ, Strugnell RA (1995) Intranasal immunization with liposomes induces strong mucosal immune responses in mice. Eur J Immunol 25:969–975
Valensi JP, Carlson JR, Van Nest GA (1994) Systemic cytokine profiles in BALB/c mice immunized with trivalent influenza vaccine containing MF59 oil emulsion and other advanced adjuvants. J Immunol 153:4029–4039
Van Hoecke C, Prikazsky V, Uto I, Menschikowski C (1996) Immunogenicity of an inactivated split influenza vaccine in institutionalized elderly patients. Gerontology 42:190–198
Villacres Eriksson M (1995) Antigen presentation by naive macrophages, dendritic cells and B cells to primed T lymphocytes and their cytokine production following exposure to immunostimulating complexes. Clin Exp Immunol 102:46–52
Wolff JA, Ludtke J, Acsadi G, Williams P, Jani A (1992) Long term persistence of plasmid DNA and foreign gene expression in mouse muscles. Hum Mol Genet 1:363–369
Wu JY, Gardner BH, Murphy CI, Seals JR, Kensil CR, Recchia J, Beltz GA, Newman GW, Newman MJ (1992) Saponin adjuvant enhancement of antigen-specific immune responses to an experimental HIV-1 vaccine. J Immunol 148:1519–1525
Yang LP, Byun DG, Demeure CE, Vezzio N, Delespesse G (1995) Default development of cloned human naive CD4 T cells into interleukin-4- and interleukin-5-producing effector cells. Eur J Immunol 25:3517–3520
Zhou F, Huang L (1993) Monophosphoryl lipid A enhances specific CTL induction by a soluble protein antigen entrapped in liposomes. Vaccine 11:1139–1144

CHAPTER 5

Vaccines Against Measles, Mumps, Rubella, and Varicella

E. NORRBY

A. Introduction

There are good reasons to draws parallels between the vaccines against the four diseases measles, mumps, rubella, and varicella. Each is a classical systemic childhood disease, and infection with wild-types of each of these viruses bequeaths a life-long protection upon renewed exposure to the infectious agent. Furthermore, live vaccines effectively preventing each of these four diseases have been developed by empirical techniques. However, there are also important differences in pathogenic events connected with the acute diseases caused by these agents.

The severity of acute disease generally decreases from measles to mumps, rubella, and varicella. Vaccination against measles and mumps aims at preventing acute disease and particularly in the case of mumps the potential complications. With rubella the acute disease in children is relatively harmless, but infection of a pregnant woman is associated with a major risk of congenital abnormalities, particularly when the infection occurs during the first trimester of pregnancy. Infection of fetal tissue may lead to prolonged carrier states, and children with such long-term infection may disseminate the virus further. However, as with measles and mumps, the infection is normally spread by a chain of consecutive contagious transfers of the virus from an acutely infected person to a susceptible person. Use of the rubella vaccine seeks to provide individual and/or herd immunity, thus preventing the infection of pregnant women.

Using the live varicella vaccine to prevent infection by wild virus has certain distinctive features. As in the case of other herpesviruses, varicella virus after causing a primary systemic infection is not effectively cleared from the body by the host's immune response. Rather, latent infection is established, and later activation of the dormant virus in the immune individual may lead to a dermatomally restricted zoster, reflecting centrifugal spread of virus in sensory neurons and deposition in epidermal tissue. The use of live varicella virus vaccine came to be considered in the United States after long deliberations, originally for use in immunocompromised children, predominantly in those with underlying leukemia, and later to protect entire cohorts of children. The vaccine was therefore first shown significantly to reduce the risk of developing varicella after household exposure of immunocompromised individuals, and evidence accumulated only later that healthy children can also profit from vaccination. The parenterally injected vaccine virus tends less than the wild virus to remain in a latent state after systemic infection, and therefore the risk of persons developing zoster who have had only a single infection with the vaccine virus is substantially reduced.

Each of the four vaccines are discussed separately below. Childhood vaccinations with these are the most cost-effective interventions available for improving health and well-being, as concluded by the World Bank (1993) and the United States National Academy of Sciences study on the future of children's vaccines (Institute of Medicine 1993). For each vaccine a decision needs to be taken as to the ultimate goal of its use: restriction or elimination of disease, or a global eradication of infection.

B. Measles Vaccine

Efforts at attenuating the measles virus began with its isolation in cell cultures by Enders and collaborators in 1954. Knowledge about its biology and particularly its molecular biology was virtually nonexistent at the time, and only rudimentary information was then available about pathogenic events. The researchers were able, however, to select viruses with reduced virulence by using empirical approaches including consecutive passaging in cells of non-primate origin and at reduced temperature.

Although it is possible today to distinguish wild-type and attenuated measles virus strains based on molecular characteristics (Bellini et al. 1994), specific molecular determinants of virulence have not yet been identified. Virulence is presumably determined by several different genes, and it is thus difficult to attribute specific characteristics to any. The identification of candidate receptor structures for the measles virus (Dorig et al. 1993; Naniche et al. 1993; Dunster et al. 1994) allowed examination of the role of early events in virus-cell interaction for virulence. Recent lymphotropic wild-type virus isolates have been show to have a lower tendency to downregulate the expression of the CD46 receptor structure than laboratory-adapted strains (Schneider-Schaulies et al. 1995). Since lack of CD46 expression at the cell surface leads to susceptibility of cells to complement lysis, it may be that cell culture adapted and attenuated virus strains are associated with a more self-limiting and less immunogenic course of disease when replicating in man.

Information has been obtained on the immunogenic potential of various structural components. This knowledge derives partially from insights into the defects of inactivated vaccine products used for field trials in the 1960s. It has long been known that passively transferred antibodies can prevent or modify a measles infection (Black 1989). The inactivated vaccine products given by repeated injections induce high titers of in vitro neutralizing antibodies, but these antibodies are only about 1% as effective in providing protection as the corresponding antibodies occurring in immunoglobulin preparations. This puzzling observation was eventually resolved when the virus was demonstrated to have two surface glycoproteins (TYRRELL and NORRBY 1978), each of which can induce in vivo protective antibodies. The immunoprotective potential of the two glycoproteins was confirmed in later studies employing purified virus components (VARSANYI et al. 1984, 1987), component-specific monoclonal antibodies (GIRAUDON and WILD 1985), and vaccinia vectors containing genes for the F and H components (BRINCKMAN et al. 1991; WILD et al. 1992). The inactivating procedures used (formalin of Tween-ether treatment) apparently had detrimental effects on the immunogenic properties of the fusion (F) component but left the attachment component, the hemagglutinin (H), unimpaired. The deficient humoral immunity in children receiving inactivated vaccine not only resulted in susceptibility to infection but also provided conditions for the evolution of atypical immunopathological disease in some cases (see NORRBY 1997). The live measles vaccine induces immunity to both surface glycoproteins (NORRBY et al. 1975), and in addition provides the bonus of inducing cell-mediated immunity.

The live measles vaccine has been widely used in both industrialized and in developing countries, and it is estimated that about 80% of today's new cohorts of children throughout the world become vaccinated. It is calculated that this immunization prevents about 90 million cases of measles and 1.5 million deaths per year. Further research is still needed, however, since annually there are still some 45 million cases of measles and more than 1 million deaths among children. New vaccine logistics (SABIN 1992) have enabled major advances to be made toward the elimination and eventual eradication of measles from the Americas. This program has given encouragement to efforts aimed at the global eradication of measles by the year 2010.

I. Safety and Efficacy of Live Measles Vaccine

When properly applied, the vaccine has a high take rate, provided no maternal antibodies are present. Seroconversion occurs in more than 90% of vaccines, but there is always a proportion of primary vaccine failure (BLACK 1989). Genetic factors on the part of the host may have some effect on the capacity of virus to replicate in a given individual. Inappropriate storage of vaccine, such as may occur in developing countries, can cause the virus infectivity titer to drop and hence reduce the take rate. However, the use of thermostabilized vaccines has markedly improved this situation. Immunization-related reactions generally do cause no inconvenience, and the vaccine infection does not spread from immunized children. In cases of vaccination-associated measles symptoms, it may be of value to isolate and characterize the virus. A Japanese study (KOBUNE et al. 1995) examined seven virus isolates, focusing particularly on the nucleotide sequence of the H gene. Only one of the virus isolates proved to be of true vaccine origin. Antibody titers are lower after vaccination than after a wild virus infection (KRUGMAN 1971). The average titer difference is about tenfold. Still, the immunity after vaccination is durable.

The possible role of continued production of measles antigen in an individual is a matter of debate among scientists. In rare cases measles virus persists and eventually, usually after many years, causes progressive and destructive brain infection. This disseminating disease, called subacute sclerosing panencephalitis (SSPE), is caused by a defective cell-associated virus spreading by some form of cell-to-cell contact. No infectious virus is produced, and consequently the afflicted individual is not contagious. However, antigen is released from disrupted cells, and a state of hyperimmunization characterizes both the general circulation and in particular the central nervous system. By contrast, antibody levels taper off gradually in the absence of exposure to virus from the outside after infection and vaccination, which suggests that the immune system can sustain durable immunity in the absence of renewed antigen exposure.

Although secondary vaccine failures do occur, overt clinical symptoms are relatively rare. Secondary vaccine failures are estimated to occur at a frequency of roughly 5% 10–15 years after immunization (MATHIAS et al. 1989). Subclinical reinfection may be relatively common when wild virus is still circulating in the community; such infections boost the level of immunity. Hence the general level of immunity in a population, as reflected in the average antibody levels, may vary depending on whether virus is continuing to circulate in the herd.

In industrialized countries the earliest injection of the live vaccine is given at the age of 12 months in order to prevent maternal antibodies interfering with the replication of vaccine virus. It has been suggested that this early age of immunization will have to be lowered in the future when the immunity of women of childbearing age relies solely on vaccination. So far only minor differences have been found in average antibody titers in populations immunized by infection with wild-type virus or by vaccination; however, the latter group under the conditions may have been reexposed to circulating wild-type virus.

A critical factor in determining the efficacy of immunity is the antigenic stability of the virus. Measles virus is the prototype of a stable monotypic virus. As with other single-stranded RNA viruses, it has a high mutation rate. It is therefore possible to trace the molecular evolution of measles virus strains over time (ROTA et al. 1992, 1994; TAYLOR et al. 1991). Furthermore, spontaneous changes can be demonstrated in epitopes identified with monoclonal antibodies (BIRRER et al. 1981; SHESHBERADARAN and NORRBY 1986; GIRAUDON et al. 1988). In spite of such changes the virus retains its stable monotypic characteristics as defined by examination with polyclonal antisera (SHESHBERADARAN and NORRBY 1986). No antigenic drift of virus of practical consequence occurs. It is therefore possible to use the existing live measles vaccine in all parts of the world and for the foreseeable future.

The nature of interaction between the measles virus and the host is often affected by the host's genetic constitution and nutritional conditions. A low percentage of children experience mild reactions such as slight fever and a transitory rash 9–11 days after immunization, depending on the strain of vaccine virus used. In addition to these acceptable side reactions, complications may also occur. For example, vaccine virus may also cause SSPE, but this complication is extremely rare, with a frequency still lower than after wild virus infection (DYKEN et al. 1989). Interestingly, there appears to be a shorter incubation time (4–5 years) for SSPE after vaccination than after wild-type measles virus infection. Congenital or acquired immunodeficiency of the host may enhance virus replication, including that of vaccine virus. For example, progressive infectious encephalitis caused by vaccine virus has been seen in children both with congenital (MONAFO et al. 1994) and with acquired immunodeficiency (VALMARI et al. 1987). Still, it is advisable for many immunosuppressed children to endure a vaccine virus infection rather than to run the risk of attempting to contain replication of wild-type virus. Therefore it is recommended that, for example, children with HIV infection receive measles vaccine, and that immunization be given as early as possible (PALUMBO et al. 1992). The take rate in these children is somewhat lower than in HIV-seronegative individuals. Children with severe immunodeficiency should not be vaccinated.

II. Epidemiological Consequences of Using Live Measles Vaccine

As noted above, the acutely infected individual is an indispensable link in the chain of epidemic transmission of measles virus. Since rinderpest virus adapted to become a human virus some 3000–4000 years ago, man has been the sole host of this infectious agent. Although the virus has evolved to be highly contagious, for an epidemic to be sustained it requires a clustered human setting of some 300000 persons to enjoy continuous access to susceptible individuals. There are two reasons for this dependence on civilization: the long generation time of humans, required to allow maturation of our unique brain cortex, and the life-long immunity established after infection.

Vaccine virus replication also induces an immunity that is durable and for practical purposes life-long. However, occasional cases of secondary vaccine failure have been recorded, and subclinical reinfections may occur more often after vaccination than after infection-induced immunity. While a symptomatic reinfection may well be contagious, it is much less certain whether an individual with subclinical infection can serve as a link in spread of infections. Epidemiological evidence argues against this (EDMONSON et al. 1990).

Practical experience has shown that herd immunity can be established solely by the use of vaccine. However, this requires an effective vaccine-induced immunity in at least 90%–95% of the individuals. Even within such an overall immunized population susceptible individuals may be clustered and thus allow a limited appearance of disease in pockets. The combined effect of primary vaccine failures and the waning of immunity in a small proportion of the vaccinees may lead to the occurrence of measles in adolescents and adults. Appreciation of this problem has led to the development of vaccination schemes comprising two injections. This is the best approach for securing an immunity level higher than 90% in the population. In industrialized countries the first immunization is given at 13–18 months of age. When a combined measles-mumps-rubella vaccine is used, a second dose can be given at an age of about 11 years to optimize conditions for appropriate establishment of rubella immunity close to sexual maturity, but the second dose may also be given earlier, at the age of about 6 years. A two-dose scheme has worked well in Nordic countries such as Sweden (CHRISTENSEN and BÖTTIGER 1994b) and Finland (PELTOLA et al. 1994) and is now being implemented also in many other industrialized countries.

The problems of vaccine use differ in developing countries and industrialized countries. The intermittent endemic circulation of virus may allow for exposure early in life. It is therefore too late to vaccinate at an age after maternal antibodies have completely disappeared. Rather, vaccination should be given as early as possible, but the earlier the vaccination is attempted, the greater is the barrier generally to virus replication mounted by maternal antibodies. The compromise age recommended for vaccination is 8–9 months, representing the optimal window of opportunity. Since, however, there is some reduction in the take rate at this age, a supplementary second immunization 6 months later is advisable. In order to completely prevent young-age measles in these milieus, vaccinations would have to be given at the age of 4–6 months.

The amount of measles virus routinely used for individual immunization is between 10^3 and 10^4 plaque-forming units. Increasing the dose 10–100 times improves seroconversion and take rates in young infants (BERRY et al. 1992; CUTTS et al. 1993; JOB et al. 1991). Best results are obtained with the Edmonston-Zagreb strain of vaccine virus, which is not associated with adverse immediate reactions to the high virus dose vaccination. However, follow-up of vaccinees shows significantly higher mortality among vaccinated children 1–2 years after immunization. This increased mortality is not due to the direct effects of measles virus replication but to other, predominantly infectious causes. Moreover, it occurred in countries with a high childhood mortality (GARENNE et al. 1991) but not in those with a relatively low rate (DIAZ-ORTEGA et al. 1992). In addition, there was a sex difference in that the increased mortality affected principally girls (HOLT et al. 1993). It has been conjectured that vaccine virus replicated somewhat more intensively in girls than in boys, and that a state of greater immunosuppression was therefore established.

It has long been known that immunosuppression follows a normal measles infection. In addition, however, a vaccine virus infection also leaves behind immunosuppressive effects, although of less intensity. For example, it has been found that the vaccine induces transient lymphopenia (BLACK and SHERIDAN 1967), suppression of delayed hypersensitivity skin test responses to recall antigens (MELLMAN and WETTON 1963), a decrease in mitogen- and antigen-stimulated proliferation of lymphocytes (FIREMAN et al. 1969; ZWEIMAN et al. 1971), and altered cytokine production (WARD and GRIFFIN 1993). Attempts at early vaccination have now been abandoned, however, as the result of the unfortunate experiences of some young infants receiving high-dose immunization. New kinds of vaccines are required before further projects at early vaccination can be pursued in the future.

III. New Kinds of Measles Vaccines

Although the empirically generated live vaccine is generally a very efficient immunogen, attempts have continued toward developing new types of products. These attempts have been spurred by remarkable advances in understanding the biology and molecular biology of measles virus (GRIFFIN and BELLINI 1996). It is now possible to use the technique of reverse genetics since Billeter and colleagues (RADECKE and BILLETER 1996) recently succeeded in establishing an infectious cDNA representing the whole measles virus genomic RNA. This may allow the development of new kinds of live vaccines. Such vaccines may also be produced by other approaches (for reviews see NORRBY 1995, 1997). In addition, various replicating vectors containing one or more measles virus genes have been characterized in animal systems. Also, other avenues towards new vaccines are being explored. These include infectious DNA representing one or more genes and various kinds of nonreplicating immunogens, including various types of whole components or other epitope-based constructs. Limitations of space prevent presentation of further details of the interesting studies that have been carried out. The reader is referred instead to recent specialized reviews (NORRBY 1995, 1997). In view of the excellent performance of the current live vaccine, it is questionable whether competing products will ever be introduced onto the market.

IV. Remaining Problems in Using Measles Vaccine and Projected Future Developments

Measles is an eradicable disease. The monotypic virus survives as a selfish gene only in an endless series of consecutive acute infections in humans. As noted above, the survival of the virus is precarious because of the long human generation time and the durable immunity following infection or vaccination. Its potential advantage as a result of expanding civilization and thus access to ever-larger cohorts of humans is counterbalanced by the increasingly effective programs of artificial vaccine-induced immunity in populations throughout the world.

The virological sage Albert SABIN (1992) recommended the institutionalization of national immunization days in which whole cohorts of children of different ages are immunized simultaneously to stamp out poliomyelitis and measles. He further advocated mucosal immunization also in the case of measles to optimize development of protection at the portal of virus entry into the body. This route of vaccination has worked well in protecting the gut by use of live polio vaccines, but less effectively in trials aimed at instituting respiratory tract immunity against measles. The reason for this is practical, namely, no satisfactory methods exist for administering live measles vaccine intranasally with an acceptable high take rate (KHANUM et al. 1987). However, the arrangements of national immunization days has delivered splendid success. We are now advancing rapidly towards the eradication of poliomyelitis by the end of the millenium, and plans call for global eradication of measles about a decade later.

To recapitulate, there is a four-pronged approach to the possible eradication of measles:

- Effective yearly vaccination of a cohort of children at a selected early age (approx. 9 months in developing countries and 13–18 months in industrialized countries) with recommendations for a second dose of vaccine
- National mass vaccination campaigns covering a broad range of age groups of children
- Regional mass vaccination campaigns ("mop-up" immunization) targeting a broad range of age groups of children
- Effective surveillance to identify suspected cases

The success in eradicating measles in the Americas has encouraged the drafting of plans for a global eradication of the disease. Figures on measles for the year 1995 illustrate the impressive achievements. There have been very few cases of the disease in South America and the English-speaking Caribbean for more than 4 years, and measles has been entirely eliminated in Brazil, Chile, and Cuba. The United States had a record low number of 301 cases, whereas 80% of all cases occurred in Canada. The latter country is now remedying this situation by improving the coverage of the yearly vaccination of cohorts of young children and by introducing a second dose of vaccination.

In conclusion, the time is now ripe to formulate definite plans for eradicating measles under the auspices of international organizations. Achieving this goal will mean the elimination of the major killer of children throughout human history.

C. Mumps Vaccine

Mumps virus, a paramyxovirus, as is measles virus, leads to a civilization-dependent acute disease. Large, interacting populations are required to sustain an endemic spread of the disease, and this afflicts primarily children. The infectiousness of the virus is somewhat lower than of measles, as illustrated by the relatively slower accumulation of mumps-seropositive individuals in groups of increasing age in an unimmunized population. Therefore, prior to the advent of vaccination, mumps was a greater problem than measles among adult populations, such as military recruits. Approximately one-third of all mumps infections occur without recognizable symptoms. Clinical cases are generally moderately severe, but the impact of the disease is accentuated by the coincidence of meningitis/encephalitis in about 10% of cases and that of orchitis. The frequency of the latter complication increases somewhat with age. The medical importance of the disease has motivated prophylactic interventions in industrialized countries, but preventive measures have not yet been given priority in developing countries. For this reason mumps is still endemic in most parts of the world.

The first live vaccine was produced in the mid 1960s by passaging of virus in chick embryo cell cultures (BUYNAK and HILLEMAN 1966). The virus strain used was recovered from the daughter of the well-known senior scientist in the team. The attenuated virus has been given in more than 100 million doses (HILLEMAN 1992, 1996) and displays a track record of safety and efficacy. The immunity established is durable.

Early attempts were also made to produce an inactivated vaccine. Such a vaccine was developed in Finland and used to protect military recruits, where a 95% reduction in mumps and complicating orchitis was achieved (PENTTINEN et al. 1968). As was mentioned above, it had been found that formalin inactivation has a detrimental effect on the F protein of measles virus. It was therefore investigated whether this type of inactivation has a similar effect on mumps virus. These studies demonstrated that the inactivated vaccine does in fact lack the capacity to induce antibodies inhibiting hemolysis by direct interaction with the F component (NORRBY and PENTTINEN 1978), but later studies have shown that antibodies to the denatured F protein are present (JULKUNEN et al. 1984). It has never been determined whether the defective humoral immunity which developed in originally seronegative individuals immunized with the inactivated vaccine predisposes to some form of atypical disease upon reexposure to the wild-type virus, in analogy to that seen after the use of inactivated measles vaccine. It is possible that atypical mumps would display such a unique clinical picture that no relationship to the traditional disease would be recognizable. The inactivated vaccine was exchanged for the live vaccine in Finland in 1982–1984.

I. Safety and Efficacy of Live Mumps Vaccine

As noted above, a large number of vaccinations against mumps have been administered, almost without exception in the form of combined measles, mumps, and rubella vaccine. The mumps vaccine virus produces no clinical symptoms when used alone, and no virus can be isolated from blood, urine, or saliva (ENNIS et al. 1969). The frequency of vaccine-associated reactions occurring within 4 weeks of immunization is low (MCDONALD et al. 1989; STRATTON et al. 1994), and it is estimated that complicating meningitis occurs in one case per 11000 (MILLER et al. 1993). Analyses of idiopathic thrombocytopenic purpura occurring after triple measles, mumps, rubella vaccine (FARRINGTON et al. 1995) has found one case in 24000, which is five times the expected rate. There was no evidence of a mumps strain specific effect.

However, such an effect was found in an analysis of convulsions occurring 15–35 days after vaccination, and this concerned only vaccine including the Urabe strain of virus. This Japanese vaccine strain has been found to be somewhat more virulent than the Jeryl Lynn strain and to cause a more disseminating infection, including meningitis (Brown et al. 1991). In many countries this virus strain has been withdrawn from the market. A similar fate has befallen a Bulgarian mumps virus vaccine strain called Sofia 6, prepared in guinea pig kidney primary cell culture (Odissev and Gacheva 1994).

Regarding vaccine-associated complications one sometimes sees mild allergic reactions such as rash, pruritus, and purpura. However, there is no increased risk of complications in egg-allergic children. Mumps vaccine virus has not been demonstrated to spread from a vaccinee to the surroundings (Ennis et al. 1969). In suspected cases occurring among vaccinees and in the event of an increase in unexpected complications, today's molecular biological techniques can distinguish whether the isolated virus is of wild-type or vaccine origin (Kunkel et al. 1995).

Measurable titers of neutralizing antibodies are induced in more than 90% of immunized children at the first vaccination (Harmsen et al. 1992; Davidkin et al. 1995). Antibody titers decline relatively rapidly during the first year, but stabilize in subsequent years. After revaccination some 5–6 years later seroconversion levels exceed 95%, and antibody titers decline more slowly. The mean mumps antibody titer 4 years after a second dose of vaccine is higher than the corresponding titer after the first dose. The protection against development of clinical disease after a two-dose vaccination scheme appears high and durable. Secondary vaccine failures are probably rare (Gut et al. 1995).

II. Epidemiological Consequences of Using Live Mumps Vaccine Use

The epidemiological effect of one or two childhood injections of mumps vaccines can be seen, for example, in the United States and Scandinavia. In the United States there was a substantial decrease in the incidence of mumps even in the first years after licensing of the vaccine in 1967. More vigorously pursued vaccination programs, still with a single dose of immunization at the age of 13–18 months, led to a record low number of mumps cases, 1692, in this country in 1993. Among these cases children in school ages and in later adolescence were overrepresented. As in the case of immunoprevention of the other infections discussed in this chapter, it has proven difficult to achieve a solid herd immunity. Thus focal outbreaks can be seen in highly immunized populations (Wharton et al. 1988; Hersch et al. 1991; Briss et al. 1994). It is important that general vaccine coverage be as effective as possible because in older children there is a higher tendency to develop orchitis (Andersson et al. 1987).

Use of the available live vaccine is currently limited to the more affluent industrialized countries. In these countries impressive results have been achieved, with the virtual elimination of the virus and the disease, as exemplified by the current situation in Sweden (BÖTTIGER et al. 1987) and Finland (PELTOLA et al. 1994).

III. New Kinds of Mumps Vaccines

There seems to be little interest at present in developing new mumps vaccines. The market today is dominated by a single strain of live virus vaccine, and this situation does not seem to change rapidly. Still, there are possibilities for the development of new kinds of mumps vaccine by using basically the same approaches as in the development of various types of new measles vaccines, as discussed above. The immunogenic shortcoming in previous nonreplicating immunogen products could probably be remedied by the development of new gene expression systems allowing a native folding also of the F protein. Fulfilling the dream of producing an equally effective and even less expensive mumps vaccine would facilitate a wider global use of vaccine against the disease.

IV. Remaining Problems in Using Mumps Vaccine and Projected Future Developments

There appear to be very few problems in the use of live mumps vaccine. Side reactions are negligible, and the immune protection appears solid. In industrialized countries cost-benefit analysis clearly favors intensive vaccination. The vaccine is now less heat labile, can be effectively lyophilized (MCALEER et al. 1980), and is available for convenient use in the combined measles-mumps-rubella vaccine. It may therefore potentially be used in developing countries, but its use there depends upon the health priorities in the respective countries. Because it is a rare killer of children, it does not rank high on a list of such priorities. However, the prospective elimination of mumps in industrialized countries by vaccination may encourage more immunization in other parts of the world.

Industrialized countries must define the optimal logistics for immunization of children. The possibility exists of eliminating mumps, but basic characteristics of the virus and the infection that it causes are such that even its eradication is feasible. The next question is therefore whether and when such an eradication should be attempted. A great bonus in the event of global eradication of mumps, as for any other virus, is of course that vaccination eventually can cease.

D. Rubella Vaccine

As with measles and mumps, rubella is a civilization-dependent acute disease, but it generally runs a rather mild course than the former two. The infection may be somewhat less contagious than measles, and the virus spreads preferably in densely populated areas. It has been considered a typically childhood disease throughout history. However, the medical impact of the infection does not concern primarily young children, although complications such as encephalitis may occur, but rather infected women during the early stage of pregnancy. The spread of virus to the developing fetus may have disastrous effects, including spontaneous abortions and fetal malformations. Children infected at an early stage in utero, with or without malformations, may carry the virus infection for some time after birth. However, from the practical point of view, such virus carriers do not represent an important source of virus for endemic spread.

In earlier times it was sometimes the practice to hold rubella parties to ensure that girls acquired immunity before reaching puberty. As soon as the virus had been isolated in tissue culture in 1962, empirical work began towards developing a live vaccine (reviewed in PLOTKIN 1988). The first vaccine strain introduced in the United States was a modified form of the strain HPV-77. The original strain proved to entail too may complications for extensive use and was abandoned. The strain became much less reactogenic after fivefold passage in duck cells. Eventually this HPV-77 variant was progressively replaced by the RA 27/3 strain, which showed somewhat fewer side reactions. The vaccine virus in the latter product originated from passaging of virus in human diploid cells at a reduced temperature, 30°C. In Europe the Cendehill vaccine strain and in Japan a number of indigenously attenuated virus strains were developed in the early phase.

Since the aim of vaccination is to prevent infection of pregnant women, several different methods of administering the vaccine have been tried. An effective reduction in virus-induced congenital malformations was achieved in a number of programs, but no coherent policy for optimally achieving this result has been developed. The World Health Organization does not include rubella vaccine in its program for expanded immunization of children, and the infection therefore remains endemic in most parts of the world (MILLER 1991). However, it seems that an effective herd immunity can be obtained by childhood immunization, preferably with a two-dose schedule as implemented in Finland (PELTOLA et al. 1994) and Sweden (CHRISTENSSON and BÖTTIGER 1994a). It is theoretically possible to eradicate rubella. Whether the eradication is ever in fact achieved depends on to what extent the disease will be prioritized relative to other pressing needs in the prevention of infectious diseases. However, it should be remembered that it can be very costly to care for children with congenital rubella syndrome.

I. Safety and Efficacy of Live Rubella Vaccine

Wild-type rubella virus has moderate to low virulence in children, and the attenuated virus produces negligible side effects in such vaccinees. However, the severity of side effects increases with age, primarily in the form of transient to sometimes prolonged conditions of arthritis and occasional neurological complications. In addition to its use in the immunization of children, the vaccine is administered to two groups of adult women, those postpartum who are seronegative, and those who may be professionally exposed to children with rubella or congenital rubella syndrome. Precautions are taken to prevent women being immunized during pregnancy; however, this does sometimes occur accidentally, and a cohort of such cases in the United States has been carefully examined. The virus, disregarding vaccine strain origin, can in some instances be recovered from the placenta and from fetal tissue (HAYDEN et al. 1980; VAHERI et al. 1972). However, it does not seem that the virus is teratogenic since none of 322 children born to the mothers accidentally immunized during early pregnancy and who were examined in this study displayed any malformations.

The dimension of the problem presented by vaccine-associated arthropathy, potentially of a more chronic nature, is a matter of dispute. Some researchers maintain that it is very important (TINGLE et al. 1985) while others estimate the risk more conservatively. A review by the Institute of Medicine of the United States National Academy of Sciences (HOWSON and FINEBERG 1992) concluded that insufficient data exist to allow any firm conclusions. It was recommended that more research be performed with careful longitudinal follow-up of cases. Use of the polymerase chain reaction technique to detect rubella virus RNA in diseased joints has produced conflicting results (MITCHELL et al. 1993; FRENKEL et al. 1996). Findings from further studies to resolve this issue should be forthcoming in the near future.

When properly administered, the vaccine yields a conversion rate exceeding 95%. Symptoms related to the vaccine virus infection are rarely seen and the vaccinee is not contagious for his surroundings (HALSTEAD and DIWAN 1971). The immunity established is durable (MILLER et al. 1995) and highly protective, but with time reinfections with wild-type virus may occur. Immunization with the RA 27/3 vaccine results in the lowest frequency of reinfections. Five years after immunization with this vaccine 10% of girls experienced subclinical infection upon exposure to wild-type virus as evidenced by serological response (CUSI et al. 1993). A limited replication of virus in the nasopharynx can be detected during reinfection, but viremia and systemic manifestations are rare events (DAVIS et al. 1971). The critical question is whether virus replicating locally in the mucosa can spread to the fetus under these circumstances. The likelihood of this appears very small (O'SHEA et al. 1983). A very limited number of cases of congenital rubella syndrome following reinfection have been reported (FREY 1994b), but these limited data do not resolve the question of the efficacy of postvaccination immunity. However, it should be emphasized that in a thoroughly immunized population, the potential problem of reinfection and fetal damage does not exist, provided circulation of virus is effectively blocked by herd immunity.

II. Epidemiological Consequences of Using Live Rubella Vaccine

In principle it is possible not only to eliminate but also to eradicate rubella infections and the virus causing them. However, the prime target in vaccine goal in its use has been the prevention of infection among pregnant women. Various approaches have been taken. In the United States a program has been employed that involves early childhood (12–15 months) immunization with rubella vaccine combined with measles and mumps vaccines supplemented with vaccination of seronegative women (reviewed in WOLINSKY 1996). A substantial spur to the development of prophylactic measures in the United States was the last major epidemic there in 1964, in which an estimated 20000 infants were born with congenital rubella syndrome, and there were at least 10000 cases of spontaneous or therapeutic abortions due to rubella. Epidemics of rubella were disrupted, but a certain number of infections still occurred, leading to about 50 cases per year of the congenital rubella syndrome. The vaccination program was intensified in the attempt to eliminate even these cases. As a result, only about two cases of congenital rubella syndrome per year occurred during the 1980s in the United States. However, there was a resurgence of rubella, as well as of measles, during 1989–1991 (MELLINGER et al. 1995), illustrating that effective, persistent herd immunity had not been established. In the immunized population there were pockets of susceptible individuals, mostly in their late teens, who could sustain an endemic spread.

As a supplementary initiative, it was decided to give two childhood vaccinations, with the second dose at 5–12 years of age. A similar vaccination schedule had already been initiated at an earlier stage in some European countries. In the United Kingdom a vaccination program was started by target only girls at the age of 12–15 years, combined with revaccination of seronegative women of childbearing age. After the rubella epidemic in 1986 (BEST et al. 1987) preschool children were included in the program of vaccination. There has since been a waning occurrence of rubella in the United Kingdom. A two-dose immunization scheme has been used in Finland (PELTOLA et al. 1994) and Sweden (CHRISTENSSON and BÖTTIGER 1994a) since the early 1980s. Rubella has been eliminated in both of these countries. In Japan early childhood a vaccination program against measles, mumps, and rubella has been in place for a number of years, but this program was stopped in 1991 when problems with the mumps component mentioned above were identified. In many countries no rubella vaccination is pursued, and in most parts of the world rubella infections still prevail (MILLER 1991).

III. New Kinds of Rubella Vaccines

Our knowledge about the biology and the molecular biology of rubella virus has advanced markedly during the past decade (FREY 1994a; WOLINSKY 1996).

It may now prove possible to develop new types of live vaccines using reverse genetics with this positive strand virus RNA of alphavirus characteristics, or some type of inactivated vaccines. Since the goal is to prevent virus spreading from the infected mother to her fetus, it might well suffice to induce durable titers of circulating neutralizing antibodies. It has been proposed that the surface glycoprotein E1 is the main component carrying immunodominant protection-inducing sites. This protein has been expressed in the baculovirus system (SETO and GILLIAM 1994), but data on immunogenicity have not been presented. In another study the two other structural proteins E2 and C were also expressed in this system (CUSI et al. 1995); when tested in a mouse model, these proteins did not provide protection, whereas replicating RA 27/3 vaccine virus did. The capacity of synthetic peptides representing either the E1 or the E2 glycoprotein have also been evaluated for their capacity to induce neutralizing antibodies (ROBINSON et al. 1995), and these were seen with some peptide immunogens representing both proteins. Further attempts to develop an inactivated vaccine may be worth pursuing. DNA immunization with a construct including one or more of the genes for structural proteins of the virus may also be of interest to study.

IV. Remaining Problems in Using Rubella Vaccine and Projected Future Developments

Major advances have been made in current applications of available live rubella vaccines. Many cases of congenital rubella syndrome have been prevented. However, there is no global consensus about how to apply the available vaccine in different settings. Because of the competition with other prioritized diseases in developing countries, the vaccine is presently used only to a small extent in this setting. Developments of new and hopefully relatively inexpensive vaccines could contribute to a change in attitude. New kinds of vaccines may also prove valuable in diminishing or eliminating the risk of transitory arthritis seen in some immunized women of childbearing age. In fact, as mentioned above, it is urgent to define the scope of the problem that such complications may represent.

Since rubella is an eradicable disease, it is tempting to define the required conditions for such an eradication in a selected setting. The Americas have lead the way in eradicating first smallpox and then poliomyelitis. In this region the attempt could also be made to eliminate rubella, and some success has already been made in this direction in the United States. An approach of this kind would include, for example, evaluating the effects of national campaign days; encouraging data have already been presented from such a campaign in Brazil using the combined measles, mumps, and rubella vaccine (MASSAD et al. 1995). In addition, effective means of surveillance must be developed for characterizing a variety of conditions ranging from clinically overt to subclinical rubella.

E. Varicella Vaccine

The basic biology of varicella virus and the pathogenic features of the infections that it causes are very distinct from the others discussed in this chapter. Measles, mumps and rubella viruses cause acute infections that are civilization dependent and have therefore become established in humans only relatively recently, but varicella-zoster virus, as other herpesviruses, have coevolved with the human species. The latter group of intelligent, selfish, viral genes have evolved means to outsmart the immune system and to remain latent in the human body. This "sleeping beauty" behavior enables the virus repeatedly to cause endemics in human populations regardless of their size and demographic composition. The relationship between virus and host is intimate and has been refined over a very long time. Only a minority of the genes in the DNA of herpesviruses are essential for the fundamental steps of virus replication as expressed in infected cell cultures; the nonessential gene products serve to tune the intensity of virus-host interaction by involvement in many operative contexts. Examples include humoral and cellular immune responses, induction, release and action of lymphokines, etc. The end result of the evolutionary balance thus established is an infection of moderate intensity, expressing itself eventually in the form of skin lesions for virus transmission, and the ability to lie in a dormant state suppressible by a normal cell-mediated immune response.

Attempts to develop a live varicella vaccine began in the 1960s. The KMcC strain was passaged in human diploid cells to a level optimally balanced between attenuation and retention of immunogenicity (NEFF et al. 1981). However, even at the minimal immunogenicity level up to 100 minimal lesions are seen in vaccinated children. The Japanese Oka strain of virus was fortunately found to provide better vaccine characteristics (TAKAHASHI et al. 1974): less reactogenicity but still satisfactory immunogenicity. It has nevertheless taken very long for this vaccine to become accepted by the microbiological and infectious disease community. For this reason less information is available on the vaccine than on those for measles, mumps, and rubella. The greatest amount of data has been gathered in Japan and South Korea, where more than 2 million children have been vaccinated. The corresponding figure in the United States is the less than 10000 children, divided among different trials. But the tide has now turned since the United States Food and Drug Administration in March 1995 licenced the use of the vaccine in varicella-susceptible children and adults. Important supplementary information is expected in the very near future. At the time of writing, about 1.5 million children have been vaccinated in the United States. The live vaccine has already proven its value for use in leukemic children (GERSHON et al. 1984), and it can be predicted that the vaccine will prove valuable for protecting not only this vulnerable group of patients but also healthy children (ARBETER et al. 1986).

There are several reasons for the long delay in acceptance of the live varicella vaccine for general use. On the technical side, the virus is difficult to propagate effectively in the laboratory, and in the early phase reliable techniques were lacking for determining seroimmunity and for distinguishing vaccine and laboratory strains. Furthermore, no representative animal model was available, and it was therefore difficult to predict whether the pathogenic potential in immunosuppressed leukemic children would be so pronounced to preclude its use, and whether the tendency to latency would be greater or less for the attenuated virus than for the wild-type virus.

I. Safety and Efficacy of Live Varicella Vaccine

Varicella caused by the wild-type virus is generally a relatively mild disease. However, pathogenetic events may become aggravated in immunosuppressed children and in newborns infected perinatally. The activated latent infection, the zoster disease, is also accentuated under conditions of immune suppression and represents one of the complicating herpes infections in transplanted individuals.

Counting the number of skin lesions is an effective means for assessing the severity of an acute infection (Tsolia et al. 1990). In fact, this information may have prognostic significance for the development of zoster, if deposition of latent virus in sensory ganglia is contingent upon a retrograde spread of virus from dermal lesions.

Since children and adolescents/adults react differently to the vaccine, the safety and efficacy of vaccination must be distinguished for three different cohorts: healthy children, leukemic children, and adolescents/adults (Krause and Klinman 1995). Healthy children react well to a single dose of vaccine, but the efficacy is dose dependent. When the dose of vaccine virus was increased from 1000–1500 plaque-forming units (pfu) to 3000–9000 pfu, the protective efficacy over a 3-year period increased from 67% to 93% (Krause and Klinman 1995). In another true double-blind study, using a somewhat higher dose of vaccine virus (17000 pfu), a 2-year efficacy of 96% was found (Weibel et al. 1984; Kuter et al. 1991).

The vaccine virus infection is very mild. A slight fever may develop in a small fraction of children, and skin lesions are seen in only 3%–4% of vaccinees. Transmissibility of vaccine virus from skin lesions may occur, but the infection is largely nonapparent or mild (Weibel et al. 1984). The level of protection against varicella upon renewed exposure to virus is high, but occasionally the immunity is broken through with an ensuing development of skin lesions (Watson et al. 1993). Clinically overt varicella cases were reported in 2% of a small group of children studied for 20 years in Japan (Asano et al. 1994). However, serological boosting is demonstrable in a larger percentage. It is important to assess the frequency of development of zoster in individuals whose primary immunity results from vaccine virus replication. The accuracy of such data will increase with as the observation time of the cohorts lengthens. However, encouraging preliminary data have already been collected. The incidence of zoster in vaccinated children is estimated at 18 per 100000 person-years, compared with a predicted value of 20–59 for the age group studied (GUESS et al. 1985; PLOTKIN et al. 1989).

Only limited data are available on the use of live varicella vaccine in adolescents and adults (WHITE et al. 1991; NADER et al. 1995). A slightly larger percentage of vaccinees in this group than healthy children develop a vaccine-associated rash, but the seroconversion rate is nevertheless lower. A second dose of vaccine has been found necessary to obtain seroconversion greater than 95%. The current recommendation is to give two vaccinations, 2 months apart, to all vaccinees aged 13 years or older. The attack rate after household exposure (estimated at 80%–90% in seronegative children) is 26% in healthy adults. At 11- to 13-year follow-up 82% of adults prove to be seropositive.

Leukemic children represent a particular cohort. They were selected as a target for vaccination in the late 1970s when it was found that 30% of these children receiving steroids develop disseminated chickenpox and 7% die from the infection (FELDMAN et al. 1975). The vaccine virus infection, as would be expected, is more intense in leukemic than in healthy children, but it is clearly manageable. At primary vaccination of the former group about 50% have skin lesions, and the seroconversion rate is about 80% (GERSHON et al. 1984; GERSHON 1995). Because of this relatively low seroconversion rate, a second vaccination is recommended after 2 months. This second vaccination occasionally causes a vaccine-associated rash (in 7%), but it increases the seroconversion rate to 95%. In the household setting vaccine virus is transmitted in 17% of cases, but only when skin lesions are present (TSOLIA et al. 1990). The expected transmission of wild-type virus is 80%–90%, as mentioned above. Interestingly, 27% of contact cases of vaccinated children have subclinical infections, compared to an expected figure of 5% for infections with wild-type virus, and in remaining contact cases the average number of skin lesions are markedly reduced. Restriction endonuclease analysis has revealed the vaccine virus nature of the agent replicating in contact cases (GERSHON et al. 1984). The protection efficacy is estimated at about 85%, and in household exposure the attack rate is 14%, compared with an expected value of 80%–90%. Follow-up after 11–13 years has shown that 87% of the children remained seropositive (GERSHON 1995). The frequency of zoster in vaccinated leukemic children requires a particular assessment (HARDY et al. 1991). The expected frequency in this group is estimated at 2460 per 100000 person-years, and the figure recorded in vaccinees is 800.

II. Epidemiological Consequences of Using Live Varicella Vaccine

Only in Japan and South Korea has varicella vaccine been used in children on a larger scale. The wider epidemiological consequences of using the vaccine can therefore not be derived from experience but must be conjectured on the basis of carefully evaluated, limited-scale field trials. Assuming that 90% vaccine coverage can be obtained, and that secondary vaccine failures involve 1%–2% of a cohort each year, mathematical modeling shows little risk of an increasing incidence of varicella in adults, provided that basically the whole population is immunized (HALLORAN et al. 1994).

Eradication of the infection is a realistic goal as regards the acute, community-dependent diseases discussed above. It is interesting to speculate whether it would be possible to eradicate herpesviruses and their diseases, which have characteristics very distinct from those of the other diseases discussed in this chapter. The key epidemiological questions concern, on one hand, the possibilities for establishing an efficient herd immunity and, on the other, means for the restrictive reintroduction of new viruses via activated latent infections, namely cases of zoster in this context. Efficient and durable herd immunity can most likely be achieved by vigorous immunization programs, potentially including repeated vaccinations and national immunization days. The relatively low frequency of vaccine-associated rash after immunization of healthy children appears to pose a limited risk for reintroduction of virus. Furthermore, the virus circulating in these cases is debilitated and therefore spreads less efficiently.

More data are required for a definitive assessment of the risk of developing zoster by the vaccine virus. Again, it should be remembered that the crippled vaccine virus is an inefficient means of transmission. It therefore seems that careful surveillance and the introduction of moderate hygienic conditions can make it possible to stop circulation of the virus. If a vigilant program aimed at eradication could be sustained over more than one human generation, the eventual outcome may in fact be eradication even of this kind of virus, with its capacity to establish latent infections. Attempts could be made to test the feasibility of eradication in a restricted geographical area before a decision (priority in relation to other herpesviruses and other infectious diseases?) can be taken on attempts at a global eradication. It can be noted that even under present conditions of proposed usage it can be shown that a routine administration of varicella vaccine in healthy children in the United States is cost-beneficial (LIEU et al. 1994).

III. New Kinds of Varicella Vaccines

Interestingly, all the current varicella vaccines produced by different manufacturers originate from the Japanese Oka strain developed by empirical methods in the early 1970s. Since that time a detailed mapping has been made of genes in the varicella virus genome. These refined molecular biological studies have shown the existence in the varicella virus genome, as in other herpesvirus genomes, of a large number of nonessential genes. The products of these genes affect pathogenetic events at the time of the acute infection and in establishing (and releasing) the state of latency. Although the absence of useful animal models limits studies on the pathogenetic role of all these different genes, attempts should be made to define their effect on pathogenetic events. Deletions of individual genes may result in either an increase or a decrease in pathogenicity. It can be assumed that there are a number of genes belonging in the latter category, and that removal of such genes would be an attractive means of diminishing virulence. Another attractive approach to developing a new kind of live vaccine would be identification and deletion of genes that are critical for establishing latency. Finding a live vaccine herpes-virus lacking the capacity to establish latency would be close to success in the quest for the "holy grail."

Finally, it should be mentioned in this section that candidate non-replicating immunogens could be of considerable use, provided that long-term immunity can be established. Cell-mediated immunity clearly plays a central role in clearing the acute infection and in control of latent virus, and the specific responder cell frequencies after vaccination are comparable to those elicited by a wild-type virus infection (HAYWARD et al. 1992; SHARP et al. 1992). However, circulating antibodies also play a central role in preventing a varicella infection. This is shown by the documented effect of zoster immunoglobulins (GERSHON et al. 1974).

IV. Remaining Problems in Using Varicella Vaccine and Priorities for the Future

More information is needed on the effects of large-scale routine use of the live varicella vaccine in healthy young children. Not only the extent to which effective protection can be established needs to be determined, but also the durability of this protection. It may be an advantage to include the varicella vaccine in the combined measles, mumps, and rubella vaccine. Trials with such a tetravalent product have produced encouraging results (BRUNELL et al. 1988; ENGLUND et al. 1989; WATSON et al. 1996).

Spread of vaccine virus from immunized children appears to be a minor problem, but a critical question concerns the risk of activating latent infection, and the risk that dermally exposed virus in these cases can be a threat as contagion. Finally, it may be important to consider what efforts should be undertaken to develop an even more effective varicella vaccine, perhaps completely devoid of the capacity to establish a state of latency.

F. Epilogue

This chapter discusses four classical childhood diseases – measles, mumps, rubella, and varicella – and the possibilities for preventing them by the use of vaccines. Remarkable advances have been made in administering the triple live measles, mumps, rubella vaccine licensed in the United States in the late 1960s, and interesting results can be expected for the more recently licensed vaccine against varicella. The measles vaccine is clearly of greatest importance in both industrialized and developing countries. The situation is somewhat different for other vaccines in developing countries. Clearly it would also be of value in these countries to eliminate mumps, rubella, and varicella, but the will to do so is hampered by the limited financial resources available for investment in health care and by the range of other, more devastating diseases. A definite program for global eradication of measles is soon to be ratified. After a certain number of years following its eradication, general vaccination will cease globally. Similar developmental scenarios can be sketched for the other diseases, including even the agent with the capacity to establish a latent state, varicella. In the trend to emphasizing global health, it will remain for representative authorities, such as the World Health Organization and the World Bank, to decide the sequence in which theoretically eradicable infectious diseases should be wiped out during the coming century, and to determine the means for realizing such transnational efforts.

References

Anderson RM, Crombie JA, Grenfell BT (1987) The epidemiology of mumps in the UK: a preliminary study of virus transmission, herd immunity and the potential impact of immunization. Epidemiol Infect 99:65–84

Arbeter AM, Starr SE, Plotkin SA (1986) Varicella vaccine studies in healthy children and adults. Pediatrics 78:748–756

Asano Y, Suga S, Yoshikawa T, Kobayashi H, Yazaki T, Shibata M, Tsuzuki K, Ito S (1994) Experience and reason: twenty year follow up of protective immunity of the Oka live varicella vaccine. Pediatrics 94:524–526

Bellini WJ, Rota JS, Rota PA (1994) Virology of measles virus. J Infect Dis 170 [Suppl 1]:S15–S23

Berry S, Hernandez H, Kanashiro R, Campos M, Azabache V, Gomez G, Gutierrez M, Weirs B, de Quadros C, Halsey N (1992) Comparison of high titer Edmonston-Zagreb, Biken-CAM and Schwarz measles vaccines in Peruvian Infants. Pediatr Infect Dis J 11:822–827

Best JM, Welch JM, Baker DA, Banatvala JE (1987) Maternal rubella at St Thomas' Hospital in 1978 and 1986: support for augmenting the rubella vaccination programme. Lancet 2:88–90

Birrer MJ, Udem S, Nathenson S, Bloom BR (1981) Antigenic variants of measles virus. Nature 293:67–69

Black FL (1989) Measles active and passive immunity in a worldwide perspective. Prog Med Virol 36:1–33

Black FL, Sheridan SR (1967) Blood leukocyte response to live measles vaccine. Am J Dis Child 113:301–304

Böttiger M, Christenson B, Romanus V, Taranger J, Strandell A (1987) Swedish experience of two dose vaccination programme aiming at eliminating measles, mumps and rubella. BMJ 295:1264–1267

Brinckman UG, Bankamp B, Reich A, ter Meulen V, Liebert UG (1991) Efficacy of individual measles virus structural proteins in the protection of rats from measles encephalitis. J Gen Virol 72:2491–2500

Briss PA, Fehrs LJ, Parker RA, Wright PF, Sanella EC, Hutcheson RH, Schaffner W (1994) Sustained transmission of mumps in a highly vaccinated population: assessment of primary vaccine failure and waning vaccine-induced immunity. J Infect Dis 169:77–82

Brown EG, Furesz J, Dimock K, Yarosh W, Contreras G (1991) Nucleotide sequence analysis of Urabe mumps vaccine strain that caused meningitis in vaccine recipients. Vaccine 9:840–842

Brunell PA, Novelli VM, Lipton SV, Pollock B (1988) Combined vaccine against measles, mumps, rubella and varicella. Pediatr 81:779–784

Buynak EB, Hilleman MR (1966) Live attenuated mumps virus vaccine. I. Vaccine development. Proc Soc Exp Biol Med 123:768–775
Christenson B, Böttiger M (1994a) Long-term follow-up study of rubella antibodies in naturally immune and vaccinated young adults. Vaccine 12:41–45
Christenson B, Böttiger M (1994b) Measles antibody: comparison of long-term vaccination titres, early vaccination titres and naturally acquired immunity to and booster effects of the measles virus. Vaccine 12(2):129–133
Cusi MG, Valosin PE, Cellesi C (1993) Possibility of reinfection after immunization with RA27/3 live attenuated rubella virus. Arch Virol 129:337–340
Cusi MG, Valassina M, Bianchi S, Wunner W, Valensin PE (1995) Evaluation of rubella virus E2 and C proteins in protection against rubella virus in a mouse model. Virus Res 37:199–208
Cutts FT, Mandala K, St Louis M, Brown C, Mayala B, Zell ER, Deforest A, Kamenga M, Davachi F, Markowitz LE (1993) Immunogenicity of high-titer Edmonston-Zagreb measles vaccine in human immunodeficiency virus-infected in Kinshasa, Zaire. J Infect Dis 167:1418–1421
Davidkin I, Valle M, Julkunen I (1995) Persistence of anti-mumps virus antibodies after a two-dose MMR vaccination. A nine-year follow-up. Vaccine 13:1617–1622
Davis WJ, Larson HE, Simsarian JP, Parkman PD, Meyer HM (1971) A study of rubella immunity and resistance to infection. JAMA 215:600–608
Diaz-Ortega JL, Luna Abascal M, Valdesping JL, Sepulveda J, Markowitz LE, Zell ER (1992) Mortality and morbidity after high titre measles vaccine in Mexico. Lancet 340:924
Dorig RE, Marcil A Chopra A Richardson CD (1993) The human CD46 molecule is a receptor for measles virus (Edmonston strain). Cell 75:295–305
Dunster LM, Schneider-Schaulies J, Löffler S, Lankes W, Schwartz-Albiez R, Lottspeich F, ter Meulen V (1994) Moesin: a cell membrane protein linked with susceptibility to measles virus infection. Virology 198:265–274
Dyken PR, Cunningham SC, Ward LC (1989) Changing character of subacute sclerosing panencephalitis in the United States. Pediatr Neurol 5:339–341
Edmonson MB, Addiss DG, McPherson JT, Berg JL, Circo SR, Davis JP (1990) Mild measles and secondary vaccine failure during a sustained outbreak in a highly vaccinated population. JAMA 263:2467–2471
Englund JA, Suarez CS, Kelly J, Tate DY, Balfour HH Jr (1989) Placebo-controlled trial of varicella vaccine given with or after measles-mumps-rubella vaccine. J Pediatr 114:37–44
Ennis FA, Douglas RD, Hopps HE, Meyer HM Jr, Brown ER, Hobbins TE, Biehusen FC (1969) Clinical studies with virulent and attenuated mumps viruses. Am J Epidemiol 89:176–183
Farrington P, Pugh S, Colville A, Flower A, Nash J, Morgan-Capner P, Rush M, Miller E (1995) A new method for active surveillance of adverse events from diphtheria/tetanus/pertussis and measles/mumps/rubella vaccines. Lancet 345:567–569
Feldman S, Hughes W, Daniel C (1975) Varicella in children with cancer: 77 cases. Pediatrics 80:388–397
Fireman P, Friday G, Kumate J (1969) Effect of measles virus vaccine on immunological responsiveness. Pediatrics 43:264–272
Frenkel LM, Nielsen K, Garakian A, Wolinsky JF, Cherry JD (1996) A search for persistent rubella infection in persons with chronic symptoms after rubella and rubella immunizations and in patients with juvenile rheumatoid arthritis. Clin Infect Dis 22:287–294
Frey TK (1994a) Molecular biology of rubella virus. Adv Virus Res 44:69–160
Frey TK (1994b) Report of an international meeting on rubella vaccines and vaccination, 9 August 1993, Glasgow, United Kingdom. J Infect Dis 70:507–509
Garenne M, Leroy O, Beau J-P, Sene I (1991) Child mortality after high-titre measles vaccines: prospective study in Senegal. Lancet 338:903–907

Gershon A (1995) Varicella-zoster virus: prospects for control. Adv Ped Infect Dis 10:93–124
Gershon A, Steinberg S Brunell P (1974) Zoster immune globulin: a further assessment. N Engl J Med 290:243–245
Gershon AA, Steinberg S, Gelb L, NIAID-Collaborative-Varicella-Vaccine-Study-Group (1984) Live attenuated varicella vaccine: efficacy for children with leukemia in remission. JAMA 252:355–362
Giraudon P, Wild TF (1985) Correlation between epitopes on hemagglutinin of measles virus and biological activities: passive protection by monoclonal antibodies is related to their hemagglutinin inhibiting activity. Virology 144:46–58
Giraudon P, Jacquier MF, Wild TF (1988) Antigenic analysis of African measles virus field isolates; identification and localisation of one conserved and two variable epitope sites on the NP protein. Virus Res 18:137–152
Griffin DE, Bellini WJ (1996) Measles virus. In: Fields BN, Knipe PM, Howley PM, Chanock RM, Melnick JL, Monath TP, Roizman B, Straus SE (eds) Fields virology, 3rd edn. Lippincott-Raven, Philadelphia, pp 1267–1312
Guess H, Broughton D, Melton L, Kurland L (1985) Epidemiology of herpes-zoster in children and adolescents: a population-based study. Pediatrics 76:512–517
Gut JP, Lablache C, Behr S, Kirn A (1995) Symptomatic mumps virus reinfections. J Med Virol 45:17–23
Halloran E, Cochi S, Lieu T, Wharton M, Fehrs L (1994) Theoretical epidemiological and morbidity effects of routine immunization of preschool children with varicella vaccine in the United States. Am J Epidemiol 140:81–104
Halstead SB, Diwan AJ (1971) Failure to transmit rubella virus vaccine: a close-contact study in adults. JAMA 215:634–636
Hardy IB, Gershon A, Steinberg S, LaRussa P, the NIAID Collaborative Varicella Vaccine Study Group (1991) The incidence of zoster after immunization with live attenuated varicella vaccine. A study in children with leukemia. N Engl J Med 325:1545–1550
Harmsen T, Jongerius MC, van der Zwan CV, Plantinga AD, Kraaijeveld CA, Berbers GAM (1992) Comparison of a neutralization enzyme immunoassay and an enzyme-linked immunosorbent assay for evaluation of children vaccinated for mumps. J Clin Microbiol 30:2193–2144
Hayden GC, Herrmann KL, Buimovici-Klein E (1980) Subclinical congenital rubella infection associated with maternal rubella vaccination in early pregnancy. J Pediatr 96:869–872
Hayward A, Villanueba E, Cosyns M, Levin M (1992) Varicella-zoster virus (VZV)-specific cytotoxicity after immunization of nonimmune adults with Oka strain attenuated VZV vaccine. J Infect Dis 166:260–264
Hersh BS, Fine PEM, Kent WK, Cochi SL, Kahn LH, Zell ER, Hays PL, Wood CL (1991) Mumps outbreak in a highly vaccinated population. J Pediatr 119:187–193
Hilleman MR (1992) Past, present and future of measles, mumps and rubella virus vaccines. Pediatrics 90:149–153
Hilleman MR (1996) The development of live attenuated mumps virus vaccine in historic perspective and its role in the evolution of combined measles-mumps-rubella. In: Plotkin S, Fantini B (eds) Vaccinia, vaccination and vaccinology: Jenner, Pasteur and their successors. Elsevier, Paris, pp 283–292
Holt A, Moulton LH, Siberry GK, Halsey NA (1993) Differential mortality by measles vaccine titer and sex. J Infect Dis 168:1087–1096
Howson CP, Fineberg HV (1992) Adverse events following pertussis and rubella vaccine. Summary of a report of the Institute of Medicine. JAMA 267:392–396
Institute of Medicine (1993) The Children's Vaccine Initiative: achieving the vision. National Academy, Washington DC
Job JS, Halsey NA, Boulos R, Holt E, Farrell D, Albrecht P, Brutus JR, Adrien M, Andre J, Chan E, Kissinger P Boulos C, Cité Soleil/JHU project team (1991)

Successful immunization of infants at 6 months of age with high dose Edmonston-Zagreb measles vaccine. Pediatr Infect Dis J 10:303–311

Julkunen I, Vaananen P, Penttinen K (1984) Antibody responses to mumps virus proteins in natural mumps infection and after vaccination with live and inactivated mumps virus vaccines. J Med Virol 14:209–221

Khanum S, Garelick H, Uddin N, Mann G, Tomkins A (1987) Comparison of Edmonston-Zagreb and Schwarz strains of measles vaccine given by aerosol or subcutaneous injection. Lancet 1:150–153

Kobune F, Funatu M, Takahashi H, Fukushima M, Kawamoto A, Iizuka S, Sakata H, Yamazaki S, Arita M, Xu W, Li-Bi Z (1995) Characterization of measles virus isolated after measles vaccination. Vaccine 13:370–372

Krause P, Klinman DM (1995) Efficacy, immunogenicity, safety and use of live attenuated chickenpox vaccine. J Pediatr 127:518–525

Krugman S (1971) Present status of measles and rubella immunization in the United States: a medical progress report. J Pediatr 78:1–16

Kunkel U, Driesel G, Henning U, Gerike E, Willers H, Schreier E (1995) Differentiation of vaccine and wild mumps viruses by polymerase chain reaction and nucleotide sequencing of the SH gene: brief report. J Med Virol 45:121–126

Kuter BJ, Weibel RE, Guess HA, Matthews H, Morton DH, Neff BJ, Provost PJ, Watson BA, Starr S, Plotkin S (1991) Oka/Merck varicella vaccine in healthy children: final report of a 2-year efficacy study and 7-year follow-up studies. Vaccine 9:643–647

Lieu T, Cochi S, Black S, Halloran ME, Shinefield HR, Holmes SJ, Wharton M, Washington E (1994) Cost-effectiveness of a routine varicella vaccination program for US children. JAMA 271:375–381

Massad E, Azevedo-Neto RS, Burattini MN, Zanetta DM, Coutinho FA, Yang HM, Moraes JC, Pannuti CS, Souza VA, Silveira AS, Strouchiner CJ, Oselka GW, Camargo MCC, Omoto TM Passos SD (1995) Assessing the efficacy of a mixed vaccination strategy against rubella in Sao Paulo, Brazil. Int J Epidemiol 24:842–850

Mathias RG, Meekison WG, Arcand TA (1989) The role of secondary vaccine failures in measles outbreaks. Am J Public Health 79:475–478

McAleer WJ, Markurs HZ, McLean AA, Buynak EB, Hilleman MR (1980) Stability on storage of various temperatures of live measles, mumps and rubella virus vaccines in new stabilizer. J Biol Stand 8:281–287

McDonald JC, Moore DL, Quennec P (1989) Clinical and epidemiological features of mumps meningoencephalitis and possible vaccine-related disease. Pediatr Infect Dis J 8:751–755

Mellinger AK, Cragan JD, Atkinson WL, Williams WW, Kleger B, Kimber RG, Tavris D (1995) High incidence of congenital rubella syndrome after a rubella outbreak. Pediatr Infect Dis J 14:573–578

Mellman WJ, Wetton R (1963) Depression of the tuberculin reaction by attenuated measles virus vaccine. J Lab Clin Med 61:453–458

Miller CL (1991) Rubella in the developing world. Epidemiol Infect 107:63–68

Miller E, Goldacre M, Pugh S, Colville A, Farrington P, Flower A, Nash J, MacFarlane L, Tettmar R (1993) Risk of aseptic meningitis after measles, mumps and rubella vaccine in UK children. Lancet 341:979–982

Miller E, Hill A, Morgan-Capner P, Forsey T, Rush M (1995) Antibodies to measles, mumps and rubella in UK children 4 years after vaccination with different MMR vaccines. Vaccine 13:799–802

Mitchell LA, Tingle AJ, Shukin R, Sangeorgan JA, McCune J, Braun DK (1993) Chronic rubella vaccine-associated arthropathy. Arch Intern Med 153:2268–2274

Monafo WJ, Haslam DB, Roberts RL, Zaki SR, Bellini WJ, Coffin CM (1994) Disseminated measles infection after vaccination in a child with a congenital immunodeficiency. J Pediatr 124:273–276

Nader S, Bergen R, Sharp M, Arvin A (1995) Comparison of cell-mediated immunity (CMI) to varicella-zoster virus (VZV) in children and adults immunized with live attenuated varicella vaccine. J Infect Dis 171:13–17
Naniche D, Varior-Krishnan G, Cervoni F, Wild TF, Rossi B, Rabourdin-Combe C, Gerlier D (1993) Human membrane cofactor protein (CD46) acts as a cellular receptor for measles virus. J Virol 67:6025–6032
Neff BJ, Weibel RE, Villarejos VM, Buynak EB, McLean AA, Morton DH, Wolanski BS, Hilleman MR (1981) Clinical and laboratory studies of KMcC strain live attenuated varicella virus (41071). Proc Soc Exp Biol Med 166:339–347
Norrby E (1995) The paradigms of measles vaccinology. In: ter Meulen V, Billeter MA (eds) Measles virus. Springer, Berlin Heidelberg New York, pp 167–180 (Current topics in microbiology and immunology, vol 191)
Norrby E (1997) Human viral vaccines – yesterday, today and tomorrow. In: Brown F et al. (eds) Vaccines 97. Cold Spring Harbor Laboratory, Cold Spring Harbor, New York, pp 219–227
Norrby E, Penttinen K (1978) Differences in antibodies to the surface components of mumps virus after immunization with formalin-inactivated and live virus vaccines. J Infect Dis 138:672–676
Norrby E, Enders-Ruckle G, ter Meulen V (1975) Differences in the appearance of antibodies to structural components of measles virus after immunization with inactivated and live virus. J Infect Dis 132:262–269
O'Shea S, Best JM, Banatvala JE (1983) Viremia, virus excretion and antibody responses after challenge in volunteers with low levels of antibody to rubella virus. J Infect Dis 148:639–647
Odisseev H, Gacheva N (1994) Vaccinoprophylaxis of mumps using mumps vaccine, strain Sofia 6, in Bulgaria. Vaccine 12:1251–1254
Palumbo P, Hoyt L, Demasio K, Oleske J, Connor E (1992) Population-based study of measles and measles immunization in human immunodeficiency virus-infected children. Pediatr Infect Dis J 11:1008–1014
Peltola H, Heinonen OP, Valle M, Paunio M, Virtanen M, Karanko V, Cantell K (1994) The elimination of indigenous measles, mumps and rubella from Finland by a 12-year, two-dose vaccination program. N Engl J Med 231 (21):1397–1402
Penttinen K, Cantell K, Somer P, Poikolainen A (1968) Mumps vaccination in the Finnish defense forces. Am J Epidemiol 88:234–244
Plotkin SA (1988) Rubella vaccine. In: Plotkin SA, Mortimer EA Jr (eds) Vaccines. Saunders, Philadelphia, pp 325–265
Plotkin SA, Starr SE, Connor K, Morton D (1989) Zoster in normal children after varicella vaccine. J Infect Dis 159:1000–1001
Radecke F, Billeter MA (1996) The nonstructural C protein is not essential for multiplication of Edmonston B strain measles virus in cultured cells. Virology 217:418–421
Robinson K, Mostrator A, Grencis RK (1995) Generation of rubella virus-neutralising antibodies by vaccination with synthetic peptides. FEMS Immunol Med Microbiol 10:191–198
Rota JS, Hummel KB, Rota PA, Bellini WJ (1992) Genetic variability of the glycoprotein genes of current wild-type measles isolates. Virology 188:135–142
Rota PA, Bloom AE, Vanchiere JA, Bellini WJ (1994) Evolution of the nucleoprotein and matrix genes of wild-type strains of measles virus. Virology 198:724–730
Sabin AB (1992) My last will and testament on rapid elimination and ultimate global eradication of poliomyelitis and measles. Paediatrics 90:162–169
Schneider-Schaulies J, Schnorr JJ, Brinckman U, Dunster LM, Baczko K, Liebert UG, Schneider-Schaulies S, ter Meulen V (1995) Receptor usage and differential downregulation of CD46 by measles virus wild-type and vaccine strains. Proc Natl Acad Sci USA 92 (9):3943–3947
Seto NO, Gilliam S (1994) Expression and characterization of a soluble rubella virus E1 envelope protein. J Med Virol 44:192–199

Sharp M, Terada K, Wilson A, Nader S, Kinchington PE, Ruyechan WT, Hay J, Arvin AM (1992) Kinetics and viral protein specificity of the cytotoxic T lymphocyte response in healthy adults immunized with live attenuated varicella vaccine. J Infect Dis 165:852–858
Sheshberadaran H, Norrby E (1986) Characterization of epitopes on the measles virus hemagglutinin. Virology 152:58–65
Stratton KR, Howe CJ, Johnston RB Jr (eds) (1994) Adverse events associated with childhood vaccines: evidence bearing on causality. National Academy, Washington DC, pp 118–186
Takahashi M, Otsuka T, Okuno Y, Asano Y, Yazaki T, Isomura S (1974) Live vaccine used to prevent the spread of varicella in children in hospital. Lancet 2:1288–1290
Taylor MJ, Godfrey E, Baczko K, ter Meulen V, Wild TF, Rima BK (1991) Identification of several different lineages of measles virus. J Gen Virol 72:83–88
Tingle AJ, Chantler JK, Pot KH, Paty DW, Ford DK (1985) Postpartum rubella immunization: association with development of prolonged arthritis, neurological sequelae, and chronic rubella viremia. J Infect Dis 152:606–612
Tsolia M, Gershon A, Steinberg S, Gelb L (1990) Live attenuated varicella vaccine: evidence that the virus is attenuated and the importance of skin lesions in transmission of varicella-zoster virus. J Pediatr 116:184–189
Tyrrell L, Norrby E (1978) The polypeptides of measles virus. J Gen Virol 39:219–229
Vaheri A, Vesikari T, Oker-Blom N, Parkman PD, Veronelli J, Robbins TC (1972) Isolation of attenuated rubella-vaccine virus from human products of conception and uterine cervix. N Engl J Med 286:1071–1074
Valmari P, Lanning M, Tuokko H, Kouvalainen K (1987) Measles virus in the cerebrospinal fluid in postvaccination immunosuppressive measles encephalopathy. Pediatr Infect Dis J 6:59–63
Varsanyi TM, Utter G, Norrby E (1984) Purification, morphology and antigenic characterization of measles virus envelope components. J Gen Virol 65:355–366
Varsanyi TM, Morein B, Löve A, Norrby E (1987) Protection agains lethal measles virus infection in mice by immune-stimulating complexes containing the hemagglutinin or fusion protein. J Virol 61:3896–3901
Ward BJ, Griffin DE (1993) Changes in cytokine production after measles virus vaccination: predominant production of IL-4 suggests induction of a Th2 response. Clin Immunol Immunopathol 67:171–177
Watson BM, Piercy SA, Plotkin SA, Starr SE (1993) Modified chickenpox in children immunized with the Oka/Merck varicella vaccine. Pediatrics 91:17–22
Watson BM, Laufer DS, Kuter BJ, Staehle B, White CJ (1996) Safety and immunogenicity of a combined measles, mumps, rubella and varicella vaccine (MMR(II)V) in healthy children. J Infect Dis 173:731–734
Weibel R, Neff BJ, Kuter BJ, Guess HA, Rothenberger CA, Fitzgerald AJ, Connor KA, McLean AA, Hilleman MR, Bunyak EB, Scolnick EM (1984) Live attenuated varicella virus vaccine: efficacy trial in healthy children. N Engl J Med 310:1409–1415
Wharton M, Cochi SL, Hutcheson RH, Bistowish JM, Schaffner W (1988) A large outbreak of mumps in the postvaccine era. J Infect Dis 158:1253–1260
White CJ, Kuter BJ, Hildebrand CS, Isganitis KL, Matthews H, Miller WJ, Provost PJ, Ellis RW, Gerety RJ, Calandra GB (1991) Varicella vaccine (VARIVAX) in healthy children and adolescents: results from clinical trials 1987 to 1989. Pediatrics 87:604–610
Wild TF, Bernard A Spehner D, Drillien R (1992) Construction of vaccinia virus recombinants expressing several measles virus proteins and analysis of their efficacy in vaccination of mice. J Gen Virol 73:359–367
Wolinsky JS (1996) Rubella. In: Fields BN, Knipe PM, Howley PM, Chanock RM, Melnick JL, Monath TP, Roizman B, Straus SE (eds) Fields virology, 3rd edn. Lippincott-Raven, Philadelphia, pp 899–929

World Bank (1993) World development report 1993: investing in health. Oxford University Press, New York

Zweiman B, Pappagianis D, Maibach H Hildreth EA (1971) Effect of measles immunization on tuberculin hypersensitivity and in vitro lymphocyte reactivity. Int Arch Allergy Appl Immunol 40:834–841

CHAPTER 6

Hepatitis and Polio Vaccines

F. SCHÖDEL and P. MINOR

A. Introduction

This chapter provides a brief overview of vaccination strategies against hepatitis viruses and poliomyelitis viruses. For other recent review see MINOR (1991), HOLLINGER and TICEHURST (1996), LEMON and THOMAS (1997), STAPLETON and LEMON (1997) and SCHÖDEL (1998).

At least four known viruses cause hepatitis in man: hepatitis A, B, C and E viruses (HAV, HBV, HBC, HBE). A fifth virus family, that of hepatitis G viruses (HGV), related to HCVs, has recently been described (SIMONS et al. 1995; LINNEN et al. 1996); its role in pathogenesis, if any, is as yet unclear. In addition, a viroid, hepatitis delta or D virus, exacerbates the clinical course of chronic HBV, which it uses as a helper virus. As none of the hepatitis viruses appears to be highly cytopathogenic, another feature that the diseases caused by the hepatitis viruses share is that the ensuing diseases are largely due to immune mechanisms. Apart from being hepatotropic and causing liver disease in man, these viruses are different. Four are positive-stranded RNA viruses (HAV, HCV and HEV, HGV) and one is a reverse transcribing DNA virus (HBV). They also differ in their respective modes of transmission: HBV and HCV are transmitted primarily through blood and blood products and sexually and vertically in the case of HBV. In fact vertical transmission from chronic carrier mother to child is the major mode of transmission in most countries (predominantly in Asia) in which HBV chronic infection is endemic. Sexual transmission appears to play a lesser role for HCV, its major known route of transmission is by contaminated blood products. There is a large cohort of HCV chronic carriers for which no route of transmission has been established. In contrast, HAV and HEV cause hepatitis epidemics: they are both primarily transmitted by the faecal-oral route.

This review focuses on the two viruses for which efficacious vaccines have been developed to date (HAV and HBV). Since hepatitis delta infection is primarily prevented by the prevention of HBV infection and a vaccine for HDV is not available, HDV is not further discussed (for a recent review see TAYLOR 1996). Experimental HCV infection has been prevented by vaccination in the chimpanzee model, but there are numerous obstacles for the development of HCV vaccines, some of which relate to the hypervariability of the viral envelope proteins and the absence of an animal model other than chimpanzees or a cell culture system. The uncertainty whether viral infection – if resolved – results in protective immunity, in turn the absence of correlates of protection in man make development of HCV vaccines difficult. We therefore summarize only very briefly some of the main features of HCV.

B. Hepatitis A Virus

I. The Virus

HAV is a non-enveloped positive-sense single-stranded RNA virus. The viral RNA consists of approximately 7478 nucleotides. HAV is classified as a hepatovirus, a unique member of the picornaviridae family. A multitude of human HAV strains have been isolated, and while they harbour distinct epitopes which can be differentiated using neutralizing monoclonal antibodies they are all crossreactive with HAV polyclonal sera and even somewhat more distantly related simian strains were shown to elicit crossprotection in chimpanzees (LEMON and BINN 1983; PING and LEMON 1992; LEMON et al. 1992). It therefore appears that at least some critical virus neutralizing epitopes are conserved throughout HAV isolates.

The viral RNA also serves as a messenger RNA and directs translation via internal ribosomal entry at *cis*-acting RNA elements in the 5'-untranslated region similar to some other positive-stranded RNA viruses (see below for e.g. poliovirus). The RNA encodes a large polyprotein of 2227 amino acids. This polyprotein is co-translationally processed by a protease domain of the large precursor protein into three or four structural proteins (1AB, 1C, 1D and a 1D precursor; or in another nomenclature VP0, VP3, VP1 and PX) and seven or more nonstructural proteins. Sixty copies of each of the structural proteins co-assemble to form the viral capsids of 27–32 nm in diameter which have an icosahedral structure and package the viral genomes.

II. The Antigens

The structural proteins forming the viral capsid, predominantly VP1 and VP3 are targets of virus neutralizing antibodies (PING and LEMON 1992; STAPLETON and LEMON 1987; PING et al. 1988; LEMON et al. 1991; NAINAN et al. 1992). Some amino acids which are involved in binding to virus neutralizing antibodies have been mapped in vitro. Most polyclonal and monoclonal virus neutralizing antibodies do not react with individual structural proteins in immunoblots. It is therefore likely that the critical neutralizing epitopes are highly conformation dependent, and in some cases it has been shown that amino acids from several neighbouring structural protein participate in antibody binding. This fact together with the difficulty of co-synthesizing viral structural proteins in recombinant heterologous expression systems has largely restricted the development of HAV vaccines to classically attenuated vaccine candidates and –

the only HAV vaccines which are currently available worldwide – to the development of inactivated whole virus vaccines. Recently it has been shown that by expressing the complete HAV open reading frame from a recombinant vacciniavirus it becomes possible to generate subviral capsid-like structures that preserve native HAV antigenicity and immunogenicity (WINOKUR et al. 1991). Such expression systems could theoretically yield alternative recombinant vaccine candidates. The crystal structure of HAV has not yet been dissolved.

III. The Disease

HAV is secreted in large amounts in the stool of infected persons (10^8 particles/ml). This is the major source of infection. After an incubation period of 15–50 days with an average of 28 days, HAV can cause hepatitis in susceptible adults. The degree of symptoms can vary from asymptomatic to acute liver failure, the latter being rare. Interestingly, HAV superinfection in HCV chronic carriers seems to be associated with a higher frequency of severe acute hepatitis (VENTO et al. 1998). HAV hepatitis is characterized by a rapid onset of symptoms in about 60% of patients. The clinical course of HAV infection has been reviewed elsewhere (KOFF 1993; HOLLINGER and TICEHURST 1996) and is not the focus of this overview. HAV replicates predominantly in hepatocytes in humans. HAV is thought to reach hepatocytes via the bile, the exact route of infection after oral ingestion and gastroduodenal passage has not been elucidated. Recent data suggest that primary HAV replication occurs in enterocytes (KARAYANNIS et al. 1986; ASHER et al. 1995) A receptor on hepatocytes has also not been characterized. Faecal shedding and viraemia are found concomitantly, approximately 2 weeks after experimental infection. It is unclear whether HAV is cytopathic; the lack of a direct correlation between cytopathic effects and the time of highest viral replication suggest that it is at least not highly cytopathic, and that much of the ensuing hepatitis is caused by immune mechanisms (for a review see SIEGL and WEITZ 1993). With the onset of symptomatic hepatitis, viraemia usually subsides rapidly while faecal shedding persists for 1–2 weeks. At the time when viraemia subsides, elevated neutralizing serum antibody titres are found, at the same time, virus-specific cytotoxic T cells can be isolated in the liver and in blood (KURANE et al. 1985; VALLBRACHT et al. 1984, 1989). Cytokines released by virus-specific T cells probably contribute to the control and eventual elimination of viral replication. HAV infection is typically self-limited; a chronic carrier state does not occur in immunocompetent individuals. HAV infection is usually asymptomatic in infants and leaves a life-long protection. The degree of severity of disease increases drastically with age (case fatality rates in reported cases range from 0.0045% in children aged 4–15 years to 2.7% in adults aged over 49; LEMON and SHAPIRO 1994; HADLER 1991). Due to the faecal-oral mode of transmission HAV infection is largely an infection of early childhood, where it is endemic. In crowded and underdeveloped parts of the world symptomatic epidemics rarely occur because persons are asymptomatically infected as children and protected as adult. As sanitary conditions improve, adult populations are no longer immune. Where susceptible adults populations exist, epidemics can spread from, for example, day-care centres and then cause symptomatic disease in adults, typically in community-wide outbreaks.

IV. Immunity and Vaccines

1. Passive Immunization

Immunoglobulin preparations containing HAV-specific antibodies given before HAV exposure can prevent HAV associated disease. This was first demonstrated in a military camp setting (STOKES and NEEFE 1945) and confirmed by later studies (see references in STAPLETON 1995). Administered after exposure, Ig can still attenuate disease. This established at least one mechanism of protection; as in HBV infection, the presence of virus neutralizing antibodies in serum suffices to prevent disease. This is especially noteworthy as HAV is an enterically transmitted virus. Yet, serum IgG is apparently sufficient for protection. The role of secretory IgA in protection, if any, is not established. Clearly, it is not required. The level and nature of serum antibodies required for protection is somewhat less firmly established than for HBV. A WHO reference IG standard has been prepared for comparing IG lots which has a virus neutralizing titre of approximately 1:800000 by radioimmunofocus assay and has been assigned a value of 100 mIU/ml. A dose of 0.06 ml/kg body weight of this IG preparation results in a titre of approx. 45 mIU/ml. Since this dose of IG protects against HAV infection at a level of approx. 90% for several months, serum levels higher than 10–20 mIU/ml are generally associated with protection (cited from STAPLETON and LEMON, 1997). The level of protection afforded by postexposure passive immunization is likely dependent on the specific anti-HAV titre of the immunoglobulin preparation, the size of the HAV inoculum and the time between HAV exposure and immunoglobulin administration.

2. Inactivated Whole Virus Vaccines

The production of HAV vaccines became possible with the successful in vivo propagation of HAV in cell culture (PROVOST and HILLEMANN 1979). The two most widely distributed commercially available vaccines are based on formalin inactivated virus preparations: Havrix® produced by SmithKline and Beecham and Vaqta® produced by Merck, Sharp and Dohme. Two other inactivated whole virus vaccines for which considerably less clinical information is available are produced by Pasteur-Mérieux (Avaxim®) and Berna (Hepacomb®), respectively. The latter also contains influenza proteins in a rather complicated mixture. Further inactivated viral vaccines have been developed, but since less information about them is available, this review focuses on some characteristics of the two most widely distributed HAV vaccines, Vaqta® and Havrix®.

Both vaccines are produced from cell-culture adapted and attenuated strains of HAV, HM175 for Vaqta® and CR326F for Havrix®. The viruses are grown in MRC-5 human diploid fibroblasts and formalin inactivated after harvesting and purification. Both are adsorbed to aluminium salts as adjuvants. Vaqta® is substantially more purified than Havrix® (the residual non-viral proteins are indicated at less than 0.1 µg per adult dose for the former and less than 5 g per adult dose for the latter). Havrix also contains more formalin than Vaqta® (less than 50 µg/infant dose compared to less than 0.1 µg/dose for Vaqta®). Inactivated HAV vaccines are typically administered as two i.m. doses spaced several months apart (e.g. at 0 and 6 months; see package inserts). Different formulations for children, adolescents and for adults are available. Both vaccines appear safe, are immunogenic and both vaccines protect against HAV infection. The use of different activity units and different systems for antibody determination make a direct comparison of the antigenicity and immunogenicity of the two vaccines impossible. The two trials conducted to demonstrate vaccine efficacy are also radically different in design and therefore do not facilitate comparison of the vaccines.

The clinical efficacy of Vaqta® was demonstrated in a double-blind placebo controlled trial in children aged 2–16 in an Orthodox Jewish community in Monroe, New York (Werzberger et al. 1992). Infectious hepatitis had been a persistent problem in this community for several years. Shortly after immunization of 1037 children with 25 U of Vaqta® or placebo a HAV epidemic occurred. Immunized children were highly protected 4 weeks after receiving a single dose of Vaqta® with a level of protection of 100% (95% confidence intervals of 87%–100% 50 days after immunization). A level of 100% protection in immunized children was maintained for at least 6 years of follow-up (Werzberger et al. 1998).

The efficacy of Havrix® was analysed in 40,119 Thai children (with a surprising age range for schoolchildren of 1–16 years) randomized to receive either Havrix® or hepatitis B vaccines at a three-dose schedule (at 0, 1, and 12 months; Innis et al. 1994). Between 138 and 386 days following two doses of 360 ELISA units of Havrix®, the children were protected against symptomatic HAV infection (a school absence of 2 days was used as the screening criterion) at a rate of 94% (79%–99%). Forty HAV infections were seen in the control group and the two infections seen in HAV vaccinated infants were brief and mild. Data on protection immediately following the primary series of vaccinations are not available.

While the efficacy of HAV vaccines has not formally been tested in adult populations, the seroconversion rates suggest that adults are also protected. In view of the two studies described above and the fact that passive immunization protects against HAV infection it is probably not necessary to conduct further efficacy trials for HAV vaccines in adults.

The duration of protection following immunization with either of the inactivated HAV vaccines is not exactly known, it can be speculated that protection may last beyond the decay of measurable antibodies as the inoculation with HAV would probably result in a secondary (memory) antibody response which might suffice for protection against disease given the long incubation period of HAV. This remains speculation and will be rather difficult to confirm clinically.

Both an experimental study in chimpanzees and the early onset of protection after a single dose of Vaqta® (indicating that some of the protected individuals were probably HAV infected before immunization) suggest that active immunization provides some level of post-exposure protection. Active immunization can therefore be used in the control of epidemics even in the absence of passive immunization.

3. Live Attenuated Vaccines

Theoretically, live virus vaccines could have the advantages of oral immunization and life long immunity after a single immunization. Passage in tissue culture attenuates HAV and both inactivated HAV vaccines are made from attenuated viruses. Apart from potential safety concerns, some life virus vaccine candidates were poorly immunogenic and had to be given at high doses (KARRON et al. 1988; SJÖGREN et al. 1992; MIDTHUN et al. 1991). A candidate live vaccine has been developed in China which appears to be more immunogenic and less attenuated (MAO et al. 1992). The available clinical information is, however, limited.

C. Hepatitis B Virus

I. The Virus

HBV is a small, enveloped DNA virus classified in the family hepadnaviridae (for hepatotropic DNA virus; for recent reviews see SCHÖDEL et al. 1989; GANEM 1996; NASSAL 1996). Mature virions or Dane particles measure approximately 42 nm in diameter. The envelope surrounding the viral nucleocapsid is composed of a unit membrane containing three species of surface antigens. The viral capsids are co-asssembled of 180–240 U of a single protein, hepatitis core antigen (CROWTHER et al. 1994), and contain both the viral genome and the transcription apparatus. The virus encoded polymerase which serves as a reverse transcriptase and probably an endonuclease and DNA-dependent DNA polymerase is covalently linked to the genome minus strand. The genome consists of a partly double-stranded DNA which is held in a circular conformation by base pairing between the 5′ end of the DNA plus strand and the 5′-end of the minus strand. The DNA minus strand is greater than unit length and the DNA plus strand is incomplete in mature virions. The replication of hepadnaviruses has been molecularly unravelled in recent years. Hepadnaviruses similar to, for example, cauliflower mosaic viruses, are DNA viruses replicating through an RNA intermediate: an RNA transcript serves both as the pregenomic RNA and as a messenger RNA. This pregenomic RNA, which is complexed with the viral polymerase is preferentially packaged into capsids during capsid assembly and virion maturation, the polymerase complex is involved in the packaging specificity. For a more detailed description of hepadnavirus replication (see NASSAL 1996; GANEM 1996).

II. The Antigens

HBV is a small enveloped DNA virus. The viral envelope consists of three species of carboxy-coterminally colinear proteins and a unit membrane. It envelopes a nucleocapsid structure which harbours the viral genome and replication machinery (polymerase). Of the at least seven proteins coded for by the viral genome we shall limit the discussion to the three surface proteins and the two forms of the nucleocapsid antigen, as they appear most relevant for eliciting protective immune responses. Every viral gene product could of course be a relevant target for anti-viral T-cell responses.

1. Envelope Proteins

The three envelope proteins of HBV are coded for by a single gene. As in the synthesis of the nucleocapsid antigen and e antigen, HBV uses 5′ heterologous RNA transcript to allow for the synthesis of more than one gene product off a single gene (for references see TIOLLAIS et al. 1985; SCHÖDEL et al. 1990b). The envelope proteins are all three colinear at the C-terminus. The smallest envelope protein, HBsAg, or major surface antigen, has 226 amino acids. It is the most abundant species both in empty spheres and in mature virions and harbours the information necessary for secretion and envelope particle assembly. HBsAg is an important target of virus neutralizing antibodies. HBsAg carries the group-specific a determinant, which is shared by all known HBV serotypes. The a epitope is an immunodominant epitope, most of the anti-HBsAg antibody after vaccination is directed against a (SZMUNESS et al. 1980). In addition, there are two pairs of subtype-specific determinants which are labelled d/y and r/w and which are found in various combinations such as ayw, adr, adw and ayr. A number of fine specificities further dividing the subtypes have been described (e.g. OHNUMA et al. 1993). Due to the presence of the common a determinant, immunization with HBsAg of various subtypes elicits crossneutralizing antibodies. Most epitopes recognized on HBsAg appear to be highly conformation dependent, and epitope mapping or induction of anti-native HBsAg antibodies using synthetic peptides has met with limited success. A single amino acid substitution at amino acid position 145 suffices to abolish recognition of HBsAg by type-specific monoclonal antibodies (WATERS et al. 1992). Recently, an escape mutant in the HBs gene has been described in a liver transplant recipient who was treated with anti-HBs monoclonal antibodies (McMahon et al. 1992). In addition, vaccine escape variants were found in children acquiring an HBV infection after immunization (Carman et al. 1990, 1995a,b; Okamoto et al. 1992; Wallace et al. 1994; Yamamoto et al. 1994). Most of these escape variants affect glycine to arginine mutations at amino acid position 145, in one instance an additional aspartic acid to threonine/isoleucine at position 126 was found which also resulted in the loss of the a determinant (Okamoto et al. 1992).

The middle (mHBsAg) and large surface antigens (lHBsAg) contain the pre-S2 and pre-S1 regions, C-terminal extensions of the major surface antigen by 55 amino acids and 108 or 119 amino acids respectively, depending on subtype. During HBV infection, two types of empty particles containing surface protein(s) are found in the blood, empty spheres of 20 nm diameter consisting of HBsAg with a variable proportion of mHBsAg and filaments consisting of lHBsAg, mHBsAg and HBsAg, also about 20 nm in diameter and of variable lengths. The mature infectious virions, or Dane particles after their discoverer (Dane et al. 1970), are double-shelled spheres of about 42 nm which contain all three species of surface proteins (Gerlich and Bruss 1993). The pre-S1 and pre-S2 amine terminal extensions carry epitopes which have been defined using synthetic peptides (for an exhaustive list see Milich 1993). Both conserved and subtype-specific epitopes are present on the pre-S sequence. Immunization of chimpanzees with synthetic peptides of the pre-S1 and pre-S2 region provides protection against HBV challenge (Emini et al. 1989; Itoh et al. 1986; Neurath et al. 1989; Thornton et al. 1987, 1989). Similarly, incubation of HBV virions with rabbit anti-pre-S2 serum abolished their infectivity in chimpanzees (Neurath et al. 1986). These experiments prove that the pre-S region carries virus neutralizing epitopes. Pre-S1 is implicated in the receptor mediated uptake of HBV by hepatocytes (Neurath et al. 1990) and appears essential for replication and infectivity. While pre-S2 is also the target of virus neutralizing antibodies, its presence appears not to be absolutely required for the viral replication cycle (Fernholz et al. 1993). Mutants with deletions of parts or the entire pre-S2 region are sometimes seen in chronic carriers and are viable upon in vitro transfection with the cloned DNA. Antibodies against pre-S belong to the early markers of HBV infection (Neurath et al. 1985).

2. Nucleocapsid Antigen

The nucleocapsid or core antigen (HBcAg) is a 21.5-kDa protein that assembles to package the viral RNA and transcription machinery. In mature virions it contains the viral polymerase and a partially double-stranded viral DNA. HBcAg can be expressed in various heterologous expression systems, where it self-assembles to form 32 nm particles. HBcAg particles exist in two forms, a larger particle composed of 240 subunits and a smaller particle composed of 180 HBcAg subunits (Crowther et al. 1994). In contrast to HBsAg, HBcAg is an extraordinarily immunogenic antigen (for a review see Schödel et al. 1994b): every HBV infected individual mounts a – mostly high titred – anti-HBc antibody response. It is both a T dependent and T cell independent antigen (MILICH and MCLACHLAN 1986a) and owns T help sites which are recognized in every species and inbred mouse strain tested so far (MILICH et al. 1987a, 1988).

Immunization of chimpanzees with HBV core antigen was shown to partly protect against HBV challenge (MURRAY et al. 1984, 1987). More recent experiments have demonstrated that woodchucks are protected against a high-titred (10^6 ID_{50}) WHV challenge by immunization with HBcAg or WHcAg (SCHÖDEL et al. 1993a). HBcAg or WHcAg are internal structural components of the virus. Antibodies against HBcAg or WHcAg do not recognize intact virions and are not protective. The observed protection is therefore likely to be T-cell mediated. Murine experiments have demonstrated that immunization with HBcAg led to a secondary type of antibody responses against surface antigen when the HBcAg primed mice were boosted with complete virions in which HBcAg was surrounded by envelope proteins present in the same particle (MILICH et al. 1987b). This intermolecular/intrastructural T help was also observed in the woodchuck model. Woodchucks immunized with WHcAg or HBcAg, responded with a very fast anti-S antibody responses when challenged with whole infectious virions (SCHÖDEL et al. 1993a). It is therefore likely that the priming of HBcAg-specific T helper cells which provide help for HBsAg-specific B cells explains why woodchucks immunized with WHcAg or HBcAg are protected. Another explanation which has not been ruled out is the induction of HBcAg-specific cytotoxic T cells which would, however, not explain the accelerated kinetic of the anti-S response. The enhanced immunogenicity of HBcAg has also been exploited to use it as a carrier for foreign epitopes (for a reviews see SCHÖDEL et al. 1994b), among them for pre-S epitopes of HBV (SCHÖDEL et al. 1992). Hybrid HBcAg particles carrying virus neutralizing epitopes of the pre-S region combine the protective capacity of HBcAg with the ability to elicit high titered virus neutralizing antibodies (SCHÖDEL et al. 1992) and could therefore be a theoretical HBV vaccine candidate.

While the inclusion of HBcAg in more complex HBV vaccines in the future might be desirable to enhance their immunogenicity, immunization with HBcAg alone would suffer from the lack of a characterized and measurable correlate of protective immunity. HBcAg is an important target of immune responses in HBV infection and might also come to play a role in therapeutic vaccines (see below).

III. The Disease

HBV is one of the major infectious agents causing acute and chronic liver disease. HBV is parenterally transmitted. The existence of two forms of infectious jaundice was recognized during World War II (references in KRUGMAN and STEVENS 1994). MacCallum suggested in 1947 the name hepatitis A for the enterically transmitted form with a shorter period of incubation and hepatitis B for the form transmitted with blood products (vaccines, transfusion) and displaying a longer incubation period, also called serum hepatitis (see references in Krugman and Stevens 1994). The clinical distinction (the agents were called MS-1 and MS-2) was confirmed and elaborated in a series of clinical observations and human transmission experiments in Willowbrook State Mental Hospital, an institution for mentally handicapped children in New York State from the 1950s to the early 1970s (Krugman et al. 1967). Later the two types of infectious hepatitis, epidemic infectious hepatitis and serum transmitted hepatitis were recognized to be caused by HAV and HBV, respectively. The first successful attempts at vaccination against hepatitis B were also made at Willowbrook: a heat inactivated infectious serum was used to actively immunize children and prevent experimental transmission of HBV in up to about 70% of immunized patients (Krugman et al. 1971). Baruch Blumberg described an antigen in the serum of leukemia patients which he called Australia antigen (Blumberg et al. 1965). Subsequently it was more aptly termed S antigen by Prince (1968) who found it in serum of patients suffering from serum hepatitis. In subsequent years the term HBsAg or S antigen came to be used for the major surface antigen of HBV, although historically speaking S antigen as observed by Prince probably corresponded to the complex of all three HBV envelope proteins. The three species of HBV envelope proteins are: HBsAg, the small or major surface antigen, the middle surface antigen, mHBsAg, and the large surface antigen, HBsAg (see below and Gerlich and Bruss 1993). HBV has two major patterns of transmission: in areas of low endemicity, it is usually transmitted via sexual contact or needle sharing in i.v. drug abusers. Plasma or blood products, which used to be a source of iatrogenic infection are now routinely tested and safe. In areas of high endemicity, the major routes of transmission are from chronic carrier mothers to their offspring in a mostly perinatal transmission, predominant in Asian countries (Stevens et al. 1975), and less frequently from chronic carriers to household members. In parts of Africa chronic carriage of HBV is common in children and sibling to sibling transmission is more frequent (Whittle et al. 1983). HBeAg positivity in chronic carrier mothers is a high risk factor for transmission of HBV to newborns.

The age at infection with HBV is a major determinant of morbidity. A high proportion of perinatally infected infants (up to 90%) develops a chronic carrier state, often associated with chronic progressive liver disease ultimately resulting in chronic liver failure or hepatocellular carcinoma. Infection later in life more often results in acute, transient liver disease. Fewer than 10% of acutely infected adults become chronically infected and fewer than 3% develop significant chronic liver disease. While this still represents a significant disease burden, the overwhelming majority of the estimated 200–300 million chronic carriers worldwide are infected early in life. It was therefore especially important to find that immunization at birth of infants born to chronic carrier mothers provides effective protection from HBV transmission.

IV. Immunity and Vaccines

1. Protective Immunity

HBV infection in nature is restricted to man. The discovery in the seventies that chimpanzees were susceptible to HBV infection (BARKER et al. 1973) has played an important role in the development of current HBV vaccines. A convenient animal model to analyse measurable correlates and molecular mechanisms of protective immunity against HBV infection is still absent. Infection of animals with related hepadnaviruses (for a review see SCHÖDEL et al. 1989), such as infection of woodchucks with woodchuck HBV, can be used to study some aspects of protective immunity with the restriction that the viruses and the resultant diseases are closely related but not identical and that woodchucks are immunologically not well characterized. Similarly, there are no cultured cells which can be efficiently and reproducibly infected with HBV in vitro. Therefore assays for the analysis of in vitro neutralizing antibodies are absent. This also implies that a classical convenient system for the production of whole cell or subunit viral vaccines which derives the viruses from cell culture is not available. HBV vaccines must purified from the plasma of infected persons or produced by recombinant or synthetic technologies.

2. Passive Immunization

Before active HBV vaccines became commercially available, the only available prophylactic measure after exposure to HBV was the administration of immunoglobulins with a high anti-HBV titre. A single passive immunization of infants with immunoglobulins containing high anti-HBV titres (HBIG) was about 50% effective in preventing the transmission of HBV from chronically infected mothers, with multiple administrations an effectiveness of about 75% was achieved (BEASLEY et al. 1983). For HBV it was therefore clearly shown that virus-neutralizing antibodies (IgG) suffice to protect against HBV infection, in newborns, in spouses of acute HBV patients and after accidental exposure (BEASLEY et al. 1983; GRADY et al. 1978; REDEKER et al. 1975).

3. Active Immunization

a) Targets

Immunization with HBsAg, peptides of the pre-S region and hepatitis core antigen have been shown to protect against HBV infection in chimpanzees or in man (HBsAg only) or against WHV infection in woodchuck.

b) Current Vaccines: Plasma-Derived Vaccines

The first vaccines successfully used to protect against HBV infection consisted of major viral surface antigen particles (HBsAg) purified from the plasma of chronically HBV infected donors (for further details see Hillemann 1993). This was possible because HBV affords the luxurious synthesis of a large surplus of empty viral spheres during infection and because these can be conveniently separated from infectious virions or Dane particles. The first demonstration that HBV positive serum can inactivated (by heat), and that immunization with such serum elicited protective immunity against experimental infection (in children) was published by Krugman et al. in 1971. Later it was published that this inactivated serum had contained about 1 μg/ml HBsAg and had indeed induced anti-HBs antibodies in the recipients (Hillemann 1993).

Subsequently HBsAg containing empty spheres (composed of pre-S2 and HBsAg) and tubules (composed of pre-S1, pre-S2 and HBsAg) were purified from plasma by a combination of ammonium sulfate precipitation, isopycnic banding in sodium bromide and rate zonal sedimentation through sucrose gradients followed by a pepsin digestion, which accidentally also removed the pre-S components. The essentially pure HBsAg was denatured and renatured using urea, further purified by gel filtration and additionally inactivated by formalin treatment (Hillemann 1993). Other purification procedures based on enrichment of HBsAg particles by size, density, adsorption to aerosil and chromatographic characteristics combined with formalin and heat inactivation procedures were developed in various laboratories and adapted by manufacturers of plasma-derived HBsAg (for a review see Sitrin et al. 1993). The resulting HBsAg preparations were adsorbed to aluminum salts as an adjuvant. The first licensed product in the United States was available by 1982. All these different plasma-derived products were safe, immunogenic and effective in protecting against HBV infection. The first clinical trial demonstrating protective efficacy in a high risk population was published by Szmuness et al. 1980. After the first two doses, 77% of the vaccinees had high titered serum anti-HBs antibodies, rising to 96% after the third dose. Within 75 days after randomization there was a significant reduction in the frequency of hepatitis B in the vaccinated group, suggesting for the first time that active vaccination is efficacious even when given after HBV exposure. The reduction in hepatitis B incidence in the vaccinated group reached up to 92.3% in this study, data which were confirmed by numerous later clinical trials (see references in for example West 1993). Efficacy and immunogenicity trials of plasma-derived HBV vaccines also led to the observation that protection from infection was anti-HBs titre correlated and that a titre of above 10 sample ration units or later milli-international units (mIU) was protective in more than 98% of vaccinees (Francis et al. 1982). The early placebo controlled trials of plasma-derived HBsAg established a serum level higher than 10 mIU anti-HBs with protection. The level of 10 mIU anti-HBs or higher has since been used as a surrogate marker and is often referred to as seroprotection. To this day plasma-derived vaccines continue to be manufactured and used.

c) Recombinant Vaccines

α) HBsAg-Based

When the HBV genome was cloned and the genes encoding the envelope proteins were characterized (GALIBERT et al. 1979; PASEK et al. 1979; VALENZUELA et al. 1979) it became possible to replace the plasma-derived vaccine preparations with recombinant material. HBsAg can be synthesized in recombinant brewer's yeast. While it is not secreted as it is in higher eukaryotes, it co-assembles to form particles containing a unit membrane which are similar to the HBsAg spheres purified from human plasma (VALENZUELA et al. 1982). Yeast expression systems for HBsAg were developed in the United States (Merck in collaboration with Chiron) (MCALEER et al. 1984; VALENZUELA et al. 1982) and in Belgium (SmithKline Beecham; HARFORD et al. 1983). The recombinant HBsAg particles purified from yeast proved to be antigenically identical to plasma-derived HBsAg and of a similar immunogenicity. A recombinant yeast-derived HBsAg of the subtype adw2 was shown to confer crossprotection against challenge with ayw and adw in chimpanzees (MCALEER et al. 1984). Formulated into vaccines similar to plasma-derived HBsAg, recombinant HBsAg was also found to be safe and well tolerated (for an overview see MCMAHON and WAINWRIGHT 1993). Formal double-blinded efficacy studies as performed with the plasma-derived vaccines were no longer considered ethical when the recombinant yeast-derived vaccines were introduced, but the combined available data suggest that the recombinant vaccines are as least as efficacious as the plasma-derived vaccines (MCMAHON and WAINWRIGHT 1993). Therefore on a worldwide scale recombinant HBsAg containing vaccines produced in bakers yeast have been replacing the earlier plasma-derived products.

Potential Problems. Vaccines that contain only HBsAg of one subtype (the most widespread licensed recombinant vaccines are of subtype adw2) are effective and provide crossprotection against all known a containing HBV subtypes. While a seroconversion rate of up to 98% after administration of three doses of the HBsAg-based vaccines in many clinical trials (see MCMAHON and WAINWRIGHT 1993) is impressive and probably enough to eradicate HBV carriage, vaccines based solely on HBsAg have some potential disadvantages: (a) nonresponders: there is a small but substantial percentage of vaccinees who do not mount a protective antibody response even after more than the recommended three doses of HBsAg; (b) escape mutants: a single point mutation causing an exchange of amino acid mutation at position 145 in HBsAg from a glycine to an arginine suffices to destroy the dominant a epitope and such mutants are not recognized by anti-a sera or monoclonal antibodies; and (c) immunogenicity: it would be desirable to develop vaccines which require less than three immùnizations to achieve the same or an enhanced seroconversion rate in all populations.

Nonresponders. HBsAg is a strictly T cell dependent antigen. To mount an anti-HBs response, the cognate interaction between T helper cells and B cells is required. It is well established in mice that several MHC class II haplotypes are non-responders to HBsAg; HBsAg does not contain T cell epitopes effectively presented by these haplotypes (Milich et al. 1983a,b, 1984, 1985a,c; Milich and Chisari 1982). HBsAg non-responsiveness in the murine model can be overcome by providing extraneous sources of T cell help in these haplotypes, for example from T cell sites of the pre-S region or the HBcAg (Milich and McLachlan 1986b; Milich et al. 1985b). It has been demonstrated that inclusion of the pre-S1 or pre-S2 region can overcome non-responsiveness against HBs in the murine model (Milich et al. 1985b). In humans there is evidence that non-responsiveness might at least be partly genetically determined (Alper et al. 1989). All the same, non-genetic factors contribute to non-responsiveness, such as gender, age, smoking and others. Dialysis patients are also known to be low responders. Low response rates can be partly ameliorated by increased vaccine dosage or repeat immunizations. A distinction between non-responsiveness and low responder status, if difficult to assess, may be of practical importance: while there is some evidence that vaccinees having once mounted an anti-HBs antibody response, albeit low, and whose antibody response has since declined below the seroprotection level of 10 mIU/ml may still be protected by immune memory against disease to some extent (for further discussion see West 1993), there is no evidence to suggest that true non-responders to HBsAg are ever protected. Efforts have also been made to formulate vaccines that include the pre-S1 and/or pre-S2 region to test whether genetic non-responsiveness can be overcome by inclusion of additional T-cell epitopes.

Escape Mutants. Viral DNA point mutants resulting in an amino acid exchange from glycine to arginine at position 145 were found in HBsAg positive babies of vaccinated mothers (Hino et al. 1995; Okamoto et al. 1992), in chronic carriers (Carman et al. 1990; Fujii et al. 1992; Wallace et al. 1994; Yamamoto et al. 1994) and in a patient who had received anti-HBs monoclonal antibodies as an adjunct therapy during a liver transplantation (McMahon et al. 1992). Since their original discovery HBsAg escape mutants have been described more and more regularly by a number of investigators. It is unclear to which extent these escape mutants are transmitted to the general population and whether they have an important impact on the HBV epidemiology under routine HBsAg vaccination. To this date it does not appear that the HBsAg escape mutants pose a major epidemiological threat. However, it is worth noting that the impact of a vaccination resistant HBV variant vertically transmitted to a small percentage of vaccinated infants (less than 3%) born to carrier mothers would take a long time to become apparent. The transmission of HBsAg escape variants could theoretically be prevented by either the inclusion of variant HBsAg in the vaccine or by inclusion of conserved pre-S epitopes eliciting HBsAg-independent virus neutralizing antibodies.

Enhanced Immunogenicity. If the observation in mice holds in man, that the immunogenicity of HBsAg is to a large extent a function of the availability of T cell help epitopes which are suboptimally provided by HBsAg itself, than the inclusion of T cell epitopes from the pre-S region and possibly from HBcAg, might lead to the formulation of more immunogenic vaccines, which in turn would allow to reduce the number of immunization, thereby affording economic and compliance advantages.

β) Pre-S Containing Vaccines

Pre-S2. The pre-S2 sequence carries several B cell epitopes and has been shown to elicit protective immune responses in chimpanzees when these were immunized with synthetic peptides containing the whole or parts of the pre-S2 sequence (see also above). In addition, it has been shown that the pre-S2 sequence contains T help sites that allow to overcome genetic non-responsiveness to HBsAg in mice immunized with middle surface antigen.

Several approaches were therefore chosen to manufacture mHBsAg containing vaccines. A mHBsAg containing recombinant vaccine based on synthesis of the M protein in Chinese hamster ovary (CHO) cells was developed (GenHevac B® Pasteur). This vaccine contains HBsAg, glycosylated HBsAg and middle surface antigen in a ratio of approximately 6:2:2 and is licensed in France (Tron 1994). The vaccine has been used in mass vaccination campaigns and was safe, well tolerated and immunogenic. Comparative studies with a plasma-derived vaccine preparation have not shown an enhanced immunogenicity of the recombinant pre-S2 containing vaccine when the seroconversion rate to anti-HBs was analysed. An anti-pre-S2 response was demonstrated in 56% of vaccinees after the first dose of 20 μg and up to 93% and 95% after the first and second boosts (cited from: Tron 1994). The anti-pre-S2 responses appeared earlier than the anti-HBs response in adult volunteers and babies but was also less long-lived. The safety of a CHO-derived vaccine was clearly demonstrated by GenHevac-B® Pasteur. With progress in tissue culture technique and experience with tissue culture derived products original theoretical safety concerns have been reduced and novel pre-S1/S2 containing products come back to expression in CHO cells. These particles contain the polymerized human albumin binding site described in the pre-S2 region and their clinical safety has invalidated theoretical concerns associated with its presence in a vaccine preparation.

An effort to produce pre-S2 containing particles in yeast was made by several laboratories (see also Rutgers et al. 1993). Production of pre-S containing HBsAg particles in yeast has met with several technical difficulties: hyperglycosylation, protease sensitivity of the pre-S2 sequence, possibly incorrect folding of the assembled polyproteins and poor yields. Some of these difficulties have been overcome by producing the vaccine candidates for example in non-glycosylating and protease deficient yeast strains (references see Rutgers et al. 1993). Another difficulty is that for the presentation of the

HBsAg epitopes it may be advantageous to have mixed particles: in particles containing only middle protein, critical HBsAg epitopes may be partly masked by the pre-S2 sequence. Two yeast-derived mHBsAg preparations resulted in somewhat conflicting data in clinical trials: one preparation elicited a lesser immune response to HBsAg in clinical studies in healthy young adults if compared to recombinant HBsAg alone (WILDE et al. 1991). A second mHBsAg preparation was tested in a dose range of 12–48 μg mHBsAg and compared to 10 μg HBsAg and fared somewhat better: a slightly higher anti-HBs mean titre and a serconversion rate of 100% for the groups receiving mHBsAg was reported (GERETY and WEST 1990). Trials in small groups of nonresponders have yielded conflicting results (references see RUTGERS et al. 1993).

A Japanese group removed six amino acids constituting the major protease site in the pre-S2 region by genetic engineering and produced mHBsAg containing particles in yeast that had an increased stability (ITOH and FUJISAWA 1986; KOBAYASHI et al. 1988; KURODA et al. 1989). This mHBsAg containing vaccine preparation was found to elicit a very high rate of seroconversion (up to 100% after two doses of 20 μg in healthy adults) and to also elicit anti-pre-S2 antibodies in 100% of vaccinated individuals (SUZUKI et al. 1994). It is noteworthy that the anti-pre-S2 antibody response appears to be more rapid than the anti-HBs response. In addition, both anti-HBs and anti-pre-S2 responses are boosted, and the mHBsAg vaccine induces seroconversion to both anti-HBs and anti-pre-S2 in a majority of nonresponders to HBsAg. In view of the HBV subtype specificity of the T-cell responses to pre-S2, at least in the murine system (MILICH et al. 1990a), it is not surprising that seroconversion of nonresponders does not reach 100% after immunization with mHBsAg of a single subtype, even if we assume that nonresponsiveness was entirely due to the absence of suitable T-cell epitopes for a given set of HLA molecules. The serum anti-pre-S2 titers of volunteers immunized with this vaccine were comparable to the serum titers in chimpanzees were protected against HBV challenge after immunization with pre-S2 synthetic peptides (KURODA et al. 1991). It can therefore be assumed that this vaccine elicits protective levels of anti-pre-S2. These data show that immunization with a stable pre-S2 containing vaccine results in an enhanced immunogenicity and the induction of anti-pre-S2 antibodies.

Pre-S1. The technical difficulties encountered in efforts to include lHBsAg in recombinant vaccine preparations were even greater than those encountered with mHBsAg. The assembly and secretion of virion envelopes requires a careful balance between HBsAg, mHBsAg and lHBsAg. The pre-S1 sequence in its transmembrane region contains a transfer stop signal and is myristilated at the N-terminus. Overexpression of lHBsAg in yeast does not lead to the assembly of native-like particles and overexpression of lHBsAg in higher eukaryotic cells suppresses secretion of mHBsAg and HBsAg and leads to the intracellular retention and accumulation of envelope gene products

(CHENG et al. 1986; CHISARI et al. 1986; MCLACHLAN et al. 1987; MOLNARKIMBER et al. 1988; PERSING et al. 1986; STANDRING et al. 1986). To secrete lHBsAg containing empty envelope particles in a eukaryotic system requires the cosynthesis of lHBsAg, mHBsAg and HBsAg in a ratio corresponding approximately to the one found during virion biosynthesis in HBV infection. One biotechnolological solution to the difficulty was to express the HBV envelope genes under control of their native promoters in CHO cells. One group has developed an expression system in which multiple copies of the HBV envelope gene are integrated into CHO cells. The resultant particles secreted at a relatively high rate contain lHBsAg (4%–7%), mHBsAg (17%–21%) and HBsAg (75%–77%) (HOURVITZ et al. 1996) in a ratio approximately corresponding to that found in virions. These mixed particles were formulated into a new vaccine (Bio-Hep-B®, Biotechnology General, Israel). The vaccine displays an enhanced immunogenicity in animal models, elicits, for example, anti-HBs seroconversion in nonresponder mice (SHOUVAL et al. 1994a) and the first clinical trials are encouraging. The vaccine is safe and well tolerated and an anti-HBs seroconversion rate of 100% after two immunizations with a 5-μg dose was recently reported in a clinical trial in which 105 healthy young adult volunteers were immunized three times (HOURVITZ et al. 1996). In this clinical trial and in several others (MADALINSKY et al. 1994; SHOUVAL et al. 1994b; YAP et al. 1995) not only the anti-HBs seroconversion rate but the geometric mean titres achieved were impressive (e.g. 12 156 mIUml 1 month after the third immunization with the 5-μg dose (HOURVITZ et al. 1996). In a recently published comparative trial, the pre-S1/2 containing vaccine (Bio-Hep-B, Biotechnology General, Israel) was found to be more immunogenic in neonates than a recombinant HBsAg-based vaccine (Engerix-B, SmithKline Biologicals, Belgium; YERUSHALMI et al. 1997).

A second similar lHBsAg, mHBsAg and HBsAg containing vaccine was developed (THOMA et al. 1991), but published information on this vaccine has been slow in forthcoming. In a study with this novel pre-S containing vaccine, two thirds of true nonresponders who had already received up to six doses of HBsAg containing vaccines seroconverted after a single dose (ZUCKERMANN et al. 1996). A high frequency of a certain HLA class II allele was found frequently in the remaining third of non-responders. All initial non-responders of another HLA class II allele known for HBsAg nonresponsiveness seroconverted. These data suggest that a different MHC restriction pattern for HBs than pre-S observed in the murine system is also found in man.

For both the pre-S1 and pre-S2 containing vaccines, no data about seroconversion against the pre-S sequences are so far available. This may be due to the difficulty of measuring anti-pre-S antibodies in human sera and/or to the relatively low amount of lHBsAg and mHBsAg present in the vaccine preparations.

Due to the efficacy of the currently licensed HBsAg containing vaccines it will be difficult to conclusively prove the superiority of novel vaccines containing pre-S2 or pre-S1 and pre-S2. Recent data suggest that they are more immunogenic. In addition, they could have the at least theoretical advantage of eliciting neutralizing antibodies against additional independent epitopes.

γ) Alternative Vaccination Approaches

The success of the currently available preventive vaccines has made improvements difficult both for scientific and for economic and proprietory reasons. A successful alternative vaccination approach must be as safe as current vaccines while offering advantages in price, route of application and/or the ease of combination with other relevant vaccines for the target population. Only a vaccine that would provide a treatment for chronic HBV infection (see also below) might be partly exempt from such considerations.

Synthetic Peptides. Since the amino acid sequences of the HBV envelope and core proteins became available, a substantial effort has been made to characterize B and T cell epitopes using synthetic peptides (reviewed in MILICH 1993). It would be beyond the scope of this article to review in detail the contributions of synthetic peptide work to the unravelling of interactions between the immune system and amino acid sequences on HBV proteins. This is done elsewhere (MILICH 1993). Epitopes recognized by antibodies (here also called B cell epitopes) can be of two kinds: those that can be represented by synthetic peptides or contiguous short stretches of amino acids heretofore called peptidic epitopes and those that are discontinuous and cannot be mimicked by conventional synthetic peptides. We use this terminology because the terminology commonly used to describe antibody binding sites tends to be misleading. Every epitope is conformation dependent as it consists of relative affinities between amino acids on two or more molecules influenced by their charge, their location in a watery milieu, in space and in time. While the majority of antibody binding sites (on proteins) found in nature cannot be adequately mimicked by short linear peptides (and are therefore often called conformation dependent), there are a number of well defined peptidic epitopes on the envelope proteins of HIVgp120, of HBV pre-S, Plasmodium falciparum CS protein to name but a few. Mapping of epitopes on HBsAg with synthetic peptides has met with limited success (MILICH 1993). The generation of cyclic peptides has been reported to enhance the HBsAg-like antigenicity and immunogenicity of such preparations (BROWN et al. 1984; IONESCU-MATIU et al. 1983).

In contrast, several peptidic epitopes, both subtype-specific and broadly crossreacting epitopes have been characterized on pre-S1 and pre-S2 (reviewed in MILICH 1993; SCHÖDEL et al. 1990b). Immunization with some of these peptides linked to suitable carrier molecules was shown to confer protective immunity against HBV challenge to chimpanzees (EMINI et al. 1989; ITOH et al. 1986; NEURATH et al. 1989; THORNTON et al. 1987, 1989) in one instance even upon immunization with the free peptide (THORNTON et al. 1989). Stimulation of antibody synthesis generally requires at least two events: interaction of the free native protein with the B cell immunoglobulin receptors and cognate T cell help, that is specific stimulation of T helper cells via peptidic fragments of the protein bound to the major histocompatibility complex of an antigen presenting cell. In contrast to B cells, which recognize sequential or complex epitopes on native proteins via their immunoglobulin receptors, T cells normally recognize small peptidic fragments of proteins presented to the T cell receptor by the proteins of the major histocompatibility complex. Peptidic fragments of proteins synthesized within the cell preferentially enter the MHC class I presentation pathway and are preferentially seen by CD8 positive cells, giving rise to cytolytic T cell responses. Peptidic fragments of proteins taken up through phagocytosis go through the lysosomal pathways and are preferentially associated with MHC class II molecules, either by loading the MHC with peptides from outside or inside the cell and stimulate preferentially $CD4^+$ T cells, which in turn provide T cell help for antibody synthesis. Several epitopes on HBV structural and non-structural proteins recognized by T helper cells associated with MHC class II or recognized by cytotoxic T cells in the context of MHC class I have been mapped with the help of synthetic peptide chemistry (see Milich 1993; Chisari and Ferrari 1995). In view of the success of currently available vaccines and the inherent restrictions of synthetic peptide technology, it is questionable whether synthetic peptides will yet come to play a role in prophylactic vaccination. The most interesting potential field of application for synthetic peptides is in treatment of chronic HBV infection by immune modulation. A possibly relevant finding comes from a transgenic mouse model; mice which are transgenic for HBeAg (see also below) are tolerant to HBcAg and HBeAg on a T cell level (Milich et al. 1990b). It has been possible to define T cell sites that escape tolerance induction because they are not dominant. Immunization with such non-dominant T cell sites as synthetic peptides elicits autoimmune responses in transgenic mice (Milich et al. 1991). If T cell tolerance, predominantly to HBeAg/HBcAg, is a major mechanism favouring chronicity in HBV infection, as suggested by the absence of cytolytic or proliferative T cell activity during asymptomatic chronic carriage and by several other observations (Ig isotype distribution, preferential perinatal transmission of HBV from $HBeAg^+$ carrier mothers), immunization with such synthetic peptides to break tolerance and induce immune responses that in chronically infected individuals might accelerate or trigger virus elimination seems a promising strategy. A more detailed discussion of the advantages and disadvantages of synthetic peptide approaches to vaccination is found elsewhere (Milich 1993).

Live Vectors. Live vectors present some advantages: they allow simultaneous immunization against the vector and one or several carried antigens, they can be cheaply manufactured, do not require extensive purification and in some instances can be given via the oral or other mucosal routes, eliciting both systemic and local immune responses. In addition, viral vectors deliver the gene coding for target antigens to the infected host cell. The gene product is thus synthesized in the host cell and presented to the immune system as the antigens during infection with wild-type pathogens (viruses and parasites). This can have the advantage – or disadvantage (see for example: OEHEN et al. 1991) – that the antigens enter the histocompatibility complex class I restricted pathway which facilitates the induction of specific cytotoxic T cell responses. The disadvantages of live vectors are: safety concerns, potential environmental release of recombinant live vectors, difficulties in standardization and the possible negative impact of preexisting anti-carrier immunity on the carried antigen immunogenicity.

Viral Vectors. The most important viral vectors suggested for the development of recombinant HBV vaccines so far are orthopoxviruses and adenoviruses. The genes coding for major, middle and large HBV surface antigens were cloned into vaccinia virus early after the cloning of HBV. These recombinant viruses expressing several HBV genes have become important tools for the molecular analysis of anti-HBV immune responses, especially cytotoxic T cell responses. A vaccinia virus expressing the HBsAg gene has been demonstrated to provide protective immunity against HBV challenge in chimpanzees in the absence of high titered antibodies against HBsAg (MOSS et al. 1984). Vaccination in spite of its relative safety and tolerance during the smallpox eradication program has caused a number of rare but occasionally severe vaccine related complications. Vaccination with the vaccine strains used in the smallpox vaccination program for the prevention of other transmittable diseases therefore meets with considerable safety concerns. More highly attenuated poxviruses have been sought as potential vaccine carriers. One possible candidate carrier vaccinia strain is modified vaccinia virus Ankara (MVA), a strain which was selected by classical serial passage in cell culture and later sequenced and found to carry several gene deletions (HOCHSTEIN-MINTZEL et al. 1972; MAYR et al. 1975, 1978; MEYER et al. 1991; STICKL et al. 1974). This strain was licensed in Germany for vaccination of adults and persons at risk for vaccination. MVA has been found safe in 120000 persons, which makes it an attractive novel carrier candidate, provided that carried heterologous gene products acquire sufficient immunogenicity. A new generation of poxvirus vectors was developed by Paoletti and coworkers, based either on complete DNA sequencing of the vaccinia genome and directed removal of virulence genes or on the use of poxviruses with a different host-range. Two different novel vectors have resulted from this work: Nyvac, a vaccinia virus mutant which carries tailored deletions disrupting 18 genes implicated or involved in the pathogenesis of vaccinia virus infection (TARTAGLIA et al. 1992) and Alvac, a canarypoxvirus which does not replicate in mammalian tissue (TAYLOR et al. 1988). Both viruses are highly attenuated, they can still carry genes to human cells but either do not replicate at all or at a very limited rate. Nyvac-based recombinant vectors expressing several HBV genes from the same vector have been engineered (PINCUS et al. 1995). Whether such novel recombinant vaccinia vectors will be capable of eliciting high-titered and long-lasting HBV neutralizing antibody responses remains to be shown.

The second recombinant virus of some potential interest is adenovirus. Adenoviruses used in the existing live adenoviral vaccines has been used to clone the HBV envelope genes (BALLAY et al. 1987; LUBECK et al. 1989; MORIN et al. 1987). The major interest in adenovirus is that they can be applied intranasally or orally and might be able to stimulate both local and systemic immunity. A recombinant adenovirus expressing the HBs gene has been shown to elicit anti-S antibodies in laboratory animals (BALLAY et al. 1987; MORIN et al. 1987). As previously with HBsAg recombinant vaccinia virus, the anti-S titres achieved in chimpanzees after vaccination with the recombinant adenovirus are rather low (MORIN et al. 1987). The same adenovirus 7 HBsAg recombinant when tested in human anti-adenovirus 7 negative human volunteers did not induce any measurable anti-HBs antibodies (TACKET et al. 1992).

Bacterial Vectors. The most widely used bacterial carriers are avirulent *Salmonella* spp., *S. typhimurium* in mice and other small animal models and *S. typhi* in man (for reviews on Salmonella spp. as vaccine carriers see CURTISS 1990; SCHÖDEL 1992). Their major attractions are that the genetics are well defined, and that they can be attenuated while preserving some degree of invasiveness and therefore immunogenicity across mucosal surfaces. It might therefore be possible to create cheap multivalent live oral vaccines based on avirulent recombinant *Salmonella* spp. It has not been possible to synthesize HBsAg in an immunogenic form in prokaryotes. However, several approaches have been used to synthesize epitopes of the pre-S region in avirulent Salmonella spp. Expression systems coding for hybrid HBcAg particles displaying epitopes of the pre-S1 and pre-S2 region have been developed in avirulent *S. typhimurium* and *S. typhi* vaccine strains (SCHÖDEL et al. 1990a, 1993a, 1994a). A single oral immunization of mice with *S. typhimurium* expressing such hybrid HBc-pre-S genes elicits high-titered anti-pre-S antibodies (SCHÖDEL et al. 1994a). A first avirulent *S. typhi cya crp cdt* candidate vaccine strain expressing the same hybrid HBc-pre-S genes did not elicit satisfactory anti-pre-S antibody responses when tested in two phase I clinical trials in human volunteers (NARDELLI-HAEFLIGER et al. 1996; TACKET et al. 1996). Appropriate safe and immunogenic *S. typhi* or other *Salmonella* spp. carrier strains remain to be developed which would make this immunization approach feasible in man.

DNA Vaccination. A promising new technology in vaccine development is the use of DNA injection for immunization (for an overview see WORLD HEALTH ORGANIZATION 1994). It had been demonstrated in the late seventies and early eighties that in vivo transfection of animals with cloned viral DNA led to productive replication and synthesis of viral proteins (CHAN et al. 1979; ISRAEL et al. 1979; WILL et al. 1982). This work was recently taken up again by groups originally interested in in vivo transfection for gene therapy purposes (e.g. NABEL et al. 1990; WOLFF et al. 1990). In a further development it was found that direct in vivo injection of nucleic acid or transfection using ballistic devices led to prolonged tissue expression of the injected genes and subsequent development of antibody and T cell responses (DAVIS et al. 1993; FYNAN et al. 1993; ROBINSON et al. 1993; TANG et al. 1992; ULMER et al. 1993; WANG et al. 1993). Apparently plasmid DNA is taken up by some eukaryotic cells by so far not understood mechanisms, is transported to the nucleus and both transcribed and translated. Again, similar to the live vector situation, it is the pathogen's genetic information that is expressed within the host cells. This fact facilitates both the generation of neutralizing antibodies against complex native antigens that have withstood the efforts of classical recombinant expression cloning and the presentation of antigens to MHC class I restricted CTLs. The genes coding for the three HBV envelope proteins have been cloned in plasmids under CMV promoter control. When mice were intramuscularly injected with these plasmids, HBsAg secretion was found off plasmids coding for HBsAg and for the middle surface antigen. The mice seroconverted to serum anti-HBs at a level that would be protective in man and those immunized with pre-S containing plasmids also mount an anti-pre-S response (DAVIS et al. 1993, 1994; MANCINI et al. 1993, 1996; MICHEL et al. 1995; MICHEL 1995). Most recently, protection of two newborn chimpanzees from HBV challenge was demonstrated after injection at birth of plasmid DNA specifying HBsAg and mHBsAg sequences (PRINCE et al. 1997). Provided all safety concerns can be overcome and efficacious human immunization would be demonstrated, purified DNA vaccination could become an attractive new vaccination approach. Plasmid DNA is well characterized, can be simply produced, the generation of multivalent vaccines should be straightforward and an identical manufacturing technology could theoretically be used for any DNA vaccine.

Edible Vaccines? Recent advances in biotechnology have made it possible to express foreign genes in plants, among them in important food crops such as potatoes or bananas. Arntzen and colleagues have recently developed an expression system for HBsAg in plants (MASON et al. 1992). The first expression system was developed for tobacco, using the transcriptional signals of the tobacco mosaic virus. Since then the group has developed expression systems for plants closer to human diet such as potatoes (THANAVALA et al. 1995), which, however, still normally require cooking before consumption. HBsAg can be synthesized in plants and similar to the synthesis in yeast assumes a native conformation and forms immunogenic particles. The dream behind such approaches is to develop expression systems in edible plants that could be transformed into very cheap and easily accessible oral vaccines. Before this dream becomes reality, several obstacles must be overcome: appropriate high level expression systems for fruits or food crops must be developed, and it will have to be seen whether any of these antigens become indeed immunogenic, that is capable of eliciting protective titres of virus neutralizing serum antibodies, after oral delivery within an eaten and digested plant.

δ) Therapeutic Vaccination?

Chronic infection with HBV reflects a balance between sustained viral replication and the ability of the host immune system to eliminate the virus. HBV replication is not cytopathogenic; both viral clearance and disease associated with viral infection are immune mediated. Therapeutic vaccination efforts rely on the hope that the immune response could be driven in favour of viral clearance. For a rational approach to therapeutic immunization it would be important to know the regulatory mechanisms of immunity to HBV and therefore of chronic infection. Despite an increasing knowledge of effector functions in HBV immunity, the regulation of immune responses to HBV remains poorly understood. Some recent observations may present important insights for the development of novel therapeutic strategies. It has been observed for a while that acute viral infection and clearance is associated with vigorous T cell responses, both proliferative responses and a high cytotoxic T cell precursor frequency (FERRARI et al. 1992, 1993, 1995; JUNG et al. 1991, 1993, 1995; PENNA et al. 1992a,b; TSAI et al. 1992). In contrast, it has been difficult if not impossible to detect T cell activation during asymptomatic chronic carriage. Two mechanisms among others might account for some of this observed phenotype: the establishment of transplacental tolerance to a circulating antigen HBeAg favouring the establishment of chronic infection after neonatal transmission (MILICH et al. 1990b) and the association of chronic carriage with a Th2-like activity (MARUYAMA et al. 1993) in contrast to acute hepatitis, in which a Th1-like immune response patter is observed. Both observations are explained in somewhat further detail below.

The HBV core gene codes for the nucleocapsid antigen (HBcAg) and for the partly colinear but confomationally and immunogenically distinct secreted HBeAg. The precursor protein of HBeAg is C-terminally colinear with HBcAg and carries an N-terminal extension of 29 amino acids which serves as a leader peptide directing the newly synthesized precore protein into the secretory pathway where it is C-terminally and N-terminally processed and released into the circulation as mature HBeAg. An unprocessed ten N-terminal amino acids stemming from the precore sequence and within this sequence specifically a conserved cysteine at position –7 are critical for the folding of HBeAg (references see SCHÖDEL et al. 1993b; ZHENG et al. 1992). The function of HBeAg, which is not a structural antigen is unknown. HBcAg and HBeAg share all known T cell recognition sites and are critical targets for the immune response against HBV. HBeAg positive, anti-HBe negative mothers transmit HBV at a high rate to their children. In a murine transgenic model, HBeAg transgenic mice are tolerized at the T helper cell level (MILICH et al. 1990b). In addition, and more importantly, the nontransgenic offspring of HBeAg transgenic mice are partly tolerized as well. It has therefore been suggested that a function of HBeAg, a small molecule that crosses the placenta, is the induction of neonatal tolerance to HBcAg/HBeAg thereby facilitating the establishment of chronic infection following in utero or post partum transmission (Milich et al. 1990b). Interestingly it was also observed in this model that Th1 cells are more readily anergized following transplacental exposure to HBeAg whereas Th2 cells to some degree escape tolerization (Milich et al. 1995a). In parallel, when reanalysing the serology of chronic HBV infection with novel techniques which permit the detection of antibody in the presence of circulating antigen several new findings were made: a, anti-HBe, anti-HBc and anti HBs are readily detected in immune complexes before and even without seroconversion in classical assays and b, the IgG isotype distribution of anti-HBc antibodies in chronic asymptomatic patients is consistent with a predominant Th2-like activity whereas in acute self-limited hepatitis a Th1-activity-like IgG isotype distribution is found (Maruyama et al. 1993). Finally, it was most recently found that HBeAg also exerts a negative regulatory effect on antigen-specific Th1 cells in the periphery; it induces Fas-dependent apoptosis (Milich et al. 1998). Together with the observation that Th2-like cells more readily escape tolerization (Milich et al. 1995a) this suggests that a Th1-like activity is critical for virus elimination (and disease) whereas a predominant Th2-like activity favours viral persistence. The virus may have genetically evolved to actively evade host immune responses by preserving a secreted non-structural antigens that skews the Th1/Th2 balance in favour of an immune response permissive for viral persistence. At least in the murine model coadministration of HBcAg with interleukin-12 can completely reverse a Th2-like activity and suppress autoantibody production (Milich et al. 1995b). Therefore it is conceivable that efforts to break tolerance or modulate the type of immune response by active immunization might be of benefit in the treatment of chronic hepatitis (Milich 1997). The apparent danger is that any active immunization regimen that modulates the immune status of chronic carriers might also exacerbate disease at least transiently, similar to episodes of acute exacerbation of hepatitis occasionally seen under interferon therapy before resolution of the infection.

Along a similar line, it was observed that HBV-specific CTLs, recognizing epitopes on various HBV antigens (Nayersina et al. 1993), are readily detected during acute disease whereas they are undetectable or at a low frequency during chronic carriage (see additional references in Chisari and Ferrari 1995). It may well be that the absence of CTL precursors in chronic diseases is a reflection of a preferential Th2-like type of T cell response whereas the cytokine pattern provided by Th1 cells which is needed for optimal stimulation of CTL activity is absent. One effort at therapeutic vaccination tries to circumvent this potential problem by an attempt at direct CTL stimulation utilizing lipopeptides with synthetically fused non-HBV T helper epitopes (Vitiello et al. 1995).

In a recent letter a French group described that active immunization of chronic HBV carriers with a commercially available recombinant pre-S2/S vaccine (GenHevac B®) was followed by virus clearance in a percentage of the patients (Pol et al. 1994). Further data are required to confirm this finding.

In several transgenic mouse models expressing parts of the HBV genome or the entire HBV genome it was found that active immunization with purified HBsAg and adjuvants, with naked DNA coding for the envelope proteins, transfer of CTLs and infusion of both tumour-necrosis factor-α and interferon-γ downregulate HBV gene expression and lead to the disappearance of circulating HBV antigen (GILLES et al. 1992; GUIDOTTI et al. 1994a,b; GUILHOT et al. 1993; MANCINI et al. 1996). In these transgenic mouse models the complete or parts of the HBV DNA are chromosomally integrated and an important intermediate in HBV replication in susceptible hosts, the closed circular DNA is absent. While these experimental findings are encouraging it is not clear what the effect of these mediators would be on HBV replication in chronic carriers.

ε) Eradication of Hepatitis B?

The sole known reservoir of HBV is man. Hepatitis B continues to be one of the most important infectious diseases. It has widely escaped public attention that HBV causes an estimated 200–300 million chronic carriers with a yearly mortality of between 1.8 and 3 million worldwide from HBV associated liver disease. Safe and effective vaccines are available. While the initial strategy in developed countries to immunize high risk groups has had a substantial impact on the prevalence of HBV disease in these at-risk groups it has not had any measurable impact on the rate of HBV carriage or infection in the general population. This has led to recommendations for universal infant immunization in many countries (for a recent United States review see ZIMMERMANN et al. 1997) and to the inclusion of HBV vaccines into the WHO program on expanded immunization (EPI, M. KANE 1995, personal communication). Both plasma-derived and recombinant HBsAg-based vaccines have been shown to prevent transmission of HBV from chronically infected mothers to babies. Routine immunization of newborns has been even more effective in reducing the rate of HBV infections in the field than previously expected (WHITTLE et al. 1995). It should therefore be possible to eradicate HBV infection. All it would predictably take is to routinely immunize the majority of infants born worldwide for about two to three generations.

D. Hepatitis C Virus

I. The Virus

HCV is a small positive-stranded RNA virus (for reviews see RICE, 1996 and HOUGHTON, 1996). It was only recently described (CHOO et al. 1989). Its description incidentally marks an important transition in virology which had already been partly apparent at the description of HBV, another virus for which no in vitro propagation system was available, when it was first discovered: this is an important virus which had not been unequivocally defined by classical virological techniques, such as electron microscopy, cell culture propagation and reaction of convalescent sera with enriched viral proteins, but was described based entirely on molecular biology techniques: the genome was cloned from NANB infected chimpanzee sera based on the reaction of patient sera with recombinant or synthetic expression products of the virus (CHOO et al. 1989). Its genome structure and organization show similarities to and some differences from flaviviruses and pestiviruses and it is classified as a separate entity in a group with these two virus families (RICE 1996). The RNA genome is probably translated similarly to, for example, the pestivirus or the HAV genome by inter-nal ribosomal entry and codes for one large precursor protein which is cleaved into several polypeptides representing structural and non-structural viral proteins. HCV is a highly variable virus, it exits in at least six related but variable groups of genotypes (for a review see COOREMANN and SCHOODERMARK-VAN DE VEN 1996).

II. The Antigens

The virus codes for several structural and non-structural antigens, among them a nucleocapsid antigen (C antigen) and two envelope proteins (E1, E2 proteins). All viral antigens could be targets for virus-specific T cell responses which might be important in the regulation of the antiviral immune response and serve effector functions. The envelope antigens probably constitute the most important targets for virus neutralizing antibody responses. A region of E2, which is implicated in virus-receptor interactions and may be an important target of virus neutralizing antibodies, is hypervariable. Some data on the evolution of the hypervariable region of E2 in the quasispecies of viruses present in a chronic infection suggest that E2 is under in vivo selection pressure by virus neutralizing antibodies (e.g. KAO et al. 1995). This in turn suggests that some antibody species are functionally neutralizing in vivo.

III. The Disease

HCV appears to be mostly parenterally transmitted (ALTER, 1993). HCV has not been found in any other species than in man, it replicates and causes hepatic disease in chimpanzees. It was originally recognized as the predominant source of non-A, non-B hepatitis associated with the use of contaminated blood products. Since the introduction and routine use of reliable tests for HCV and anti-HCV antibodies blood products, have ceased to be the predominant source of infection. It should be noted, however, that the source or mode of transmission of HCV is unknown in a majority of HCV cases that have been diagnosed since diagnostic reagents have become readily available. There must be a major mode of community transmission not based on sharing needles, blood products or other medical interventions contributing to an estimated 500 million chronic carriers worldwide. HCV replicates primarily in hepatocytes. Additional extrahepatic sites of replication are likely. HCV infection causes an acute hepatitis which remains subclinical in the majority of cases. In experimentally infected chimpanzees HCV RNA is detected 3 days after inoculation in the serum. HCV infection leads to chronic persistent infection at a high frequency (>70%) (Choo et al. 1989) which can subsequently cause cirrhosis and hepatocellular carcinoma.

IV. Vaccine Development

Currently no vaccine against HCV is available. It has been shown that chimpanzees can be protected against homologous challenge by immunization with recombinant E1, E2 protein (Choo et al. 1994). This finding as well as the evolutionary pressure on the E2 sequence in the host probably exerted by virus neutralizing antibodies suggest that pre-existing antibodies of the right specificity and avidity are able to prevent HCV infection. The functional importance of viral neutralizing antibodies is underlined by the finding that the sequence diversity of E2 remained stable in a agammaglobulinaemic patient, (Kumar, 1994) whereas it rapidly evolves in patients with a normal antibody response. On the other hand, there are data indicating that primary infection does not protect against subsequent infection (Farci et al. 1992; Prince et al. 1992; Jarvis et al. 1992), confounded by the chronic nature of primary infection. Superinfection with heterologous HCV isolates, or rather, heterologous quasispecies swarms of viruses, in chronic carriers have been recorded. Finally, the variable nature of the HCV envelope proteins makes development of suitable immunogens a challenge even if it can be confirmed that pre-existing antibodies can prevent viral infection. Other vaccine development strategies against HCV are being pursued, for example, relying on generating virus-specific CTL responses (for a recent review see Inchauspé 1997). DNA vaccination has also been successfully used to elicit anti-E2 antibodies in mice (Nakano et al. 1997).

E. Hepatitis E Virus

I. The Virus

HEV is a faecal-orally transmitted virus responsible for NANB-hepatitis mostly in large epidemics caused by contaminated water (Bradley and Balayan 1988). HEV is similar to HAV in size if not in structure. A positive-stranded 3′-polyadenylated RNA, of 7.5 kba is contained in a viral capsid of approximately 27–32 nm in diameter. HEV has been tentatively grouped into the family of caliciviridae (for a review see Purcell 1996). The genome contains three open reading frames: the largest open reading frame contains non-structural proteins involved in viral replication, the second largest reading frame codes for the structural proteins forming the viral capsid and the smallest ORF codes for proteins which are immunogenic during infection but so far of unknown function for the virus.

II. The Antigens

ORF 2 encodes a 72-kDa protein believed to represent the viral capsid antigen. Upon expression in recombinant baculovirus, a truncated form of the ORF2 protein product self-assembles to form virus-like particles (TIAN-CHENG et al. 1997). The capsid antigen is a likely target for virus neutralizing antibodies. So far, only one serotype has been identified.

III. The Disease

HEV is the causative agent of up to 50% of NANBH in countries of the developing world. It is primarily spread by faecally contaminated drinking water. The incubation period of HEV ranges from 14 to 60 days with an average of 40 days. The highest rate of symptomatic infection is observed in young adults aged from 15 to 40 years. In some epidemics an excess rate of infected females over males is observed. HEV usually causes transient acute hepatitis with a case fatality rate of 0.5%–1% in the general population. In pregnant women in the second and third trimester average case fatality rates of 20% and acute hepatic failure rates of up to 32% have been described. The reason for this excess morbidity is unknown.

IV. Candidate Vaccines

ORF2 products have been expressed in a number of recombinant expression systems (PURDY et al. 1992; TSAREV et al. 1993; LI et al. 1994; PANDA et al. 1995; TSAREV et al. 1996; MCATEE et al. 1996a,b; TIAN-CHENG et al. 1997). At least two of these recombinant products from two different HEV strains (Mexico and Burma) have been shown to protect cynomolgous monkeys from HEV challenge (TSAREV et al. 1994, 1996; FUERST et al. 1996). The recent development of a rat model for HEV infection may help to facilitate HEV vaccine development (MANEERAT et al. 1996).

F. Poliovirus

I. The Virus

Polioviruses are members of the picornavirus family and the enterovirus genus (MINOR et al. 1991). Other genera of the family include the rhinoviruses, the aphthoviruses, the cardioviruses of mice, the hepatoviruses and a sixth genus incorporating the human enteroviruses echovirus 22 and 23 which has yet to be named. The virus particle is approximately 27 nm in diameter in negatively stained electron micrographs and is comprised of sixty copies of each of the main capsid proteins, designated VP1–VP4 in descending order of size. The capsid proteins are arranged with icosahedral symmetry such that VP1 surrounds the pentameric apex and VP2 and VP3 alternate about the

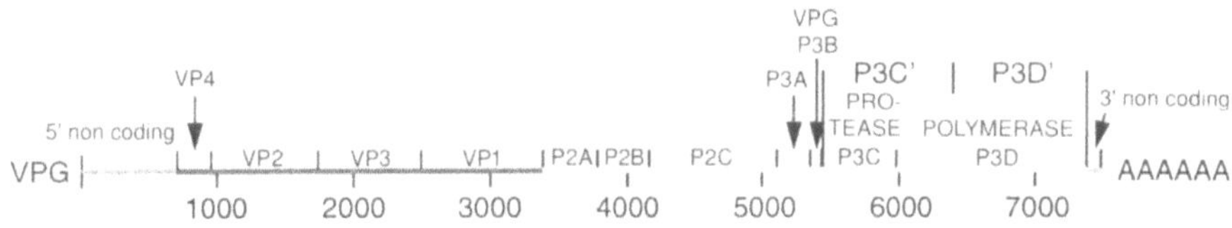

Fig. 1. The organization of the poliovirus genome

pseudosixfold axis of symmetry. Five copies of the fourth protein, VP4 are arranged about the internal surface of the pentameric apex. The atomic structure of the virus has been solved (Hogle et al. 1985). The three larger capsid proteins range in M_r from 25 to 30 kDa and VP4 is about 8 kDa. The genome consists of approximately 7500 nucleotides and includes a 3′ terminal polyadenylate tract and a covalently bound virus encoded protein, VPG at its 5′ terminus.

The genome is organized into four main portions. The 5′ noncoding region is about 750 bases in length and includes signals for RNA replication and an internal initiation site for protein translation. It is believed to be highly structured and it has been shown to be a key region in the attenuation of the virulence of the live vaccine strains in general use. It precedes the single large open reading frame in which the capsid proteins are encoded by approximately the first third and non-structural proteins by the remainder. The nascent polypeptide is cleaved by virus encoded proteases, notably 2A, which cleaves the structural proteins en bloc from the rest of the polyprotein and protease 3C which executes the remaining cleavages to generate the functional proteins. The genome terminates in a 3′ noncoding region of about 70 bases. The organization is summarized in Fig. 1.

Polioviruses effect their entry into the host cell by the use of a specific receptor, a protein of the immunoglobulin superfamily termed CD155 whose natural function is currently unknown (Mendelsohn et al. 1989). The functional receptor is found only on cells of old world and higher primates, although homologues exist in other species. Transfection of nonsusceptible cells such as those of murine origin with the gene for CD155 is sufficient to render them sensitive to infection (Pipkin et al. 1993). There is no natural host for poliovirus other than humans, a factor which makes its eradication theoretically simpler.

II. The Antigens

Polioviruses exist in three distinct serotypes such that infection with a virus of one serotype does not confer complete protection against another. Polyclonal sera in general neutralize all strains of a given type although strain-specific sera can be obtained by adsorption and it is possible to distinguish strains within a serotype by the use of monoclonal antibodies. The reason why only three serotypes are recognized is not known. It is considered likely that the antigenic properties of the virus are linked in some way to its receptor site usage, so that if the antigenic properties change sufficiently to generate a new serotype, it will be unable to recognize CD155 and therefore have a different pathological effect. No human enterovirus other than polio uses CD155. Sequence comparisons of various genomic regions suggest that polioviruses are very closely related to some of the coxsackie A group of enteroviruses which however use distinct receptor sites (HYPPIA et al. 1997).

The capsid of poliovirus is composed of closely interacting proteins, which individually elicit a very poor antibody response to the native virus. Moreover relatively mild treatments, such as UV irradiation or heating at 56°C for 30 min cause a major change in the conformation of the capsid which is recognized by polyclonal sera as a conversion from D or N antigenicity to C or H antigenicity. As acute phase sera react predominantly with C antigen and convalescent sera with D antigen it has been considered for some time that the potency of inactivated vaccines should be standardized in terms of their D antigen content. Immunization with C antigen-specific particles can induce neutralizing antibodies, however, and some neutralizing monoclonal antibodies react with both antigenic forms of the virus particle.

While the antigenic sites are conformational in nature they can be defined by the isolation of mutants resistant to neutralization by monoclonal antibodies. Such mutations can be shown to map to virtually every prominent surface feature in the three dimensional structure (MINOR 1990). Distinct sites can be identified, one surrounding the pentameric apex of the icosahedron, composed of sequences from VP1, a second composed of sequences from VP2 and VP1 near the three fold axis of symmetry, and a third involving sequences from VP3 and VP1 also near the three fold axis of symmetry. A fourth site, which may be linked to the third in some cases, is composed of sequences from VP3 which span adjacent pentamers and is therefore the only one specific for the icosahedral capsid and not reacting with the pentamers which form it. Antibodies specific for particular strains of virus tend to react with the third and fourth sites, and have been used in strain identification.

III. The Disease

The development of the understanding of the pathogenesis of poliomyelitis has recently been reviewed (MINOR 1996). Paralytic poliomyelitis is characterized by the selective destruction of the motor neurons, usually of the spine although polioencephalitis also occurs. The basis of this selectivity remains unknown; in particular the receptor site is widely expressed in a range of tissues including those where virus replication has not been noted, such as the kidney. Paralytic poliomyelitis is in fact a rare consequence of infection with polioviruses.

Infection is most often by the faecal-oral route and is usually silent, replication occurring in the gut. In a proportion of infections a mild febrile illness follows 3–8 days post-infection, termed the minor illness, and in a small proportion of these the virus invades the central nervous system causing the major disease, either meningitis (non-paralytic poliomyelitis), or paralytic poliomyelitis if the neurones are infected. The frequency of poliomyelitis depends on the strain and serotype involved. Type 1 poliovirus causes disease in about 1% of infections, type 3 in about one infection per 1000, and type 2 in even fewer. Bulbar poliomyelitis occurs where the upper part of the spinal cord including regions involved in breathing are affected and is the most dangerous other than polioencephalitis. About 10% of cases recover completely, about 5%–10% die, and the remainder have some residual paralysis. The period between infection and disease ranges from 8 to 30 days. Because of the comparative rarity of the disease asymptomatic carriers are a major source of infection in a way that is not true of a disease such as small-pox. This affects strategies to eradicate the disease, and was recognized in the early years of the twentieth century when the disease first became a major public health problem.

In areas with poor hygiene most individuals are exposed to poliovirus infection while still protected by maternal antibody. In such cases an immunizing infection of the gut occurs without any disease. As standards of hygiene improve, however, the first exposure is delayed, and poliomyelitis appears in epidemics in infants, hence the alternative term infantile paralysis. The site of infection in the gut is still not firmly established. According to the lymphatic model of Bodian the primary sites of infection are the lymphatic tissues of the pharynx or gut, either the tonsils of the pharynx or the Peyer's patches associated with the walls of the small intestine. An alternative view proposed by Sabin is that the mucosal layer of the gut or throat is infected, and that infectious virus is found in the draining lymphatic tissue but does not necessarily replicate there. A viraemic phase occurs within 1 week of infection as shown by the effectiveness of passive antibodies in preventing disease if given within 1 week of infection, but the source of the viraemia is not known. Invasion of the CNS may involve transport of the virus along neurones, or via the blood-brain barrier.

IV. Inactivated Poliovaccines

The first vaccines against poliomyelitis were devised and tested in the 1930's, and consisted of extracts of infected monkey spinal cord. One preparation was believed to be inactivated and another attenuated in its virulence. Both caused paralytic poliomyelitis in clinical trials. The first effective and safe vaccine used was developed by Salk, and was based on tissue culture grown virus treated with formalin. The treatment was sufficiently mild to preserve the delicate antigenicity of the virus particle, but sufficiently stringent to inactivate infectivity. The vaccine, currently manufactured in a highly potent form by the use of large scale cell culture fermenter technology has been used in Scandinavian countries, in the Netherlands and increasingly in France to great effect. The earlier less potent form had a major impact on the incidence of disease in the United States, reducing it by about 90% before the introduction of the live vaccines developed by Sabin.

Inactivated poliovaccine has been shown to be able to interrupt transmission of poliovirus under the appropriate conditions. In Finland in 1984 there was an outbreak of poliomyelitis in which eight persons were affected, with evidence that the virus infected a substantial proportion of the rest of the population without causing disease. The outbreak was attributed in large part to declining public interest in vaccination due to the success of public health programmes in eliminating the disease and the use of old style low potency preparations. However, during the preceding 25 years no poliovirus had been isolated from any source in Finland, implying that transmission had been interrupted by earlier effective immunization strategies which had exclusively used inactivated vaccines. Similarly there have been occasional outbreaks of poliomyelitis in the Netherlands in communities who refuse vaccination on religious grounds. The disease and virus isolations were in all cases essentially confined to the particular community, and did not appear as silent infections in the surrounding population where a high level of immunity had been generated by the use of high potency inactivated poliovaccine.

The basis for the interruption of transmission has been hotly debated for many years and is of great importance in the efforts to eradicate poliomyeltis from the world entirely, as is described below. It has been considered for some time that inactivated vaccine does not generate gut immunity and does not therefore prevent infection of the intestine of the immune individual. In fact it has been reported in several studies that previous immunization with the inactivated vaccine can inhibit immunization with the live vaccine to some extent, but it seems likely that the effect is not strong. The interruption of transmission has therefore been attributed to the effect of the vaccine in preventing viraemia, and therefore the secondary infection of the pharynx from the gut. By this reasoning, transmission could only be broken where the standards of hygiene are sufficient to prevent significant faecal oral spread, and where most infections are from the pharynx as in the Netherlands or Scandinavia. It is not clear if this is the case, but if it were, the general value of inactivated vaccine world wide in interrupting transmission would be questionable. It is therefore a question of great importance in attempts to eradicate all polioviruses.

V. Oral Poliovaccines

The relative merits of inactivated and attenuated vaccines against poliomyelitis have been argued continuously even before their development. Both are clearly highly effective and safe, and the extensive use of the live attenuated vaccine strains developed by Sabin in preference to the inactivated vaccines of Salk probably originated as much in their cost and the fact that they could be manufactured in sufficient quantities as in any intrinsic clinical merit. However, it is widely accepted that epidemics are best controlled by the use of the live vaccine, probably because the infection of available individuals means that the epidemic virus has no unoccupied host to infect rather than through the induction of gut immunity in the usually accepted sense. Clearly if wild-type poliovirus is to be eradicated, therefore the wide use of the live attenuated strains seems the most suitable strategy, as described below.

The origin of the live attenuated strains developed by Sabin has been recorded (SABIN and BOULGER 1973). The type 1 strain LSc originated from the Mahoney isolate and had been passaged through monkey testes culture. The type 2 strain P712 was isolated from an asymptomatic carrier and the type 3 strain. Leon 12a1b from an isolate from a fatal case in 1937 which was extensively passaged in vitro in tissue culture. Over the past 15 years or so the molecular basis of the attenuated phenotypes has been studied by a number of workers (MINOR 1992). Purified genomic RNA from poliovirus is infectious and full length cDNA clones can be produced, from which it is possible to transcribe infectious RNA. Where clones are available from attenuated and closely related virulent viruses it is possible to generate chimaeric viruses which can be tested for virulence in animal models, thus identifying the differences responsible for the different phenotypes. Despite the differences in passage history and origin of the Sabin vaccine strains it appears that there are similar factors involved. All have mutations in the 5′ noncoding region which reduce the efficiency of initiation of translation apparently by weakening without disrupting a particular domain in the predicted secondary structure. All also have attenuating mutations in the capsid regions. The type 3 strain has a mutation in capsid protein VP3 which affects the formation of the pentameric subunit of the capsid. The type 2 strain has a mutation in capsid protein VP1 which may have an effect in attachment, but it may also have a mutation in VP4. So far as is known these are the only mutations which affect the virulence of the vaccine strains of these two serotypes. While the type 1 strain also has a mutation in the 5′ noncoding region and mutations in VP1 and probably VP4 the full basis of attenuation is believed to be more complicated and more mutations may be involved. However, the similarities between the strains are striking.

In rare cases vaccine associated poliomyelitis can occur. The rate is estimated at one per 530000 primary vaccinees and one per 2.5 million overall recipients (NKOWANE et al. 1987). Some 90% of cases are attributable to types 2 or 3, with cases due to type 1 making up the remainder. The molecular and biological properties of isolates from cases in recipients or their contacts are consistent with the molecular basis of virulence given above. However other changes also occur, including recombination between serotypes, mutations in antigenic sites and mutations having the effect of fine tuning the optimal temperature of growth to that prevailing in the gut, as well as mutations whose selective advantage if any is unknown. Precisely the same mutations are observed at high frequency and extremely rapidly in healthy infants after immunization when isolates are made. For example the mutations in the 5′ noncoding region revert to the virulent form within at most six days in all instances for type 3, and usually much sooner, within 1 week in all instances for type 2 and in half of the instances within 1 week for type 1. Moreover type 1 isolates may either revert the attenuating base, or change the base with which it is predicted to pair in the secondary structure to restore a strong base pair, or another base pair in the same stem which is mismatched in the vaccine strain but strongly paired in the revertant. These changes not only suggest that the predicted structures have a real physiological significance but also indicate the viruses (and by implication possibly all viruses) may be able to adapt rapidly and with great subtlety to the environment in which they are growing. Excretion of viruses after vaccination continues for 5–6 weeks in about half of the recipients, although 1% continue to excrete virus for more than 10 weeks. In hypogammaglobulinaemic individuals, however, there is evidence that excretion of viruses can continue for at least 3 years, and reason to believe that it may continue for much longer. In normal recipients the type 2 strain shows the highest take when all three serotypes are given, followed by type1 and then type 3.

VI. Eradication of Poliomyelitis

By the early seventies poliomyelitis was well controlled in developed countries, including Western Europe and the United States. The incidence of poliomyelitis in the latter is shown in Fig. 2 in numbers of cases per 100000 inhabitants and had declined to a low base line level explained by occasional importation of virus from endemic countries and rare vaccine associated cases as described above. In Scandinavia and the Netherlands inactivated vaccine was used exclusively and effectively eliminated the disease, although occasional outbreaks occurred when wild-type viruses were imported into groups refusing vaccination for religious reasons.

At the same time, live poliovaccine was considered to have very little impact in tropical countries, for which a variety of explanations were put forward, including the occurrence of gastrointestinal infections at high frequency, and a more endemic pattern of infection with wild-type poliovirus because of poor standards of hygiene. It is likely that the availability of vaccine of high quality correctly handled and administered to the appropriate age of children who in any event received a far higher standard of health care played a major role in the effectiveness of vaccination in developed countries. However, poliomyelitis was controlled in Cuba and later to a large extent in Brazil by adopting a strategy of immunization campaigns, in which all children below a relevant age were immunized more or less simultaneously, rather than as they reached a certain age, which was the strategy used in other countries. This led to the strategy of National Immunization Days and has the effect of colonizing all susceptible individuals with the live vaccine virus, so that the wild-type virus is unable to spread. Transmission is thus interrupted and the wild virus dies out. Following the success of early versions of this strategy in South America the World Health Organization declared the goal of eradicating poliomyelitis due to wild-type poliovirus by the year 2000. Figure 3 shows

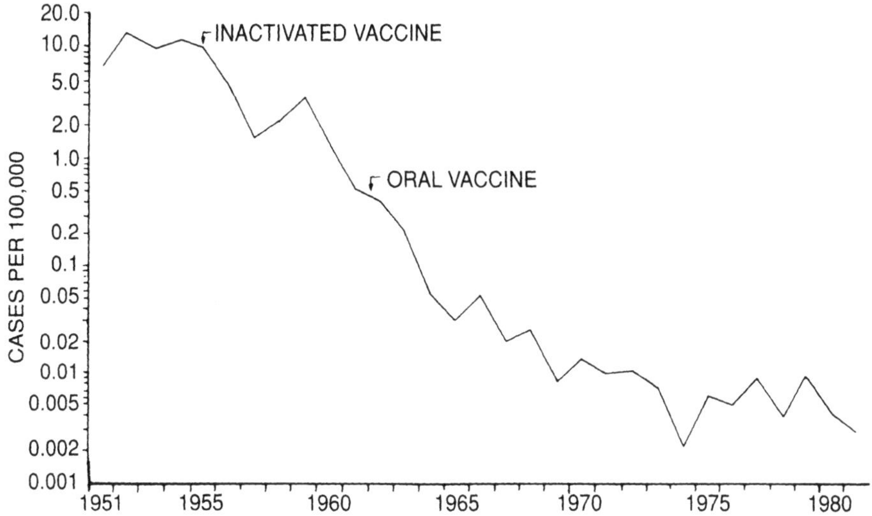

Fig. 2. Incidence of poliomyelitis in the United States from 1951 to 1980 in cases per 100000 head of population, showint the effects of the introduction of inactivated vaccine in 1955 and oral vaccine in 1961

the countries considered polio free in 1988 and the progress made by 1994. Currently there has been no poliomyelitis attributable to wild-type poliovirus in the Americas since 1992, and the whole region has been declared polio free. Major progress has been made in China where there have been only a few cases in 1996, attributable to imported virus, and there is every reason to think that transmission will be broken in India in 1997, if it has not already ceased. This extraordinary achievement has involved the use of the live vaccines developed by Sabin to eradicate the wild virus. Pockets of transmission persist in areas of conflict or among populations who are difficult to reach for some reason, and it is clear that further effort is needed, but it seems increasingly likely that the WHO goal will be met at or soon after the target date.

In the immediate future therefore poliomyelitis will be caused only by strains derived from the vaccines used in immunization programmes, and considerable thought is currently going into strategies for cessation of vaccination. One view is that vaccination can simply cease. There is a case to be made that the Sabin vaccine strains are less readily transmitted from person to person than wild-type strains. This is based in part on early studies on transmission among family members in the United States, partly on experience in Finland, where live poliovaccine was used extensively in a campaign to control the outbreak in 1984, and the virus was isolated from sewage only for 6 months after the campaign, and partly on the observation that in susceptible populations such as those refusing vaccination for religious reasons, outbreaks have so far always been caused by wild-type not vaccine related viruses. None of these arguments is conclusive. For example, in Finland the population had been previously exposed to a type three wild-type strain which had infected

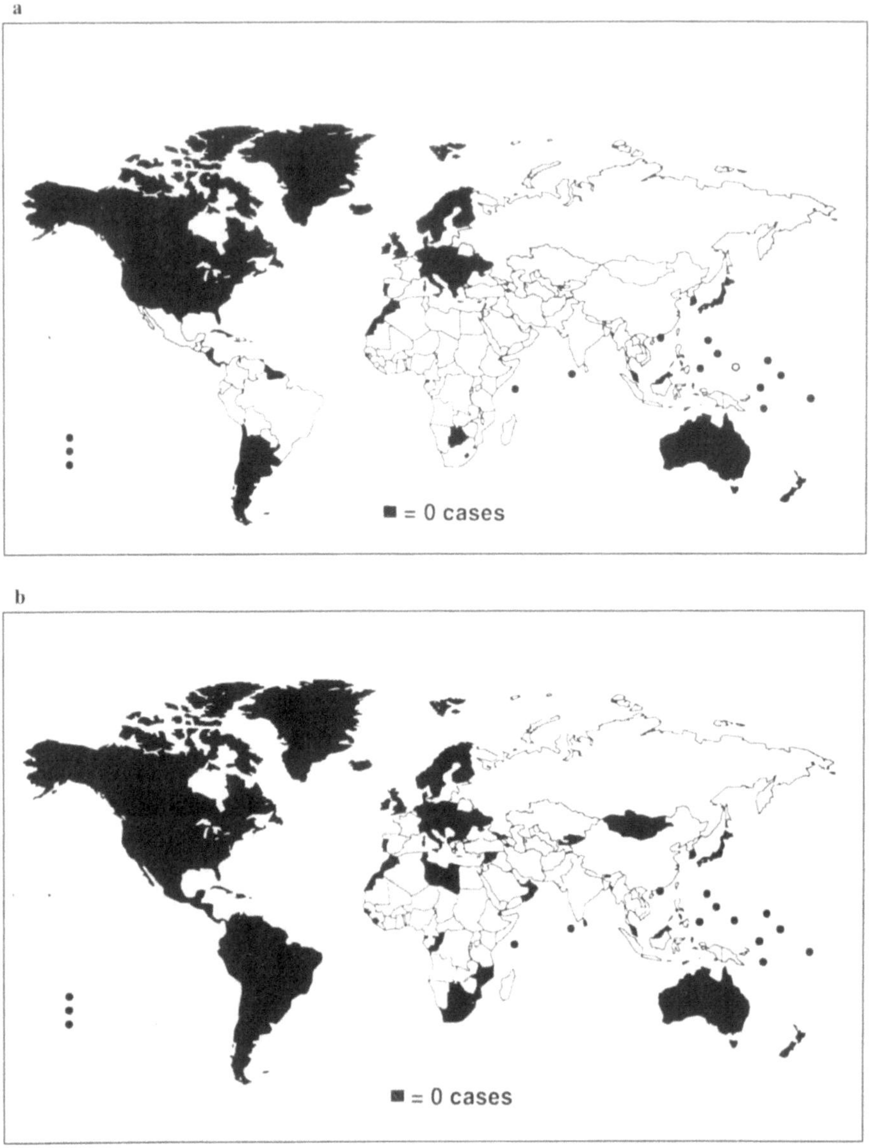

Fig. 3. Global distribution of countries with no indigenous polio (**a**) in 1988 and 1994 (**b**)

much of the population, and at the end of the live vaccine campaign use of high potency inactivated vaccine was resumed. A high level of immunity and awareness was therefore likely.

Alternative strategies could include the use of inactivated poliovaccine for some time until strains derived from the live vaccine are no longer isolated.

It is not clear how long this would be, and the question of hypogammaglobulinaemic individuals, or AIDS victims acting as a reservoir of infection is of some interest. The final strategy remains to be decided. The fact that one is required is a tribute to the amazing achievements of those who have conceived and helped to execute the eradication programme to date.

References

Alper CA, Kruskall MS, Marcusbagley D, Craven DE, Katz AJ, Brink SJ, Dienstag JL, Awdeh Z, Yunis EJ (1989) Genetic prediction of nonresponse to hepatitis-B vaccine. N Engl J Med 321:708–712

Asher LV, Binn LN, Mensing TL, Marchwicki RH, Vassell RA, Young GD (1995) Pathogenesis of hepatitis A in orally inoculated owl monkeys (Aotus trivirgatus). J Med Virol 47:260–268

Ballay A, Levrero M, Tiollais P, Perricaudet M (1987) Hepatitis B adenovirus recombinants as a potential live vaccine. In: Robinson W, Koike K, Will H (eds) Hepadna viruses, vol 70. Liss, New York, pp 481–493

Barker L, Chisari F, McGrath P et al (1973) Transmission of type B viral hepatitis to chimpanzees. J Infect Dis 127:648

Beasley R, Hwang L, Stevens C et al (1983) Efficacy of hepatitis B immune globulin for prevention of perinatal transmission of the hepatitis B virus carrier state: final report of a randomized double-blind placebo-controlled trial. Hepatology 3:135

Blumberg BS, Alter HJ, Visnich S (1965) A 'new' antigen in leukemia sera. J Am Med Assoc 191:101–106

Bradley DW, Balayan MS (1988) Virus of enterically transmitted non-A non-B hepatitis. Lancet I:819

Brown S, Howard C, Zuckerman A, Steward M (1984) Affinity of antibody response in man to hepatitis B vaccines determined with synthetic peptides. Lancet II:184

Carman WF, Zanetti A, Karayiannis P, Manzillo G, Tanzi E, Zuckerman AJ, Thomas HC (1990) A vaccine induced surface mutant of Hbv that is replication competent. Gut 31:A591–A591

Carman WF, Korula J, Wallace L, Macphee R, Mimms L, Decker R (1995a) Fulminant reactivation of hepatitis-B due to envelope protein mutant that escaped detection by monoclonal Hbsag Elisa. Lancet 345:1406–1407

Carman WF, Thursz M, Hadziyannis S, McIntyre G, Colman K, Gioustoz A, Fattovich G, Alberti A, Thomas HC (1995b) Hepatitis-B E-antigen negative chronic active hepatitis – hepatitis-B. J Viral Hepatitis 2:77–84

Chan H, Israel M, Garon C, Rowe W, Martin M (1979) Molecular cloning of polyoma virus DNA in Escherichia coli: lambda phage vector system. Science 203:887–892

Cheng KC, Smith GL, Moss B (1986) Hepatitis-B virus large surface protein is not secreted but is immunogenic when selectively expressed by recombinant vaccinia virus. J Virol 60:337–344

Chisari FV, Ferrari C (1995) Hepatitis-B virus immunopathogenesis. Annu Rev Immunol 13:29–60

Chisari FV, Filippi P, McLachlan A, Milich DR, Riggs M, Lee S, Palmiter RD, Pinkert CA, Brinster RL (1986) Expression of hepatitis-B virus large envelope polypeptide inhibits hepatitis-B surface-antigen secretion in transgenic mice. J Virol 60:880–887

Choo QL, Kuo G, Weiner AJ, Overby LR, Bradley DW, Houghton M (1989) Isolation of a cDNA clone derived from a blood-borne non-A non-B viral hepatitis genome. Science 244:359–362

Choo QL, Kuo G, Ralston R, Weiner A, Chien D, Van Nest G, Han J, Berger K, Thudium K, Kuo C (1994) Vaccination of chimpanzees against infection by the hepatitis C virus. Pro Natl Acad Sci USA 91:1294–1298

Cooremann MP, Schoodermark-Van de Ven EME (1996) Hepatitis C virus: biological and clinical consequence of genetic heterogeneity. Scand J Gastroenterol 31:106–115
Crowther RA, Kiselev NA, Bottcher B, Berriman JA, Borisova GP, Ose V, Pumpens P (1994) 3-dimensional structure of hepatitis-B virus core particles determined by electron cryomicroscopy. Cell 77:943–950
Curtiss R (1990) Attenuated salmonella strains as live vectors for the expression of foreign antigens. In: Woodrow GC, Levine MM (eds) New generation vaccines. Dekker, New York, pp 161–188
Dane D, Cameron C, Briggs M (1970) Virus-like particles in serum from patients with Australia antigen-associated hepatitis. Lancet I:695–698
Davis H, Michel M-L, Whalen R (1993) DNA based immunization for hepatitis N induces continuous secretion of antigen and high levels of circulating antibody. Hum Mol Genet 2:1847–1851
Davis HL, Michel ML, Mancini M, Watkins SC, Schleef M, Whalen RG (1994) Direct gene-transfer in muscle for genetic vaccination against hepatitis-B virus (Hbv). J Cell Biochem 527–527
Emini E, Larson V, Eichberg J, Cinard P, Garsky V, Lee D, Ellis R, Miller W, Anderson C, Gerety R (1989) Protective effect of a synthetic peptide comprising the complete preS2 region of the hepatitis B virus surface protein. J Med Virol 28:7–12
Farci P, Alter HJ, Govindarajan S, Wong DC, Engle R, Lesniewski RR, Mushahwar IK, Desai SM, Miller RH, Ogata N et al (1992) Lack of prective immunity against reinfection with hepatitis C virus. Science 258:135–140
Fernholz D, Galle PR, Stemler M, Brunetto M, Bonino F, Will H (1993) Infectious hepatitis-B virus variant defective in pre-S2 protein expression in a chronic carrier. Virology 194:137–148
Ferrari C, Cavalli A, Penna A, Valli A, Bertoletti A, Pedretti G, Pilli M, Vitali P, Neri TM, Giuberti T, Fiaccadori F (1992) Fine specificity of the human T-cell response to the hepatitis-B virus pres1 antigen. Gastroenterology 103:255–263
Ferrari C, Penna A, Bertoletti A, Fiaccadori F (1993) Cell-mediated immune-response to hepatitis-B virus nucleocapsid antigen. Arch Virol 8 [Suppl]:91–101
Ferrari G, Place C, Bartlett J, Moody D, Tartaglia J, Paoletti E, Weinhold K (1995) Antigen-specific activation/expansion of human anti-Hiv Ctl using recombinant alvac (canarypox)-based vectors. J Cell Biochem 219–219
Francis D, Hadler S, Thompson S, Maynard J, Ostrow D, Altman N, Braff E, Omalley P, Hawkins D, Judson F, Penley K, Nylund T, Christie G, Meyers F, Moore J, Gardner A, Doto I, Miller J, Reynolds G, Murphy B, Schable C, Clark B, Curran J, Redeker A (1982) The prevention of hepatitis B with vaccine. Report of the Centers for Disease Control multi-center efficacy trial among homosexual men. Ann Intern Med 97:362–366
Fuerst TR, Yarbough PO, Zhang Y, McAtee P, Tam AW, Lifson J, McCaustland K, Spellbring J, Bradley D, Margolis HS et al (1996) Prevention of hepatitis E using a novel ORF2 subunit vaccine. In: Buison Y, Coursaget P, Kane M (eds) Enterically transmitted hepatitis viruses. France, La Simarre Joue-les-Tours, pp 384–392
Fujii H, Moriyama K, Sakamoto N, Kondo T, Yasuda K, Hiraizumi Y, Yamazaki M, Sakaki Y, Okochi K, Nakajima E (1992) Gly^{145} to Arg substitution in HBs antigen of immune escape mutant of hepatitis B virus. Biochem Biophys Res Commun 184:1152–1157
Fynan E, Webster R, Fuller D, Haynes J, Santoro J, Robinson H (1993) DNA vaccines protective immunization by parenteral mucosal and gene gun inoculations. Proc Natl Acad Sci USA 90:11478–11482
Galibert F, Mandart E, Fitoussi F, Tiollais P, Charnay P (1979) Nucleotide sequence of the hepatitis B virus genome (subtype ayw) cloned in E coli. Nature 281:646–650

Ganem D (1996) Hepadnavridae and their replication. In: Fields BN, Knipe DM, Howley PM et al (eds) Fields virology, 3rd edn. Lippincott-Raven, Philadelphia, pp 2703–2737
Gerety R, West D (1990) Current and future hepatitis B vaccines. In: Coursaget P, Tong M (eds) Progress in hepatitis B immunization. John Libbey Eurotext, Paris, pp 215–225 (Colloque INSERM, vol 194)
Gerlich W, Bruss V (1993) Functions of hepatitis B virus proteins and molecular targets for protective immunity. In: Ellis R (ed) Hepatitis B vaccines in clinical practice. Dekker, New York, pp 41–82
Gilles P, Fey G, Chisari F (1992) Tumor necrosis factor-alpha negatively regulates hepatitis B virus gene expression in transgenic mice. J Virol 66:3955–3960
Grady G, Lee V, Prince A et al (1978) Hepatitis B immune globulin for accidental exposures among medical personnel: final report of a multicenter controlled trial. J Infect Dis 138:625
Guidotti LG, Ando K, Hobbs MV, Ishikawa T, Runkel L, Schreiber RD, Chisari FV (1994a) Cytotoxic T-lymphocytes inhibit hepatitis-B virus gene-expression by a noncytolytic mechanism in transgenic mice. Proc Natl Acad Sci USA 91:3764–3768
Guidotti LG, Guilhot S, Chisari FV (1994b) Interleukin-2 and alpha/beta interferon down-regulate hepatitis-B virus gene-expression in-vivo by tumor necrosis factor-dependent and factor-independent pathways. J Virol 68:1265–1270
Guilhot S, Guidotti LG, Chisari FV (1993) Interleukin-2 down-regulates hepatitis-B virus gene-expression in transgenic mice by a posttranscriptional mechanism. J Virol 67:7444–7449
Hadler SC (1991) Global impact of hepatitis A virus infection changing patterns. In: Hollinger FB, Lemon SM, Margolis HS (eds) Viral hepatitis and liver disease. Williams and Wilkins, Baltimore, pp 14–20
Harford N, Cabezon T, Crabeel M, Simoen E, Rutgers A, De Wilde M (1983) Expression of hepatitis B surface antigen in yeast. Dev Biol Stand 54:125–130
Hillemann MR (1993) Plasma-derived hepatitis B vaccine: a breakthrough in preventive medicine. In: Ellis RW (ed) Hepatitis B vaccines in clinical practice. Dekker, New York, pp 17–39
Hino K, Okuda M, Hashimoto O, Ishiko H, Okazaki M, Fujii K, Hanada H, Okita K (1995) Glycine-to-arginine substitution at codon-145 of Hbsag in 2 infants born to hepatitis-B E-antigen-positive carrier. Dig Dis Sci 40:566–570
Hochstein-Mintzel V, Huber H, Stickl H (1972) Virulenz und Immunogenität eunes modifizierten Vakzinia-Virus (Stamm MVA). Z Immun Forsch 144:140–145
Hogle JM, Chow M, Filman DJ (1985) The three dimensional structure of poliovirus at 29 A resolution. Science 229:1358–1365
Hollinger FB, Ticehurst JR (1996) Hepatitis A virus fields. Virology 24:735–782
Houghton M (1996) Hepatitis C Viruses. In: Fields BN, Knipe DM, Howley PM et al (eds) Fields Virology. Lippincott-Raven, Philadelphia, pp. 1035–1058
Hourvitz A, Mosseri R, Solomon A, Yehezkelli Y, Atsmon J, Danon Y, Koren R, Shouval D (1996) Reactogenicity and immunogenicity of a new recombinant hepatitis B vaccine containing preS antigens: a preliminary report. J Viral Hepatitis 3:37–42
HyppiaT, Hovi T, Knowles NJ, Stanway G (1997) Classification of enteroviruses based on molecular and biological properties. J Gen Virol 78:1–11
Inchauspé G (1997) Gene vaccination for hepatitis C. Springer Semin Immunopathol 19:211–221
Innis BL, Snitbhan R, Kunasol P, Laorakpongse T, Poopatanakool W, Kozik CA, Suntayakorn S, Suknuntapong T, Safary A, Tang DB et al (1994) Protection against hepatitis A by an inactivated vaccine. JAMA 271:1328–1334
Ionescu-Matiu I, Kennedy R, Sparrow J, Culwell A, Sanchez J, Melnick L, Dressman G (1983) Epitopes associated with a synthetic hepatitis B surface antigen peptide. J Immunol 130:1947

Israel M, Chan H, Rowe W, Martin M (1979) Molecular cloning of polyoma virus DNA in Eschrichia coli: plasmid vector system. Science 203:883–887
Itoh Y, Fujisawa Y (1986) Synthesis in yeast of hepatitis B virus surface antigen modified P31 particles by gene modification. Biochem Biophys Res Commun 141:942–948
Itoh Y, Takai E, Ohnuma H, Kitajima K, Tsuda F, Machida A, Mishiro S, Nakamura T, Miyakawa Y, Mayumi M (1986) A synthetic peptide vaccine involving the product of the pre-S (2) region of hepatitis B virus DNA: protective efficacy in chimpanzees. Proc Natl Acad Sci USA 83:9174–9178
Jarvis LM, Watson HG, MsOmish F, Peutherer JF, Ludlam CA, Simmonds P (1992) Frequent reinfection and reactivation of hepatitis C virus genotypes in multitransfused hemophiliacs. J Infect Dis 170:1018–1022
Jung MC, Stemler M, Weimer T, Spengler U, Dohrmann J, Hoffmann R, Eichenlaub D, Eisenburg J, Paumgartner G, Riethmuller G, Will H, Pape GR (1991) Immune-response of peripheral-blood mononuclear-cells to Hbx-antigen of hepatitis-B virus. Hepatology 13:637–643
Jung MC, Schraut W, Santantonio T, Spengler U, Eichenlaub D, Eisenburg J, Zachoval R, Hoffmann R, Paumgartner G, Pastore G, Will H, Riethmuller G, Zieglerheitbrock HWL, Pape GR (1993) Increased frequency of Cd8+ Cd45r0+ memory T-lymphocytes in acute hepatitis-B virus-infection. J Hepatol 18:295–300
Jung MC, Diepolder HM, Spengler U, Wierenga EA, Zachoval R, Hoffmann RM, Eichenlaub D, Frosner G, Will H, Pape GR (1995) Activation of a heterogeneous hepatitis-B (Hb) core and E-antigen- specific Cd4 (+) T-cell population during seroconversion to anti-Hbe and anti-Hbs in hepatitis-B virus-infection. J Virol 69:3358–3368
Kao JH, Chen PJ, Lai MY, Wang TH, Chen DS (1995) Quasispecies of hepatitis C virus and genetic drift of the hypervariable region in chronic type C hepatitis. J Infect Dis 172:261–264
Karayiannis P, Jowett T, Enticott M et al (1986) Hepatitis A virus replication in tamarins and host immune response in relation to pathogenesis of liver cell damage. J Med Virol 18:261–276
Karron RA, Daemer RJ, Ticehurst JR et al (1988) Evaluation of attenuation of cell-culture-adapted hepatitis A virus in nonhuman primates. In: Zuckerman AJ (ed) Viral hepatitis and liver disease. Liss, New York, pp 74–76
Kobayashi M, Asano T, Utsunomiya M, Itoh Y, Fujisawa Y, Nishimura O, Kato K, Kakinuma A (1988) Recombinant hepatitis B virus surface antigen carrying the pre-S2 region derived from yeast: purification and characterization. J Biotechnol 8:1–22
Koff RS (1993) Viral hepatitis. In: Schiff L, Schiff ER (eds) Diseases of the liver. pp 492–577
Krugman S, Stevens CE (1994) Hepatitis B vaccine. In: Plotkin S, Mortimer E (eds) Vaccines. Saunders, Chicago, pp 419–437
Krugman S, Giles JP, Hammond J (1967) Infectious hepatitis. J Am Med Assoc 200:365–373
Krugman S, Giles JP, Hammond J (1971) Viral hepatitis type B (MS-2 strain). Studies on active immunization. J Am Med Assoc 217:41–45
Kumar U, Monjardino J, Promise HC (1994) Hypervariable region of the hepatitis C virus envelope glycoprotein "E2-NS1" in an agammaglobulinemic patient. Gastroenterology 106:1072–1075
Kurane I, Binn LN, Bancroft WH, Ennis FA (1985) Human lymphocyte responses to hepatitis A virus-infected cells: interferon productio and lysis of infected cells. J Immunol 135:2140–2144
Kuroda S, Itoh Y, Miyazaki T, Otaka-Imai S, Fujisawa Y (1989) Efficient expression of genetically engineered hepatitis B virus surface antigen P31 proteins in yeast. Gene 78(2):297–308
Kuroda S, Fujisawa Y, Iino S, Akahane Y, Suzuki H (1991) Induction of protection level of anti-pre-S2 antibodies in humans immunized with a novel hepatitis B

vaccine consisting of M (pre-S2+S) protein particles (a third generation vaccine). Vaccine 9:163–169
Lemon SM, Binn LN (1983) Antigenic relatedness of two strains of hepatitis A virus determined by cross-neutralization. Infect Immun 42:418–420
Lemon SM, Binn LN (1993) Serum neutralizing of hepatitis A virus in vitro due to lipid-associated virus. J Gen Virol 148:1033–1039
Lemon SM, Shapiro CN (1994) The value of immunization against hepatitis A. Infect Agents Dis 3:38–49
Lemon SM, Thomas DL (1997) Vaccines to prevent viral hepatitis. N Engl J Med 336 (3):196–204
Lemon SM, Ping LH, Day S et al (1991) Immunobiology of hepatitis A virus. In: Hollinger FB, Lemon SM, Margolis HS (eds) Viral hepatitis and liver disease. Williams and Williams, Baltimore, pp 20–24
Lemon SM, Jansen RW, Brown EA (1992) Genetic antigenic and biological differences between strains of hepatitis A virus. Vaccine 10 [Suppl 1]:S40–S44
Li F, Zhuang H, Kolivas S, Locarnini SA, Anderson DS (1994) Persistent and transient antibody responses to hepatitis E virus detected by Western immunoblot using open reading frame 2 and 3 and glutathione S-transferase fusion protiens. J Clin Microbiol 32:2060–2066
Linnen J, Wages J Jr, Zhang-Keck ZY, et al (1996) Molecular cloning and disease association of hepatitis G virus: a transfusion-transmissible agent. Science 271: 505–508
Lubeck M, Davis A, Chengalvala M, Natuk R, Morin J, Molnar-Kimber K, Mason B, Bhat B, Mizutani S, Hung P, Purcell R (1989) Immunogenicity and efficacy testing of an oral hepatitis B vaccine based on live recombinant adenovirus. Proc Natl Acad Sci USA 86:6763–6767
MacCallum FO (1947) Homologous serum jaundice. Lancet 2:228
Madalinsky K, Zychowicz C, Gornicki J, Czubkowska I, Skarpetowsk A, Gorecki M, Koren R (1994) Immunogenicity in children of recombinant hepatitis B vaccine containing S pre-S2 and pre-S1 antigens. II. Persistence of antibodies. Polish J Immunol 19:333–340
Mancini M, Hadchouel M, Tiollais P, Pourcel C, Michel ML (1993) Induction of antihepatitis-B surface-antigen (Hbsag) antibodies in Hbsag producing transgenic mice – a possible way of circumventing nonresponse to Hbsag. J Med Virol 39:67–74
Mancini M, Hadchouel M, Davis HL, Whalen R, Tiollais P, Michel M-L (1996) DNA-mediated immunization breaks tolerance in a transgenic mouse model of hepatitis B surface antigen chronic carriers. Proc Natl Acad Sci USA 93:12496–12501
Maneerat Y, Clayson ET, Myint KSA, Young GD, Innis BL (1996) Experimental infection of the laboratory rat with the hepatitis E virus. J Med Virol 48:121–128
Mao JS, Chen NL, Huang HY, Chai SA, Dong DX, Coa YY, Zhang HY, Wu DM, Zhang SY (1992) Development of live attenuated hepatitis A vaccine (H2-strain). Chin Med J 105:189–193
Maruyama T, McLachlan A, Iino S, Koike K, Kurokawa K, Milich DR (1993) The serology of chronic hepatitis-B infection revisited. J Clin Invest 91:2586–2595
Mason H, Lam D-K, Arntzen C (1992) Expression of hepatitis B surcace antigen in transgenic plants. Proc Natl Acad Sci USA 89:11745–11749
Mayr A, Hochstein-Mintzel V, Stickl H (1975) Abstammung Eigenschaften und Verwendung des attenuierten Vakzinia-Stammes MVA. Infektion 6–14
Mayr A, Stickl H, Müller H, Danner K, Singer H (1978) Pockenimpfstamm MVA: Marker genetische Struktur Erfahrungen mit der parenteralen Schutzimpfung und Verhalten im abwehrgeschwächten Organismus. Zentralbl Bakt Hyg I Abt Orig B 167:365–390
McAleer W, Buynak E, Maigetter R, Wampler D, Miller W, Hilleman M (1984) Human hepatitis B vaccine from recombinant yeast. Nature 307:178–180
McAtee CP, Zhang Y, Yarbough PO, Bird T, Fuerst TR, Stone KL, Samander S, Williams KR (1996a) Purification and characterization of a recombinant hepatitis

E protein vaccine candidate by liquid chromatography-mass spectrometry. J Chromatogr B 685:91–104
McAtee CP, Zhang Y, Yarbough PO, Bird T, Fuerst TR (1996b) Purification of a soluble hepatitis E open reading frame 2-derived protein with unique antigenic properties. Protein Express 8:262–270
McLachlan A, Milich DR, Raney AK, Riggs MG, Hughes JL, Sorge J, Chisari FV (1987) Expression of hepatitis-B virus surface and core antigens – influences of pre-S and precore sequences. J Virol 61:683–692
McMahon B, Wainwright R (1993) Protective efficacy of hepatitis B vaccines in infants children and adults. In: Ellis R (ed) Hepatitis B vaccines in clinical practice. Dekker, New York, pp 243–261
McMahon G, Ehrlich P, Moustafa Z, McCarthy L, Dottavio D, Tolpin M, Nadler P, Östberg L (1992) Genetic alterations in the gene encoding the major HBsAg: DNA and immunological analysis of recurrent HBsAg derived from monoclonal antibody-treated liver transplant patient. Hepatology 15:757–766
Mendelsohn CL, Wimmer E, Racaniello VR (1989) Cellular receptor for poliovirus: molecular cloning nucleotide sequence and expression of a new member of the immunoglobulin superfamily. Cell 56:855–865
Meyer A, Sutter G, Mayr A (1991) Mapping of deletions in the genome of the highly attenuated vaccinia virus MVA and their influence on virulence. J Gen Virol 72:1031–1038
Michel M-L (1995) Dna-mediated immunization – prospects for hepatitis-B vaccination. Res Virol 146:261–265
Michel M-L, Davis HL, Schleef M, Mancini M, Tiollais P, Whalen R (1995) DNA-mediated immunization to the hepatitis B surface antigen in mice: aspects of the humoral response mimic hepatitis B viral infection in humans. Proc Natl Acad Sci USA 92:5307–5311
Midthun K, Ellerbeck E, Gershman K, Calandra G, Krah D, McCaughtry M, Nalin D, Provost P (1991) Safety and immunogenicity of a live attenuated hepatitis A virus vaccine in seronegative volunteers. J Infect Dis 163:735–739
Milich D (1993) Application of synthetic peptide technology to experimental HBV vaccine design. In: Ellis R (ed) Hepatitis B vaccines in clinical practice. Dekker, New York, pp 351–381
Milich DR (1997) Influence of T-helper subsets and crossregulation in hepatitis B virus infection. J Viral Hepat 4 [Suppl 2]:48–59
Milich DR, Chisari FV (1982) Genetic-regulation of the immune-response to hepatitis-B surface-antigen (Hbsag) 1 H-2-restriction of the murine humoral immuneresponse to the a and D determinants Of Hbsag. J Immunol 129:320–325
Milich DR, McLachlan A (1986a) The nucleocapsid of hepatitis B virus is both a T-cell-independent and a T-cell-dependent antigen. Science 234:1398–1401
Milich DR, McLachlan A (1986b) The nucleocapsid of hepatitis-B virus is both a T-cell independent and a T-cell-dependent antigen. Science 234:1398–1401
Milich DR, Alexander H, Chisari FV (1983a) Genetic-regulation of the immune-response to hepatitis-B surface- antigen (Hbsag) 3 circumvention of non-responsiveness in mice bearing Hbsag nonresponder haplotypes. J Immunol 130:1401–1407
Milich DR, Lerouxroels GG, Chisari FV (1983b) Genetic-regulation of the immune-response to hepatitis-B surface-antigen (Hbsag) 2. Qualitative characteristics of the humoral immune-response to the a-determinant D-determinant and Y-determinant of Hbsag. J Immunol 130:1395–1400
Milich DR, Lerouxroels GG, Louie RE, Chisari FV (1984) Genetic-regulation of the immune-response to hepatitis-B surface-antigen (Hbsag) 4. Distinct H-2-linked ir genes control antibody-responses to different Hbsag determinants on the same molecule and map to the I-a and I-C subregions. J Exp Med 159:41–56

Milich DR, Louie RE, Chisari FV (1985a) Genetic-regulation of the immune-response to hepatitis-B surface-antigen (Hbsag) 5. T-cell proliferative response and cellular interactions. J Immunol 134:4194–4202
Milich DR, McNamara MK, McLachlan A, Thornton GB, Chisari FV (1985b) Distinct H-2-linked regulation of T-cell responses to the pre-S-region and S-region of the same hepatitis-B surface-antigen polypeptide allows circumvention of non-responsiveness to the S-region. Proc Natl Acad Sci USA 82:8168–8172
Milich DR, Peterson DL, Lerouxroels GG, Lerner RA, Chisari FV (1985c) Genetic-regulation of the immune-response to hepatitis-B surface- antigen (Hbsag) 6 T-cell fine specificity. J Immunol 134:4203–4211
Milich DR, MacLachlan A, Moriarty A, Thornton GB (1987a) Immune responses to hepatitis B virus core antigen (HBcAg): localization of T cell recognition sites within HBcAg/HBeAg. J Immunol 139:1223–1231
Milich DR, McLachlan A, Thornton GB, Hughes JL (1987b) A synthetic T-cell site primes antibody production to both the nucleocapsid and the envelope of hepatitis B virus. Nature 330:547–549
Milich DR, McLachlan A, Stahl S, Wingfield P, Thornton GB, Hughes JL, Jones JE (1988) Comparative immunogenicity of hepatitis B virus core and e antigens. J Immunol 141:3617–3624
Milich DR, Hughes JL, McLachlan A, Langley KE, Thornton GB, Jones JE (1990a) Importance of subtype in the immune-response to the pre-S(2) region of the hepatitis-B surface-antigen 1 T-cell fine specificity. J Immunol 144:3535–3543
Milich DR, Jones JE, Hughes JL, Price J, Raney AK, McLachlan A (1990b) Is a function of the secreted hepatitis-B antigen to induce immunological-tolerance inutero. Proc Natl Acad Sci USA 87:6599–6603
Milich DR, McLachlan A, Raney AK, Houghten R, Thornton GB, Maruyama T, Hughes JL, Jones JE (1991) Autoantibody production in hepatitis-B E-antigen transgenic mice elicited with a self T-cell peptide and inhibited with nonself peptides. Proc Natl Acad Sci USA 88:4348–4352
Milich DR, Schödel F, Peterson DL, Jones JE, Hughes JL (1995a) Characterization of self-reactive T-cells that evade tolerance in hepatitis-B E-antigen transgenic mice. Eur J Immunol 25:1663–1672
Milich DR, Wolf SF, Hughes JL, Jones JE (1995b) Interleukin-12 suppresses autoantibody production by reversing helper T-cell phenotype in hepatitis-B E-antigen transgenic mice. Proc Natl Acad Sci USA 92:6847–6851
Milich DR, Chen MK, Hughes JL, Jones JE (1998) The secreted hepatitis B eantigen preferentially depletes Th1 cells via Fas: a mechanisms for persistent infection. J Immunol (in press)
Minor PD (1990) Antigenic structure of picornaviruses. In: Compans Rw, Cooper M, Koprowski H et al (eds) Current topics in microbiology and immunology, vol 161. Springer, Berlin Heidelberg New York
Minor PD (1991) Picornaviridae. In: Francki RIB, Fauquet CM, Knudson DL, Brown F (eds) Classification and nomenclature of viruses. Springer, Vienna New York, pp 320–326 (Archives of virology, suppl 2)
Minor PD (1992) The molecular biology of poliovaccines. J Gen Virol 73:3065–3077
Minor PD (1996) Poliovirus. In: Nathanson N et al (eds) Viral pathogenesis. Lippincott-Raven, Philadelphia, pp 555–574
Minor PD, Brown F, King A, Knowles NJ, Lemon S, Martin S, Melnick J, Moore N, Palmenberg A, Rueckert RR, Yin Murphy M (1991) Picornaviridae. In: Francki RIB, Fauquet CM, Knudson DL, Brown F (eds) Classification and nomenclature of viruses; fifth report of the International Committee on Taxonomy of Viruses. Springer, Vienna New York, pp 320–326 (Archives of virology, suppl 2)
Molnarkimber KL, Jarockiwitek V, Dheer SK, Vernon SK, Conley AJ, Davis AR, Hung PP (1988) Distinctive properties of the hepatitis-B virus envelope proteins. J Virol 62:407–416

Morin J, Lubeck M, Barton J, Conley A, Davis A, Hung P (1987) Recombinant adenovirus induces antibody response to hepatitis B virus surface antigen in hamsters. Proc Natl Acad Sci USA 84:4626–4630
Moss B, Smith GL, Gerin JL, Purcell RH (1984) Live recombinant vaccinia virus protects chimpanzees against hepatitis-B. Nature 311:67–69
Murray K, Bruce SA, Hinnen A, Wingfield P, van Eerd PMCA, de Reus A, Schellekens H (1984) Hepatitis B virus antigens made in microbial cells immunise against viral infection. EMBO J 3:645–650
Murray K, Bruce SA, Wingfield P, van Eerd P, de Reus A, Schellekens H (1987) Protective immunisation against hepatitis B with an internal antigen of the virus. J Med Virol 23:101–107
Nabel E, Plautz G, Nabel G (1990) Site-specific gene expression in vivo by direct gene transfer into the arterial wall. Science 249:1285–1288
Nainan OV, Brinton MA, Margolis HS (1992) Identification of amino acids located in the antibody binding sites of human hepatitis A virus. Virology 191:984–987
Nakano I, Maertens G, Major ME, Vitvitski L, Dubuisson J, Fournillier A, Martynoff G, de Trepo C, Inchauspe G (1997) Immunization with plasmid DNA encoding hepatitis C virus envelope E2 antigenic domains induces antibodies whose immune reactivity is linked to the injection mode. J Virol 71:7101–7109
Nardelli-Haefliger D, Kraehenbuhl J-P, Curtiss R III, Schödel F, Potts A, Kelly S, De Grandi P (1996) Oral and rectal immunization of adult female volunteers with a recombinant attenuated Salmonella typhi vaccine strain. Infect Immun 64:5219–5224
Nassal M (1996) Hepatitis B virus morphogenesis. In: Capron A, Compans RW, Cooper M et al (eds) Current topics in microbiology and immunology, vol 214. Springer, Berlin Heidelberg New York, pp 297–337
Nayersina R, Fowler P, Guilhot S, Missale G, Schlicht H-J, Vitielo A, Chesnut R, Person J, Redeker A, Chisari F (1993) HLA-A2 restricted cytotoxic T lymphocyte responses to multiple hepatitis B surface antigen epitopes during hepatitis B virus infection. J Immunol 150:4659–4671
Neurath AR, Kent SBH, Strick N, Taylor P, Stevens CE (1985) Hepatitis-B virus contains pre-S gene-encoded domains. Nature 315:154–156
Neurath AR, Kent SBH, Parker K, Prince AM, Strick N, Brotman B, Sproul P (1986) Antibodies to a synthetic peptide from the pres 120–145 region of the hepatitis-B virus envelope are virus-neutralizing. Vaccine 4:35–37
Neurath AR, Seto B, Strick N (1989) Antibodies to synthetic peptides from the pres1 region of the hepatitis-B virus (Hbv) envelope (Env) protein are virus-neutralizing and protective. Vaccine 7:234–236
Neurath AR, Strick N, Sproul P, Ralph HE, Valinsky J (1990) Detection of receptors for hepatitis-B virus on cells of extrahepatic origin. Virology 176:448–457
Nkowane BM, Wassilak SG, Oversteen WA, Bart KJ, Schonberger LB, Hinman AR, Kew OM (1987) Vaccine associated paralytic poliomyelitis in the United States: 1973 through 1984. J Am Med Assoc 257:1335–1340
Oehen S, Hengartner H, Zinkernagel R (1991) Vaccination for disease. Science 251:195–198
Ohnuma H, Machida A, Okamoto H, Tsuda F, Sakamoto M, Tanaka T, Miyakawa Y, Mayumi M (1993) Allelic subtypic determinants of hepatitis B surface antigen (I and t) that are distinct from d/y or w/r. J Virol 67:937–932
Okamoto H, Yano K, Nozaki Y, Matsui A, Miyazaki H, Yamamoto K, Tsuda F, Machida A, Mishiro S (1992) Mutations within the S gene of hepatitis-B virus transmitted from mothers to babies immunized with hepatitis-B immune globulin and vaccine. Pediatric Res 32:264–268
Panda SK, Nanda SK, Zafrullah M, Ansari I, Ozdenar MH, Jameel S (1995) An Indian strain of hepatitis E virus (HEV): cloning sequence and expression of structural region and antibody responses in sera from individuals from an area of high-level HEV endemicity. J Clin Microbiol 33:2653–2659

Pasek M, Goto T, Gilbert W, Zink B, Schaller H, MacKay P, Leadbetter G, Murray K (1979) Hepatitis B virus genes and their expression in E coli. Nature 282:575–579
Penna A, Bertoletti A, Cavalli A, Valli A, Missale G, Pilli M, Marchelli S, Giuberti T, Fowler P, Chisari FV, Fiaccadori F, Ferrari C (1992a) Fine specificity of the human T-cell response to hepatitis-B virus core antigen. Arch Virol 4 [Suppl]:23–28
Penna A, Fowler P, Bertoletti A, Guilhot S, Moss B, Margolskee RF, Cavalli A, Valli A, Fiaccadori F, Chisari FV, Ferrari C (1992b) Hepatitis-B virus (Hbv)-specific cytotoxic T-cell (Ctl) response in humans – characterization of Hla class-Ii-restricted Ctls that recognize endogenously synthesized Hbv envelope antigens. J Virol 66:1193–1198
Persing DH, Varmus HE, Ganem D (1986) Inhibition of secretion of hepatitis-B surface-antigen by a related presurface polypeptide. Science 234:1388–1391
Pincus S, Tartaglia J, Paoletti E (1995) Poxvirus-based vectors as vaccine candidates. Biologicals 23:159–164
Ping LH, Lemon SM (1992) Antigenic structure of human hepatitis A virus defined by analysis of escape mutants selected against murine monoclonal antibodies. J Virol 66:2208–2216
Ping LH, Jansen RW, Stapleton JT, Cohen JI, Lemon SM (1988) Identification of an immunodominant antigenic site involoving the capsid protein VP3 of hepatitis A virus. Proc Natl Acad Sci USA 85:8281–8285
Pipkin PA, Wood DJ, Racaniello VR, Minor PD (1993) Characterisation of L cells expressing the human poliovirus receptor for the specific detection of polioviruses in vitro. J Virol Methods 41:333–340
Pol S, Driss F, Michel M-L, Nalpas B, Berthelot P, Brechot C (1994) Specific vaccine therapy in chronic hepatitis B infection (letter to the editor). Lancet 344:342
Prince AM (1968) An antigen detected in the blood during the incubation period of serum hepatitis. Proc Natl Acad Sci USA 60:814–821
Prince AM, Brotman B, Huima T, Pascual D, Jaffrey M, Inchauspé G (1992) Immunity in hepatitis C virus infection. J Infect Dis 165:438–443
Prince AM, Whalen R, Brotman B (1997) Successful nucleic acid based immunization of newborn chimpanzees against hepatitis B virus. Vaccine 15:916–919
Provost PJ, Hilleman MR (1979) Propagation of human hepatitis A virus in cell culture in vitro. Proc Soc Exp Biol Med 160:213–221
Purcell RH (1996) Hepatitis E virus. In: Fields BN, Knipe DM, Howley PM, Chanock RM, Melnich JL, Monath TP, Roizman B, Straus SE (eds) Fields virology, 3rd edn. Lippincott-Raven, Philadelphia, pp 2831–2843
Purdy MA, McCaustland KA, Krawczynski K et al (1992) Expression of a hepatitis E virus (HEV)-trpE fusion protein containing epitopes recoganized by antibodies in sera from human cases and experimentally infected primates. Arch Virol 123:335–349
Redeker A, Mosley J, Gocke D, McKee A, Pollack W (1975) Hepatitis B immune globulin as a prophylactic measure for spouses exposed to acute type B hepatitis. N Engl J Med 293:1055–1059
Rice CM (1996) Flaviviridae: the viruses and their replication. In: Fields BN, Knipe DM, Howley PM et al (eds) Fields virology, 3rd edn. Lippincott-Raven, Philadelphia, pp 931–1034
Robinson H, Hunt L, Webster R (1993) protection against a lethal influenza virus challenge by immunization with a haemagglutinin-expressing plasmid. Vaccine 11:967–960
Rutgers T, Hauser P, De Wilde M (1993) Potential future recombinant vaccines. In: Ellis R (ed) Hepatitis B virus vaccines in clinical praxis. Dekker, New York, pp 383–407
Sabin AB, Boulger LR (1973) History of the Sabin attenuated poliovirus oral live vaccine strains. J Biol Stand 1:115–118

Schödel F (1992) Prospects for oral vaccination using recombinant bacteria expressing viral epitopes. Adv Vir Res 41:409–446
Schödel F (1998) Hepatitis B virus: molecular mechanisms in disease and novel strategies for therapy. In: Koshy R, Caselmann WH (eds) Hepatitis B virus, molecular mechanisms in disease and novel strategies for therapy. Imperial College Press, London (in press)
Schödel F, Sprengel R, Weimer T, Fernholz D, Schneider R, Will H (1989) Animal hepatitis B viruses. Adv Viral Oncol 8:73–102
Schödel F, Milich DR, Will H (1990a) Hepatitis B virus nucleocapsid/pre-S2 fusion proteins expressed in attenuated Salmonella for oral vaccination. J Immunol 145:4317–4321
Schödel F, Weimer T, Will H (1990b) HBV: molecular biology and immunology. Biotest Bull 4:63–83
Schödel F, Moriarty AM, Peterson DL, Zheng J, Hughes JL, Will H, Leturcq DJ, McGee JS, Milich DR (1992) The position of heterologous epitopes inserted in hepatitis-B virus core particles determines their immunogenicity. J Virol 66:106–114
Schödel F, Neckermann G, Peterson D, Fuchs K, Fuller S, Will H, Roggendorf M (1993a) Immunization with recombinant woodchuck hepatitis virus nucleocapsid antigen protects woodchucks from woodchuck hepatitis virus infection. Vaccine 11:624–628
Schödel F, Peterson D, Zheng J, Jones JE, Hughes JL, Milich DR (1993b) Structure of hepatitis-B virus core and E-antigen – a single precore amino-acid prevents nucleocapsid assembly. J Biol Chem 268:1332–1337
Schödel F, Kelly SM, Peterson DL, Milich DR, Curtiss R (1994a) Hybrid hepatitis B virus core-pre-S proteins synthesized in avirulent Salmonella typhimurium andSalmonella typhi for oral vaccination. Infect Immun 62:1669–1776
Schödel F, Peterson D, Hughes J, Milich D (1994b) Hepatitis B virus core particles as a vaccine carrier moiety. Int Rev Immunol 11:153–165
Shouval D, Ilan Y, Adler R, Deepen R, Panet A, Evenchen Z, Gorecki M, Gerlich WH (1994a) Improved immunogenicity in mice of a mammalian cell-derived recombinant hepatitis-B vaccine containing pre-S-1 and pre-S-2 antigens as compared with conventional yeast-derived vaccines. Vaccine 12:1453–1459
Shouval D, Ilan Y, Hourvitz A, Mosseri R, Solomon A, Zychowicz A, Gornicki J, Czubkowska I, Madalinski K, Burczynska B, Adler R, Gorecki M, Koren R (eds) (1994b) Immunogenicity of a mammalian cell-derived recombinant hepatitis B vaccine containing preS2 and preS1 antigens: a preliminary report. Viral Hepatitis Liver Dis
Siegl G, Weitz M (1993) Pathogenesis of hepatitis A: persistent viral infection as basis of an acute disease? Microb Pathog 14:1–8
Simons JN, Leary TP, Dawson GJ, et al (1995) Isolation of novel virus-like sequences associated with human hepatitis. Nat Med 1:564–569
Sitrin R, Wampler D, Ellis R (1993) Survey of licensed hepatitis B vaccines and their production processes. In: Ellis R (ed) Hepatitis B vaccines in clinical practice. Dekker, New York, pp 83–101
Sjögren MH, Purcell RH, McKee K, Binn L, Macarthy P, Ticehurst J, Feinstone S, Caudil J, See A, Hoke C et al (1992) Clinical and laboratory observations following oral or intramuscular administation of a live attenuated hepatitis A vaccine candidate. Vaccine 10 [Suppl 1]:S135–S137
Standring D, Ou J, Rutter W (1986) Assembly of viral particles in Xenopus oocytes: pre-surface-antigens regulate secretion of the hepatitis B viral surface envelope particle. Proc Natl Acad Sci USA 83:9338–9342
Stapleton JT (1995) Host immune response to hepatitis A virus. J Infect Dis 171 [Suppl 1]:S9–S14
Stapleton JT, Lemon SM (1987) Neutralization escape mutants define a dominant immunogenic neutralization site on hepatitis A virus. J Virol 61:491–498

Stapleton JT, Lemon SM (1997) New Vaccines Against Hepatitis A. In: Levine MM, Woodrow GC, Kaper JB, Cobon GS (ed) New generation vaccines. Marcel Dekker, New York, pp 571–585

Stevens C, Beasley R, Tsuji J, Lee W (1975) Vertical transmission of hepatitis B in Taiwan. N Engl J Med 292:771–774

Stickl H, Hochstein-Mintzel V, Mayr A, Huber H, Schäfer H, Holzner A (1974) MVA-Stufenimpfung gegen Pocken Klinische Erprobung des attenuierten Pocken-Lebendimpfstoffes Stamm MVA. Dtsch Med Wochenschr 99:2386–2392

Stokes J Jr, Neefe J (1945) The preventio and attenuation of infectious hepatitis by gamma globulin. JAMA 127:144–145

Suzuki H, Iino S, Shiraki K, Akahane Y, Okamoto H, Domoto K, Mishiro S (1994) Safety and efficacy of a recombinant yeast-derived pre-S2+S containig hepatitis B vaccine (TGP-943): phase 1 phase 2 and 3 clinical testing. Vaccine 12:1090–1096

Szmuness W, Stevens C, Harley E, Zang E, Oleszko W, William D, Sadovshy R, Morrison J, Kellner A (1980) Hepatitis B vaccine: demonstration of efficacy in a controlled clinical trial in a high risk population in the United States. N Engl J Med 303:833–841

Tacket C, Kelly S, Schödel F, Losonsky G, Nataro J, Edelman R, Levine M, Curtiss R III Safety and immunogenicity in humans of an attenuated Salmonella typhi vaccine vector strain expressing plasmid-encoded hepatitis B antigens stabilized by the ASD balanced-lethal system (in preparation)

Tacket CO, Losonsky G, Lubeck M, Davis A, Mizutani S, Horwith G, Hung P, Edelman R, Levine MM (1992) Initial safety and immunogenicity studies of an oral recombinant adeno-hepatitis B vaccine. Vaccine 10:673–676

Tang D-C, DeVit M, Johnston S (1992) Genetic immunization is a simple method for eliciting an immune response. Nature 356:152–154

Tartaglia J, Perkus ME, Taylor J, Norton EK, Audonnet JC, Cox WI, Davis SW, Vanderhoeven J, Meignier B, Riviere M, Languet B, Paoletti E (1992) Nyvac – a highly attenuated strain of vaccinia virus. Virology 188:217–232

Taylor JM (1996) Hepatitis delta virus and its replication. In: Fields BN, Knipe DM, Howley PM et al (eds) Fields virology, vol 2, , 3rd edn. Lippincott-Raven, Philadelphia; pp 2809–2818

Taylor J, Weinberg R, Languet B, Desmettre P, Paoletti E (1988) Recombinant fowlpox virus inducing protective immunity in non-avian species. Vaccine 6:497–503

Thanavala Y, Yang Y, Lyons P, Mason F, Arntzen C (1995) Immunogenicity of transgenic plant-derived hepatitis B surface antigen. Proc Natl Acad Sci USA 92:3358–3361

Thoma H, Hemmerling A, Koller E, Kaper G (1991) Does preS2 have the same effect in improving the HBV immune response as preS1. In: Hollinger F, Lemon S, Margolis H (eds) Viral hepatitis and liver disease. Williams and Wilkins, Baltimore, pp 736–741

Thornton G, Milich D, Chisari F, Mitamura K, Kent S, Neurath R, Purcell R, Gerin J (1987) Immune responses in primates to the pre-S2 region of hepatitis-B surface antigen: Identification of a protective determinant. In: Chanock R, Lerner R, Brown F, Ginsberg H (eds) Vaccines 87. Cold Spring Harbor Laboratory, Cold Spring Harbor

Thornton G, Moriarty A, Milich D, Eichberg J, Purcell R, Gerin J (1989) Protection of chimpanzees from hepatitis-B virus infection after immunization with synthetic peptides: identification of protective epitopes in the pre-S region. In: Brown F, Chanock R, Ginsberg H, Lerner R (eds) Vaccines 89. Cold Spring Harbor Laboratory, Cold Spring Harbor

Tian-Cheng L, Yoshio Y, Kenji S, Masashi T, Mohammed A, Azeez R, Toshikazu U, Naokazu T, Tatsuo M (1997) Expression and self-assembly of empty virus-like particles of hepatitis E virus. J Virol 71:7207–7213

Tiollais P, Pourcel C, Dejean A (1985) The hepatitis B virus. Nature 317:489–495
Tron F (1994) Hepatitis B virus recombinant vaccines: achievement and progress. In: Kurstak E (ed) Modern vaccinology. Plenum, New York, pp 153–167
Tsai SL, Chen PJ, Lai MY, Yang PM, Sung JL, Huang JH, Hwang LH, Chang TH, Chen DS (1992) Acute exacerbations of chronic type B-hepatitis are accompanied by increased T-cell responses to hepatitis-B core and E-antigens – implications for hepatitis-B E-antigen seroconversion. J Clin Invest 89:87–96
Tsarev SA, Emerson SU, Tsareva TS et al (1993) Variation in course of hepatitis E in experimentally infected cynomolgus monkeys. J Infect Dis 167:1302–1306
Tsarev SA, Tsareva TS, Emerson SU et al (1994) Successful passive and active immunization of cynomolgous monkeys against hepatitis E. Proc Natl Acad Sci USA 91:10198–10202
Tsarev SA, Tsareva TS, Emerson SU, Govindarajan S, Shapiro M, Gerin JL, Robinson R, Gorbalenya AE, Purcell RH (1996) Prospects for prevention of hepatitis E. In: Buisson Y, Coursaget P, Kane M (eds) Enterically-transmitted hepatitis viruses. France, La Simarre Tours, pp 373–383
Ulmer J, Donnelly J, Parker S, Rhodes G, Felgner P, Gromkowski S, Deck R, DeWitt C, Friedman A, Hawe L, Leander K, Martinez D, Perry H, Shiver J, Montgomery D, Liu M (1993) Heterologous protection against influenza by injection of DNA encoding a viral protein. Science 259:1745–1749
Valenzuela P, Gray P, Quiroga M, Zaldivar J, Goodman H, Rutter W (1979) Nucleotide sequence of the gene coding for the major protein of hepatitis B virus surface antigen. Nature 280:815–819
Valenzuela P, Medina A, Rutter W, Ammerer G, Hall B (1982) Synthesis and assembly of hepatitis B virus surface antigen particles in yeast. Nature 298:347–350
Vallbracht A, Hoffmann L, Wurster KG, Flehmig B (1984) Persistend infection of human fibroblasts by hepatitis A virus. J Gen Virol 65:609–615
Vallbracht A, Maier K, Stierhof YD, Wiedmann KH, Flehmig B, Fleischer B (1989) Liver-derived cytotoxic T cells in hepatitis A virus infection. J Infect Dis 160:209–217
Vento S, Garofano T, Renzini C, Cainelli F, Casali F, Ghironzi G, Ferraro T, Concia E (1998) Fulminant hepatitis associated with hepatitis A virus superinfection in patients with chronic hepatitis C. N Engl J Med 338 (5):286–290
Vitiello A, Ishioka G, Grey HM, Rose R, Farness P, Lafond R, Yuan LL, Chisari FV, Furze J, Bartholomeuz R, Chesnut RW (1995) Development of a lipopeptide-based therapeutic vaccine to treat chronic Hbv infection 1. Induction of a primary cytotoxic T-lymphocyte response in humans. J Clin Invest 95:341–349
Wallace LA, Echevarria JE, Echevarria JM, Carman WF (1994) Molecular characterization of envelope antigenic variants of hepatitis-B virus from Spain. J Infect Dis 170:1300–1303
Wang B, Ugen K, Srikantan V, Agadjanyan M, Dang K, Refaeli Y, Sato A, Boyer J, Williams W, Weiner D (1993) Gene inoculation generates immune responses against human immunodeficiency virus type 1. Proc Natl Acad Sci USA 90:4156–4160
Waters JA, Kennedy M, Voet P, Hauser P, Petre J, Carmann W, Thomas HC (1992) Loss of the common "A" determinant of hepatitis B antigen by a vaccine-induced escape mutant. J Clin Invest 90:2543–2547
Werzberger A, Mensch B, Kuter B, Brown L, Lewis J, Sitrin R, Miller W, Shouval D, Wiens B, Calandra G et al (1992) A controlled trial of a formalin-inactivated hepatitis A vaccine in healthy children. N Engl J Med 327 453–457
Werzberger A, Kuter B, Nalin D (1998) N Engl J Med (publication pending)
West D (1993) Scope and design of hepatitis B vaccine clinical trials. In: Ellis R (ed) Hepatitis B vaccines in clinical practice. Dekker, New York, pp 159–177
Whittle H, Bradley A, McLauchlan K, Ajdukiewicz A, Howard C, Zuckerman A, McGregor I (1983) Hepatitis B virus infection in two Gambian villages. Lancet i:1203–1206

Whittle HC, Maine N, Pilkington J, Mendy M, Fortuin M, Bunn J, Allison L, Howard C, Hall A (1995) Long-term efficacy of continuing hepatitis-B vaccination in infancy in 2 Gambian villages. Lancet 345:1089–1092
Wilde MD, Rutgers T, Cabezon T, Hauser P, Opstaal OV, Harford N, Wijnendaele FV, Desmons P, Comberbach MPR, Safary A, Wiedermann G, Ambrusch FP, Voet Delem A, Petre J (1991) PreS containing HBsAg particles from Saccharomyces cerevisae: production antigenicity and immunogenicity. In: Hollinger F, Lemon S, Margolis H (eds) Viral hepatitis and liver disease. Williams and Wilkins, Baltimore, pp 732–736
Will H, Cattaneo R, Hock H, Darai G, Schaller H, Schellekens H (1982) Cloned HBV DNA causes hepatitis in chimpanzees. Nature 299:740–742
Winokur PL, McLinden JH, Stapleton JT (1991) The hepatitis A virus polyprotein expressed by a recombinant vaccinia virus undergoes proteolytic processing and assembly into viruslike particles. J Virol 65:5029–5036
Wolff J, Malone R, Williams P, Chong W, Ascadi G, Jani A, Felgner P (1990) Direct gene transfer into mouse muscle in vivo. Science 247:1465–1468
World Health Organization (1994) WHO meeting on nucleic acid vaccines. Vaccine 12:1491–1567
Yamamoto K, Horikita M, Tsuda F, Itoh K, Akahane Y, Yotsumoto S, Okamoto H, Miyakawa Y, Mayumi M (1994) Naturally-occurring escape mutants of hepatitis-B virus with various mutations in the S-gene in carriers seropositive for antibody to hepatitis-B surface-antigen. J Virol 68:2671–2676
Yerushalmi B, Raz R, Blondheim O, Shumov E, Koren R, Dagan R (1997) Safety and immunogenicity of a novel mammalian cell-derived recombinant hepatitis B vaccine containing Pre-S1 and Pre-S2 antigens in neonates. Pediatr Infect Dis J 16:587–592
Yap I, Guan R, Chan SH (1995) Study on the comparative immunogenicity of a recombinant-Dna hepatitis-B vaccine containing pre-S components of the Hbv coat protein with non pre-S containing vaccines. J Gastroenterol Hepatol 10:51–55
Zheng J, Schödel F, Peterson DL (1992) The structure of hepadnaviral core antigens – identification of free thiols and determination of the disulfide bonding pattern. J Biol Chem 267:9422–9429
Zimmerman RK, Ruben FL, Ahwesh ER (1997) Hepatitis B virus infection hepatitis B vaccine and hepatitis B immune globulin. J Fam Pract 45:295–315
Zuckermann JN, McDermott A, Craig FM, Williams A, Madrigal J, Zuckerman A (1996) Immunogenicity of a novel pre-S1/pre-S2? S hepatitis B vaccine in non-responders and high resolution analysis of HLA haplotypes. In: IXth Triennal international symposium on viral hepatitis and liver disease, Rome Italy

CHAPTER 7
Herpes

A. VAHLNE, T. BERGSTRÖM, and B. SVENNERHOLM

A. *Herpesviridae*

Of the nearly 100 herpes viruses that have been described, host species include primates, other mammals such as swine and horse as well as birds and fish (ROIZMAN and SEARS 1996), indicating a long coevolution of this virus family and animal life. A typical herpesvirus consists of a core containing a linear, double-stranded DNA of 150–230 kb, an icosahedral capsid of 162 capsomeres (100–110 nm in diameter), a surrounding amorphous material (the tegument), and an envelope derived from the cellular membrane and containing viral glycoprotein spikes (ROIZMAN and SEARS 1996).

Eight herpesviruses have so far been isolated from humans: herpes simplex virus type 1 (HSV-1), herpes simplex virus type 2 (HSV-2), varicella zostervirus (VZV), cytomegalovirus (CMV), Epstein-Barr virus (EBV), and human herpesvirus 6 (HHV-6) human herpesvirus 7 (HHV-7), and human herpesvirus 8 (HHV-8). This chapter reviews HSV-1, HSV-2, CMV, and EBV. VZV is reviewed in Chap. 5. No vaccines have as yet been developed for HHV-6, HHV-7, or HHV-8 (Kaposi's sarcoma-associated herpesvirus), and these are therefore not included in this review.

HSV types 1 and 2 together with VZV are classified in the *Herpesviridae* subfamily of *Alphavirinae* (MURPHY 1996) based on similar biological properties (ROIZMAN and SEARS 1996) such as variable host range, relatively short replicative cycle, rapid and lytic spread in cell culture, and capacity to establish latent infections in sensory ganglia.

HSV-1 and HSV-2 initially infects and replicates in the epithelial cells of the mucosal membranes of the body orifices. Replicated virus is taken up by the sensory nerve endings and is transported retrogradely intra-axonally (KRISTENSSON et al. 1971, 1978, 1982a; COOK and STEVENS 1973; THOMANDER et al. 1988; LYCKE et al. 1984) to the corresponding ganglia, where a latent reactivable infection is established (STEVENS 1975; STEVENS et al. 1972; BARINGER and SWOVELAND 1973; WHITLEY 1996). HSV-1 infection is associated primarily with infection of the upper part of the body, in particular the mouth and lips. Hence the latent HSV-1 infection is carried primarily by the trigeminal ganglia. HSV-2 infection is a sexually transmitted infection, thus affecting the genital tracts, and the latent HSV-2 infection is in the neurons of the sacral sensory ganglia.

CMV belongs to the *Betavirinae* subfamily of *Herpesviridae* owing to its biological characteristics such as restricted host range, long replication cycle, slow progression in culture, and the induction of latency in secretory glands, kidneys, and lymphoreticular cells. The other virus of this subfamily is murine cytomegalovirus (ROIZMAN and SEARS 1996).

Primary CMV infection is usually subclinical in otherwise healthy individuals. It is probably transmitted through close contact and takes place during childhood and adolescence. After primary infection the virus establishes a latent infection. Which cells harbor the latent CMV infection is not clear. Recently, however, SÖDERBERG NAUCLER et al. (1997) managed to retrieve latent CMV from peripheral blood monocytes obtained from healthy CMV-seropositive individuals.

Based on properties such as its ability to establish latent infection in lymphocytes and its association with proliferation EBV has been classified in the *Gammavirinea* subfamily of *Herpesviridae.* This subfamily includes the *lymphocryptovirus* and *rhadinovirus* genera. EBV is the prototype *lymphocryptovirus* (KIEFF 1996). There are two subtypes of EBV, referred to as type 1 and type 2. These two subtypes differ in only few genes but have some biological differences (KIEFF 1996). EBV infects only B-lymphocytes and epithelial cells of the oropharynx, uterine cervix, and male genital tract. Horizontal oral spread of EBV is from infectious virus shed in the oropharynx (MORGAN 1992). After the initial infection of the epithelial cells the virus infection is spread to the B-lymphocytes where a latent infection is established (MORGAN 1992; RICKINSON and KIEFF 1996). As with the other herpes viruses, primary EBV infection occurs in the early years and is usually subclinical. Sometimes, particularly in adolescent individuals experiencing a primary EBV infection, infectious mononucleosis is seen.

B. Historical Background

I. HSV

The first known clinical record of lesions possibly caused by HSV was the description by the Roman scholar Herodotus around 100 AD of "the herpetic eruptions which appeared about the mouth at the crisis of simple fevers" (for review, see NAHMIAS and DOWDLE 1968). The Greek word *herpes* means to creep or crawl, and was used already by Hippocrates to describe spreading lesions of the skin in general (BESWICK 1962).

In the early twentieth century herpes febrilis was associated with the nervous system by the observation that oral herpetic lesions occurring during acute pneumonia is related to trigeminal ganglionitis (HOWARD 1905). Furthermore, facial herpetic lesions surrounding the area of anesthesia appeared after surgical section of the proximal trigeminal root (CUSHING 1905). Using the rabbit as a model, GOODPASTURE and TEAGUE (1923a,b) showed in the early 1920s that: (a) herpes febrilis can invade the CNS and cause encephalitis after peripheral infection, (b) the virus travels along the axis cylinders of the corresponding nerves, and (c) ganglion cells and neuroglial cells are particularly easily penetrated by the virus.

II. CMV

In the early 1920s it was suggested that cytoplasmic inclusion disease, at that time thought to be a protozoan disease, is caused by a virus (LIPSCHUTZ 1921) and more particularly a herpes virus (VON GLAHN and PAPPENHEIMER 1925; BRITT and AHLFORD 1996; SÖDERBERG 1995). This disease was initially described in 1904 (SÖDERBERG 1995; RIBBERT 1904). However, it was not possible to isolate the virus until the early 1950s when human cells could be cultured (SÖDERBERG 1995). The virus nature of the pathogen and isolation of the virus was established independently by SMITH (1956), ROWE and coworkers (1956) and WELLER and coworkers (1957; BRITT and AHLFORD 1996). Initially the virus was called salivary gland virus, but the name was changed in the mid-1950ies to cytomegalovirus (SÖDERBERG 1995; WELLER et al. 1960).

III. EBV

EBV was first described in 1964 by EPSTEIN and coworkers as herpeslike particles in tumor tissue and in cell lines from Burkitt's lymphomas (BL) (KIEFF 1996). EPSTEIN had been encouraged to search for viruses in BL tissue by a missionary physician working in East Africa, Denis Burkitt. Burkitt had described a new childhood sarcoma involving the jaws of children, found the disease to be endemic in equatorial Africa, and suggested that the tumor disease is etiologically related to a possibly arthropod-borne infectious agent (KIEFF 1996). The virus described by Epstein an coworkers was found to be biologically distinct from the other known human herpes viruses at that time (HSV, VZV, and CMV), and therefore it was considered to be new human herpes virus. Since the discovery of the virus was ascribed to Tony Epstein and his Ph.D. student Yvonne Barr it was named Epstein-Barr virus (KIEFF 1996). EBV as the causal pathogen in infectious mononucleosis (IM) was discovered in 1968 by the HENLE couple at Baylor University in Houston (HENLE et al. 1968) when a technician who had been seronegative for EBV, and whose serum was used as a negative control in EBV antibody assays, after IM became EBV seropositive.

C. The Glycoproteins

Virus-encoded envelope glycoproteins are involved in several facets of virus-host cell interactions and are regarded as important determinants of viral pathogenicity. They are also the targets for virus neutralizing antibodies.

The glycoproteins of HSV are those that have been best characterized as yet. The 11 HSV-1 glycoproteins identified hitherto are designated gB1, gC1, gD1, gE1, gG1, gH1, gI1, gJ1, gK1, gL1, and gM1 (ROIZMAN and SEARS 1996). The HSV-1 glycoproteins seem to have their HSV-2 counterpart (ROIZMAN and SEARS 1996). None of them is identical between the two types, but the degree of homology varies.

The use of gG1 and gG2 in type-selective seroassays have allowed the serological distinction between HSV-1 and HSV-2 infection (SVENNERHOLM et al. 1984; OLOFSSON et al. 1986; LEE et al. 1986).

Five of the HSV glycoproteins – gB, gD, gH, gL, and gM – are regarded as essential for replication in cells based on deletion mutagenesis, while the others are dispensable (ROIZMAN and SEARS 1996). A plausible reason for HSV to carry so many dispensable genes is the complex life cycle of the virus, with a wide host cell range.

Two glycoproteins act as receptors for host molecules: gC binds the complement factor C3b and functions as inhibitor of the complement cascade; while gE (both gE1 and gE2) functions as IgG Fc receptor, probably in complex with gI (SPEAR 1985; JOHNSON et al. 1988). The role of both these functions is probably to counteract immune responses of the host (see below). Since HSV neutralizing antibodies readily can be induced with the gB and gD, efforts in subunit vaccine development have focused on these two glycoproteins (WHITLEY 1997).

CMV particles contain at least eight different glycoproteins arranged in three or four complexes (MOCARSKI 1996). However, the CMV genome contains as many as 55 open reading frames encoding potential glycoproteins (CHEE et al. 1990). Four of the CMV glycoproteins have functional similarities with the HSV glycoproteins gB, gH, gL, and gM (MOCARSKI 1996). The major glycoprotein of CMV, as well as the major target for CMV neutralizing antibodies, is gB, also designated gpUL55 (MOCARSKI 1996). This 150-kDa protein is cleaved into two parts that remain complexed by disulfide bridging (PEREIRA 1994). The CMV gB promotes virion penetration into cells, transmission of infection from cell to cell, and fusion of infected cells (NAVARRO et al. 1993). The protein is also a target for the cell mediated ($CD4^+$ and $CD8^+$) immune responses (BORYSIEWICZ et al. 1988; LIU et al. 1993). Another disulfide-linked complex is gH or gpUL75 with a molecular weight of 245 kDa. Antibodies to gH may neutralize CMV infection by preventing fusion of the infecting virion envelope with the host cell (KEAV and BALDWIN 1991) and cell-to-cell transmission (RASMUSSEN et al. 1991). The gH is hence a candidate protein for subunit vaccine development.

A CMV glycoprotein complex (gC-II), which is encoded for in the US6 family of open reading frames, binds heparin (KARI and GEHRZ 1992) in analogy with what has been found for gC of HSV-1 (see below). Another biological property encoded for by the same open reading frame family is a function that downregulates MHC class I expression in infected cells (MOCARSKI 1996).

The glycoproteins of EBV are those that are the least characterized. The virion contains at least four major glycoproteins, i.e., gp350/220, gp250/200, gp85, and gp78/55 (MACKETT et al. 1990). The major EBV outer envelope glycoprotein 350/220 is transcribed from the same reading frame, where the transcript for gp350 is not spliced, and the transcript for gp220 is spliced in frame (KIEFF 1996). The pg350/220 is the major target for virus neutralizing antibodies (Moss et al. 1996). Neutralizing determinants have also been found on gp250/200 and gp85 (Moss et al. 1996).

D. Early Events of Human Herpes Virus Replication

Since one important function by which neutralizing antibodies work is by inhibiting the early advents of the virus infectious cycle, i.e., the adsorption and penetration of the infecting virus into the host cell, these steps are discussed here.

Recent work has clarified some of the steps involved in the early interaction between HSV and the cell surface. The initial phase of infection is the attachment of virus to heparan sulfate (HS) present on most cell surfaces (WUDUNN and SPEAR 1989; LYCKE et al. 1991). HS is required for infection by both HSV-1 and HSV-2 (WUDUNN and SPEAR 1989; LYCKE et al. 1991). However, type differences in attachment and type-specific receptors have been found in nonneuronal cells, in neuronal synaptosomes, and in glial cells (VAHLNE et al. 1979, 1980; LANGELAND et al. 1990). That different pathways of attachment exist is supported by the finding that HSV-1 but not HSV-2 can be blocked by the polycations neomycin and polylysine (LANGELAND et al. 1987, 1988; CAMPADELLI et al. 1990).

The HSV glycoprotein gC (LANGELAND et al. 1990; KUHN et al. 1990; WU et al. 1990; HEROLD et al. 1991; SVENNERHOLM et al. 1991) has been suggested to play a role in attachment. The HS-binding domain of HSV-1 gC (gC1) has been mapped to an arginine-rich stretch in the N-terminal part of the protein which overlaps with a conserved epitope for an HSV-1 type-specific monoclonal antibody (TRYBALA et al. 1994). The possible role of HSV gB and gD in attachment (KUHN et al. 1990) as well as the role of other HSV glycoproteins needs more clarification. Recent findings suggest that the step during the attachment process following HS-binding is mediated by the interaction of gD to an orphan member of the tumor necrosis factor/nerve growth factor receptor family designated as herpesvirus entry mediator, which in turn utilizes as its cellular ligands LIGHT, a new member of the tumor-necrosis factor superfamily, and lymphotoxin-α (MONTGOMERY et al. 1996; MAURI et al. 1998).

The process of cell entry is believed to be mediated by fusion of the viral envelope and plasma membranes rather than by endocytosis (MORGAN et al. 1968; FULLER and SPEAR 1987). The fusion step seems to be dependent on the glycosylation of viral proteins (SVENNERHOLM et al. 1982). The essential role of gB in this step has been shown by the use of temperature-sensitive mutants which fail to express mature gB or gB-null viruses, which adsorb to but do not enter into cells (MANSERVIGI et al. 1977; CAI et al. 1988). The responsible domain of gB has been mapped by monoclonal antibodies to amino acid residues 241–441, centrally located in the external part (HIGHLANDER et al. 1988). This domain is also essential for cell fusion, whereby HSV might spread from cell to cell.

The essential role of gD for fusion has been shown in a similar way as for gB (ROIZMAN and SEARS 1996). In addition, gD might sequester cell membrane proteins necessary for fusion of viral and cellular membranes (CAMPADELLI et al. 1988). Also gE and gH may be active in the cell fusion process, although data are scarce (GOMPELS and MINSON 1986; CHATTERJEE et al. 1989).

As for HSV the initial binding of CMV with its host cell is believed to be mediated by the interaction of viral envelope glycoproteins (gCII and possibly gB) and HS proteoglycans on the cell surface (COMPTON et al. 1993; KARI and GEHRZ 1992, 1993). Subsequent events leading to fusion of the virus envelope with the cell membrane are not yet fully understood but seem to involve the gH binding to a 92.5-kDa cellular protein (KEAY and BALDWIN 1991; KEAY et al. 1989). Also CD13 (SÖDERBERG 1995) and cellular proteins of 30-kDa (fibroblasts; TAYLOR and COOPER 1990) and 32- to 34-kDa (lymphoid cells) (ADLISH et al. 1990) have been suggested to be involved in the early events of CMV replication.

The receptor for EBV on both B-lymphocytes and epithelial cells is the type 2 complement receptor CR2 or CD21 (KIEFF 1996). The virus attachment protein binding to CD21 is gp350/220 (KIEFF 1996). After EBV adsorption to lymphocytes the virus is internalized into cytoplasmic vesicles. This step involves the coaggregation of CD21 and sIg (KIEFF 1996). The EBV gp85 seems to be important for the subsequent virus envelope fusion with the host cellular membrane and has sequence homologies with HSV-1 gH (KIEFF 1996).

E. Latency

Following primary infection HSV infects sensory nerve endings and is transported retrogradely to trigeminal and dorsal root ganglia where the virus establishes a latent infection (KRISTENSSON et al. 1971, 1978, 1986; COOK and STEVENS 1973; THOMANDER et al. 1988; LYCKE et al. 1984; GOODPASTURE and TEAGUE 1923a,b). The sensory ganglia offer a sanctuary to the virus, where it is protected from attacks of the immune defense system of the host and from antiviral therapy (WHITLEY 1996; SVENNERHOLM et al. 1981).

Whether latency is restricted to ganglionic sensory neurons is unknown. HSV isolation has never been accomplished from human brain tissue without signs of active infection, but hybridization techniques have shown positive findings of HSV DNA (FRASER et al. 1981). Furthermore, after PCR amplification HSV-1 genomes have been found to be present in cerebral regions of the human brain such as the medulla, the olfactory bulbs and the hippocampus (BARINGER and PISANI 1994). Although shown experimentally (CLEMENTS and SUBAK SHARPE 1988; CLEMENTS and JAMIESON 1989), HSV latency in peripheral tissue has been sought for but not found in humans (RUSTIGAN et al. 1966).

In most individuals HSV remain clinically silent throughout the life of the host but can still be isolated from sensory ganglia after death (BARINGER and SWOVELAND 1973; BASTIAN et al. 1972). Within the ganglia the cell type harboring latent HSV is the neuron (COOK et al. 1974; MCLENNAN and DARBY 1980). The role of LAT-1, the viral transcript expressed during latency (WECHSLER et al. 1988) is debated, since negative mutants also are capable of establishing latency (JAVIER et al. 1988).

Little is known about the CMV latency. Although white blood cells have been considered to be those that harbor the latent CMV infection, the true identity of the cell carrying the latent virus infection has eluded scientists. A breakthrough, however, seems to be the recent finding that infectious CMV can be rescued from peripheral blood mononuclear cells obtained from healthy CMV antibody positive individuals by allogenic stimulation (SÖDERBERG et al. 1997)

There are three different latency programs of EBV described. The latency I program is found in BL cells (MOSS et al. 1996; ROWE et al. 1987) and lately also in a subpopulation of EBV infected B-lymphocytes in healthy individuals (CHEN et al. 1995). This type of EBV latency is characterized by the expression of only the EBV nuclear antigen 1 (EBNA-1) and two nontranslated transcripts EBER1 and EBER2 (RYMO 1979; HOWE and SHU 1989). Latency program I is found in cells of the EBV associated tumors nasopharyngeal carcinoma and Hodgkin's disease (RICKINSON and KIEFF 1996; MOSS et al. 1996). In latency program II in addition to EBNA1 and the EBER1 and EBER2 also the latency membrane proteins LMP1 and/or LMP2 A and/or LMP2B are expressed (MOSS et al. 1996). In latency program III all six EBV nuclear antigens (EBNA1 through EBNA6) as well as LMP1, LMP2A, and LMP2B are expressed. The two diseases associated with this latency program are IM and posttransplantation lymphoproliferative disease (RICKINSON and KIEFF 1996; MOSS et al. 1996).

F. Reactivation

Numerous factors may induce HSV reactivation in the human host: trauma, UV irradiation, emotional stress, menstruation, infections such as pneumococcal pneumonia, and administration of immunosuppressive agents (WHITLEY 1996). It is unknown whether these different factors have any common denominator in form of a specific alteration of nerve cell function. Studies have shown that deprivation of nerve growth factor by antiserum induces HSV reactivation of latently infected neurons in vitro (WILCOX and JOHNSON 1987).

In the ganglia only a minor fraction of the neurons are believed to contain latent HSV, and often in multiple copies (ROIZMAN and SEARS 1996; NICHOLLS and BLYTH 1989). It has been suggested that HSV replication is restricted by a neuronal inhibitor, which can be overcome by a large enough number of HSV genomes in a single neuron by so called multiplicity activation (VAHLNE et al. 1981b; NILHEDEN et al. 1985; THOMAS et al. 1985). HSV receptors are found only on the nerve endings (synaptosomes) and not on the neuronal cell somas (perikarya) of central nervous system neurons (VAHLNE et al. 1978). If this is the case also for peripheral nervous system neurons, it would constitute a physiological barrier that might explain the demonstrated lack of interneural spread of HSV in the sensory ganglia after reactivation of latent virus infection. Instead, the virus dose in a single neuron would have to be increased by reseeding into new neurons from the peripheral lesions caused during reactivation, hereby creating prerequisites for future reactivations as suggested in the "round trip" hypothesis (VAHLNE et al. 1981; NILHEDEN et al. 1985; THOMAS et al. 1985; KLEIN 1976).

Virus might also spread to a neighboring sensory neuron centrally in the brainstem and spinal cord and than be transported back to the ganglia, "the back door route" (NICHOLLS and BLYTH 1989; TULLO et al. 1982; BLYTH et al. 1984). This hypothesis provides a mechanism whereby a great number of neurons might be latently infected, even in absence of peripheral lesions.

An HSV reactivation leading to clinical symptoms is termed recurrence. HSV type differences exist in inclination to cause recurrent disease. In a study of genital HSV infection, the risk of recurrence was two- to fourfold greater after HSV-2 infection than after HSV-1 infection (REEVES et al. 1981; COREY et al. 1983). In addition, the rate of HSV recurrences may be strain dependent, since isolates from patients with high frequency of recurrences retained this biological property in a rabbit model of keratitis (GERDES and SMITH 1983).

The finding of multiple strains of HSV-1 in ganglia from one individual indicate the possibility of reinfection (LEWIS et al. 1984). Exogenous infection with homologous HSV type in already infected individuals has been described in singular cases both for HSV-2 (BUCHMAN et al. 1979) and HSV-1 (MANZELLA et al. 1984). Although reinfection may be a rare event, the possibility of intratypic recombination in future reactivations is of interest from pathogenetic aspects.

It has been known since the early 1970s that allogenic stimulation induces the reactivation of latent murine CMV in blood cells of mice (OLDING et al. 1976). Recently SÖDERBERG NAUCLER and coworkers (1997) were able to reactivate also human CMV from the peripheral blood mononuclear cells of EBV latently infected healthy individuals by allogenic stimulation. Thus this immune stimulation is most probably the reason for the CMV reactivation often seen after blood transfusion and allogenic organ transplantation (BRITT and AHLFORD 1996).

It is not clear what mechanisms induce reactivation of the latent EBV infection in vivo. In EBV latently infected cultured cells the infection is reactivated by the stimulation with phorbol esters with or without calcium ionophore, by sodium butyrate, 5′ azacytidine, and by cross-linking with surface IgM (RICKINSON and KIEFF 1996; MOSS et al. 1996).

G. Clinical Infections

I. HSV

Seroepidemiological data indicate that a majority of the adult population are infected with HSV-1, and up to 50% with HSV-2 in women of certain age groups (WHITLEY 1996; ADES et al. 1989; LOWHAGEN et al. 1990). Seroprevalence is highly related to socioeconomic conditions (the poorer the country or population group studied, the higher the seroprevalence), and a sharp decrease in HSV-1 prevalence has been described among students in the United States in recent years (WHITLEY 1996). The protective effect of oral HSV-1 infection to later genital HSV-2 infection has also been indicated in experimental studies (STURN and SCHNEWEIS 1978; BERNSTEIN et al. 1989), hence giving a rational for a prophylactic HSV vaccine.

Neonates appear to have the highest frequency of CNS infection of any HSV-infected population. Of a calculated incidence of 14 cases/100000 live births 70% develop disseminated (including brain engagement) or CNS disease, with a mortality in untreated cases of 65% (COREY and SPEAR 1986). Neonatal CNS infections are caused by HSV-2 in more than 70% (WHITLEY 1996), and the infection is probably acquired during the passage through the birth canal.

Clear intertypic differences exist in the clinical patterns of HSV-induced CNS infection. In adults HSV-2 preferably induces a meningitis while HSV-1 almost exclusively is isolated from brains of patients with encephalitis (WHITLEY 1996; CRAIG and NAHMIAS 1973; SKOLDENBERG et al. 1973; WHITLEY et al. 1982; BERGSTROM et al. 1990). However, immunocompromised subjects seem to be an exception from this rule since brain engagement has been described, for example, in cases of lymphadenopathy (DIX et al. 1985).

One explanation to why HSV-1 but not HSV-2 causes encephalitis in the immunocompetent host might be the presence of HSV-1 receptors but not HSV-2 receptors on the CNS glial cells (VAHLNE et al. 1980; WILCOX and JOHNSON 1987).

The ability of HSV to induce demyelination is evident by experimental animal studies (KRISTENSSON et al. 1978, 1979, 1982; MARTIN 1982; MARTIN and STONER 1984; SOFFER and MARTIN 1988). Furthermore, a herpeslike virus (later typed as HSV-2) was derived from the brain of a patient with a 27-year history of multiple sclerosis (MS) (GUDNADOTTIR et al. 1964). In addition, HSV-1 was isolated from the CSF of a patient during the first attack of MS (BERGSTROM et al. 1989). Thus singular cases of HSV-related demyelination might be included within the diagnostic framework of MS criteria (MARTIN et al. 1988). Also the peripheral type of facial palsy known as Bell's palsy has been found to be caused by HSV infection, and probably be owing to a transient demyelination in the brainstem (MCCORMICK 1972; ADOUR et al. 1978, 1980; VAHLNE et al. 1981, 1985; ROSENHALL et al. 1983; EDSTROM et al. 1987; BARINGER 1996).

The most severe complication of HSV infection in humans is the HSV-1 induced focal encephalitis (HSE). Although the frequency is low, HSE is still considered to be the most common cause of sporadic encephalitis (WHITLEY 1996).

II. CMV

In otherwise healthy individuals primary CMV infection passes without overt symptoms. As with the other herpes viruses, it is an ubiquitous infection. Among children aged under 1 year 10% transmit CMV in their urine, and about 50% of women at fertile age are seropositive for CMV (ROSENTHAL et al. 1997). In pregnant women who experience a primary infection the risk of transmission of the virus to the fetus is 40% (STAGNO et al. 1986). Of these, 25% have serious sequelae from the intrauterine infection. In fact, CMV is the major cause of damage to the unborn child due to intrauterine infection, and it has been estimated that in the United States up to 10000 infants per year suffer significant consequences from such an intrauterine infection (STAGNO et al. 1977; DOBBINS et al. 1992). The classical fetal CMV syndrome presented at birth includes microcephaly, enlarged spleen and liver, and a thrombocytopenic purpura (AHLFORS et al. 1978). However, also those children not presenting any symptoms at birth may develop late sequelae such as deafness (MARSHALL 1997). Women infected with CMV before the pregnancy may also transmit the infection to the unborn child. However, both the risk of transmission and the consequences of the intrauterine CMV infection are drastically reduced as compared to a primary infection (STAGNO et al. 1977; FOWLER et al. 1992).

In rare cases of immunocompetent individuals CMV infection has been associated with diseases such as pneumonia, encephalitis, myocarditis, pericarditis, epidermolysis, Guillain-Barré syndrome, and a mononucleosislike syndrome (BRITT and AHLFORD 1996). However, in the immunocompromised host, particularly in AIDS patients and bone marrow transplantation (BMT) recipients CMV is a major cause of morbidity and mortality (GRIFFITHS 1993; GALLANT et al. 1992, 1995). In AIDS patients CMV affects not only the retina, leading to blindness but also to the gastrointestinal tract with colitis (GALLANT et al. 1995; HOOVER et al. 1993). CMV infection in BMT patients is manifested as fever, leukopenia, thrombocytopenia, hepatitis, arthralgias, gastrointestinal disease, and interstitial pneumonia (BRITT and AHLFORD 1996). In organ transplant recipients CMV has been implicated in graft rejection, glomerulopathy after renal transplantation, and vanishing bile duct syndrome after liver transplantation (BRUGGEMAN and VAN DAM MIERAS 1991; ARNOLD et al. 1992). Also bone marrow failure and chronic graft versus host disease in BMT patients may be caused by CMV (Söderberg 1995). The latter has been suggested to be caused by an autoimmunity induced to CD13 as a consequence to its incorporation into the CMV virion (Söderberg 1995).

III. EBV

EBV is one of the most ubiquitous virus infections in humans. Over 95% of the population worldwide is infected with EBV before the age of 5 years (Rickinson and Kieff 1996). In the vast majority this infection is subclinical. However, in an individual first infected in adolescence or as a young adult, EBV infection may cause IM in as many as 50% (Morgan 1992). Although IM usually is a benign self-limiting disease, some complication may cause concern (Andersson 1996). These are: meningoencephalitis with seizures, splenic rupture, upper airway obstruction, bilateral interstitial infiltrates causing severe hypoxia, severe hepatitis and liver failure, and thrombocytopenia and hemorrhages secondary to disseminated intravascular coagulopathy (Andersson 1996).

In the immunocompromised host EBV may cause a severe lymphoproliferative disorder. A rare and inherited form of fatal IM-like illness caused by EBV is called X-linked lymphoproliferative syndrome or Duncan's syndrome (Rickinson and Kieff 1996; Andersson 1996). This X-linked immunodeficiency selectively predisposes affected boys to EBV-associated disease (Rickinson and Kieff 1996), usually overwhelming pneumonia-hepatitis and death (Andersson 1996). Other clinical manifestations of EBV infection in the immunocompromised include oral hairy leukoplakia and immunoblastic lymphoma in AIDS patients, and posttransplantation lymphoproliferative disorder (Rickinson and Kieff 1996).

Malignancies that to varying certainties have been associated with EBV are BL, nasopharyngeal carcinoma, Hodgkin's disease, T-cell lymphoma, and immunoblastic lymphoma (Rickinson and Kieff 1996).

H. Immunology

Only the adaptive immune response are discussed here. Both humoral and cell-mediated immunity most probably play a role in recovery and protection from herpes viruses. The relative importance of these two facets of the immune response is under fairly constant debate. It seems that cell-mediated immunity is more important for the recovery of the primary infection, but administration of neutralizing antibodies has been shown to be protective for all four herpes viruses discussed in this chapter. For vaccine purposes an immune response entailing both neutralizing antibodies and cytotoxic T-lymphocytes would be desired – the former to reduce the infectious dose and thereby possibly ameliorating the clinical symptoms of the infection and a CTL response to help to clear the body from virus replicating cells.

Interestingly, in a recent study KHANNA and coworkers (1997) found an inverse relationship between genetic variations within certain CTL epitopes and the frequency distribution of HLA alleles that presented the CTL epitopes, suggesting that the evolutionary pressure on the CTL epitopes of the virus may be toward their conservation rather than their inactivation.

An important aspect to be considered in immune intervention of herpes virus infections and recurrences and in prophylactic vaccine development is that these viruses have adapted a number of mechanisms for the evasion of the host immune response.

The cells that harbor the latent HSV infection, i.e., the neurons, do not express HLA class I molecules and hence are not targets for $CD8^+$ cytotoxic T-cells. Furthermore, HSV encodes for a protein that shuts off host cellular macromolecular synthesis (ROIZMAN and SEARS 1996) and also for an α-protein (ICP47) that downregulates HLA class I by preventing its maturation in the endoplasmic reticulum (YORK et al. 1994; TOMAZIN et al. 1996; AHN et al. 1996). The mechanism by which ICP47 downregulates the MHC class I molecules is by inhibiting the transporter associated with antigen presentation (TAP) mediated translocation of peptides from the cytosol into the endoplasmic reticulum (FRUH et al. 1995; HILL et al. 1995). Hence the formation of stable MHC class I heavy-chain β_2-microglobulin peptide complexes are not formed and transported to the cell membrane. This inhibition seems to involve the direct binding of ICP47 to the TAP outcompeting the binding of peptide to TAP (WIERTZ et al. 1997). In addition, both gC1 and gC2 may utilize binding of the complement factor C3b as a means of immune evasion (HUNG et al. 1994). Furthermore, purified gC1, but not gC2, inhibits the binding of properdin to C3b, which accelerates the decay of the alternative pathway C3 convertase and thereby affects later steps of the complement cascade (HUNG et al. 1994). These circumstances may help to explain why recurrences of HSV occur despite a functional immune response to the virus.

Also CMV has developed complex strategies to evade the immune response, and both HLA class I and class II molecules on the surface of infected cells are downregulated by CMV encoded gene products (BARNES and GRUNDY 1992; WARREN et al. 1994; SEDMAK et al. 1994; BEERSMA et al. 1993; JONES et al. 1995, 1996; FISH et al. 1996; WIERTZ et al. 1996a,b; GILBERT et al. 1996). The two CMV gene-products US2 and US11 induce a rapid degradation of HLA class I molecules, probably by translocating the heavy-chain to the cytosol directly after its translation (WARREN et al. 1994; BEERSMA et al. 1993; JONES et al. 1995; FISH et al. 1996; WIERTZ et al. 1996b). The CMV gene product US3 induces retention of HLA class I molecules in the endoplasmic reticulum and thereby also a downregulation of the expression of these molecules on the cell membrane (WIERTZ et al. 1997; JONES et al. 1996).

In the subset of B-lymphocytes that latently harbor the EBV and express only the EBNA-1 protein (CHEN et al. 1995) this herpes virus has also developed a mechanism for evading the immune surveillance by inhibiting antigen presentation of the EBNA-1 protein. The mechanisms for this are not known but depend on the internal repeat of Ala-Gly in EBNA-1 (CHEN et al. 1995; WIERTZ et al. 1997). It has been suggested that this amino acid repeat prevents the EBNA-1 protein to bind to and be cleaved by the proteasome complex, and hence there would be no formation of EBNA-1 peptides that could bind to the HLA class I complex (WIERTZ et al. 1997).

I. Therapeutic Vaccines

Efforts aimed at developing a vaccine that could affect the clinical recurrences of reactivated virus infection in latently infected individuals who experience multiple recurrences and/or severe clinical symptoms upon reactivation of the latent infection have focused on HSV-1 and HSV-2. HSV-specific noninfectious vaccines, for example, using inactivated (killed) virus vaccines or HSV subunit vaccines have been tried. Since HSV latently infected individuals have already developed both humoral and cell-mediated immune responses and uphold this anti-HSV immune reactivity owing to the recurrences of the infection, the vaccine should do better than the natural infection with HSV does. Furthermore, the patients and/or the physicians should appreciate the vaccine to be superior to available antiviral treatments.

Decisive studies on therapeutic HSV vaccines are difficult to perform. In placebo-controlled clinical trials studying the putative therapeutic effect of HSV vaccines a varying degree of beneficial effect is obtained with the placebo vaccine alone concerning both number of recurrences and severity of each recrudence (WHITLEY 1997). Indeed, in one study up to 76% improvement was registered with placebo alone (KERN and SCHIFF 1964).

I. Killed Vaccines

Killed vaccines were the first to be tested. Numerous studies have been performed since 1938 (WHITLEY 1997). Most have either yielded disappointing results, or when more beneficial effects have been claimed from the vaccination, the studies have been hampered by shortcomings in study design. They have either been open trials, or when they have been placebo controlled, they have been difficult to interpret properly due to differences in the control and study group populations (KERN and SCHIFF 1964; FRANK 1938; WEITGASSER 1977). Hence a recent study comparing the efficacy of killed virus vaccine (142 patients) with intermittent oral acyclovir (ACV) treatment (50 patients) that claimed the vaccinee reduced their annual active disease days by a factor 5 (from 65 to 12 days) as compared to an only slightly more than twofold reduction (from 72 to 30 days) in the ACV-treated group. However this trial had a diverse study group of patients, entailing both those having facial – presumably HSV-1 – infections, and genital – presumably HSV-2 – infections (MASTROLORENZO et al. 1995).

II. Subunit Vaccines

The possible cancer-inducing potential of HSV prompted the development of subunit vaccines not containing any HSV genome. Subunit vaccines may be produced from HSV-infected cells with production steps including DNase treatment and glycoprotein purification. The most studied subunit HSV vaccine is the so-called Skinner vaccine, which is composed of mixed HSV-1 glycoproteins obtained from infected cells after detergent extractions and formalin inactivation. In a recent report (SKINNER et al. 1997) of a randomized, placebo-controlled, multicenter trial using this vaccine on herpes genital patients, all anti-HSV immune parameters improved in the vaccine group. However, although the frequency of recurrences was statistically reduced in the vaccinated female patients, the overall reduction when both sexes were included was not significantly different from that of the placebo group (SKINNER et al. 1997). A decrease in severity of the recurrences in the vaccinated population just reached statistical significance ($p = 0.04$).

HSV subunit candidate vaccines based on virus glycoproteins produced by recombinant DNA technology have focused on gB and gD. Both these glycoproteins are required for infectivity, and gD in particular has been shown to be a strong inducer of neutralizing antibodies. Such vaccines have given very promising results in a guinea pig model of HSV reactivation. A critical factor for obtaining a good immune response when using a subunit vaccine is the adjuvant. The only adjuvant approved for human use is aluminum hydroxide (alum). However, better adjuvants have been tested in humans. Using a recombinant HSV-2 gB and gD vaccine with the more potent squalene oil based adjuvant (MF59) Straus and coworkers (STRAUS et al. 1997) in a placebo-controlled trial of genital herpes obtained a reduction in both the duration and severity of symptoms, but the rate of recurrences was not significantly improved.

Biological properties of glycoproteins are determined by the sequence of the polypeptide backbone, creating specific folding patterns, and also defining glycosylation sites of the virus. However, the process of glycosylation is regulated by host cell enzymes. Important for the production of subunit vaccines employing virus glycoproteins is that the structure of the carbohydrate trees might vary depending on the cell type replicating the virus (LUNDSTROM et al. 1987). The degree of glycosylation has important implications for properties such as antigenicity (SJOBLOM et al. 1987). Thus, since HSV in its normal life cycle exploits both neuronal and epithelial cells, the sensitivity of the progeny virus to a subunit gC vaccine induced immune response may vary depending upon the cell type in which the virus was replicated in.

In summary, a number of clinical HSV vaccination trials using either killed virus vaccine or subunit vaccine have been shown to ameliorate the recurrences of reactivated HSV infections. However, the results are probably not impressive enough for any of the vaccines tested to constitute a realistic therapeutic alternative to the antiviral treatments now available.

J. Prophylactic Vaccines

Ideally a prophylactic vaccine should protect the vaccinated individual from becoming infected. However, as yet no viral vaccine has met this objective. What is expected of a vaccine is therefore that it should at least ameliorate the clinical outcome of an infection or even render the infection subclinical. Another a beneficial effect of a prophylactic vaccine would be a reduction in virus spread at subclinical reactivation from latently infected to noninfected individuals, in particular mother to child horizontal neonatal transmission of HSV and vertical intrauterine transmission of CMV. For the prophylactic vaccines most attention has been given to the development of a vaccine that would replicate in the vaccinee. Reasons for this are, among others, that such vaccines would be expected to give a longer lasting immunity than subunit or killed vaccines. Also, antigenic peptides from proteins expressed by foreign genes in the cells of the immunized individual are more likely to bind to HLA class I molecules and hence induce a cell mediated immunity than are peptides from external proteins taken up by the cells.

I. HSV

A vaccine effective in inhibiting the clinical symptoms of the primary infection with HSV most probably would not inhibit the establishment of a latent infection in the sensory ganglia. However, if the virus replication is considerably reduced by a vaccine induced preinfection immunity, the viral load of the ganglia may be reduced, and thus possibly the likelihood of reactivation, the asymptomatic shedding of virus and the recurrence of clinical symptoms. If this were achieved, the spread of HSV in the population might be reduced. The effect such a scenario in the long-term would have on the incidence of possibly HSV-related diseases such as multiple sclerosis, Bell's palsy and arteriosclerosis would be interesting to see. Apart from the reduction in clinical recurrences, the more short-term and realistic benefits of a prophylactic HSV vaccine, however, would include reduction in the incidence of the severe diseases of neonatal HSV infection and HSV-1 encephalitis, and fewer cases of HSV-2 induced recurrent meningitis and HSV keratoconjunctivitis. The four following types of HSV vaccines are being developed.

1. HSV replication competent live virus that has been genetically modified to give a less virulent phenotype (Mocarski et al. 1980; Post and Roizman 1981; Meignier et al. 1990). An attenuated virus was obtained by making deletions in the thymidine kinase (TK) gene and in the junction region between UL and US of the HSV-1 strain F. To ensure an HSV-1/HSV-2 cross-reactive immune response a gene fragment encoding gD2, gG2, and gI2 was inserted at the site of the deleted internal repeat. Since this thus modified HSV (designated R7017) was lacking the TK gene, it was resistant to acyclovir. To obtain a variant of R7017 that was acyclovir sensitive (R7020) the TK gene was inserted next to the inserted HSV-2 gene fragment. Both the R7017 and the

R7020 given at 10^7 plaque-forming units were virtually avirulent in owl monkeys in which 100 pfu of wild-type HSV-1 causes a fatal infection (MEIGNIER et al. 1988). However, human studies have yielded disappointing results as only weak immune responses have been obtained (WHITLEY 1997). A new HSV construct in which the two copies (UL 55 and 56) of the g1 34.5 neurovirulence gene (CHOU et al. 1990) have been deleted is under evaluation by the same group (WHITLEY 1997).

2. HSV in which genes essential for the production of infectious progeny virus have been deleted will be replication incompetent but still able to infect cells if the viruses are produced in cells carrying the deleted gene. Such replication-incompetent viruses, termed disabled infectious single cycle (DISC) virus, encodes for viral proteins in the infected host and thus induce an immune response. However, since new infectious viruses are not produced, vaccine based on such modified viruses should be safe. Genes that have been deleted for this purpose are the UL αa-gene coding for ICP27 (MORRISON and KNIPE 1994) and the UL g-gene coding for gH (FORRESTER et al. 1992). The former gene is necessary for late gene expression, and the latter appears to play a role in *inter alia* the egress and cell spread of the virus (ROIZMAN and SEARS 1996). Such vaccines have produced promising results in animal models, inducing a $CD8^+$ cytotoxic T-lymphocyte response, and conferred a protection similar to that of wild-type HSV, but their efficacy in humans still waits to be seen (BREHM et al. 1997).

3. Recombinant viruses where HSV genes under the control of a strong promoter have been inserted into a carrier virus, i.e., infectious subunit vaccines, are also being tested. Genes that have been inserted into the genome of a carrier virus include inter *alia* gB in adenovirus (McDERMOTT et al. 1989) and gD and gC in vaccinia virus (ALLEN et al. 1990). In the latter recombinant vaccinia virus also the gene for interleukin-2 was incorporated to augment the immune response (ALLEN et al. 1990). Such prototype vaccines have been shown to induce good immune responses in animal models. However, safety concerns have been raised for these type of vaccines, in particular for vaccinia virus for use as a vaccine carrier virus.

4. Vaccines employing the observation that an immune response to gene products may be obtained after injection of DNA as such provided that the gene is under the control of an expression promoter. Genes used in such so-called naked DNA prototype vaccines and that have been shown to induce an immune response are the α-gene coding for ICP27 (MANICKAN et al. 1995), gB (KUKLIN et al. 1997), and the genes coding for gD2 (KRIESEL et al. 1996) and gD1 (BOURNE et al. 1996).

II. CMV

The major objectives for a prophylactic CMV vaccine would be to prevent the mother to fetus transmission of the virus that occurs in about 40% of primary infected mothers and to ameliorate the disease after CMV infection in CMV naive bone marrow and organ transplant recipients. A complicating factor for the development of a human CMV vaccine is that there is no animal model for testing a prototype vaccine since the virus is species specific. Another problem, as discussed by MARSHALL (1997), is that it is difficult to set realistically achievable endpoints for evaluating the efficacy of the vaccine in clinical trials on relatively rare diseases such as CMV-induced symptomatic congenital disease.

1. Live Vaccine

A CMV vaccine that has been tested vigorously in clinical trials is a live attenuated vaccine based on the Towne strain of CMV attenuated by passaged 125 times in the WI-38 human diploid fibroblast cell line (PLOTKIN and HUYGELEN 1976; PLOTKIN et al. 1976, 1984, 1990, 1991; GLAZER et al. 1979; FRIEDMAN et al. 1982; MARSHALL and PLOTKIN 1990). The Towne vaccine virus seems to be well attenuated and apparently does not even establish a latent infection (PLOTKIN and HUANG 1985). The vaccine virus induces both humoral and cell-mediated immune response (GEHRZ et al. 1980; STARR et al. 1981; QUINNAN et al. 1984). In placebo-controlled trials of kidney transplant recipients (PLOTKIN et al. 1991, 1994) the vaccine did not prevent CMV infection and had only a slight effect on CMV disease at large. It did, however, significantly reduce severe CMV disease (0%–10% in the vaccine groups versus 17%–36% in the placebo groups). Although the Towne vaccine does not seem to protect from infection (ADLER 1995), it might be justified for prophylactic use before transplantation of CMV naive individuals.

Recombinant vaccinia virus (BRITT et al. 1988; GONCZOL et al. 1991), adenovirus (MARSHALL et al. 1990) and recombinant canarypox virus (GONCZOL et al. 1995) expressing the major envelope glycoprotein (gB) of CMV all induce neutralizing antibodies in animals. None of them has as yet been tested on humans. A candidate protein to be incorporated in live recombinant vaccines should be the CMV phosphoprotein pp65. This matrix protein is a major target for cytotoxic $CD8^+$ T-lymphocytes (MCLAUGHLIN TAYLOR et al. 1994).

So-called naked DNA or polynucleotide vaccines are also being developed for human CMV. Immunization in mice with naked DNA plasmid constructs expressing CMV pp65 give rise to neutralizing antibodies (PANDE et al. 1995), and injection with MCMV IE1 expressing plasmid provide protection in the mice (GONZALEZ et al. 1996).

2. Subunit Vaccine

The major candidate CMV protein for a subunit vaccine is gB (MARSHALL 1997). A prototype vaccine consisting of a modified CMV glycoprotein gB produced by Chiron Vaccines (Emersville, CA), employing recombinant DNA technology in Chinese hamster ovary cells combined with the MF59 adjuvant is reported to give a good immune response with CMV crossneutralizing antibodies without major side effects (MARSHALL 1997; WANG et al. 1996) and is currently being tested in more extensive human trials. Another CMV protein to be a candidate for inclusion in a subunit CMV vaccine is gH (MARSHALL 1997). A CMV vaccine based on HLA restricted nonapeptide is also under development (DIAMOND et al. 1997).

III. EBV

The EBV vaccines developed have aimed at preventing latency program III associated clinical disease, for example, IM and posttransplant lymphoproliferative disorder. There is in effect only one good animal model for EBV infection and immune prevention of EBV disease. This is the cottontop tamarin (*Saquinus oedipus oedipus*) in which EBV induces a tumor disease resembling that induced by EBV in immunosuppressed individuals as seen in organ graft recipients (posttransplant lymphoproliferative disorder) (MORGAN 1992).

1. Subunit Vaccines

The first vaccines tested in the tamarin model utilized gp340 purified from cell membrane of EBV-infected cultured cells. Such protein denatured with sodium dodecyl sulfate and incorporated into liposomes has been shown to induce a protective immunity for a tumorogenic dose of EBV in the animals (EPSTEIN et al. 1985). However, a vaccine based on EBV-infected cells would probably not be accepted because of its potential contamination with putative oncogenic EBV DNA.

Another way of producing gp340 for vaccine purposes is by recombinant DNA technology. Recombinant EBV gp340 produced in eukaryotic induce EBV-neutralizing antibodies (EMINI et al. 1988; FINERTY et al. 1992; MADEJ et al. 1992) and protect marmosets from EBV challenge (FINERTY et al. 1992; EMINI et al. 1989). By removing the membrane anchor sequence a secreted form of gp340 can be produced in large quantities (FINERTY et al. 1992; MADEJ et al. 1992).

2. Live Vaccines

The few vaccines tested also have focused on the major membrane antigen gp340/220. A number of such EBV gp340 live vector vaccines have been constructed using vaccinia virus (MORGAN 1992; MORGAN et al. 1988), varicella (LOWE et al. 1987), and replication incompetent adenovirus (see DISC for HSV above). The latter type of vaccine has recently been tested in the tamarin model and been shown to induce protective immunity against EBV lymphoma (RAGOT et al. 1993).

Recently a vaccine based on a vaccinia vector expressing gp350/220 was tested in humans (GU et al. 1995). Three groups were tested: vaccinia virus and EBV-exposed adults, EBV-seropositive non-vaccinia-virus exposed juveniles, and vaccinia virus naive infants. In the first group no significant changes in serum titers to EBV were observed. The juveniles of the second group developed neutralizing antibodies. In the third group all of nine vaccinated infants developed neutralizing antibodies to EBV. Of these nine vaccinated infants three became naturally infected with EBV during a 16-month observation period, whereas all of ten unvaccinated control infants had become infected during that same period.

K. Conclusions

Vaccines to herpes viruses preferably should induce humoral and cell-mediated immune responses. This might not be feasible with only a single type of vaccine; hence, one should be open to vaccine regiments including, for example, priming with a live vaccine and boosting with a subunit nonreplicating vaccine. In the development of such vaccines consideration should be given the strategies these virus have developed for immune evasion.

References

Ades AE, Peckham CS, Dale GE, Best JM, Jeansson S (1989) Prevalence of antibodies to herpes simplex virus types 1 and 2 in pregnant women, and estimated rates of infection. J Epidemiol Community Health 43:53–60

Adler SP (1995) Immunoprophylaxis against cytomegalovirus disease. Scand J Infect Dis Suppl 99:105–109

Adlish JD, Lahijani RS, St Jeor SC (1990) Identification of a putative cell receptor for human cytomegalovirus. Virology 176:337–345

Adour KK, Byl FM, Hilsinger RL Jr, Kahn ZM, Sheldon MI (1978) The true nature of Bell's palsy: analysis of 1000 consecutive patients. Laryngoscope 88:787–801

Adour KK, Hilsinger Jr RL, Byl FM (1980) Herpes simplex polyganglionitis. Otolaryngol Head Neck Surg 88:270–274

Ahlfors K, Ivarsson SA, Johnsson T, Svensson I (1978) Congenital and acquired cytomegalovirus infections. Virological and clinical studies on a Swedish infant population. Acta Paediatr Scand 67:321–328

Ahn K, Meyer TH, Uebel S, Sempe P, Djaballah H, Yang Y, Peterson PA, Fruh K, Tampe R (1996) Molecular mechanism and species specificity of TAP inhibition by herpes simplex virus ICP47. EMBO J 15:3247–3255

Allen EM, Weir JP, Martin S, Mercadal C, Rouse BT (1990) Role of coexpression of IL-2 and herpes simplex virus proteins in recombinant vaccinia virus vectors on levels of induced immunity. Viral Immunol 3:207–215

Andersson J (1996) EBV, Everybody's virus: the management of Epstein-Barr virus infection. Herpes 3:55–62

Arnold JC, Portmann BC, O'Grady JG, Naoumov NV, Alexander GJ, Williams R (1992) Cytomegalovirus infection persists in the liver graft in the vanishing bile duct syndrome. Hepatology 16:285–292

Baringer JR (1996) Herpes simplex virus and Bell palsy (editorial; comment). Ann Intern Med 124:63–65

Baringer JR, Pisani P (1994) Herpes simplex virus genomes in human nervous system tissue analyzed by polymerase chain reaction. Ann Neurol 36:823–829

Baringer JR, Swoveland P (1973) Recovery of herpes-simplex virus from human trigeminal ganglions. N Engl J Med 288:648–650

Barnes PD, Grundy JE (1992) Down-regulation of the class I HLA heterodimer and beta 2-microglobulin on the surface of cells infected with cytomegalovirus. J Gen Virol 73:2395–2403
Bastian FO, Rabson AS, Yee CL, Tralka TS (1972) Herpesvirus hominis: isolation from human trigeminal ganglion. Science 178:306–307
Beersma MF, Bijlmakers MJ, Ploegh HL (1993) Human cytomegalovirus down-regulates HLA class I expression by reducing the stability of class I H chains. J Immunol 151:4455–4464
Bergstrom T, Andersen O, Vahlne A (1989) Isolation of herpes simplex virus type 1 during first attack of multiple sclerosis. Ann Neurol 26:283–285
Bergstrom T, Vahlne A, Alestig K, Jeansson S, Forsgren M, Lycke E (1990) Primary and recurrent herpes simplex virus type 2-induced meningitis. J Infect Dis 162:322–330
Bernstein DI, Lee FK, Echler G, Nahmias AJ (1989) Clinical and serological outcome of genital herpes simplex virus (HSV) type 2 inoculation following oral HSV type 1 infection in guinea-pigs. J Gen Virol 70:2365–2372
Beswick TSL (1962) The origin and the use of the word herpes. Med Hist 6:214–232
Blyth WA, Harbour DA, Hill TJ (1984) Pathogenesis of zosteriform spread of herpes simplex virus in the mouse. J Gen Virol 65:1477–1486
Borysiewicz LK, Hickling JK, Graham S, Sinclair J, Cranage MP, Smith GL, Sissons JG (1988) Human cytomegalovirus-specific cytotoxic T cells. Relative frequency of stage-specific CTL recognizing the 72-kD immediate early protein and glycoprotein B expressed by recombinant vaccinia viruses. J Exp Med 168:919–931
Bourne N, Milligan GN, Schleiss MR, Bernstein DI, Stanberry LR (1996) DNA immunization confers protective immunity on mice challenged intravaginally with herpes simplex virus type 2. Vaccine 14:1230–1234
Brehm MA, Bonneau RH, Knipe DM, Tevethia SS (1997) Immunization with a replication-deficient mutant of herpes simplex virus type 1 (HSV-1) induces a CD8+ cytotoxic T-lymphocyte response and confers a level of protection comparable to that of wild-type HSV-1. J Virol 71:3534–3544
Britt WJ, Ahlford CA (1996) Cytomegalovirus. In: Fields BN (ed) Fields virology. Lippincott-Raven, New York, pp 2493–22524
Britt WJ, Vugler L, Stephens EB (1988) Induction of complement-dependent and -independent neutralizing antibodies by recombinant-derived human cytomegalovirus gp55–116 (gB). J Virol 62:3309–3318
Bruggeman CA, van Dam Mieras MC (1991) The possible role of cytomegalovirus in atherogenesis. Prog Med Virol 38:1–26
Buchman TG, Roizman B, Nahmias AJ (1979) Demonstration of exogenous genital reinfection with herpes simplex virus type 2 by restriction endonuclease fingerprinting of viral DNA. J Infect Dis 140:295–304
Cai WH, Gu B, Person S (1988) Role of glycoprotein B of herpes simplex virus type 1 in viral entry and cell fusion. J Virol 62:2596–2604 [published erratum appears in J Virol 1988, 62(11):4438]
Campadelli Fiume G, Arsenakis M, Farabegoli F, Roizman B (1988) Entry of herpes simplex virus 1 in BJ cells that constitutively express viral glycoprotein D is by endocytosis and results in degradation of the virus. J Virol 62:159–167
Campadelli Fiume G, Stirpe D, Boscaro A, Avitabile E, Foa Tomasi L, Barker D, Roizman B (1990) Glycoprotein C-dependent attachment of herpes simplex virus to susceptible cells leading to productive infection. Virology 178:213–222
Chatterjee S, Koga J, Whitley RJ (1989) A role for herpes simplex virus type 1 glycoprotein E in induction of cell fusion. J Gen Virol 70:2157–2162
Chee MS, Bankier AT, Beck S, Bohni R, Brown CM, Cerny R, Horsnell T, Hutchison CA 3d, Kouzarides T, Martignetti JA et al (1990) Analysis of the protein-coding content of the sequence of human cytomegalovirus strain AD169. In: McDougall JK (ed) Cytomegaloviruses. Springer, Berlin Heidelberg New York, pp 125–169 (Current topics in microbiology and immunology, vol 154)

Chen F, Zou JZ, di Renzo L, Winberg G, Hu LF, Klein E, Klein G, Ernberg I (1995) A subpopulation of normal B cells latently infected with Epstein-Barr virus resembles Burkitt lymphoma cells in expressing EBNA-1 but not EBNA-2 or LMP1. J Virol 69:3752–3758
Chou J, Kern ER, Whitley RJ, Roizman B (1990) Mapping of herpes simplex virus-1 neurovirulence to gamma 134.5, a gene nonessential for growth in culture. Science 250:1262–1266
Clements GB, Subak Sharpe JH (1988) Herpes simplex virus type 2 establishes latency in the mouse footpad. J Gen Virol 69:375–383
Clements GB, Jamieson FE (1989) Reactivation of latent herpes simplex virus-1 (HSV) from mouse footpad cells demonstrated by in situ hybridization. Arch Virol 104:95–106
Compton T, Nowlin DM, Cooper NR (1993) Initiation of human cytomegalovirus infection requires initial interaction with cell surface heparan sulfate. Virology 193:834–841
Cook ML, Stevens JG (1973) Pathogenesis of herpetic neuritis and ganglionitis in mice: evidence for intraaxonal transport of infection. Infect Immun 7:272–288
Cook ML, Bastone WB, Stevens JG (1974) Evidence that neurons harbor latent herpes simplex virus. Infect Immun 9:946–951
Corey L, Spear PG (1986) Infections with herpes simplex viruses (2). N Engl J Med 314:749–757
Corey L, Adams HG, Brown ZA, Holmes KK (1983) Genital herpes simplex virus infections: clinical manifestations, course, and complications. Ann Intern Med 98:958–972
Craig CP, Nahmias AJ (1973) Different patterns of neurological involvement with herpes simplex virus types 1 and 2: isolation of herpes simplex virus type 2 from the buffy coat of two adults with meningitis. J Infect Dis 127:365–372
Cushing H (1905) Surgical aspect of major neuralgia of trigeminal nerve: report of 20 cases of operation upon the gasserian ganglion with anatomic and physiological notes on the consequence of its removal. JAMA 44:1002–1008
Diamond DJ, York J, Sun JY, Wright CL, Forman SJ (1997) Development of a candidate HLA A*0201 restricted peptide-based vaccine against human cytomegalovirus infection. Blood 90:1751–1767
Dix RD, Waitzman DM, Follansbee S, Pearson BS, Mendelson T, Smith P, Davis RL, Mills J (1985) Herpes simplex virus type 2 encephalitis in two homosexual men with persistent lymphadenopathy. Ann Neurol 17:203–206
Dobbins JG, Stewart JA, Demmler GJ (1992) Surveillance of congenital cytomegalovirus disease, 1990–1991. Collaborating Registry Group. MMWR CDC Surveill Summ 41:35–39
Edstrom S, Hanner P, Andersen O, Rosenhall U, Vahlne A, Karlsson B (1987) Elevated levels of myelin basic protein in CSF in relation to auditory brainstem responses in Bell's palsy. Acta Otolaryngol (Stockh) 103:198–203
Emini EA, Schleif WA, Armstrong ME, Silberklang M, Schultz LD, Lehman D, Maigetter RZ, Qualtiere LF, Pearson GR, Ellis RW (1988) Antigenic analysis of the Epstein-Barr virus major membrane antigen (gp350/220) expressed in yeast and mammalian cells: implications for the development of a subunit vaccine. Virology 166:387–393
Emini EA, Schleif WA, Silberklang M, Lehman D, Ellis RW (1989) Vero cell-expressed Epstein-Barr virus (EBV) gp350/220 protects marmosets from EBV challenge. J Med Virol 27:120–123
Epstein M, Achong B, Barr Y (1964) Virus particles in cultured lymphoblasts from Burkitt's lymphoma. Lancet 1:702–703
Epstein MA, Morgan AJ, Finerty S, Randle BJ, Kirkwood JK (1985) Protection of cottontop tamarins against Epstein-Barr virus-induced malignant lymphoma by a prototype subunit vaccine. Nature 318:287–289

Finerty S, Tarlton J, Mackett M, Conway M, Arrand JR, Watkins PE, Morgan AJ (1992) Protective immunization against Epstein-Barr virus-induced disease in cottontop tamarins using the virus envelope glycoprotein gp340 produced from a bovine papillomavirus expression vector. J Gen Virol 73:449–453
Fish KN, Britt W, Nelson JA (1996) A novel mechanism for persistence of human cytomegalovirus in macrophages. J Virol 70:1855–1862
Forrester A, Farrell H, Wilkinson G, Kaye J, Davis Poynter N, Minson T (1992) Construction and properties of a mutant of herpes simplex virus type 1 with glycoprotein H coding sequences deleted. J Virol 66:341–348
Fowler KB, Stagno S, Pass RF, Britt WJ, Boll TJ, Alford CA (1992) The outcome of congenital cytomegalovirus infection in relation to maternal antibody status. N Engl J Med 326:663–667
Frank AB (1938) Formalized herpes virus therapy and the neutralizing substance in herpes simplex. J Invest Dermatol 1:267–282
Fraser NW, Lawrence WC, Wroblewska Z, Gilden DH, Koprowski H (1981) Herpes simplex type 1 DNA in human brain tissue. Proc Natl Acad Sci USA 78:6461–6465
Friedman AD, Furukawa T, Plotkin SA (1982) Detection of antibody to cytomegalovirus early antigen in vaccinated, normal volunteers and renal transplant candidates. J Infect Dis 146:255–259
Fruh K, Ahn K, Djaballah H, Sempe P, van Endert PM, Tampe R, Peterson PA, Yang Y (1995) A viral inhibitor of peptide transporters for antigen presentation. Nature 375:415–418
Fuller AO, Spear PG (1987) Anti-glycoprotein D antibodies that permit adsorption but block infection by herpes simplex virus 1 prevent virion-cell fusion at the cell surface. Proc Natl Acad Sci USA 84:5454–5458
Gallant JE, Moore RD, Richman DD, Keruly J, Chaisson RE (1992) Incidence and natural history of cytomegalovirus disease in patients with advanced human immunodeficiency virus disease treated with zidovudine. The Zidovudine Epidemiology Study Group. J Infect Dis 166:1223–1227
Gallant JE, Moore RD, Keruly J, Richman DD, Chaisson RE (1995) Lack of association between acyclovir use and survival in patients with advanced human immunodeficiency virus disease treated with zidovudine. Zidovudine Epidemiology Study Group. J Infect Dis 172:346–352
Gehrz RC, Christianson WR, Linner KM, Groth KE, Balfour HH Jr (1980) Cytomegalovirus vaccine. Specific humoral and cellular immune responses in human volunteers. Arch Intern Med 140:936–939
Gerdes JC, Smith DS (1983) Recurrence phenotypes and establishment of latency following rabbit keratitis produced by multiple herpes simplex virus strains. J Gen Virol 64:2441–2454
Gilbert MJ, Riddell SR, Plachter B, Greenberg PD (1996) Cytomegalovirus selectively blocks antigen processing and presentation of its immediate-early gene product. Nature 383:720–722
Glazer JP, Friedman HM, Grossman RA, Starr SE, Barker CF, Perloff LJ, Huang ES, Plotkin SA (1979) Live cytomegalovirus vaccination of renal transplant candidates. A preliminary trial. Ann Intern Med 91:676–683
Gompels U, Minson A (1986) The properties and sequence of glycoprotein H of herpes simplex virus type 1. Virology 153:230–247
Gonczol E, deTaisne C, Hirka G, Berencsi K, Lin WC, Paoletti E, Plotkin S (1991) High expression of human cytomegalovirus (HCMV)-gB protein in cells infected with a vaccinia-gB recombinant: the importance of the gB protein in HCMV immunity. Vaccine 9:631–637
Gonczol E, Berensci K, Pincus S, Endresz V, Meric C, Paoletti E, Plotkin SA (1995) Preclinical evaluation of an ALVAC (canarypox)-human cytomegalovirus glycoprotein B vaccine candidate. Vaccine 13:1080–1085

Gonzalez Armas JC, Morello CS, Cranmer LD, Spector DH (1996) DNA immunization confers protection against murine cytomegalovirus infection. J Virol 70:7921–7928
Goodpasture EW, Teague O (1923a) Surgical aspect of major neuralgia of trigeminal nerve: report of 20 cases of experimental production of herpetic lesions in organs and tissues of the rabbit. J Med Res 44:121–137
Goodpasture EW, Teague O (1923b) Surgical aspect of major neuralgia of trigeminal nerve: report of 20 cases of transmission of the virus of herpestic febrilis along nerves in experimentally infected rabbits. J Med Res 44:139–184
Griffiths PD (1993) Current management of cytomegalovirus disease. J Med Virol 1:106–111
Gu SY, Huang TM, Ruan L, Miao YH, Lu H, Chu CM, Motz M, Wolf H (1995) First EBV vaccine trial in humans using recombinant vaccinia virus expressing the major membrane antigen. Dev Biol Stand 84:171–177
Gudnadottir M, Helgadottir H, Bjarnason O, Jonsdottir K (1964) Virus isolated from the brain of a patient with multiple sclerosis. Exp Neurol 9:85–95
Henle G, Henle W, Diehl V (1968) Relation of Burkitt's tumour associated herpes-type virus to infectious mononucleosis. Proc Natl Acad Sci USA 59:94–101
Herold BC, WuDunn D, Soltys N, Spear PG (1991) Glycoprotein C of herpes simplex virus type 1 plays a principal role in the adsorption of virus to cells and in infectivity. J Virol 65:1090–1098
Highlander SL, Cai WH, Person S, Levine M, Glorioso JC (1988) Monoclonal antibodies define a domain on herpes simplex virus glycoprotein B involved in virus penetration. J Virol 62:1881–1888
Hill A, Jugovic P, York I, Russ G, Bennink J, Yewdell J, Ploegh H, Johnson D (1995) Herpes simplex virus turns off the TAP to evade host immunity. Nature 375:411–415
Hoover DR, Saah AJ, Bacellar H, Phair J, Detels R, Anderson R, Kaslow RA (1993) Clinical manifestations of AIDS in the era of pneumocystis prophylaxis. Multicenter AIDS Cohort Study. N Engl J Med 329:1922–1926
Howard WT (1905) Further observation on the relation of lesions of the Gasserian and posterior root ganglia to herpes occurring in pneumonia and cerebrospinal meningitis. Am J Med Sci 130:1012–1019
Howe JG, Shu MD (1989) Epstein-Barr virus small RNA (EBER) genes: unique transcription units that combine RNA polymerase II and III promoter elements. Cell 57:825–834
Hung SL, Peng C, Kostavasili I, Friedman HM, Lambris JD, Eisenberg RJ, Cohen GH (1994) The interaction of glycoprotein C of herpes simplex virus types 1 and 2 with the alternative complement pathway. Virology 203:299–312
Javier RT, Stevens JG, Dissette VB, Wagner EK (1988) A herpes simplex virus transcript abundant in latently infected neurons is dispensable for establishment of the latent state. Virology 166:254–257
Johnson DC, Frame MC, Ligas MW, Cross AM, Stow ND (1988) Herpes simplex virus immunoglobulin G Fc receptor activity depends on a complex of two viral glycoproteins, gE and gI. J Virol 62:1347–1354
Jones TR, Hanson LK, Sun L, Slater JS, Stenberg RM, Campbell AE (1995) Multiple independent loci within the human cytomegalovirus unique short region downregulate expression of major histocompatibility complex class I heavy chains. J Virol 69:4830–4841
Jones TR, Wiertz EJ, Sun L, Fish KN, Nelson JA, Ploegh HL (1996) Human cytomegalovirus US3 impairs transport and maturation of major histocompatibility complex class I heavy chains. Proc Natl Acad Sci USA 93:11327–11333
Kari B, Gehrz R (1992) A human cytomegalovirus glycoprotein complex designated gC-II is a major heparin-binding component of the envelope. J Virol 66:1761–1764

Kari B, Gehrz R (1993) Structure, composition and heparin binding properties of a human cytomegalovirus glycoprotein complex designated gC-II. J Gen Virol 74:255–264
Keay S, Baldwin B (1991) Anti-idiotype antibodies that mimic gp86 of human cytomegalovirus inhibit viral fusion but not attachment. J Virol 65:5124–5128
Keay S, Merigan TC, Rasmussen L (1989) Identification of cell surface receptors for the 86-kilodalton glycoprotein of human cytomegalovirus. Proc Natl Acad Sci USA 86:10100–10103
Kern AB, Schiff BL (1964) Vaccine therapy in recurrent herpes simplex. Arch Dermatol 89:844–845
Khanna R, Slade RW, Poulsen L, Moss DJ, Burrows, SR, Nicholls J, Burrows JM (1997) Evolutionary dynamics of genetic variation in Epstein-Barr virus isolates of diverse geographical origins: evidence for immune pressure-independent genetic drift. J Virol 71:8340–8346
Kieff E (1996) Epstein-Barr virus and its replication. In: Fields BN (ed) Fields virology. Lippincott-Raven, New York, pp 2343–2396
Klein RJ (1976) Pathogenetic mechansims of recurrent herpes simplex virus infections. Arch Virol 51:1–13
Kriesel JD, Spruance SL, Daynes RA, Araneo BA (1996) Nucleic acid vaccine encoding gD2 protects mice from herpes simplex virus type 2 disease. J Infect Dis 173:536–541
Kristensson K, Lycke E, Sjostrand J (1971) Spread of herpes simplex virus in peripheral nerves. Acta Neuropathol (Berl) 17:44–53
Kristensson K, Vahlne A, Persson LA, Lycke E (1978) Neural spread of herpes simplex virus types 1 and 2 in mice after corneal or subcutaneous (footpad) inoculation. J Neurol Sci 35:331–340
Kristensson K, Svennerholm B, Persson L, Vahlne A, Lycke E (1979) Latent herpes simplex virus trigeminal ganglionic infection in mice and demyelination in the central nervous system. J Neurol Sci 43:253–263
Kristensson K, Nennesmo L, Persson L, Lycke E (1982a) Neuron to neuron transmission of herpes simplex virus. Transport of virus from skin to brainstem nuclei. J Neurol Sci 54:149–156
Kristensson K, Svennerholm B, Vahlne A, Nilheden E, Persson L, Lycke E (1982b) Virus-induced demyelination in herpes simplex virus-infected mice. J Neurol Sci 53:205–216
Kristensson K, Lycke E, Roytta M, Svennerholm B, Vahlne A (1986) Neuritic transport of herpes simplex virus in rat sensory neurons in vitro. Effects of substances interacting with microtubular function and axonal flow [nocodazole, taxol and erythro-9-3-(2-hydroxynonyl)adenine]. J Gen Virol 67:2023–2028
Kuhn JE, Kramer MD, Willenbacher W, Wieland U, Lorentzen EU, Braun RW (1990) Identification of herpes simplex virus type 1 glycoproteins interacting with the cell surface. J Virol 64:2491–2497
Kuklin N, Daheshia M, Karem K, Manickan E, Rouse BT (1997) Induction of mucosal immunity against herpes simplex virus by plasmid DNA immunization. J Virol 71:3138–3145
Langeland N, Holmsen H, Lillehaug JR, Haarr L (1987) Evidence that neomycin inhibits binding of herpes simplex virus type 1 to the cellular receptor. J Virol 61:3388–3393
Langeland N, Moore LJ, Holmsen H, Haarr L (1988) Interaction of polylysine with the cellular receptor for herpes simplex virus type 1. J Gen Virol 69:1137–1145
Langeland N, Oyan AM, Marsden HS, Cross A, Glorioso JC, Moore LJ, Haarr L (1990) Localization on the herpes simplex virus type 1 genome of a region encoding proteins involved in adsorption to the cellular receptor. J Virol 64:1271–1277
Lee FK, Pereira L, Griffin C, Reid E, Nahmias A (1986) A novel glycoprotein for detection of herpes simplex virus type 1-specific antibodies. J Virol Methods 14:111–118

Lewis ME, Leung WC, Jeffrey VM, Warren KG (1984) Detection of multiple strains of latent herpes simplex virus type 1 within individual human hosts. J Virol 52:300-305

Lipschutz B (1921) Untersuchungen über die Atiologie der Krankenheit der Herpesgruppe (herpes zoster, herpes genitalis, herpes febrilis). Arch Derm Syphilol 136:428–482

Liu YN, Curtsinger J, Donahue PR, Klaus A, Optiz G, Cooper J, Karr RW, Bach FH, Gehrz RC (1993) Molecular analysis of the immune response to human cytomegalovirus glycoprotein B. I. Mapping of HLA-restricted helper T cell epitopes on gp93. J Gen Virol 74:2207–2214

Lowe RS, Keller PM, Keech BJ, Davison AJ, Whang Y, Morgan AJ, Kieff E, Ellis RW (1987) Varicella-zoster virus as a live vector for the expression of foreign genes. Proc Natl Acad Sci USA 84:3896–3900

Lowhagen GB, Jansen E, Nordenfelt E, Lycke E (1990) Epidemiology of genital herpes infections in Sweden. Acta Derm Venereol (Stockh) 70:330–334

Lundstrom M, Olofsson S, Jeansson S, Lycke E, Datema R, Mansson JE (1987) Host cell-induced differences in O-glycosylation of herpes simplex virus gC-1. I. Structures of nonsialylated HPA- and PNA-binding carbohydrates. Virology 161:385–394

Lycke E, Kristensson K, Svennerholm B, Vahlne A, Ziegler R (1984) Uptake and transport of herpes simplex virus in neurites of rat dorsal root ganglia cells in culture. J Gen Virol 65:55–64

Lycke E, Johansson M, Svennerholm B, Lindahl U (1991) Binding of herpes simplex virus to cellular heparan sulphate, an initial step in the adsorption process. J Gen Virol 72:1131–1137

Mackett M, Conway MJ, Arrand JR, Haddad RS, Hutt Fletcher LM (1990) Characterization and expression of a glycoprotein encoded by the Epstein-Barr virus BamHI I fragment. J Virol 64:2545–2552

Madej M, Conway MJ, Morgan AJ, Sweet J, Wallace L, Qualtiere LF, Arrand JR, Mackett M (1992) Purification and characterization of Epstein-Barr virus gp340/220 produced by a bovine papillomavirus virus expression vector system. Vaccine 10:777–782

Manickan E, Yu Z, Rouse RJ, Wire WS, Rouse BT (1995) Induction of protective immunity against herpes simplex virus with DNA encoding the immediate early protein ICP 27. Viral Immunol 8:53–61

Manservigi R, Spear PG, Buchan A (1977) Cell fusion induced by herpes simplex virus is promoted and suppressed by different viral glycoproteins. Proc Natl Acad Sci USA 74:3913–3917

Manzella JP, McConville JH, Valenti W, Menegus MA, Swierkosz EM, Arens M (1984) An outbreak of herpes simplex virus type I gingivostomatitis in a dental hygiene practice. JAMA 252:2019–2022

Marshall GS (1997) Cytomegalovirus vaccines – two decades of progress. Herpes 4:20–24

Marshall GS, Plotkin SA (1990) Progress toward developing a cytomegalovirus vaccine. Infect Dis Clin North Am 4:283–298

Marshall GS, Ricciardi RP, Rando RF, Puck J, Ge RW, Plotkin SA, Gonczol E (1990) An adenovirus recombinant that expresses the human cytomegalovirus major envelope glycoprotein and induces neutralizing antibodies. J Infect Dis 162:1177–1181

Martin JR (1982) Spinal cord and optic nerve demyelination in experimental herpes simplex virus type 2 infection. J Neuropathol Exp Neurol 41:253–266

Martin JR, Stoner GL (1984) The nervous system in genital herpes simplex virus type 2 infections in mice. Lethal panmyelitis or nonlethal demyelinative myelitis or meningitis. Lab Invest 51:556–566

Martin JR, Holt RK, Webster HD (1988) Herpes-simplex-related antigen in human demyelinative disease and encephalitis. Acta Neuropathol (Berl) 76:325–337

Mastrolorenzo A, Tiradritti L, Salimbeni L, Zuccati G (1995) Multicentre clinical trial with herpes simplex virus vaccine in recurrent herpes infection. Int J STD AIDS 6:431–435
Mauri DN, Ebner R, Montgomery RI, Kochel KD, Cheung TC, Yu GL, Ruben S, Murphy M, Eisenberg RJ, Cohen GH, Spear PG, Ware CF (1998) LIGHT, a new member of the TNF superfamily, and lymphotoxin alpha are ligands for herpesvirus entry mediator. Immunity 8:21–30
McCormick DP (1972) Herpes-simplex virus as a cause of Bell's palsy. Lancet 1:937–939
McDermott MR, Graham FL, Hanke T, Johnson DC (1989) Protection of mice against lethal challenge with herpes simplex virus by vaccination with an adenovirus vector expressing HSV glycoprotein B. Virology 169:244–247
McLaughlin Taylor E, Pande H, Forman SJ, Tanamachi B, Li CR, Zaia JA, Greenberg PD, Riddell SR (1994) Identification of the major late human cytomegalovirus matrix protein pp65 as a target antigen for CD8+ virus-specific cytotoxic T lymphocytes. J Med Virol 43:103–110
McLennan JL, Darby G (1980) Herpes simplex virus latency: the cellular location of virus in dorsal root ganglia and the fate of the infected cell following virus activation. J Gen Virol 51:233–243
Meignier B, Longnecker R, Roizman B (1988) In vivo behavior of genetically engineered herpes simplex viruses R7017 and R7020: construction and evaluation in rodents. J Infect Dis 158:602–614
Meignier B, Martin B, Whitley RJ, Roizman B (1990) In vivo behavior of genetically engineered herpes simplex viruses R7017 and R7020. II. Studies in immunocompetent and immunosuppressed owl monkeys (Aotus trivirgatus). J Infect Dis 162:313–321
Mocarski ES, Post LE, Roizman B (1980) Molecular engineering of the herpes simplex virus genome: insertion of a second L-S junction into the genome causes additional genome inversions. Cell 22:243–255
Mocarski ES Jr (1996) Cytomegaloviruses and their replication. In: Fields BN, Knipe DM, Howley PM (eds) Fields virology. Lippincott-Raven, New York, pp 2447–2492
Montgomery RI, Warner MS, Lum BJ, Spear PG (1996) Herpes simplex virus-1 entry into cells mediated by a novel member of the TNF/NGF receptor family. Cell 87:427–436
Morgan C, Rose HM, Mednis B (1968) Electron microscopy of herpes simplex virus. I. Entry. J Virol 2:507–516
Morgan AJ, Mackett M, Finerty S, Arrand JR, Scullion FT, Epstein MA (1988) Recombinant vaccinia virus expressing Epstein-Barr virus glycoprotein gp340 protects cottontop tamarins against EB virus-induced malignant lymphomas. J Med Virol 25 189–195
Morgan JA (1992) Epstein-Barr virus vaccines. Vaccine 10:563–571
Morrison LA, Knipe DM (1994) Immunization with replication-defective mutants of herpes simplex virus type 1: sites of immune intervention in pathogenesis of challenge virus infection. J Virol 68:689–696
Moss DJ, Schmidt C, Elliott S, Suhrbier A, Burrows S, Khanna R. (1996) Strategies involved in developing an effective vaccine for EBV-associated diseases. Adv Cancer Res 69:213–245
Murphy FA (1996) Virus taxonomy. In: Fields BN, Knipe DM, Howley PM (eds) Fields virology. Lippincott-Raven, New York, pp 15–58
Nahmias AJ, Dowdle WR (1968) Antigenic and biological differences in herpesvirus hominis. Prog Med Virol 10:110–159
Navarro D, Paz P, Tugizov S, Topp K, La Vail J, Pereira L (1993) Glycoprotein B of human cytomegalovirus promotes virion penetration into cells, transmission of infection from cell to cell, and fusion of infected cells. Virology 197:143–158

Nicholls SM, Blyth WA (1989) Quantification of herpes simplex virus infection in cervical ganglia of mice. J Gen Virol 70:1779–1788
Nilheden E, Jeansson S, Vahlne A (1985) Herpes simplex virus latency in a hyperresistant clone of mouse neuroblastoma (C1300) cells. Arch Virol 83:319–325
Olding LB, Kingsbury DT, Oldstone MB (1976) Pathogenesis of cytomegalovirus infection. Distribution of viral products, immune complexes and autoimmunity during latent murine infection. J Gen Virol 33:267–280
Olofsson S, Lundstrom M, Marsden H, Jeansson S, Vahlne A (1986) Characterization of a herpes simplex virus type 2-specified glycoprotein with affinity for N-acetylgalactosamine-specific lectins and its identification as g92 K or gG. J Gen Virol 67:737–744
Pande H, Campo K, Tanamachi B, Forman SJ, Zaia JA (1995) Direct DNA immunization of mice with plasmid DNA encoding the tegument protein pp65 (ppUL83) of human cytomegalovirus induces high levels of circulating antibody to the encoded protein. Scand J Infect Dis [Suppl] 99:117–120
Pereira L (1994) Function of glycoprotein B homologues of the family herpesviridae. Infect Agents Dis 3:9–28
Plotkin SA, Huang ES (1985) Cytomegalovirus vaccine virus (Towne strain) does not induce latency. J Infect Dis 152:395–397
Plotkin SA, Huygelen C (1976) Cytomegalovirus vaccine prepared in WI-38. Dev Biol Stand 37:301–305
Plotkin SA, Farquhar J, Horberger E (1976) Clinical trials of immunization with the Towne 125 strain of human cytomegalovirus. J Infect Dis 134:470–475
Plotkin SA, Smiley ML, Friedman HM, Starr SE, Fleisher GR, Wlodaver C, Dafoe DC, Friedman AD, Grossman RA, Barker CF (1984) Prevention of cytomegalovirus disease by Towne strain live attenuated vaccine. Birth Defects 20:271–287
Plotkin SA, Starr SE, Friedman HM, Gonczol E, Brayman K (1990) Vaccines for the prevention of human cytomegalovirus infection. Rev Infect Dis 12 [Suppl 7]:S827–S838
Plotkin SA, Starr SE, Friedman HM, Brayman K, Harris S, Jackson S, Tustin NB, Grossman R, Dafoe D, Barker C (1991) Effect of Towne live virus vaccine on cytomegalovirus disease after renal transplant. A controlled trial. Ann Intern Med 114:525–531
Plotkin SA, Higgins R, Kurtz JB, Morris PJ, Campbell DA Jr, Shope TC, Spector SA, Dankner WM (1994) Multicenter trial of Towne strain attenuated virus vaccine in seronegative renal transplant recipients. Transplantation 58:1176–1178
Post LE, Roizman B (1981) A generalized technique for deletion of specific genes in large genomes: alpha gene 22 of herpes simplex virus 1 is not essential for growth. Cell 25:227–232
Quinnan GV Jr, Delery M, Rook AH, Frederick WR, Epstein JS, Manischewitz JF, Jackson L, Ramsey KM, Mittal K, Plotkin SA et al (1984) Comparative virulence and immunogenicity of the Towne strain and a nonattenuated strain of cytomegalovirus. Ann Intern Med 101:478–483
Ragot T, Finerty S, Watkins PE, Perricaudet M, Morgan AJ (1993) Replication-defective recombinant adenovirus expressing the Epstein-Barr virus (EBV) envelope glycoprotein gp340/220 induces protective immunity against EBV-induced lymphomas in the cottontop tamarin. J Gen Virol 74:501–507
Rasmussen L, Resta S, Merigan T (1991) Human cytomegalovirus glycoprotein-receptor interactions. Transplant Proc 23:60–63
Reeves WC, Corey L, Adams HG, Vontver LA, Holmes KK (1981) Risk of recurrence after first episodes of genital herpes. Relation to HSV type and antibody response. N Engl J Med 305:315–319
Ribbert H (1904) Über protozoanartige Zellen in der Niere eines syphilitischen Neugeboren und in der Parotis von Kindern. Zentralbl Allg Path Path Anat 15:945–948

Rickinson AB, Kieff E (1996) Epstein-Barr Virus. In: Fields BN, Knipe DM, Howley PM (eds) Fields virology. Lippincott-Raven, New York, pp 2397–2446
Roizman B, Sears AE (1996) Herpes simplex viruses and their replication. In: Fields BN, Knipe DM, Howley PM (eds) Fields virology. Lippincott-Raven, New York, pp 2231–2295
Rosenhall U, Edstrom S, Hanner P, Badr G, Vahlne A (1983) Auditory brain stem response abnormalities in patients with Bell's palsy. Otolaryngol Head Neck Surg 91:412–416
Rosenthal SL, Stanberry LR, Biro FM, Slaoui M, Francotte M, Koutsoukos M, Hayes M, Bernstein DI. (1997) Seroprevalence of herpes simplex virus types 1 and 2 and cytomegalovirus in adolescents. Clin Infect Dis 24:135–139
Rowe WP, Hartley JW, Waterman S, Turner HC, Huebner RJ (1956) Cytopathogenic agents resembling human salivary gland virus recovered from tissue cultures of human adenoids. Proc Soc Exp Biol Med 92:418–424
Rowe M, Rowe DT, Gregory CD, Young LS, Farrell PJ, Rupani H, Rickinson AB (1987) Differences in B cell growth phenotype reflect novel patterns of Epstein-Barr virus latent gene expression in Burkitt's lymphoma cells. EMBO J 6:2743–2751
Rustigan R, Smulow JB, Tye M, Gibson WA, Shindell E (1966) Studies on latent infection of skin and oral mucosa in individuals with recurrent herpes simplex. J Invest Dermatol 47:218–221
Rymo L (1979) Identification of transcribed regions of Epstein-Barr virus DNA in Burkitt lymphoma-derived cells. J Virol 32:8–18
Sedmak DD, Guglielmo AM, Knight DA, Birmingham DJ, Huang EH, Waldman WJ (1994) Cytomegalovirus inhibits major histocompatibility class II expression on infected endothelial cells. Am J Pathol 144:683–692
Sjoblom I, Lundstrom M, Sjogren Jansson E, Glorioso JC, Jeansson S, Olofsson S (1987) Demonstration and mapping of highly carbohydrate-dependent epitopes in the herpes simplex virus type 1-specified glycoprotein C. J Gen Virol 68:545–554
Skinner GR, Turyk ME, Benson CA, Wilbanks GD, Heseltine P, Galpin J, Kaufman R, Goldberg L, Hartley CE, Buchan A (1997) The efficacy and safety of Skinner herpes simplex vaccine towards modulation of herpes genitalis; report of a prospective double-blind placebo-controlled trial. Med Microbiol Immunol (Berl) 186:31–36
Skoldenberg B, Jeansson S, Wolontis S (1973) Herpes simplex virus type 2 in acute aseptic meningitis. BMJ 2:611
Smith MG (1956) Propagation in tissue cultures of a cytopathogenic virus from human salivary gland virus disease. Proc Soc Exp Biol Med 92:424–430
Söderberg C (1995) Thesis. Karolinska institutet, Stockholm
Söderberg Naucler C, Fish KN, Nelson JA (1997) Reactivation of latent human cytomegalovirus by allogeneic stimulation of blood cells from healthy donors. Cell 91:119–126
Soffer D, Martin JR (1988) Remyelination of central nervous system lesions in experimental genital herpes simplex virus infection. J Neurol Sci 86:83–95
Spear PG (1985) Glycoproteins specified by herpes simplex virus. In: Roizman B (ed) The herpesviruses. Plenum, New York, pp 315–356
Stagno S, Reynolds DW, Huang ES, Thames SD, Smith RJ, Alford CA (1977) Congenital cytomegalovirus infection. N Engl J Med 296:1254–1258
Stagno S, Pass RF, Cloud G, Britt WJ, Henderson RE, Walton PD, Veren DA, Page F, Alford CA (1986) Primary cytomegalovirus infection in pregnancy. Incidence, transmission to fetus, and clinical outcome. JAMA 256:1904–1908
Starr SE, Glazer JP, Friedman HM, Farquhar JD, Plotkin SA (1981) Specific cellular and humoral immunity after immunization with live Towne strain cytomegalovirus vaccine. J Infect Dis 143:585–589
Stevens JG (1975) Herpes simplex viral latency – a review. IARC Sci Publ 11(2):67–72

Stevens JG, Newburn AB, Cook ML (1972) Latent herpes simplex virus from trigeminal ganglia of rabbits with recurrent eye infection. Nat New Biol 235:216–217
Straus SE, Wald A, Kost RG, McKenzie R, Langenberg AG, Hohman P, Lekstrom J, Cox E, Nakamura M, Sekulovich R, Izu A, Dekker C, Corey L (1997) Immunotherapy of recurrent genital herpes with recombinant herpes simplex virus type 2 glycoproteins D and B: results of a placebo-controlled vaccine trial. J Infect Dis 176:1129–1134
Sturn B, Schneweis KE (1978) Protective effect of an oral infection with herpes simplex virus type 1 against subsequent genital infection with herpes simplex virus type 2. Med Microbiol Immunol (Berl) 165:119–127
Svennerholm B, Vahlne A, Lycke E (1981) Persistent reactivable latent herpes simplex virus infection in trigeminal ganglia of mice treated with antiviral drugs. Arch Virol 69:43–48
Svennerholm B, Olofsson S, Lunden R, Vahlne A, Lycke E (1982) Adsorption and penetration of enveloped herpes simplex virus particles modified by tunicamycin or 2-deoxy-D-glucose. J Gen Virol 63:343–349
Svennerholm B, Olofsson S, Jeansson S, Vahlne A, Lycke E (1984) Herpes simplex virus type-selective enzyme-linked immunosorbent assay with Helix pomatia lectin-purified antigens. J Clin Microbiol 19:235–239
Svennerholm B, Jeansson S, Vahlne A, Lycke E (1991) Involvement of glycoprotein C (gC) in adsorption of herpes simplex virus type 1 (HSV-1) to the cell. Arch Virol 120:273–279
Taylor HP, Cooper NR (1990) The human cytomegalovirus receptor on fibroblasts is a 30-kilodalton membrane protein. J Virol 64:2484–2490
Thomander L, Aldskogius H, Vahlne A, Kristensson K, Thomas E (1988) Invasion of cranial nerves and brain stem by herpes simplex virus inoculated into the mouse tongue. Ann Otol Rhinol Laryngol 97:554–558
Thomas E, Lycke E, Vahlne A (1985) Retrieval of latent herpes simplex virus type 1 genetic information from murine trigeminal ganglia by superinfection with heterotypic virus in vivo. J Gen Virol 66:1763–1770
Tomazin R, Hill AB, Jugovic P, York I, van Endert P, Ploegh HL, Andrews DW, Johnson DC (1996) Stable binding of the herpes simplex virus ICP47 protein to the peptide binding site of TAP. EMBO J 15:3256–3266
Trybala E, Bergstrom T, Svennerholm B, Jeansson S, Glorioso JC, Olofsson S (1994) Localization of a functional site on herpes simplex virus type 1 glycoprotein C involved in binding to cell surface heparan sulphate. J Gen Virol 75:743–752
Tullo AB, Easty DL, Hill TJ, Blyth WA (1982) Ocular herpes simplex and the establishment of latent infection. Trans Ophthalmol Soc UK 102:15–18
Vahlne A, Nystrom B, Sandberg M, Hamberger A, Lycke E (1978) Attachment of herpes simplex virus to neurons and glial cells. J Gen Virol 40:359–371
Vahlne A, Svennerholm B, Lycke E (1979) Evidence for herpes simplex virus type-selective receptors on cellular plasma membranes. J Gen Virol 44:217–225
Vahlne A, Svennerholm B, Sandberg M, Hamberger A, Lycke E (1980) Differences in attachment between herpes simplex type 1 and type 2 viruses to neurons and glial cells. Infect Immun 28:675–680
Vahlne A, Edstrom S, Arstila P, Beran M, Ejnell H, Nylen O, Lycke E (1981a) Bell's palsy and herpes simplex virus. Arch Otolaryngol 107:79–81
Vahlne A, Nilheden E, Svennerholm B (1981b) Multiplicity activation of herpes simplex virus in mouse neuroblastoma (C1300) cells. Arch Virol 70:345–356
Vahlne A, Edstrom S, Hanner P, Andersen O, Svennerholm B, Lycke E (1985) Possible association of herpes simplex virus infection with demyelinating disease. Scand J Infect Dis Suppl 47:16–21
von Glahn WC, Pappenheimer AM (1925) Intranuclear inclusions in visceral disease. Am J Pathol 1:445–465
Wang JB, Adler SP, Hempfling S, Burke RL, Duliege AM, Starr SE, Plotkin SA (1996) Mucosal antibodies to human cytomegalovirus glycoprotein B occur following

both natural infection and immunization with human cytomegalovirus vaccines. J Infect Dis 174:387–392

Warren AP, Ducroq DH, Lehner PJ, Borysiewicz LK (1994) Human cytomegalovirus-infected cells have unstable assembly of major histocompatibility complex class I complexes and are resistant to lysis by cytotoxic T lymphocytes. J Virol 68:2822–2829

Wechsler SL, Nesburn AB, Watson R, Slanina SM, Ghiasi H (1988) Fine mapping of the latency-related gene of herpes simplex virus type 1: alternative splicing produces distinct latency-related RNAs containing open reading frames. J Virol 62:4051–4058

Weitgasser H (1977) Controlled clinical study of the herpes antigens (Lupidon H, Lupidon G). Z Hautkr 52:625–628

Weller TH, Macaulay JC, Craig JM, Wirth P (1957) Isolation of intranuclear inclusion producing agents from infants with illnesses resemling cytomegalic inclusion disease. Proc Soc Exp Biol Med 94:4–12

Weller TTH, Hanshaw JB, Scott DE (1960) Cytopathogenic agents resembling human salivary gland virus recovered from tissue cultures of human adenoids. Virology 12:130–132

Whitley RJ (1996) Herpes simplex virus. In: Fields BN, Knipe DM, Howley PM (eds) Fields virology. Lippincott-Raven, New York, pp 2297–2342

Whitley RJ (1997) Herpes simplex vaccines. In: Levine MM, Woodrow GC, Kaper JB, Cobon GS (eds) New generation vaccines. Dekker, New York, pp 727–747

Whitley RJ, Soong SJ, Linneman C Jr, Liu C, Pazin G, Alford CA (1982) Herpes simplex encephalitis. Clinical assessment. JAMA 247:317–320

Wiertz EJ, Jones TR, Sun L, Bogyo M, Geuze HJ, Ploegh HL (1996a) The human cytomegalovirus US11 gene product dislocates MHC class I heavy chains from the endoplasmic reticulum to the cytosol. Cell 84:769–779

Wiertz EJ, Tortorella D, Bogyo M, Yu J, Mothes W, Jones TR, Rapoport TA, Ploegh HL (1996b) Sec61-mediated transfer of a membrane protein from the endoplasmic reticulum to the proteasome for destruction. Nature 384:432–438

Wiertz EJ, Mukherjee S, Ploegh HL (1997) Viruses use stealth technology to escape from the host immune system. Mol Med Today 3:116–123

Wilcox CL, Johnson EM Jr (1987) Nerve growth factor deprivation results in the reactivation of latent herpes simplex virus in vitro. J Virol 61:2311–2315

Wu CT, Levine M, Homa F, Highlander SL, Glorioso JC (1990) Characterization of the antigenic structure of herpes simplex virus type 1 glycoprotein C through DNA sequence analysis of monoclonal antibody-resistant mutants. J Virol 64:856–863

WuDunn D, Spear PG (1989) Initial interaction of herpes simplex virus with cells is binding to heparan sulfate. J Virol 63:52–58

York IA, Roop C, Andrews DW, Riddell SR, Graham FL, Johnson DC (1994) A cytosolic herpes simplex virus protein inhibits antigen presentation to CD8+ T lymphocytes. Cell 77:525–535

CHAPTER 8

Toxin-Based Vaccines (Diphtheria, Tetanus, Pertussis)

R. RAPPUOLI and M. PIZZA

A. Introduction

Corynebacterium diphtheriae, *Clostridium tetani*, and *Bordetella pertussis* produce potent toxins that are entirely or in part responsible for the severe diseases caused by these micro-organisms. The clinical manifestations of diphtheria and tetanus can be thoroughly reproduced by the systemic administration of the toxins. Both diseases can be prevented by administration of toxin-neutralizing antibodies or by immunization with partially purified and detoxified toxins (toxoids). Diphtheria and tetanus vaccines are commonly considered as composed only of the detoxified form of the toxins. However, it should be emphasized that diphtheria and tetanus vaccines usually contain approximately only 75% and 50% of detoxified toxins, respectively. The remaining impurities are antigens produced by the bacteria that may play a role in controlling colonization.

Pertussis is a more complex disease since the pertussis toxin is the main, but not the only, molecule contributing to toxicity induced by the bacterium. Administration of the pertussis toxin by the systemic route reproduces the systemic symptoms of the disease, but not the local toxicity and cough. Administration of toxin-neutralizing antibodies has been shown to reduce the duration and the severity of the disease. Immunization with detoxified toxin either alone or combined with other purified antigens from *B. pertussis* has been shown to prevent the disease.

In order to be safely administered as vaccines toxins need to be detoxified. Detoxification of bacterial toxins was a milestone in vaccine development. Consequently, in this chapter we propose to trace the past and present of the vaccines that have been developed to defeat these diseases and conclude that a bright future lies ahead.

The starting point of such a story is definitely the discovery by RAMON (1924) in France and GLENNY and HOPKINS (1923) in the United Kingdom of an effective method to detoxify tetanus and diphtheria toxins based on formaldehyde treatment. The detoxification of these two toxins allowed their use as vaccines and the introduction of mass immunization, which led to the almost complete eradication of both diseases from developed countries. Diphtheria (D) and tetanus (T) vaccines, which are used mostly for infant immunization, combined with whole cell pertussis (P) as DPT, are still prepared using the method described by Ramon. New generation vaccines, such as the acellular pertussis vaccines, may contain either the toxoid detoxified by the old method of Ramon, or nontoxic mutants produced by the genetically manipulated of the micro-organism. The latter molecules have been reported to be better immunogens than the classical chemically detoxified toxoids, and to provide excellent protection from disease.

Programs to provide diphtheria and tetanus vaccinations have been among the safest and most successful campaigns against infectious diseases. Both diseases have almost disappeared from those countries in which appropriate immunization is carried out (PAPPENHEIMER 1984; BIZZINI 1984).

In developing countries where vaccination is not yet fully practiced both diseases are still responsible for a remarkably high number of deaths (RAY et al. 1983; BIZZINI 1984): it is estimated that between 300000 and one million cases of tetanus occur annually, 80% of them being newborn children (neonatal tetanus).

In the early 1990s a massive epidemic of diphtheria occurred in the former Soviet Union, causing more than 50000 cases and several thousand deaths over 4 years. This outbreak of diphtheria showed in a definitive way that diphtheria vaccination of infants and booster of the adults is still a must for controlling the disease.

B. Diphtheria Toxin

Diphtheria toxin is a 58.35 kDa protein molecule that is released into the supernatant by toxinogenic *C. diphtheriae* strains (PAPPENHEIMER 1977; COLLIER 1982). The toxin is synthesized as a single polypeptide chain of 58.35 kDa that, following mild trypsin treatment and reduction of a disulfide bond, can be divided into two functionally different moieties: fragment A and fragment B of 21.15 and 37.2 kDa, respectively (PAPPENHEIMER 1977; RATTI et al. 1983; GREENFIELD et al. 1983).

The three-dimensional structure of the molecule has shown that the toxin contains three different domains: the C (catalytic) domain, corresponding to fragment A; the T (transmembrane or translocation) domain, composed of nine α-helices (TH1–TH9); and the carboxyterminal R (receptor binding) domain (CHOE et al. 1992; BENNETT et al. 1994).

Fragment A is a NAD^+-binding enzyme that catalyzes the transfer of the ADP-ribosyl group of the NAD molecule, to a posttranslationally modified histidine residue (diphthamide) of elongation factor 2 (EF2), a GTP-binding protein involved in protein synthesis in eukaryotic cells (BROWN and BODLEY 1979; VAN NESS et al. 1980). The EF2-ADP-ribose complex is inactive, and consequently diphtheria toxin causes inhibition of protein synthesis and cell death. It has been shown that in vitro a single molecule of fragment A is enough to kill one eukaryotic cell (YAMAIZUMI et al. 1978). This observation is correlated with the findings of in vivo studies showing that diphtheria toxin is one of the most potent bacterial toxins. In fact, the minimal lethal dose is below 0.1 μg/kg body weight (PAPPENHEIMER 1984).

Although the entire lethal activity of diphtheria toxin resides in fragment A, fragment B is required for receptor binding and for translocation of fragment A across the cell membrane. The COOH-terminal domain involved in receptor binding (R) and the NH_2-terminal hydrophobic domain (T), which is involved in the interaction with the cell membrane, mediate this process. The toxin receptor is the heparin-binding EGF-like growth factor precursor (NAGLICH et al. 1992; BROWN et al. 1993; IWAMOTO et al. 1994; HOOPER and EIDELS 1995), that is present on most mammalian cells but is not found in murine cells.

The diphtheria toxin gene is carried by a family of closely-related bacteriophages (corynebacteriophages), that are able to integrate into the bacterial chromosome and convert nontoxinogenic, nonvirulent *C. diphtheriae* strains into toxinogenic, highly virulent species (FREEMAN 1951; UCHIDA et al. 1971).

The chromosome of *C. diphtheriae* contains two primary attachment sites (attB1 and attB2) (RATTI et al. 1983; RAPPUOLI and RATTI 1984). The presence of the sites on the chromosome allows the stable integration of two copies of corynephage DNA into the bacterial chromosome (RAPPUOLI et al. 1983). Because each phage carries one copy of the *tox* gene, double lysogens produce twice the toxin produced by monolysogens. This property has been very important for obtaining strains hyperproducing CRM 197, a nontoxic form of diphtheria toxin that is used as a carrier for *Haemophilus influenzae* type B (ANDERSON et al. 1987; EGAN et al. 1995; ROTHBROCK et al. 1995) and for meningococcal conjugate vaccines (COSTANTINO et al. 1992). CRM 197 has also been proposed as a new vaccine against diphtheria (RAPPUOLI 1990; PODDA and RAPPUOLI, unpublished data).

C. Tetanus Toxin

Tetanus toxin is a potent neurotoxin that is synthesized intracellularly by *Clostridium tetani* as a single polypeptide chain of 150.5 kDa. After cell lysis the toxin is released into the medium and cleaved by endogenous proteases generating an NH_2-terminal light chain of 52.3 kDa (fragment A) and a COOH-terminal heavy chain of 98.3 kDa (fragment B–C). A disulfide bridge holds together light and heavy chains. The intact, unnicked toxin can be recovered from the washed bacteria by extraction with a neutral, hypertonic solution. Treatment of the toxin with papain yields an NH_2-terminal fragment A-B of 99 kDa and a COOH-terminal fragment C of 51.5 kDa (BIZZINI 1978, 1984; FAIRWEATHER et al. 1986).

Tetanus toxin binds to gangliosides and proteins of peripheral nerve terminals through the COOH-terminal of fragment C; it is then internalized and translocated through retrograde transport to the spinal cord neurones, where it produces a spastic paralysis accompanied by convulsions (BIZZINI 1978).

Purified fragments A and B–C are nontoxic, but if they are reassociated, toxicity is restored.

The toxin can be neutralized by monoclonal antibodies binding the A, B, or C fragments, suggesting that in theory any of the three nontoxic fragments could be used as a vaccine (VOLK et al. 1984). Fragments A, B, and C have been successfully used to immunize mice against tetanus (BIZZINI 1984).

The gene encoding tetanus toxin, contained within a large plasmid of 75 kb, has been cloned and sequenced (EISEL et al. 1986; FAIRWEATHER and LYNESS 1986; NIEMAN et al. 1988). The knowledge of the sequence of this gene led to the cloning of genes coding for the related botulinum toxin produced by A–G serotypes. Despite the lack of high homology within the amino acid sequences of these neurotoxins, a conserved "HEXXH" motif was identified in the light chain. This motif is typically present in the active site of metalloproteases. On the basis of this evidence it was demonstrated that the light chain of tetanus toxin is a zinc-metalloprotease able to cleave with remarkable specificity synaptobrevin, an integral membrane protein of the presynaptic vesicle (SCHIAVO et al. 1992; MONTECUCCO and SCHIAVO 1994), thus interfering with the release of neurotransmitters.

D. Production, Effectiveness, and Problems of Conventional Diphtheria and Tetanus Vaccines

I. Production and Detoxification of Diphtheria and Tetanus Toxoids

To produce a conventional diphtheria vaccine, the hypertoxinogenic *C. diphtheriae* strain PW8 (RAPPUOLI et al. 1983; PARK and WILLIAMS 1986) is grown in fermentors for 36–48 h until the concentration of the toxin in the supernatant reaches 150–250 Lf/ml (1 flocculation unit, or Lf, corresponds to 2.5 μg diphtheria toxin) (BIZZINI 1978). Although several semisynthetic media for the industrial growth of diphtheria have been described so far, many of the manufacturers still grow the PW8 strain in a medium containing enzymatic digests of beef (LINGOOD 1941). This method of growing bacteria is likely to change in the near future due to the need to remove any substance of bovine origin from the production media of biologicals, following the epidemic of bovine spongiform encephalopathy in the United Kingdom in the 1990s. Once the fermentation is complete, the bacteria are removed by centrifugation or filtration, and formaldehyde is added to the supernatant to a final concentration of 0.75%. The supernatant is then stored for 4–6 weeks at 37°C to allow complete detoxification of diphtheria toxin.

Similarly, to produce the conventional tetanus vaccine the highly toxinogenic Harvard strain of *C. tetani* is grown in a semisynthetic medium (usually modified from MUELLER and MILLER 1945) in a fermentor for about 1 week until the bacteria lyse and release tetanus toxin into the supernatant. The average yield obtained under these conditions is approximately 60–80

Lf/ml (BIZZINI 1984). The culture is then filtered, and the filtrate containing the toxin is detoxified by adding formaldehyde to a final concentration of 0.5%. The pH is adjusted to 7.6 and the supernatant is then stored at 37°C for 4 weeks to allow the complete detoxification of the tetanus toxin.

During the detoxification process formaldehyde reacts with the toxin molecules, peptones, and other proteins present in the medium (BLASS 1964; BLASS et al. 1967; BIZZINI and RAYNAUD 1974; SINKOVIC 1975; BIZZINI 1984; PAPPENHEIMER 1984). The formulation proceeds in two steps: the first step involves the reaction, which is very rapid and reversible, mainly with the ε-amino groups of lysine; the second step involves a slower reaction of the unstable product generated in the first step, with another amino acid side chain containing a reactive hydrogen (indole ring of Trp, amide group of Asn and Gln, phenol ring of Tyr, ε-NH_2 group of Lys, imidazole group of His, and guanidine group of Arg) (FRAENKEL-CONRAT and OLCOTT 1948; BIZZINI and RAYNAUD 1974; SINKOVIC 1975).

The final result of the formaldehyde treatment can be the formation of an internal link within the protein molecule, a cross-link between different protein molecules, which then form dimers or oligomers, or a cross-link between the protein molecule and peptides or amino acids present in the medium (RAPPUOLI 1990). When the formaldehyde is added to the crude culture supernatant containing the toxin molecules, this results in the coating of the toxoid by peptides deriving from the beef meat used to prepare the medium. These peptides are unnecessary antigenic determinants of bovine origin which might be responsible for some of the side effects associated with diphtheria and tetanus vaccination, especially in the adult population, which has been repeatedly exposed to bovine antigens.

To overcome this problem and to produce better vaccines, the toxins may be purified before the addition of formaldehyde. This process, which is now used by very few vaccine manufacturers, is likely to become of general use in the near future.

II. Production and Detoxification of Purified Diphtheria and Tetanus Toxoids

The conventional method of preparation used for diphtheria and tetanus vaccines entails a large amount of impurities, some of which are covalently linked to the toxoids and therefore impossible to remove. An easy way to avoid this problem is to treat with formaldehyde the purified proteins instead of the crude culture supernatant. In fact, when purified proteins are formaldehyde-treated, only intra- and intermolecular bridging can occur.

Several methods have been described for the purification of diphtheria and tetanus toxins, and most of them are based on a diafiltration of the culture supernatant, precipitation by ammonium sulfate, and if necessary a further purification step on a gel filtration or on an ion-exchange chromatography (BIZZINI 1984; PAPPENHEIMER 1984). Using these methods, diphtheria and tetanus toxins can be obtained at a purity ranging from 85% to 95%. The partially or highly purified toxin preparations can then be detoxified by the conventional formaldehyde treatment.

The purity of the diphtheria toxoid prepared by detoxification of purified toxin is much higher than that of the conventional diphtheria vaccines. Purity ranges from 85% to 95%, whereas the purity in the conventional vaccines does not exceed 60%.

Early studies on the formulation of purified diphtheria toxin showed that the toxoids can revert to toxicity if diluted, or if the excess of formaldehyde is removed (LINGOOD et al. 1963; STAINER 1968; AKAMA et al. 1971; PAPPENHEIMER 1984). At a later stage it was shown that if 0.025 *M* lysine is used in the reaction, the toxoids obtained were stable and immunogenic. An alternative way to detoxify toxins has been developed by Relyveld and is based on the use of glutaraldehyde (STAINER 1968; RELYVELD 1969; PAPPENHEIMER 1984). Diphtheria and tetanus toxins can be completely and irreversibly inactivated by treatment for a short time with 0.025 *M* or 0.0025 *M* glutaraldehyde, respectively. The presence of lysine is still necessary for the irreversible detoxification of diphtheria toxin, but not for tetanus toxin. The vaccines obtained by glutaraldehyde detoxification of purified tetanus and diphtheria toxins (called POLAN for polymerized antigen) (RELYVELD et al. 1983; RELYVELD and BEN EFRAIM 1983; BIZZINI 1984) have been successfully used for the immunization of large groups of infants and children in France without noticeable adverse reactions.

E. Future Prospects

I. Engineered Live-Attenuated Strains

Infection with pathogens can efficiently stimulate both humoral and cellular immunity. In recent years it has been possible to construct fully defined attenuated micro-organisms and to test them as potential live vaccines. An example of this approach is the use of genetically defined attenuated *Salmonella* strains as oral vaccines. Such strains undergo a limited subclinical infection in mice and provide solid immunity to salmonellosis (CHATFIELD et al. 1992a). These strains have been also used as carriers for heterologous antigens and the fragment C of tetanus toxin has been an excellent model antigen. In fact, FAIRWEATHER et al. (1990) reported that fragment C could be expressed within a *Salmonella typhimurium aro* A strain under the control of the *tac* promoter, and that this strain could be used to immunize mice both orally and intravenously. After two oral immunizations with the recombinant *Salmonella* strain, mice developed high anti-tetanus antibody titers and were 100% protected by challenge with tetanus toxin. This approach was further refined by the use of the *nir*B promoter, which is repressed under aerobic conditions and is activated under anaerobic conditions. With this promoter, a single oral dose of *S. typhimurium aroA aroD* expressing fragment C was sufficient to completely protect mice from tetanus toxin challenge (CHATFIELD et al. 1992b). Fragment C has been also expressed in attenuated strains of *S. typhi* (CHATFIELD et al. 1992a), the human pathogen responsible for typhoid fever, increasing the possibility of obtaining an oral typhoid-tetanus vaccine a step closer to reality.

II. Recombinant Molecules

Ever since the initial cloning and sequencing of the genes encoding diphtheria and tetanus toxins, parts of the two molecules have been expressed in *Escherichia coli* using a wide variety of vectors (LEONG et al. 1983; GREENFIELD et al. 1984; EISEL et al. 1986; FAIRWEATHER et al. 1986, 1987; CABIAUX et al. 1988; NIEMAN et al. 1988). For diphtheria these molecules have been tested for enzymatic activity of fragment A (LEONG et al. 1983), functionality of fragment B (CABIAUX et al. 1988), the possibility of using them as specific tumor-killing agents (MURPHY et al. 1986; BACHA et al. 1988; MURPHY and VANDERSPEK 1995; NEGRO and SKAPER 1995), and their ability to induce protective antitoxic antibodies. So far, the attempts to induce neutralizing antibodies against diphtheria using recombinant molecules have met a poor success.

In marked contrast, for tetanus the possibility of using recombinant molecules as vaccines has been successful. Fragment C of the tetanus toxin has been produced in a soluble form using *E. coli* and *Pichia pastoris* expression systems (FAIRWEATHER et al. 1987; MAKOFF et al. 1989; CLARE et al. 1991). Recombinant fragment C produced in this way retains all the biological properties of native fragment C, including the internal disulfide bridge structure and the ganglioside binding activity. Moreover, it is as effective as native fragment C derived from tetanus toxin in inducing protective antibodies in mice. No tetanus molecules analogous to the diphtheria toxin CRMs have been produced from *C. tetani*, mostly because of the absence of efficient genetic mechanisms to manipulate the bacterium. However, recombinant CRMs have been generated by site-directed mutagenesis of the glutamic acid in position 234 located within the HEXXH motif in the light chain. The Glu 234→Ala substitution renders the light chain nontoxic and, in combination with native heavy chain, a completely nontoxic molecule able to induce protective immunity in mice can be generated (LI et al. 1994). The use of recombinant molecules would solve some of the problems associated with the production of the conventional vaccine as, for instance, the containment necessary to grow *C. tetani*, the long fermentation periods (1 week), the low yield of purified toxin, the low purity of the toxoid, and the handling of large volumes of highly toxic materials.

III. Mucosal Vaccination

It is clear that the development of mucosal (oral or intranasal) tetanus and diphtheria vaccines would be a significant advantage in global immunization against both diseases. Nevertheless, it is known that nonliving antigens are poor mucosal immunogens, unable to induce local and systemic immune response when delivered at mucosal sites. The coadministration of nonliving antigens with cholera toxin (CT) or heat-labile toxin (LT), induce in the host the ability to mount both a mucosal and systemic immune response. Conversely, the toxicity of both LT and CT has precluded their use as adjuvants in humans. Using computer modeling to predict the mutations that would generate nontoxic derivatives of LT (PIZZA et al. 1994), LT mutants able to retain the adjuvant properties of the wild-type molecule have been produced. These mutants have been tested in combination with purified recombinant fragment C of tetanus toxin as a mucosal vaccine in intranasally immunized mice. In this experiment fragment C induced high levels of mucosal and systemic antibodies that were able to protect all mice from challenge with tetanus toxin (DOUCE et al. 1995, 1997). This result shows that mucosal vaccines against tetanus have become more feasible by the use of newly developed mucosal adjuvants.

To improve vaccination in developing countries, where there is the logistic difficulty of delivering the two or three doses necessary to give protection, the development of single dose vaccines against diphtheria and tetanus has been proposed. In order to achieve this goal efforts are being focused on developing controlled-release formulations which can deliver the necessary vaccine in a single dose. Most of the work on controlled-release vaccines has concentrated on the use of poly(lactide-coglycolide) (PLG) polymers (COHEN et al. 1991).

Single-dose vaccines may become a reality in the future since many of the technical problems of the stability of micro-encapsulated antigens are being solved (SCHWENDEMAN et al. 1995).

IV. CRM 197

Following nitrosoguanidine mutagenesis of corynephage βtox+, several phages have been isolated that encode for nontoxic proteins, which are immunologically related to diphtheria toxin (UCHIDA et al. 1971, 1973a–c; LAIRD and GROMAN 1976). These proteins are generally called cross-reacting materials (CRMs).

CRM 197 contains a single glycine to glutamic acid substitution in position 52 that makes the fragment A unable to bind NAD^+ and therefore enzymatically inactive (PAPPENHEIMER et al. 1972; GIANNINI et al. 1984). Being enzymatically inactive and therefore nontoxic but otherwise identical to diphtheria toxin, CRM 197 has been proposed as a natural candidate to develop a new vaccine against diphtheria. However, structural differences between CRM 197 and diphtheria toxin make this mutant more susceptible to proteases and less immunogenic than diphtheria toxoid (PAPPENHEIMER et al. 1972). After stabilization with formalin, CRM 197 becomes more immunogenic and able to induce protective antibody titers against diphtheria (PORRO et al. 1980); however, its potency per microgram of protein is still slightly lower than that of diphtheria toxoid. A formalin-stabilized form of CRM 197 has been successfully used in a clinical trial combined with tetanus toxoid, as a booster in young adults (Podda and Rappuoli, unpublished data).

F. Diphtheria and Tetanus Toxoids as Carriers for Polysaccharide Vaccines

Invasive bacterial diseases caused by pneumococci, meningococci, and *H. influenzae* type b can be prevented by serum antibodies against the capsular polysaccharide of these organisms (MacLeod et al. 1945; Robbins et al. 1984; Robbins and Robbins 1986). These polysaccharides are high molecular mass polymers of repeating oligosaccharide units that induce only a primary antibody response in adults but are not immunogenic at all in infants younger than 2 years of age (MacLeod et al. 1945; Robbins et al. 1984). The reason probably lies in the nature of the major histocompatibility complex, which binds peptides but not sugars and polysaccharide molecules. Consequently, the immunogenicity of the polysaccharide is mostly independent of helper T cells. This problem has been overcome by conjugation of carrier proteins containing T-cell epitopes to the polysaccharides. Conjugates between CRM 197, tetanus, or diphtheria toxoids, and capsular polysaccharides have been developed. They are immunogenic in infants and induce a memory response (MacLeod et al. 1945; Robbins et al. 1984).

Today they are among the most successful vaccines, recently introduced for immunization of all infants.

G. Pertussis

I. The Disease

Pertussis, or whooping cough, is a disease caused by *B. pertussis,* a gram-negative bacterium which adheres to the cilia of the upper respiratory tract of humans, colonizes this tissue and releases a number of virulence factors responsible for the local and systemic damages associated to the disease (Weiss and Hewlett 1986; Rappuoli 1994). The disease is characterized by long-lasting paroxysmal cough, accompanied by whoops, vomiting, cyanosis, and apnea. The most common complications are pneumonia, seizures, encephalopathy, and death. Antibiotic treatment is very effective in clearing the bacteria but has little consequence on the disease, mostly because the disease is usually diagnosed when the bacteria have already released the toxins. *B. pertussis* infects virtually all children who are not immune; however, the disease can occur at any age. Mortality is frequent under 1 year of age and occurs mostly in developing countries, where incidence can be as high as 1/100. In developed countries the frequency is between 1/1000 and 1/10000; of the infants in Western countries who contract the disease in the first year of life 50% are hospitalized and 1% die. After the first year of life only 4% of children with the disease are hospitalized. It has been estimated worldwide that whooping cough causes approximately 300000 deaths each year (MULLER et al. 1986; SUTTER and COCHI 1992).

II. History of Acellular Vaccines

Vaccination is the only way to control pertussis. Mass vaccination using killed bacteria (cellular vaccine) was introduced in the 1950s and reduced by 99% the incidence of diseases in infants.

In spite of its efficacy, the vaccine composed of whole, inactivated *B. pertussis* cells is not widely used, mostly because of the fear of the side effects which have been associated with it. Most of these are mild reactions such as redness, swelling, and fever and occur in 30%–70% of vaccinees. However, the vaccine has also been associated with more severe reactions, such as neurological damages and death, which have been reported with a frequency of 1/100000 and 1/300000, respectively (MILLER et al. 1981). Although never proven to be caused by the vaccine, these reactions have caused a drop in vaccine uptake and stressed the need for a new vaccine, devoid of side effects and possibly composed of well-defined and purified antigens.

Many molecules produced by *B. pertussis* and involved in the virulence of the bacterium, have been identified as candidates for inclusion in a vaccine against whooping cough. These include molecules involved in the adhesion of the bacteria to the eukaryotic cells and to the cilia of the upper respiratory tract, and molecules which cause local and systemic damage of the host. Using a combination of pertussis toxin, filamentous hemagglutinin, and an outer membrane protein that has been subsequently characterized, named pertactin or 69K, partially purified from culture supernatants of *B. pertussis* and detoxified with formaldehyde, Y. and H. Sato developed a new vaccine, which has been used in Japan since 1981 in children over 2 years of age. Following the Japanese example, many laboratories developed acellular vaccines containing pertussis toxin, filamentous hemagglutinin, and other purified *B. pertussis* antigens, such as the fimbriae or agglutinogens, and 69 K (MOXON and RAPPUOLI 1990; RAPPUOLI 1996).

In order to be included in vaccines pertussis toxin needs to be detoxified. Y. Sato used formalin treatment to make the toxin inactive, but this procedure, although effective for diphtheria and tetanus, was not entirely satisfactory for pertussis toxin, which in some cases showed some reversion to toxicity (STORSAETER et al. 1990). To obtain completely detoxified molecules, a number of other chemical methods were used: these included stronger formaldehyde treatments, hydrogen peroxide (SEKURA et al. 1988; KRANTZ et al. 1990), tetranitromethane (WINBERRY et al. 1988), and glutaraldehyde. In our laboratory we have opted to use genetic tools instead of chemical reagents to detoxify pertussis toxin.

III. Pertussis Toxin

Pertussis toxin is a complex bacterial protein toxin of 105 kDa composed of five noncovalently linked subunits termed S1–S5 (TAMURA et al. 1982). It is organized into two functional domains called A and B. The A domain, which is composed of the S1 subunit, is an enzyme of 26.22 kDa with ADP-ribosyltransferase activity. It binds NAD and transfers the ADP-ribose group to a cysteine residue present in a ... XCGLX motif, located at the carboxyterminal region of the α-subunit of many GTP-binding proteins, such as G_1 and G_0, that are involved in signal transduction in eukaryotic cells, altering their response to extracellular stimuli. The B domain is a nontoxic oligomer formed by four distinct subunits, termed S2–S5, of 21.92 kDa, 21.86 kDa, 12.06 kDa and 11.77 kDa, respectively, which are present in a 1:1:2:1 ratio (TAMURA et al. 1982). The B oligomer binds the receptor on the surface of eukaryotic cells and facilitates the translocation of the enzymatically active subunit across the cell membrane so that it can reach the target G proteins.

The crystal structure of the protein has been solved (STEIN et al. 1994). The genes coding for the five subunits of pertussis toxin are clustered in a 3.2-kb fragment of the chromosomal DNA, with the typical organization of a bacterial operon. The PT operon was sequenced by two different research groups in 1986 and was described to contain the genes in the following order: S1, S2, S4, S5, S3 (LOCHT and KEITH 1986; NICOSIA et al. 1986). Each gene contains a sequence coding for a signal peptide. The five subunits are cotranslationally secreted into the bacterial periplasm where the protein is assembled into the holotoxin. Secretion of the toxin occurs only after assembly of the holotoxin, or of the B domain, and requires the product of the nine-genes *ptl* operon, which is located downstream from the PT operon and is regulated by the same promoter (COVACCI and RAPPUOLI 1993; WEISS et al. 1993).

IV. Genetic Detoxification

Following the cloning and sequencing of the 3.5-kb DNA fragment containing the five PT genes (LOCHT and KEITH 1986; NICOSIA et al. 1986), the attempt to express the operon in *E. coli* failed. The individual subunits of the pertussis toxin were successfully expressed in *E. coli* and *Bacillus subtilis*; however, these recombinant molecules were unable to induce protective immunity because protection was found to be mediated by conformational epitopes which are present in the entire, assembled pertussis toxin but absent in the individual recombinant subunits (BARTOLONI et al. 1988). The failure to induce a good protective immunity with recombinant subunits expressed in *E. coli* and *B. subtilis* suggested that the ideal vaccine should be a PT molecule whose toxicity had been eliminated by the genetic manipulation of the gene coding for the S1 subunit. Therefore, to obtain a molecule suitable for vaccine use we decided to modify the PT genes in the chromosome of *B. pertussis* in order to obtain a molecule fully assembled but devoid of toxicity.

For this purpose, we and other investigators generated a number of recombinant Sl molecules containing amino acid substitutions and tested their enzymatic activity (PIZZA et al. 1988; BARBIERI and CORTINA 1988; CIEPLAK et al. 1988; LOCHT et al. 1989).

The mutations which render the S1 subunit nontoxic were then introduced into the chromosome of *B. pertussis* using homologous recombination (PIZZA et al. 1989). These new *B. pertussis* strains were found to produce molecules indistinguishable from PT in sodium dodecyl sulfate–polyacrylamide gel electrophoresis, but that had a toxicity ranging from 0.1% to 10% of wild-type PT. Since even 0.1% of the toxicity is by far too high for a molecule to be used in a vaccine, we combined some of the above mutations and obtained several PT double mutants. One of these, the PT 9K/129G mutant, containing mutations of arginine in position 9 to lysine and glutamic acid in position 129 to Gly, resulted to have lost all toxic properties of PT, and maintained all the antigenic properties.

Once the safety and antigenic properties of the PT 9K/129G mutant were demonstrated, the molecule was produced in large scale and formulated in an anti-diphtheria, tetanus, and pertussis (DTaP) vaccine. The anti-pertussis component was either made by the PT 9K/129G alone (10 μg), or by the PT 9K/129G in combination with filamentous hemagglutinin (FHA) and 69 K (5, 2.5, and 2.5 μg each, respectively). The vaccines were tested for safety and immunogenicity in phase I and phase II studies (PODDA et al. 1990, 1991, 1992, 1993, 1994) with optimal results before being used for larger comparative international studies.

V. Acellular Vaccines Proposed

During the period 1980–1989 many acellular vaccines were developed by vaccine manufacturers. All of them contained either detoxified pertussis toxin alone or combined with FHA, with pertactin or 69 K, and with the agglutinogens. Some of the acellular vaccines developed are reported in Table 1. As shown, the pertussis toxin included in the vaccines has been detoxified with a variety of chemical methods, including formaldehyde, glutaraldehyde, tetranitromethane and hydrogen peroxide. The Chiron vaccines were the only ones containing genetically detoxified pertussis toxin.

VI. Clinical Trials

Following the Japanese experience, an efficacy study was organized in 1986 in Sweden to test two vaccines: one containing formalin-treated PT, and the other containing formalin-treated PT and FHA. Although infants were immunized only with two doses, both vaccines were able to protect infants from disease (54% and 69%, respectively) (AD HOC GROUP 1988). At the end of this

Table 1. List of acellular vaccines and their composition

Manufacturer	Vaccine composition	Antigen quantity (μg)	PT detoxification
Chiron (C-1)	PT	10	Genetic
North American Vaccine (NAV)	PT	40	Hydrogen peroxide
Swiss Serum & Vaccine Institute (SSVI-1)	PT	50	Tetranitromethane
Connaught (US)/Biken (CB-2)	PT, FHA	23.4, 23.4	Formaldehyde
Michigan Dept Public Health (Mich-2)	PT, FHA	25.0, 25.0	Formaldehyde
Pasteur Merieux (PM-2)	PT, FHA	25.0, 25.0	Glutaraldehyde
SmithKline Beecham Biologicals (SKB-2)	PT, FHA	25.0, 25.0	Formaldehyde + glutaraldehyde
Chiron (C-3)	PT, FHA, 69K	5.0, 2.5, 2.5	Genetic
Lederle-Praxis Biologicals (LPB-3)	PT, FHA, 69K	10.0, 20.0, 5.0	Formaldehyde
SmithKline Beecham Biologicals (SKB-3)	PT, FHA, 69K	25.0, 25.0, 8.0	Formaldehyde + glutaraldehyde
Connaught Laboratories (Canada) (CLL-3)	PT, FHA, FIM2, FIM3	10.0, 5.0, −5.0[a]	Glutaraldehyde
Porton International (Por-3)	PT, FHA, FIM2, FIM3	10.0, 10.0, −10.0[a]	Formaldehyde
Lederle Praxis/Takeda (LPT-4)	PT, FHA, 69K, FIM2	3.5, 35.0, 2.0, 0.8	Formaldehyde
Connaught Labs. (Canada) (CLL-4)	PT, FHA, 69K, FIM2, FIM3	20.0, 20.0, 3.0, 5.0[a,b]	Glutaraldehyde

[a] FIM2 + FIM3.
[b] The composition of the vaccine used in the Sweden I trial was 10.0, 5.0, 3.0, 5.0.

trial acellular vaccines were not licensed because the observed efficacy was not considered to be high enough, and the absence in the study of an arm immunized with the whole cell vaccine did not allow to compare the efficacy of acellular vaccines with that of the vaccine already in use. In addition, during the study the formalin detoxified pertussis toxin showed some reversion to toxicity and this suggested that further development of acellular vaccines was still necessary (Storsaeter et al. 1990).

In 1990 the National Institute of Allergy and Infectious Diseases, performed a large scale phase II trial in the United States to compare the safety and the immunogenicity of most of the acellular vaccines available at that time in order to select the vaccines to be subsequently used in new efficacy studies. Thirteen acellular vaccines (Table 1) were tested and compared with two whole cell vaccines. In each arm 120 infants were immunized with three doses of each vaccine. Eleven of the acellular vaccines contained chemically detoxified PT, and two vaccines contained genetically detoxified PT. The results showed that all acellular vaccines were much safer (Decker et al. 1995) and more immunogenic than the whole cell vaccines (Edwards et al. 1995).

A comparison of the immunogenicity of the pertussis toxin used in the vaccines in the study is reported in Table 2. This trial demonstrated unequivocally that the genetically detoxified pertussis toxin induced anti-PT levels that in enzyme-linked immunosorbent assay (ELISA) and toxin neutralization were 5- to 20-fold higher than those induced by chemically detoxified forms of PT present in the other vaccines.

Table 2. Immunogenicity of genetically (Chiron) versus chemically (all others) inactivated PT in phase II and phase III trials (from Edwards et al. 1995; Greco et al. 1996; Gustafsson et al. 1996). See Table 1 for vaccine identification

Vaccine	Immunogenicity PT (phase II NIH)				Immunogenicity (phase III efficacy studies)			
	Absolute value		Units/μg protein		Absolute value		Units/μg protein	
	ELISA	CHO	ELISA	CHO	ELISA	CHO	ELISA	CHO
C-1	180	1035	18.0	103.5	–	–	–	–
C-3	99	487	19.8	97.4	94.4	787	18.8	157
CB-2	127	841	5.4	36.0	–	–	–	–
SKB-2	104	530	4.1	21.2	59.9	ND	2.3	ND
SKB-3	54	205	2.1	8.2	51.3	230	2.0	9.2
SSVI-1	99	259	1.9	5.1	–	–	–	–
PM-2	68	432	2.7	17.2	–	–	–	–
Mich-2	66	327	2.6	13.0	–	–	–	–
LPB-3	39	163	3.9	16.3	–	–	–	–
CLL-3	38	158	3.8	15.8	–	–	–	–
CLL-4	36	142	3.6	14.2	49.4	ND	4.9	ND
Por-3	29	118	2.9	11.8	–	–	–	–
LPT-4	14	116	4.0	33.1	–	–	–	–

ND = not done

After the comparative phase II study described above four vaccines were selected to be tested in efficacy trials. Two of them, the Connaught five-component vaccine and the SmithKline Beecham Biologicals two-component vaccine, were tested in an efficacy trial performed in Sweden (Sweden I in Table 3), while the Chiron vaccine was tested in Italy, in parallel with the three-component vaccine produced by SmithKline Beecham Biologicals (Italy in Table 3). This was a unique opportunity to test two vaccines containing exactly the same components (PT, FHA, and 69K) but differing in the method used to detoxify PT and in the amount of antigens present (25, 25, and 8 μg in the SmithKline Beecham Biologicals vaccine, and 5, 2.5, and 2.5 μg in the Chiron vaccine, respectively). The results of the Italian and Swedish trials clearly showed that all acellular vaccines have a greatly superior safety to that of the whole cell vaccines (e.g., fever occurred in 40% of infants vaccinated with whole cell vaccines and in 5%–8% of infants vaccinated with acellular vaccines). All acellular vaccines showed efficacy; however, the most efficacious vaccines were the three-component vaccine of Chiron, containing the genetically inactivated PT (84.2%), the SmithKline Beecham Biologicals vaccine containing the formalin detoxified PT (83.9%), and the five-component of Connaught Laboratories, also containing formalin-inactivated PT (85.2%). The SmithKline Beecham Biologicals vaccine containing the PT and FHA alone showed a surprisingly low efficacy (58.9%). The whole cell vaccine performed very poorly (only 36.1%–48.3% protection) (GRECO et al. 1996; GUSTAFSSON et al. 1996). As previously observed in phase II studies, the genetically detoxified PT was the most immunogenic in these studies, inducing a superior ELISA and toxin neutralizing antibody titers (Table 2; GRECO et al. 1996). This superior immunogenicity may be responsible for other important features observed in the study: the vaccine containing genetically detoxified PT was able to protect starting from the first vaccination dose, and showed a longer lasting protective immunity (SALMASO et al. 1998). In addition, it showed a lower reactogenicity.

A subsequent study was then organized in Sweden (Sweden II, Table 3) to compare the relative efficacy of the two-component acellular vaccine from SmithKline Beecham Biologicals, the recombinant three-component from Chiron, the five-component from Connaught, and a whole-cell vaccine produced by Wellcome. The study was not placebo-controlled, and therefore no absolute efficacy could be calculated. Given the large population samples involved (nearly 90000 children), the study had the statistical power to discriminate even very small differences in efficacy. The results confirmed that the two-component SmithKline Beecham Biologicals vaccine was poorly efficacious, and therefore the arm containing this vaccine was discontinued during the trial. The remaining vaccines (Chiron's recombinant three-component, Connaught's five-component, and Wellcome's whole-cell) were all very efficacious in protecting from disease. The small but statistically significant differences observed indicated that among the very efficacious vaccines the efficacy was superior in the arm vaccinted with the whole-cell, five- and three-component vaccines, respectively.

Table 3. Efficacy trials sponsored by the National Institutes of Health (United States) performed in Sweden and Italy, and other clinical studies of pertussis vaccines

Trial	Design	Schedule	Vaccines used	Efficacy
NIH studies				
Sweden I	Prospective, cohort, randomized, double-blind, placebo-controlled	3 doses: 2, 4, 6 months	DT (SBL)	Control
			DTa2P (SKB-2)	58.9%
			DTa5P (CLL-4)	85.2%
			DTwP (CLI)	48.3%
Sweden II	Prospective, cohort, randomized, double-blind	3 doses: 3, 5, 12 months; 2–4, 6 months	DTa3P (C-3)	0.31[a]
			DTa5P (CLL-4)	0.25[a]
			DTa2P (SKB-2)	1.0[a]
			DTwP (Well)	0.13[a]
Italy	Prospective, cohort, randomized, double-blind, placebo-controlled	3 doses: 2, 4, 6 months	DT (C)	control
			DTa3P (C-3)	84.2%
			DTa3P (SKB-3)	83.9%
			DTwP (CLI)	36.1%
Other studies				
Gothenburg, Sweden	Prospective, double-blind, placebo-controlled	3 doses: 3, 5, 12 months	DT (SS)	Control
			DTa1P (NAV)	71%
Erlangen, Germany	Prospective, cohort, randomized, double-blind, household contact study	4 doses: 2, 4, 6 months with booster at 18 months	DT	Control
			DTa4P (LPT-4)	81.5%
			DTwP (L)	90.8%
Senegal	Prospective, cohort, randomized, double-blind	3 doses: 2, 4, 6 months	DTa2P (PM-2)	86%
			DTwP (PM)	96%
Mainz, Germany	Prospective, blinded, household contact study	4 doses: 3, 4, 5 months with booster before 25 months	DT (BW)	Control
			DTa3P (SKB-3)	88.7%
			DTwP (BW)	97%
Munich, Germany	Case-control	9, 16, 24 months with booster at 15–25 months	DT	Control
			DTa2P (CB-2)	82%
			DTwP (BW)	96%

[a] Relative risk after third dose using the 3, 5, and 12 month schedule.
Abbreviations: SBL, Swedish Bactercology Laboratories; CLI, Connaught Laboratories Inc.; Well, Wellcome; SS, Statens Seruminstitut; L, Lederle; BW, Behringwerke.

The overall conclusion from these trials confirm that while three-component vaccines are excellent, the addition of other antigens may increase slightly the protective effect, and that the best protection can still be achieved with the whole-cell vaccine. However, the small differences found between the efficacious vaccines tested in the Sweden II trial are only a statistical curiosity, and they have no relevance for the selection of an efficacious vaccine to be used. Other criteria such as safety, duration of protection, etc., should be also considered. The design, schedule of immunizations, and results of the efficacy trials performed in Italy and Sweden are summarized in Table 3.

VII. Other Clinical Studies

In addition to the above trials which were sponsored and performed under the supervision of the National Institutes of Health (United States), using consistent criteria of disease definition and protocol, several other clinical studies have been performed using different standards, different study designs, and often different schedules of immunization. The additional studies are also reported in Table 3. These trials provide some information about the vaccines used, but the values obtained cannot be compared to those obtained in the Italian and Swedish trials reported in Table 3. For example, one obvious discrepancy between the two trials is the different efficacy between the two-component vaccine of SmithKline Beecham Biologicals in Sweden (58.9%), and the two-component vaccine of Pasteur Merieux in Senegal (86%, Table 3). Given the difference in trial design, it is not clear whether the manufacturing of the Pasteur Merieux vaccine produces a two-component vaccine superior to that manufactured by SmithKline Beecham Biologicals, or whether the absolute efficacy reported for the trial in Senegal has been overestimated.

The development of an acellular vaccine containing genetically detoxified PT represents a new milestone in vaccine development and proves several new concepts that should be taken into account for future vaccine development and for selection of vaccine usage: (a) The native conformation of the genetically detoxified pertussis toxin results in a superior immunogenicity. In addition to the higher antibody titers achieved with a very low dose of PT, in the phase III trial in Italy, the vaccine containing the genetically detoxified PT was able to confer protective immunity starting after the first vaccine dose, allowing the protection of infants in the first few months of life, when the disease is most dangerous. (b) The native conformation of the genetically pertussis toxin induced a better immunological priming against the natural molecule that resulted in a longer lasting protective immunity. (c) The superior immunogenicity of genetically detoxified PT allowed to use a lower antigen dose in the vaccine, which resulted in a lower frequency of common side effects such as fever, redness and swelling. (d) Only genetic detoxification can guarantee the absolute absence of active pertussis toxin, a molecule present in whole-cell vaccines. The absence, even in minimal amounts, of active pertussis toxin is crucial in novel vaccines, because this toxin has been shown to cause anaphylaxis and permanent modification of the nerve-mediated permeability of the intestine (MUNOZ et al. 1987; KOSECKA et al. 1994). Chemical detoxification is less reliable, and active pertussis toxin has been reported in several acellular vaccine preparations (MILLER et al. 1995), while on another occasion, reversion to toxicity has been reported (STORSAETER et al. 1990).

References

Ad hoc group for the study of pertussis vaccine (1988) Placebo-controlled trial of two acellular pertussis vaccines in Sweden. Protective efficacy and adverse events. Lancet i:955–960

Akama K, Kameyama S, Otani S, Sadahiro S, Murata R (1971) Reversion of toxicity of diphtheria toxoid. Jpn J Med Sci Biol 24:183–187

Anderson P, Pichichero M, Edwards K, Porch CR, Insel R (1987) Priming and induction of *Haemophilus influenzae* type b capsular antibodies in early infancy by Dpo20, an oligosaccharide-protein conjugate vaccine. J Pediatr 111:644–650

Bacha P, Williams DP, Waters C, Williams JM, Murphy JR (1988) Strom TB. Interleukin 2 receptor-targeted cytotoxicity. Interleukin 2 receptor-mediated action of a diphtheria toxin-related interleukin 2 fusion protein. J Exp Med 167:612–622

Barbieri JT, Cortina G (1988) ADP-ribosyltransferase mutations in the catalytic S-subunit of pertussis toxin. Infect Immun 56:1934–1941

Bartoloni A, Pizza M, Bigio M, Nucci D, Ashworth LA, Irons LI, Robinson A, Burns D, Manclark C, Sato H, Rappuoli R (1988) Mapping of a protective epitope of pertussis toxin by in vitro refolding of recombinant fragments. Biotechnology 6:709–712

Bennett MJ, Choe S, Eisenberg D (1994) Refined structure of dimeric diphtheria toxin at 2.0 angstrom resolution. Protein Sci 3(9):1444–1463

Bizzini B (1978) Tetanus toxin structure as a basis for elucidating its immunological and neuropharmacological activities. In: Cuatrecasas P (ed) The specificity and action of animal, bacterial and plant toxins. Receptors and recognition, series B, vol 1. Chapman and Hall, London, p 177

Bizzini B (1984) Tetanus. In: Germanier R (ed) Bacterial vaccines. Academic, New York, pp 37–68

Bizzini B, Raynaud M (1974) La detoxification des toxines protéiques par le formol: mécanismes supposés et nouveaux développements. Biochemie 56:297–303

Blass J (1964) Etat actuel de nos connaissances sur le mécanisme de la detoxification par le formol. Biol Med 1964 53:202–234

Blass J, Bizzini B, Raynaud M (1967) Etudes sur le mécanisme de la detoxification des toxines protéiques par le formol. Bull Soc Chim Fr 10:3957–3964

Brown BA, Bodley JW (1979) Primary structure at the site in beef and wheat elongation factor 2 of ADP-ribosylation by diphtheria toxin. FEBS Lett 103:253–255

Brown JG, Almond BD, Naglich JG, Eidels L (1993) Hypersensitivity to diphtheria toxin by mouse cells expressing both diphtheria toxin receptor and CD9 antigen. Proc Natl Acad Sci USA 90:8184–8188

Cabiaux V, Phalipon A, Wattiez R, Falmagne P, Ruysschaert JM, Kaczorek M (1988) Expression of a biologically active diphtheria toxin fragment B in *Escherichia coli.* Mol Microbiol 2:339–346

Chatfield SN, Fairweather NF, Charles I, Pickard D, Levine M, Hone D, Posada M, Strugnell RA, Dougan G (1992a) Construction of a genetically defined *Salmonella typhi* Ty2 aroA, aroC mutant for the engineering of a candidate oral typhoid-tetanus vaccine. Vaccine 10(1):53–60

Chatfield SN, Charles IG, Makoff AJ, Oxer MD, Dougan G, Pickard D, Slater D, Fairweather NF (1992b) Use of NirB promoter to direct the stable expression of heterologous antigens in Salmonella oral vaccine strains: development of a single dose oral tetanus vaccine. Biotechnology 10(8):888–892

Choe S, Bennett MJ, Fujii G, Curmi PM, Kantardjieff KA, Collier RJ, Eisenberg D (1992) The crystal structure of diphtheria toxin. Nature 357:216–222

Cieplak W, Burnette WN, Mar VL, Kaljot KT, Morris CF, Chen KK, Sato H, Keith JM (1988) Identification of a region in the S1 subunit of pertussis toxin that is required for enzymatic activity and that contributes to the formation of a neutralizing antigenic determinant. Proc Natl Acad Sci USA 85:4667–4671

Clare JJ, Rayment FB, Ballantine SP, Srekrshna K, Romanos MA (1991) High level expression of tetanus toxin fragment C in *Pichia pastoris* strains containing multiple tandem integrations of the gene. Biotechnology 9(5):455–460

Cohen S, Yoshioka T, Lucareli M, Hwang LH, Langer R (1991) Controlled delivery systems for proteins based on poly(lactic/glycolic acid) microspheres. Pharm Res 8(6):713–720

Collier RJ (1982) Structure and activity of diphtheria toxin. In: Hayashi D, Ueda K (eds) ADP-ribosylation reactions. Academic, New York, pp 575–592

Costantino P, Viti S, Podda A, Velmonte MA, Nencioni L, Rappuoli R (1992) Development and phase I clinical testing of a conjugate vaccine against meningococcus A and C. Vaccine 10:691–698

Covacci A, Rappuoli R (1993) Pertussis toxin export requires accessory genes located downstream from the pertussis toxin operon. Mol Microbiol 8:429–434

Decker MD, Edwards KM, Steinhoff MC, Rennels MB, Pichichero ME, Englund JA, Anderson EL, Deloria MA, Reed GF (1995) Comparison of 13 acellular pertussis vaccines: adverse reactions. Pediatrics 96(3):557–566

Douce G, Turcotte C, Cropley I, Roberts M, Pizza M, Domenghini M, Rappuoli R, Dougan G (1995) Mutants of *Escherichia coli* heat-labile toxin lacking ADP-ribosyltransferase activity act as nontoxic, mucosal adjuvants. Proc Natl Acad Sci USA 92(5):1644–1648

Douce G, Fontana MR, Pizza M, Rappuoli R, Dougan G (1997) Intranasal immunogenicity and adjuvanticity of site-directed mutant derivatives of cholera toxin. Infect Immun 65(7):2821–2828

Edwards KE, Meade BD, Decker MD, Reed GF, Rennels MB, Steinhoff MC, Anderson EL, Englund JA, Pichichero ME, Deloria MA, Deforest A (1995) Comparison of 13 acellular pertussis vaccines: overview and serologic response. Pediatrics 96(3):548–557

Egan W, Frasch CE, Anthony BF (1995) Lot-release criteria, postlicensure quality control, and the *Haemophilus influenzae* type b conjugate vaccines. JAMA 273(11):888–889

Eisel U, Jarausch W, Goretzki K et al (1986) Tetanus toxin: primary structure, expression in *E. coli*, and homology with botulinum toxins. EMBO J 5:2495–2502

Fairweather NF, Lyness VA (1986) The complete nucleotide sequence of tetanus toxin. Nucleic Acids Res 14(19):7809–7812

Fairweather NF, Lyness VA, Pickard DJ, Allen G, Thomson OR (1986) Cloning, nucleotide sequencing and expression of tetanus toxin fragment C in *Escherichia coli*. J Bacteriol 165(1):21–27

Fairweather NF, Lyness VA, Maskell DJ (1987) Immunization of mice against tetanus with fragments of tetanus toxin synthesized in *Escherichia coli*. Infect Immun 55(11):2541–2545

Fairweather NF, Chatfield SN, Makoff AJ, Strugnell RA, Bester J, Maskell DJ, Dougan G (1990) Oral vaccination of mice against tetanus by use of a live attenuated Salmonella carrier. Infect Immun 58(5):1323–1326

Fraenkel-Conrat H, Olcott HS (1948) Reaction of formaldehyde with proteins. J Biol Chem 174:827–848

Freeman VJ (1951) Studies on the virulence of bacteriophage-infected strains of *Corynebacterium diphtheriae*. J Bacteriol 61:675–688
Giannini G, Rappuoli R, Ratti G (1984) The amino acid sequence of two nontoxic mutants of diphtheria toxin: CRM 45 and CRM 197. Nucleic Acids Res 12:4063–4069
Glenny AT, Hopkins BE (1923) Diphtheria toxoid as an immunizing agent. Br J Exp Pathol 4:283–288
Greco D, Salmaso S, Mastrantonio P, Giuliano M, Tozzi AE, Ciofi ML, Giammanco A, Panei P, Blackwelder WC, Klein DL, Wassilack SGF, The Progetto Pertosse Working Group (1996) A controlled trial of two acellular vaccines and one whole-cell vaccine against pertussis. N Engl J Med 334(6):341–348
Greenfield L, Bjorn MJ, Horn G, Font D, Buck GA, Collier RJ, Kaplan DA (1983) Nucleotide sequence of the structural gene for diphtheria toxin carried by corynebacteriophage data. Proc Natl Acad Sci USA 80:6853–6857
Greenfield L, Dovey HF, Lawyer FC, Gelfand DH (1984) High-level expression of diphtheria toxin peptides in *Escherichia coli*. Biotechnology 4:1006–1011
Gustafsson L, Hallander HO, Olin P, Reizenstein E, Storsaeter J (1996) Marked differences in efficacy, immunogenicity and reactogenicity between a 5-component acellular, a 2-component acellular, and a US licenced whole-cell pertussis vaccine. N Engl J Med 334(6):349–355
Hooper KP, Eidels L (1995) Localization of a critical diphtheria toxin-binding domain to the C-terminus of the mature heparin-binding EGF-like growth factor region of the diphtheria toxin receptor. Biochem Biophys Res Commun 206(2):710–717
Iwamoto R, Higashiyama S, Mitamura T, Taniguchi N, Klagsbrun M, Mekada E (1994) Heparin-binding EGF-like growth factor, which acts as the diphtheria toxin receptor, forms a complex with membrane protein DRAP27/CD9, which up-regulates functional receptors and diphtheria toxin sensitivity. EMBO J 13(10):2322–2330
Kosecka U, Marshall JS, Crowe SE, Bienenstock J, Perdue MH (1994) Pertussis toxin stimulates hypersensitivity and enhances nerve-mediated antigen uptake in rat intestine. Am J Physiol Gastrointest L 30:G745–G753
Krantz I, Sekura R, Trollfors B, Taranger J, Zackrisson G, Lagergard T, Schneerson R, Robbins J (1990) Immunogenicity and safety of a pertussis vaccine composed of pertussis toxin inactivated by hydrogen peroxide, in 18 to 23 month old children. J Pediatr 116:539–543
Laird W, Groman N (1976) Isolation and characterization of tox mutants of corynebacteriophage beta. J Virol 19(1):220–227
Leong D, Coleman KD, Murphy JR (1983) Cloned fragment A of diphtheria toxin is expressed and secreted into the periplasmic space of *Escherichia coli* K 1 2. Science 220:515–517
Li Y, Foran P, Fairweather NF, Depaiva A, Weller U, Dougan G, Dolly JO (1994) A single mutation in the recombinant light chain of tetanus toxin abolishes its proteolytic activity and removes the toxicity seen after reconstitution with native heavy chain. Biochemistry 33(22):7014–7020
Lingood FV (1941) Purification of diphtheria toxin and toxoid made from tryptic digest broths. Br J Exp Pathol 22:255–261
Lingood FV, Stevens MF, Fulthorpe AJ, Woiwod AJ, Pope CG (1963) The toxoiding of purified diphtheria toxin. Br J Exp Pathol 44:177–188
Locht C, Keith JM (1986) Pertussis toxin gene: nucleotide sequence and genetic organization. Science 232:1258–1264
Locht C, Capiau C, Feron C (1989) Identification of amino acid residues essential for the enzymatic activities of pertussis toxin. Proc Natl Acad Sci USA 86:3075–3079
MacLeod CM, Hodges RG, Heidelberger M, Bernhard WG (1945) Prevention of pneumococcal pneumonia by immunization with specific capsular polysaccharides. J Exp Med 83:445–465
Makoff AJ, Ballantine SP, Smallwood AE, Fairweather NF (1989) Expression of tetanus toxin fragment C in *E. coli*: its purification and potential use a s a vaccine. Biotechnology 7:1043–1046

Miller DL, Ross EM, Alderslade R, Bellman MH, Rawson NSB (1981) Pertussis immunization and serious acute neurological illness in children. Br Med J 282:1595–1599
Miller E, Waight P, Ashworth E, Thornton C, Redhead K (1995) Summary of clinical and laboratory data for acellular and whole cell pertussis vaccines from UK studies. In: Salisbury M (ed) Compatibility of acellular pertussis with other vaccines in UK primary immunization and boosting schedules. PHLS, CAMR and NIBSC, London, pp 1–16
Montecucco C, Schiavo G (1994) Mechanism of action of tetanus and botulinum neurotoxins. Mol Microbiol 13(1):1–8
Moxon ER, Rappuoli R (1990) *Haemophilus influenzae* infections and whooping cough. Lancet 335:1324–1329
Mueller JH, Miller PA (1945) Production of tetanal toxin. J Immunol 50 (6):377–384
Muller AS, Leeuwenburg J, Pratt DS (1986) Pertussis: epidemiology and control. Bull WHO 64:321–331
Munoz JJ Peacock MG, Hadlow WJ (1987) Anaphylaxis or so-called encephalopathy in mice sensitized to an antigen with the aid of pertussigen (pertussis toxin). Infect Immun 55:1004–1008
Murphy JR, Vanderspek JC (1995) Targeting diphtheria toxin to growth factor receptors. Semin Cancer Biol 6(5):259–267
Murphy JR, Bishai W, Borowski M, Miyanohara A, Boyd J, Nagle S (1986) Genetic construction, expression, and melanoma-selective cytotoxicity of a diphtheria toxin related a-melanocyte stimulating hormone fusion protein. Proc Natl Acad Sci USA 83(21):8258–8262
Naglich JG, Metherall JE, Russell DW, Eidels L (1992) Expression cloning of a diphtheria toxin receptor: identity with a heparin-binding EGF-like growth factor precursor. Cell 69:1051–1061
Negro A, Skaper SD (1995) Production, characterization and cytotoxic properties of a diphtheria toxin ciliary neurotrophic factor fusion protein. Protein Eng 8(2):175–183
Nicosia A, Perugini M, Franzini C, Casagli MC, Borri MG, Antoni G, Almoni M, Neri P, Ratti G, Rappuoli R (1986) Cloning and sequencing of the pertussis toxin genes: operon structure and gene duplication. Proc Natl Acad Sci USA 83:4631–4635
Nieman H, Beckh BA, Binz T, Demotz S, Mayer T, Widman C (1988) Tetanus toxin: evaluation of the primary sequence and potential applications. In: Fehrenbach FJ, Alouf JE, Falmagne P, Goebel W, Jeljaszewicz, Jürgens D, Rappuoli R (eds) Bacterial protein toxins. Fischer, Stuttgart, p 29
Pappenheimer AM Jr (1977) Diphtheria toxin. Annu Rev Biochem 46:69–93
Pappenheimer AM Jr (1984) Diphtheria. In: Germanier R (ed) Bacterial vaccines. Academic, New York, pp 1–16
Pappenheimer AM Jr, Uchida T, Harper AA (1972) An immunological study of the diphtheria toxin molecule. Immunochemistry 9:891–906
Park WH, Williams AW (1986) The production of diphtheria toxin. J Exp Med 11:164–185
Pizza M, Bartoloni A, Prugnola A, Silvestri S, Rappuoli R (1988) Subunit S1 of pertussis toxin: mapping of the regions essential for ADP-ribosyltransferase activity. Proc Natl Acad Sci USA 85:7521–7525
Pizza M, Covacci A, Bartoloni A, Perugini M, Nencioni L, De Magistris MT, Villa L, Nucci D, Manetti R, Bugnoli M, Giovannoni F, Olivieri R, Barbieri JT, Sato H, Rappuoli R (1989) Mutants of pertussis toxin suitable for vaccine development. Science 246:497–500
Pizza MG, Domenighini M, Hol W, Giannelli V, Fontana MR, Giuliani M, Magagnoli C, Peppoloni S, Manetti R, Rappuoli R (1994) Probing the strucutre-activity relationships of *E. coli* LT-A by site-directed mutagenesis. Mol Microbiol 14(1):51–60
Podda A, Nencioni L, De Magistris MT, Di Tommaso A, Bossù P, Nuti S, Pileri P, Peppoloni S, Bugnoli M, Ruggiero P, Marsili I, D'Errico A, Tagliabue A,

Rappuoli R (1990) Metabolic, humoral and cellular responses in adult volunteers immunized with the genetically inactivated pertussis toxin mutant PT-9 K/129G. J Exp Med 172:861–868
Podda A, Nencioni L, Marsili I, Peppoloni S, Volpini G, Donati D, Di Tommaso A, De Magistris MT, Rappuoli R (1991) Phase I clinical trial of an acellular pertussis vaccine composed of genetically detoxified pertussis toxin combined with FHA and 69 kDa. Vaccine 9:741–745
Podda A, De Luca EC, Titone L, Casadei AM, Cascio A, Peppoloni S, Volpini G, Marsili I, Nencioni L, Rappuoli R (1992) Acellular pertussis vaccine composed of genetically inactivated pertussis toxin: safety and immunogenicity in 12- to 24- and 2- to 4-month-old children. J Pediatr 120:680–685
Podda A, De Luca EC, Titone L, Casadei AM, Cascio A, Bartalini M, Volpini G, Peppoloni S, Marsili I, Nencioni L, Rappuoli R (1993) Immunogenicity of an acellular pertussis vaccine composed of genetically inactivated pertussis toxin combined with filamentous hemagglutinin and pertactin in infants and children. J Pediatr 123:81–84
Podda A, De Luca EC, Contu B, Furlan R, Maida A, Moiraghi A, Stramare D, Titone L, Uxa F, Di Pisa F, Peppoloni S, Nencioni L, Rappuoli R, Bartalini M, Bona G, Budroni M, Pistilli AMC, Cascio A, Cascio G, Cossu M, Dall'Orto P, Dileo G, Furlan A, Macagno F, Marsili I, Meloni T, Regoli D, Rigo G, Trappan A, Vargiu G, Volpini G (1994) Comparative study of a whole-cell pertussis vaccine and a recombinant acellular pertussis vaccine. J Pediatr 124:921–926
Porro M, Saletti M, Nencioni L, Tagliaferri L, Marsili I (1980) Immunogenic correlation between cross-reacting material (CRM 197) produced by a mutant of *Corynebacterium diphtheriae* and diphtheria toxoid. J Infect Dis 142:716–724
Ramon G (1924) Sur la toxine et surranatoxine diphtheriques. Ann Inst Pasteur 38: 1–10
Rappuoli R (1990) New and improved vaccines against diphtheria and tetanus. In: Woodrow GC, Levine MM (eds) New generation vaccines. Dekker, New York, pp 251–267
Rappuoli R (1994) Pathogenicity mechanisms of Bordetella. In: Dangl JL (ed) Bacterial pathogenesis of plants and animals. Springer, Berlin Heidelberg New York, pp 319–336
Rappuoli R (1996) Acellular pertussis vaccines: a turning point in infant and adolescent vaccination. Infect Agents Dis 5(1):21–28
Rappuoli R, Ratti G (1984) Physical map of the chromosomal region of *Corynebacterium diphtheriae* containing corynephage attachment sites attB1 and attB2. J Bacteriol 158(1):325–330
Rappuoli R, Michel JL, Murphy JR (1983) Restriction endonuclease map of corynebacteriophage ω ctox+ isolated from the Park Williams no 8 strain of *Corynebacterium diphtheriae*. J Virol 45(2):524–530
Ratti G, Rappuoli R, Giannini G (1983) Complete nucleotide sequence of the gene coding for diphtheria toxin in the corynephage omegatox+ genome. Nucleic Acids Res 11:6589–6595
Ray K, Dutta KK, Mukherjee AK, Anand BR, Rai Chowdhuri AN (1983) Hospital based surveillance of faucial diphtheria in and around Delhi. Indian J Med Res 78:776–779
Relyveld EH (1969) Studies on the detoxification of purified diphtheria toxin. Prog Immunobiol Stand 3:258–268
Relyveld EH, Ben-Efraim S (1983) Preparation of vaccines by the action of glutaraldehyde on toxins, bacteria, viruses, allergens and cells. Methods Enzymol 93:24–60
Relyveld EH, Girard O, Désormeau-Bedot JP (1983) Procede de fabrication de vaccins à l'aide dy glutaraldehyde. Ann Immunol Hung 17:21–32
Robbins JB, Schneerson R, Pittman M (1984) Haemophilus influenzae type b. In: Germanier R (ed) Bacterial vaccines. Academic, New York, p 289

Robbins FC, Robbins JB (1986) Current status and prospects for some improved and new bacterial vaccines. Annu Rev Public Health 7:105–125

Rothbrock G, Smithee L, Rados M, Baughman W (1995) Progress toward elimination of *Haemophilus influenzae* type b disease among infants and children – United States, 1993-1994. (Reprinted from MMWR 44:545–550, 1995.) JAMA 274:1334

Salmaso S, Mastrantonio P, Wassilak SG, Giuliano M, Anemona A, Giammanco A, Tozzi AE, Ciofi Degli Atti ML, Greco D (1998) Persistence of protection through 33 months of age provided by immunization in infancy with two three-component a cellular pertussis vaccines. Stage II Working Group. Vaccine 16:1270–1275

Schiavo G, Benfenati F, Poulain B, Rossetto O, Polverino de Laureto P, DasGupta BR, Montecucco C (1992) Tetanus and botulinum-B neurotoxins block neurotransmitter release by proteolytic cleavage of synaptobrevin [see comments]. Nature 359:832–835

Schwendeman S, Costantino HR, Gupta RK et al (1995) Stabilization of tetanus and diphtheria toxoids against moisture-induced aggregation. Proc Natl Acad Sci USA 92(24):11234–11238

Sekura RD, Zhang YL, Roberson R, Acton B, Trollfors B, Tolson N, Shiloach J, Bryla D, Muir-Nash J, Koeller D et al (1988) Clinical, metabolic, and antibody responses of adult volunteers to an investigational vaccine composed of pertussis toxin inactivated by hydrogen peroxide. J Pediatr 113:806–813

Sinkovic D (1975) Effect of formaldehyde on diphtheria toxin protein molecule. Rad Imunol Zadova Zagrab 18:85–90

Stainer DW (1968) Preparation and properties of diphtheria toxoids in submerged culture. II. Purification, detoxification, antigenicity and stability. Can J Microbiol 14(4):327–330

Stein PE, Boodhoo A, Armstrong GD, Cockle SA, Klein MH, Read RJ (1994) The crystal structure of pertussis toxin. Structure 2:45–57

Storsaeter J, Hallander H, Farrington CP, Olin P, Mollby R, Miller E (1990) Secondary analyses of the efficacy of two acellular pertussis vaccines evaluated in a Swedish phase III trial. Vaccine 8:457–461

Sutter RW, Cochi SL (1992) Pertussis hospitalizations and mortality in the United States, 1985-1988. Evaluation of the completeness of national reporting. JAMA 267:386–391

Tamura M, Nogimori K, Murai S, Yajima M, Ito K, Katada T, Ui M, Ishii S (1982) Subunit structure of the islet-activating protein, pertussis toxin, in conformity with the A-B model. Biochemistry 21:5516–5522

Uchida T, Gill DM, Pappenheimer AM Jr (1971) Mutation in the structural gene for diphtheria toxin carried by temperate phage β. Nature New Biol 233:8–11

Uchida T, Pappenheimer AM Jr, Greany R (1973a) Diphtheria toxin and related proteins. I. Isolation and properties of mutant proteins serologically related to diphtheria toxin. J Biol Chem 248:3838–3844

Uchida T, Pappenheimer AM Jr, Harper AA (1973b) II. Kinetic studies on intoxication of HeLa cells by diphtheria toxin and related proteins. J Biol Chem 248:3845–3850

Uchida T, Pappenheimer AM Jr, Harper AA (1973c) Diphtheria toxin and related proteins. III. Reconstruction of hybrid "diphtheria toxin" from nontoxic mutant proteins. J Biol Chem 248:3851–3854

Van Ness BG, Howard JB, Bodley JW (1980) ADP-ribosylation of elongation factor 2 by diphtheria toxin. J Biol Chem 255:10710–10716

Volk WA, Bizzini B, Snyder RM, Bernhard E, Wagner RR (1984) Neutralization of tetanus toxin by distinct monoclonal antibodies binding to multiple epitopes on the toxin molecule. Infect Immun 45(3):604–609

Weiss A, Hewlett EL (1986) Virulence factors of *Bordetella pertussis*. Annu Rev Microbiol 40:661–686

Weiss AA, Johnson FD, Burns DL (1993) Molecular characterization of an operon required for pertussis toxin secretion. Proc Natl Acad Sci USA 90:2970–2974

Winberry L, Walker R, Cohen N, Todd C, Sentissi A, Siber G (1988) Evaluation of a new method for inactivating pertussis toxin with tetranitromethane. In: Keith J (ed) Abstracts of the international workshop of *B. pertussis*. Rocky Mountains Laboratory, Hamilton, Montana, Hamilton

Yamaizumi M, Mekada E, Uchida T, Okada Y (1978) One molecule of diphtheria toxin fragment A introduced into a cell can kill the cell. Cell 15:245–250

CHAPTER 9

Outer Membrane Protein Vaccines

J. POOLMAN

A. Introduction

Bacterial vaccines that are used on a large-scale are composed of killed or attenuated whole cells, toxoids, and polysaccharide-protein conjugate vaccines. Examples of such vaccines for human use include diphtheria, pertussis, tetanus, bacille Calmette-Guérin, *Haemophilus influenzae* type b conjugates, cholera, and typhoid. New subunit pertussis vaccines have recently been registered which in addition to pertussis toxoid may contain surface proteins such as filamentous hemagglutinin (FHA) and pertactin. FHA and pertactin are the first two bacterial surface proteins to become part of widely accepted vaccines.

This chapter describes the ongoing efforts, including large-scale field trials, to prevent bacterial diseases for which currently no accepted vaccine formulations exist. Vaccine research and development efforts are summarized, which focus on *Neisseria gonorrhoeae*, *N. meningitidis*, nontypable *H. influenzae*, *Moraxella catarrhalis*. Many aspects of bacterial pathogenesis and bacterial surface characteristics are shared among these species to allow for a combined description. For of this reason other bacterial surface protein vaccine development efforts are not described in this chapter.

B. *Neisseria gonorrhoeae*

I. Pili

It was recognized early that the presence of pili on the gonococcus is associated with virulence, and that nonselective passage in vitro selects for nonpiliation (KELLOGG et al. 1963). The abundance, easy accessibility of pili, and the critical adhesion function (SWANSON 1973) to eukaryotic cells led to the development of a parenteral pilus vaccine (TRAMONT et al. 1985; BRINTON et al. 1982). The vaccine was first tested in a human challenge model using the homologous organism and shown highly protective. In volunteers antibodies were induced that demonstrated opsonic and adhesion-blocking activities with broad cross-reactivity. A double-blinded, placebo-controlled field trial was performed in a military population with a high incidence of gonorrhea. Although high levels of antibodies were produced in the vaccine group, the number of cases with gonorrhea was nearly identical in the control and vaccine groups. As opposed to the pretrial studies, the trial volunteers did not show increases in antibody titers that inhibited gonococcal attachment in vitro (BOSLEGO et al. 1991).

Over time it became evident that gonococcal pili demonstrate a huge intra- and interstrain heterogeneity. The gene *pilE* encoding the pilin subunit PilE was shown to undergo reciprocal exchange of genetic information with silent copies of pilin genes, *pilS*, leading to intrastrain antigenic variability (SPARLING et al. 1986; HAAS et al. 1992; KOOMEY et al. 1987).

Transformation mechanisms may further extend the range of variability. At some point it was hoped to be possible to induce opsonic and/or adhesion-blocking antibodies against conserved epitopes present within the pilin subunit (ROTHBARD et al. 1984; SCHOOLNIK et al. 1984). No successful studies with this approach have been published. The explanation for this comes from the recent elucidation of the three-dimensional structure at 2.6 Å resolution of the gonococcal fiber (PARGE et al. 1995). It appears that the hypervariable C-terminal region is exposed on the surface, and that the conserved amino acid sequences are covered inside the fiber. Some epitopes become exposed at the tip of the fiber because the polymerization stops (FOREST et al. 1996).

At one point it was believed that the pilus-associated protein PilC is located at the top of the fiber; however, recent findings show PilC to be located close to the outer membrane (RUDEL et al. 1995; JONSSON 1996).

The recent discovery that the neisserial fiber is covered with O-linked trisaccharides on one side is intriguing (STIMSON et al. 1995).

II. Outer Membrane Protein PI or Por

The most abundant protein from the gonococcal outer membrane is PI or Por, the gonococcal porin. Por has limited variability in its serotype determinants (BUCHANAN and HILDEBRANDT 1981). Anti-Por antibodies induced in man were found to exhibit bactericidal activity (HOOK et al. 1984). Por vaccines, designed to cover most serotypes were developed but found invariably unsuccessful (GULATI et al. 1991; TRAMONT 1989). The Por vaccines contained 85%–90% Por, the impurities mostly being PIII or reduction-modifiable protein (Rmp) and lipopolysaccharide (LPS) (WETZLER et al. 1992a).

Antibody induction to Rmp blocks antibody activity directed against porin (RICE et al. 1986), which partly explains the disappointing results of the Por vaccine trials. Much later the most likely conformation and topology of Por in the gonococcal outer membrane was predicted (VAN DER LEY et al. 1991). The conformation of Por explains why a proper vaccine formulation within membranous material such as liposomes or blebs are an essential feature of vaccines containing integral outer membrane protein (OMPs) (WETZLER et al. 1992a). However, an additional issue surrounding Por as a vaccine candidate, is the observation that LPS sialylation prevents Por-mediated bactericidal antibody killing (WETZLER et al. 1992b).

III. PII or Opacity-Associated Proteins

A strain of *Neisseria gonorrhoeae* can express no or multiple PIIs, or opacity-associated proteins (Swanson 1982). The presence of PII gives rise to the formation of opaque colonies due to interbacterial associations. PIIs can mediate adhesion properties of gonococci in addition to pili (King and Swanson 1978; Lambden et al. 1979). The expression of *opa* genes by *N. gonorrhoeae*, up to 12 different ones in a single gonococcal strain, is regulated in response to addition or deletion of CTCTT pentanucleotide coding repeat units (Stern et al. 1986). The total repertoire of antigenic variation among gonococci is unknown, but discourages vaccine development efforts. A model for Opa has been proposed assuming four cell surface exposed, variable regions (van der Ley 1988). Different receptors on different cell types are likely recognized by different Opa's (Virji 1996). The possibility of gonococci to switch on and off 12 different Opa adhesins allows the bacterium to stick to (and let loose) a variety of human cells being encountered. The natural competence of gonococci may further broaden the variability by way of transformation mechanisms.

IV. Lipopolysaccharides

Although not OMPs, the discussion of gonococcal LPS fits well in this chapter. The antibody response to gonococcal infection is directed mostly against pili, Rmp, Opa's, LPS, and to a lesser extend Por (Hook et al. 1984) (At the time of the study, the importance of OMPs induced by Fe limitation was yet not appreciated). Antibodies against LPS demonstrate bactericidal and opsonic activities (Apicella et al. 1986; Ross and Densen 1985). The heterogeneity of gonococcal LPS is limited (Mandrell et al. 1986).

A complicating factor, however, is the observed structural identity between gonococcal LPS and human paragloboside and globotriose (Mandrell 1992). A major finding in gonococcal research was the observation that the terminal lacto-*N*-neotetraose structure can be sialylated in vivo, thereby explaining the long-known gonococcal mechanism of serum resistance (Smith et al. 1995). Recently the gene locus for the biosynthesis of the terminal, variable portion of *N. gonorrhoeae* lipo-oligosaccharide has been identified (Gotschlich 1994). The mechanism controlling the variable expression of the terminal structure operates through slipped strand mispairing of a homopolymeric track of 14 guanosine residues (Jennings et al. 1995). The knowledge about this genetic locus opens the way for the construction of LPS mutants devoid of the human tissue cross-reactive epitopes. Another possibility for vaccine development is the use of monoclonal antibodies against LPS with no immunochemical similarity to human glycosphingolipids (Gulati et al. 1996). The recognized epitope can then be mimicked by anti-idiotypic monoclonal antibodies or by peptide or other mimotopic structures.

V. Fe Limitation Inducible OMPs

The availability of free iron in the host is limited and pathogenic bacteria in vivo grow under iron limitation.

The gonococcus expresses a number of surface OMPs allowing the binding and removal of iron from human transferrin, lactoferrin and hemoglobin/haptoglobin (SPARLING et al. 1996) Not all gonococcal proteins expressed under iron limitation are known; however, a number of these are understood reasonably well. The gonococcal human transferrin receptor is made up of two transferrin-binding proteins (Tbp). TbpA is an integral outer membrane protein and essential for the entry of Fe (CORNELISSEN et al. 1992). TbpB is a lipoprotein exposed on the outside which facilitates transferrin binding but is not essential for Fe entry (ANDERSON et al. 1994). TbpA and TbpB occur together in patches on the membrane, which is energized by a TonB-ExbB-ExbD system (SPARLING et al. 1996; CORNELISSEN and SPARLING 1996). TbpB is more surface exposed than TbpA and also appears to be much more antigenically variable (SPARLING et al. 1996). About 50% of the clinical gonococcal isolates produce a functional human lactoferrin receptor (MICKELSEIN et al. 1982).

A two-protein system LbpA-LbpB appears to exist very similar to the transferrin receptor system (BISWAS and SPARLING 1995). Gonococci also produce iron limitation inducible receptors that bind hemoglobin/haptoglobin (LEWIS and DYER 1995).

Another major iron-repressed OMP is found called Fe-repressed protein (FrpB), a member of the TonB-dependent integral OMP family (BEUCHER and SPARLING 1995). To date no function of FrpB has been found; it may be involved in the uptake of siderophores (SPARLING et al. 1996).

VI. Miscellaneous

Over the years in addition to the pili, major OMPs, LPS, and Fe limitation inducible OMPs a number of other vaccine candidates have been investigated. As far can now be determined, the adhesion/invasion properties of gonococci depend mostly upon pili and PIIs, whereby the on-and-off mechanisms are critical. The on-and-off mechanisms of the LPS sialylation also plays a crucial role during the invasion process (VAN PUTTEN 1993).

It cannot be excluded that as yet unknown gonococcal adhesins exist (BRAUDE et al. 1985). Anaerobic growth of gonococci results in the expression of several OMPs, of which the Pan1 protein is recognized by patient sera (HOEHN and CLARK 1992a). Pan1 appears to be a lipoprotein (HOEHN and CLARK 1992b). Another two lipoproteins, H8 and lipid-modified azurin which are well conserved among pathogenic *Neisseria*, have attracted considerable attention (CANNON 1989). Antibodies to H8 have failed to demonstrate bactericidal activity (BHATTACHARJEE et al. 1990) Another interesting class of OMPs are those involved in the secretion of exoproteins. The gonococcal OMP-macromolecular complex (Omc) was characterized molecularly (TSAI et al. 1989) and found to share homologies with enterobacterial OMPs involved in protein excretion (POOLMAN et al. 1995).

Omc is a surface-exposed, antigenically conserved major OMP of approximately 800 kDa and composed of 10–12 subunits of a 76-kDa subunit (NEWHALL et al. 1980). Antibodies to Omc appear to be bactericidal for serum-resistant gonococci (CORBETT et al. 1987). The 76-kDa subunit has recently been recognized as crucial for pilin excretion and has been renamed PilQ (DRAKE and KOOMEY 1995). Gonococci do not seem to excrete exotoxins, and the toxic effect seems to be caused mostly by LPS (VIRJI 1996).

However, one excreted protein has attracted considerable attention: IgA1 protease, assuming a possibility to cleave mucosal IgA. Two distinct types of IgA1 protease are produced by gonococci (MULKS et al. 1987). An isogenic gonococcal variant devoid of IgA1 protease demonstrates identical behavior in the fallopian tube organ model (COOPER et al. 1984).

The IgA1 hinge region susceptible to bacterial IgA1 protease also is present in lysosome-associated-macrophage protein 1 (LAMP1) (MULKS and SHOBERG 1994). In line with this pathogenic *Neisseriae* increase. LAMP 1 turnover, while IgA1 protease negative mutants are negative in this respect and fail to survive within epithelial cells (LIN et al. 1996).

C. *Neisseria meningitidis*

The development of polysaccharide and conjugate vaccines against diseases caused by serogroup A, C, Y, W-135 are described in this volume. The development of a group B meningococcal vaccine has focused on OMP vaccines. However, interesting work on the B polysaccharide is still continuing. Because of its very special character, the menB polysaccharide vaccine research and development activities are described in this chapter.

I. The Serogroup B Capsular Polysaccharide

It was recognized early that the α-(2→8)-sialic acid homopolymer which composes the menB capsular polysaccharide is nonimmunogenic in man (WYLE et al. 1972). The basis for this immunotolerance is related to the structural mimicry between the menB polysaccharide and human tissue antigens, particularly neural-cell-adhesion-molecule (NCAM), present mostly during vertebrate organogenesis (HÄYRINEN et al. 1995; LACKIE et al. 1993).

The nonimmunogenicity of the menB polysaccharide cannot be overcome by conjugation to protein (JENNINGS and LUGOWSKI 1982). However, chemical alteration by replacing the *N*-acetyl for *N*-propionyl at the C-5 atom of sialic acid and consecutive conjugation to a carrier protein results in an immune response to group B meningococci (JENNINGS et al. 1986). The *N*-propionyl-BPS conjugate induces two distinct population of antibodies, one of which reacts with BPS while the other population does not but is still able to kill group B meningococci in a group-specific manner (JENNINGS et al. 1987).

Both families of antibodies react with conformational epitopes present on the extended helical form of the MenB PS and/or the *N*-propionyl-MenB PS (JENNINGS et al. 1996). The latter epitope is present on the BPS in its form on group B meningococci. One possible explanation for this paradox is the difference in real-time availability of the two major families of epitopes on menB PS, *N*-propionyl menB PS, and living group B meningococci. It needs to be investigated how the two families of epitopes are presented by NCAM-expressing neonatal human cells. Monoclonal antibodies against the two major families of epitopes are being developed to better understand the paradox (BARTOLINI et al. 1996; DEVI et al. 1995). The first attempts to induce a rise in bactericidal antibodies in adult volunteers with the *N*-propionylated menB PS conjugate have been disappointing (DANVE et al. 1996).

The occurrence of natural bactericidal antibodies against group B meningococci with age and absence of a correlation with antibodies to menB PS, can be taken as promising for a noncapsular vaccine (GOLDSCHNEIDER et al. 1969). Antibody measurements against the *N*-propionyl menB PS-associated epitope(s) to explain natural immunity have not been performed yet.

II. OMP Vaccines

The correlation between the presence of bactericidal antibodies and susceptibility to meningococcal disease has been documented (GOLDSCHNEIDER et al. 1969). This strong correlation has been underscored by the observation that individuals with terminal complement component deficiencies have an increased risk for meningococcal disease (DENSEN 1989). The presence of opsonic antibodies, however, can have an additional beneficial effect (ROSS et al. 1987). Meningococcal carriage or disease results into a bactericidal antibody response to OMP antigens (JONES and ELDRIDGE 1979; MANDRELL and ZOLLINGER 1989). Because of these findings menB-OMP vaccines have been developed (Table 1) (SIERRA et al. 1991; DE MORAES et al. 1992; BJUNE et al. 1991; BOSLEGO et al. 1995).

The three vaccine tested differed considerably in their composition and formulation (POOLMAN 1996); they were tested in a number of large-scale field trials. The common antigens, although having different serosubtype specificity, in these vaccines are PorA (class 1 OMP), Por B (class 2/3 OMP), and Rmp (Class 4 OMP). Sometimes additional OMPs are present, such as Opc and certain high molecular weight OMPs (Table 2). The efficacy of vaccines ranges from 57% to 83%; however, in the very young (<4 years), very poor or negligible protection rates are encountered. Although it is complicated to pinpoint the relevant antigens for the protective, bactericidal antibodies in these multicomponent vaccines, the Chilean and Brasilean studies revealed evidence for the importance of antibodies against PorA, the class 1 OMP (MILAGRES et al. 1994; ZOLLINGER et al. 1991).

Table 1. Efficacy trials with meningococcal OMP vaccines

Vaccine	Study	Efficacy
Purified total OMP in proteoliposome with added high MW OMP (Finlay) 50 μg (B:4:P1.19, 15)	Cuba, teenagers, randomized at school level, double-blinded, placebo control, two immunization, 16-month study period	83%
Same vaccine	Brazil, case-control, two immunizations, 12-month follow-up	−37% (3–23 mo) 47% (24–47 mo) 74% (48–93 mo)
Total OMP in vesicle (OMV) formulation (SIFF) 25 μg (B:15:P1.7, 16)	Norway, teenager, randomized at school level, double-blinded, placebo control, two immunizations, almost 3-year study period	57.4%
Purified class 1, 3, 4 containing OMP (WRAIR) 100 μg (B:15:P1.3)	Chile, ages 1–21 years, randomized, double-blinded, placebo control, two immunizations, 20-month follow-up	−39% (1–4 yr) 70% (5–21 yr)

The Norwegian study demonstrated the importance of Opc and PorA with respect to the induction of bactericidal antibodies (ROSENQVIST et al. 1995). The critical importance of the vaccine formulation is evident from the recent finding that the Norwegian outer membrane vesicle (OMV) vaccine is able to induce high levels of bactericidal antibodies in infants in a Chilean immunogenicity study (Perkins and Carlone, CDC, personal communication). The Norwegian vaccine has not been tested for efficacy in infants so far. The Cuban and Walter Reed Army Institute's vaccines contain proteoliposomes or purified proteins (Table 1) and do not induce bactericidal antibodies in infants (MILAGRES et al. 1994; BOSLEGO et al. 1995).

The OMV vaccine formulation resulting from excreted/extracted native outer membrane blebs (POOLMAN 1995) appear to present the integral OMPs in its native configuration and allows proper priming of infants. Although additional high molecular weight OMPs are sometimes present in the OMP vaccines, little evidence has been gathered so far to demonstrate their potential to induce bactericidal antibodies. Also, in those who respond to Rmp or class 4 OMP no evidence has been obtained for a blocking effect (MUSACCHIO et al. 1996). There may exist a parallel between the gonococcal PI-PIII (Por-Rmp) and meningococcal class 2/3–class 4 (PorB-Rmp) interactions; the bactericidal antibodies against the meningococcal class 1 OMP or PorA appear to act in a different way. This also appears true for the effect of LPS sialylation and its blocking activity on bactericidal antibodies against Por and PorB (WETZLER et al. 1992b; POOLMAN et al. 1987).

The results of clinical studies with respect to the importance of PorA are in line with those obtained with a collection of monoclonal antibodies when

Table 2. MenB, NTHi, Mcat vaccine candidates

	MenB		NTHi		Mcat	
	Presence	Possible protection	Presence	Possible protection	Presence	Possible protection
Capsule	+	+	–		+/–	ND
High MW adhesins	+/–	ND	+	+	–	
High MW OMP involved in protein excretion PilQ; UspA	+	+/–	ND may be D15		+/–	ND
Tbps	+	+	+	+	+	ND
Hbps	+	ND	+	ND	ND	
Lbps	+	ND	ND		ND	
FrpB/CopB	+	+	ND		+	+
Porins	+	+	+	+	+	ND
Low MW adhesins	+	+	ND		ND	
Pili	+	ND	+	+	+	ND
LPS	+	+	+	+	+	ND
IgA1 protease	+	ND	+	ND	–	
Exotoxins	+	ND	ND		ND	
H8	+	–	+/–			
NspA	+	+				
LPD			+	+		
P6			+	+		

ND = not determined

tested for in vitro bactericidal activity and passive protection in the infant rat meningitis model (SAUKKONEN et al. 1989).

The observed worldwide limited heterogeneity for PorA (ABDILLAHI and POOLMAN 1988) has provided the justification for developing a multivalent PorA vaccine in OMV formulation (VAN DEN LEY et al. 1995). Studies in adults indicate the possibility to induce PorA-specific bactericidal antibodies (PEETERS et al. 1996; ROUPPE VAN DER VOORT et al. 1996). The results of an infant immunogenicity study appear very promising (POOLMAN 1996).

III. Opa and Opc

A single meningococcal strain encodes fewer (≤4) Opa's than gonococci (≤12), and some functions for adhesion/invasion are related to meningococcal Opc, a protein not present among gonococci (VIRJI et al. 1993). However, the expression of Opc is low or absent in most case isolates, making it a less attractive vaccine candidate for inducing systemic immunity (ROSENQVIST et al. 1993). It does make it an attractive candidate, however, for the induction of mucosal immunity, which is currently being explored using intranasal immunization (WEDEGE et al. 1996) with OMV vaccines. The extent of antigenic variability of meningococcal Opa's (or class 5 OMPs) is unknown but may be limited; Opa's do induce bactericidal antibodies (DANELLI et al. 1996).

IV. Pili

Two structurally distinct classes of pili occur in meningococci, and both undergo antigenic variations and function as adhesins (VIRJI et al. 1992). The disappointing results obtained with gonococcal pili and the observed variability of meningococcal pili has prevented substantial vaccine research from being carried out. The recent findings of pilin glycosylation may reopen the case (STIMSON et al. 1995, 1996).

V. Lipopolysaccharides

Data on patient sera indicate that meningococcal LPS is a target for bactericidal and opsonic antibodies (VERHEUL et al. 1993; ESTABROOK et al. 1993; PRESTON et al. 1996) The most prevalent L2 and L3 immunotypes among case isolates contain lacto-*N*-neotetraose, are sialylated, and determine the serum resistance properties. The elucidation of the gene cluster involved in the terminal saccharide synthesis opens ways to construct mutants devoid of epitopes cross-reactive with human tissue (JENNINGS et al. 1995).

Group B meningococcal carrier isolates generally do not express the lacto-*N*-neotetraose part, in contrast to case isolates (JONES et al. 1992). A variation in the expression of pili, Opa, Opc, lacto-*N*-neotetraose, and B capsule allows the meningococcus to attach to and penetrate epithelial cells and resist complement attack in the subepithelial tissues. A slipped-strand mispairing system for the expression of the amount of B polysaccharide has recently been discovered (HAMMERSCHMIDT et al. 1996).

VI. Fe Limitation Inducible OMPs

As with gonococci, meningococci express a receptor for human transferrin composed of two bacterial OMPs, the integral OMP TbpA, and the more exposed lipoprotein TbpB (IRWIN et al. 1993). Particularly TbpB induces bactericidal and animal-protective antibodies (LISSOLO et al. 1995). TbpB reveals a significant degree of antigenic variability, which may hamper vaccine development (ROKBI et al. 1995, 1997). The meningococcal lactoferrin receptor appears to follow the same two-component system by combining LbpA and LbpB (PETTERSSON et al. 1993, 1994). The meningococcal hemoglobin/haptoglobin receptors encompass two different receptors, Hpu and HmbR (LEWIS and DYER 1995; STOJILKOVIC et al. 1995), of which Hpu follows the system of an integral OMP in association with an associated lipoprotein. Meningococci also express the integral OMP FrpB, which function remains undefined (PETTERSSON et al. 1995). The sequence variation of FrpB is confined mostly to one exposed loop which binds highly bactericidal antibodies (VAN DER LEY et al. 1996; PETTERSSON et al. 1990).

VII. Miscellaneous

The conserved H8 and lipid-modified azurin are common among pathogenic *Neisseria*; however, no bactericidal antibodies could be induced (BHATTACHARJEE et al. 1990). The PilQ protein, formerly known as Omc, was also found among meningococci and may be of interest as a vaccine antigen (TØNJUM et al. 1996) In general the excretion systems for proteins are ill-defined among meningococci. Meningococci also express IgA1 protease (PLAUT et al. 1975). A very intriguing property of meningococci is the possibility to excrete the putative exotoxins FrpA and FrpC being expressed under iron limitation (THOMPSON et al. 1993), which share homology with *B. pertussis* adenylate cyclase.

The search for conserved OMPs still continues and may yet produce pleasant surprises such as NspA, a low molecular weight OMP (MARTIN et al. 1996).

D. Nontypable *Haemophilus influenzae*

As with gonococci and meningococci, *H. influenzae* is a highly competent bacterial species, and antigenic variability is the major obstacle to the development of a NTHi vaccine. Although oral immunization with formalin-killed nontypable *H. influenzae* enhances nasopharyngeal clearance in a mouse model (KURONO et al. 1996), a definition of protective antigens is required to enable a multivalent vaccine. During persistence of NTHi in patients with chronic bronchitis, antigenic drift can be observed with respect to OMPs (GROENEVELD et al. 1988). NTHi has a set of defined surface components: antigenic variation suggest a role for antibody-mediated immune mechanisms; however, in the lung clearance rat model immunity is correlated better with $CD4^+$ T cells (KYD and CRIPPS 1996).

I. P1, P2, P4

Original studies using *H. influenzae* type b strains showed that antibodies specific for P1 and P2 protect infant rats from subsequent challenge (LOEB 1987; MUNSON et al. 1983). P2 was found to be able to induce bactericidal and opsonic antibodies which are directed against the variable regions present within surface exposed loop structures of this integral OMP (HAASE et al. 1994; TROELSTRA et al. 1994) The lipoprotein P4 can also induce bactericidal antibodies (GREEN et al. 1991).

II. P6 and PCP

P6 is a conserved peptidoglycan associated lipoprotein making up 1%–5% of the outer membrane (NELSON et al. 1991). Later a lipoprotein of about the same molecular weight was recognized, called P6 cross-reactive protein (PCP) (DEICH et al. 1990). A mixture of the conserved lipoproteins P4, P6, and PCP did not reveal protection, as measured in the chinchilla otitis media model (GREEN et al. 1993). P6 alone appears to induce protection in the chinchilla model (DEMARIA et al. 1996).

III. Pili, Fimbriae, P5

Pili and fimbriae are used as synonyms in the bacterial literature. The gene cluster of NTHi has been characterized (HAM et al. 1994). The pilin appears to have a molecular weight of 22 kDa, typical for pili (COLEMAN et al. 1991). The extent of antigenic variability of NTHi pili is ill defined. Another fimbrin is described with homology to P5, which in itself has sequence homology to the integral *E. coli* OmpA (MIYAMOTO and BAKALETZ 1996; MUNSON et al. 1993). This paradox requires further investigations to clarify the nature and role of pilin, pilin-associated proteins, pilin-excreting proteins, and P5. It has been shown, however, that NTHi adhere to mucus by way of fimbriae (MIYAMOTO and BAKALETZ 1996). P5 appears to undergo antigenic drift during persistent infections with NTHi (DUIM et al. 1997).

IV. High Molecular Weight Adhesins

Three high molecular weight adhesins have recently been described: Hmw1, Hmw2 and Hia (ST. GEME et al. 1993). Hmw1 and Hmw2 occur among ±75% of NTHi strains, are conserved, and reveal homology with Proteus and Serratia OMPs involved in export of exoproteins (BARENKAMP and ST. GEME

1994). Recently *H. influenzae* adhesin (Hia) was recognized, a 114-kDa surface adhesin present in Hmw1/2 negative strains and absent in Hmw1/2 positive strains (BARENKAMP and ST. GEME 1996). The adhesion properties to epithelial cells for Hmw1/2 and Hia have been demonstrated. Immunization with Hmw1/2 modifies NTHi otitis media in the chinchilla model (BARENKAMP 1996). Hmw1/2 share epitopes with *B. pertussis* FHA and show morphological similarities with FHA (BAKALETZ and BARENKAMP 1994).

V. Lipopolysaccharide or Lipo-oligosaccharide

LPS from NTHi contains a limited number of epitopes, some of which cross-react with human tissue. A rapid change in the epitope composition has been encountered. The elucidation of the genes involved could lead to LPS mutants suitable for vaccine development (PRESTON et al. 1996). One possible vaccine approach is the construction of conjugates from detoxified LPS (GU et al. 1996).

VI. Fe Limitation Inducible OMPs

In line with the observations made with gonococci and meningococci, NTHi expresses dual human transferrin receptors composed of TbpA and TbpB when grown under iron limitation. Anti-TbpB protects infant rats (LOOSMORE et al. 1996). A hemoglobin/haptoglobin receptor has also been described for NTHi (MACIVER et al. 1996; JIN et al. 1996). A receptor for Hem:Hemopexin has also been identified (COPE et al. 1994). A lactoferrin receptor is also present among NTHi, but has not yet been characterized (SCHRYVERS 1989). No protein resembling neisserial FrpB protein has been described among NTHi.

VII. Miscellaneous

A 80-kDa OMP, the D15 surface antigen, provides protection against NTHi in a mouse challenge model (FLACK et al. 1995). A 42-kDa outer membrane lipoprotein, LPD, is conserved among *H. influenzae* and induces bactericidal antibodies (AKKOYUNLU et al. 1996). A minor 98-kDa OMP (KIMURA et al. 1985) was found to be a protective antigen; this OMP may very well be one of the Fe limitation inducible OMPs or high molecular weight adhesins that have since been characterized. *H. influenzae* produces IgA1 protease activity (MULKS and SHOBERG 1994). IgA1 proteases of NTHi reveals a high degree of antigenic variability (LOMHOLT et al. 1993).

E. *Moraxella catarrhalis*

Moraxella catarrhalis is now well accepted as a human pathogen responsible for respiratory tract infections. The following vaccine candidates can be identified.

I. OMP CD, OMP E

OMP CD is a 46-kDa OMP with homology to *Pseudomonas aeruginosa* OprF, which is a porin. OMP CD is conserved among strains (HSIAO et al. 1995; MURPHY et al. 1993). OMP E reveals sequence homology to the OmpF family of porin proteins (BHUSHAN et al. 1994) and appears to contain surface-exposed conserved epitopes (MURPHY 1996). Two other major OMPs in the range of 30–40 kDa have been recognized (MURPHY 1996).

II. Pili/Fimbriae/High Molecular Weight OMPs

M. catarrhalis produces pili, but no correlation has been established with the adherence to epithelial cells (AHMED et al. 1994; RIKITOMI et al. 1991). A high molecular weight OMP of 350–700 kDa composed of subunits (UspA) has been found among *Moraxella* stains (KLINGMAN and MURPHY 1994). These characteristics suggest a similarity with neisserial PilQ (or Omc) (DRAKE and KOOMEY 1995). A monoclonal antibody has been found to enhance clearance of *M. catarrhalis* in a mouse colonization model (HELMINEN et al. 1994). The possibility of UspA being a vaccine candidate has recently been strengthened by active immunization showing pulmonary clearance in the mouse model (CHEN et al. 1996).

III. Lipopolysaccharide or Lipo-oligosaccharide

M. catarrhalis expresses LPS without O-repeat units, such as gonococci, meningococci, NTHi (FOMSGAARD et al. 1991; EDEBRINK et al. 1994; PRESTON et al. 1996). Little antigenic variation exists; a protective effect of antibodies has not yet been described (VANEECHOUTTE et al. 1990).

IV. Fe Limitation Inducible OMPs

A human transferrin receptor expressed by *M. catarrhalis* has recently been described; OMP B1 is involved in this receptor function (CAMPAGNARI et al. 1996). Other receptors have not yet been described. An approximately 80-kDa OMP, CopD, is expressed by *M. catarrhalis* under Fe limitation, which clearly is a virulence factor and a monoclonal antibody against CopB-enhanced pulmonary clearance in the mouse model (HELMINEN et al. 1993a,b). CopB has been found to have sequence homology with neisserial FrpB (PETTERSSON et al. 1995; BEUCHER and SPARLING 1995).

F. Animal Models

The intraperitoneal challenge of infant rat is being used to measure the protective immunity against systemic infection (SAUKKONEN et al. 1989). This model can evaluate the effect of antibodies against *H. influenzae* type b and meningococcal bacteremia and meningitis by passive immunization.

Various models are being used to evaluate protective immunity against respiratory infection by NTHi, *M. catarrhalis*, and meningococci (Kyd et al. 1989; Unhanand et al. 1992; Mackinnon et al. 1993). Infant and adult mice and rat are infected by intranasal or endobronchial challenge, and lung and/or nasopharyngeal colonization are used as endpoints. A model for persistent lung infection with NTHi has also been developed (Slater 1990). The availability of iron during challenge by way of added mucin or iron containing compounds such as Fe dextran are critical for achieving low challenge doses (Holbein et al. 1979; Holbein 1980).

For NTHi and *M. catarrhalis* a chinchilla model is being used to evaluate protective immunity against otitis media (Suzuki and Bakkaletz 1994; Bakaletz et al. 1995) by transbullar challenge. For NTHi a preinfection with adenovirus followed by intranasal NTHi challenge also results in otitis media. A transbullar challenge in a rat model for NTHi is also being used (Melhus et al. 1995).

References

Abdillahi H, Poolman JT (1988) Neisseria meningitidis group B serosubtyping using monoclonal antibodies in a whole-cell ELISA. Microbial Pathog 7:27–32

Ahmed K, Masaki H, Dai TC, Ichinose A, Utsunomiya Y, Tao M, Nagatake T, Matsumoto K (1994) Expression of fimbriae and host response in Branhamella catarrhalis respiratory infections. Microbiol Immunol 38:767–771

Akkoyunlu M, Janson H, Ruan M, Forsgren A (1996) Biological activity of serum antibodies to a nonacylated form of lipoprotein D of Haemophilus influenzae. Infect Immun 64:4586–4592

Anderson JE, Sparling PF, Cornelissen C (1994) Gonococcal transferrin-binding protein 2 facilitates but is not essential for transferrin utilization. J Bacteriol 176:3162–3170

Apicella MA, Westerink MAJ, Morse SA, Schneider H, Rice PA, Griffiss JM (1986) Bactericidal antibody response of normal human serum to the lipooligosaccharide of Neisseria gonorrhoeae. J Infect Dis 153:520–526

Bakaletz LO, Barenkamp SJ (1994) Localization of high-molecular-weight adhesion properties of nontypable Haemophilus influenzae by immunoelectron microscopy. Infect Immun 62:4460–4468

Bakaletz LO, Murwin DA, Billy JM (1995) Adenovirus serotype 1 does not act synergistically with Moraxella (Branhamella) catarrhalis to induce otitis media in the chinchilla. Infect Immun 63:4188–4190

Barenkamp SJ (1996) Immunization with high-molecular-weight adhesion proteins of nontypable Haemophilus influenzae modifies experimental otitis-media in chinchillas. Infect Immun 64:1246–1251

Barenkamp SJ, St Geme JW (1994) Genes encoding high-molecular-weight adhesion proteins of nontypable Haemophilus influenzae are part of gene clusters. Infect Immun 62:3220–3328

Barenkamp SJ, St Geme JW (1996) Identification of a 2nd family of high-molecular-weight-adhesion proteins expressed by nontypable Haemophilus influenzae. Mol Microbiol 19:1215–1223

Bartolini A, Ricci S, Gallo E, Rosa D, Ravenscrof N, Guarnieri V, Seid R, Shan A, Usinger W, Mc Hugh YE, Granoff DM (1996) Murine monoclonal antibodies to an N-propionylated meningococcal B polysaccharide exhibit heterogeneity with respect to cross-reactivity with N-acetylated meningococcal B polysaccharide and autoreactivity to host polysialylated glycoproteins. In: Zollinger WD, Frasch CE, Deal CD (eds) Pathogenic Neisseria, the Tenth International Pathogenic Neisseria Conference, pp 218–219

Beucher M, Sparling PF (1995) Cloning, sequencing and characterization of the gene encoding FrpB, a major iron-regulated outer membrane protein of Neisseria gonorrhoeae. J Bacteriol 177:2041–2049

Bhattacharjee AK, Moran EE, Zollinger WD (1990) Antibodies to meningococcal H8 (Lip) Antigen fail to show bactericidal activity. Can J Microbiol 36:117–122

Bhushan R, Craigie R, Murphy TF (1994) Molecular cloning and characterization of outer membrane protein E of Moraxella (Branhamella) catarrhalis. J Bacteriol 176:6636–6643

Biswas GD, Sparling PF (1995) Characterization of LpbA, the structural gene for lactoferrin receptor in Neisseria gonorrhoeae. Infect Immun 63:2958–2967

Bjune G, Høiby EA, Grønnesby JK, Arnesen O, Frederiksen JH, Halstensen A, Holten E, Lindbak AK, Nøkleby H, Rosenqvist E, Solberg LK, Class O, Eng J, Frøholm LO, Lystad A, Bakketeig LS, Hareide B (1991) Effect of outer membrane vesicle vaccine against group B meningococcal disease in Norway. Lancet 338:1093–1096

Boslego J, Garcia J, Cruz C (1995) Efficacy, safety, and immunogenicity of a meningococcal group B:15:P1.3 outer membrane protein vaccine in Iquique, Chile. Vaccine 13:821–829

Boslego JW, Tramont EC, Chung RC (1991) Efficacy trial of a parenteral gonococcal pilus vaccine in men. Vaccine 9:154–162

Braude A, Wunderlich A, Chimaki G, Douglas H, McCutchan J (1985) Generation of adherence-blocking immunoglobulin G(IgG) by vaccinating women with nonpiliated boiled gonococci observation on IgG binding to a high-molecular-weight protein in the outer membrane fraction. In: Schoolrik GK, Brooks GF, Falkow S et al (eds) The pathogenic Neisseria. American Society for Microbiology, pp 323–328

Brinton CC Jr, Wood SW, Brown A, Labik AM, Lee JR, Polen SW, Tramont EC, Sadoff JC, Zollinger WD (1982) The development of a neisserial pilus vaccine for gonorrhoeae and meningococcal meningitis. In: Robbins JB, Hill JH, Sadoff JC (eds) Bacterial vaccines. Thieme-Stratton, New York, pp 140–150

Buchanan TM, Hildebrandt JF (1981) Antigen-specific serotyping of Neisseria gonorrhoeae characterization based upon principal outer membrane protein. Infect Immun 32:985–994

Campagnari AA, Ducey TF, Rebmann CA (1996) Outer membrane protein B1, an iron-repressible protein conserved in the outer membrane of Moraxella (Branhamella) catarrhalis, binds human trasferrin. Infect Immun 64:3920–3924

Cannon J (1989) Conserved lipoproteins of pathogenic Neisseria species bearing the H.8 outer membrane protein. Clin Microbiol Rev 2:S1–S4

Chen D, Mc Michael JC, Vandermeid KR, Hahn D, Mininni T, Cowell J, Eldridge J (1996) Evaluation of UspA from Moraxella catarrhalis as a vaccine in a murine model after active immunization. Infect Immun 64:1900–1905

Coleman T, Grass S, Munson R (1991) Molecular cloning, expression and sequence of the pilin gene from nontypable Haemophilus influenzae M37. Infect Immun 59:1716–1722

Cooper MD, Mc Gee ZA, Mulks MH Koomey JM, Hindman TL (1984) Attachment to and invasion of human fallopian tube mucosa by an IgA1 protease-deficient mutant of Neisseria gonorrhoeae and its wild-type parent J Infect Dis 150:737–744

Cope LD, Thomas SE, Latimer CA, Slaughter A, Muller-Eberhard U, Hansen EJ (1994) The 100 kDa heme: hemopexin-binding protein of Haemophilus influenzae: structure and localization. Mol Microbiol 13:868–873

Corbett MJ, Black JR, Wilde CE II (1987) Antibodies to outer membrane protein-macromolecular complex (OMP-MC) are bactericidal for serum-resistant gonococci. In: Poolman JT, Zanen HC, Meyer TF et al (eds) Gonococci and meningococci. Kluwer, London, pp 685–691

Cornelissen CN Sparling PF (1996) Binding and surface exposure of the gonococcal transferrin receptor are dependent on both transferrin-binding proteins. J Bacteriol 178:1437–1444

Cornelissen CN, Biswas GD, Tsai J, Paruchuri DK, Thompson SA, Sparling PF (1992) Gonococcal transferrin-binding protein 1 is required for tranferrin utilization and is homologous to Ton B-dependent outer membrane receptors. J Bacterial 174:5788–5797

Danelli MGM, Alves CMA, Bastos RC, Batoren NM, Barroso DE, Peralta JM, Frasch CE (1996) Human immune response to epitope on the meningococcal outer membrane class 5 protein following natural infection. FEMS Immunol Med Microbiol 15:159–168

Danve B, Bruge J, Bouveret-Le Cam N, Chassard D, Rougon G, Schultz D (1996) Safety and immunogenicity of an N-propionylated group B meningococcal polysaccharide conjugate vaccine in adult volunteers. In: Zollinger WD, Frasch CE, Deal CD (eds) Pathogenic Neisseria, the Tenth International Pathogenic Neisseria Conference, pp 225–226

Deich RM Anilionis A, Fulginiti J, Metcalf BJ, Quataert S, Quinn-Dey T, Zlotnick GW, Green BA (1990) Antigenic conservation of the 15,000 dalton outer membrane lipoprotein PCP of Haemophilus influenzae and biologic activity of anti-PCP antisera. Infect Immun 58:3388–3393

Densen P (1989) Interaction of complement with Neisseria meningitidis and Neisseria gonorrhoeae. Clin Microbiol Rev 2 [Suppl]:511–517

Demaria TF, Murwin DM, Leake ER (1996) Immunization with outer membrane protein P6 from nontypable Haemophilus influenzae induces bactericidal antibody and affords protection in the chinchilla model of otitis media. Infect Immun 64:5187–5192

De Moraes JC, Perkins BA, Camargo MCC, Rossetto Hidalgo NT, Aparecida Barbosa H, Tavares Sacchi C, Land Gral IM, Gatta VL, Vasconcelos, H de G, Plikaytis BD, Wenger JD, Broome CV (1992) Protective efficacy of a serogroup B meningococcal vaccine in São Paulo, Brazil. Lancet 340:1074–1078

Devi SJN, Karpas A, Frasch CE (1996) Binding deversity of monoclonal antibodies to α-(2→8) polysialic acid conjugates do outer membrane vesicle via adipic acid dihydrazide. FEMS Immunol Med Micorbiol 14:211–220

Drake SL, Koomey M (1995) The product of the pilQ gene is essential for the biogenesis of type IV pili in Neisseria gonorrhoeae. Mol Microbiol 18:975–986

Duim B, Bowler LD, Eijk PP, Ruiter P, Jansen HM, Dankert J, van Alphen L (1997) Molecular variation in the MOMP P5 gene of nonencapsulated Haemophilus influenzae during chronic infections. Infect Immun 65:1351–1356

Edebrink P, Jansson PE, Rahman MM, Widmalm G, Holme T, Rahman M, Weintraub A (1994) Structural studies of the O-polysaccharide from the lipopolysaccharide of Moraxella (Branhamella) catarrhalis serotype A (strain ATCC 25238). Carbohydr Res 257:269–284

Estabrook MM, Baker CJ, Griffiss JM (1993) The immune response of children to meningococcal lipooligosaccharides during disseminated disease is directed primarily against two monoclonal antibody-defined epitopes. J Infect Dis 167:966–970

Flack FS, Loosmore S, Chong P, Thomas WR (1995) The sequencing of the 80 kDa D15 protective surface antigen of Haemophilus influenzae. Gene 156:97–99

Fomsgaard JS, Fomsgaard A, Hoiby N, Bruun B, Galanos C (1991) Comparative immunochemistry of lipopolysaccharides from Branhamella catarrhalis strains. Infect Immun 59:3346–3349

Forest KT, Berstein SL, Getzoff ED, So M, Tribbick G, Geysen HM, Deal CD, Tainer JA (1996) Mapping of the gonorrhoeae pilus with antibodies. Infect Immun 64:644–652

Goldschneider, Gotschlich EC, Artenstein MS (1969) Human immunity to the meningococcus. II Delvelopment of natural immunity. J Exp Med 129:1327–1348

Gotschlich EC (1994) Genetic locus for the biosynthesis of the variable portion of Neisseria gonorrhoeae. Lipooligosaccharide. J Exp Med 180:2181–2190

Green BA, Farley JE, Quinndey T, Deich RA, Zlotnick GW (1991) The E(P4) outer membrane protein of Haemophilus influenzae. Biological activity of anti-e serum and cloning and sequencing of the structural gene. Infect Immun 59:3191–3198

Green BA, Vasquez ME, Zlotnick GW, Quigleyreape G, Swarts JD, Green I, Cowell JL, Bluestone CD, Doyle WJ (1993) Evaluation of mixtures of purified Haemophilus influenzae outer membrane proteins in protection against challenge with nontypable Haemophilus influenzae in the chinchilla otitis-media model. Infect Immun 61:1950–1957

Groeneveld K, Alphen LV, Eijk PP, Jansen HM, Zanen HC (1988) Changes in outer membrane proteins of nontypable Haemophilus influenzae in patients with chronic obstructive pulmonary disease. J Infect Dis 158:360–365

Gu XX, Tsai CM, Keyama T, Barenkamp SJ, Robbins JB, Lim DJ (1996) Synthesis, characterization, and immunologic properties of detoxified lipooligosaccharide from nontypable. Haemophilus influenzae conjugated to proteins. Infect Immun 64:4047–4053

Gulati S, Mc Quillen DP, Mandrell RE, Jani DB, Rice PA (1996) Immunogenicity of Neisseria gonorrhoeae lipooligossacharide epitope 2C7, widely expressed in vivo with no immunochemical similarity to human glycosphingolipids. J Infect Dis 174:1223–1237

Gulati S, Rice PA, Blake MS, Sarafian SK, Morse SA, Quentin-Millet MJ, Arminjon F (1991) Antibody responses in six volunteers immunized with a gonococcal protein I vaccine. In: Achtman MA, Kohl P, Marchal C et al (eds) Neisseria 1990. De Gruyter, Berlin, pp 229–234

Haase EM, Kyungcheol Y, Morse GD, Murphy TF (1994) Mapping of bactericidal epitopes on the P2 porin protein of nontypable Haemophilus influenzae. Infect Immun 62:3712–3722

Ham SML, von Alphen L, Mooi FR, Putten JPM (1994) The fimbrial gene cluster of Haemophilus influenzae type b. Mol Microbiol 13:673–384

Hammerschmidt S, Müller A, Sillman M (1996) Capsule phase variation in Neisseria meningitidis serogroup B by slipped-strand mispairing in the polysialyltransferase (sia D): correlation with bacterial invasion and the outbreak of meningococcal disease. Mol Microbiol 20:1211–1220

Häyrinen J, Jennings HJ, Raff HV (1995) Antibodies to polysialic acid and its N-propyl derivative: binding properties and interaction with human embryonal brain glycopeptides. J Infect Dis 171:1481–1490

Helminen ME, Maciver I, Latiner JL, Cope LD, Mc Cracken GH JR, Hansen EJ (1993a) A major outer membrane protein of Moraxella catarrhalis is a target for antibodies that enhance pulmonary clearance of the pathogen in an animal model. Infect Immun 61:2003–2010

Helminen ME, Maciver I, Paris M, Latiner JL, Lumbley SL, Cope LD, McCracken GH Jr, Hansen EJ (1993b) A mutation affecting expression of a major outer membrane protein of Moxarella catarrhalis alters serum resistance and survival in vivo. J Infect Dis 168:1194–1201

Helminen ME, Maciver I, Latimer JL, Klesney-Tait J, Cope LD, Paris M, McCracken GH Jr, Hansen EJ (1994) A large antigenically conserved protein on the surface of Moraxella catarrhalis is a target for protective antibodies. J Infect Dis 170:867–872

Hoehn GT, Clark VL (1992a) Isolation and nucleotide sequence of the gene (ani A) encoding the major anaerobically induced outer membrane protein of Neisseria gonorrhoeae. Infect Immun 60:4695–4703

Hoehn GT, Clark VL (1992b) The major anaerobically induced outer membrane protein of Neisseria gonorrhoeae, Pan 1, is a lipoprotein. Infect Immun 60:4704–4708

Holbein BE (1980) Iron-controlled infection with Neisseria meningitidis in mice. Infect Immun 29:886–891

Holbein BE, Jericho KWF, Likes GC (1979) Neisseria meningitidis infection in mice: influence of iron, variations in virulence among strains, and pathology. Infect Immun 24:545–551

Hook EW III, Olsen DA, Buchanan TM (1984) Analysis of antigen specificity of the human serum immunoglobulin G immune response to complicated gonococcal infection. Infect Immun 43:706–709

Hsiao CB, Sethi S, Murphy TF (1995) Outer membran protein CD of Branhamella catarrhalis: sequence conservation in strains recovered from the human respiratory tract. Microb Pathog 19:215–225

Irwin SW, Averill N, Cheng CY, Schryvers AB (1993) Preparation and analysisi of isogenic mutants in the transferrin receptor protein genes, tbp1 and tbp2, from Neisseria meningitidis. Mol Microbiol 8:1125–1133

Jennings HJ, Lugowski C (1982) Immunochemistry of group A, B and C meningococcal polysaccharide-tetanus toxoid conjugates. J Immunol 127:1011–1018

Jennings HJ, Roy R, Gamian A (1986) Induction of meningococcal group B polysaccharide-specific IgG antibodies in mice by using an N-propionylated B polysaccharide-tetanus toxoid conjugate vaccine. J Immunol 137:1708–1713

Jennings HJ, Gamian A, Ashton FE (1987) N-propionylated group B meningococcal polysaccharide mimics a unique epitope on group B Neisseria meningitidis. J Exp Med 165:1207–1211

Jennings MP, Hood D, Peak IRA (1995) Molecular analysis of a locus which controls the biosynthesis and phase variable expression of the lacto-N-neotetraose terminal LPS structure in Neisseria meningitidis. Mol Microbiol 18:724–740

Jennings HJ, Pon R, Lussier M, Lang QL (1996) Protective epitope of N-propionylated group B meningococcal polysaccharide. In: Zollinger WD, Frasch CE, Deal CD (eds) Pathogenic Neisseria, the Tenth International Pathogenic Neisseria Conference, pp 212–213

Jin H, Ren Z, Pozsgay JM, Elhins C, Whitby PW, Morton DJ, Stull TL (1996) Cloning of a DNA fragment encoding a heme-repressible hemoglobin-binding outer membrane protein from Haemophilus influenzae. Infect Immun 64:3134–3141

Jones DM, Eldridge J (1979) Development of antibodies to meningococcal protein and lipopolysaccharide serotype antigens in healthy carriers. J Med Microbiol 12:107–111

Jones DM, Borrow R, Fox AJ, Gray S, Cartwright KA, Poolman JT (1992) The lipooligosaccharide immunotype as a virulence determinant in Neisseria meningitidis. Microb Pathog 13:219–224

Jonsson AB (1996) Molecular interaction of Neisseria gonorrhoeae and Neisseria meningitidis to host cell receptors. In: Zollinger WD, Frasch CE, Deal CD (eds) The Tenth International Pathogenic Neisseria Conference, pp 264–265

Kellogg DS, Peacock WL, Deacon WE, Brown L, Pirkle C (1963) Neisseria gonorrhoeae I. Virulence genetically linked to clonal variation. J Bacteriol 85:1274–1279

Kimura A, Gulig PA, McCracken GH Jr, Loftus TA, Hansen EJ (1985) A minor high-molecular-weight outer membrane protein of Haemophilus influenzae type b is a protective antigen. Infect Immun 47:253–259

King GJ, Swanson J (1978) Studies on gononcoccus infection. XV. Identification of surface proteins of Neisseria gonorrhoeae correlated with leukocyte association. Infect Immun 21:575–584

Klingman KL, Murphy TF (1994) Purification and characterization of a high-molecular-weight outer membrane protein of Moxarella (Branhamella) catarrhalis. Infect Immun 62:1150–1155
Koomey M, Gotschlich EC, Robbins K, Bergstrom S, Swanson JL (1987) Opacity genes in Neisseria gonorrhoeae: control of phase and antigenic variation. Genetics 117:391–398
Kurono Y, Shigemi H, Kodama S, Mogi G (1996) Effects of oral and systemic immunization on nasopharyngeal clearance of nontypable Haemophilus influenzae in BaLB/c mice. Laryngoscope 106:614–618
Kyd JM, Cripps AW (1996) Modulation of antigen-specific T-cell and B-cell responses influence bacterial clearance of nontypable Haemophilus influenzae from the lung in a rat model. Vaccine 14:1471–1478
Kyd JM, Dunkley ML, Cripps AW (1989) Enhanced respiratory clearance of nontypeable Haemophilus influenzae following mucosal immunization with P6 in a rat model. Infect Immun 63 2931–2940
Lackie P, Zuber C, Roth J (1993) Expression patterns of polysialic acid during vertebrate organogenesis. In: Roth J, Rutishauer U, Troy FA (eds) Polysialic acid, from microbes to man, pp 263–278
Lambden PR, Heckels JE, James LT, Watt PJ (1979) Variation in surface protein composition associated with virulence properties in opacity types of Neisseria gonorrhoeae. J Gen Microbiol 114:305–312
Lewis LA, Dyer DW (1995) Identification of an iron-regulated outer membrane protein of Neisseria meningitidis involved in the utilization of hemoglobin complexed to haptoglobin. J Bacteriol 177:1299–1306
Lin L, Ayala P, Mulks M, Enns C, So M (1996) The IgA1 proteases of pathogenic Neisseria increase LAMP 1 turnover and promotes survival of bacteria in epithelial cells. In: Zollinger WD, Frasch CE, Deal CD (eds) Pathogenic Neisseria, the Tenth International Pathogenic Neisseria Conference, p 293
Lissolo L, Maitre-Wilmotte G, Dumas P, Mignon M, Danve B, Quentin-Millet MJ (1995) Evaluation of transferrin-binding protein 2 within the transferrin-binding protein complex as a potential antigen for future meningococcal vaccines. Infect Immun 63:884–890
Loeb MR (1987) Protection of infant rats from Haemophilus influenzae type b infection by antiserum to purified outer membrane protein. Infect Immun 55:2612–2618
Lomholt H, Alphen Lv, Kilian M (1993) Antigenic variation of IgA1 proteases among sequential isolates of Haemophilus influenzae from healthy children and patients with chronic obstructive pulmonary disease. Infect Immun 61:4575–4581
Loosmore SM, Yang YP, Coleman DC Shortreed JM, England DM, Harkness RE, Chong PSC, Klein MH (1996) Cloning and expression of the Haemophilus influenzae transferrin receptor genes. Mol Microbiol 19:575–586
Maciver I, Latimer JL, Liem HH, Muller-Eberhard U, Hrkal Z, Hansen EJ (1996) Identification of an outer membrane protein involved in utilization of hemoglobin-haptoglobin complexes by nontypable Haemophilus influenzae. Infect Immun 64:3703–3712
Mackinnon FG, Borrow R, Gorringe AR, Fox AJ, Jones DM, Robinson A (1993) Demonstration of lipooligosaccharide immunotype and capsule as virulence factors for Neisseria meningitidis using an infant mouse intranasal infection model. Microb Pathog 15:359–366
Mandrell RE (1992) Further antigenic similarities of Neisseria gonorrhoeae lipooligosaccharides and human glycosphingolipids. Infect Immun 60:3017–3020
Mandrell RE, Zollinger WD (1989) Human immune response to meningococcal outer membrane protein epitopes after natural infection or vaccination. Infect Immun 57:1590–1598
Mandrell RE, Schneider H, Apicella M, Zollinger W, Rice PA, Griffis JM (1986) Antigenic and physical diversity of Neisseria gonorrhoeae lipooligosaccharides. Infect Immun 54:63–69

Martin D, Cadieux N, Hamel J, Rioux CR, Brodeur BR, (1996) Immunization with a low molecular weight meningococcal outer membrane protein protects against lethal experimental infection. In: Zollinger WD, Frasch CE, Deal CD (eds) Pathogenic Neisseria, the Tenth International Pathogenic Neisseria Conference, pp 127–128

Haas RS, Veit S, Meyer TF (1992) Silent pilin genes of Neisseria gonorrhoeae. ms 11 and occurrence of related hypervariant sequences among other gonococcal isolates. Mol Microbiol 6:197–208

Melhus Å, Hermansson A, Akkoyunlu M, Prellner K, Forsgren A (1995) Experimental recurrent otitis media induced by Haemophilus influenzae. Protection and serum antibodies. Am J Otolaryngol 16:383–390

Mickelsein PA, Blackman E, Sparling PF (1982) Ability of Neisseria gonorrhoeae, Neisseria meningitidis and commensal Neisseria species to obtain iron from lactoferrin. Infect Immun 35:915–920

Milagres LG, Ramos SR, Sacchi CT, Melles CEA, Viera VSD, Sato H, Brito GS, Moraes JC, Frasch CE (1994) Immune response of Brazilian children to a Neisseria meningitidis serogroup B outer membrane protein vaccine: comparison with efficacy. Infect Immun 62:4419–4424

Miyamoto N, Bakaletz LO (1996) Selective adherence of nontypable Haemophilus influenzae (NTHi) to mucus or epithelial cells in the chichilla Eustachian tube and middle ear. Microb Pathog 21:343–356

Mulks MH, Shoberg RJ (1994) Bacterial immunoglobulin A1 proteases. Methods Enzymol 235:543–554

Mulks MH, Simpson DA, Shoberg RJ (1987) Restriction site polymorphism in genes encoding type 2 but not type 1 gonococcal IgA1 proteases. In: Poolman JT, Zanen HC, Meyer TF et al (eds) Gonococci and meningococci. Kluwer, London, pp 419–426

Murphy TF (1996) Branhamella catarrhalis: epidemiology, surface antigenic structure, and immune response. Microbiol Rev 60:267–279

Murphy TF, Kirkham C, Lesse AJ (1993) The major heat-modifiable outer membrane protein CD is highly conserved among strains of Branhamella catarrhalis. Mol Microbiol 10:87–98

Musacchio A, Aase A, Dalseg R, Delwig AA, Høiby EA, Kolberg J, Michaelson T, Wedege E, Rosenqvist E (1996) Epitope specificity and functional activities of human and murine antibodies against class 4 outer membrane protein from Neisseria. In: Zollinger WD, Frasch CE, Deal CD (eds) Pathogenic Neisseria, the Tenth International Pathogenic Neisseria Conference, pp 165–167

Munson RS Jr, Shenep J, Barenkamp SJ, Granoff DM (1983) Purification and comparison of outer membrane protein P2 from Haemophilus influenzae type b isolates. J Clin Invest 72:677–684

Munson RS Jr, Grass S, West R (1993) Molecular cloning and sequence of the gene for outer membrane P5 of Haemophilus influenzae. Infect Immun 61:1017–4020

Newhall WJ, Wilde III CE, Sawyer D, Haak RA (1980) High molecular weight antigenic protein complex in the outer membrane of Neisseria gonorrhoeae. Infect Immun 27:475–482

Nelson MB, Munson RS Jr, Apicella MA, Sikkema DJ, Molleston JP, Murphy TF (1991) Molecular conservation of the P6 outer membrane protein among strains of Haemophilus influenzae: analysis of antigenic determinants, gene sequences, and restriction fragment length polymorphisms. Infect Immun 59:2658–2663

Parge HE, Forest KT, Hickey MJ, Christensen DA, Getzoff ED, Tainer JA (1995) Structuire of the fibre-forming protein pilin at 2.6Å resolution. Nature 378:32–38

Peeters CCAM, Rümke HC, Sundermann LC, Rouppe van der Voort EM, Meulenbelt J, Schuller M, Kuipers AJ, van der Ley P, Poolman JT (1996) Phase I clinical trial with a hexavalent PorA containing meningococcal outer membrane vesicle vaccine. Vaccine 14:1009–1015

Pettersson A, Kuipers B, Pelzer M, Verhagen E, Tiesjema RM, Tommassen J, Poolman JT (1990) Monoclonal antibodies against the 70-kilodalton iron-regulated protein of Neisseria meningitidis are bacterial and strain specific. Infect Immun 58:3036–3041

Pettersson A, van der Ley P, von Poolman J, Tommassen J (1993) Molecular characterization of the 70-kilodalton iron-regulated outer membrane protein of Neisseria meningitidis. Infect Immun 61:4724–4733

Pettersson A, Klarenbeek V, von Deurzen J, Poolman JT, Tommassen J (1994) Molecular characterization of the structural gene for the lactoferrin receptors of the meningococcal strain H44/76. Microb Dathog 17:395–408

Pettersson A, Maas A, von Wassenaar D, van der Ley P, Tommassen J (1995) Molecular characterization of FrpB, the 70-kilodalton iron-regulated outer membrane protein of Neisseria meningitidis. Infect Immun. 63:4181–4184

Plaut AG, Gilbert JV, Artenstein MS, Capra JD (1975) Neisseria gonorrhoeae and Neisseria meningitidis extracellular enzyme cleaves human immunoglobulin A. Science 190:1103–1105

Preston A, Mandrell RE, Gibson BW, Apicella MA (1996) The lipopolysaccharide of pathogenic gram-negative bacteria. Crit Rev Microbiol 22:139–180

Poolman JT (1995) Development of a meningococcal vaccine. Infect Ag Dis 4:13–28

Poolman JT (1996) Clinical trials with outer membran protein vaccines and PorA recombinant vaccines. In: Zollinger WD, Fransch CE, Deal CD (eds) Pathogenic Neisseria, the Tenth International Pathogenic Neisseria Conference, pp 117–122

Poolman JT, Timmermans JAM, Hopman CTP, Teerlink T, von Vught PAM, Beuvery EC (1987) Comparison of meningococcal outer membrane protein vaccines, solubilized with detergents or C polysaccharide. Antonie Van Leeuwenhoek 53:413–419

Poolman JT, van der Ley PA, Tommassen J (1995) Surface structures and secreted products of meningococci. In: Cartwright K (ed) Meningococcal disease. Wiley, New York, pp 21–34

Rice PA, Vayo HE, Tam MR, Blake MS (1986) Immunoglobulin G antibodies directed against protein III block killing of serum-resistant Neisseria gonorrhoeae by immune serum. J Exp Med 164:1735–1748

Rikitomi N, Anderson B, Matsumoto K, Lindstedt R, Svanborg C (1991) Mechanism of adherence of Moraxella (Branhamella) catarrhalis. Scand J Infect Dis 23:559–567

Rokbi B, Maitre-Wilmotte G, Mazarin L, Fourrichon L, Lissolo L, Quentin-Millet MJ (1995) Variable sequences in a mosaic-like domain of meningococcal tbp2 encode immunoreactive epitopes FEMS Microbiol Lett 132:277–283

Rokbi B, Mignon M, Maitre-Wilmotte G, Lissolo L, Danve B, Caugant D, Quentin-Millet MJ (1997) Evaluation of recombinant transferrin-binding protein B variants from Neisseria meningitidis for their ability to induce cross-reactive and bacterial antibodies against a genetically diverse collection of serogroup B strains. Infect Immun 65:55–63

Rosenqvist E, Høiby EA, Wedege E, Kusecek B, Achtman M (1993) The 5 C protein of Neisseria meningitidis is highly immunogenic in humans and induces bactericidal antibodies. Infect Immun J Infect Dis 167:1065–1073

Rosenqvist E, Høiby EA, Wedege E, Bryn K, Kolberg J, Klem A, Rønnild E, Bjune G, Nøkleby H (1995) Human antibody responses to meningococcal outer membrane antigens after three doses of the Norwegian group B meningococcal vaccine. Infect Immun 63:4642–4652

Ross SC, Densen P (1985) Opsonophagocytosis of Neisseria gonorrhoeae: interaction of local and disseminated isolates with complement and neutrophils. J Infect Dis 151:33–41

Ross SC, Rosenthal PJ, Berberic HM, Densen P (1987) Killing of Neisseria meningitidis by human neutrophils: implications for normal and complement deficient individuals. J Infect Dis 155:1266–1275

Rothbard JB, Fernandez R, Schoolnik GK (1984) Strain specific and common epitopes of gonococcal pili. J Exp Med 160:208–221
Rouppe van der Voort E, van der Ley P, van den Biezen J, Poolman JT (1996) Specificity for human bactericidal antibodies induced with a hexavalent meningococcal PorA outer membrane vesicle vaccine, as determined using isogenic target strains. Infect Immun 64:2745–2751
Rudel T, Scheurpflug I, Meyer T (1995) Neisseria PilC protein identified as type-4 pilus tip-located adhesin. Nature 373:357–359
Saukkonen K, Leinonen M, Abdillahi H, Poolman JT (1989) Comparative evaluation of potential components for group B meningococcal vaccine by passive protection in the infant rat and in vitro bactericidal assay. Vaccine 7:325–328
Schoolnik GK, Fernandez R, Tai JY, Rothbard JB, Gotschlich EC (1984) Gonococcal pili: primary structure and receptor binding domain. J Exp Med 159:1351–1370
Schryvers AB (1989) Indentification of the transferrin-and lactoferrin-binding proteins in Haemophilus influenzae. J Med Microbiol 29:121–130
Sierra VG, Campa C, Garcia L, Sotolongo F, Izquierdo L, Valcarcél M, Casunueva V, Baró M, Leguen F, Rodriguéz R, Terry H (1991) Efficacy evaluation of the Cuban vaccine VA-mengoc-BC against disease caused by serogroup B Neisseria meningitidis. In: Achtman M, Kohl P, Marchal C et al (eds) Neisseria 1990. De Gruyter, pp 129–134
Slater LN (1990) A rat model of prolonged pulmonary infection due to nontypable Haemophilus influenzae. Am Rev Respir Dis 142:1429–1435
Smith H, Parsons NJ, Cole JA (1995) Sialylation of neisserial lipopolysaccharide: a major influence on pathogenicity. Microb Pathog 19:365–377
Sparling PF, Cannon JG, So M (1986) J Infect Dis 153:196–201
Sparling SF, Cornelissen CN, Biswas GD, Anderson JE, Carson SB, Elkins C, Chen CJ, Kashari M (1996) Neisseria iron regulated stress proteins: what do they do? What can we do with them? In: Zollinger WD, Frasch CE, Deal CD (eds) Pathogenic Neisseria, the Tenth Pathogenic Neisseria Conference, pp 545–551
Stern A, Brown M, Nickel P, Meyer TF (1986) Cell 47:61–71
St Geme JW, Falkow S, Barenkamp SJ (1993) High-molecular-weight proteins of nontypable Haemophilus influenzae mediate attachment to human epithelial cells. Proc Natl Acad Sci USA 90:2875–2879
Stimson E, Virji M, Makepeace K, Dell A, Morsis H, Payne G, Saunders J, Jennings M, Barker S, Parico M, Blench I, Moxon ER (1995) Meningococcal pilin: a glycoprotein substituted with digalactosyl 2, 4-diacetamido-2, 4, 6- trideoxyhexose. Mol Microbiol 17:1201–1214
Stimson E, Virji M, Baker S, Danico M, Blench I, Saunders J, Payne G, Moxon ER, Dell A, Morris MR (1996) Discovery of a novel protein modification: α-glycerophosphate is a substituent of meningococcal pilin. Biochem J 316:29–33
Stojilkovic I, Hwa V, Martin LdS, O'Gaora P, Nassif X, Heffon F, So M (1995) The Neisseria meningitidis hemoglobin receptor: its role in iron utilization and virulence. Mol Microbiol 15:531–541
Suzuki K, Bakaletz LO (1994) Synergistic effect of adenovirus type 1 and nontypable Haemophilus influenzae in chinchilla model of experimental otitis media. Infect Immun 62:1710–1718
Swanson J (1973) Studies of gonococcal infection. IV. Pili: their role in attachment of gonococci to tissue culture cells. J Exp Med 137:571–589
Swanson J (1982) Colony opacity and protein II composition of gonococci. Infect Immun 37:359–368
Thompson SA, Wang LL, West SE, Sparling PF (1993) Neisseria meningitidis produces iron-regulated proteins related to the RTX family to exoproteins. J Bacteriol 175:811–818
Tønjum T, Caugant DA, Koomey M (1996) A repetitive sequence element cause polymorphism in the PilQ protein of Neisseria meningitidis. In: Zollinger WD,

Frasch CE, Deal CD (eds) Pathogenic Neisseria, the Tenth International Pathogenic Neisseria Conference, pp 402–403
Tramont EC (1989) Gonococcal vaccines. Clin Microbiol Rev 2 [Suppl]:S74–S77
Tramont EC, Boslego JW, Chung R, Mc Chesney, D; Ciak J, Sadoff J, Piziak M, Brinton CC, Wood S, Bryan J (1985) Parenteral gonococcal pilus vaccine. In: Schoolnik GK et al (eds) The pathogenic Neisseriae. American Society for Microbiology, pp 316–322
Troelstra A, Vogel L, Alphen Lv, Eijk P, Jansen H, Dankert J (1994) Opsonic antibodies ot outer membrane protein P2 of monencapsulated Haemophilus influenzae are strain specific. Infect Immun 62:779–784
Tsai WM, Larsen SH, Wilde CE III (1989) Cloning and DNA sequence of the omc gene encoding the outer membrane protein-macromolecular complex from Neisseria gonorrhoeae. Infect Immun 57:2653–2659
Unhanand M, Maciver I, Ramilo O, Arencibia-Mireles O, Argyle JC, McCracken GH Jr, Hansen EJ (1992) Pulmonary clearance of Moraxella catarrhalis in an animal model. J Infect Dis 165:644–650
van der Ley PA (1988) Three copies of a single protein II encoding sequence in the genome of Neisseria gonorrhoeae JS3: evidence for gene conversion and gene duplication. Mol Microbiol 2:797–806
van der Ley PA, Heckels JE, Virji M, Hoogerhout P, Poolman JT (1991) Topology of outer membrane porins in pathogenic Neisseria pp. Infect Immun 59:2963–2971
van der Ley PA, van den Biezen J, Poolman JT (1995) Construction of Neisseria meningitidis strains carrying multiple chromosomal copies of the PorA gene for use in the production of a multivalent outer membrane vesicle vaccine. Vaccine 13:401–407
van der Ley P, van den Biezen J, Stutmuller R, Hoogerhout P, Poolman JT (1996) Sequence variability of FrpB, a major iron-regulated outer-membrane protein in the pathogenic neisseriae. Microbiology 142:3269–3274
Vaneechoutte M, Verschraegen G, Claeys G, Van Den Abeele AM (1990) Serological typing of Branhamella catarrhalis strains on the basis of lipopolysaccharide antigens. J Clin Microbiol 28:182–187
Verheul AFM, Snippe H, Poolman JT (1993) Meningococcal lipopolysaccharides: virulence factor and potential vaccine component. Microbiol Rev 57:34–49
Virji M (1996) Virulence determinants of meningococci and factors that may determine between the carrier state and invasive disease. In: Zollinger WD, Frasch CE, Deal CD (eds) Pathogenic Neisseria, the Tenth Pathogenic Neisseria Conference, pp 247–255
Virji M, Alexandrescu C, Ferguson DJP, Saunders JR, Moxon ER (1992) Variation in the expression of pili: the effect on adherence of Neisseria meningitidis to human epithelial and endothelial cells. Mol Microbiol 6:1271–1279
Virji M, Makepeace K, Ferguson DJP, Achtman M, Moxon ER (1993) Meningococcal Opa and Opc proteins: role in colonisation and invasion of human epithelial and endothelial cells. Mol Microbiol 10:499–510
van Putten JPM (1993) Phase variation of lipopolysaccharide directs interconversion of invasive and immuno-resistant phenotypes of Neisseria gonorrhoeae. EMBO J 12:4043–4051
Wedege E, Høiby EA, Andersen SR, Bolstad K, Haugen IL, Haneberg B (1996) Immunoblot analyses of antibody responses in sera, saliva and nasal fluids from volunteers immunized nasally with the Norwegian group B meningococcal vacccine. In: Zollinger WD, Frasch CE, Deal CD (eds) Pathogenic Neisseria, the Tenth Pathogenic Neisseria Conference, pp 192–193
Wetzler LM, Blake MS, Barry K, Gotschlich EC (1992a) Gonococcal porin vaccine evaluation. Comparison of Por proteosomes, liposomes, and blebs isolated from rmp deletion mutants. J Infect Dis 166:551–555

Wetzler LM, Barry K, Blake MS, Gotschlich EC (1992b) Gonococcal lipoolisaccharide sialylation prevents complement-dependent killing by immune sera. Infect Immun 60:39–43
Wyle FA, Artenstein MS, Brandt BL, Tramont EC, Kasper DL, Alteri PL, Berman SL, Lowenthal JP (1972) Immunological response of man to group B. meningococcal polysaccharide vaccines. J Infect Dis 126:514–522
Zollinger WD, Boslego J, Moran E (1991) Meningococcal serogroup B vaccine protection trial and follow-up studies in Chile. NIPH Ann 14:211–213

CHAPTER 10

Carbohydrate-Based Bacterial Vaccines

J.T. POOLMAN, C. LAFERRIÈRE, and S.B. SVENSON

A. Introduction

It was recognized early in the twentieth century that pneumococcal capsular polysaccharides have potential as vaccine components. Chemical conjugation of the polysaccharides to a carrier protein improves immunogenicity (AVERY and GOEBEL 1969). The development of methods to purify high molecular weight meningococcal capsular polysaccharides led the administration of vaccines to humans (GOTSCHLICH et al. 1969).

Meningococcal, pneumococcal, and *Hemophilus influenzae* type b (Hib) polysaccharide vaccines have been found to be protective in the field (ARTENSTEIN et al. 1970; PELTOLA et al. 1977a,b). Field experience with polysaccharide vaccines have demonstrated that infants mount no or only a poor immune response to polysaccharides (PELTOLA et al. 1977a). The age limit of responsiveness varies with the capsular polysaccharide, meningococcal group A polysaccharide being an example of early responsiveness of as low as 3 months of age (KÄYHTY et al. 1980). With respect to the Hib polysaccharide, significant immune responses and clinical protection can be induced only above the age of 18 months (PELTOLA et al. 1977b).

Another issue of major concern is the duration of protection, i.e., immunological memory (REINGOLD et al. 1985). The lack of induction of immunological memory with plain polysaccharides can also be demonstrated by the absence of a booster response and the absence of antibody isotype switching (KÄYHTY et al. 1984, 1987). The poor response in infants and the lack of immunological memory have prevented the use of polysaccharide vaccines on a large scale.

The mechanisms explaining the peculiar immunological properties of polysaccharides relate to three phenomena:

- Polysaccharides contain a number of repeating epitopes within one molecule which can result in the cross-linking of antigen receptors on the surface of B lymphocytes.
- Many bacterial polysaccharides activate the alternative pathway leading to a deposition of C3d onto the polysaccharide (GRIFFIOEN et al. 1991). C3d can bind to complement receptor 2 (CR2 or CD21) on the surface of B lymphocytes. Coactivation of CR2 and surface immunoglobulins can explain the induction of an antibody response. CR2 is present at low levels on the

surface of neonatal B lymphocytes, which may explain the poor responsiveness of infants to polysaccharides (RIJKERS et al. 1996).
- The results described under the two above points explain why polysaccharide-specific B lymphocytes can be activated in a T-independent manner leading to absence of immunological memory, isotype switching, and affinity maturation. However, T cell regulation of the antipolysaccharide response in man does exist (RIJKERS et al. 1996), although it has not been studied in great detail.

Both $CD4^+$ $\alpha\beta$ as well as $\gamma\delta$ T cell receptor cells can affect the B cell response in a non-MHC-restricted fashion. The recognition of idiotypic determinants on B cells is a possible explanation for these observations while non-MHC binding of polysaccharides to T cells requires further attention. Polysaccharide-specific B cells are present at birth (RIJKERS et al. 1988). The activation of such cells has been found to be possible by phorbol esters and ionomycin. A clinically relevant solution to the immunological shortcomings of polysaccharides has been provided by chemical conjugation to carrier proteins (AVERY and GOEBEL 1969; SCHNEERSON et al. 1980). T cells specific for the carrier protein become involved in the activation of B cell specific for the polysaccharide. Such conjugate vaccines can induce clinically relevant immune responses in infants, including the induction of immunological memory (ESKOLA and KÄYHTY 1996).

B. The Hib Experience

The Hib polysaccharide vaccine poly-ribose-ribitol-phosphate (PRP) has been found to be 90% efficacious at the age of 18–24 months or older in preventing Hib infectious diseases (PELTOLA et al. 1977b). This study provided population data that indicated 0.15 μg/ml anti-PRP to be associated with short-term and 1.0 μg/ml for long-term protection (KÄYHTY et al. 1983). These levels correspond to protective levels achieved after passive immunization with polyclonal anti-PRP (AMBROSINO et al. 1986). The plain PRP vaccine has come into use in the United States to some extent but has only had a marginal effect on the incidence of Hib infectious diseases.

Because of the immunological shortcomings of polysaccharide vaccines Hib conjugate vaccines were developed and initially tested in Finland (ESKOLA et al. 1985). The PRP–diphtheria toxoid vaccine was found 90% efficacious after a primary immunization series of three doses (ESKOLA et al. 1987) at the age of 3, 4, and 6 months. In addition to PRP-D, other PRP conjugate vaccines were developed and shown to be efficacious in infants (BLACK et al. 1991; SANTOSHAM et al. 1991; BOOY et al. 1994). Diphtheria cross-reacting material, atoxic mutant CRM197, tetanus toxoid, or meningococcal outer membrane proteins (OMPs) were used as carrier proteins. Two major kinds of conjugate vaccine principles were thus used and tested clinically: (a) high molecular weight polysaccharide cross-linked to the carrier protein (PRP-D, PRP-T,

PRP–outer membrane preparation) and (b) low molecular weight oligosaccharide end-linked to the carrier protein (HbOC).

In general PRP-D appeared to be the least immunogenic and demonstrated a poor efficacy in a field trial in Alaska (WARD et al. 1990) (DECKER et al. 1992). A combination of low immunogenicity, high infection pressure, and other factors explain this low efficacy. In other studies the experience have been positive (ESKOLA and KÄYHTY 1996). It appears that both kinds of conjugate vaccines, i.e., the cross-linked (high molecular weight) and the end-linked (low molecular weight) are effective for the induction of relevant immune responses. This finding is in line with the immunological character of the polysaccharide, containing many repetitive epitopes. As long as the relevant epitope(s) are presented to the immune system the appropriate response can be induced, as demonstrated by the use of chemically synthesized, short-chain ribose-ribitol-phosphate (PEETERS et al. 1992b).

An important conclusion can be drawn from the efficacy trials with conjugate vaccines: the antibody concentrations that are correlated with protection on the basis of polysaccharide vaccine studies and after passive immunization with immunoglobulins do not seem relevant for conjugate vaccines. Lower levels of antibodies as obtained with conjugate vaccines still afford protection as compared with the polysaccharide immunization experience. The induction of immunological memory appears to be a second major mechanism of protection in vaccinated individuals (INSEL and ANDERSON 1986; GRANOFF et al. 1993; ESKOLA and KÄYHTY 1996). This memory leads to a vigorous, IgG-dominated antibody response after the primary series with conjugate vaccine, which can also be achieved by the plain polysaccharide (KÄYHTY et al. 1992). It remains to be seen how long such memory persists, but it appears to be at least 8–10 years (ESKOLA and KÄYHTY 1996). Interestingly, T cells can be isolated from conjugate-primed individuals which can be activated by the plain polysaccharide (PEETERS et al. 1992c).

In addition to the induction of memory because of the T cell dependent immune response, antibody avidity appears to be different (KÄYHTY et al. 1988; SCHLESINGER et al. 1992; LUCAS and GRANOFF 1995). Antibodies with higher avidity, bactericidal activity, and animal protection capacity have been found to be induced by the conjugates as compared to the plain polysaccharide. Differences among the various conjugate vaccines can also be found, although no relationship to clinical protection is evident. The IgGl/G2 ratios found are generally in the same range when the conjugates are compared, and even with the plain polysaccharide. The HbOC vaccine appears to induce higher IgGl/G2 ratios (SÄPPÄLÄ et al. 1988). No apparent differences in functional activities among IgGl or G2 anti-PRP were found in one study (WEINBERG et al. 1986); however, in another study IgGl antibodies were clearly superior in bactericidal and opsonic activities (AMIR et al. 1990).

It has been demonstrated that the allotype of IgG2 heavy chains G2m (n) is correlated with a high, while the $\varkappa$-light chain allotype Km(1) corresponds to a low level of the IgG anti-polysaccharide antibody response (GRANOFF and

HOLMES 1992). In adults the IgG2 subclass dominates among anti-polysaccharide immunoglobulins. An interesting biallelic polymorphism has been described among human granulocyte Fc receptors for IgG: FcγRII, which is the only FcγR that binds IgG2 expresses polymorphism at amino acid 131, being either histidine or arginine. In Dutch patients with recurrent respiratory tract infections from encapsulated bacteria a significantly lower frequency of H131 homozygous individuals was encountered. Only H131 FcγRII receptors can interact with IgG2 (SANDERS et al. 1994). Both the immune response related to avidity and functional activities and the immunological memory (and thus a rapid response after exposure to infection) differ between polysaccharide (or natural infection) and conjugate immunizations. This implies that conjugate immunization changes observed susceptibilities with respect to Km(1) and FcγRII.

The outcome of Hib conjugate immunization is virtually total elimination of Hib disease (ESKOLA and KÄYHTY 1996). The rapid disappearance of Hib diseases is related not only to relevant antibody levels and memory in vaccinees, but a reduction in carriage among vaccinees is also observed (MURPHY et al. 1993; BARBOUR 1996). Such a reduced carriage has an impact on the spread of Hib and causes a concomitant reduction in disease in nonvaccinated individuals (ESKOLA and KÄYHTY 1996). Infants and children vaccinated with Hib conjugates have been shown to possess IgG and IgA antibodies in saliva (KAUPPI et al. 1995).

C. Meningococcal Polysaccharide and Conjugate Vaccines

It was discovered in the 1940s that antibodies to the group-specific capsular polysaccharide protect mice from lethal challenge (SCHERP and RAKE 1945). The development of purification procedures by GOTSCHLICH et al. (1969) of high molecular weight bacterial polysaccharides led to efficacy trials and eventually registered tetravalent ACYW-135 vaccines (LEPOW et al. 1986; KING et al. 1996). The protective efficacies of A and C polysaccharide vaccines have been demonstrated (ARTENSTEIN et al. 1970; PELTOLA et al. 1977a).

Group C vaccine demonstrates an immunological pattern resembling that of Hib-PRP vaccine: poor response in children under 2 years of age and no evidence of booster responses (GOLD et al. 1975; LEPOW et al. 1977). Booster immunizations within 1 year after primary immunization may lead to hyporesponsiveness (GOLD et al. 1975). Some heterogeneity is found among group C meningococci; ±15% of clinical isolates appear to produce a polysaccharide devoid of O-acetyl groups (APICELLA 1974); O-acetyl-negative and O-acetyl-positive group C polysaccharides produce similar immune responses when tested in a tetravalent ACYW135 formulation (PELTOLA et al. 1985). In contrast to the results obtained with menC and Hib-PRP polysaccharides, the menA polysaccharide induces immune responses in infants around 6 months of age, and the response can be boosted (GOLD et al. 1975; PELTOLA et al. 1977a).

Despite these encouraging results the field experiences in Burkina Faso indicate a decline in protection rates over 2–3 years (REINGOLD et al. 1985). It remains unknown, despite the possibility of boosting the menA-PS response, what would happen if a multidose schedule is implemented, particularly in regions with high menA incidences.

The meningococcal polysaccharide vaccines as a whole have similar immunological shortcomings to those of Hib-PRP, with some exception for menA-PS, and the vaccines are not being used in the general immunization programs. The vaccines are recommended for high-risk groups such as soldiers, travellers to epidemic areas, individuals with high susceptibility such as those with complement deficiencies, and asplenic persons. Occasionally the vaccines are used during outbreaks and epidemics (LEPOW 1992).

The meningococcal B polysaccharide appears nonimmunogenic in man, most likely because the α-(2→8)-polysialic acid is identically present on human glycoproteins such as the neural cell adhesion molecules (WYLE et al. 1972). Because of its special character the efforts towards a menB vaccine are described in Chap. 9.

After the success of the Hib conjugate vaccines, vaccine development towards meningococcal conjugates was initiated. Early investigations demonstrated the possibility of improving the immunogenicity of meningococcal polysaccharides by way of conjugation to carrier proteins (JENNINGS and LUGOWSKI 1981; BEUVERY et al. 1983). Recently immunogenicity studies in children and infants have been initiated with men A/C end-linked oligosaccharide CRM 197 conjugates. Clearly better immune responses were obtained with the conjugate vaccines than with the plain polysaccharides, and further developments towards application of these vaccines are anticipated (FAIRLEY et al. 1996; LIEBERMAN et al. 1996; LEACHK et al. 1997).

D. Pneumococcal Polysaccharide and Conjugate Vaccines

Streptococcus pneumoniae causes an overall bacteremic infection rate of 8 per 100000 in the United States. It is carried in the nasopharynx in 18%–60% of preschool children (FEDSON 1988) and is responsible for between 9% and 21% of childhood meningitis and between 25% and 50% of cases of acute bacterial *Otitis media* (KLEIN 1994). The capsular polysaccharide of pneumococcus is its major virulence factor (ALONSO DE VELASCO et al. 1995; WATSON et al. 1995), and antisera to it have demonstrated protection against disease and infection. The chemical structure of the capsular polysaccharide is responsible for serum specificity (VAN DAM et al. 1990), and 90 different serotypes have now been identified (HENRICHSEN 1995).

Serotypes 6B, 14, 18C, 19F, and 23F are the most strongly associated with disease in infants and children; this is probably related to the nonimmunogenicity of their capsules in children younger than 2 years since their relative prevalence in causing disease decreases with age. Other immunogenic serotypes such as 3 and 7F have increasing relative prevalence with age for unknown reasons (SCOTT et al. 1996). A hexavalent polysaccharide vaccine was first licensed in the United States in the late 1940s (WATSON et al. 1993), but it was withdrawn from the market for lack of use since antibiotics were seen as more cost effective; however, the vaccine was reintroduced because of the continued high fatality rate in certain populations at risk even with antibiotic intervention (AUSTRIAN 1984). Indeed, antibiotic resistance is now a major factor for the advancement of pneumococcal vaccines (APPLEBAUM 1992; MUNFORD and MURPHY 1994; BUTLER et al. 1996).

In 1977 a 14 valent vaccine was licensed, followed in 1983 by the 23 valent. The selection of serotypes included was based on their relative frequency in causing invasive disease and on cross-reaction between related serotypes (ROBBINS et al. 1983). Recently there has been a revaluation of the vaccine efficacy (FEDSON et al. 1994; HIRSCHMAN and LIPSKY 1994), and case controlled versus double-blinded studies have reported values between 56% and 79% (MILLS and RHOADS 1996), although it is estimated to be lower in groups at risk such as patients with renal transplant or failure, hematological or metastatic cancer, nephrotic syndrome, systemic lupus erythematosus, sickle cell anemia, or alcoholism (FINE et al. 1994). A protective antibody level against invasive disease has been estimated at 200–300 ng/ml N Ab (1.3–2.0 μg/ml Ig) (SCHIFFMAN 1983). Likewise, in passive immunization trials, IgGl and IgG2 levels of 2.3 and 4.2 μg/ml were associated with a reduction in the number of *Otitis media* episodes per year (ISHIZAKA et al. 1994).

It was quickly recognized that the polysaccharide vaccine is not immunogenic in infants (SELL et al. 1981; KLEIN 1995), and this has been explained by reasons described above. Hence efforts have since focused on increasing the immunogenicity of the polysaccharides via chemical conjugation with a protein carrier.

Pneumococcal conjugate vaccine development has been hampered by the diversity in chemistry for the many different polysaccharide structures of the serotypes causing disease. There was a proliferation of preclinical work in the early 1980s and continuing into this decade, which is now culminating in extensive clinical testing.

The optimization of polysaccharide-protein conjugates depends upon many variables, including: the choice of the protein carrier, the size of the polysaccharide, the chemistry used in linkage, including the use of small spacer molecules, the ratio of polysaccharide to protein and ultimately the adjuvant formulation. Each of these are considered briefly below.

I. Protein Carrier

Protein carriers used in preclinical studies have included: *Salmonella typhi* flagella (VAN DE WIJGERT et al. 1991), keyhole limpit hemocyanin (SNIPPE et al. 1983; ALONSO DE VELASCO et al. 1993, 1994a,b), pneumolysin (VERHEUL et al. 1989; PATON et al. 1991; LEE et al. 1994; KUO et al. 1995), pertussis toxin

(SCHNEERSON et al. 1992), diphtheria toxoid (ANDERSON and BETTS 1989), tetanus toxoid (PEETERS et al. 1991a,b, 1992a,b; SCHNEERSON et al. 1986; LAFERRIÈRE et al. 1997), meningococcal group B outer membrane preparation (VELLA et al. 1992; GIEBINK et al. 1993, 1996), and CRM197 (ARNDT and PORRO 1991; VAN DEN DOBBELSTEEN et al. 1995). Of these, the latter four have advanced to clinical studies by Connaught Laboratories, Pasteur Merieux, Merck, and Lederle, respectively (CVI FORUM 1996).

II. Polysaccharide Size

The degree of polymerization (dp) of the pneumococcal polysaccharide in the conjugate has not been studied extensively, and it is difficult to draw conclusions from results presented in the literature. LAFERRIÈRE et al. (1997) found no effect of size on the immunogenicity in rabbits of serotypes 3, 6A, 18C, and 23F, but ARNDT and PORRO (1991) found conjugates of 19F and 23F of dp 10 to be nonimmunogenic. LAFERRIÈRE (1997) found that serotype 19F has less immunogenicity at 40 dp than at 10 dp. For types 4 and 14 higher molecular weight polysaccharides have been found to be more immunogenic (PEETERS et al. 1991a; LAFERRIÈRE and JENNINGS 1997). Very few data are available in humans, but it appears that higher molecular weight polysaccharides are more immunogenic in conjugates with 6B and 23F (STEINHOFF et al. 1994). The major advantage of using smaller polysaccharides is that nonconjugated polysaccharide, which significantly reduces the immunogenicity of the conjugate (PEETERS et al. 1992a), can be removed by size exclusion chromatography.

III. Coupling Chemistry

The most common chemistry used to make the linkage between the polysaccharide and protein carrier can be divided into three classifications based on the way in which the polysaccharide is activated for coupling. Conjugates made by Merck use carbonyldiimidazole to activate the polysaccharides which are then reacted with outer membrane preparations through a complicated series of reactions (MARBURG et al. 1986). Pasteur Merieux Connaught have used the chemistry originally developed by SCHNEERSON et al. (1980) employing cyanogen bromide to activate the polysaccharide which is then coupled to a small linker molecule and used to couple to protein. The third method, used by Lederle and North American Vaccine activates the polysaccharide by oxidation with periodate, resulting in the formation of aldehydes which are coupled directly with the amino of the lysines in the protein by reductive amination (JENNINGS and SOOD 1994).

IV. Polysaccharide-Protein Ratio

While theoretical considerations exist for the optimal polysaccharide-protein ratio (PEETERS et al. 1991a; JENNINGS 1992), little work has been reported towards optimizing pneumococcal polysaccharide-protein ratio. PEETERS et al. (1991a) found that a serotype 4 saccharide-tetanus toxoid ratio less than 1 is more immunogenic for polysaccharides, but that the inverse is the case for oligosaccharides. LAFERRIÈRE et al. (1997) found no significant effect of the ratio on serotype 3-tetanus toxoid conjugate immunogenicity. Merck selected their ratio based on results from the Hib vaccine (VELLA et al. 1992), using rather low saccharide protein ratios.

V. Adjuvant

The classical adjuvants of $Al(OH)_3$ (VELLA et al. 1992; SCHNEERSON et al. 1992; LU et al. 1994; GIEBINK et al. 1996; LAFERRIÈRE et al. 1997) and $ALPO_4$ (STEINHOFF et al. 1994; KUO et al. 1995; SHELLY 1997) have been used with pneumococcal conjugate vaccines. There have also been experiments with new adjuvants such as Quil A (VERHEUL et al. 1989; ALONSO DE VELASCO et al. 1993, 1994b) and MPL (VAN DER WIJGERT et al. 1991; GARG and BONDADA 1992; KEYSERLING et al. 1996).

VI. Animal Models and Correlates of Protection

Pneumococcal conjugate vaccines have been tested most often in mice (MUSHER et al. 1990; LEE et al. 1994, 1996; PATON et al. 1991; PEETERS et al. 1991a,b, 1992a,b; VAN DE WIJGERT et al. 1991; VELLA et al. 1992; ALONSO DE VELASCO et al. 1993; EBY et al. 1994; LU et al. 1994; KUO et al. 1995; VAN DEN DOBBELSTEEN et al. 1995; AABERGE et al. 1996), followed by monkeys (VELLA et al. 1992; KEYSERLING et al. 1996) and rabbits (ALONSO DE VELASCO et al. 1994; EBY et al. 1994; LEE et al. 1996; LAFERRIÈRE et al. 1997). GIEBINK et al. (1993, 1996) have developed a chinchilla model of *Otitis media.*

The use of an enzyme-linked immunosorbent assay to quantitate the humoral response to the vaccine in humans has undergone several revisions since it was discovered that residual cell wall polysaccharide can adversely effect the results (KOSKELA 1987; MUSHER et al. 1990; VIÔARSSON et al. 1994). A standard protocol now exists (QUATAERT et al. 1995), and it is hoped that this will make the results between different laboratories more comparable (GO and BALLAS 1996). This same assay can also be used in animal models.

Antibody isotype response is correlated with bacterial disease in humans; hence the measurement of isotype response is encouraged. For example, chronic or recurrent pneumococcal infection is associated with IgG2 deficiency (FRENCH et al. 1995; SANDERS et al. 1995), and this subclass has been correlated with pneumococcal opsonization (LORTON et al. 1993; VIÔARSSON et al. 1994), although opsonization has also been associated with IgGl and IgG4 (BARDARDOTTIR et al. 1990). Extending the analysis further, high antibody levels are correlated with G2m (n) allotype (FEDSON 1988; KONRADSEN et al. 1994), whereas responsiveness in HIV-infected whites associated with the Km(l) negative allotype (RODRIGUEZ-BARRADAS et al. 1996). This correlation does not exist with conjugate vaccines; hence the association is probably with the T-independent mechanism of antibody production. It is tempting to speculate that this is related to the idiotypic determinant of T-amplifier cells (CAULFIELD and STANKO 1995; RIJKERS et al. 1996).

Phagocytosis is the dominant defense mechanism against pneumococcal disease. The in vitro opsonophagocytic assay has been employed in many different configurations to determine correlates with serology. Most frequently human polymorphonuclear leukocytes (neutrophils) isolated from willing donors have been used as the phagocytic cell (CHUDWIN et al. 1985; BARDARDOTTIR et al. 1990; ESPOSITO and CLARK 1990; LORTAN et al. 1993; VIÔARSSON et al. 1994; ANDERSON et al. 1996; LEE et al. 1996; OBARO et al. 1996), but murine J774 A.1 (ALONSO DE VELASCO et al. 1994b) and HL60 cells (WHITIN and ANDERSON 1993; LAFERRIERE et al. 1997) have also been used. A standardized protocol will also be useful for interlaboratory comparison (ROMERO-STEINER et al. 1997).

Opsonophagocytosis is a direct biological test of serum activity, and preliminary results indicate that it may be a strong indicator of protection (WENGER et al. 1996). Because pneumococci contain a complement C3 degrading enzyme (ANGEL et al. 1994), it will have to be considered whether to use human complement in the assay, as was deemed necessary in the case of the meningococcal bactericidal assay (ZOLLINGER and MANDRELL 1983). Indeed, IgG2 may be the evolutionarily favored subclass against pneumococcal polysaccharide because this subclass can opsonize pneumococci in the absence of complement (LORTON et al. 1993).

Active protection experiments in mice have been reported for the important types 4 (MUSHER et al. 1990), 9V (LEE et al. 1996), 18C (KUO et al. 1994), and 19F (PATON et al. 1991; LEE et al. 1994). BLANK et al. (1994) reported a guinea pig model of meningitis with type 3. Giebink has developed a chinchilla model of *Otitis media* using types 6A and 6B, 14, 19A and 19F (GIEBINK et al. 1993, 1996). It should be noted that the different virulence observed between different strains of serotype 6B pneumococci in a (murine) passive protection model (AABERGE et al. 1996) indicates the existence of unknown virulence factors.

VII. Clinical Trials of Pneumococcal Polysaccharide-Protein Conjugate Vaccines

Postlicensure testing of the 23 valent vaccine is still being performed in populations at risk, such as alcoholics (MCMAHON et al. 1993), splenectomy patients (REINERT et al. 1994), AIDS patients (RODRIGUEZ-BARRADAS et al. 1996), children (LEE et al. 1995), and the elderly (HEDLUND et al. 1994; KONRADSEN 1995). Added to this now are phase 1 and phase 2 clinical trials with the new conjugate vaccines, and Table 1 summarizes the clinical reports published to date. None of the reports indicates symptoms of vaccination that were considered dangerous. The studies have examined the dose response or

Table 1. Clinical trials with new conjugate vaccines

Vaccine	Steinhoff et al. (1994) Lederle	Käyhty et al. (1995) Merck	Ahmed et al. (1996) Lederle	Dagan et al. (1996) Merck	Anderson et al. (1996) Merck	Chan et al. (1996) Merck	Åhman et al. (1996a) Lederle	Powers et al. (1996) Lederle	Åhman et al. (1996b) Pasteur TT	Åhman et al. (1996c) Connaught	Käyhty et al. (1996) Lederle	Pichichero et al. (1997) Lederle	Shelly et al. (1997) Lederle
Adjuvant	$ALPO_4$	$AL(OH)_3$	?	$AL(OH)_3$	?		$AL(OH)_3$	None	?	?	?	$ALPO_4$	$ALPO_4$
Population Schedule	18–30 mo 1 dose, oligo. vs. native	Infants 2, 4, 6, vs. 4, 6, 14 mo	AIDS 1 dose vs. 23 valent	12–18 mo 1 vs. 2 dose 3 mo apart	Infants 2, 4, 6 mo + 23 valent at 12 mo	Hodgkins 1 dose + 23 valent at 12 mo	Infants 2, 4, 6 mo 23 valent	Elderly 1 dose + 23 valent vs. 23 V	Infants 2, 4, 6 mo + 23 valent at 14 mo	Infants 2, 4, 6 mo + boost 23 at 14 mo	Infants 2, 4, 6 mo valent at 24 mo	21–27 mo 2 doses 2 mo apart	Elderly 1 dose + boost 23 valent at 2 mo vs. 23 valent
Dose (μg, type PS)													
4				1	1	1							
6	2 vs. 10	1	10	3.5	3.5	2.5	10	10		1 vs. 3 vs. 10	10	10	10
9				1.5	1.5	1							
14		1	10	1	1	1	10	10		1 vs. 3 vs. 10	10	10	10
18			10	1	1	1	10	10			10	10	10
19		1	10	2.0	2.0	1	10	10		1 vs. 3 vs. 10	10	10	10
23	2 vs. 10	1	10	1	1	1	10	10		1 vs. 3 vs. 10	10	10	10
Endpoint	Safety, IG after 1 mo	Safety, IG after 1 mo	Safety, IG after 1 mo	Safety, IG and carriage after 12 mo	Safety, IG after 1 mo	IG after 3–6 weeks	Safety, IG after 1 mo	Safety, IG after 1 mo	IG after 1 mo	IG after 1 mo	IG after 1 Mo	Safety, IG after 1 mo	Safety, IG after 1, 3 mo

IG Immunogenicity.

immunization schedule. The Lederle conjugate appears most immunogenic at 10 μg per polysaccharide, while the Pasteur and Connaught vaccines have dose-response curves that decrease significantly from 1 to 10 μg. This kind of tolerizing effect has been attributed to the presence of unconjugated polysaccharide (PEETERS et al. 1992a). KÄYHTY et al. (1995) studied two different schedules and found 3 doses at 2, 4, and 6 months to give significantly higher geometric mean concentrations of specific antibody than 2 doses at 4 and 6 months.

An interesting approach in the strategy in infants has been to prime with the conjugate vaccine and to boost with the 23 valent vaccine. The value of this approach has been confirmed in several studies (ANDERSON et al. 1996; KÄYHTY et al. 1996), and it is likely to be incorporated into the childhood vaccination recommendations. The conjugate vaccine may hold some benefits for certain populations at risk, such as those with Hodgkins disease (CHAN et al. 1996), but not for other populations, such as AIDS patients (AHMED et al. 1996) and the elderly (POWERS et al. 1996; SHELLEY et al. 1997), although the latter may benefit from longer duration of protection. Longitudinal studies must be performed to verify this. At least 20 other studies are planned or now ongoing (CVI FORUM 1994), and there are thus great expectations for a pneumococcal conjugate vaccine.

VIII. The Future of Pneumococcal Vaccines

New ideas are required to increase the efficacy of the pneumococcus vaccine, and in this light innovations such as maternal immunization (LEE et al. 1994; LU et al. 1994; SHAHID et al. 1995), mucosal immunization (Van Den Dobbelsteen 1994; VAN COTT et al. 1996) and inclusion of pneumococcal proteins (PATON et al. 1991; KUO et al. 1995) are currently under investigation.

E. *Salmonella*, *Shigella*, and *E. coli*

I. *Salmonella:* O-Antigen Specific Conjugates

The preparation of a *Salmonella* serotype BO specific O-antigen-protein conjugate was described as early as 1977 (SVENSON and LINDBERG 1977). This conjugate was based on oligosaccharides of the O-polysaccharide chain obtained by the use of specific bacteriophage-associated endo-alpharhamnosidase. The resulting conjugate was shown to yield substantial anti-serotype BO antibody titers upon injection in rabbits (SVENSON and LINDBERG 1977). This first conjugate used a coupling procedure based on the conversion of the oligosaccharide to its corresponding aldonic acid and conjugation to the carrier protein by a water soluble carbodiimide. In 1979 the same authors reported the construction of a number of conjugates with varying carrier proteins by a newly developed technique [2-(4-isothiocyanotophenyl)-ethylamine derivatives] for covalent conjugation (SVENSON and LINDBERG

1979) (JÖSBECK et al. 1979). This new technique allowed the preparation of non-cross-linked conjugates as opposed to the aldonate method previously used (SVENSON and LINDBERG 1977). Also these conjugates proved to be highly immunogenic in rabbits – an octasaccharide-diphtheria conjugate increased the titer in the range of 20000–50000 against both the oligosaccharide hapten and the carrier protein (SVENSON and LINDBERG 1979; JÖSBECK et al. 1979). The same year this group of researchers also showed in the mouse typhoid model that such a conjugate vaccine confers protection after both active immunization and passive transference of antibodies elicited by the conjugate vaccine (SVENSON et al. 1979).

During the 1990s in particular Robbins and coworkers at Bethesda have been very active in the establishment of many bacterial carbohydrate antigen protein conjugates intended for vaccine purposes (WATSON et al. 1992; KONADU et al. 1996; POLOTSKY et al. 1994; COHEN et al. 1996; SZU et al. 1994; GUPTA et al. 1994; KLUGMAN et al. 1996; KONADU et al. 1994). As regards *Salmonella* O-antigen specific conjugate vaccine candidates these researchers have developed methods for detoxification of the intrinsic endotoxic activities of O-antigenic lipopolysaccharides and devised methods for the covalent attachment of such detoxified lipopolysaccharides to proteins. In 1996 they reported the synthesis and immunological properties in mice of conjugates composed of detoxified lipopolysaccharide of *Salmonella paratyphi* A bound to tetanus toxoid (KONADU et al. 1996). With their developed detoxification procedure (treatment with acetic acid or hydrazin) they detoxified, isolated, and purified lipopolysaccharides from *Salmonella paratyphi* A serogroup A and one of the most important pathogens among *Salmonella* (KONADU et al. 1996). The detoxified O-polysaccharide was then activated with cyanogen bromide or with 1-cyano-4-dimethylaminopyridinium tetrafluoroborate. After activation of the O-polysaccharide it was covalently linked to tetanus toxoid, either using a spacer-arm or directly to the tetanus toxoid molecule itself (KONADU et al. 1996). It should be noted that the conjugation method that these authors used results in a cross-linked type of conjugate. Even using as low doses as 2.5 µg of the resulting conjugates (based on saccharide content) they were able upon subcutaneous injections in young mice to evoke substantial titers of O-antigenic as well as anti-tetanus toxoid specific antibodies of IgG subclass. They found that a conjugate synthesized with higher molecular weight detoxified lipopolysaccharide elicited the highest anti-O-antigen titers and that these antibodies also had bactericidal activity (KONADU et al. 1996). Interestingly, using their conjugation methods they found no immunological difference between the conjugate synthesized with or without a spacer arm.

Recently, also SAXENA and DIFABIO (1994) reported the preparation of a *Salmonella typhi* O-polysaccharide tetanus toxoid conjugate vaccine. They used a method involving introduction of reducing groups into the O-antigenic polysaccharide side-chain of the *S. typhi* by selective periodate oxidation, after which they conjugated it to tetanus toxoid by reductive amination. The immunogenicity of the resulting cross-linked conjugate was evaluated in mice using microgram doses in Alhydrogel adjuvant. They found that the elicited anti-O-polysaccharide specific antibodies exhibited complement mediated bactericidal activity against *S. typhi*. Moreover, immunized mice were fully protected against challenge with 10 LD_{50} of *S. typhi* Ty2 and partially protected against challenge with 100 LD_{50}.

II. *S. typhi:* Vi-Antigen Specific Conjugates

S. typhi – in contrast to other *Salmonella* spp. – often carries a capsular polysaccharide called Vi-polysaccharide (Vi for virulence) It was long debated whether the Vi-polysaccharide has anything at all to do with the virulence of *S. typhi*. However, the partial protection (55% protective capacity) seen in trials using isolated and purified Vi-polysaccharide as vaccine at least in endemic areas seems promising (KLUGMAN et al. 1996).

To further improve the current Vi-polysaccharide based vaccine attempts have been made to enhance the immunogenicity by preparation of a conjugate vaccine (COHEN et al. 1996). These authors conjugated (cross-linked) the Vi-polysaccharide to the B subunit of the heat-labile toxin of *Escherichia coli* and also to a recombinant exoprotein A from *Pseudomonas aeruginosa* (COHEN et al. 1996). Both of the conjugates elicited higher antibody titers against Vi in mice and in guinea pigs than did the nonconjugate type of Vi-polysaccharide vaccine.

In limited preliminary studies in adult volunteers both Vi-conjugates were shown to elicit higher levels of antibodies than Vi-polysaccharide vaccine alone after the first injection. Interesting – and somewhat surprising – is that upon boosting 26 weeks later no booster responses were seen in any of the volunteers (COHEN et al. 1996).

III. *Shigella:* O-Antigen Specific Conjugates

Robbins and coworkers have utilized their above detoxification method or slight modifications thereof (mildly acidic treatment) also on *Shigelia flexneri* type 2a (POLOTSKY et al. 1994) and *S. sonnei* (COHEN et al. 1996) lipopolysaccharide. Initially they used tetanus toxoid as carrier protein with *S. flexneri* type 2a and found that such conjugates upon repeated injections in mice elicited antibodies with specificity for the O-antigen epitopes of *S. flexneri* type 2a (POLOTSKY et al. 1994). In a subsequent study the safety and immunogenicity of investigational vaccines, composed of the O-specific polysaccharide of *S. sonnei* and *S. flexneri* type 2a covalently bound to *P. aeruginosa* recombinant exoprotein A, were evaluated in 192 Israeli soldiers. As early as 14 days after immunization up to 90% of those given the *S. sonnei* vaccine and over 70% of those receiving the *S. flexneri* type 2a vaccine showed a fourfold or greater increase in serum IgG and IgA titers (COHEN et al. 1996). As expected from conjugate vaccines, the predominant subclass was IgG. It is interesting that after a 2-year follow-up these titers still were significantly higher than the prevaccination titers, showing a rather long-lasting immune response towards both of these investigational *Shigella* vaccines.

IV. *E. coli:* O-Antigen Specific Conjugates

Several groups have prepared various conjugate type candidate vaccines using the different O-antigens from *E. coli*. Among these can be mentioned that GUPTA et al. (1994) using detoxified serotype O111 lipopolysaccharide constructed a series of conjugates using tetanus toxoid as carrier protein. The rational was that some clones of *E. coli* O111 cause enteritis. They reported that the conjugate which they prepared by using mildly acidic treatment and linkage via a spacer arm (apidic acid dihydrazide) was in mice the most immunogenic conjugate, although the mild acidic treatment resulted in partial loss of colitosyl groups, and although these are the most important sugar residues of the major epitope of the *E. coli* O111 O-antigen.

KONADU et al. (1994) from the same laboratory prepared an *E. coli* serotype O157 conjugate. Some clones of this serotype may result not only in severe enteritis but also disseminated infections including hemolytic uremic syndrome. The techniques used in preparation of these conjugates were essentially the same as for the O111 conjugate, detoxification with mildly acidic treatment and use of different carrier protein. The authors reported that the conjugates were able to evoke IgG responses with bactericidal activity.

ROBERTS et al. (1993) showed in the nonhuman primate model of pyelonephritis that an *E. coli* serotype O8-conjugate vaccine confers significant protection against renal damage and dysfunction upon experimental challenge with O-antigenically homologous bacteria. The conjugate used in this study was prepared using O8-antigen specific oligosaccharide derived after partial specific enzyme degradation of the O-polysaccharide chain of *E. coli* O8. These oligosaccharides were then conjugated by the techniques developed by SVENSON et al. (COHEN 1996) As expected, the highest antibody titers were seen in subclass IgG antibodies. As most cases of pyelonephritis are caused by clones of P-fimbriated *E. coli* of a limited number of O-antigen serotypes, it should be possible to construct a multivalent conjugate vaccine that could be combined with a subunit P-fimbriae vaccine.

References

Aaberge IS, Eng J, Lermark G, Lvik M (1995) Virulence of *Streptococcus pneumoniae* in mice: a standardized method for preparation and frozen storage of the experimental bacterial inoculum. Microb Pathog 18:141–152

Aaberge IS, Hvalbye B, Lvik M (1996) Enhancement of *Streptococcus pneumoniae* serotype 6B infection in mice after passive immunization with human serum. Microb Pathog 21:125–137

Åhman F, Käyhty H, Tamminen P, Vuorela A, Malinoski F, Eskola J (1996a) Pentavalent pneumococcal oligosaccharide conjugate vaccine PncCRM is well-tolerated

and able to induce an antibody response in infants. Pediatr Infect Dis J 15:134–139
Åhman F, Käyhty H, Leroy O, Eskola J (1996b) Booster responses to polysaccharide vaccine at 14 months after immunisation with tetravalent pneumococcal (Pnc) conjugate vaccine PncT in infancy is dose dependent (abstact G109). In: Abstracts of the 36th Interscience Conference on Antimicrobial Agents and Chemotherapy. New Orleans, 15–18 September
Åhman F, Kayhty H, Leroy O, Froeschle J, Eskola J (1996c) Booster responses to polysaccharide and conjugate vaccine at 14 months after immunisation with tetravalent pneumococcal conjugate vaccine PncD in infancy (abstract Gl10). In: Abstracts of the 36th Interscience Conference on Antimicrobial Agents and Chemotherapy. New Orleans, 15–18 September
Åhmed H, Steinhoff MC, Rodriguez-Barradas MC, Hamilton RG, Musher DM, Nelson KE (1996) Effect of human immunodeficiency virus type 1 infection on the antibody response to a glycoprotein conjugate pneumococcal vaccine: results from a randomized trial. J lnf Dis 173:83–90
Alonso de Velasco E, Verheul AFM, Veeneman GH, Gomes LJF, van Boom JH, Verhoef J, Snippe H (1993) Protein-conjugated synthetic di- and trisaccharides of pneumococcal type 17F exhibit a different immunogenicity and antigenicity than tetrasaccharide. Vaccine 11:1429–1436
Alonso de Velasco E, Verheul AFM, van Steijn AMP, Dekker HAT, Feldman RG, Fernandez IM, Kamerling JP, Viiegenthart JFG, Verhoef J, Snippe H (1994a) Epitope of rabbit immunoglobulins G (IgG) elicited by pneumococcal type 23F synthetic oligosaccharide- and native polysaccharide-protein conjugate vaccines: comparison with human antipolysaccharide 23F IgG. Infect Immun 63:799–808
Alonso de Velasco E, Dekker HA Th, Antal P, Jalink KP, van Strijp JAG, Verheul AFM, Verhoef J, Snippe H (1994b) Adjuvant quil A improves protection in mice and enhances opsonic capacity of antisera induced by pneumococcal polysaccharide conjugate vaccines. Vaccine 12:1419–1422
Alonso de Velasco E, Verheul FM, Verhoef J, Snippe H (1995) *Streptococcus pneumoniae:* virulence factors, pathogenesis, and vaccines. Microbiol Rev 59:591–603
Ambrosino DM, Landesman SH, Gorham GC, Siber GR (1986) Passive immunization against disease due to *Haemophilus influenzae* type b: concentrations of antibody to capsular polysaccharide in high-risk children. J Infect Dis 153:1–7
Amir J, Scott MG, Nahm MH, Granoff DM (1990) Bactericidal and opsonic activity of IgGl and IgG2 anticapsular antibodies to *Haemophilus influenzae* type b. J Infect Dis 162:163–171
Anderson P, Betts R (1989) Human adult immunogenicity of protein-coupled pneumococcal capsular antigens of serotypes prevalent in *Otitis media.* Pediatr Infect Dis J 8:S50–S53
Anderson EL, Kennedy DJ, Geldmacher KM, Donnelly J, Mendelman P (1996) Immunogenicity of heptavalent pneumococcal conjugate vaccine in infants. J Pediatr128:649–653
Angel CS, Ruzek M, Hostetter MK (1994) Degradation of C3 by *Streptococcus pneumoniae.* J Inf Dis 170:600–608
Apicella MA (1974) Identification of a subgroup antigen on the *Neisseria meninaitidis* group C capsular polysaccharide. J Infect Dis 129:147
Appelbaum PC (1992) Antimicrobial resistance in *Streptococcus pneumoniae:* an overview. Clin Inf Dis 15:77–83
Arndt B, Porro M (1991) Strategies for type-specific glycoconjugate vaccines of *Streptococcus pneumoniae.* In: Atassi MZ (ed) Immunology of proteins and peptides VI. Plenum, New York, pp 129–148
Artenstein MS, Gold R, Zimmerly JG, Wyle FA, Schneider H, Harkins C (1970) Prevention of meningococcal disease by group C polysaccharide vaccine. N Engl J Med 282:417–420

Austrian R (1984) A reassessment of pneumococcal vaccine. N Engl J Med 310:651–653

Avery OT, Goebel WF (1929) Chemo-immunological studies on conjugated carbohydrate-proteins. II. Immunological specificity of synthetic sugar-protein antigens. J Exp Med 50:533–550

Bardardottir E, Jonsson S, Jonsdottir I, Sigfusson A, Vaidimarsson H (1990) IgG subclass response and opsonization of *Streptococcus pneumoniae* after vaccination of healthy adults. J lnf Dis 162:482–488

Barbour ML (1996) Conjugate vaccines and the carriage of *Haemophilus influenzae* type b. Peteatr Infect Dis 7:149–156

Beuvery EC, Miedema F, van Delft R, Haverkamp J (1983) Preparation and immunochemical characterization of meningococcal serogroup C polysaccharide-tetanus toxoid conjugates as a new generation of vaccines. Infect Immun 40:39–45

Black SB, Shinefield HR, Fireman B, Hiatt R, Polen M, Vittinghof E, and the Northern California Kaiser Permanente Vaccine Stuudy Center Pediatrics Group (1991) Efficacy in infancy of oligosaccharide conjugate *Haemophilus influenzae* type b (HBOC) in a United States population of 61,080 children. Pediatr Infect Dis J 10:97–104

Blank AL, Davis GL, VanDeWater TR, Ruben RJ (1994) Acute *Streptococcus pneumoniae* meningogenic labyrinthitis. An experimental guinea pig model and literature review. Arch Otolaryngol Head Neck Surg 120:1342–1346

Booy R, Hodgson S, Carpenter L Mayon-White, RT, Slack S, St. Clair Roberts J et al (1975) Efficacy of *Haemophilus influenzae* type b conjugate vaccine RPT-T. Lancet 344:362–366

Butler JC, Hoffmann J, Cetron MS, Elliott JA, Facklman RF, Breiman RF (1996) The continued emergence of drug-resistant *Streptococcus pneumoniae* in the United States: an update from the centers for disease control and prevention's pneumococcal sentinel surveillance system. J Inf Dis 174:986–993

Caulfield MJ, Stanko D (1995) T-cell dependent response to immune complexes abrogates B-cell unresponsiveness to pneumococcal cell wall polysaccharide. lmmunol 86:331–335

Chan CY, Molrine DC, George S, Tarbell HJ, Mauch P, Diller L, Shamberger RC, Phillips NR, Goorin A, Ambrosino DM (1996) Pneumococcal conjugate vaccine primes for antibody responses to polysaccharide pneumococcal vaccine after treatment of Hodgkin's disease. J Inf Dis 173:256–258

Chudwin DS, Artrip SG, Korenblit A, Schiffman G, Rao S (1985) Correlation of serum opsonis with in vitro phagocytosis of *streptococcus pneumoniae.* Infect Immun 50:213–217

Cohen D, Ashkenazi S, Green M, Lerman Y, Slepon R, Robin G, Orr N, Taylor DN, Sadoff JC, Chu C, Shiloach J, Schneerson R, Robbins JB (1996) Safety and immunogenicity of investigational Shigelia conjugate vaccines in Israeli volunteers. Infect Immun 64:4074–4077

CVI Forum (1996) A pneumococcal vaccine to save children of all ages nears final testing. WHO Publication, December, pp 3–11

Dagan R, Melamed R, Muallem M, Pigiansky L, Greenberg D, Abramson O, Mendelman PM, Bohidar N, Yagupsky P (1996) Reduction of nasopharyngeal carriage of pneumococci during the second year of life by a Heptavalent conjugate pneumococcal vaccine. J Infect Dis 174:1271–1278

Decker MD, Edwars KM, Bradley R, Palmer P (1992) Comparative trial in infants of four conjugate *Haemophilus influenzae* type b vaccines. J Pediatr 120:184–189

Eby R, Koster M, Hogerman D, Malinoski F (1994) Pneumococcal conjugate vaccines. In: Norrby E, Brown F, Chanock RM, Ginsberg MS (eds) Vaccines 1994. Cold Spring Harbor Laboratory, New York, pp 119–123

Eskola J, Käyhty H (1996) Ten years experience with *Haemophilus influenzae* type b (Hib) conjugate vaccines in Finland. Rev Med Microbiol 7:231–241

Eskola J, Käyhty H, Peltola H (1985) Antibody levels achieved in infants by course of *Haemophilus influenzae* type b polysaccharide/diphtheria toxoid conjugate vaccine. Lancet I:1184–1186

Eskola J, Peltola H, Takala AK (1987) Efficacy of *Haemophilus influenzae* type b polysaccharide diphtheria toxoid conjugate vaccine in infancy. N Engl J Med 317:717–722

Esposito AL, Clark CA (1990) The effect of capsular polysaccharide on the capacity of serum to support the killing of type 3 *Streptococcus pneumoniae* by human neutrophils. APMIS 98:812–822

Fairley CF, Begg N, Borrow R, Fox AJ, Jones DM, Cartwright K (1996) Conjugate meningococcal serogroup A and C vaccine: reactogenicity and immunogenicity in United Kingdom infants. J Infect Dis 174:1360–1363

Fedson DD (1988) Pneumococcal vaccine. In: Plotkin S, Mortimer E (eds) Vaccines. Saunders, Philadelphia, pp 271–299

Fedson DS, Shapiro ED, Laforce EM, Mufson MA, Musher DM, Spika JS, Breiman RF, Broome CV et al (1994) Pneumococcal vaccine after 15 years of use. Arch Intern Med 154:2531–2535

Fine MJ, Smith MA, Carson CA, Meffe F, Sankey SS, Weissfeld LA, Detsky AS, Kapoor WN (1994) Efficacy of pneumococcal vaccination in adults. Arch Intern Med 154:2666–2677

French MA H, Denis KA, Dawkins R, Peter JB (1995) Severity of infections in IgA deficiency: correlation with decreased serum antibodies to pneumococcal polysaccharides and decreased serum IgG2 and/or IgG4. Clin Exp Immunol 100:47–53

Garg M, Bondada S (1992) Immune responses of systemic and mucosal lymphoid organs to pnu-immune vaccine as a function of age and the efficacy of monophosphoryl lipid A as an adjuvant. Infect Immun 60:2329–2336

Giebink GS, Koskela M, Vella PP, Harris M, Le CT (1993) Pneumococcal capsular polysaccharide-meningococcal outer membrane protein complex conjugate vaccines: immunogenicity and efficacy in experimental pneumococcal otitis media. J Infect Dis 167:347–355

Giebink GS, Meier JD, Quartey MK, Liebeler CL, Le CT (1996) Immunogenicity and efficacy of Streptococcus pneumoniae polysaccharide-protein conjugate vaccines against homologous and heterologous serotypes in the chinchilla otitis media model. J Infect Dis 173:119–127

Go ES, Ballas ZK (1996) Anti-pneumococcal antibody response in normal subjects: a meta-analysis. J Allergy Clin Immunol 98:205

Gold R, Lepow, ML, Goldschneider I, Draper TL, Gotschlich EC (1975) Clinical evaluation of group A and group C meningococcal polysaccharide vaccines in infants. J Clin Invest 56:1536–1547

Gotschlich EC, Liu, TY, Artenstein MS (1969) Human immunity to the meningococcus. III. Preparation and immunochemical properties of the group A, group B and group C meningococcol polysaccharides. J Exp Med 129:1349–1365

Granoff DM, Holmes SJ (1992) G2m(23) immunoglobulihn allotype and immunity to Haemophilus influenzae type b. J Infect Dis 165:S66–S69

Granoff DM, Holmes SJ, Osterholm MT, McHugh JE, Lucas AM, Anderson EL, Belshe RB, Jacobs JL, Medley F, Murphy TVl (1993) Induction of immunologic memory in infants primed with Haemophilus influenzae type b conjugate vaccines. J Infect Dis 168:663–671

Griffioen AW, Rijkers GT, Janssens-Korpela P, Zegers BJM (1994) Pneumococcal polysaccharides complexed with C3d bind to human B lymphocytes via complement receptor type 2. Infect Immun 59:1839–1845

Gupta RK, Egan W, Bryla DA, Robbins JB, Szu SC (1994) Comparative immunogenicity of conjugates composed of Escherichia coli O111 O-specific polysaccharide, prepared by treatment with acetic acid or hydrazine, bound to tetanus toxoid by two synthetic schemes. Infect Immun 63:2805–2810

Hedlund JU, Kalin ME, Örtqvist AB, Henrichsen J (1994) Antibody response to pneumococcal vaccine in middle-aged and elderly patients recently treated for pneumonia. Arch Intern Med 154:11961–11965

Henrichsen H (1995) Six newly recognized types of *Streptococcus pneumoniae*. J Clin Microbiol 33 n'10:2759–2762

Hirschmann JV, Lipsky BA (1994) The pneumococcal vaccine after 15 years of use. Arch Intern Med 154:373–377

lnsel D (1986) Oligosaccharide-protein conjugate vaccines induce and prime for oligoctonal IgG antibody responses to the *Haemophilus influenzae* type b capsular polysaccharide in human infants. J Exp Med 163:262–269

Insel RA, Anderson PW (1986) Response to oligosaccharide-protein conjugate vaccine against Haemophilus b in two patients with IgG2 deficiency unrespsonsive to capsular polysaccharide vaccine. N Engl J Med 315:499–503

Ishizaka A, Sakiyama Y, Otsu M, Ozutsumi K, Matsumoto S (1994) Successful intravenous immunoglobulin therapy for recurrent pneumococcal otitis media in young children. Eur J Pediatr 153:174

Jennings HJ, Lugowski C (1981) Immunochemistry of serogroups A, B and C polysaccharide-tetanus toxoid conjugates. J Immunol 127:1011–1018

Jennings H (1992) Further approaches for optimizing polysaccharideprotein conjugate vaccines for prevention of ivasive bacterial disease J Inf Dis 165:5156–5159

Jennings HJ, Sood RK (1994) Synthetic glycoconjugates as human vaccines. In: Lee YC, Lee RT (eds) Neoglycoconjugates: preparation and applications. Academic, New York, pp 325–371

Jösbeck HJA, Svenson SB, Lindberg AA (1979) Immunochemistry of Salmonella 0-antigens: specificity of rabbit antibodies against the 0antigen 4 determinant elicited by whole bacteria and 0-antigen 4 specific saccharide-protein conjugates. J lmmunol 123:1376–1381

Karma P, Luotonen J, Timonen M, Pontynen S, Pukander J, Herva E, Gronroos P, Leinonen M, Sibakov M, Makela PH (1980) Efficacy of pneumococcal vacillation against recurrent *Otitis media*. Ann Otol Rhinol Laryngol Suppl 89:357–362

Kauppi M, Eskola J, Kayhty H (1995) Anti-capsular polysaccharide antibodies reduce nasopharyngeal colonization by *Haemophilus influenzae* type b in infant rats. J Infect Dis 167:365–371

Käyhty H, Karanko V, Peltola H, Sarna S, Mäkelä PH (1980) Serum antibodies to capsular polysaccharide vaccine of group A *Neisseria meningitides* followed for 3 years in infants and children. J Infect Dis 142:861–868

Käyhty H, Peltola H, Karanko V (1983) The protective level of serum antibodies to the capsular polysaccharide of *Haemophilus influenzae* type b. J Infect Dis 147:1100

Käyhty H, Karanko V, Peltola H, Mäkelä PH (1984) Serum antibodies after vaccination with *Haemophilus influenzae* type b capsular polysaccharide and responses to reimmunization: no evidence of immunologic tolerance or memory. Pediatrics 74:857–865

Käyhty H, Eskola J, Peltola H, Stout MG, Samuelson JS, Gordon LK (1987) Immunogenicity in infants of a vaccine composed of *Haemophilus influenzae* type b capsular polysaccharide mixed with DTP or conjugated to diphteria toxoid. J Infect Dis 155:100–106

Käyhty H, Mäkelä O, Eskola J, Saarinen L, Seppala 1 (1988) Isotype distribution and bactericidal activity of antibodies after immunization with *Haemophilus influenzae* type b vaccines at 18–24 months of age. J Infect Dis 158:973–982

Käyhty H, Eskola J, Peltola H, Saarinen L, Mäkelä PH (1992) High antibody responses to booster doses of either *Haemophilus influenzae* type b capsular polysaccharide or conjugate vaccine after primary immunization with conjugate vaccine. J Infect Dis 165 (Suppl) 165–166

Käyhty H, Ahman H, Rönnberg PR, Tillikainen R, Eskola J (1995) Pneumococcal polysaccharide–meningococcal outer membrane protein complex conjugate vaccine is immunogenic in infants and children. J Infect Dis 172:1273–1278

Käyhty H, Ahman H, Vuorela A, Malinkoski F, Eskola J (1996) Response at 23 months to booster dose to pneumococal (PnC) polysaccharide (PS) vaccine in children immunised with pentavalent Pnc conjugate vaccine (PncCRM) in infancy (abstract G108). In: 36th Interscience Conference on Antimicrobial Agents and Chemotherapy, New Orleans, 15–18 September

Kelly T, Dillard JP, Yother J (1994) Effect of genetic switching of capsular type on virulence of *Streptococcus pneumoniae.* Infect Immun 62:1813–1819

Keyserling H, Kwan B, Romero-Steiner S, Pais L, McClure H, Carlone G (1996) Immunogenicity of a pneumococcal polysaccharideprotein conjugate vaccine with adjuvants in infants rhesus monkeys (abstract G36). In: 36th Interscience Conference on Antimicrobial Agents and Chemotherapy, New Orleans, 15–18 September

King WJ, MacDonald NE, Wells G, Huang J, Allen U, Chan F, Ferris W, Diaz-Mitoma F, Ashton F (1996) Total and functional antibody response to a quadrivalent meningococcal polysaccharide vaccine among children. J Pedriatr 128:196–202

Klein JO (1994) Otitis media. Clin Infect Dis 19:823–833

Klugman KP, Koornhof HJ, Robbins JB, Le Cam NN (1996) Immunogenicity, efficacy and serological correlate of protection of *Salmonella typhi* Vi capsular polysaccharide vaccine three years after immunization. Vaccine 14:435–438

Konadu E, Robbins JB, Shiloach J, Bryla DA, Szu SC (1994) Preparation, characterization, and immunological properties in mice of *Eschrichia coli* O1 57-specific polysaccharide-protein conjugate vaccines. Infect Immun 62:5048–5055

Konadu E, Shiloach J, Bryla DA, Robbins JB, Szu SC (1996) Synthesis, characterization, and immunological properties in mice of conjugates composed of detoxified lipopolysaccharide of *Salmonella paratyphi* A bound to tetanus toxoid with emphasis on the role of O-acetyls. Infect Immun 64:2709–2715

Konradsen HB (1995) Quantity and avidity of pneumococcal antibodies before and up to five years after pneumococcal vaccination of elderly persons. Clin Infect Dis 21:616–620

Konradsen HB, Oxelius VA, Hahn-Zoric M, Hanson LA (1994) The importance of Gl m and 2 allotypes for the IgG2 antibody levels and avidity against pneumococcal polysaccharide type 1 within mono- and dizygotic twin-pairs. Scand J Immunol 40:251–256

Koskela M (1987) Serum antibodies to pneumoccal C polysaccharide in children: response to acute pneumococcal otitis media or to vaccination. Pediatr Infect Dis J 6:519–526

Kuo J, Douglas M, Ree HK, Lindberg Alf A (1995) Characterization of a recombinant pneumolysin and its use as a protein carrier for pneumococcal type 18C conjugate vaccines. Infect Immun 63(7):2706–2713

Laferrière CA, Sood RK, de Muys JM, Michon F, Jennings HJ (1997a) The synthesis of *Streptococcus pneumoniae* polysaccharidetetanus toxoid conjugates and the effect of chain length on immunogenicity. Vaccine 15(2):179–186

Laferrière C, Jennings HJ et al (1997b) Infect Immun (in press)

Leachk A, Twumasi PA, Kumah S, Banya WS, Jaffar S, Forrest BD, Granoff DM, Libutti DE, Carione GM, Paris LB, Broome CV, Greenwood BM (1997) Induction of immunologic memory in Gambian children by vaccination infancy with a group A plus group C meningococcal polisaccharide-protein conjugate vaccine. J Infect Dis 175:200–204

Lee C-J, Lock RA, Andrew PW, Mitchell TJ, Hansman D, Paton JC (1994) Protection of infant mice from challenge with *Streptococcus pneumoniae* type 19F by immunization with a type 19F polysaccharidepneumolysoid conjugate. Vaccine 12:875–878

Lee H-J, Kang JH, Henrichsen J, Konradsen HB, Jang S-H, Shin HY, Ahn H-S, Choi Y, Hessel L, Nam S-W (1995) Immunogenicity and safety of a 23-valent pneumococcal polysaccharide vaccine in healthy children and in' children at increased risk of pneumococcal infection. Vaccine 13(16):1533–1538

Lee C-J, Karpas A, Kosaka T, Kaio Koizumi (1996) Production, binding characteristics and protective immunity of monoclonal antibody to pneumococcal type-9 V conjugate. Microbiol Immunol 40(11):857–865
Lepow M (1992) Meningococcal vaccines. In: Plotkin S, Mortimer EA Jr (eds) Meningococcal vaccines. Saunders, Philadephia, pp 503–515
Lepow M, Beeler J, Randolph M, Samuelson JS, Hankins WA (1986) Reactogenicity and immunogenicity of a quadravalent combined meningococcal polysaccharide vaccine in children. J Infect Dis 154:1033–1036
Lepow ML, Goldschneider 1, Gold R, Randolph M, Gotschlich EC (1977) Persistence of antibody following immunization of children with groups A and C meningococcal polysaccharide vaccines. Pediatrics 60:673–680
Lieberman JM, Chiu SS, Wong UK, Partridge S, Chang S-J, Chiu C-Y, Gheesling LL, Carlone GM, Ward JL (1996) Safety and immunogenicity of a serogroup A/C *Neisseria meninaitidis* oligosaccharideprotein conjugate vaccine in young children JAMA 275:1499–1503
Lortan JE, Kaniuk StC, Monteil M (1993) Relationship of in vitro phagocytosis of serotype 14 *Streptococcus pneumoniae* to specific class and IgG subclass antibody levels in healthy adults. Clin Exp Immunol 91:54–57
Lu C-H, Lee C-J, Kind P (1994) Immune response of young mice to pneumococcal type 9 V polysaccharide-tetanus toxoid conjugate. Infect Immun 62(7):2754–2760
Lucas AH, Granoff DM (1995) Functional differences in idiotypically defined IgGl anti-polysaccharide antibodies elicited by vaccination with *Haemophilus influenzae* type b polysaccharide-protein conjugates J Immunol 154:4195–4202
Marburg S, Jorn D, Tolman RL, Arison B, McCauley J, Kniskern PJ, Hagopian A, Vella PP (1986) Bimolecular chemistry of macromolecules: synthesis of bacterial polysaccharide conjugates with Neisseria meningitides membrane protein. J Am Chem Soc 108:5282–5287
McMahon BJ, Parkinson AJ, Bulkow L, Davidson M, Wainwright K, Wolfe P, Schiffman G (1993) Immunogenicity of 23-valent pneumococcal polysaccharide vaccine in Alaska native chronic alcoholics compared with nonalcoholic native and non-native controls. Am J Med 95:589–594
Mills OF, Rhoads GG (1996) The contribution of the case-control approach to vaccine evaluation: pneumococcal and haemophilus influenzae type B PRP vaccines. J Clin Epidemiol 49(6):631–636
Munford RS, Murphy TV (1994) Antimicrobial resistance in streptococcus pneumoniae: Can immunization prevent its spread? J Invest Med 42 (4):613–621
Murphy TV, Pastor P, Medley F, Osterholm T, Granoff DM (1993) Decreased *Haemophilus* colonization in children vaccinated with *Haemophilus influenzae* type b conjugate vaccine. J Pediatr 122:517
Musher DM, Johnson B, Watson DA (1990) Quantitative relationship between anticapsular antibody measured by enzyme-linked immunosorbent assay or radioimmunoassay and protection of mice against challenge *with Streptococcus pneumoniae* serotype 4. Infect Immun 58(12):3871–3876
Obaro SK, Henderson DC, Monteil MA (1996) Defective antibodymediated opsonisation of *S. pneumoniae* in high risk patients detected by flow cytometry. Immunol Lett 49:83–89
Paton JC, Lock RA, Lee C-J, Li JP, Berry AM, Mitchell TJ, Andrew PW, Hansman D, Boulnois GJ (1991) Purification and immunogenicity of genetically obtained pneumolysin toxoids and their conjugation to *Streptococcus pneumoniae* type 19F polysaccharide. Infect Immun 59(7):2297–2304
Peeters CCAM, Tenbergen-Meekes A-M, Evenberg DE, Poolman JT, Zegers BJM, Rijkers GT (1991a) A comparative study of the immunogenicity of pneumococcal type 4 polysaccharide and oligosaccharide tetanus toxoid conjugates in adult mice. J Immunol 146(12):4308–4314
Peeters CCAM, Tenbergen-Meekes AM, Poolman JT,, Beurret M, Zegers BJM, Rijkers T (1991b) Effect of carrier priming on immunogenicity of saccharide-protein conjugate vaccines. Infect Immun 59(10):3504–3510

Peeters CCAM, Tenbergen-Meekes A-M, Poolman JT, Zegers BJM, Rijkers GT (1992a) Immunogenicity of a *Streptococcus pneumoniae* type 4 polysaccharide-protein conjugate vaccine is decreased by admixture of high doses of free saccharide. Vaccines 10(12):833

Peeters CCAM, Evenberg DE, Hoogerhout P, Käyhty H, Saarinen L, Van Boeckel CCA, van der Marel GA, Ban Boon JH, Poolman JT (1992b) Synthetic timer and tetramer of 3b-D-ribose-(l→l)-D-ribitol-5-phosphate conjugated to protein induce antibody responses to *Haemophilus influenzae* type b polysaccharide in mice and monkeys. Infect Immun 60:1826–1833

Peeters CCAM, Tenbergen-Meekes AM, Heÿnen CJ, Poolman JT, Zegers BJM, Rijkers GT (1992c) In vitro activation of human T lymphocytes by *Haemophilus influenzae* type b. Scand J Immunol 35:137–148

Peeters CAM, Lagerman PR, de Weers L, Oomen LA, Hoogerhout P, Beurret M, Poolman JT (1996) Polysaccharide-conjugate vaccines. In: Robinson A, Farrar G, Wilblin C (eds) Polysaccharide conjugate vaccines. Methods in molecular medicine: vaccine protocols. Humana, Totowa, pp 111–133

Peltola H, Käyhty H, Sivonen A (1977a) Haemophilus influenzae type b capsular polysaccharide vaccine in children: a double-blind field trial of 100000 vaccines 3 months to 5 years of age in Finland. Pediatrics 60:730–737

Peltola H, Mäkelä PH, Käyhty H, Jousimies H (1977b) Clinical efficacy of meningococcus group A capsular polysaccharide vaccine in children three months to five years of age. N Engl J Med 297:686–691

Peltola H, Safety A, Käyhty H, Karanko V, André FE (1985) Evaluation of two tetravalent (ACYW135) meningococcal vaccines in infants and small children, a clinical study comparing immunogenicity of O-acetyl-negative and O-acetyl-positive group C polysaccharides. Pediatrics 76:91–96

Pichichero ME, Shelly MA, Treanor JJ (1997) Evaluation of a pentavalent conjugated pneumococcal vaccine in toddlers. Pediatr Infect Dis J 16(1):72–74

Polotsky VY, Robbins JB, Bryla D, Schneerson R (1994) Comparison of conjugates composed of lipopolysaccharide frorm *Shigelia flexner* type 2a detoxified by two methods and bound to tetanus toxoid. Infect Immun 6:210–214

Powers DC, Anderson EL, Lottenbach K, Mink CM (1996) Reactogenicity and immunogenicity of a protein-conjugated pneumococcal oligosaccharide vaccine in older adults. J Infect Dis 173:1014–1018

Quataert SA, Kirch CS, Quackenbush Wiedl LJ, Phipps DC, Strohmeyer S, Cimino CO, Skuse J, Madore DV (1995) Assignment of weight-based antibody units to a human antipneumococcal standard reference serum, lot 89-S. Clin Diagn LabImmunol 2(5):590–597

Reinert RR, Kaufhold A, Kühnemud O, Lutticken R (1994) Serum antibody responses to vaccination with 23-valent pneumococcal vaccine in splenectomized patients. Zentralbl Bakteriol 281:481–490

Reingoid AL, Broome CV, Hightower AW (1985) Age-specific differences in duration of clinical protection after vaccination with meningococcal polysaccharide A vaccine. Lancet 2:114–118

Rijkers GT, Dollekamp I, Zegers BJM (1988) 8-mercaptoguanosine overcomes unresponsiveness of human neonatal B cells to polysaccharide antigens. J Immunol 141:2313–2316

Rijkers GT, Sanders EAM, Breukels MA, Zegers BJM (1996) Responsiveness of infants to capsular polysaccharides: implications for vaccine development. Rev Med Microbiol 7:3–12

Robbins JB, Austrian R, Lee C-J, Rastogi SC, Schiffman G, Henrichsen J, Makela PH, Broome CV, Facklam RR, Tiesjema RH, Parke JC (1983) Consideration for formulating the secondgeneration pneumococcal capsular polysaccharide vaccine with emphasis on the cross-reactie types. J Infect Dis 148:1136–1159

Roberts JA, Kaacks B, Baskin G, Svenson SB (1993) Vaccination with a synthetic E. coli serotype O8 specific oligosaccharideprotein conjugate prevents renal scarring from pyelonephritis in non-human primates. Infect Immun 61:5214–5218

Rodriguez-Barradas MC, Groover JE, Lacke CE, Gump DW, Lahar CJ, Pandey JP, Musher DM (1996) IgG antibody to pneumococcal capsular polysaccharide in human immunodeficiency virusinfected subjects: persistence of antibody in responders, revaccination in nonresponders, and relationship of immunoglobulin allotype to response. J Infecti Dis 173:1347–1353

Romero-Steiner S, LiButti, D, Pais LB, Dykes J, Anderson P, Whitin JC, Keyserling HL, Carlone GM (1997) Standardization of an opsonophagocytic assay for the measurement of functional antibody activity against *Streptococcus pneumoniae* using differentiated HL-60 cells. Clin Diagn Lab Immunol (submitted)

Sanders EAM, van de Winkel JGJ, Rijkers GT (1994) Fcg receptor Ila (CD32) heterogeneity in patients with recurrent bacterial respiratory tract. infections. J Infect Dis 170:854–861

Sanders LAM, Rijkers GT, Tenbergen-Meekes A-M, VoorhorstOgink MM, Zegers BJM (1995) Immunoglobulin isotype-specific antibody responses to pneumococcal polysaccharide vaccine in patients with recurrent bacterial respiratory tract infections. Pediatr Res 37:812–819

Santosham M, Wolff M, Reid R (1991) The efficacy in Navajo infants of a conjugate vaccine consisting of *Haemophilus influenzae* type b polysaccharide and *Neisseria meningitides* outer-membrane protein complex. N Engl J Med 324:1767–1772

Saxema M, Di Fabio JL (1994) *Salmonella typhi* O-polysaccharide tetanus toxoid conjugated vaccine. Vaccine 12:879–884

Schiffman G (1983) Pneumococcal vaccine: a tool for the evaluation of the B-cell function of the iummune system. Proc Soc Exp Biol Med 174:309–315

Scherp HW, Rake G (1945) Studies on meningococcal infection. XIII. Correlation between antipolysaccharide and the antibody which protects mice against infection with type I meningococci. J Exp Med 8:85–92

Schlesinger Y,'Granoff the Vaccine Study Group (1992) Avidity and bactericidal activity of antibody elicited by different *Haemophilus influenzae* type b conjugate vaccines. JAMA 267:1489–1494

Schneerson R, Barrera O, Sutton A, Robbins JB (1980) Preparation, characterisation and immunogenicity of *Haemophilus influenzae* type b polysaccharide-protein conjugates. J Exp Med 152:361–376

Schneerson R, Robbins JB, Parke JC et al (1986) Quantitative and qualitative analyses of serum antibodies elicited in adults by *Haemophilus influenzae* type b and pneumococcus type 6A capsular polysaccharide tetanus toxoid conjugates. Infect Immun 519–528

Schneerson R, Levi L, Robbins JB et al (1992) Synthesis of a conjugate vaccine composed of pneumococcus type 14 capsular polysaccharide bound to pertusis toxin. Infect Immun 60:3528–3532

Scott JAG, Hall AJ, Dagan R, Dixon JMS et al (1996) Serogroup specfic epidemiology of *Streptococcus pneumoniae:* associations with age, sex and geography in 7,000 episodes of invasive disease. Clin Infect Dis 22:973–981

Sell SH, Wright PF, Vaughn WK, Thompson J, Schiffman G (1981) Clinical studies of pneumococcal vaccines in infants: Reactogenicity and immunogenicity of two polyvalent polysaccharide vaccines. Rev Infect Dis 3:S97–107

Seppala I, Sarvas H, Mäkelä O, Mattila P, Eskola J, Käyhty H (1988) Human antibody responses to two conjugate vaccines of *Haemophilus infuenzae* type b saccharides and diphteria toxin. Scand J lmmunol 28:471–479

Shahid NS, Steinhoff MC, Hoque SS, et al (1995) Serum, breast milk, and infant antibody after maternal immunisation with pneumococcal vaccine. Lancet 346

Shelly MA, Jacoby H, Riley GJ, Graves BT, Pichichero M, Treanor JJ (1997) Comparison of Pneumococcal polysaccharide and CRM, conjugated pneumococcal oligosaccharide vaccines in young and elderly adults. Infect Immun 65:242–247

Snippe H, Van Houtte A-J, Van Dam JEG, De Reuver MJ, Jansze M, Willers JMN (1983) Immunogenic properties in mice of hexasaccharide from the capsular polysaccharide of *Streptococcus pneumoniae* type 3. Infect Immun 40:856–861

Steinhoff MC, Edwards K, Keyserling H, Thoms ML, Johnson C, Madore D, Hogerman D (1994) A randomized comparison of three bivalent streptococcus pneumoniae glycoprotein conjugate vaccines in young children: effect of polysaccharide size and linkage characteristics. Pediatr Infect Dis J 13:368–372

Svenson SB, Lindberg AA (1977) Oligosaccharide-protein conjugate a novel approach for making *Salmonella* O-antigen immunogens. FEMS Microbiol Lett 1:145–148

Svenson SB, Lindberg AA (1979) Coupling of acid labile *Salmonella* specific oligosaccharides to macromolecular carriers. J Immunol Methods 25:323–335

Svenson SB, Nurminen M, Lindberg AA (1979) Artificial vaccines: O-antigenic oligosaccharide-protein conjugates induce protection against infection with *Salmonella thypimurium*. Infect Immun 25(3):863–872

Szu SC, Taylor DN, Tofa AC, Clements JD, Shiloach J, Sadoff JC, Bryla DA, Robbins JB (1994) Laboratory and preliminary clinical characterization of Vi capsular polysaccharide-protein conjugate vaccines. Infect Immun 62:4440–4444

Van Cott JL, Kobayashi T, Yamamoto M, Pillai S, McGhee JR, Kiyono H (1996) Induction of pneumococcal polysaccharide-specific mucosal immune responses by oral immunization. Vaccines 145:392–398

Van Dam JEG, Fleer A, Snippe H (1990) Immunogenicity and immunochemistry of *Streptococcus pneumoniae* capsular polysaccharides. Antonie van Leeuwenhoek 58:1–47

Van Den Dobbelsteen G (1994) Mucosal immune response to pneumococcal polysaccharides in the rat. Thesis, Drukkerij Elinkwijk, Utrecht

Van Den Dobbeisteen G, Kroes H, Van Rees EP (1995) Characteristics of immune responses to native and protein conjugated pneumococcal polysaccharide type 14. Scand J Immunol 41:273–280

van de Wijgert JHHM, Verheul AFM, Snippe H, Check IJ, Hunter RL (1991) lmmunogenicity of *Streptococcus pneumoniae* type 14 capsular polysaccharide: influence of carriers and adjuvants on isotype distribution. Infect Immun 59:8:2750–2757

Vella PP, Marburg S, Staub JM, Kniskern PJ, Miller W, Hagopian A, Ip C, Tolman RL, Rusk CM, Chupak LS, Ellis RW (1992) Immunogenicity of conjugate vaccines consisting of pneumococcal capsular polysaccharide type 6B, 14, 19F, and 23F and a meningococcal outer membrane protein complex. Infect Immun 60:4977–4983

Verheul AFM, Versteeg AA, De Reuver MJ, Jansze M, Snippe H (1989) Modulation of the immune response to pneumococcal type 14 capsular polysaccharide-protein conjugates by the adjuvant quil A depends on the properties of the conjugates. Infect Immun 59:1078–1083

Viôarsson G, Jónsdóttir I, Jónsson S, Vaidimarsson H (1994) Opsonization and antibodies to capsular and cell wall polysaccharides of *Streptococcus pneumoniae*. J Infect Dis 170:592–599

Ward J, Breneman G, Letson GW, Heyward WL (1990) The Alaska *Haemophilus influenzae* Vaccine Study Group. Limited efficacy of a *Haemophilus influenzae* type b conjugate vaccine in Alaska native infants. N Engl J Med 323:1393–1401

Watson DC, Robbins JB, Szu SC (1992) Infect Immun 60:4679–4686

Watson DA, Musher DM, Jacobson JW, Verhoef J (1993) A brief history of the pneumococcus in biomedical research: a panoply of scientific discovery. Clin Infect Dis 17:913–924

Watson DA, Musher DM, Verhoef J (1995) Pneumococcal virulence factors and host immune responses to them. Eur J Clin Microbiol Infect Dis 14:479–490

Wenger JD, Steiner SR, Pais LB, Butler JC, Perkins B, Carione GM, Broome CV (1996) Laboratory correlates for protective efficacy of pneumococcal vaccines: how can they be identified and validated? In: Abstracts of the 36th Interscience Conference on Antimicrobial Agents and Chemotherapy. New Orleans, 15–18 September

Weinberg GA, Granoff DM, Nahm MH, Shackerford PG (1986) Functional activity of different IgG subclass antibodies against type b capsular polysaccharide of *Haemophilus influenzae*. J Immunol 136:4232–4236

Whitin JC, Anderson P (1993) Dimethyl formamide (DMF) activated HL60 cells in a convenient assay for antibodies opsonic of pneumococcus. Pediatr Res 4:187A

Wyle FA, Artenstein MS, Brandt BL Tramondt EC (1972) Immunologic response of man to group B. Meningococcal vaccines. J Infect Dis 126:514–522

Zollinger W, Mandrell R (1983) Importance of complement source in bactericidal activity of human antibody and murine antibody to meningococcal group B polysaccharide. Infect Immun 40:257–264

CHAPTER 11

Mycobacteria

R. JANSSEN, D. YOUNG, and J. THOLE

A. Introduction

Mycobacterium tuberculosis, the causative agent of tuberculosis, is expected to cause 30 million deaths in the coming 10 years – more than any other single infectious agent. The current vaccine against tuberculosis, bacille Calmette-Guérin (BCG), has a highly variable record of protection, and a widely accepted need exists for improved vaccines to control this disease. Infection with tuberculosis triggers a broad spectrum of immune responses which can contribute to elimination of the pathogen and to disease-associated pathology. In an ideal world a complete understanding of these responses would have provided an optimal rational design for improved prophylactic, or therapeutic, vaccines against tuberculosis and other mycobacterial diseases such as leprosy. However, the complexity of this immune response has thus far confronted us with a formidable obstacle, and novel vaccine strategies currently are still based on more pragmatic arguments. In the first part of this chapter we summarise the current knowledge of the host response to mycobacterial pathogens in order to highlight its complex role in protection and disease. The second part describes current approaches taken to produce novel, improved vaccines against mycobacterial pathogens.

B. The Cellular Response to Mycobacteria

While mycobacteria are potent inducers of humoral immunity, there is no evidence that it plays a role in providing protection, and we therefore confine the discussion to the various interactions of mycobacteria with the cellular arm of the immune response.

I. Macrophage Interactions

Mycobacteria typically infect the lung (or perhaps nasal mucosa in the case of leprosy) after entering the airways in the form of aerosolised droplets, and are probably taken up by alveolar macrophages (DANNENBERG and ROOK 1994). These cells represent the key effector cells in a protective immune response and may be responsible for killing of mycobacteria. On the other hand, they probably provide a niche for mycobacterial replication during progressive infection. Many studies have addressed the viability of mycobacteria in macrophages, and the diverse observations that have been made most probably reflect the diversity of macrophage populations in their ability to kill mycobacteria (reviewed by CHAN and KAUFMANN 1994). Killing of bacteria by macrophages generally involves uptake into a phagosome which is made toxic by acidification, generation of oxygen and nitrogen intermediates, and exposure to hydrolytic enzymes released by lysosomal fusion. It is generally believed that reactive nitrogen intermediates play a key role in the killing of mycobacteria. This activity is enhanced in murine macrophages (DENIS 1991; FLESCH and KAUFMANN 1991; CHAN et al. 1992) by exposure to a combination of cytokines, such as interferon-γ (IFNγ) and tumour necrosis factor-α (TNFα) (DING et al. 1988; NUSSLER and BILLIAR 1993). The role of reactive nitrogen intermediates in human macrophages is less clear, although a recent study showed that these effector molecules may be important in the killing of mycobacteria by human cells as well (NICHOLSON et al. 1996).

While macrophages are the key effector cells in restricting mycobacterial growth, they may also play an important role in the presentation of mycobacterial antigens to T cells. Mycobacteria are very effective in inducing responses by a variety of T cell subsets (see below) that are stimulated by presentation of mycobacterial antigens via MHC class I, class II or CD1 molecules. Morphological and biochemical studies have indicated, however, that mycobacteria in general do not gain access to the cytoplasm of macrophages and occupy a phagosomal compartment that remains relatively separated from the endosomal-lysosomal pathway (ARMSTRONG and HART 1975; STURGILL-KOSZYCKI et al. 1994; CLEMENS and HORWITZ 1995; BARKER et al. 1997; HASAN et al. 1997). It is therefore currently unclear how mycobacterial antigens from this location enter into the various antigen processing pathways. There is an abundance of evidence that mycobacteria stimulate CD4 T cells, and therefore gain access to the class II pathway.

The class II pathway, however, seems to be 'endocytic' in nature and may not be easily accessible from the mycobacterial phagosome, providing a possible explanation for the observation that live mycobacteria impair antigen presentation via class II molecules (PANCHOLI et al. 1993). Killed mycobacteria, in contrast, are rapidly transported to the endosomal-lysosomal pathway (ARMSTRONG and HART 1975; STURGILL-KOSZYCKI et al. 1994; CLEMENS and HORWITZ 1995; BARKER et al. 1997; HASAN et al. 1997), and killing or crippling of mycobacteria is perhaps an important factor in promoting access of mycobacterial antigens to the class II pathway. It is possible that the ability of different macrophages to kill or disable mycobacteria may influence the efficiency with which mycobacterial antigens gain access to the class II pathway. The finding that CD8 T cells recognise mycobacterial antigens suggests that some antigens also gain access to the MHC class I pathway.

Although some researchers have suggested that virulent mycobacteria can escape from the phagosome (MCDONOUGH et al. 1993) and in that way enter the MHC class I pathway, most studies have indicated that mycobacteria remain inside phagosomes indicating that other mechanisms may be involved. The recent finding that viable, but not killed, mycobacteria facilitate TAP-dependent MHC class I restricted presentation of admixed proteins suggests that antigens derived from viable mycobacteria within a segregated phagosome may be somehow translocated to the cytoplasm to gain access the MHC class I pathway (MAZZACCARO et al. 1996). It is therefore possible that particularly antigens secreted by metabolically active mycobacteria inside the segregated compartment are amenable for entering the class I processing pathway. The recent finding that culture filtrate antigens are a target for CD8 responses in mice infected with live BCG supports this notion (DENIS et al. 1997). While mycobacterial protein antigens are presented via MHC class I and class II molecules, (glyco)lipid antigens seem to be presented via CD1 molecules (BECKMAN et al. 1994; SIELING et al. 1995). The processing and presentation pathway of antigens via CD1 molecules is largely undefined; the recent observation that CD1 molecules are co-localised in the MHC class II loading compartment may suggest similarity with the MHC class II pathway (SUGITA et al. 1996).

In addition to the macrophage, other host cells may act as antigen-presenting cells. Dendritic cells are attractive candidates as they are specialised in inducing a primary immune response, and as they can take up mycobacteria (INABA et al. 1993). Dendritic cells may act as presenting cells either after taking up whole bacteria or after taking up antigens synthesised by extracellular bacteria or by taking up bacterial antigens released following death of other infected cells such as macrophages. While the role of dentritic cells in the processing and presentation of mycobacterial antigens remains to be established, the characterisation of antigen processing pathways in various host cell types might provide important novel insights into mycobacterial immunogenicity.

II. Recognition by CD4 T Cells

Antigen recognition by CD4 T cells, followed by the production of macrophage-activating cytokines such as IFNγ, represents a key element in the immune response to mycobacteria. The importance of this response in protective immunity in humans is demonstrated by the enhanced susceptibility to tuberculosis in HIV infected individuals where the CD4 response is diminished (DE-COCK 1994). In addition, individuals with mutations in the gene encoding the IFNγ receptor are highly susceptible to mycobacterial infections (NEWPORT et al. 1996). Similarly, in experimental animal models disruption of the genes encoding CD4 or IFNγ (COOPER et al. 1993; FLYNN et al. 1993) both result in a strong enhancement of the susceptibility to mycobacterial challenge. In the simplest conceptual model for protective immunity against mycobacteria therefore one would propose a direct correlation between the level of IFNγ production by T cells and the efficiency of mycobacterial killing by macrophages.

In response to an antigen-specific challenge T cells may produce a variety of cytokines that differ in their effector role. Based on the cytokines produced, T cells can be classified into distinct subsets, among which two subsets have been particularly prominent; Th1 cells which stimulate activation of macrophages and produce cytokines such as interleukin (IL) 2 and IFNγ, and Th2 cells which promote B cell differentiation and produce cytokines such as IL4 and IL5. Cytokines produced by each of these T cell subset counteract the alternative subset, and immune responses generally tend to polarise towards one or the other subset. Mycobacteria, like other intracellular bacterial pathogens, stimulate predominantly Th1 responses (HSIEH et al. 1993). This is probably a reflection of their adjuvant activity as this is the case for both live and dead mycobacteria. One potential mechanism is the production of IL12 by macrophages infected by mycobacteria, which subsequently triggers production of Th1 cytokines by T cells (FULTON et al. 1996). From the mycobacterial perspective such a response may not be very desirable, and there is evidence of a generalised reduction in the Th1 response of patients with active tuberculosis; manifested by transient skin test anergy, and by low levels of IFNγ production by lymphocytes obtained from peripheral blood or pleural effusions (ZHANG et al. 1995). The diminished Th1 response appears to be associated with an increase in expression of certain macrophage products, such as transforming growth factor (TGF) β in humans and IL6 in mice, rather than by activation of Th2 T cells (TOOSSI et al. 1995; LIN et al. 1996; VANHEYNINGEN et al. 1997). Lesions from patients with the lepromatous leprosy are characterised by the presence of Th2 cytokines, whereas predominantly Th1 cytokines are found at tuberculoid pole of the leprosy spectrum (YAMAMURA et al. 1991). Thus, while there it is clear that a Th1 response seems essential for protection against mycobacterial disease, there is no simple association of a Th1 response with protection and Th2 response with disease. Additional factors are important in determining the outcome of mycobacterial infection.

III. The CD8 Response

Observations from mice models shows that both depletion of CD8 T cells and disruption of the β_2-microglobulin gene (FLYNN et al. 1992) results in an increased susceptibility to a challenge with *M. tuberculosis*. It is therefore likely that CD8 T cells play an important role in the protective immune response generated following a mycobacterial infection. The contribution of CD8 T cells may be through their cytotoxic activity (SILVA et al. 1994; TURNER and DOCKRELL 1996). Some macrophages may not be effective in killing a phagocytosed mycobacterium, and killing of this cell by CD8 T cells may be advantageous for the host, as bacteria become available to other more activated phagocytes. Although the cytotoxic mechanisms of CD8 T cells may make them well-suited to this task, CD4 T cells might also display this function (OTTENHOFF and MUTIS 1990). Alternatively CD8 T cells may function through the production of specific cytokines. As with CD4 T cells, there is evidence for type1 and type 2 CD8 subsets, and production of IFNγ (particularly by type1 cells) may be a key element of the CD8 response (COOPER et al. 1997). It is probable that immunity to mycobacteria involves the coordinated interaction of multiple cell types in a process involving recruitment and activation of new macrophages together with the disposal of lethally infected cells.

While many target antigens for CD4 responses have been identified, the characterisation of mycobacterial antigens involved in CD8 responses has been difficult. It may be that mycobacterial antigens do not gain access to the class I processing pathway very easily, and there have been relatively few descriptions of the detection of CD8 responses to mycobacteria. The viability of mycobacteria seems a prerequisite, and BCG either used alone or used as a delivery system for recombinant antigens can induce a detectable CD8 response both in humans cells in vitro or in mice in vivo (STOVER et al. 1991; TURNER and DOCKRELL 1996). Another indication that viability is an important factor is the observation that viable *M. tuberculosis* is more efficient in the facilitation of class I presentation of an admixed peptide than BCG, which survives less well in macrophages (MAZZACCARO et al. 1996). The finding that the efficiency of CD8 responses induced in mice can be dependent on genetic background indicates that different hosts may differ in their ability to mount a CD8 responses (DENIS et al. 1997). As is mentioned below, the injection of DNA encoding appropriate antigens seems to stimulate CD8 T cells, and this may be an extremely useful approach to investigate the role of CD8 T cells in mycobacterial infection.

IV. Other Subsets

In addition to CD4 and CD8 T cells, mycobacteria trigger several other T cell subsets. Although the mechanism is unknown, natural killer (NK) cells are induced, and they may provide an important source of IFNγ for macrophage activation at the early stage of infection. Another subset that is triggered is composed of T cells with an antigen-specific T cell receptor (TCR) comprised of γ and δ chains (as opposed to the $\alpha\beta$ TCR of the majority CD4 and CD8 subsets). At least two mycobacterial antigens stimulate these so-called $\gamma\delta$ T cells. The first is a phosphorylated nucleotide derivative that stimulates the dominant $\gamma\delta$ subset (Vγ2/Vδ2) present in human peripheral blood (TANAKA et al. 1994, 1995; BEHR et al. 1996). The second triggers a subset of murine hybridomas and has been identified as a peptide from the hsp60 protein antigen (O'BRIEN et al. 1989). Although the diversity of the molecules recognised by $\gamma\delta$ T cells is unknown, these findings suggest that various subsets with different $\gamma\delta$ TCRs exist that may be directed to peptide and non-peptide antigens. The functional role of $\gamma\delta$ T cells is unclear. Similar to CD4 and CD8 T cells, type1 and type 2 $\gamma\delta$ T cells have been identified, and type 1 cells may provide an early source of IFNγ for macrophage activation. The increased bacterial growth in $\gamma\delta$ knock out mice infected with high dose of mycobacteria suggests an essential role for $\gamma\delta$ T cells in protection (LADEL et al. 1995). However, challenge with low doses of bacteria does not result in increased bacterial growth but in different pathology, and an anti-inflammatory role for $\gamma\delta$ cells has been postulated by others (D'Souza et al. 1997). Thus, while it is clear that mycobacterial infection stimulates recruitment and activation of $\gamma\delta$ T cells, the exact role and the functional significance of this response remains to be fully defined.

Another minor T cell subset that may express $\alpha\beta$ or $\gamma\delta$ TCR but are negative for both the CD4 and the CD8 co-ligand have been identified more recently. These so-called double negative T cells seem to be directed to mycobacterial (glyco)lipid antigens in the context of the cell surface protein CD1 (Beckman et al. 1994; Sieling et al. 1995). CD1-restricted responses have been shown to be triggered by mycolic acids and lipoarabinomannan, both of which are important components of the mycobacterial cell wall. Both molecules have long been recognised as dominant targets of the antibody response, and from the host perspective it seems advantageous to recognise this class of molecules by the T cell mediated arm of the immune response as well. As with the $\gamma\delta$ response, the functional role of these CD1-restricted T cells remains to be clarified.

C. Mycobacterial Antigens

Considerable effort has been invested in identification of the target molecules of the immune response to mycobacteria. A diverse range of mycobacterial protein antigens have been characterised by a variety of biochemical and recombinant DNA techniques, and a number of these have been extensively studied in terms of antibody and cellular immune responses (for reviews see Young et al. 1992; Thole et al. 1995). Although the immune response to mycobacterial proteins seems to vary from individual to individual, three broad categories of antigens have been identified that are referred to as "immunodominant". The first class of proteins was identified based on homology to proteins of the highly conserved heat shock families, in particular of the hsp60 and hsp70 families (Young et al. 1988). Such antigens are prominent components of mycobacterial cell extracts, and were found to be particularly immunodominant during early attempts to generate monoclonal antibodies against mycobacterial proteins in BALB/c mice. The high degree of sequence conservation amongst such proteins has stimulated discussion as to their possible role in autoimmune responses (van Eden 1991). However, while the mycobacterial hsp60 has been shown to be involved in autoimmune responses in animal models of autoimmunity, it is also clear that the majority of the immune response to hsp proteins in humans infected with mycobacteria are directed towards epitopes that are not shared with the corresponding self proteins (Adams et al. 1997).

A second class of protein antigens that has received much attention recently are those that are secreted from live mycobacteria. It may be that in particular these proteins become available for immune recognition at the early stages of an infection – prior to release of cytoplasmic proteins from killed bacteria – and therefore induce an early protective immune response (ANDERSEN et al. 1991; ORME et al. 1993). The antigen 85 complex, a family of secreted proteins now known to function as mycolate transferase enzymes involved in the biosynthesis of cell wall components (G. Besra, personal communication), is a very prominent example of this category of proteins. ESAT6 is another (low molecular weight) secreted antigen that has been identified as a an important target in the memory immune response against mycobacteria (SORENSEN et al. 1995). Interestingly, the gene encoding ESAT6 is absent from BCG, and therefore this protein has been considered as a candidate antigen for an improved BCG vaccine (MAHAIRAS et al. 1996).

Finally, cell-associated lipoproteins constitute a third class of prominent protein antigens (YOUNG and GARBE 1991). Some members, such as the 19 kDa antigen, have a proteolytically sensitive site near the N-terminus, allowing them to be released into the culture medium as soluble secreted antigens (HERRMANN et al. 1996). Proteolysis is regulated by the presence or absence of O-linked sugars on threonine residues within the cleavage site. It is possible that the various classes of antigen differ in their access to different antigen-processing pathways and may therefore differ in their recognition by the various T cell subsets activated by mycobacterial infection.

D. Protection Versus Disease

The central issue in development of vaccines against mycobacterial disease lies in the distinction between a protective and pathological response after a mycobacterial infection. None of the pathogenic mycobacteria have thus far been shown to produce toxin molecules, and pathology is thought to be mediated mainly by the immune response itself. It is therefore attractive to investigate the hypothesis that mechanisms leading to pathology are different from those that render protection. However, as discussed above, while there is some evidence to link Th1 and Th2 responses with two different disease forms (tuberculoid and lepromatous) of leprosy, the pathology associated with tuberculoid leprosy and tuberculosis may be mediated by a similar Th1 response that may mediate protection. Some of the pathology observed in mycobacterial diseases may therefore be mediated by a slightly different version of the protective immune response, rather than by a distinct immunological mechanism.

The effects exerted by TNFα may illustrate this. TNFα together with IFNγ forms a key component in the generation of a protective immune response, by stimulating macrophages to express nitric oxide intermediates and thus promoting mycobacterial killing (CHAN et al. 1992; CHAN and KAUFMANN 1994; FLYNN et al. 1995). Macrophages infected with mycobacteria are stimulated to produce TNFα, in part triggered by LAM from the mycobacterial cell wall.

Interestingly, nonpathogenic mycobacterial species carry a structural modification in the LAM molecule as compared to some pathogenic species, and these are less efficient in stimulation of TNFα production (CHATTERJEE et al. 1992). During some stages in the mycobacterial disease process extremely high levels of TNFα have been implicated to play role in pathology, and strategies to reduce these high levels of TNFα expression are currently undergoing clinical evaluation in both leprosy and tuberculosis. Such interventions – by the use of thalidomide, for example (SAMPAIO et al. 1993; TURK et al. 1996) – are aiming to reduce TNFα production to a level sufficient to carry out its protective function. It may therefore be that an appropriate balance of different immunological activities may provide the key to protective immunity. The identification of the key factors that determine the outcome of this balance will enable an improved understanding of our concept of protective immunity and will form an important component in the rationale design of improved vaccines against mycobacterial disease.

E. Mycobacterial Vaccines

I. Live Vaccines

The traditional BCG vaccine has been and is still extensively used for vaccination against tuberculosis. It is a safe vaccine that can be given very early in life. Although the results of major trials show that the average protective efficacy of BCG is about 50%, there is huge variability in this efficacy, ranging from –57% to over 75% (COLDITZ et al. 1994). Many explanations have been suggested for this observed variability (for a review see BLOOM and FINE 1994). Different BCG strains and numerous vaccine preparations have been used in trials and the viability of these preparations has been shown to vary significantly. Furthermore, the vaccination studies have been carried out in many parts of the world and in very diverse populations. For instance, populations have been used that are either homogeneous or heterogeneous in age, BCG status before vaccination, and exposure to *M. tuberculosis* before and after vaccination. The low incidence of tuberculosis resulting into low numbers of actual cases observed adds to this variability. Another explanation may lie in the prevalence of environmental mycobacteria. A high prevalence of environmental mycobacteria is thought to be associated with partial protection against tuberculosis and the hypothesis that BCG vaccination cannot add to this protection could explain low protection levels found in some trials.

In trials in which the efficacy of BCG vaccination against leprosy was determined BCG invariably was shown to protect, and in some trials protection against leprosy was greater than protection against tuberculosis (PONNIGHAUS et al. 1992). These results taken together with the observation that environmental mycobacteria may induce protection against tuberculosis and leprosy has led to the hypothesis that protective immune responses are directed partly against conserved antigens. It has therefore been postulated that one of the ways to improve the efficacy of BCG is by the addition of more specific antigens from other mycobacteria. Thus far results from trials in which heat-killed *M. leprae* was added to BCG have been disappointing, and no extra protection was observed using such a mixtures (CONVIT et al. 1992; KARONGA PREVENTION TRIAL GROUP 1996). However, the use of killed preparations of some mycobacteria such as *M. vaccae* and *Mycobacterium w* may have immuno-modulatory effects, and such preparations are now being tried as immuno-therapeutic vaccines in combination with traditional antibiotic therapy (ZAHEER et al. 1993; ROOK and HERNANDEZ-PANDO 1996).

With a rising incidence of tuberculosis there is a great need for improved vaccines. Current attempts to develop improved vaccines for prevention of tuberculosis follow different general strategies. The first strategy involves generation of live attenuated mycobacterial strains that retain full immunogenicity without the ability to cause disease. Genetic studies of mycobacteria have been greatly hampered because of the difficulties associated with the introduction of specific genomic mutations in mycobacteria. Although there is certainly room for improvement, a growing number of studies now indicate that it is possible to introduce specific mutations in mycobacteria. Electroporation with very long linear DNA fragments resulted in allelic exchange (BALASUBRAMANIAN et al. 1996), whereas the development of suicide vectors with *sac*B as a counter-selectable marker in addition allowed efficient introduction of mutations in the mycobacterial genome (PELICIC et al. 1996). These developments have led to the isolation of auxotrophic strains of *M. tuberculosis* that are immunogenic but unable to survive in animals and, by analogy with other intracellular bacterial pathogens, provide a promising route for vaccine development (MCADAM et al. 1995; REYRAT et al. 1995; GULERIA et al. 1996). In addition, these strains could possibly be used safely in immuno-compromised hosts. Although the protective efficacy in such hosts is expected to be low, complications arising from vaccination that can sometimes be observed after BCG vaccination are less likely to occur. As information is gathered about the molecular mechanisms underlying mycobacteria-host cell interactions, a new generation of attenuated strains can be envisaged in which specific virulence-associated genes have been inactivated.

Another class of new live vaccines is based on improving the efficacy of the existing BCG vaccine. Different approaches have been followed to accomplish this. One way forward may be the administration of a second 'boost' vaccination. A recent study in Malawi indicated that a second BCG vaccination improves protective efficacy against leprosy, but not against tuberculosis (KARONGA PREVENTION TRIAL GROUP 1996). However, latter finding may be attributable partly to an excess in HIV-positive cases, which could be due to a synergism between persistence of mycobacteria and the immunosuppressive effects of HIV (WELTMAN and ROSE 1993; WHALEN et al. 1995). The possibilities of this approach therefore needs further study, and subunit vaccines based on mycobacterial antigens should perhaps also be considered for use in booster vaccinations.

An alternative way to improve the vaccine potential of BCG is by introducing cytokine genes, either to further improve its Th1 inducing potential, for instance, by the introduction of IFNγ or IL2 (O'DONNELL et al. 1994; MURRAY et al. 1996), or to reduce possible inflammatory responses arising as a result of such strong Th1 responses, for instance, by the introduction of the IL10 gene (MARSHALL et al. 1997). The in vivo protective or anti-inflammatory effects of such recombinant BCG strains await further experimentation, and such studies may provide useful information about immunological correlates of protection. A third way to improve the existing BCG vaccine has been based on adding *M. tuberculosis* genes that are absent in BCG. The genetic difference between the virulent *M. bovis, M. tuberculosis* and attenuated BCG has recently been analysed by subtractive genomic hybridisation, and revealed genomic regions that are absent from BCG (MAHAIRAS et al. 1996; PHILIPP et al. 1996). By characterising the genes present in these regions not only new insights into the virulence of pathogenic mycobacteria can be obtained, but also genes with potential protective efficacy can be identified. These genes can subsequently be reintroduced in BCG to improve its protective efficacy. One candidate antigen identified is ESAT6 (SORENSEN et al. 1995), and experiments to evaluate the protective efficacy of recombinant BCG expressing this gene in animal models are currently underway. This approach could in addition generate BCG strains with an improved viability, which could improve their protective potential, for example, because of a more efficient induction of CD8 responses.

II. Subunit Vaccines

A second general strategy for the development of new vaccines involves identification of key antigenic determinants of *M. tuberculosis* and their delivery in the form of a subunit vaccine. It has been postulated that early in infection when the bacteria are not yet killed, preferentially secreted antigens are seen by the host immune system as opposed to somatic antigens that are presented later in infection (ANDERSEN et al. 1991; ORME et al. 1993). This has led to the assumption that secreted antigens are important in induction of the initial response and therefore research has mainly concentrated on the use of such secreted antigens for vaccination. Some success has been achieved by immunisation with proteins isolated from the culture filtrate or with complete culture filtrate of *M. tuberculosis* mixed with appropriate adjuvant (ANDERSEN 1994).

In a recent study it was shown that the adjuvant used for the delivery of such antigen preparation has a great impact on the type and magnitude of immune response that is induced and the level of protection that is subsequently observed (LINDBLAD et al. 1997). Whereas Th1-inducing adjuvants such as DDA mixed with this antigen preparation induced significant levels of protection, mixing with Th2 inducing adjuvants such as $Al(OH)_3$ did not induce protection at all and even increased susceptibility to tuberculosis in a mouse model system. The amounts of IFNγ induced with the different adjuvants was generally low and did not vary significantly, suggesting no direct correlation between IFNγ production and protection. There, however, seemed to be an inverse correlation between the presence of IL4 mRNA and the level of protection. Inclusion of cytokines such as IL12 and IL2 in these vaccine preparations enhanced the IFNγ response; however, in the case of IL12 this seemed to have a transient effect on protection. Also in a guinea pig model for pulmonary tuberculosis a selected extracellular protein preparation of *M. tuberculosis*, which contained mainly a 67-kDa protein reactive with hsp71 monoclonal antibodies, administered in IFA, induced significant levels of protection against subsequent aerosol challenge (Pal and Horwitz 1992).

Recently a promising new vaccine candidate for subunit vaccines, ESAT6, has been identified (Sorensen et al. 1995). The corresponding gene is present in one of the genomic regions that is deleted in BCG, and indeed BCG vaccine strains do not express ESAT6 (Mahairas et al. 1996). ESAT-6 appears to be a very immunodominant antigen for which high precursor frequencies of T cells are present in memory immune mice (Andersen and Heron 1993). These mice produce high amounts of IFNγ upon subsequent in vitro stimulation with ESAT6. Using the mouse foodpad model, immunisation with a soluble protein fraction derived from the cell wall of *M. leprae* as well as with purified 10-kDa and 65-kDa hsp proteins resulted in a limited amount of protection against *M. leprae* (Gelber et al. 1992, 1994). Although in this model the efficacy of secreted antigens has not been evaluated, these findings suggest that also nonsecreted antigens should be considered as candidate antigens for subunit vaccines.

III. DNA Vaccines

An exciting novel development is nucleic acid vaccination. One of the major advantages of DNA vaccination is the fact that also CD8 T cells can be stimulated in this way, and if protection in humans relies on the presence of CD8 T cells, as has been suggested from studies in animal models using knock-out mice, this approach may prove to be very valuable. Recently several DNA vaccines have been successfully used in mouse model systems. A DNA vaccine based on the 30/31-kDa secreted antigen induced antibody responses, CD4 T cell and indeed also CD8 T cell responses (Huygen et al. 1996). Also, significant levels of protection were obtained in this study. A DNA vaccine based on the 65-kDa heat shock protein induced similar responses and protection levels whereas the purified hsp65 did not induce these responses in the same study (Tascon et al. 1996). Also good protective responses have been observed using a DNA vaccine based on the 36-kDa proline-rich antigen (Tascon et al. 1996). These DNA vaccination studies were preceded by the observation that a mouse macrophage-like cell-line that was transfected with the hsp65 gene induced protection in mice, suggesting that expression of the antigen by "host" cells is a powerful way for induction of adequate immune responses (Silva and

LOWRIE 1994). These results with genes encoding selected mycobacterial antigens have demonstrated the feasibility of the nucleic acid vaccination approach, and attempts are now underway to screen the entire genome of *M. tuberculosis* to identify optimally protective antigens.

F. Future Prospects

While it seems likely that one of these strategies will result in identification of novel vaccine candidates with efficacy in experimental animal models, there are formidable logistic problems associated with evaluation of tuberculosis vaccine efficacy in man. It has taken more than 70 years of clinical trials with BCG to establish that its efficacy varies markedly in different geographical areas, and considerable controversy remains as to its overall contribution to tuberculosis control efforts. Evaluation of new vaccine candidates will depend on the identification of suitable correlates of protection – immunological parameters that can be used to give an initial idea of vaccine efficacy. The diversity of the immune response to mycobacteria, and the overlap of protective and pathological responses, imposes considerable problems on identification of appropriate correlates of protection. Progress in disentangling the complex web of cytokines and T cell subsets will be an essential element of tuberculosis vaccine development.

Generation and testing of novel vaccine candidates in protection models will provide a powerful tool in this process over next years, with alternative animal model systems to the mouse model that allow a larger window for measuring protective immunity becoming increasingly more important for accurate evaluation of novel vaccines. In addition, to the possibility of developing an effective vaccine for tuberculosis, the potential reward for such work is that, by understanding the fundamental mechanisms of mycobacterial immunogenicity we may be able to exploit these organisms in designing immunomodulatory strategies from other diseases.

References

Adams E, Basten A, Rodda S, Britton WJ (1997) Human T-cell clones to the 70-kilodalton heat shock protein of Mycobacterium leprae define Mycobacterium-specific epitopes rather than shared epitopes. Infect Immun 65:1061–1070

Andersen P (1994) Effective vaccination of mice against Mycobacterium tuberculosis infection with a soluble mixture of secreted mycobacterial proteins. Infect Immun 62:2536–2544

Andersen P, Heron I (1993) Specificity of a protective memory immune response against Mycobacterium tuberculosis. Infect Immun 61:844–851

Andersen P, Askgaard D, Ljungqvist L, Bennedsen J, Heron I (1991) Proteins released from Mycobacterium tuberculosis during growth. Infect Immun 59:1905–1910

Armstrong JA, Hart PD (1975) Phagosome-lysosome interactions in cultured macrophages infected with virulent tubercle bacilli. Reversal of the usual nonfusion pattern and observations on bacterial survival. J Exp Med 142:1–16

Balasubramanian V, Pavelka M Jr, Bardarov SS, Martin J, Weisbrod TR, McAdam RA, Bloom BR, Jacobs W Jr (1996) Allelic exchange in Mycobacterium tuberculosis with long linear recombination substrates. J Bacteriol 178:273–279

Barker LP, George KM, Falkow S, Small PLC (1997) Differential Trafficking of live and dead Mycobacterium marinum organisms in macrophages. Infect Immun 65:1497–1504

Beckman EM, Porcelli SA, Morita CT, Behar SM, Furlong ST, Brenner MB (1994) Recognition of a lipid antigen by CD1-restricted alpha beta+ T cells. Nature 372:691–694

Behr C, Poupot R, Peyrat MA, Poquet Y, Constant P, Dubois P, Bonneville M, Fournie JJ (1996) Plasmodium falciparum stimuli for human gammadelta T cells are related to phosphorylated antigens of mycobacteria. Infect Immun 64:2892–2896

Bloom BR, Fine PEM (1994) The BCG experience: implications for future vaccines against tuberculosis. In: Bloom BR (ed) The BCG experience: implications for future vaccines against tuberculosis. ASM, Washington DC, pp 531–558

Chan J, Kaufmann SHE (1994) Immune mechanisms of protection. In: Bloom, BR (ed) Immune mechanisms of protection. ASM, Washington DC, pp 389–416

Chan J, Xing Y, Magliozzo RS, Bloom BR (1992) Killing of virulent Mycobacterium tuberculosis by reactive nitrogen intermediates produced by activated murine macrophages. J Exp Med 175:1111–1122

Chatterjee D, Roberts AD, Lowell K, Brennan PJ, Orme IM (1992) Structural basis of capacity of lipoarabinomannan to induce secretion of tumor necrosis factor. Infect Immun 60:1249–1253

Clemens DL, Horwitz MA (1995) Characterization of the Mycobacterium tuberculosis phagosome and evidence that phagosomal maturation is inhibited. J Exp Med 181:257–270

Colditz GA, Brewer TF, Berkey CS, Wilson ME, Burdick E, Fineberg HV, Mosteller F (1994) Efficacy of BCG vaccine in the prevention of tuberculosis. Meta-analysis of the published literature. JAMA 271:698–702

Convit J, Sampson C, Zuniga M, Smith PG, Plata J, Silva J, Molina J, Pinardi ME, Bloom BR, Salgado A (1992) Immunoprophylactic trial with combined Mycobacterium leprae/BCG vaccine against leprosy: preliminary results. Lancet 339:446–450

Cooper AM, Dalton DK, Stewart TA, Griffin JP, Russell DG, Orme IM (1993) Disseminated tuberculosis in interferon gamma gene-disrupted mice. J Exp Med 178:2243–2247

Cooper AM, D'Souza C, Frank AA, Orme IM (1997) The course of Mycobacterium tuberculosis infection in the lungs of mice lacking expression of either perforin- or granzyme-mediated cytolytic mechanisms. Infect Immun 65:1317–1320

Dannenberg AM Rook GAW (1994) Pathogenesis of pulmonary tuberculosis: an interplay of tissue-damaging and macrophage-activating immune responses – dual mechanisms that control bacillary multiplication. In: Bloom BR (ed) Pathogenesis of pulmonary tuberculosis: an interplay of tissue-damaging and macrophage-activating immune responses – dual mechanisms that control bacillary multiplication. ASM, Washington, pp 459–484

de-Cock KM (1994) Impact of interaction with HIV. In: Porter JDH, McAdam KPWJ (eds) Impact of interaction with HIV. Wiley, Chichester, pp 35–49

Denis M (1991) Interferon-gamma-treated murine macrophages inhibit growth of tubercle bacilli via the generation of reactive nitrogen intermediates. Cell Immunol 132:150–157

Denis O, Lozes E, Huygen K (1997) Induction of cytotoxic T-cell responses against culture filtrate antigens in Mycobacterium bovis bacillus Calmette-Guerin-infected mice. Infect Immun 65:676–684

Ding AH, Nathan CF, Stuehr DJ (1988) Release of reactive nitrogen intermediates and reactive oxygen intermediates from mouse peritoneal macrophages. Comparison

of activating cytokines and evidence for independent production. J Immunol 141:2407–2412
D'Souza CD, Cooper AM, Frank AA, Mazzaccaro RJ, Bloom BR, Orme IM (1997) An anti-inflammatory role for $\gamma\delta$ T lumphocytes in acquired immunity to Mycobacterium tuberculosis. J Immunol 158:1217–1221
Flesch IEA, Kaufmann SHE (1991) Mechanisms involved in mycobacterial growth inhibition by gamma interferon-activated bone marrow macrophages: role of reactive nitrogen intermediates. Infect Immun 59:3213–3218
Flynn JL, Goldstein MM, Triebold KJ, Koller B, Bloom BR (1992) Major histocompatibility complex class I-restricted T cells are required for resistance to Mycobacterium tuberculosis infection. Proc Natl Acad Sci USA 89:12013–12017
Flynn JL, Chan J, Triebold KJ, Dalton DK, Stewart TA, Bloom BR (1993) An essential role for interferon gamma in resistance to Mycobacterium tuberculosis infection. J Exp Med 178:2249–2254
Flynn JL, Goldstein MM, Chan J, Triebold KJ, Pfeffer K, Lowenstein CJ, Schreiber R, Mak TW, Bloom BR (1995) Tumor necrosis factor-alpha is required in the protective immune response against Mycobacterium tuberculosis in mice. Immunity 2:561–572
Fulton SA, Johnsen JM, Wolf SA, Sieburth DS, Boom WH (1996) Interleukin-12 production by human monocytes infected with Mycobacterium tuberculosis: role of phagocytosis. Infect Immun 64:2523–2531
Gelber RH, Murray L, Siu P, Tsang M (1992) Vaccination of mice with a soluble protein fraction of Mycobacterium leprae provides consistent and long-term protection against M. leprae infection. Infect Immun 60:1840–1844
Gelber RH, Mehra V, Bloom B, Murray LP, Siu P, Tsang M, Brennan PJ (1994) Vaccination with pure Mycobacterium leprae proteins inhibits M. leprae multiplication in mouse footpads. Infect Immun 62:4250–4255
Guleria I, McAdam RA, Kalpana G, Jacobs WR, Bloom BR (1996) Auxotrophic vaccines for tuberculosis. Nature Med 2:334–337
Hasan Z, Schlax C, Kuhn L, Lefkovits I, Young D, Thole J, Pieters J (1997) Isolation and characterization of the mycobacterial phagosome: segregation from the endosomal/lysosomal pathway. Mol Microbiol
Herrmann JL, O'Gaora P, Gallagher A, Thole JER, Young DB (1996) Bacterial glycoproteins: a link between glycosylation and proteolytic cleavage of a 19 kDa antigen from Mycobacterium tuberculosis. EMBO 15:3547–3554
Hsieh CS, Macatonia SE, Tripp CS, Wolf SF, O'Garra A, Murphy KM (1993) Development of TH1 CD4+ T cells through IL-12 produced by Listeria-induced macrophages. Science 260:547–549
Huygen K, Content J, Denis O, Montgomery DL, Yawman AM, Deck RR, DeWitt CM, Orme IM, Baldwin S, D'Souza C, Drowart A, Lozes E, Vandenbussche P, Van Vooren J-P, Liu MA, Ulmer JB (1996) Imunogenicity and protective efficacy of a tuberculosis DNA vaccine. Nat Med 2:893–898
Inaba K, Inaba M, Naito M, Steinman RM (1993) Dendritic cell progenitors phagocytose particulates, including bacillus Calmette-Guerin organisms, and sensitize mice to mycobacterial antigens in vivo. J Exp Med 178:479–488
Karonga Prevention Trial Group (1996) Randomised controlled trial of single BCG, repeated BCG, or combined BCG and killed Mycobacterium leprae vaccine for prevention of leprosy and tuberculosis in Malawi. Lancet 348:17–24
Ladel CH, Blum C, Dreher A, Reifenberg K, Kaufmann SH (1995) Protective role of gamma/delta T cells and alpha/beta T cells in tuberculosis. Eur J Immunol 25:2877–2881
Lin Y, Zhang M, Hofman FM, Gong J, Barnes PF (1996) Absence of a prominent Th2 cytokine response in human tuberculosis. Infect Immun 64:1351–1356
Lindblad ER, Elhay MJ, Silva R, Appelberg R, Andersen P (1997) Adjuvant modulation of immune responses to tuberculosis subunit vaccines. Infect Immun 65:623–629

Mahairas GC, Sabo PJ, Hickey MJ, Singh DC, Stover CK (1996) Molecular analysis of genetic differences between Mycobacterium bovis BCG and virulent M. bovis. J Bacteriol 178:1274–1282

Marshall BG, Chambers MA, Wangoo A, Shaw RJ, Young DB (1997) Production of tumor necrosis factor and nitric oxide by macrophages infected with live and dead mycobacteria and their suppression by an interleukin-10-secreting recombinant. Infect Immun

Mazzaccaro RJ, Gedde M, Jensen ER, van Santen HM, Ploegh HL, Rock KL, Bloom BR (1996) Major histocompatibility class I presentation of soluble antigen facilitated by Mycobacterium tuberculosis infection. Proc Natl Acad Sci USA 93:11786–11791

McAdam RA, Weisbrod TR, Martin J, Scuderi JD, Brown AM, Cirillo JD, Bloom BR, Jacobs WR (1995) In vivo growth characteristics of leucine and methionine auxotrophic mutants of Mycobacterium bovis BCG generated by transposon mutagenesis. Infect Immun 63:1004–1012

McDonough KA, Kress Y, Bloom BR (1993) Pathogenesis of tuberculosis: interaction of Mycobacterium tuberculosis with macrophages. Infect Immun 61:2763–2773

Murray PJ, Aldovini A, Young RA (1996) Manipulation and potentiation of antimycobacterial immunity using recombinant bacille Calmette-Guerin strains that secrete cytokines. Proc Natl Acad Sci USA 93:934–939

Newport MJ, Huxley CMP, Huston S, Hawrylowycz C, Oostra BA, Williamson R, Levin M (1996) Mutation in the interferon-γ receptor and mycobacterial susceptibility in man. N Engl J Med 335:1941–1949

Nicholson S, da Bonecini Almeida MG, Lapa e Silva JR, Nathan C, Xie QW, Mumford R, Weidner JR, Calaycay J, Geng J, Boechat N, Linhares C, Rom W, Ho JL (1996) Inducible nitric oxide synthase in pulmonary alveolar macrophages from patients with tuberculosis. J Exp Med 183:2293–2302

Nussler AK, Billiar TR (1993) Inflammation, immunoregulation, and inducible nitric oxide synthase. J Leukocyte Biol 54:171–178

O'Brien RL, Happ MP, Dallas A, Palmer E, Kubo R, Born WK (1989) Stimulation of a major subset of lymphocytes expressing T cell receptor gammadelta by an antigen derived from Mycobacterium tuberculosis. Cell 57:667–674

O'Donnell MA, Aldovini A, Duda RB, Yang H, Szilvasi A, Young RA, DeWolf WC (1994) Recombinant Mycobacterium bovis BCG secreting functional interleukin-2 enhances gamma interferon production by splenocytes. Infect Immun 62:2508–2514

Orme IM, Andersen P, Boom WH (1993) T cell response to Mycobacterium tuberculosis. J Infect Dis 167:1481–1497

Ottenhoff THM, Mutis T (1990) Specific killing of cytotoxic T cells and antigen presenting cells by cytotxic CD4+ T cell clones. A novel potentially immunoregulatory T-T cell interaction in man. J Exp Med 171:2011–2024

Pal PG, Horwitz MA (1992) Immunization with extracellular proteins of Mycobacterium tuberculosis induces cell-mediated immune responses and substantial protective immunity in a guinea pig model of pulmonary tuberculosis. Infect Immun 60:4781–4792

Pancholi P, Mirza A, Bhardwaj N, Steinman RM (1993) Sequestration from immune CD4+ T cells of mycobacteria growing in human macrophages. Science 260:984–986

Pelicic V, Reyrat JM, Gicquel B (1996) Expression of the Bacillus subtilis sacB gene confers sucrose sensitivity on mycobacteria. J Bacteriol 178:1197–1199

Philipp WJ, Nair S, Guglielmi G, Lagranderie M, Gicquel B, Cole ST (1996) Physical mapping of Mycobacterium bovis BCG Pasteur reveals differences from the genome maps of Mycobacterium tuberculosis H37Rv and M. bovis. Microbiology 142:3135–3145

Ponnighaus JM, Fine PE, Sterne JA, Wilson RJ, Msosa E, Gruer PJ, Jenkins PA, Lucas SB, Liomba NG, Bliss L (1992) Efficacy of BCG vaccine against leprosy and tuberculosis in northern Malawi. Lancet 339:636–639
Reyrat JM, Berthet FX, Gicquel B (1995) The urease locus of M. tuberculosis and its utilisation for the demonstration of allelic exchange in M. bovis BCG. Proc Natl Acad Sci USA 92:8768–8772
Rook GAW, Hernandez-Pando R (1996) The pathogenesis of tuberculosis. Annu Rev Microbiol 50:259–284
Sampaio EP, Kaplan G, Miranda A, Nery JA, Miguel CP, Viana SM, Sarno EN (1993) The influence of thalidomide on the clinical and immunologic manifestation of erythema nodosum leprosum. J Infect Dis 168:408–414
Sieling PA, Chatterjee D, Porcelli SA, Prigozy TI, Mazzaccaro RJ, Soriano T, Bloom BR, Brenner MB, Kronenberg M, Brennan PJ (1995) CD1-restricted T cell recognition of microbial lipoglycan antigens. Science 269:227–230
Silva CL, Lowrie DB (1994) A single mycobacterial protein (hsp 65) expressed by a transgenic antigen-presenting cell vaccinates mice against tuberculosis. Immunology 82:244–248
Silva CL, Silva RC, Pietro RCLR, Lowrie DB (1994) Protection against tuberculosis by passive transfer with T-cell clones recognising mycobacterial heat-shock protein 65. Immunology 83:341–346
Sorensen AL, Nagai S, Houen G, Andersen P, Andersen AB (1995) Purification and characterisation of a low-molecular-mass T-cell antigen secreted by Mycobacterium tuberculosis. Infect Immun 63:1710–1717
Stover CK, de la Cruz VF, Fuerst TR, Burlein JE, Benson LA, Bennett LT, Bansal GP, Young JF, Lee MH, Hatfull GF et al (1991) New use of BCG for recombinant vaccines. Nature 351:456–460
Sturgill-Koszycki S, Schlesinger PH, Chakraborty P, Haddix PL, Collins HL, Fok AK, Allen RD, Gluck SL, Heuser J, Russell DG (1994) Lack of acidification in Mycobacterium phagosomes produced by exclusion of the vesicular proton-ATPase. Science 263:678–681
Sugita M, Jackman RM, van-Donselaar E, Behar SM, Rogers RA, Peters PJ, Brenner MB, Porcelli SA (1996) Cytoplasmic tail-dependent localization of CD1b antigen-presenting molecules to MIICs. Science 273:349–352
Tanaka Y, Sano S, Nieves E, DeLibero G, Rosa D, Modlin R, Brenner MB, Bloom BR, Morita CT (1994) Nonpeptide ligands for human gammadelta T cells. Proc Natl Acad Sci USA 91:8175–8179
Tanaka Y, Morita CT, Tanaka Y, Nieves E, Brenner MB, Bloom BR (1995) Natural and synthetic non-peptide antigens recognized by human gamma delta T cells. Nature 375:155–158
Tascon RE, Colston MJ, Tagno S, Stavropoulos E, Gregory D, Lowrie DB (1996) Vaccination against tuberculosis by DNA injection. Nature Med 2:888–892
Thole JER, Wieles B, Clark-Curtiss JE, Ottenhoff THM, Rinke de Wit TF (1995) Immunological and functional characterisation of Mycobacterium leprae protein antigens: an overview. Mol Microbiol 18:791–800
Toossi Z, Gogate P, Shiratsuchi H, Young T, Ellner JJ (1995) Enhanced production of TGF-beta by blood monocytes from patients with active tuberculosis and presence of TGF-beta in tuberculous granulomatous lung lesions. J Immunol 154:465–473
Turk BE, Jiang H, Liu JO (1996) Binding of thalidomide to alpha1-acid glycoprotein may be involved in its inhibition of tumor necrosis factor alpha production. Proc Natl Acad Sci USA 93:7552–7556
Turner J, Dockrell H (1996) Stimulation of human peripheral blood mononuclear cells with live Mycobacterium bovis BCG activates cytolytic CD8+ T cells in vitro. Immunology 8:339–342
van Eden W (1991) Heat shock proteins as immunogenic bacterial antigens with the potential to induce and regulate autoimmune arthritis. Immunol Rev 121:5–28

VanHeyningen TK, Collins H, Russell DG (1997) IL-6 produced by macrophages infected with Mycobacterium species supresses T cell responses. J Immunol 158:330–337
Weltman AC, Rose DN (1993) The safety of Bacille Calmette-Guerin vaccination in HIV infection and AIDS. Aids 7:149–157
Whalen C, Horsburgh CR, Hom D, Lahart C, Simberkoff M, Ellner J (1995) Accelerated course of human immunodeficiency virus infection after tuberculosis. Am J Respir Crit Care Med 151:129–135
Yamamura M, Uyemura K, Deans RJ, Weinberg K, Rea TH, Bloom BR, Modlin RL (1991) Defining protective responses to pathogens: cytokine profiles in leprosy lesions. Science 254:277–279
Young D, Lathigra R, Hendrix R, Sweetser D, Young RA (1988) Stress proteins are immune targets in leprosy and tuberculosis. Proc Natl Acad Sci USA 85:4267–4270
Young DB, Garbe TR (1991) Lipoprotein antigens of Mycobacterium tuberculosis. Res Microbiol 142:55–65
Young DB, Kaufmann SH, Hermans PW, Thole JE (1992) Mycobacterial protein antigens: a compilation. Mol Microbiol 6:133–145
Zaheer SA, Mukherjee R, Ramkumar B, Misra RS, Sharma AK, Kar HK, Kaur H, Nair S, Mukherjee A, Talwar GP (1993) Combined multidrug and Mycobacterium w vaccine therapy in patients with multibacillary leprosy. J Infect Dis 167:401–410
Zhang M, Lin Y, Iyer DV, Gong J, Abrams JS, Barnes PF (1995) T-cell cytokine responses in human infection with Mycobacterium tuberculosis. Infect Immun 63:3231–3234

CHAPTER 12

Vaccines Against Diarrheal Diseases

J. HOLMGREN and A.-M. SVENNERHOLM

A. Introduction

Diarrheal disease remains one of the leading global health problems. It has been estimated that 3–5 billion episodes of diarrhea, resulting in 3–5 million deaths, occur annually in developing countries with the highest incidence and severity in children below the age of 5 years (FARTHING and KEUSCH 1989; WHO 1996). The vast majority of these diseases are caused by enteric infections. The etiological agents include a large number of bacteria, some viral agents and limited number of parasites, which give rise to diarrhea in different clinical presentation patterns. The following are the major causes and clinical presentation patterns of gastrointestinal infection associated with diarrhea (modified from FARTHING and KEUSCH 1989):

- Acute watery diarrhea
  - Enterotoxigenic *Escherichia coli*
  - *Vibrio cholerae* O1 and O139
  - Noncholera vibrios
  - Aeromonas
  - Plesiomonas
  - Rotavirus
  - Calicivirus norwalk agent
  - *Cryptosporidium* spp.
- Dysentery
  - *Shigella* spp.
  - Enteroinvasive *E. coli* (EIEC)
  - Enterohemorrhagic *E. coli* (EHEC)
  - *Salmonella* spp.
  - *Campylobacter jejuni*
  - *Entamoeba histolytica*
- Chronic diarrhea/malabsorption
  - *Giardia lamblia*
  - *Strongyloides stercoralis*

About half of all diarrheas are due to bacteria that cause watery diarrheal disease by producing one or more enterotoxins. Cholera, resulting from infection with *Vibrio cholerae*, is the most severe of these "enterotoxic enteropathies" (CRAIG 1980), whereas infection with enterotoxigenic *Escherichia coli* (ETEC) is causing the largest number of cases. The other most important bacterial pathogens include various *Shigella* spp., *Campylobacter jejuni*, and various non-ETEC *E. coli* that cause disease by different mechanisms. Among the viral pathogens, rotaviruses are without comparison the most common cause of diarrheal disease although essentially limited to children below the age of 3 years, where they together with ETEC account for more than 50% of all hospital admittances for diarrhea (KAPIKIAN and CHANOCK 1996).

Diarrhea is also a leading cause of illness and incapacitation among international travelers visiting countries with a high incidence of diarrhea. It has been estimated that ca. 50% of persons traveling to developing countries experience diarrheal disease, accounting for at least 10 million cases per year. ETEC causes one-third to one-half of these diseases, and other important agents include *C. jejuni*, *Shigella* spp. and *Salmonella* spp. Common causes of traveler's diarrhea are (modified from PELTOLA and GORBACH 1997):

- Bacteria
  - Enterotoxigenic *E. coli*
  - *Campylobacter* spp.
  - *Shigella* spp.
  - *Salmonella* spp.
  - *Aeromonas*
  - *Plesiomonas*
- Viruses
  - Rotavirus
  - Calicivirus
  - Enteroviruses
  - Protozoa
  - *Giardia lamblia*
  - *Entamoeba histolytica*
  - *Cryptosporidium* spp.

Based on the great health impact of infections with these different enteropathogens, both in children in developing countries and in travelers to these areas there has been a great interest in providing and/or developing effective vaccines against the most important diarrheal diseases. Until recently vaccines against most of these diseases have been lacking. Recently, however, effective vaccines against cholera and rotavirus diarrhea have been developed and have been and/or are being licensed in different countries. Vaccines against disease caused by ETEC, *Shigella* and *Campylobacter jejuni* have reached the stage of being tested for their protective efficacy in humans. In this review we provide a brief description of the most important diarrheal pathogens and discuss current knowledge on the mechanisms of disease and immunity in these different enteric infections and the state of vaccine development against the most important diarrheal pathogens.

B. Overview of the Main Diarrheal Pathogens

I. *Vibrio cholerae*

V. cholerae of serogroup O1 is the prototype for the enterotoxin-producing bacteria. It was first isolated and (under the initial name of *Vibrio comma)* identified as the etiological agent of cholera by Robert Koch in 1884. Until the beginning of the 20s all examined *V. cholerae* O1 isolates were of the so-called classical, nonhemolytic biotype. In 1906, however, vibrios of a new hemolytic biotype, El Tor, were recognized and for many years vibrios of either the classical or El Tor biotype were isolated from cholera cases (Blake 1994). Both classical and El Tor *V. cholerae* (now usually distinguished mainly by the polymixin resistance of the El Tor biotype) have the same O-antigen (lipopolysaccharide) antigen, referred to serologically as O group 1, dividable in two main different serotypes, Inaba and Ogawa. At the beginning of the nineteenth century cholera started to spread from Bengal, and since then seven large cholera pandemics have been described. The latest "global" pandemic, which began in Celebes in Indonesia in 1961, spread in the 60s and 70s to a large number of countries in first Asia and then Africa. Since 1991 it has also given rise to major epidemics and become endemic in South and Central America (Blake 1994). The causative agent in Latin America has proved to be identical to the seventh pandemic O1 El Tor organisms isolated from Asia and Africa. In addition, from 1992 *V. cholerae* of a previously unknown serogroup, referred to as O139 Bengal, emerged as an additional "new" cause of cholera in India and Bangladesh (Morris 1994). Cholera caused by *V. cholerae* O139 Bengal has since then also been reported from a number of neighboring countries, for example, Thailand and Nepal, but has as yet been restricted to Southeast Asia. Also there it has declined in significance (whether permanently or transiently remains to be seen) and at present *V. cholerae* O139 only accounts for a small percentage of cholera isolates.

Although the highest incidence of cholera in endemic areas is seen in children below 5 years (Mosley et al. 1968), still and in contrast to most other diarrheal pathogens approximately two thirds of all *V. cholerae* O1 cases occur in older children and adults in these areas. When, on the other hand, cholera has spread to new countries, all age groups have had a similar incidence, probably reflecting that acquired immunity which normally develops by increasing age in endemic countries due to natural exposure, has been missing (Tauxe et al. 1994). Consistent with this, cholera caused by the serogroup O139 Bengal has been recorded with similar incidence in adults and in children (Morris 1994). The total number of cholera cases annually is uncertain since several affected countries do not monitor and for a variety of reasons underreport the disease. Therefore recent figures from WHO (1996) of ca 5.5 million cases and 120000 deaths from cholera annually are almost certainly significant underestimates.

The diseases caused by the enterotoxin-producing bacteria are characterized by watery stools without blood and mucus (Farthing and Keusch 1989).

In cholera, which has the highest frequency of dehydrating disease, the most severe cases can purge as much as 15–25 l water and electrolytes per day and the mortality rate in severe, nontreated cholera may be as high as 30%–50%.

II. Enterotoxigenic *Escherichia coli*

ETEC is the most common cause of diarrhea in developing countries accounting for an estimated 400 million cases a year. Although many episodes are relatively mild, disease caused by ETEC is responsible for 300000–700000 deaths a year among children under 5 years in these areas, corresponding to almost 10%–20% of the global total deaths from diarrheal diseases in this age group (WHO 1996). As noted above, ETEC is also the most common cause of traveler's diarrhea being isolated in one-third to one-half of these episodes (BLACK 1990).

Although most episodes of ETEC disease are relatively mild, the diarrhea may result in moderate to severe dehydration that is sometimes fatal. Indeed, ETEC disease may vary from mild diarrhea to a severe choleralike disease and is often accompanied by nausea, vomiting, abdominal cramps, anorexia and more often than in cholera also with significant fever.

III. *Shigella* spp.

Among the children who die from diarrheal diseases each year in developing countries, almost 600000 die from infection with different *Shigella* species causing the serious disease known as bacillary dysentery. The disease is endemic throughout the world and can occur in major epidemics in developing countries. Children suffering from malnutrition are especially prone to develop severe *Shigella* infection (WHO 1996). Most cases of life-threatening disease are caused by members of *S. dysenterie* and *S. flexneri*, whereas *S. sonnei* is the most common cause of endemic disease in industrialized countries (WHO 1996).

The *Shigella* pathogens typically cause dysentery, which is a serious form of enteric infection characterized clinically by fever, blood and/or mucus in the stool. Localized mainly in the large intestine, the acute inflammation of dysentery may involve the entire intestinal wall; in addition, dysentery may be accompanied by extraintestinal diseases, including arthritis and the hemolytic uremic syndrome. In general, dysentery is rare in infants up to 6 months of age, has its highest incidence and worst severity in children and continues to cause infection and disease at a lower level throughout adult life (FARTHING and KEUSCH 1989). Although a number of different bacterial enteric pathogens and also parasites such as *Entamoeba histolytica* may cause the clinical syndrome of dysentery, *Shigella* is the most common cause in developing countries (TAYLOR et al. 1986).

IV. *Campylobacter jejuni*

After having first been identified as a significant diarrheal pathogen in the early 1970s, the global health impact of *Campylobacter jejuni* has been well documented. *Campylobacter* is currently thought to be the most common cause of bacterial diarrhea in the United States and is probably next to ETEC the leading cause of travelers' diarrhea (SCOTT et al. 1997). However, the incidence of *Campylobacter*-induced diarrhea in the industrialized world is low in comparison to the impact of this disease in young children in developing countries. Thus it has been estimated that the incidence of *Campylobacter* disease in children under 5 years of age in Mexico is as high as 0.4 episodes per child per year (CALVA et al. 1988).

The microaerophilic *Campylobacter* bacteria are commensals of many wild and domestic birds and mammals, and humans usually acquire the infection by eating contaminated and insufficiently heated meat or poultry. Diarrhea is the most common human illness and *C. jejuni* is the predominant species associated with disease. *Campylobacter* disease can present with a range of symptoms from watery diarrhea to dysentery, and in some cases the abdominal pain can be mistaken for acute appendicitis. Usually there are moderate volumes of watery diarrhea that in about 20%–50% of the cases also contain mucus and blood, and there is often also fever and abdominal cramps. Untreated, acute infection usually lasts for 3–5 days, but in about 10%–20% of the patients illness lasts up to 2 weeks.

V. Rotavirus

Rotavirus is the primary cause of severe dehydrating diarrhea in children below 3 years of age throughout the world. Nearly every child is infected in the first few years of life. In developing countries rotaviruses are estimated to account for as many as 870000 deaths a year in children under 5 years of age (WHO 1996). Although rotavirus is found in less than 10% of all cases of diarrhea in infants in developing countries, it accounts for a much higher proportion (20%–60% in different studies) of all hospital admittances for diarrhea (GLASS et al. 1997). Because rotaviruses affect children in industrialized and developing countries with similar frequency it is believed that improvements in hygiene would probably only have a limited effect on the incidence of disease. The dehydration caused by rotavirus diarrhea can usually be successfully treated with standard rehydration therapy, but much interest has been focused on prevention by vaccination.

C. Mechanisms of Disease and Immunity in Diarrheal Diseases

Rational vaccine development requires knowledge, not only of the causative agents of disease, but also of their main pathogenic mechanisms and of the protective immune mechanisms that can be mounted by the host. The identification of the mechanisms of disease and the underlying microbial virulence factors can directly guide the identification of the most important antigens to target vaccination against and thus to be included in a vaccine. Similarly, knowledge about the immune mechanisms, both protective and potentially immunopathogenic ones, is important in guiding which type of vaccine, for example, oral versus parenteral or killed versus live, should be most likely to induce protection while avoiding adverse immune reactions or other side-effects.

As is discussed below, the various diarrheal pathogens may use a range of different mechanisms for both infecting the host and inducing disease, and the host defense mechanisms may also differ substantially between the different types of infection.

I. Enterotoxins and Antitoxic Immunity

The major pathogenic mechanisms of enterotoxigenic bacteria include initial bacterial colonization of the small intestine followed by production of one or more enterotoxins that through various mechanisms can induce electrolyte and water secretion resulting in diarrhea (GUERRANT 1985; HOLMGREN and SVENNERHOLM 1992). These enterotoxins stimulate secretion, primarily from the small intestinal crypt cells, probably mainly by inducing increased formation of cyclic AMP and/or cyclic GMP in the epithelial cells. However, there is increasing evidence that in addition to the fluid secretion mediated via these cyclic AMP and/or cyclic GMP pathways there may also be a local neurogenic component in the diarrheal secretory process, triggered primarily via the action of enterotoxins on intestinal enterochromaffin cells (JODAL and LUNDGREN 1995).

The prototype enterotoxin is cholera toxin (CT), which is produced in an almost identical form by *V. cholerae* O1 as well as O139 bacteria (HOLMGREN 1981; WALDOR and MEKALANOS 1994). CT consists of five identical binding (B) subunits associated in a ring into which a single toxic-active (A) subunit is noncovalently inserted; the binding receptor for CT on cells is a specific glycolipid, the ganglioside GM1 (HOLMGREN 1981).

ETEC bacteria may produce either or both of a heat-labile enterotoxin (LT) and a heat-stable enterotoxin (ST; GUERRANT 1985; HOLMGREN and SVENNERHOLM 1992). While both geographic and age-related variations may occur, approximately one-third of all clinical ETEC isolates produce LT alone, one-third ST alone and one-third LT in combination with ST (SVENNERHOLM and HOLMGREN 1995). LT is structurally, functionally and immunologically closely related – although not identical – to CT. Thus, LT also consists of five B subunits and one A subunit and both of these proteins cross-react strongly immunologically with the corresponding CT subunits, even though there are also specific A- and B-subunit epitopes on both toxins (GUERRANT 1985; HOLMGREN and SVENNERHOLM 1992). Both anti-CT and anti-LT immune responses are mainly directed against the B-subunit portions of the molecules and prevent the toxins from binding to the intestinal cell receptors (SVENNERHOLM et al. 1986a).

Experimental studies during the 70s and 80s indicated that the purified CT or LT B-subunits (CTB or LTB) are suitable toxoid candidates for inclusion in vaccines against enterotoxic enteropathies (HOLMGREN et al. 1977). B-subunits are particularly well suited as oral immunogens because they are stable in the intestinal milieu and are capable of binding to the intestinal epithelium, including the M-cells of the Peyer's patches, properties which are important for stimulating mucosal immunity including local immunological memory (NEUTRA and KRAEHENBÜHL 1992). Furthermore, studies in animals have suggested that locally formed secretory IgA (SIgA) antibodies are of major importance for providing antitoxic immunity in the gut (HOLMGREN and SVENNERHOLM 1992). Antibodies against CTB may cross-protect against *E. coli* LT disease and vice versa, anti-LT immunity may be effective against experimental cholera, although protection against the homologous toxin may be somewhat stronger (SVENNERHOLM et al. 1986a). As a practical vaccine consequence of this substantial degree of immunological cross-reactivity between CT and LT, studies both in endemic areas and in travelers have shown that peroral vaccination with a cholera vaccine containing CTB induces a significant degree of protection also against diarrhea caused by LT-producing *E. coli* (CLEMENS et al. 1988a; PELTOLA et al. 1991; SCERPELLA et al. 1995).

E. coli ST has very distinct properties from the heat-labile enterotoxins. The important form of ST for diarrhea in humans, STa (another structurally related ST called STb may be produced by ETEC isolated form piglets with diarrhea) is a small, tightly packed molecule consisting of 18 or 19 amino acids, which is not immunogenic unless coupled to a carrier protein, either chemically or by recombinant technology (FRANTZ and ROBERTSON 1981; SVENNERHOLM et al. 1986b; SANCHEZ et al. 1988). Accordingly, STa that is released during infection does not induce any immune response and it is still unclear whether sufficiently strong anti-ST immunity can be induced by vaccination with STa-carrier protein conjugates to provide practically significant protection against disease caused by ST-producing *E. coli* in humans (SVENNERHOLM and HOLMGREN 1995).

II. Colonization and Antibacterial Immunity in Cholera and ETEC Infections

Enterotoxin-producing bacteria need to colonize the small intestine to cause diarrhea. The colonization is dependent on receptor-ligand interactions between the bacteria and the host cells, which usually are specific for the species. Colonization is promoted by so called adhesins or colonization factors, that may be fimbrial or fibrillar in nature (EVANS and EVANS 1989; GAASTRA and SVENNERHOLM 1996).

In *V. cholerae* O1 bacteria a toxin-coregulated pilus (TCP) has been shown to be of importance for colonization of the small intestine (Voss et al. 1996). In addition, *V. cholerae* bacteria may express a number of other fimbrial structures, for example, a mannose-sensitive hemagglutinin (MSHA), which can mediate bacterial attachment to epithelial cells (JONSON et al. 1991). In experimental systems it has been shown that antibodies against TCP as well as MSHA can protect against infection and disease with *V. cholerae* expressing the corresponding fimbriae (OSEK et al. 1992). However, it remains to be defined whether mucosal immune responses against TCP and MSHA can add significantly to the strong protective action mediated by antibodies to the *V. cholerae* LPS antigens. Thus it has been well established that *V. cholerae* O1 LPS is the predominant target antigen for antibacterial immunity against experimental O1 classical or El Tor cholera (SVENNERHOLM 1980), and recent studies have suggested that antibacterial immunity against *V. cholerae* O139 to a large extent is provided by antibodies against the corresponding LPS (JONSON et al. 1996). In accordance with this and what was said above about antitoxic cholera immunity, the major protective antibodies against cholera have been identified as those directed against LPS and CTB, respectively (HOLMGREN et al. 1977). Either of these two types of antibodies can confer strong protection against disease by inhibiting bacterial colonization and toxin binding, respectively, and when present together in the gut they can have a strongly synergistic protective effect (SVENNERHOLM and HOLMGREN 1976; HOLMGREN et al. 1977).

In ETEC isolated from humans as many as 20 different colonization factors (CFs) have been identified to date (GAASTRA and SVENNERHOLM 1996), and different from the situation in cholera, protective antibacterial immunity against experimental ETEC infection has been found to be specific for the different CFs rather than for the LPS O antigen (SVENNERHOLM et al. 1990). Fortunately from the point of vaccine development, 50%–80% of the clinical ETEC isolates express either of three distinct CF antigens (CFA), referred to as CFA/I, CFA/II or CFA/IV (EVANS and EVANS 1989). CFA/I is a homogenous protein made up of hundreds of copies of an identical 15-kDa subunit protein, which determines the immunological characteristics of the fimbriae. CFA/II and CFA/IV, on the other hand, consist of three different subcomponents each, the coli surface (CS) antigens CS1, CS2 and CS3 and CS4, CS5 and CS6 (EVANS and EVANS 1989). Usually, the fibrillar CS3 is expressed alone or together with the fimbrial CS1 and CS2 and CS6, which neither has a fimbrial or fibrillar structure, is found together with the fimbrial CS4 or CS5. All of these CFs have been shown to promote ETEC colonization and to induce specific immune responses following infection (ÅHRÉN and SVENNERHOLM 1985; SVENNERHOLM et al. 1990; STOLL et al. 1986; QADRI et al. 1998). In addition, a number of other fimbriated CFs, for example, PCFO159, PCFO166, CS7, CS17 and CFA/III, have been identified (MCCONNELL 1991; GAASTRA and SVENNERHOLM 1996). The latter fimbriae and the additional CFs identified to date, have usually been found in relatively low frequencies on clinical ETEC isolates. However, in some areas CS17 (Sommerfelt et al. 1996) or PCFO166 (Paniagua et al. 1997) have been identified on a substantial proportion of ETEC strains isolated from children with diarrhea.

The three predominant CFAs (CFA/I, CFA/II and CFA/IV) are found in especially high frequencies on ETEC strains expressing either ST alone or ST in combination with LT (Svennerholm and Holmgren 1997). There is still, however, a high proportion of LT-only producing *E. coli* strains in which a specific CF has not yet been identified.

The great diversity of O-groups associated with human ETEC (Black 1986) together with the relatively less protective importance of anti-LPS antibodies as indicated by experimental studies (Åhrén and Svennerholm 1985; Svennerholm et al. 1990), limits the utility of LPS antigens as a basis for vaccine development against ETEC. Antibodies against different CFs, on the other hand, have been highly effective in protecting against diarrhea caused by ETEC expressing the homologous CF. Furthermore, similar to the synergistic cooperation observed between anti-LPS and anti-CTB antibodies against cholera, antibodies against different CFAs have been found to cooperate synergistically with anti-LT antibodies in protecting against disease caused by LT-producing *E. coli* expressing the corresponding CFA (Åhrén and Svennerholm 1985). These results suggest that an effective ETEC vaccine should ideally provide anti-CF as well as anti-LT immunity and should thus contain the most prevalent CFs in combination with a suitable toxoid. Since in both cholera and ETEC infections the bacteria as well as the toxins they produce are noninvasive and reside at the mucosal surfaces of the small intestine and in the gut lumen, locally produced mucosal antibodies are of prime importance for protection. Such antibodies are most efficiently induced by intestinal or, more practically, by oral antigen administration (Holmgren and Svennerholm 1983), which has guided most vaccine development efforts against cholera and ETEC diarrhea over the past 10–20 years.

III. Pathogenic and Immune Mechanisms in *Shigella* Infections

At variance with the enterotoxin-producing bacteria, which mainly affect the small intestine, the *Shigella* pathogens mainly infect the large intestine where they have developed a sophisticated process of invading the colonic epithelium and by using the host cells' cytoskeletal proteins can spread from cell to cell without being exposed to the extracellular environment. The dysenteric disease, associated with bloody and mucus-containing diarrhea, is caused by a combination of the resulting destruction of the colonic epithelium and the intense inflammation evoked in response to the infection. Before entering the enterocytes in the colon the bacteria probably adhere to the surface of the intestinal mucosa by as yet nonidentified adhesins.

The genes associated with virulence in *Shigella* are found on large plasmids, often referred to as virulence plasmids or invasion plasmids. In vitro studies have shown that *Shigella* that carry a virulence plasmid can adhere considerably better to epithelial cells than isogenic mutants lacking this plasmid (PAL and HALE 1989). Following attachment, virulent *Shigella* bacteria can enter the epithelial cells by inducing a micropinocytotic process (SANSONETTI and PHALIPON 1996). This process results in massive rearrangement of the subcortical cytoskeleton of the cell, which is caused by the secretion of bacterial so-called invasion plasmid antigens, Ipas, following contact of the bacteria with the host cell surface and is a critical event for invasion to occur. Once intracellular, *Shigellae* can lyse the phagocytic vesicle by using the three Ipa proteins A, B and C and escape into the cytoplasm where they grow rapidly. Furthermore, the bacteria can move intracellularly towards the intercellular junctions where they push the host cell membrane into the adjacent colonic cells thus allowing for a direct cell-to-cell spread (GOLDBERG et al. 1993).

Recent studies have suggested that the machinery required for bacterial internalization is only present on the basolateral side of the colonic cells (MOUNIER et al. 1992). In animal models it has been shown that *Shigella* largely is taken up via M-cells covering the lymphoid follicles in the intestinal mucosa. It has therefore been proposed that the initial step of invasion by *Shigella* may be an essentially passive uptake and transport by M-cells which is then followed by the described Ipa-dependent colonic cell attachment and invasion steps taking place from the basolateral back side of the epithelium (SANSONETTI and PHALIPON 1996). This is an attractive hypothesis, especially as it has also been shown that the uptake via M-cells is followed by a significant inflammation, which is characteristic of shigellosis (SANSONETTI and PHALIPON 1996). However, the significance of this M-cell uptake pathway for infection in humans remains to be defined.

Shigella bacteria can also produce various toxins, among which Shiga toxin produced by *S. dysenteriae* strains is best known. This is an oligomeric protein, which similar to cholera toxin and *E. coli* LT is composed of 5 B-subunits and a single toxic-active A-subunit. However, different from the latter enterotoxins the Shiga toxin A-subunit makes this toxin cytotoxic rather than cytotonic by blocking protein synthesis by exposed cells. Current knowledge about both the Shiga toxin and the related Shiga-like toxins produced by other *Shigella* species suggest that the primary targets for these toxins are blood vessel endothelia rather than the intestinal epithelial cells, and that these toxins may rather be involved in the pathogenesis of the more rare systemic complications of *Shigella* infection including the hemolytic-uremic syndrome (WENNERÅS and SANSONETTI, to be published). Immunity against *Shigella* infection appears to be mainly of the Th-2 type, and humoral as well as cell-mediated immune mechanisms may contribute to protection, although humoral immune responses appear to be most important at least for protection (SANSONETTI and PHALIPON 1996).

It is well accepted that protective immunity against *Shigella* species is primarily directed against the O-antigens. Since there are more than 45 O-antigen serotypes and subtypes it is not realistic to attempt to develop vaccines protecting against all these *Shigella* types. However, fortunately a majority of clinicially significant *Shigella* infections fall in three serotypes: *S. dysenteriae* type I, *S. sonnei* (which is a single serotype) and *S. flexneri* 2a. These types are therefore the primary targets for vaccine development (possibly combined with *S. flexneri* 1 and 3; NORIEGA et al. 1997). It has been assumed that mucosal – rather than systemic – immune responses play the major role in protection against *Shigella*, since the bacteria remain localized to the colonic mucosa. However, a protective role of systemic antibodies against *Shigella* LPS is also suggested since parenteral vaccination with an *S. sonnei* LPS toxoid conjugate, which predominantly induced a systemic immune response, conferred protection against severe *S. sonnei* infection in Israeli soldiers (COHEN et al. 1997). Conceivably, these antibodies could have transudated from the circulation into the colonic submucosal tissues, and there been able to interfere with bacterial invasion of the colonic cells from the basolateral side if this is an important infection mechanism. At the same time, it cannot be excluded that the parenteral vaccination also induced some mucosal anti-LPS responses, since it has been shown that parenteral vaccination (in this case with an inactivated cholera vaccine) can induce a mucosal immune response in persons who had been immunologically primed via previous natural exposure to antigen (HOLMGREN and SVENNERHOLM 1983).

IV. Protective Immunity in Rotavirus Infections

Rotaviruses are small rheoviruses, circa 70 nm in diameter. The viruses are nonenveloped and possess a double-shelled outer capsid (KAPIKIAN and CHANOCK 1996). Within the double-capsid is a third layer, the core, which contains the virus genome comprising 11 segments of double-stranded RNA; the segmented genome often undergoes genetic reassortment during coinfection. Rotaviruses have three major types of protein antigens, referred to as group-, subgroup-, and serotype-specific antigens (HOSHINO and KAPIKIAN 1994). The group specificity of rotaviruses is predominantly defined by VP6, which is the major structural protein encoded by gene number 6. Seven different subgroups of VP6 have been defined (A–G), but almost all human rotaviruses of epidemiological importance belong to group A. Against this background rotavirus vaccine development has focused on inclusion of VP6, group A antigen. The serotype specificity of rotaviruses is defined by the two major neutralization antigens that are located on the outer capsid, i.e., VP7 and VP4. VP7, which is the most abundant protein on the outer capsid, has been considered to be the only determinant of serotype specificity (KAPIKIAN 1996). Among the 14 different types of VP7 identified, only serotypes 1, 2, 3 and 4 are of epidemiological importance. Overall, serotype 1 occurs most frequently, but the other serotypes may predominate occasionally or in certain areas. VP4, which protrudes from the outer surface in the form of discrete spikes of 10–12 nm in length, has also been divided into different serotypes based on neutralization as well as genotyping (ESTES and COHEN 1989; HOSHINO and KAPIKIAN 1994).

Antibodies against VP4 and VP7 are independently associated with protection against rotavirus as studied in different animal models (HOSHINO and KAPIKIAN 1994). However, the mechanisms of immune protection have not been clearly established. Locally produced antibodies have been shown to be of importance in some studies and a role of cellular immunity has also been suggested (KAPIKIAN 1996). Recently, DONG et al. (1997) demonstrated the capacity of rotavirus to produce an enterotoxin, but this toxin seems to be poorly immunogenic following infection in humans, and its role as a protective antigen has not been established.

D. Cholera Vaccines

Until the middle 1980s the cholera vaccines available were injectable killed whole cell preparations, that usually contained vibrios of the two serotypes Inaba and Ogawa, but no toxin antigen. These vaccines only provided limited protection of short duration, at most approx. 50% protection for 3–6 months, and only then limited to adults and older children in cholera endemic areas (FEELEY and GANGAROSA 1980; HOLMGREN and SVENNERHOLM 1992). Following the identification of CT as a major virulence factor in cholera, the documentation of the strictly mucosal nature of the cholera disease and the growing insight that protective immunity depends on stimulation of gut mucosal rather than systemic antibody production (HOLMGREN and SVENNERHOLM 1983), intensive work was initiated in many laboratories during the 70s and 80s to develop new cholera vaccines that induce strong immune responses locally in the gut and that provided antitoxic in addition to antibacterial immunity. Based on this both oral inactivated vaccines and live genetically engineered vaccines against cholera have been developed and recently licensed in several countries.

I. Oral Inactivated Vaccines

An oral cholera vaccine consisting of the nontoxic, highly immunogenic CTB protein in combination with heat- and formalin-killed *V. cholerae* O1 vibrios of the classical and El Tor biotypes (Table 1) was developed in Sweden in the early 80s (HOLMGREN and SVENNERHOLM 1992). This CTB-whole cell (B-WC) vaccine, which is produced and marketed by SBL Vaccin, Stockholm, Sweden was first licensed in 1991, and in a second-generation form based on the use of recombinantly produced CTB licensed since 1993. The vaccine is given together with a bicarbonate buffer to preserve the CTB pentameric structure necessary for good oral mucosal immunogenicity. In extensive clinical trials,

Table 1. Oral B subunit (B) whole cell (WC) cholera and ETEC vaccines

	Composition per dose	Clinical evaluation
Cholera	A. B-O1 WC (Bangladesh field trial formulation) 1 mg CTB (purified from CT) + 1×10^{11} killed bacteria 2.5×10^{10} heat-killed Inaba vibrios (strain Cairo 48) 2.5×10^{10} heat-killed Ogawa vibrios (strain Cairo 50) 2.5×10^{10} formalin-killed classical vibrios (strain Cairo 50) 2.5×10^{10} formalin-killed E1 Tor vibrios (strain Phil 6973)	Safe, immunogenic and protective in both volunteers and in large field trials in Bangladesh: 85% efficacy first 6 months; ca 60% first 3 years. Cross-protection against ETEC: ca 70% short-term efficacy. Licensed 1991. Replaced by (B) in 1993.
	B. rB-O1 WC (currently licensed formulation) 1 mg recombinant CTB + same WC composition as in A	Same safety, immunogenicity and protective efficacy as for (A). Field trial in Peru showed 86% protection against O1 E1 Tor cholera. Licensed in 1993.
ETEC	C. rB-CF ETEC vaccine 1 mg rCTB + 1×10^{11} formalin-killed ETEC $\sim 2 \times 10^{10}$ CFA/I (O78) $\sim 2 \times 10^{10}$ CS1 (O6) $\sim 2 \times 10^{10}$ CS2 + CS3 (OR) $\sim 2 \times 10^{10}$ CS4 + CS6 (O25) $\sim 2 \times 10^{10}$ CS5 + CS6 (O167)	Safe and immunogenic in phase I and phase II clinical trials. Protective efficacy trials initiated in travelers.

including large field trials, the B-WC vaccine has proved to be completely safe and to provide good protection against cholera and also partial protection against diarrhea caused by LT-producing ETEC.

The B-WC vaccine was designed to evoke antitoxic as well as antibacterial intestinal immunity, since both of these types of immunity have been shown to protect and when present together to provide synergistic cooperative protection against experimental cholera (Svennerholm and Holmgren 1976; Holmgren et al. 1977). Phase I and phase II clinical studies in different endemic and nonendemic countries established that the vaccine does not cause any detectable side-effects and that, after either two or three doses, it stimulates a gut mucosal IgA antitoxic and antibacterial immune response (including memory) comparable to that induced by cholera disease itself (Svennerholm et al. 1984; Jertborn et al. 1988; Quiding et al. 1991; Sanchez et al. 1993). Furthermore, immunization with either the complete B-WC vaccine or the WC component alone was found to protect American volunteers against challenge with a dose of live cholera vibrios of biotype El Tor, that caused disease in 100% of concurrently tested unvaccinated controls (Black et al. 1987).

On this basis, a large, double-blind placebo-controlled field trial with more than 90000 participants was undertaken in Bangladesh. In this trial both the B-WC vaccine and the WC component alone were found to confer substantial and long-lasting protection against cholera. The B-WC vaccine had a higher initial ($p < 0.05$) efficacy level than the WC vaccine (85% versus 58% for the initial 4- to 6-month period) that was the same in all age groups studied (Clemens et al. 1986). Indeed, the protection induced by the combined vaccine was 64% higher ($p < 0.05$) than that provided by the WC component alone. The B-WC vaccine continued to be significantly more protective than the WC component alone for the first 8 months after vaccination. Thereafter, the efficacy was similar, approximately 60% for both vaccines, if calculated for the whole population, for a 3-year follow-up period (Clemens et al. 1990a). Protection was of similar magnitude after two or three doses of vaccine; even higher long-term protective efficacy, ca 70%, was seen in vaccinees older than 5 years. It is likely that also children below 5 years, in which immunity rapidly waned after the high-level protection induced in the first 6 months, could also be provided with long-lasting high-level protection by a booster immunization after 1 year.

The initial B-WC vaccine formulation tested in Bangladesh contained a CTB component that was prepared by chemical isolation from CT produced by the high level-expression wild-type strain, 569B (Tayot et al. 1981), which made the preparation of CTB relatively laborious and expensive. It was therefore a significant improvement when an efficient recombinant overexpression system for the large-scale production of CTB (Sanchez and Holmgren 1989) was constructed; the production and purification of recombinant CTB for industrial vaccine production purposes has since then been further improved (J. Holmgren and SBL Vaccin, unpublished data). Since extensive clinical testing of a vaccine formulation containing such recombinantly produced CTB (rB) (Table 1) has shown the same degree of safety and immunogenicity as the Bangladesh trial formulation, rB-WC has become the currently produced and licensed vaccine formulation under the tradename Dukoral. In a recent field trial in Peruvian militaries the rB-WC vaccine was shown to induce strong protection (SANCHEZ et al. 1994). Thus, this vaccine formulation conferred 86% protection against cholera in Peru, i.e., the same high level of protection as previously seen during the first 6-month postvaccination period in Bangladesh, in spite of the fact that the rB-WC vaccine was only given in two doses and that it protected against cholera of exclusively the O1 El Tor biotype.

In Vietnam, a locally produced WC only vaccine, being closely similar to the Swedish WC formulation was recently shown to give rise to 67% protection against O1 El Tor cholera during the first year of follow-up (TRACH et al. 1997). A large field trial of a bivalent such O1/O139 WC vaccine has recently been initiated in Hue, Vietnam.

Both the B-WC and the WC vaccines have been shown to substantially reduce the overall diarrhea morbidity among those vaccinated, for example, there was a 50% reduction in admissions for life-threatening diarrhea in the vaccinated as compared with the placebo group during 3 years of follow-up, and there was even an apparent significant impact on overall mortality during the first year postvaccination cholera season (CLEMENS et al. 1988b; J. Clemens, unpublished data). There was also a significant, ca 50%, reduction in cholera incidence in unvaccinated breast-feeding children below the age of 2 years whose mothers had received either of the oral B-WC or WC alone cholera vaccines as compared with those whose mothers had received *E. coli* K12 placebo vaccination (CLEMENS et al. 1990b). These findings support the public health potential of the B-WC cholera vaccine in settings where cholera, ETEC and additional enterotoxic diarrheal diseases account for a large number of life-threatening watery diarrheas and where adequate treatment facilities in rural areas are still scarce (SACK and FREIJ 1990; SACK et al. 1991).

II. Oral Live Vaccines

As an alternative approach, recombinant DNA techniques have been applied to construct various attenuated *V. cholerae* O1 strains for testing of their possible utility as live oral vaccines. In the early 80s a number of vaccine candidates were constructed from wild-type *V. cholerae* O1 strains by introducing deletions in the chromosomal gene(s) encoding the A-subunit or both the A- and B-subunits of CT (KAPER et al. 1984, 1997). Although these first-generation recombinant live vaccine strains, such as JBK 70 and CVD 101, were markedly attenuated compared with their wild-type parents, they still caused unacceptable adverse reactions, i.e., mild-to-moderate diarrhea often associated with additional reactions such as malaise, headache and vomiting in up to 50% of immunized volunteers. Thus, the removal of even the whole CT gene from pathogenic *V. cholerae* O1 wild-type strains represented an insufficient attenuation for providing a safe, nonreactogenic vaccine strain. However, it was evident that even a single dose immunization with such strains could induce significant protection against later challenge with the wild-type parent strain in volunteers (Levine and Tacket 1995; Kaper et al. 1997).

CVD103-HgR. To overcome the safety problems observed with the initially prepared live attenuated strains, attempts were made to use a less pathogenic and relatively poorly colonizing strain, 569B (classical biotype, Inaba serotype) as the parenteral strain for vaccine construction. This resulted in the first truly promising, engineered vaccine strain, CVD103-HgR, which was prepared by deletion of more than 90% of the gene encoding the A-subunit of CT together with the insertion of a gene encoding resistance to Hg^{2+} to readily differentiate this vaccine strain from wild-type vibrios (Kaper et al. 1997). CVD103-HgR has now been extensively tested in human subjects for its safety and immunogenicity both in industrialized and in developing countries (Levine and Tacket 1995). The results have shown that a single dose of CVD103-HgR containing 5×10^8 colony-forming units (CFU) is safe and gives vibriocidal seroconversion in 90% or more of adult volunteers living in industrialized countries. Furthermore, vaccination also conferred significant protection against challenge with wild-type *V. cholerae* O1 strains when tested in adult American volunteers. The overall protection was high (80%–100%) against challenge with vibrios of the classical biotype and moderate (49%–67%) against challenge with El Tor biotype (Levine and Tacket 1995). Protection was evident as early as 8 days after vaccination and was shown to last for at least 6 months (Tacket et al. 1992). Similar to previous findings with the inactivated B-WC cholera vaccine, which gave 64% protection against any type of cholera but 100% protection against clinically significant cholera after challenge with vibrios of the El Tor biotype (Black et al. 1987), vaccination with CVD103-HgR provided complete protection against severe cholera even after El Tor challenge. Based on these results CVD103-HgR, which is manufactured by the Swiss Serum and Vaccine Institute in Bern, has been licensed for use in travelers in some European and Latin American countries under the trade name Orochol (Levine and Svennerholm 1997).

However, when studies of CVD103-HgR began in less-developed countries where cholera is endemic, it was found that the 5×10^8 CFU dose that previously had been used successfully in the United States, elicited vibriocidal antibody seroconversions in less than 25% of the vaccinees. Apparently the “take” of the vaccine, at least in this dose, was relatively poor in populations already having some immunity based on previous natural exposure to *V. cholerae* organisms. To overcome this problem higher doses of vaccine were tested (Levine and Tacket 1995) and when a dose of CVD103-HgR containing 5×10^9 CFU was used, a markedly increased vibriocidal titer conversion rate was seen (>75%). A large-scale double-blind, placebo-controlled field trial of CVD103-HgR was initiated in North Jakarta, Indonesia in 1993 to determine the protective efficacy of a single 5×10^9 CFU dose in preventing cholera in an endemic area (KAPER et al. 1997). Recently, the results of this trial have been analyzed, and have shown that disappointingly the vaccine did not give any significant protection in any of the 4 years of surveillancc (M. Levine, personal communication). The lack of protection is surprising in view of the good protection seen in North-American volunteers given this vaccine. The results with CVD103-HgR in Indonesia are also in sharp contrast to the behaviour of the oral inactivated cholera whole cell or whole cell plus B subunit vaccines which as mentioned above gave the same (or higher) levels of protection in the field as in volunteers.

Peru 15. Mekalanos and coworkers have also prepared a series of interesting live vaccine candidates generated from wild-type El Tor strains, which have been tested for safety and immunogenicity in volunteers (TAYLOR et al. 1994; COSTER et al. 1995). These various vaccine constructs have in common a deletion of the whole 4.5 kb core element ("virulence cassette") containing the genes encoding CT, the associated virulence factors Zot (zonula occludence toxin), Ace (auxiliary cholera enterotoxin) and Cep (core encoded pilus, believed to be an accessory colonization factor) as well as the RS1 element and attRS1; the latter sequences are flanking one core element and are used in site-specific and homologous recombination. Similar to the previous observations with attenuation of strains of the classical biotype, also these first-generation El Tor vaccine candidate strains gave rise to unacceptable side-reactions with diarrhea and usually also additional gastrointestinal adverse reactions. More recently, however, a motility-deficient mutant strain, designated Peru-15 was selected from one of the previous vaccine constructs and this less well colonizing strain has given much more promising results in clinical studies, with little reactogenicity and yet good protective immunogenicity (TAYLOR et al. 1994; COSTER et al. 1995).

III. Combined Vaccines Against O1 and O139 Cholera

Inactivated Vaccines. Based on the appearance in late 1992 of *V. cholerae* O139 as an additional cause of cholera in Southeast Asia, much attention has been focused on the possibility to develop a cholera vaccine that also affords protection against this "new" type of cholera. In our laboratory we have studied the immune mechanisms and protective antigens of *V. cholerae* O139 in animal models as a basis for vaccine development (JONSON et al. 1996). Together with SBL Vaccin, Stockholm, we are working on the development of an oral bivalent B subunit – O1/O139 whole cell (B-O1/O139 WC) cholera vaccine by adding formalin-killed O139 vibrios to the recently licensed oral rB-O1 WC vaccine. When tested in Swedish volunteers, such an rB-O1/O139 WC vaccine was found to be safe and immunogenic and induced comparable vibriocidal antibody responses against both *V. cholerae* O1 and O139 (JERTBORN et al. 1996).

Live Vaccines. Following similar strategies as used for developing live O1 cholera vaccine candidates, live oral vaccines based on *V. cholerae* O139 strains have also been constructed. Thus KAPER et al. (1997) engineered an attenuated *V. cholerae* O139 vaccine strain (CVD112) by deleting the entire "virulence cassette" chromosomal region together with introducing a deletion mutation in *recA* in order to diminish the ability of the vaccine strain to recombine foreign DNA into its chromosome. WALDOR and MEKALANOS (1994) constructed an attenuated vaccine candidate (Bengal-3) from another wild-type strain of *V. cholerae* O139 by deleting the whole virulence cassette plus the RS1 and attRS1 factors involved in virulence cassette recombination, and in addition they also inserted a recombinant gene encoding CTB into *recA* to both inactivate the latter gene and to provide overexpression of CTB. Both CVD112 and Bengal-3 have in initial volunteer studies been found to have low reactogenicity and to induce protective immunity against challenge with wild-type homologous *V. cholerae* O139 organisms. However, different from the situation with the bivalent B-O1/O139 vaccine for which there was no indication of any competition between different vaccine components with regard to immunogenicity, there is evidence that it may be difficult to combine live vaccines against O1 and O139 cholera without the risk of reducing the immunogenicity of either or both vaccine strains (J. J. Mekalanos, personal communication). Furthermore, the interest in access to a bivalent cholera vaccine providing protection against both *V. cholerae* O1 and O139 cholera has diminished considerably. This is due to the decreased prevalence of *V. cholerae* O139 in Southeast Asia and to the lack of spread of vibrios of this O-group to new countries – although O139 cholera returned to Calcutta in the fall of 1996 (KAPER et al. 1997).

E. ETEC Vaccines

The findings of drastically decreased rates of ETEC diarrhea in children in developing countries with increasing age (BLACK 1986) and the decreased disease-to-infection rates observed in highly endemic areas (CRAVIOTO et al. 1988; LÓPEZ-VIDAL et al. 1990), have suggested that significant protective immunity may develop naturally against ETEC. Studies in human volunteers have also shown that strain-specific immunity follows ETEC infection (LEVINE 1990). Furthermore, travelers from industrialized countries who remain in less-developed countries for at least one year, and who therefore most likely are repeatedly exposed to ETEC and may also experience one or more episodes of ETEC diarrhea, exhibit significantly lower incidence rates of ETEC disease than newly arrived travelers, supporting protection by acquired immunity (LEVINE and SVENNERHOLM 1997).

Another strong indication of the potential of inducing effective protective immunity against ETEC disease in humans is the finding in the cholera vaccine trial in Bangladesh that the oral B-WC cholera vaccine through its CTB component, which cross-reacts immunologically with *E. coli* LTB, afforded significant protection against diarrhea caused by LT-producing ETEC (CLEMENS et al. 1988a). The protection observed was about 67% for 3 months and was similar against bacteria producing LT and against ETEC producing

LT in combination with ST. The oral B-WC cholera vaccine has also afforded significant protection, 50%–60%, against LT-producing *E. coli* (LT or LT+ST strains) in placebo-controlled trials in Finnish travelers going to Morocco (PELTOLA et al. 1991) and in American college students studying in Mexico for limited time periods (SCERPELLA et al. 1995).

Thus, there is strong support for the potential of developing effective ETEC vaccines for use in humans. A broad and strong protective efficacy is to be expected if the mucosal immunity against LT achieved by immunization with either LTB or CTB can be combined with antibacterial immunity directed against the predominant CFs on human ETEC strains. Different ETEC vaccine candidates have recently been considered based on these premises, for example, live or inactivated vaccines that may provide both anti-colonization and antitoxic immunities.

I. Oral Inactivated Vaccines

Purified Fimbriae. Attempts to use purified CF fimbriae for immunization have yielded disappointing results. Such antigens are relatively expensive to prepare and they have also proven to be very sensitive to proteolytic degradation in the human gastrointestinal tract, which probably explains their poor immunogenicity in human volunteer studies (EVANS et al. 1984; LEVINE 1990). To protect the ETEC fimbriae from degradation in the stomach, immunization with purified CFs incorporated into biodegradable microspheres has been attempted (TACKET et al. 1994). However, although repeated high doses of purified CFs were provided, no significant protection was induced against subsequent challenge with a diarrheagenic dose of ETEC expressing the homologous CF.

Enterotoxoids. Since purified CTB and LTB, which cross-react immunologically, have been shown to be strongly immunogenic and lack toxicity, it has been suggested that they may be suitable candidate antigens to provide anti-LT immunity. Furthermore, both these types of B subunits are particularly well suited as oral immunogens, because they are stable in the gastrointestinal milieu and are capable of binding to the intestinal epithelium, including M-cells (NEUTRA and KRAEHENBÜHL 1992). Even though CTB has provided significant protection against *E. coli* LT disease in animals as well as in humans (CLEMENS et al. 1988a; PELTOLA et al. 1991; HOLMGREN and SVENNERHOLM 1992), we cannot exclude that an LT toxoid may be slightly more effective than CTB in inducing protective anti-LT immunity. At present, however, suitable methods for large-scale production of LTB are not yet available. Therefore, rB is presently used for immunoprophylaxis against ETEC, but may later be replaced by a more LT-like B-subunit, since the structural gene for CTB can be modified to encode B subunits that also contain LTB specific epitopes (LEBENS et al. 1996).

The significance of anti-ST immunity for protection against ST-producing *E. coli* is unclear. By coupling *E. coli* STa either chemically or recombinantly to different carrier proteins, for example, CTB, chimeric proteins that are capable of inducing ST-neutralizing antibodies have been achieved (FRANTZ and ROBERTSON 1981; SVENNERHOLM et al. 1986b; SANCHEZ et al. 1988; AITKEN and HIRST 1993). However, all of these ST conjugates have had retained toxic activity. Although it has been possible to prepare nontoxic ST peptides, either by protein synthesis or by recombinant methods (SVENNERHOLM et al. 1988; SANCHEZ et al. 1988; AITKEN and HIRST 1993) immunization with such peptides coupled to different carrier proteins has failed to induce ST-neutralizing antibodies. However, even though it may be possible to induce high levels of ST-neutralizing antibodies it is uncertain that such a toxoid might play a significant role in an ETEC vaccine, since comparatively large amounts of anti-ST antibodies will be required to provide neutralization of the small ST molecule.

B-CF ETEC vaccines. Based on such considerations we have concluded that a practical way to construct an inactivated ETEC vaccine is to prepare killed ETEC bacteria that express the most important CFs on their surface and combine these organisms with an appropriate B subunit component, i.e., rCTB. In collaboration with SBL Vaccin we have developed a CTB-containing CF whole cell ETEC (B-CF ETEC) vaccine with the potential of providing broad protective coverage against ETEC diseases in different countries.

In initial studies a prototype vaccine consisting of a mixture of killed *E. coli* expressing CFA/I and the different CS-components of CFA/II and CTB was evaluated. The B subunit component was provided as conventionally purified CTB in the oral B-WC cholera vaccine. Strains that express the different fimbriae in high concentrations had been selected for preparation of the whole cell component. The bacteria were inactivated by mild formalin-treatment, which resulted in complete killing of the bacteria without significant losses in antigenicity of the different CFs; furthermore, the CF-antigens of the inactivated bacteria were stable after incubation in human gastrointestinal secretions containing acid and proteolytic enzymes (SVENNERHOLM et al. 1989).

The safety and immunogenicity of the prototype B-CF ETEC vaccine were studied in adult Swedish volunteers given 2 or 3 oral doses with 2-week intervals. Surveillance for side-effects revealed that the vaccine was safe, i.e., it did not give rise to any significant side-effects as tested in more than 100 volunteers (WENNERÅS et al. 1992; ÅHRÉN et al. 1993). Determination of specific immune responses in the intestine, i.e., in intestinal lavage fluid, showed that the vaccine induced significant IgA antibody responses against CFA/I, CFA/II as well as CTB in a majority of the vaccinees (ÅHRÉN et al. 1993). The prototype ETEC vaccine also gave rise to significant increases in peripheral blood antibody secreting cells (ASCs) with specificities for CFA/I, CFA/II and CTB in 85%–100% of the volunteers (WENNERÅS et al. 1992). Responses were predominantly in IgA-producing cells, but high frequencies of IgM ASC responses against CFAs and of IgG ASCs against CTB were also seen. Two oral immunizations seemed to be optimal in inducing specific immune responses, since neither specific IgA levels in intestinal lavages nor numbers of CFA-specific ASCs increased after administration of a third dose of vaccine (WENNERÅS et al. 1992; ÅHRÉN et al. 1993).

Based on the promising results from studies of the prototype ETEC vaccine in adult Swedish volunteers, a modified, more definitive formulation of the ETEC vaccine was developed and produced by SBL Vaccin (Table 1). This vaccine contains recombinantly produced CTB (the same as in the rB-WC cholera vaccine) in combination with five different *E. coli* strains expressing CFA/I and the different subcomponents of CFA/II and CFA/IV, i.e., CS1-CS6 (SVENNERHOLM and HOLMGREN 1995). Based on a large number of epidemiological studies of ETEC CFs in different geographic areas, this modified ETEC vaccine has a potential protective coverage of at least 70%–80% (SVENNERHOLM et al. 1997).

The rB-CF ETEC vaccine has been evaluated for safety and immunogenicity in several phase I and phase II trials in different countries. Peroral immunization with one, or in most instances, two doses of the vaccine 2 weeks apart of more than 500 Swedish, Bangladeshi, Egyptian, Israeli, Swiss, Austrian or American adult volunteers, and recently also in more than one hundred Egyptian children in the age group 2–15 years, has shown that the vaccine is safe. The capacity of the rB-CF vaccine to induce a mucosal immune response has predominately been assessed by determining ASC responses in peripheral blood against CTB as well as the different CFs in the vaccine, since our recent studies have suggested that peripheral blood ASC responses may be good proxy measures of intestinal immune responses. This is based on studies in Sweden in which immune responses against CFs and CTB in intestinal lavage fluids have been compared with corresponding responses in serum, and stool extracts as well as with ASC responses in peripheral blood. These analyses have shown a strong relation between specific IgA responses in intestine and peripheral blood IgA ASC responses (ÅHRÉN et al. 1998). As shown in Table 2, the rB-CF vaccine induced high frequencies of ASC responses against CFA/I, the different CS components of CFA/II and CFA/IV as well as against CTB in a majority of the immunized Swedish, Egyptian, and Bangladeshi adult volunteers (JERTBORN et al. 1998; SAVARINO et al. 1998a; QADRI et al. 1998). Interestingly, cumulative frequencies of ASC responses in peripheral blood to CTB and the different CS components were very similar in the different groups of adult volunteers, but there was a clear trend that ASC responses were considerably higher after the first than the second vaccine dose in volunteers living in ETEC endemic countries such as Egypt, whereas responses in Swedish volunteers often were comparable after the first and second vaccine dose. Similar to the prototype vaccine (ÅHRÈN et al. 1993) the rB-CF ETEC vaccine has been considerably less efficient in inducing a serum antibody responses than mucosal immune responses against the different CFs. Thus serum antibody responses against some of the CS components have been seen in less than 50% of the vaccinees, whereas ASC responses or intestinal lavage responses were observed in 70%–90% of them (JERTBORN et al. 1998).]

Table 2. Immune responses against oral rB-CF ETEC vaccine in Swedish, Egyptian, and Bangladeshi volunteers: frequency (%) with significant responses

	Sweden[a]			Egypt[b]	Bangladesh[c]	
	Blood ASC IgA	Serum IgA	Serum IgG	Blood ASC IgA	Blood ASC IgA	Serum IgA
CFA/I	23/28 (82%)	19/32	14/32	15/16 (94%)	22/27 (81%)	13/27 (48%)
CS1 + CS3	25/32 (78%)	NT	NT	11/16 (69%)	22/26 (85%)	NT
CS2	9/11 (82%)	9/32 (28%)	7/32 (22%)	13/16 (81%)	NT	NT
CS4	22/32 (69%)	3/32 (9%)	1/32 (3%)	16/16 (100%)	NT	NT
CS5	27/31 (87%)	5/32 (16%)	3/32 (9%)	NT	16/27 (59%)	NT
CTB	32/32 (100%)	29/32 (90%)	27/32 (84%)	16/16 (100%)	25/27 (93%)	24/27 (89%)

NT, Not tested.
[a] JERTBORN et al. 1997.
[b] SAVARINO et al. 1998a.
[c] QADRI et al. 1998.

The promising results obtained from the studies of the rB-CF ETEC vaccine in different countries have encouraged the initiation of several phase III trials of the vaccine, both in children in Egypt and in travelers to ETEC endemic areas. Thus, a study was recently initiated in European travelers going for vacation to Kenya (R. Steffen et al., Zürich), and another one was started in American students going to Guatemala for summer courses (D. Sack et al., John's Hopkins, Baltimore) during the spring of 1998. In these different studies adult volunteers are given two doses of the vaccine or an *E. coli* K12 placebo in a double-blind fashion and the incidence of ETEC disease in the two study groups evaluated during follow-up. Studies are also planned, initially in Egypt and then in other ETEC endemic countries, i.e. Bangladesh and Nicaragua, to test the vaccine for capacity to protect against ETEC diarrhea in children less than five years. The phase III trials are preceded by extensive phase II trials to confirm that the vaccine is safe and immunogenic also in young children. Thus, the ETEC vaccine has been tested for safety and immunogenicity in Egyptian school children, 6–10 years, and subsequently in younger children 2–5 years with encouraging results. Safety has been ascertained by showing similar, low frequencies of adverse reactions in vaccine and placebo recipients, and good immunogenicity as reflected by high ASC responses against CTB and different CFs in a majority of the vaccinated children. Based on these results a phase II trial has been initiated in the actual target group for an ETEC vaccine in Egypt, i.e., infants aged 6–18 months (SAVARINO et al. to be published). Since the vaccine proved to be safe and capable of eliciting immune responses also in this young age group, it will be tested for protective efficacy in Egyptian infants within short.

The results from these different phase III trials may reveal the possibility to use an inactivated ETEC vaccine for immunoprophylaxis against traveler's diarrhea caused by ETEC as well as for use as a public health tool to control the most prevalent form of diarrhea in children in developing countries.

II. Oral Live Vaccines

Live bacteria expressing the major CFs and producing LTB may also be considered as ETEC vaccine candidates. If such strains could effectively colonize and multiply in the gut, they might provide a sustained antigen stimulation for the local intestinal immune system (LEVINE 1990). However, since the different CFs are normally not expressed on the same strains and it has not yet been possible to successfully clone the genes for different CFs in the same host organisms to allow stable surface expression of the different fimbriae, such vaccines must, at least for the time being, be based on a cocktail of several different strains. However, with any mixed vaccines there is a risk of overgrowth of one of the included vaccine strains with suppression of the others. Other potential drawbacks of live vaccines may be the risk that they might revert to toxicity by uptake of toxin-encoding plasmids, that they would only provide insufficient low-level production of LTB during growth in vivo, and that the vaccine strains might have poor survival during storage.

Attenuated live ETEC. The potential of live ETEC vaccines was demonstrated when a prototype live vaccine strain expressing CS1 and CS3 fimbriae, but lacking genes that encode LT and ST, was given to human volunteers (LEVINE 1990). By providing a single dose of this attenuated strain, SIgA antifimbrial antibodies were induced locally in the intestine and 75% protection against experimental challenge with wild-type ETEC elaborating CS1 and CS3 fimbriae as well as LT and ST was induced. Thus one strategy involves administration of a collection of attenuated *E. coli* strains expressing the major fimbrial CFs and LTB subunit.

Recombinant Strains. Another approach to prepare a live ETEC vaccine may be to introduce CF-encoding plasmids into heterologous bacteria, such as attenuated *Salmonella*, which, because of their invasive properties, can reside in the bowel for long periods (LEVINE and SVENNERHOLM 1997). It has, for example, been possible to express CFA/I fimbriae on the surface of the Ty21a typhoid vaccine strain. However, since such attenuated typhoid strains are relatively poor intestinal immunogens, of which multiple doses are required to elicit significant protective immunity, they are probably not ideal vectors for heterologous mucosal antigens. Instead the recently developed attenuated *S. typhi* strains, that are immunogenic when administered as a single dose, may be used; for example, cloned genes for the expression of CFA/I and CS3 have been introduced on stable plasmids into such an *S. typhi* strain (CVD 908) and a high level of coexpression of these fimbriae has been demonstrated (GIRÓN et al. 1995).

Other attenuated bacterial strains, such as of *Shigella* or *V. cholerae*, have also been considered as live vectors for the expression of ETEC CFs. Thus, expression of CFA/I and CS3 fimbriae in an attenuated *S. flexneri* 2a live vaccine candidate has been reported (NORIEGA et al. 1997). However, a problem with these various approaches appears to be the introduction of several CFs together with CTB into the same host organism.

F. *Shigella* Vaccines

It has been deemed increasingly important to prevent *Shigella* infections because of the high mortality caused by these infections in children in developing countries and because *Shigella* tends to cause clinical illness in travelers in which severe complications are not uncommon (WHO 1996; LEVINE and SVENNERHOLM 1997). Moreover, antibiotic resistant strains of *Shigella* are prevalent, particularly in developing countries, often rendering effective treatment difficulties. The fact that the majority of bacillary dysentery cases in endemic areas are seen in children less than 5 years, suggests that infections that are experienced in early childhood may have induced a certain level of protection (PAL and LINDBERG 1996). This assumption has gained support from challenge studies in human volunteers. However, the incomplete knowledge of the pathogenic and immune mechanisms against shigellosis has hampered the efforts to develop effective anti-*Shigella* vaccines.

During the course of bacillary dysentery strong antibody responses develop against the major surface antigenic complexes of *Shigella*, directed against both the LPS molecule and several of the Ipas (SANSONETTI and PHALIPON 1996; PÁL and LINDBERG 1996). However, most epidemiological and experimental data indicate that the immunity noted after natural *Shigella* infection is serotype-specific, suggesting that the LPS O-antigen is the main protective antigen. *Shigella* infection or vaccination is usually followed by antibody responses against LPS and the Ipas, both systemically and at various mucosal sites. The intensity of humoral immune responses in peripheral blood as well as at mucosal surfaces has been shown to be correlated with the severity of infection (ISLAM et al. 1995), and a protective effect of SIgA antibodies in patients as well as in breast-fed babies has been suggested based on studies in endemic areas. The immunological mechanisms by which these secretory antibodies would interfere with the infective organisms is, however, still unclear. The severity of shigellosis in immunodeficient patients has also suggested the active involvement of cell-mediated effector mechanisms in counter-acting *Shigella*.

In the 1960s and early 70s, live attenuated strains of *Shigella*, including streptomycin-dependent mutants, were shown to be safe and protective when administered as live oral vaccines (LEVINE 1995). However, those early vaccines had (or were at least perceived by competing laboratories to have) shortcomings such as requirements for multiple doses, the necessity to administer frequent doses to maintain protection, occasional genetic reversions, and moderate side-effects. Therefore considerable efforts have been made during recent years to develop improved vaccines. These vaccines include parenteral vaccines inducing immunity against *Shigella* O-antigens as well as oral attenuated *Shigella* strains and attenuated live vectors expressing potentially protective *Shigella* antigens. However, the early vaccines might have been dismissed too soon based on the hope of rapid replacement with novel improved vaccines.

I. Parenteral Vaccines

Killed Whole Cells. The earliest *Shigella* vaccine candidates consisted of killed whole cell preparations for parenteral administration. Although those vaccines stimulated high levels of serum antibody responses, no protection was recorded either in challenged human volunteers or in field studies (FORMAL et al. 1967). In continued studies, Russian scientists prepared purified ribosomal vaccines against *S. sonnei* and *S. flexneri* 2a, that were administered subcutaneously (LEVENSON et al. 1991). In a preliminary clinical trial in Russia, low doses of the ribosomal vaccine were well tolerated. *S. sonnei* ribosomal extracts have also been highly protective against homologous challenge in animal models.

Conjugate Vaccines. Based on the observation that there was a positive correlation between anti-LPS antibody titers in serum and the level of protection (COHEN et al. 1991), lipid A free O-specific polysaccharides of *S. flexneri* 2a, *S. sonnei* or *S. dysenteriae* type 1 conjugated to various protein carriers, for example, tetanus toxoid, have been studied as parenteral, single dose vaccine candidates (ROBBINS and SCHNEERSON 1990). These vaccines have been shown to be safe in humans and to induce high levels of serum antibody responses against the homologous LPS antigen (TAYLOR et al. 1993). In a recent controlled field trial in military recruits in Israel, a parenteral *S. sonnei* O-polysaccharide conjugate vaccine was shown to confer approximately 70% serotype specific protection for at least 3 months (COHEN et al. 1997).

II. Oral Live Vaccines

Live Noninvasive Vaccines. The only *Shigella* vaccine currently in use is the Isterati T32 *S. flexneri* 2a strain (Badizen, Bukarest, Romania). This vaccine strain, which was selected after 32 passages of a wild isolate on artificial media (MEITERT et al. 1984) has repeatedly been shown to be safe and avirulent in humans. The safety and efficacy of this *S. flexneri* vaccine has only been tested in field trials in Rumania with promising results; further documented studies are, however, required to evaluate the safety and protective efficacy of this vaccine in other settings.

In the early 1980s an *S. sonnei* vaccine candidate was constructed by inserting into the attenuated *S. typhi* Ty21a strain the plasmid of *S. sonnei* that contains genes for synthesis of the O-antigen (FORMAL et al. 1981). The resulting hybrid strain, 5076-1C, expressed both *S. typhi* and *S. sonnei* O-antigens, but not the Ipas. Although this strain was well tolerated and some lots of this vaccine provided significant protection against *S. sonnei* shigellosis in human volunteers (HERRINGTON et al. 1990), controlled field trials were not undertaken because of marked variations in different lots of vaccine, resulting in large differences in efficacy, which was later associated with the presence or not of pili and flagellar antigens (PÁL and LINDBERG 1996).

Live Invasive Vaccines. Another group of live vaccines that has been developed, has been designed with the goal to maintain the invasive nature of *Shigella*. Examples of such vaccine candidates is the strain EcSf2a-2, which consists of an *E. coli* strain modified by introducing the *S. flexneri* invasiveness plasmid and the chromosomal genes of *S. flexneri* 2a that encode the type- and group-specific O-antigens. This resultant *E. coli* strain expressing *S. flexneri* 2a O-antigen has been shown to confer a low to moderate level of protection in challenged volunteers when given in two to four doses with approx. 7×10^8–2×10^9 CFU/dose (KOTLOFF et al. 1995). However, this vaccine was recently evaluated in large-scale clinical trials in the field with less promising result (COHEN et al. 1997).

A number of other *S. flexneri* vaccine candidates have been developed in Sweden, France and the United States; all these strains have been impaired in their capacity to proliferate in vivo after invasion or in their ability to spread from enterocyte to enterocyte. These vaccine candidates includes a thymine – requiring, temperature-sensitive double mutant of *S. flexneri* Y (strain TSF21; AHMED et al. 1990). This strain was shown to be safe in monkeys and to protect against homologous challenge in animals, but no data from human trials are available. Other such candidate vaccine strains that have been developed include a transposon Tn10 inactivated *aroD* mutant (SFL114) of a virulent *S. flexneri* Y strain SFL1 (PÁL and LINDBERG 1996). This strain had significantly reduced intracellular growth capacity, was well tolerated and highly protective in monkeys. Subsequently SFL114 was further improved by deleting the *aroD*-gene resulting in strain SFL124 (KÄRNELL et al. 1992). When administering this strain to Swedish volunteers it was shown to be safe and highly effective in stimulating systemic as well as local LPS-specific immune responses (PÁL and LINDBERG 1996). However, data from protective efficacy studies in humans are not available.

Other invasive *Shigella* vaccine candidates are strains with a mutation or deletion into the *virG* gene, harbored by the invasion plasmid and necessary for the intracellular mobility, such a strain SC560, which is derived from an *S. flexneri* 5 strain; this strain only causes focal lesions in the intestinal mucosa due to its inability to spread from cell to cell (SANSONETTI et al. 1991). Although this strain has proved to be protective against subsequent challenge in monkeys, it is still causing mild dysentery in this animal model. Attempts to further attenuate an *S. flexneri* 5 strain by making double mutants has resulted in strains that were well tolerated by the vaccinated animals, but with decreased protective efficacy, suggesting that they were too attenuated.

In general all these live vaccine candidates have proven to be well tolerated at low doses, and to induce modest immune responses, but they are reactogenic at high-dose levels, that stimulate more potent immune responses (KÄRNELL et al. 1995). Administration of two or three doses of vaccines containing lower, well tolerated numbers of vaccine organisms have been successful in eliciting acceptable immune responses. Very recently a vaccine strain, SC602, which is a double mutant of *S. flexneri* 2a and that is affected both in its capacity to spread intra- and intercellularly and in its tissue survival, was developed at Institut Pasteur (BARZU et al. 1996; SANSONETTI and PHALIPON 1996). A single dose (10^4CFU) of this vaccine strain has proved to be safe and immunogenic and to confer 100% protection against severe shigellosis and 50% protection against any diarrhea after an experimental challenge with *S. flexneri* 2a bacteria (World Health Organization, 1998).

G. *Campylobacter jejuni* Vaccines

Data from epidemiological and volunteer studies suggest that the development of an effective vaccine against *C. jejuni* is feasible. Thus studies in developing countries have provided evidence that multiple *Campylobacter* infections early in life provide lasting protective immunity. Although infection rates remain high throughout childhood, symptomatic infection rates are highest during the first year of life and by the age of 5 years most infections are asymptomatic (CALVA et al. 1988; PAZZAGLIA et al. 1991). Development of a vaccine against *C. jejuni* may be complicated by the large antigenic diversity of the bacteria. However, it appears that, similar to the situation for *Shigella*, a limited number of serotypes predominate in different regions of the world (LIOR et al. 1982). Furthermore, there are some conserved virulence determinants of *Campylobacter* that may be targets for a vaccine, for example, the single polar flagellum which is composed of two subunit proteins, i.e., FlaA and FlaB; the motility provided by these organellae seems to be necessary for colonization. Other putative protective antigens may be piluslike appendages (DOIG et al. 1996).

Different strategies for development of *C. jejuni* vaccines have been considered. They include development of live attenuated strains, subunit vaccines and killed whole cell preparations. Among these different approaches, only killed whole cell vaccines have reached the stage of clinical trials in humans. Using a genetic approach to develop a living attenuated *Campylobacter* vaccine (YAO et al. 1994), nonmotile mutants of *C. jejuni* were prepared as putative vaccine candidates. However, more information on the pathogenic mechanisms in *Campylobacter* infections are required before such a vaccine could be introduced in humans, due to the risk of inducing reactive arthritis (SCOTT et al. 1997).

Subunit Vaccines. Flagellin and a protein named PEB1 have been suggested as subunit vaccine candidates for use either as purified recombinant proteins or expressed in a carrier vaccine strain, for example, live attenuated *Salmonella* or *Shigella* (Scott et al. 1997). However, the antigenic diversity of *Campylobacter* flagellins as well as the fact that these proteins are glycosylated makes the approach of using flagellin subunit-based vaccines less attractive. Thus, the major immune response against *C. jejuni* flagellin seems to be directed against the highly conserved amino- and carboxy-ends of the protein, which is not surface exposed in the flagellar filament (Power et al. 1994). The other potential vaccine target antigen is the highly conserved, immunogenic protein PEB1, which has been suggested to function as an adhesin to eukarotic cells (Pei and Blaser 1993).

Killed Whole Cell Vaccines. Based on the promising approach in developing inactivated whole cell vaccines against cholera and ETEC (see above), it has been suggested that a killed whole cell vaccine against *Campylobacter* may be safe, immunogenic and protect against disease, particularly if combined with an effective mucosal adjuvant (Scott et al. 1997). In recent studies in monkeys two doses of a prototype oral killed whole cell *Campylobacter* vaccine administered together with *E. coli* LT as an adjuvant were well tolerated and induced *Campylobacter* specific IgA and IgG ASCs in peripheral blood of most of the vaccinated animals (Baqar et al. 1995). These results suggested that killed whole cells were sufficiently promising as an oral candidate vaccine, to be tested in humans. Against this background a monovalent, formalin-inactivated whole cell vaccine (CWC) made from strain *C. jejuni* 81–176 (Lior serogroup 5) was prepared. The bacteria used for vaccine preparation were grown under conditions that attempted to maximize motility, i.e. flagellar expression, and ability to invade eukaryotic cells in vitro: the preparation obtained was shown to have intact flagellae and was agglutinated by specific anti-Lior 5 antisera. Phase I safety and immunogenicity studies that currently are in progress in the United States indicate that the vaccine is well tolerated and at least moderately immunogenic (Scott et al. 1997).

H. Rotavirus Vaccines

Efforts to develop effective rotavirus vaccines were initiated shortly after rotaviruses were recognized as a cause of severe dehydrating diarrhea in animals (Connor et al. 1996). Indeed, development of rotavirus vaccines for children initially focused on development of live oral rotaviruses for nonhuman species, i.e. a Jennerian vaccine approach. Subsequently the most promising approach has been to develop reassortant vaccines consisting of animal rotaviruses that express human rotavirus surface protein (Table 3).

Bovine Rotaviruses. The first human rotavirus vaccines evaluated were of nonhuman origin. A bovine rotavirus vaccine, RIT4237, was shown to be

Table 3. Protection by live oral rotavirus vaccines in clinical trials

	Protection against rotavirus diarrhea	
	Any	Severe
Bovine rotavirus		
RIT 4237:		
Finland	50%–58%	82%–86%
Africa	NS	NS
WC3:		
United States	76%	100%
Central Africa	NS	
China	50%	ND
Simian rotavirus		
RRV	38%–48%	67%–80%
Animal and human reassortants		
Quadrivalent RRV reassortants:		
United States	49%–50%	64%–80%
Finland	68%	91%
Venezuela	48%	88%

NS, Not significant; ND, not determined.

totally nonreactogenic and efficient in inducing serum antibodies when given to Finnish infants (Vesikari et al. 1983). However, the strain specificity of the immune response was exclusively homeotypic. In different protective efficacy trials in Finland, RIT4237 was shown to induce protective efficacies of 50%–58% against all rotavirus disease and 82%–88% against clinically significant rotavirus disease (Vesikari et al. 1990). However, studies of RIT4237 vaccine in Africa revealed little efficacy (Hanlon et al. 1987). This vaccine has subsequently been abandoned as a candidate for further use in humans (Connor et al. 1996).

The other bovine rotavirus vaccine candidate extensively tested is WC3, which was shown to be totally nonreactogenic and efficient in inducing a homeotypic serum neutralizing antibody response. In an initial efficacy trial this vaccine was shown to give 76% protection against all rotavirus disease and 100% protection against moderate to severe diarrhea (Clark et al. 1988). However, in subsequent trials in the United States and in Central Africa no or only minimal protection was observed (George-Courbou et al. 1991; Connor et al. 1996) and this vaccine candidate is not further developed.

Simian Rotaviruses. As an alternative approach, a rhesus rotavirus (RRV) vaccine strain that had been isolated from the rhesus monkey and passaged several times was developed. The protective efficacy of this vaccine has varied widely in clinical trials between no protection at all in Arizona and Rochester in the United States and 38%–48% against all rotavirus disease and 67%–80% against severe disease in Finland and Sweden (Connor et al. 1996). Protection seemed to be considerably higher against the homeotypic VP7 serotype G3, than against other serotypes, for example, G1. Against this background RRV vaccine is no longer considered as a single candidate vaccine but is a component of a candidate tetravalent vaccine.

Second Generation Reassortant Virus Vaccines. Since monovalent vaccines composed of a single animal origin rotavirus failed to consistently protect infants in repeated clinical trials, rotavirus vaccine development work focused on exploring the possibility of constructing animal rotavirus reassortants, which express human rotavirus surface antigens. In early experimental studies it was suggested that the VP7 (G-type) surface protein is a predominant neutralization antigen (HOSHINO et al. 1984). Therefore, animal rotavirus reassortants containing genes encoding human G serotype specific proteins have gained most interest. Thus, a modified Jennerian strategy was adopted in an attempt to achieve broader antigenic coverage (for a review see KAPIKIAN 1996). For this approach, three single gene substitution reassortant strains that possess ten genes from the RRV and a single gene that encodes the antigenic specificity of VP7 serotype 1, 2, or 4 were generated. The goal was to incorporate these three reassortants with RRV into a single quadrivalent vaccine to achieve antigenic coverage against all the four epidemiologically important human rotavirus VP7 serotypes. The reassortant strains were combined into a quadrivalent formulation, which was evaluated for reactogenicity and immunogenicity in different phase I studies. It was found that the vaccine was safe but failed to induce sufficient neutralizing antibody responses against individual serotypes (FLORES et al. 1993). Based on these results the vaccine was adjusted in dose, and also it was recommended that the vaccine should be given in a two-dose instead of a single-dose regimen.

Field Trials with the Quadrivalent Vaccine. The quadrivalent rhesus rotavirus based vaccine has been evaluated in numerous different phase II and phase III trials in different countries (KAPIKIAN 1996; GLASS et al. 1997). Oral administration of the quadrivalent vaccine in three doses of 10^4–4×10^5 plaque-forming units (pfu) to infants at approximately 2, 4, and 6 months of age showed that the vaccine was safe but mildly reactogenic, inducing some febrile episodes on the fourth and fifth days after vaccination (KAPIKIAN 1996). Using the three-dose regimen with 4×10^5 pfu in each in four different trials in the United States, Finland, and Venezuela it has been shown that the quadrivalent vaccine gives approximately 50% protection (68% in Finland) against all rotavirus diarrhea and 64%–91% protection against severe rotavirus disease (KAPIKIAN et al. 1996; PEREZ-SHAEL et al. 1997; GLASS et al. 1997). Based on these promising results the quadrivalent rotavirus vaccine constitutes the first generation of licensed rotavirus vaccines. The incomplete protection of the vaccine against mild disease as well as a modest rate of adverse side effects observed after vaccination will probably result in further development of a second generation of improved rotavirus vaccines. Such candidate second generation vaccines include rotavirus strains of human and animal origin, inactivated strains and microencapsulated products (GLASS et al. 1997). Even if the protective efficacy of the quadrivalent vaccine is only partial, it has been estimated that a 50% effective rotavirus vaccine would prevent over 1 million cases of rotavirus diarrhea and almost 60,000 hospitalizations in the United States each year.

References

Ahmed ZU, Sarker MR, Sack DA (1990) Protection of adult rabbits and monkeys from lethal shigellosis by oral immunization with a thymine-requiring and temperature-sensitive mutant of Shigella flexneri Y. Vaccine 8:153–158

Åhrén CM, Svennerholm A-M (1985) Experimental enterotoxin-induced Escherichia coli diarrhea and protection induced by previous infection with bacteria of the same adhesin or enterotoxin type. Infect Immun 50:255–261

Åhrén C, Wennerås C, Holmgren J, Svennerholm A-M (1993) Intestinal antibody response after oral immunization with a prototype enterotoxigenic Escherichia coli vaccine. Vaccine 11:929–934

Åhrén C, Jertborn M, Svennerholm A-M (1998) Intestinal immune responses to an enterotoxigenic Escherichia coli vaccine and associated immunoglobulin A responses. Infect Immun 66:3311–3316

Aitken R, Hirst TR (1993) Recombinant enterotoxin as vaccines against Escherichia coli-modified diarrhoea. Vaccine 11:227–233

Baqar S, Bourgeois AL, Schultheiss PJ, Walker R, Rollins DM, Haberger RL, Pavlovskis OR (1995) Safety and immunogenicity of a prototype oral whole-cell killed Campylobacter vaccine administered with a mucosal adjuvant in non-human primates. Vaccine 13:22–28

Barzu S, Fontaine A, Sansonetti PJ, Phalipon A (1996) Induction of a local anti-IpaC antibody response in mice by use of a *Shigella flexneri* 2a vaccine candidate: implications for use of IpaC as a protein carrier. Infect Immun 64:1190–1196

Black RE (1986) The epidemiology of cholera and enterotoxigenic E. coli diarrheal disease. In: Holmgren J, Lindberg A, Möllby R (eds) Development of vaccines and drugs against diarrhea. 11th Nobel conference, Studentlitteratur, Stockholm, Lund, pp 23–32

Black RE (1990) Epidemiology of traveller's diarrhea and relative importance of various pathogens. Rev Infect Dis 12:S73–79

Blake PA (1994) Historical perspectives on pandemic cholera. In: Wachsmuth IK, Blake PA, Olsvik Ø (eds) Vibrio cholerae and cholera: molecular to global perspectives. American Society for Microbiology, Washington, pp 293–295

Black RE, Levine MM, Clemens ML, Young CR, Svennerholm A-M, Holmgren J (1987) Protective efficacy in man of killed whole vibrio oral cholera vaccine with and without the B subunit of cholera toxin. Infect Immun 77:1116–1129

Calva JJ, Ruiz-Palacios GM, Lopez-Vidal AB, Ramos A, Bojalil R (1988) Cohort study of intestinal infection with campylobacter in Mexican children. Lancet 1:503–506

Clark HF, Borian FE, Bell LM, Modesto K, Gouvea V, Plotkin SA (1988) Protective effect of WC3 vaccine against rotavirus diarrhea in infants during a predominantly serotype 1 rotavirus season. J Infect Dis 158:570–587

Clemens J, Sack DA, Harris JR, Chakraborty J, Khan MR, Stanton BF, Kay BA, Khan MU, Yunus MD, Atkinson W, Svennerholm A-M, Holmgren J (1986) Field trial of oral cholera vaccines in Bangladesh. Lancet i:124–127

Clemens J, Sack DA, Harris JR, Chakraborty J, Neogy PK, Stanton BF, Huda N, Khan MU, Kay BA, Khan MR, Ansurazzaman M, Yunus M, Rao MR, Svennerholm A-M, Holmgren J (1988a) Cross-protection by B subunit-whole cell cholera vaccine against diarrhea associated with heat-labile toxin-producing enterotoxigenic Escherichia coli:Results of a large-scale field trial. J Infect Dis 158:372–377

Clemens JD, Sack DA, Harris JR, Chakraborty J, Khan MR, Stanton BF, Ali M, Ahmed F, Yunus M, Kay BA, Khan MU, Rao MR, Svennerholm A-M, Holmgren J (1988b) Impact of B subunit killed whole-cell and killed whole-cell-only oral vaccines against cholera upon treated diarrhoeal illness and mortality in an area endemic for cholera. Lancet i:1375–1379
Clemens JD, Sack DA, Harris JR, van Loon F, Chakraborty J, Ahmed F, Rao MR, Khan MR, Yunus MD, Huda N, Stanton BF, Kay BA, Walter S, Eeckels R, Svennerholm A-M, Holmgren J (1990a) Field trial of oral cholera vaccines in Bangladesh:Results from three-year follow-up. Lancet 355:270–273
Clemens JD, Sack DA, Chakraborty J, Rao MR, Ahmed F, Harris RJ, Van Loon F, Khan MR, Yunus MD, Huda N, Stanton BF, Kay BA, Svennerholm A-M, Holmgren J (1990b) Field trial of oral cholera vaccines in Bangladesh: evaluation of anti-bacterial and anti-toxic breast-milk immunity in response to ingestion of the vaccines.Vaccine 8:469–472
Cohen D, Green MS, Block C, Slepon R, Ofek I (1991) Prospective study of the association between serum antibodies to lipopolysaccharide O antigen and the attack rate of shigellosis. J Clin Microbiol 29:386–389
Cohen D, Ashkenazi S, Green M, Gdalevich M, Robin G, Slepan R, Yovzori M, Orr N, Block C, Ashkenazi J, Shemer J, Taylor D, Hale T, Sadoff J, Pavliakova D, Schneerson R, Robbins JB (1997) Double-blind vaccine-controlled randomized efficacy trial of an investigational Shigella sonnei conjugate vaccine in young adults. Lancet 349:155–159
Connor ME, Estes MK, Offit PA, Clark HF, Franco M, Feng N, Greenberg HB (1996) Development of a mucosal rotavirus vaccine. In: Kiyono H, Ogra PL, McGhee JR (eds) Mucosal vaccines. Academic, San Diego, pp 325–344
Coster TS, Killeen KP, Waldor MK, Beattie DT, Spriggs DR, Kenner JR, Trofa A, Sadoff JC, Mekalanos JJ, Taylor DN (1995) Safety, immunogenicity, and efficacy of live attenuated Vibrio cholerae O139 vaccine prototype. Lancet 345:949–952
Craig JP (1980) A survey of the enterotoxic enteropathies. In: Ouchterlony Ö, Holmgren J (eds) Cholera and related diarrheas, molecular aspects of a global health problem, 43rd Nobel symposium, Stockholm, 6–11 Aug 1978, pp 15–25
Cravioto A, Reyes RE, Ortega R, Fernandéz G, Hernandez R, López D (1988) Prospective study of diarrhoeal diseases in a cohort of rural Mexican children: incidence and isolated pathogens during the first two years of life. Epidemiol Infect 101:123–134
Doig P, Yao R, Burr DH, et al. (1996) An environmentally regulated pilus-like appendage involved in Campylobacter pathogenesis. Mol Microbiol 20:885–894
Dong Y, Zeng CQ, Ball JM, Estes MK, Morris AP (1997) The rotavirus enterotoxin NSP4 mobilizes intracellular calcium in human intestinal cells by stimulating phospholipase C-mediated inositol 1, 4, 5-trisphosphate production. Proc Natl Acad Sci USA 94:3960–3965
Estes MK, Cohen J (1989) Rotavirus gene structure and function. Microbial Rev 53:410–449
Evans DJ, Evans DG (1989) Determinants of microbial attachment and their genetic control. In: Farthing MJG, Keusch GT (eds) Enteric infection. Mechanisms, manifestations and management. Chapman and Hall, London, pp 31–40
Evans DG, Graham DY, Evans DJ Jr, Opekun AR (1984) Administration of purified colonization factor antigens (CFA/I, CFA/II) or enterotoxigenic Escherichia coli to volunteers. Response to challenge with virulent enterotoxigenic Escherichia coli. Gastroenterology 87:934–940
Farthing MJG, Keusch GT (eds) (1989) Enteric infection. Mechanisms, manifestations and management. Chapman and Hall, London
Feeley JC, Gangarosa EJ (1980) Field trials of cholera vaccine. In: Ouchterlony Ö, Holmgren J (eds) Cholera and related diarrheas. 43rd Nobel symposium, Stockholm 1978. Karger, Basel, pp 204–210
Flores J, Perez-Schael I, Blanco M, Rojas AM, Alfonzo E, Crespo I, Cunto W, Pittman A, Kapikian AZ (1993) Reactogenicity and immunogenicity of a high-titered

rhesus rotavirus-based quadrivalent rotavirus vaccine. J Clin Microbiol 31:2439–2445

Formal SB, Maenza RM, Austin S, LaBrec EH (1967) Failure of parenteral vaccines to protect monkeys against experimental shigellosis. Proc Soc Exp Biol Med 125:347–349

Formal SB, Baron LS, Kopecko DJ, Washington O, Powell C, Life CA (1981) Construction of a potential bivalent vaccine strain: introduction of Shigella sonnei form I antigen genes into the galE Salmonella typhi Ty21a typhoid vaccine strain. Infect Immun 34:746–750

Frantz JC, Robertson DC (1981) Immunological properties of Escherichia coli heat-stable enterotoxins: development of a radioimmunoassay specific for heat-stable enterotoxins with suckling mouse activity. Infect Immun 33:193–198

Gaastra W, Svennerholm A-M (1996) Colonization factors of enterotoxigenic Escherichia coli (ETEC). Trends Microbiol 4:444–452

Georges-Courbot MC, Monges J, Siopathis MR, Roungou JP, Gresenguet A, Bellec L, Bouquety IC, Lanckviet C, Cadoz M, Hesel L, Gouvea G, Clark HF, Georges AI (1991) Evaluation of the efficacy of a low passage bovine rotavirus (strain WC3) vaccine in children in Central Africa. Res Virol 142:405–411

Girón JA, Xu J-G, González CR, Hone DM, Kaper JB, Levine MM (1995) Simultaneous expression of CFA/I and CS3 colonization factor antigens of enterotoxigenic Escherichia coli by ΔaroC, ΔaroD Salmonella typhi vaccine strain CVD 908. Vaccine 13:939–946

Glass RI, Bresee JS, Parashar U, Miller M, Gentsch JR (1997) Rotavirus vaccines at the threshold. Nature Med 3:1324–1325

Goldberg M, Barzu O, Parsot C, Sansonetti PJ (1993) Unipolar localization and ATPase activity of IcsA, a Shigella flexneri protein involved in intracellular movement. J Bacteriol 175:2189–2196

Guerrant RL (1985) Microbial toxins and diarrhoeal disease:introduction and overview. In: Evered D, Whelan J (eds) Microbial toxins and diarrhoeal disease. Pitman, London, pp 1–13 (Ciba Foundation symposium 112)

Hanlon P, Hanlon L, Marsh V, Byass P, Shenton F, Hassan-King M, Jobe O, Sillah H, Hayes R, M'Boge BH, Whittle HC, Greenwood BM (1987) Trial of an attenuated bovine rotavirus vaccine (RIT 4237) in Gambian infants. Lancet 1:1342–1345

Herrington DA, Van der Berg L, Formal SB, Hale TL, Tall BD, Cryz SJ, Tramont EC, Levine MM (1990) Studies in volunteers to evaluate candidate Shigella vaccines: further experience with a bivalent Salmonella typhi-Shigella sonnei vaccine and protection conferred by previous Shigella sonnei disease. Vaccine 8:353–357

Holmgren J (1981) Actions of cholera toxin and the prevention and treatment of cholera. Nature 292:413–417

Holmgren J, Svennerholm A-M (1983) Cholera and the immune response. Progr Allergy 33:106–119

Holmgren J, Svennerholm A-M (1992) Bacterial enteric infections and vaccine development. In: McDermott RP, Elson CO (eds) Mucosal immunology, Gastroenterology Clinics of North America. Saunders, Philadelphia, pp 283–302

Holmgren J, Svennerholm A-M, Lönnroth I, Fall-Persson M, Markman B, Lundbäck H (1977) Development of improved cholera vaccine based on subunit toxoid. Nature 269:602–604

Holmgren J, Jertborn M, Svennerholm A-M (1997) New and improved vaccines against cholera: oral B subunit killed whole-cell cholera vaccine. In: Levine MM, Woodrow GC, Kaper JB, Cobon GS (eds) New generation vaccines, 2nd edn. Dekker, New York, pp 459–468

Hoshino Y, Kapikian AZ (1994) Prospects for development of a rotavirus vaccine in infants and young children. Trends Microbiol 2:242–249

Hoshino Y, Wyatt RG, Greenberg HB, Flores J, Kapikian AZ (1984) Serotypic similarity and diversity of rotaviruses of mammalian and avian origin as studied by plaque-reduction neutralization. J Infect Dis 149:694–702

Islam D, Wretlind B, Ryd M, Lindberg AA, Christenson B (1995) Immunoglobulin subclass distribution and dynamics of Shigella-specific antibody responses in serum and stool samples in shigellosis. Infect Immun 63:2054–2061
Jertborn M, Svennerholm A-M, Holmgren J (1988) Five-year immunologic memory in Swedish volunteers after oral cholera vaccination. J Infect Dis 157:374–377
Jertborn M, Svennerholm A-M, Holmgren J (1996) Intestinal and systemic immune responses in humans after oral immunization with a bivalent B subunit-O1/O139 whole cell cholera vaccine. Vaccine 14:1459–1465
Jertborn M, Åhrén C, Holmgren J, Svennerholm A-M (1998) Safety and immunogenicity of an oral inactivated enterotoxigenic Escherichia coli vaccine. Vaccine 16:255–260
Jodal M, Lundgren O (1995) Neural reflex modulation of intestinal epithelial transport. In: Gaginella TS (ed) Regulatory mechanisms in gastrointestinal function. CRC Press, Boca Raton, pp 99–144
Jonson G, Holmgren J, Svennerholm A-M (1991) Identification of a mannose-binding pilus on Vibrio cholerae El Tor. Microb Pathog 11:433–441
Jonson G, Osek J, Svennerholm A-M, Holmgren J (1996) Immune mechanisms and protective antigens of Vibrio cholerae serogroup O139 as a basis for vaccine development. Infect Immun 64:3778–3785
Kaper JB, Lockman H, Baldini MM, Levine MM (1984) Recombinant nontoxigenic Vibrio cholerae strains as attenuated cholera vaccine candidates. Nature 308:655–658
Kaper JB, Tacket CO, Levine MM (1997) New and improved vaccines against cholera: attenuated Vibrio cholerae O1 and O139 strains as live oral cholera vaccines. In: Levine MM, Woodrow GC, Kaper JB, Cobon GS (eds) New generation vaccines, 2nd edn. Dekker, New York pp 447–458
Kapikian AZ (1996) Rotavirus vaccine: the clinical experience with the rhesus rotavirus-based vaccines. In: Kiyono H, Ogra PL, McGhee JR (eds) Mucosal Vaccines, San Diego, pp 345–356
Kapikian AZ, Chanock RM (1996) Rotaviruses. In: Fields BN, Knipe DM, Howley PM et al (eds) Virology. Lippincott-Raven, Philadelphia, pp 1657–1708
Kapikian AZ, Hishino Y, Chanock RM, Perez-Schael I (1996) Efficacy of a qadrivalent rhesus rotavirus-based human rotavirus vaccine aimed at preventing severe rotavirus diarrhea in infants and young children. J Infect Dis 174:S65–S72
Kaplan JE, Feldman R, Campbell DS, Lookabaugh C, Gary GW (1982) The frequency of a Norwalk-like pattern of illness in outbreak of acute gastroenteritis. Am J Public Health 72:1329–1332
Kärnell A, Stocker BAD, Katakura S, Reinholt FP, Lindberg AA (1992) Live oral auxotrophic Shigella flexneri SFL124 vaccine with a deleted aroD gene: characterization and monkey protection studies. Vaccine 10:389–394
Kärnell A, Li A, Zhao CR, Karlsson K, Nguyen BM, Lindberg AA (1995) Safety and immunogenicity of the auxotrophic Shigella flexneri 2a vaccine SFL1070 with a deleted aroD gene in Swedish volunteers. Vaccine 13:88–99
Kotloff KL, Losonsky GA, Nataro JP, Wasserman SS, Hale TL, Taylor DN, Newland JW, Sadoff JC, Formal SB, Levine MM (1995) Evaluation of the safety, immunogenicity and efficacy in healthy adults of four doses of live oral hybrid Escherichia coli-Shigella flexneri 2a vaccine strain EcSf-2a. Vaccine 13:495–502
Lebens M, Shahabi V, Bäckström M, Houze T, Lindblad M, Holmgren J (1996) Synthesis of hybrid molecules between heat-labile entertoxin and cholera toxin B subunits: potential for use in broad spectrum vaccine. Infect Immun 64:2144–2150
Levenson VI, Egorova TP, Belkin ZP, Fedosova VG, Subbotina JL, Rukhadze EZ, Dzhikidze EK, Stassilevich ZK (1991) Protective ribosomal preparation from Shigella sonnei as a parenteral candidate vaccine. Infect Immun 59:3610–3618
Levine MM (1990) Vaccines against enterotoxigenic Escherichia coli infections. In: Woodrow GC, Levine MM (eds) New generation vaccines. Dekker, New York, pp 649–660

Levine MM (1995) Development of bacterial vaccines. In: Blaser MJ, Smith PD, Ravdin JI, Greenberg HB, Guerrant RL (eds) Infections of the gastrointestinal tract. Raven, New York, pp 1441–1470
Levine MM, Svennerholm A-M (1997) Future enteric vaccines. In: DuPont H, Steffen R (eds) Textbook of travel medicine and health. Dekker, Ontario, pp 169–177
Levine MM, Tacket CO (1995) Live oral vaccines against cholera. In: Ala'Aldeen DAA, Hormaeche CE (eds) Molecular and clinical aspects of bacterial vaccine development. Wiley, Chichester, pp 233–258
Lior H, Woodward DK, Edgar JA, Laroche LJ, Gill P (1982) Serotyping of Campylobacter jejuni by slide agglutination based on heat-labile antigenic factors. J Clin Microbiol 15:761–768
López-Vidal Y, Calva JJ, Trujillo A, de León AP, Ramos A, Svennerholm A-M, Ruiz-Palacios G (1990) Enterotoxins and adhesins of enterotoxigenic Escherichia coli: are they risk factors for acute diarrhea in the community? J Infect Dis 162:442–447
McConnell MM (1991) Newly characterized putative colonization factors of human enterotoxigenic Escherichia coli. In: Wadström T, Mäkelä PH, Svennerholm A-M, Wolf-Watz H (eds) Molecular pathogenesis of gastrointestinal infections. Plenum, New York, pp 79–85 (FEMS symposium, no 58)
Meitert T, Pencu E, Ciudin L, Tonciu M (1984) Vaccine strain Sh. flexneri T32-Istrati. Studies in animals and in volunteers. Antidysentery immunoprophylaxis and immunotherapy by live vaccine Vadizen (Sh. flexneri T32-Istrati). Arch Roum Pathol Exp Microbiol 43:251–278
Morris JG (1994) Vibrio cholerae O139 Bengal. In: Wachsmuth IK, Blake PA, Olsvik Ø (eds) Vibrio cholerae and cholera:molecular to global perspectives. American Society for Microbiology, Washington, pp 95–102
Mosley WH, Ahmed S, Benenson AS, Ahmed A (1968) The relationship of vibriocidal antibody titre to susceptibility to cholera in family contacts of cholera patients. Bull WHO 38:777–785
Mounier J, Vasselon T, Hellio R, Lesourd M, Sansonsetti PJ (1992) Shigella flexneri enters human colonic Caco-2 epithelial cells through the basolateral pole. Infect Immun 60:237–248
Neutra MR, Kraehenbühl J-P (1992) Transepithelial transport and mucosal defence I: the role of M cells, vol 2. Elsevier Science, New York, pp 134–138
Noriega F, Formal SB, Kotloff KL, Lindberg AA (1997) Vaccines against Shigella infections:engineered attenuated mutants of Shigella as live oral vaccines. In: Levine MM, Woodrow GC, Kaper JB, Cobon GS (eds) New generation vaccines, 2nd edn. Dekker, New York, pp 853–863
Osek J, Svennerholm A-M, Holmgren J (1992) Protection against Vibrio cholerae El Tor infection by specific antibodies against mannose-binding hemagglutinin pili. Infect Immun 60:4961–4964
Pal T, Hale TL (1989) Plasmid-associated adherence of Shigella flexneri in a HeLa cell model. Infect Immun 57:2580–2582
Pál T, Lindberg AA (1996) Oral vaccines for Shigella. In: Kiyono H, Ogra PL, McGhee JR (eds) Mucosal vaccines. Academic Press, San Diego, pp 213–228
Paniagua M, Espinoza F, Ringman M, Reizenstein E, Svennerholm A-M, Hallander H (1997) ETEC in a prospective cohort study on infant diarrhea in Nicaragua. J Clin Microbiol 35:1404–1410
Pazzaglia G, Bourgeois AL, El Diwany K, Nour N, Badran N, Hablas R (1991) Campylobacter diarrhoea and an association of recent disease with asymptomatic shedding in Egyptian children. Epidemiol Infect 106:77–82
Pei Z, Blaser MJ (1993) PEB1, the major cell-binding factor of Campylobacter jejuni, is a homolog of the binding component in gram-negative nutrient transport systems. J Biol Chem 268:18717–18725
Peltola H, Gorbach SL (1997) Travelers' diarrhea: epidemiology and clinical aspects. In: DuPont HL, Steffen R (eds) Textbook of travel medicine and health. Decker, Ontario, pp 78–86

Peltola H, Siitonen A, Kyrönseppä H, Simula I, Mattila L, Oksanen P, Kataja MJ, Cadoz M (1991) Prevention of travellers' diarrhoea by oral B-subunit/whole cell cholera vaccine. Lancet 338:1285–1289

Perez-Schael, I, Guntinas MJ, Perez M, Pagone V, Rojas AM, Gonzalez R, Cunto W, Hoshino Y, Kapikian AZ (1997) Efficacy of the rhesus rotavirus-based quadrivalent vaccine in infants and young children in Venezuela. N Engl J Med 337:1181–1187

Power ME, Guerry P, McCubbin WD, Kay CM, Trust TJ (1994) Structural and antigenic characteristics of Campylobacter coli FlaA flagellin. J Bacteriol 176:3303–3313

Qadri F, Wennerås C, Bardhan PK, Hossain J, Sack RB, Svennerholm A-M (1998) B cell responses to enterotoxigenic Escherichia coli (ETEC) induced by oral vaccination and natural disease (submitted for publication)

Quiding M, Nordström I, Kilander A, Andersson G, Hanson L-Å, Holmgren J, Czerkinsky C (1991) Intestinal immune responses in humans. Oral cholera vaccination induces strong intestinal antibody responses, gamma-interferon production, and evokes local immunological memory. J Clin Invest 88:143–148

Robbins JB, Schneerson R (1990) Polysaccharide-protein conjugates: a new generation of vaccines. J Infect Dis 161:821–832

Sack DA, Freij L (1990) Prospects for public health benefits in developing countries from new vaccines against enteric infections. SAREC documentation, conference report 1990:2, SAREC symposium, Gothenburg, Sweden, 28–29 May

Sack DA, Freij L, Holmgren J (1991) Prospects for public health benefits in developing countries from new vaccines against enteric infections. J Infect Dis 163:503–506

Sanchez J, Holmgren J (1989) Recombinant system for overexpression of cholera toxin B subunit in Vibrio cholerae as a basis for vaccine development. Proc Natl Acad Sci USA 86:481–485

Sanchez J, Svennerholm A-M, Holmgren J (1988) Genetic fusion of a non-toxic heat-stable enterotoxin-related decapeptide antigen to cholera toxin B-subunit. FEBS Lett 241:110–114

Sanchez JL, Trofa AF, Taylor DN, Kuschner RA, DeFraites RF, Craig SC, Rao MR, Clemens JD, Svennerholm A-M, Sadoff JC, Holmgren J (1993) Safety and immunogenicity to the oral, whole cell/recombinant B subunit cholera vaccine in North American volunteers. J Infect Dis 167:1446–1449

Sanchez JL, Vasques B, Begue RE, Meza R, Castellares G, Cabezas C, Watts DM, Svennerholm, A-M, Sadoff JC, Taylor DN (1994) Protective efficacy of the oral, whole cell/recombinant B subunit cholera vaccine in Peruvian military recruits. Lancet 344:1273–1276

Sansonetti P, Phalipon A (1996) Shigellosis:from molecular pathogenesis of infection to protective immunity and vaccine development. Res Immunol 147:595–602

Sansonetti PJ, Arondel J, Fontaine A, d'Hauteville H, Bernardini ML (1991) OmpB (osmo regulation) and icsA (cell-to-cell spread) mutants of S. flexneri: vaccine candidates and probes to study the pathogenesis of shigellosis. Vaccine 9:416–422

Savarino SJ, Brown FM, Hall E, Bassily S, Youssef F, Wierzba T, Peruskil, El-Masry NA, Safwat M, Rao M, Bourgeois AL, Jertborn M, Svennerholm A-M, Lee YJ, Clemens JD (1998a) Safety and immunogenicity of an oral, killed enterotoxigenic Escherichia coli-cholera toxin B subunit vaccine in Egyptian adults. J Infect Dis 177:796–799

Savarino SJ, Hall ER, Bassily S, Brown M, Youssef F, Wierzba TF, Peruskil, El-Masry NA, Safwat M, Rao M, Eng M, Svennerholm A-M, Lee YJ, Clemens JD (1998b) Oral inactivated whole cell enterotoxigenic Escherichia coli plus cholera toxin B subunit vaccine: results of the initial evaluation in children. J Infect Dis (in press)

Scerpella EG, Sanchez JL, Mathweson JJ, Torres-Cordero JV, Sadoff JC, Svennerholm A-M, DuPont HL, Taylor DN, Ericsson CD (1995) Safety, immunogenicity and protective efficacy of the whole-cell/recombinant B subunit (WC/rBS) oral cholera vaccine against travelers' diarrhea. J Travel Med 2:22–27

Scott DA, Baqar S, Pazzaglia G, Guerry P, Burr DH (1997) Vaccines against Campylobacter jejuni. In: Levine MM, Woodrow GC, Kaper JB, Cobon GS (eds) New generation vaccines, 2nd edn. Dekker, New York, pp 885–896

Sommerfelt H, Steinsland H, Grewal HMS, Viboud GI, Bhandarin, Gaastra W, Svennerholm A-M and Bhan MK (1996) Colonization factors of enterotoxigenic Escherichia coli isolated from children in North India. J Infect Dis 174:768–776

Stoll BJ, Svennerholm A-M, Gothefors L, Barua D, Huda S, Holmgren J (1986) Local and systemic antibody responses to naturally acquired enterotoxigenic Escherichia coli diarrhea in an endemic area. J Infect Dis 153:527–534

Svennerholm A-M (1980) The nature of protective immunity in cholera. In: Ouchterlony Ö, Holmgren J (eds) Cholera and related diarrheal disease. 43rd Nobel symposium, Stockholm 1978, Karger, Basel, pp 171–184

Svennerholm A-M, Holmgren J (1976) Synergistic protective effect in rabbits of immunization with Vibrio cholerae lipopolysaccharide and toxin/toxoid. Infect Immun 13:735–740

Svennerholm A-M, Holmgren J (1995) Oral B-subunit whole-cell vaccines against cholera and enterotoxigenic Escherichia coli diarrhoea. In: Ala'Aldeen DAA, Hormaeche CE (eds) Molecular and clinical aspects of bacterial vaccine development. Wiley, Chichester, pp 205–232

Svennerholm A-M, Åhrén C, Jertborn M (1997) Oral inactivated vaccines against enterotoxigenic Escherichia coli. In: Levine MM, Woodrow GC, Kaper JB, Gabon GS (eds) New Generation Vaccines, edition II. Marcel Dekker Inc. New York, pp 865–874

Svennerholm A-M, Jertborn M, Gothefors L, Karim M, Sack DA, Holmgren J (1984) Mucosal antitoxic and antibacterial immunity after cholera disease and after immunization with a combined B subunit-whole cell vaccine. J Infect Dis 149:884–893

Svennerholm A-M, Wikström M, Lindblad M, Holmgren J (1986a) Monoclonal antibodies to Escherichia coli heat-labile enterotoxins:neutralizing activity and differentiation of human and porcine LTs and cholera toxin. Med Biol 64:23–30

Svennerholm A-M, Wikström M, Lindblad M, Holmgren J (1986b) Monoclonal antibodies against E. coli heat-stable toxin (STa) and their use in diagnostic ST ganglioside GM1-enzyme-linked immunosorbent assay. J Clin Microbiol 24:585–590

Svennerholm A-M, Lindblad M, Svennerholm B, Holmgren J (1988) Synthesis of nontoxic, antibody-binding Escherichia coli heat-stable enterotoxin (ST_a) peptides. FEMS Microbiol Lett 55:23–28

Svennerholm A-M, Holmgren J, Sack DA (1989) Development of oral vaccines against enterotoxigenic Escherichia coli diarrhoea. Vaccine 7:196–198

Svennerholm A-M, Wennerås C, Holmgren J, McConnell MM, Rowe B (1990) Roles of different coli surface antigens of colonization factor antigen II in colonization by and protective immunogenicity of enterotoxigenic Escherichia coli in rabbits. Infect Immun 58:341–346

Svennerholm A-M, McConnell MM, Wiklund G (1992) Roles of different putative colonization factor antigens in colonization of human enterotoxigenic Escherichia coli in rabbits. Microb Pathog 13:381–389

Tacket CO, Hone DM, Losonsky G, Guers L, Edelman R, Levine MM (1992) Clinical acceptability and immunogenicity at CVD 908 Salmonella typhi vaccine strain. Vaccine 10:443–446

Tacket CO, Reid RH, Boedeker EC, Losonsky G, Nataro JP, Bhagat H, Edelman R (1994) Enteral immunization and challenge of volunteers given enterotoxigenic E. coli CFA/II encapsulated in biodegradable microspheres. Vaccine 14:1270–1274

Tauxe R, Seminario L, Tapia R, Libel M (1994) The Latin American epidemic. In: Wachsmuth IK, Blake PA, Olsvik Ø (eds) Vibrio cholerae and cholera: molecular to global perspectives. American Society for Microbiology, Washington, pp 321–344
Taylor D, Echeverria P, Pal T, Sethabutr O, Saiborisuth S, Sricharmorn S, Rowe B, Cross J (1986) The role of Shigella spp, enteroinvasive Escherichia coli, and other enteropathogens as causes of childhood dysentery in Thailand. J Infect Dis 153:1132–1138
Taylor DN, Trofa AC, Sadoff J, Chu C, Bryla D, Shiloach J (1993) Synthesis, characterization, and clinical evaluation of conjugate vaccines composed of the O-specific polysaccharide of Shigella flexneri 2a, and Shigella sonnei (Plesiomonas shigelloides) bound to bacterial toxoids. Infect Immun 61:3678–3687
Taylor DN, Killeen KP, Hack DC, Kenner JK, Coster TS, Beattie DT, Ezzell J, Hyman T, Trofa A, Sjogren MH, Friedlander A, Mekalanos JJ, Sadoff JC (1994) Development of a live, oral and attenuated vaccine against El Tor cholera. J Infect Dis 170:1518–1523
Tayot J-L, Holmgren J, Svennerholm L, Lindblad M, Tardy M (1981) Receptor-specific large scale purification of cholera toxin on silica beads derivatized with lyso-GM1 ganglioside. Eur J Biochem 113:249–258
Trach DD, Clemens JD, Ke NT, Thuy HT, Son ND, Canh DG, Hang PV, Rao MR (1997) Field trial of a locally produced, killed, oral cholera vaccine in Vietnam. Lancet 349:231–235
Vesikari T, Isolauri E, Delem A, D'Hondt E, Andre FE, Zissis G (1983) Immunogenicity and safety of live oral attenuated bovine rotavirus vaccine strain RIT 4237 in adults and young chiuldren. Lancet 2:807–811
Vesikari T, Rautanen T, Varis T, Beards GM, Kapikian AZ (1990) Clinical trial in children vaccinated between 2 and 5 months of age. Am J Dis Child 144:285–289
Voss E, Manning PA, Attridge SR (1996) The toxin-coregulated pilus is a colonization factor and protective antigen of Vibrio cholerae El Tor. Microb Pathog 20:141–153
Waldor MK, Mekalanos JJ (1994) Emergence of a new cholera pandemic: molecular analysis of virulence determinants in Vibrio cholerae O139 and development of a live vaccine prototype. J Infect Dis 170:278–283
Wennerås C, Svennerholm A-M, Åhrén C, Czerkinsky C (1992) Antibody-secreting cells in human peripheral blood after oral immunization with an inactivated enterotoxigenic Escherichia coli vaccine. Infect Immun 60:2605–2611
World Health Organization (1996) State of the world's vaccines and immunization. Davey S (ed) WHO, Geneva
Yao R, Burr DH, Doig P, Trust TJ, Niu H, Guerry P (1994) Isolation of motile and non-motile insertional mutants of Campylobacter jejuni: the role of motility in adherence and invasion of eukaryotic cells. Mol Microbiol 14:883–893

CHAPTER 13
Sexually Transmitted Diseases

E.G. Sandström

A. Introduction

Sexually transmitted diseases (STDs) are characterized by their mode of transmission, but they also have profound similarities and differences. Most STDs tend to be asymptomatic as a sign of adaptation to the human host. This is also reflected in their species specificity, i.e., humans, which in the context of vaccines hampers our ability to study them in animal models. They are common worldwide and pose significant health problems, principally due to the late sequelae that affect a significant proportion of those infected. The combination of asymptomatic disease and late sequelae is particularly troublesome.

The nature of spread, which implies sexual infidelity, associates them with shame and may hamper rational diagnosis and treatment efforts. This is compounded by the fact that they are associated with early sexual debut, many partners, and situations of social vulnerability that often are associated with lack of resources and moral attitudes that hinder their rational handling. Provision of knowledge of sexual matters, contraception, and STD prevention are influenced by these moral values. Diagnosis and treatment are often inadequate, leading to increased suffering and risk of development of antibiotic resistance for bacterial STDs. These difficulties will all become important when the use of a potential STD vaccine is contemplated and may profoundly affects its practical application.

STDs are caused by a variety of organisms which are favored by human sexual behavior. These include bacteria, viruses, and insects:

- Bacterial
  - Gonorrhea (see also Chap. 9)
  - Genital chlamydial infections (see also Chap. 9)
  - Syphilis
  - Chancroid
- Viral
  - Genital herpes simplex infections (see also Chap. 7)
  - Genital papillomavirus infections, *Condyloma acuminata*
  - Cytomegalovirus infections (see also Chap. 14)
  - Hepatitis B virus infections (see Chap. 6)
  - *Molluscum contagiosum*

- Protozoal
  - Trichomoniasis
  - Amoebiasis
- Insects
  - Scabies
  - Pubic lice infestation

The asymptomatic phase may be due to nonapparent local colonization or to infection of distant organs from which they are intermittently or continuously produced. Infection might be limited to the epithelial surfaces or be rapidly disseminated. The sequelae may be limited to local infection or extend to the destruction of distant organs. Immunity may confer protection or be the cause of the late sequelae. Thus there is no single approach to STD vaccines. Each agent and disease must be evaluated as to its own properties.

This review deals with four bacterial diseases – gonorrhea, chlamydial infection, chancroid, and syphilis – and two viral diseases – human genital papillomavirus (HPV) infection and herpes simplex virus (HSV) infection. The bulk of these infections affect persons in the developing nations who cannot afford the high cost of modern drugs. Affordable vaccines that could be used in the existing health care infrastructure could therefore have a profound effect on the spread of and clinical consequences of these diseases. For detailed description of the pathogens and the associated diseases the reader is referred to the standard text by HOLMES et al. (1990). Excellent reviews of vaccines for bacterial sexually transmitted infections and HPV have recently been published (SPARLING et al. 1994; FRAZER 1996).

B. Special Challenges

It is important to define the goal of a vaccine. Is it possible to prevent infection by agents which exist primarily on or in epithelial surfaces, such as *Haemophilus ducreyi*, gonococci, *Chlamydia*, and HPV, or should the target here be to affect the sequelae, i.e., ulcers, salpingitis, and cancerous cells? It is essential not to stimulate immune responses which may be effective in early events, such as adhesion, but may augment late immune mediated sequelae. A similar reasoning applies to diseases that spread rapidly beyond the genital area such as syphilis, HSV, and HIV. If, on the other hand, it is feasible only to affect sequelae, the vaccine may in addition have a role as an immunotherapeutic agent.

The goal of the vaccine determines how much emphasis is put on the generation of mucosal immunity in vaccine development. Specific efforts, for instance, with mucosal immunization with cholera toxin B moieties, may induce tolerance to cell-mediated responses while achieving appropriately trafficked humoral responses (LYCKE et al. 1989). In addition, the relevance of humoral responses are greater if true prevention of infection is sought, while there is more interest in cell-mediated responses if containment of an established infection is the goal. These decisions in turn apply not only to the nature of the antigens that are used but also to an equal degree on their presentation with adjuvants in vectors or as DNA plasmids. Finally, these decisions are reflected in the types of trials designed to evaluate efficacy of the vaccines. Since the immunological correlates to protection from neither infection nor disease are known for these diseases, laboratory markers cannot be used as surrogate endpoints in trials.

C. Gonorrhea

I. Epidemiology

Gonorrhea is a worldwide disease, which has undergone a remarkable decline in many industrialized countries. However, it remains an important pathogen in North America and many developing nations. It is associated with genital symptoms such as discharge in about 50% of men but in considerably fewer women. The sequelae of salpingitis and epididymitis lead to suffering and infertility and infection of the newborn to a blinding ophthalmia neonatorum if silver nitrate prophylaxis is not used. Gonococci rapidly develop resistance to new antibiotics, especially when suboptimal doses are used, and when tests of cure are not feasible. An easy and reliable typing system exists that relies on the prime candidate for a gonococcal vaccine, reactions to the gonococcal major outer membrane protein, the porin (see below) (Sandström and Danielsson 1980; Tam et al. 1992).

II. Natural Infection

Gonococci attach to mucosal cells via several mechanisms involving pili and opacity-associated proteins (Sparling et al. 1994). They are subsequently transported through the epithelium, and some emerge from the basal layer of the cell (Sparling 1990). The porin may promote invasion by translocating into eucaryotic cells (Lynch et al. 1984) and impair neutrophil function (Sparling 1990). Whether invasion of PMN and intracellular replication is a privileged site is still a matter of debate (Sparling 1990).

In the preantibiotic era symptomatic urogenital gonorrhea eventually subsided into a nonsymptomatic state, and most patients became noninfectious (Wharton 1937); however, a significant number later developed complicating infections such as pelvic inflammatory disease, Fitz-Hugh-Curtis syndrome, epididymitis, and infections of other genital structures (Hook and Handsfield 1990). Even if resistance to infection at the mucosal level cannot be achieved, protection from these latter complications would be worthwhile in themselves.

Immunity induced by infection does not, however, protect from reinfection, which may be due to extensive strain variation. A strain-specific protection has been claimed in a study of Nairobi prostitutes in whom reinfection was common but was less likely to be caused by strains of the same serovar (PLUMMER et al. 1989). Whether this was due to immunity to the porin (see below) or to covariation with some associated molecule is not clear.

III. Antigenic Variation

Most of the molecular candidates for a gonococcal vaccine undergo extensive phase and antigenic variation, which may indicate a potential natural sensitivity to the human immune response. The least variable potential vaccine candidate, the porin, is apparently associated with two antigenic variable structures, the reduction-modifiable protein (Rmp) and the lipo-oligosaccharide (LOS) in a complex that limit the exposure of the porin (RICE et al. 1986).

IV. Whole-Cell Vaccines

Although whole-cell vaccines were evaluated in the preantibiotic era, it is not possible to draw any conclusions as to their efficacy due to the study designs used at the time. More recently there has been some support for this approach from a chimpanzee model in which a killed whole-cell vaccine conferred some resistance to urethral challenge with the same isolate (ARKO et al. 1976).

V. Subcomponent Vaccines

1. Porin

The major outer membrane protein is immunogenic and has been used for typing purposes. A limited number of serovars are present in any given area, of which a few tend to dominate over extended periods of time (KNAPP et al. 1984). None of the epitopes identified with the monoclonal antibodies used in this typing has been associated with protection so far. Gonococci carry one of two closely related porin molecules, PorA and PorB (SANDSTRÖM et al. 1982). They are variants of the same basic gene, GOTSCHLICH et al. (1987), CARBONETTI et al. (1988). The amino acid composition suggests multiple membrane-spanning domains, and the model predicts eight surface-exposed loops (VAN DER LEY et al. 1991). The porin undergoes very slow antigenic variation (KNAPP et al. 1984). It is the most abundant surface protein and is expressed constitutively.

The possibility of a porin-based vaccine was suggested by VIRJI et al. (1986, 1987a) when they demonstrated that certain anti-porin monoclonal antibodies are bactericidal, induce chemoluminescence of neutrophils, and reduce toxicity to epithelial cells. Some of these antibodies react widely with strains harboring one or the other of the two porin molecules. Broadly reactive PorA and PorB monoclonal antibodies have been defined, SM101 and YSIP, by VIRJI et al. (1987a) and BUTT et al. (1990), respectively. Polyclonal rabbit antisera proved bactericidal and cross-reactive for several serovars of PorA and PorB, respectively, when raised against synthetic peptides representing the most N-terminal surface exposed loop (ELKINS et al. 1992).

However, the association of porins with Rmp and LOS may severely hamper its utility as a vaccine. Rmp is highly immunogenic and stimulates complement-fixing antibodies which are generally nonbactericidal and block the bactericidal effects of anti-porin and anti-LOS antibodies (RICE et al. 1986; VIRJI et al. 1987b). Rmp may even increase the susceptibility to infection (PLUMMER et al. 1993). In addition, sialylation of LOS partially masks the porin on whole bacteria and can thus abolish bactericidal and opsonizing effects of anti-porin antibodies (ELKINS et al. 1992; WETZLER et al. 1992b). It was discovered early that gonococci must be boiled to be rendered reactive to antibodies in serovar determination by coagglutination (SANDSTRÖM and DANIELSSON 1980). Techniques have been developed to produce gonococcal strains or recombinant porins without Rmp contamination (WETZLER et al. 1989; SPARLING et al. 1994). When formulated in liposomes, purified porins elicit bactericidal and opsonic antibodies (WETZLER et al. 1992a). High levels of anti-porin antibodies may, however, be able to overcome these blocking effects (SPARLING et al. 1994).

To optimize the presentation it has been possible to construct *Salmonella typhimurium* strains containing the gonococcal porins. Such vectors would allow presentation of porin antigens to the gut mucosa and hopefully induce both mucosal and systemic immunity (ELKINS et al. 1994).

2. Lipo-oligosaccharides

Gonococci do not have a carbohydrate capsule. The LOS have common sugar components and lack the O-antigen. Several factors may hinder their usefulness as vaccines. The terminal core sugars mimic certain host glycolipids (MANDRELL et al. 1988). Both phase and antigenic variation is as frequent as with pili and opacity proteins (APICELLA et al. 1987). Sialylation of the terminal core sugar can result in masking of the LOS (ELKINS et al. 1992). Phase variation of the core sugar can result in the transient loss of the ability to undergo sialylation. These nonsialylated phase variants have an invasive propensity but are serum susceptible, whereas the sialylated phase variants are noninvasive and remain serum resistant (VAN PUTTEN 1993).

3. Pili

Pili function as filamentous adherence structures. Variants without pili have proved noninfectious in experimental infection in male volunteers (KELLOGG et al. 1963). Gonococcal pili undergo high frequency phase and antigenic variation, involving internal recombination of various genes involved in pilin production (SEGAL et al. 1986; KOOMEY et al. 1987; HAAS and MEIER 1987).

An early study reported partial protection from challenge with a homologous strain in experimental infection of human male volunteers (BRINTON et al. 1982). A field trial of a vaccine based on a single antigenic pili type showed no evidence of efficacy (BOSLEGO et al. 1991). In view of the extensive variability of pili these results are not surprising.

Although certain regions of pili are relatively conserved, they do not seem to be exposed in vivo (Hagblom et al. 1985; Robinson et al. 1989). Furthermore, it has been shown that protective monoclonal anti-pili antibodies are all directed at highly variable epitopes (Virji and Heckels 1985).

4. Opacity Proteins

Opacity proteins derive their name from the effect they have on colony morphology, and they serve as adherence ligands. In this and in their high-frequency phase and antigenic variation they resemble pili. They are essential for gonococci in order to establish genital infection (Swanson et al. 1988). Some may promote invasion of eucaryotic cells (Makino et al. 1991). Their expression is regulated by translational frameshifting (Stern et al. 1986). At any time 1–4 of the 11 constitutively transcribed genes are translated into proteins (Bhat et al. 1991). Presence of antibodies to opacity proteins is negatively correlated to gonococcal salpingitis (Plummer et al. 1994). However, no cross-reactive antibody has been described which blocks opacity protein mediated adherence.

5. Transferrin Binding Proteins

There are two unrelated transferrin binding proteins, Tbp1 and Tbp2, involved in the binding of transferrin and scavenging iron from the host, which is essential for gonococcal growth (Blanton et al. 1990; Cornelissen et al. 1992). Antibodies against analogous molecules on meningococci cross-react broadly, inhibit transferrin binding, and are bactericidal (Ala'Aldeen et al. 1993; Stevenson et al. 1992).

6. IgA Protease

In analogy to meningococci, antibodies against the IgA1 protease may cross-react broadly and block the proteolytic activity and thus increase the protective effect of the immune response to other antigens (Lomholt et al. 1995).

VI. Prospects

The lack of a good animal model has severely hampered progress in gonococcal vaccine research. There are some leads as to the design of vaccines aimed at the more rapid clearance of the organism. The new *Salmonella* vector may be able to present porins in an better way in the absence of LOS and Rmp in order to enhance mucosal immunity, and the level of local antibody response may then be vigorous enough to handle naturally occurring inocula. This may be facilitated by the recent development of a human infection model that has shown that experiments on male urethral gonorrhea can be safely performed (Cohen et al. 1994).

A mouse model using estradiol primed mice has been developed for the evaluation of gonococcal vaccines (Arko et al. 1997).

Protection against infection may be a difficult task to accomplish (COHEN and SPARLING 1992). However, protection from the most important complications that arise from the inflammation and subsequent scarification of structures such as the fallopian tubes and the epididymis would be worthwhile. The gonococcal infection is characterized by a neutrophil response which in turn leads to tissue destruction. To what degree this response is due to chemotactic factors produced by the gonococci (SANDSTRÖM et al. 1983) or by the host immune response against the gonococci is not clear. In addition, cell-mediated responses may be of importance in the progression of uncomplicated disease to complications, although most studies have focused on the humoral responses, which are more accessible to investigation. The role of the immunological responses remains to be elucidated in the process of the design of a vaccine.

The addition other immunogens than the porin, i.e., transferrin-binding molecules that would induce an immune response that impairs the uptake of iron, may be beneficial in a vaccine aimed primarily at reducing complications.

D. *Chlamydia* Infection

I. Epidemiology

Infections with *Chlamydia trachomatis* is one of the most common sexually transmitted bacterial pathogens worldwide. It often causes few and mild genital symptoms in both men and women. Approximately 10% of those infected develop complications such as salpingitis leading to ectopic pregnancy or tubal infertility often without having experienced genital symptoms (WESTRÖM and MÅRDH 1993). The declining rates of gonorrhea in the industrialized nations has made it the most important cause of tubal infertility.

It is not clear to what degree the cited findings apply to the protection against infections caused by serovars L1–3 causing lymphogranuloma venereum (LGV), which is characterized by an initial often transient genital sore and subsequent scarring of local draining lymph nodes leading to lymphstasis, elephantiasis, and fistulae. LGV is most often diagnosed in developing nations.

II. Natural Infection

C. trachomatis is an obligate intracellular bacterial pathogen. It exists either as extracellular infectious elementary bodies (EB) or in intracellular vacuoles as replicative forms called reticular bodies (RB). It causes trachoma and genital tract infections in humans. Infections tend to be asymptomatic and chronic and to give rise to complicating scarring reactions. There are two human biological variants, the trachoma and LGV biovars, and one murine biovar. The former are subdivided in 15 serovars based on the reactivity with major outer membrane protein (MOMP) antigens, of which A, B, and C give rise to trachoma, D–K to genital infections, and L1, L2, and L3 to LGV. Antigenically, serovars A, C, H, I, and J are members of the C complex and serovars B,

Ba, D, E, L1, and L2 are members of the B complex (F and G are related), while K and L3 are thought to bridge these complexes (FITCH et al. 1993).

There is at least partial immunity to infection by the same strain and an inverse correlation between specific IgA level and presence of *Chlamydia* in cervical secretions (BRUNHAM et al. 1983). It also has been shown that humans who are allowed to recover from ocular disease are immune to the infecting serovar (see review by BRUNHAM and PEELING 1994).

III. Pathogenesis

EB are exposed to the immune pressure as evidenced by the high variability of the MOMP (see below). *Chlamydia* in privileged reticulate bodies, however, does not seem to present peptides through either the MHC class I or class II antigen-presentation pathways.

A fascinating series of investigations has teased apart the immune responses responsible for protection from those involved in the immunopathogenesis of the disease. A 57-kDa chlamydial antigen was found to be associated with tubal infertility (BRUNHAM et al. 1985). The protein has subsequently been cloned and sequenced and found to be a 57-kDa heat shock protein related to the GroEL product (MORRISON et al. 1989). This is a member of the hsp60 family and 48% homologous to the human hsp60 (MORRISON et al. 1992). A number of studies have confirmed a strong correlation between the presence of antibodies directed against hsp60 and salpingitis, ectopic pregnancy, and infertility (BRUNHAM et al. 1992; WAGAR et al. 1990; TOYE et al. 1993). It is proposed that the high degree of homology between the chlamydial and human hsp60 homologues are responsible for the breakdown of tolerance in susceptible individuals and thus induction of an autoimmune response (BRUNHAM and PEELING 1994).

Although the exact mechanism of immunopathogenesis of tubal damage is not known, it has been observed that women with complications are more likely to have a delayed-type hypersensitivity reaction to the hsp than *Chlamydia*-infected women without signs of salpingitis (TOTH et al. 1992). Interferon-γ may induce *Chlamydia* to express normal amounts of the hsp, but markedly decrease the expression of (MOMP) (BEATTY et al. 1993). These responses may be involved in Reiter's syndrome, which is induced in individuals of certain HLA types (ZHONG and BRUNHAM 1992). Topical application of the 57-kDa hsp elicits an ocular or fallopian tube hypersensitivity reaction (MORRISON et al. 1989; PATTON et al. 1994). Thus any vaccine against *Chlamydia* needs to avoid antigens that can arouse these unwanted immune reactions.

IV. Whole-Cell Vaccines

In analogy to the genital infection, ocular chlamydial infection is often chronic and leads to scarring. Whole-cell vaccines may actually have potentiated this reactivity after reexposure to ocular chlamydial infection (reviewed in

SPARLING et al. 1994). However, it has been shown that protective immunity can be achieved with parenteral immunization with whole EB in humans. The poor immunogenicity and risk of hypersensitivity has hampered this line of investigation (GRAYSTONE and WANG 1978). However, in a new interesting model it has recently been shown that intranasal immunization with the *C. trachomatis* mouse pneumonitis strain EB can preserve fertility in mice challenged with inoculation with the same strain in the ovarian bursa. As expected, a strong IgA response was obtained, but no reactivity to the hsp60 was seen after the intranasal immunization in these BALB/c mice (PAL et al. 1994).

V. Subcomponent Vaccines

1. Major Outer Membrane Protein

MOMP is the leading vaccine candidate. Similarly to the gonococcus, it constitutes the basis for the typing system and may also function as a porin (BAVOIL et al. 1984) and/or mediate attachment to eucaryotic cells (SU et al. 1990). MOMP is a transmembrane protein with four surface-expressed loops (VD1–4) (BAEHR et al. 1988) defined by the gene *omp-1*. Proteolytic cleavage of two of these loops reduces infectivity (SU et al. 1988). A seven amino acid long sequence from loop VD4 coupled to a common MOMP T-helper cell epitope elicits polyclonal antibodies in animal systems that bind to *Chlamydia* of serovars D, E, and F, which causes the majority of human infections (SU and CALDWELL 1993). The variability of MOMP was investigated in 60 prostitutes in Nairobi presumably heavily exposed to many different *Chlamydial* strains (BRUNHAM et al. 1994). Frequent point mutations and mosaic *omp-1* sequences relative to reference strains were seen, indicating that MOMP may be the target for selected mutation and recombination. In addition to the allelic polymorphism of MOMP, *C. trachomatis* has thus evolved mechanisms to escape herd immunity against dominant serovars. This results in the maintenance of a constant prevalence of infection is maintained in a host population determined by behavioral factors, by successive waves of allelic variants (BRUNHAM and PEELING 1994).

Both polyclonal and monoclonal antibodies against MOMP are neutralizing and protect tissue cultures against infection by elementary bodies (CALDWELL and PERRY 1982; ZHANG et al. 1987, 1989; QU et al. 1993).

Since it has been found that neutralizing antibodies that differentiate *C. trachomatis* isolates into serogroups and serovars recognize the variable regions on the MOMP (VD1–VD4), oligopeptides representing these regions have become the focus of attention (MURDIN et al. 1995). The VD4 region seems particularly attractive since it contains a sequence that is conserved in all but one (K) of the human serovars (FITCH et al. 1993).

High-titer neutralizing antibodies can be obtained in mice immunized with an oligopeptide that corresponds to antigenically common T-helper and neutralizing epitopes on *C. trachomatis* MOMP, A8-VDIV (SU and CALDWELL 1993). Parenteral immunization with this oligopeptide raises high neutralizing IgG titers but not IgA titers and has a limited protection against vaginal or intrauterine challenge with serovar D in mice (SU et al. 1995). In view of the inverse correlation between chlamydial presence and IgA observed in humans (BRUNHAM et al. 1983) and the lack of IgA response to this immunization approach, the authors are attempting to couple the oligopeptide to cholera toxin subunit B to achieve such response. In a similar study of parenteral immunization with serovar L1 recombinant MOMP, chlamydial colonization was not reduced, but there was a reduction in number of mice developing severe salpingitis after challenge with *Chlamydia* of serovar F (TUFFREY et al. 1992).

In mice immunized with a phage presentation system VD1 peptides in a loop formation give rise to antibodies specific for chlamydial EB while a linear presentation give rise only to antibodies specific for the immunizing peptide (ZHONG et al. 1994).

Poliovirus hybrids that expressed complex B (VD1) or complex C (VD4) epitopes give rise to broadly cross-reactive neutralizing antibodies. The further development of these hybrids may enable induction of broadly reactive immune responses that include mucosal genital mucosal immunity after oral challenge (MURDIN et al. 1995).

DNA vaccination with chlamydial MOMP genes generated delayed type hypersensitivity, serum antibodies and resitance in the mouse pneumonitis model (ZHANG et al. 1997).

Most experience has been obtained with antibody correlates to immunity. There is, however, good reason to believe that cellular immunity is as important. In mice adoptive transfer of $CD4^+$, but not $CD8^+$, T-lymphocytes from mice that have recovered from genital infection with the mouse biovar of *C. trachomatis* showed protection against genital challenge in naive animals (SU and CALDWELL 1995).

2. Heat Shock Protein 75 kDa

Another chlamydial outer membrane protein, 75-kDa, which is a member of the heat shock 70 family, have given rise to polyclonal antibodies that are protective in vitro (DANILITON et al. 1990). Antibodies to this protein show an inverse correlation with salpingitis (BRUNHAM et al. 1987). This hsp does not cause ocular hypersensitivity in infected experimental animals (TAYLOR et al. 1990).

VI. Prospects

It seems that the understanding of the pathogenesis of genital chlamydial infection and its protective immunity is at a stage where rapid progress can be anticipated. However, the role of cell-mediated immunity remains to be elucidated as well as the most appropriate immunogens, i.e., MOMP subcomponents as well as their optimal administration (BRUNHAM et al. 1981).

E. Genital Ulcers Caused by *Haemophilus ducreyi*

I. Epidemiology

Chancroid remains a major global health care problem (RONALD and PLUMMER 1985; MORSE 1989). Genital ulceration may facilitate infection with human immunodeficiency disease (CAMERON et al. 1989). Although uncommon in most industrialized countries, *H. ducreyi* may be the cause of up to 80% of all genital ulcers in sub-Saharan Africa (PLOURDE et al. 1992). Treatment failures due to resistance to drugs and concomitant HIV infection are increasing problems (TYNDALL et al. 1993).

II. Natural Infection

Ducrey (1889) was able to serially propagate the infectious agent at least 15 times by autoinoculations in the same individual, indicating little or no immunity in natural infection. Little is know about the human immune response to *H. ducreyi* infection, although there is evidence for both a cell-mediated and a humoral response (ABECK et al. 1990). After an incubation period of 4–7 days *H. ducreyi*, which enters a break in the skin, usually at the preputium in men and introitus in women, forms a papule and then a pustule which soon ruptures to form a sore. The sore is usually irregular, soft, and painful. Local lymph nodes usually become fluctuant and rupture spontaneously. Dissemination is unusual. It has not been established whether there is a true asymptomatic carriage in contrast to other STDs.

The diagnosis is usually based on clinical findings since Gram stains are difficult to interpret, and the organism is difficult to culture (MORSE 1989).

III. Subcomponent Vaccines

Natural infection yields anti-cytotoxin antibodies, but they do not protect from disease (LAGERGÅRD and PURVEN 1993). Other potential targets would be the outer membrane proteins and pili (MORSE 1989). These proteins may be variable during natural infection (ABECK and JOHNSON 1992). Protection against heterologous challenge of crude *H. ducreyi* envelops has been observed in a temperature-dependent rabbit model (HANSEN et al. 1994). Using the same model, a pilus preparation from the same strain that previously only gave homologous protection was found to give rise to protection against an epidemiologically distinctly heterologous strain, while a LOS preparation showed no protection even against the homologous strain (DESJARDIN et al. 1995).

IV. Prospects

Recently a safe and reproducible human model has been developed in which live *H. ducreyi* are applied to abraded skin with papule and pustule formation (SPINOLA et al. 1994).

F. Syphilis

I. Epidemiology

Syphilis has declined dramatically in the industrialized world since the Second World War, with the advent of effective and cheap antibiotics, i.e., penicillin. In some countries such as the United States a resurgence has been noted in recent years concomitant with the HIV pandemic. In developing nations, however, syphilis remains a significant health hazard to the neonate and adult in spite of relatively inexpensive diagnosis and treatment.

II. Natural Infection

Treponema pallidum is one of a small group of highly related pathogenic spirochetes confined to infection of man. No culture system is available besides testicular infection in rabbits, and this has an in vivo doubling time of about 30 h. The organism is remarkably stable; for instance, there has been no development of resistance to penicillin despite its widespread use as therapy. This may be due to its comparatively small genome of 1000 kbp (WALKER et al. 1991).

A typically painless chancre is formed at the site of infection, which heals spontaneously in 3–6 weeks concomitant with dissemination of the spirochetes to regional lymph nodes and subsequently to many distant organs. During the early stages of infection there is protection to reinfection, i.e., "chancre immune," in which humans become resistant to intradermal challenge during the early stages of infection (reviewed in SELL and NORRIS 1983). This can be duplicated in the rabbit model TURNER and HOLLANER 1957).

A vigorous cellular and humoral response is correlated with the clearance of treponemas from primary and secondary lesions (WICHER et al. 1991). However, treponemas in privileged sites such as lymph nodes may persist for a long time, but syphilis in humans is not considered infectious beyond 2 years after infection.

One-third of infected individuals undergo apparent "cure," defined as loss of clinical manifestations and serological reactivity, indicating the possibility of some kind of protective immunity to the systemic infection, while another one-third go on to develop complications. Most of the syphilis pathology is due to host inflammatory reactions stimulated by the persistent infection (SELL and NORRIS 1983).

In a study of the immunity to natural infection against rabbit propagated *T. pallidum* in prisoners with various experiences of the natural infection, all 8 individuals without experience of syphilis infection developed dark-field positive lesions (spirochetes visible with dark-field microscopy). The 11 men who had been treated during early infection developed lesions, of whom 9 were dark-field positive. Of the 32 men who had been treated for late or congenital syphilis only 2 developed dark-field positive lesions, and an additional 15 developed lesions without microscopic signs of spirochetes. The remaining 14 showed no lesions. Correspondingly none of the 5 men with untreated latent syphilis developed lesions (MAGNUSON et al. 1956). For ethical reasons all individuals were treated with penicillin when lesions developed or at 4 months, thus the critical question of the effect on systemic spread of treponemas to privileged sites responsible for late disease could not be addressed in this study.

III. Antigenic Variation

No evidence exists of phase or antigenic variation during the passage of *T. pallidum*. However, remarkably few *Treponema*-specific outer membrane proteins are exposed on the surface of the organism, and there is also a lack of specific LPS exposed in the outer membrane.

IV. Whole-Cell Vaccines

Whole-cell killed antigens gives limited protection in animals, but in doses or regiments that were not suitable for human trials (MILLER 1973; FITZGERALD 1991).

V. Subcomponent Vaccines

1. Cardiolipin

Although most recent work has focused on treponemal proteins, a recent study has shown that antibodies raised against the cardiolipin reagent used in the screening test for syphilis, VDRL, can mediate opsonization but not macrophage-mediated killing. Immunization with this antigen reduces the infectivity of dermal challenge by the Nichols strain in rabbits (BAKER-ZANDER et al. 1993). However, cardiolipin is derived from mammalian cells and not a treponemal product, which raises some concerns over the possibility of autoimmune reactions.

2. Treponemal Proteins

Extensive progress has been made in the molecular characterization of the genome in search for membrane spanning proteins. There is an extensive review of humoral and cellular response in NORRIS et al. (1993). Recently published studies focus on recombinant *T. pallidum* proteins. TpN19 has evoked partial protection in rabbits (BORENSTEIN et al. 1988). Its function in the pathogenesis in syphilis in unknown. A recombinant endoflagellar protein has also resulted in partial protection in animals (FEHNIGER et al. 1986). More recently TpN36 has also shown some protection (CHAMPION et al. 1990).

VI. Prospects

It has been clearly recognized that the vigorous immune response is not sufficient for clearance of *T. pallidum*, and that further research needs to focus on the cellular immunology (LUKEHART et al. 1980).

G. Human Genital Papilloma Virus Infection

I. Epidemiology

The prevalence of genital HPV infection increases with the initiation of sexual activity. With the aid of PCR techniques it has been estimated that 20- to 24-year-old sexually active women have a genital HPV prevalence of 20%–25%. In Finland the life-time risk of genital HPV infection has been estimated at 80% for women. However, women over 35 years of age have a prevalence of only 1%–2% indicating a loss of viral genomes with age. Coinfection with several types is common. Sexual partners are frequently found to be infected with different types. HPV infection can cause local genital discomfort and psychological problems.

Individuals with infection with so-called "high-risk" HPV types run the risk of developing squamous cell carcinomas, especially of the cervix, 10–15 years after the infection. Screening for cervical dysplasia and cervical intraepithelial neoplasia can detect the majority of these lesions before manifest cancer develops.

II. Natural Infection

HPV is a small icosahedral virus without an envelope. It cannot be cultured in vitro. HPV genomes exist as episomes that replicate as squamous epithelial cells and start to keratinize. In the basal cell layer only genes E1 and E2 are expressed and facilitate HPV episome replication. In the suprabasal layer the HBV E5, E6, and E7 genes are expressed. The latter counteract the effects of differentiation and turn on essential parts of the dormant cellular nuclear metabolic machinery. This causes nuclear atypia in the affected cells typically in the epidermal intermediate zone, the so-called koilocytes. In the terminally differentiated cells the L1 and L2 proteins are abundant and associated with mature virus production. There exist more than 70 types of HPV. The "types" are distinguished by their DNA homology as determined under stringent hybridization conditions. A number of types are found most often in the genitalia. Two types, 6 and 11, are considered "low-risk" and another 10 "medium- to high-risk" types, notably types 16 and 18.

The former tend to be most common and are associated with condyloma acuminata, i.e., acuminate or papular lesions, less dysplasia, and shorter persistence. The latter with subclinical lesions, dysplasia of various degrees, and persistence. The latter forms are associated with some squamous cell genital cancers, most importantly as a cofactor for more than 90% of the cervical cancers (BOSCH et al. 1995). Cervical cancer is the second most common cause of cancer deaths worldwide, affecting about 1% of all women, and may develop after 5–25 years (PARKIN 1994).

The clearance of virus with age can be due to epithelial shedding, as the virus fails to infect the reproducing cells in the basal layer or to an activation of a specific immune response. However, it has been reported that prostitutes are first found to carry common HPV types, which are replaced by more and more uncommon types with increased sexual exposure, indicating effective genotype-specific immunity (FRAZER 1996). In certain diseases with immune dysfunction HPV can cause significant problems (PALEFSKY 1995; PETRY et al. 1994).

III. Whole-Viral Vaccines

Attempts to use autologous homogenates as therapeutic vaccines has a long history, with unsubstantiated claims of efficacy (BIBERSTEIN 1944; ABCARIAN and SHARON 1982; EFTAHIA et al. 1982). Calves have been protected by intramuscular injection of whole bovine papilloma virus (JARRETT et al. 1990). The canine oral papillomavirus infection is of particular interest since this virus infects the mucosal surface as HPV. Vaccination with formalin-inactivated virus can protect against disease and transmission but not against infection (BELL et al. 1994).

IV. Subcomponent Vaccines

Advances have been made in expressing HPV proteins that assemble into virus-like particles (VLP) in various systems (KIRNBAUER et al. 1992; ROSE et al. 1993; ZHOU et al. 1991). The cottontail rabbit can be vaccinated to induce long-lived protection against its species specific papillomavirus infection, CHRISTENSEN et al. (1996). Antibodies against L1 HPV VLPs is induced at a median 8.3 months after detection of type 16 HPV by PCR (CARTER et al. 1996).

While L1 might be of great importance in attempts to protect against infection, it seems that early proteins such as E7 which is expressed on basal layer celles and cervical cancer lesions are essential for the ellimination of tumerous cells. Inclusion of the E7 protein in VLP have given raise immune response to cancereous lesions in mice in support of this concept (GREENSTONE et al. 1998). Recombinant vaccinia vectors have been previously been constructed that present E6 and E7 (BOURSNELL et al. 1996). Since mucosal immunity is desirable in order to prevent infection it is important that nasal immunozation with Salmonella typhimurium expressing L1 (HPV16) has been shown to give rise to specific IgA and IgG in vaginal secretions (NARDELLI-HAEFLIGER et al. 1997). Salmonella vectors have also been constucted to express E6 and E7 of HPV16 (KRUL et al. 1996).

V. Prospects

Although HPV certainly is a common STD, its diagnosis and treatment are currently under debate due to the difficulties in reliably diagnosing and eradicating the types most likely to cause complicating disease, i.e., cervical cancer. In spite of the recent developments of the PCR technique there is a lack of simple methods for the confirmation of protection from subclinical HPV infection that would be pertinent to efficacy trials for a human HPV vaccine. The long interval between the infection and the development of cancer preclude using this relevant event as an endpoint except in large population based studies. The trial strategies involved in studies of both prophylactic and therapeutic HPV vaccines have recently been discussed (FRAZER 1996). Should immunity play a role in immunotherapy, there may be a role for a vaccine as an adjuvant to existing therapies in the management of symptomatic HPV disease or cervical dysplasia.

H. Herpes Simplex Infection

I. Epidemiology

There are two variants of HSV. Type 1 has classically been associated with oral and type 2 with genital infections. This distinction has started to blur with an increased incidence of genital type 1 infections at least in some industrialized countries either due to a decreasing incidence of HSV-1 infections during childhood or altered sexual practices (BARTON et al. 1982; SCOULER et al. 1990; ROSS et al. 1993). It is estimated that 40–60 million individuals in the United States alone are infected by HSV-2 (WHITLEY and GNANN 1993). The rate infection there is related to female gender and black race. Similar rates are present in other areas of the world, for example, 22% of a cohort of Swedish women developed HSV-2 antibodies between ages of 15 and 30 years (CHRISTENSSON et al. 1992). Only one-third of those infected are aware of their infection, but they may be able to transmit the disease by subclinical shedding of the virus (WALD et al. 1995). The most severe complications are neonatal infections acquired at birth and encephalitis when infected later in life, FORSGREN and MALM (1996), SKÖLDENBERG (1996). HSV is no longer considered to be a cofactor in cervical neoplasia.

II. Natural Infection

HSV enters the skin where initial local replication takes place. Subsequently HSV is transported along sensory nerve axons to the local ganglia where it establishes itself in a number of episomal genomes (ROIZMAN and SEARS 1993). The events that trigger production of new virions are not fully understood but lead to transport of infectious virions back to the skin. In the skin further replication may take place with the production of local inflammation and a pustule. Genital infection with HSV-1 has a lower tendency to recur than infection with HSV-2. It has long been observed that previous infection with HSV-1 reduces the frequency and severity of subsequent episodes of genital HSV-2 infection (REEVES et al. 1981; COREY et al. 1983; MERTZ et al. 1992). Immunity may operate at least three levels. It may function at the time of first infection, protecting nerves from infection and subsequent latency, operate via cytokines or other mechanisms to repress reactivation in the nerves, or ameliorate the epithelial viral production subsequent to reactivation.

III. Whole-Viral Vaccines

Both live and killed whole-viral particles have been used. A live attenuated HSV-1 strain was constructed to contain HSV-2 glycoproteins D, G, and I as well as a reinserted thymidine kinase gene to make it sensitive to acyclovir. It appears safe after serial passages in the mouse brain (MEIGNIER et al. 1988). Variants which have emerged with enhanced replication capabilities are currently under investigation (WHITLEY et al. 1993). Mice immunized with a HSV-1 mutant that lacks glycoprotein H and is thus capable of only one round of replication are protected from subsequent challenge with high titers of wild-type HSV (FARREL et al. 1994). This defective virus can establish latency and in a very low frequency rescue the deleted gene from the cell line in which it is grown. Whether latency with this virion is a drawback or source of continuous immune stimulation remains to be determined. This so-called disabled infection single cycle, DISC, virus can protect against HSV-2 challenge in the guinea pig model after scarification or intravaginal immunization. However, recurrences do occur albeit at a significantly lower rate in immunized animals, even in those without signs of a primary infection (McLEAN et al. 1994; BOURSNELL et al. 1997). Other HSV-1 mutants that are unable to replicate due to deletion or nonsense mutation in genes for the immediate early proteins ICP8 and ICP27, respectively, have protected mice from latent infection of ganglia after corneal challenge subsequent to subcutaneous immunization with the defective virus (MORRISON and KNIPE 1994).

IV. Subcomponent Vaccines

Purified components of HSV are of interest since they are free of genetic material that can integrate with the host and can be produced in bulk with modern techniques. Two of the HSV glycoproteins, gB and gD, are the focus of attention as subcomponent vaccines as they play a major role during natural infection.

Glycoprotein D is the target of neutralizing antibodies, antibody-dependent cellular toxicity, and $CD4^+$- and $CD8^+$-mediated cellular responses. They are highly conserved and similar in HSV-1 and HSV-2. Purified dD2 given with alum to human volunteers greatly increases total and neutralizing antibody levels in HSV naive and HSV-1 or HSV-2 experienced subjects. There is a minimal effect on cell-mediated immunity with this adjuvant (STRAUSS et al. 1993).

Success in reducing the recurrence rate was subsequently achieved with gD2, with alum as adjuvant, concomitant with rises in neutralizing antibodies (NA). However, rises in NA are not correlated to a reduction in recurrences (STRAUSS et al. 1994). The poor immunogenicity of alum adjuvants has prompted investigation of other adjuvants. One such adjuvant in clinical trials is the oil-in-water microfluidized formulation MF59, with or with out the muramyl peptide MTE-PE, investigated in the guinea pig model (BURKE et al. 1994). Responses to gD2 and gB2 in this adjuvant are well tolerated and induce humoral and cellular responses comparable to those of the natural infection (LANGENBERG et al. 1995).

In an effort to circumvent the tolerance induced by mucosal immunization a truncated HSV-1 glycoprotein D has been coupled to interleukin-2. This complex given to mice by the nasal route gave rise not only to good nasal IgA titers, which may also reflect genital trafficking, but also protection against intraperitoneal challenge indicative of a cellular response (HAZAMA et al. 1993).

Since commonly used adjuvants are poor inducers of cell-mediated immunity, which at least in animal models is very important in disease modification, several techniques have been employed to boost such immunity.

Coadministration of imiquimod, which upregulates some cell-mediated responses, with a crude HSV-2 glycoprotein preparation significantly reduced recurrences in latently infected guinea pigs given glycoprotein alone or with Freund's complete adjuvant. The specific effect on cell-mediated immunity of this regiment however, was, not documented (BERNSTEIN et al. 1995).

The recently developed genetic immunization technique has been utilized in the mouse zosteriform HSV model. Intramuscular immunization with a DNA plasmid containing the gene for HSV-1 glycoprotein B conferred protection to HSV-1 challenge. This protection has been adoptively transferred to nude mice through splenic $CD4^+$ lymphocytes with a TH-1 cytokine pattern (MANICKAN et al. 1995). Similar results have been obtained with vaccinations with a plasmid containing the gD2 gene in a mouse HSV vaginitis model (KREISEL et al. 1996). The simultaneous administration of vitamin D in the latter trial could indicate that modification of the cytokine response might be of additional benefit.

Intranasal DNA plasmid immunization with genes for gB induced a distal mucosal IgA response which was increased by simultaneous cholera toxin administration. However in these mouse experiments the immune response was not sufficient to protect from viral mucosal vaginal challenge (KUKLIN et al. 1997). A Sindbis virus derived plasmid expressing gB has been shown to induce virus specific antibody, cytotoxic T-cells and protection in mice (HAIHARAN et al. 1998).

V. Prospects

Clinical trials of gD with or without gB in different adjuvants are underway both as preventive vaccines and as immunotherapy. The results from these trials has so far been disapointing (STRAUSS et al. 1997). Whether the live attenuated vaccines will pass into clinical trials depends on the risk-benefit deliberations with respect to the possibility that they can revert to wild type and establish latency or other unknown consequences. Nucleic acid vaccine development is underway and may address these safety problems.

I. Conclusion

There seems to be good hope for a gonococcal vaccine using gonococcal porins and vectors to stimulate mucosal immunity. The aim of such a vaccine would be to shorten the infectious time and thus reduce transmission. However, if the vaccine simply turns the infection into an asymptomatic state that does not abolish infectiousness, spread of gonorrhea would increase as individuals with symptomatic disease would not seek medical care. If such a vaccine does not affect the rate of complications, matters could be even worse. A gonococcal vaccine might be acceptable by society as the level of awareness of gonorrhea and its complications is high in most areas of the world, and adequate antibiotic treatment is a practical and financial problem. A possible target population are individuals seeking health care for any STD-related problem.

Fewer research issues seem to surround a chlamydial vaccine using some presentation of the MOMP variable domains, free of hsp60 proteins, in a delivery system aimed at mucosal immunity. Also, with this vaccine the aim should be to help the host to clear the infection as fast as possible, preferentially before an immune response to hsp60 has had time to develop. The acceptance of such a vaccine might, however, be lower, as public awareness is not as great as with gonorrhea. This is due to the typically asymptomatic presentation, the difficulties and cost of diagnosis, and subsequent lack of information on its epidemiology, especially in developing nations. Complications do, however, develop in spite of the lack of genital symptoms and are often not recognized until long after the genital infection has resolved.

The prospects of a vaccine against chancroid are further removed, and it is not easy to see how such a vaccine should be handled in the health care situation. Would it be acceptable in military recruits in areas with high prevalence of genital ulcer disease and HIV?

The vaccine against syphilis raises a number of issues. Although some molecular targets have been identified, the task is formidable. Since it is aimed at the prevention of long-term sequelae that are induced by the immune system, it must be proven that all organisms are really cleared by the vaccine, and that there are no long-term safety problems in human studies of natural protection. Given this and the wide availability of relatively inexpensive diagnosis and cure that are not utilized, it is difficult to envision a vaccine campaign that could not be undertaken with mass treatment with penicillin analogous to the eradication of endemic syphilis, if there is a public health policy to do so.

A vaccine against genital HPV infection faces similar challenges. Molecular target have been established, but the desired effect on long-term sequelae would be difficult to establish. This is confounded by the difficulties in diagnosis and the presence of successful screening programs for cervical cancer in many countries. However large trials are planned to assess the VLP based vaccines in protection against cervical cancer.

The vaccine that has advanced furthest is that aimed at herpes simplex. The molecular targets are defined and the presentation by various modes are ongoing, even in large clinical trials. The goal is moderate, to reduce transmission and number of recurrences. However further planned trials have been but on hold pending analysis of the data from these trials. There is an awareness of the disease in the population, at least in the industrialized countries, and the burden of discomfort and cost is considerable. Depending on the efficacy of the vaccine the populations that could potentially be offered the vaccine vary from STD clinic populations, military recruits, to a general health care measure.

References

Abcarian H, Sharon N (1982) Long term effectiveness of the immunotherapy of anal condyloma acuminata. Dis Colon Rectum 25:648–651

Abeck D, Johnson AP (1992) Pathophysiological concept of Haemophilus ducreyi infection (chancroid). Int J AIDS 3:319–323

Abeck D, Korting HC, Zaba R, Dangor Y, Fehler G, Ballard RC (1990) Soluble interleukin-2 receptors in serum and urine of patients with chancroid and their respomse to therapy. Int J STD AIDS 1:282–284

Ala'Aldeen DA, Powell NB, Wall RA, Borriello SP (1993) Localization of the meningococcal receptors to human transferrin. Infect Immun 61:751–559

Apicella MA, Shero M, Jarvis GA, Griffis JML, Mandrell RE, Schneider H (1987) Phenotypic variation in epitope expression of the Neisseria gonorrhoeae lipoologosaccharide. Infect Immun 55:1755–1761

Arko R, Duncan W, Brown W, Peacock W, Tomizawa T (1976) Immunity in infection with Neisseria gonorrhoeae: duration and serological response in the chimpanzee. J Infect Dis 133:441–447

Arko RJ, Smith SD, Chen CY (1997) *Neisseria gonorrhoeae*: vaginal clearance and its correlation with resistance to infection in subcutaneous chambers in orally immunized estradiol treated mice. Vaccine 15:1344–1348

Baehr W, Zhang Y-X, Joseph T, Su H, Nano FE, Everett KDE, Caldwell HD (1988) Mapping antigenic domains expressed by Chlamydia trachomatis major outer membrane protein genes. Proc Natl Acad Sci USA 85:4000–4004

Baker-Zander, Shaffer JM, Lukehart SA (1993) VDRL antibodies enhance phagocytosis of Treponema pallidum by macrophages. J Infect Dis 167:1100–1105
Barton IG, Kinghorn GR, Najem S, Al-Omar LS, Potter CW (1982) Incidence of herpes simplex type 1 and 2 isolated in patients with herpes genitalis in Sheffield. Br J Vener Dis 58:44–47
Bavoil P, Ohlin A, Scachter J (1984) Role of the disulfide bonding in the outer membrane structure and permeability in Chlamydia trachomatis. Infect Immun 44:479–485
Beatty W, Byrne GI, Morrison RP (1993) Morphologic and antigenic characterization of interferon γ-mediated persistent Chlamydia trachomatis infection in vitro. Proc Natl Acad Sci USA 90:3998–4002
Bell JA, Sundberg JP, Ghim S, Newsome J, Jenson AB, Schegel R (1994) A formalin-inactivated vaccine protects against mucosal papillomavirus infection: a canine model. Pathobiology 62:194–198
Bernstein DI, Harrison CJ, Tepe ER, Shahwan A, Miller RL (1995) Effect of immiquimod as an adjuvant for immunotherapy of genital HSV in guinea-pigs. Vaccine 13:72–75
Bhat KS, Gibbs CP, Barrera O, Morrison SG, Jänhig F, Stern A, Kupsch E-M, Meyer TF, Swanson J (1991) The opacity proteins of Neisseria gonorrhoeae strain MS11 are encoded by a family of 11 complete genes. Mol Microbiol 5:1889–1901
Biberstein H (1944) Immunization therapy of warts. Arch Dermatol 50:12–22
Blanton K, Biswas GD, Tsai J, Adams J, Dyer D, Davis S, Koch G, Sen P, Sparling PF (1990) Genetic evidence the Neisseria gonorrhoeae produces specific receptors for transferrin and lactoferrin. J Bacteriol 172:5225–5235
Borenstein LA, Radolf JD, Fehniger TE, Blanco DR, Miller JN, Lovett MA (1988) Immunization of rabbits with recombinant Treponema pallidum surface antigen 4D alters the course of experimantal syphilis. J Immunol 140:2415–2421
Bosch FX, Manos MM, Munoz N, Sherman M, Jansen AM, Peto J, Schiffman MH, Moreno V, Kurman R, Shah KV (1995) Prevalence of human papillomavirus in cervical cancer: a worldwide perspective. J Natl Cancer Inst 87:796–802
Boslego JW, Tramont EC, Chung RC, McChesney DG, Ciak J, Sadoff JC, Piziak MV, Brown JD, Brinton CC, Wood SW, Bryan JW (1991) Efficacy trial of a parenteral gonococcal vaccine in men. Vaccine 9:154–152
Boursnell ME, Rutherford E, Hickling JK, Rollinson EA, Munro AJ, Rolley N, McLean CS, Borysiewics LK, Vousden K, Inglis SC (1996) Construction and characterization of a recombinant vaccinia virus expressing human papillomavirus proteins for immunotherapy of cervical cancer. Vaccine 14:1485–1494
Boursnell ME, Entwisle C, Blakeley D, Roberts C, Duncan IA, Chisholm SE, Martin GM, Jennings R, Ni Challanian D, Sobaek I, Inglis SC, McLean CS (1997) A genetically inactivated herpes simplex virus type 2 (HSV-2) vaccine provices effective protection against primary and recurrent HSV-2 disease. J Infect Dis 175:16–25
Brinton CC, Wood SW, Brown A, Labik AM, Bryan JR, Lee SW, Polen SE, Tramont EC, Sadoff J, Zollinger W (1982) The development of a neisserial pilus vaccine for gonorrhea and menigoccal meningitisIn: Weinstein L, Fields BN (eds) Seminars in infectious diseases, vol 4. Thieme-Stratton, New York, pp 140–159
Brunham RC, Peeling RW (1994) Chlamydia trachomatis antigens: role in immunity and pathogenesis. Infect Agents Dis 3:218–233
Brunham RC, Martin DH, Kuo C-C, Wang S-P, Stevens CE, Hubbard T, Holmes KK (1981) Cellular immune response during uncomplicated genital infection with Chlamydia trachomatis. Infect Immun 34:98–104
Brunham RC, Kua C-C, Cles L, Holmes KK (1983) Correlation of host immune response with quantitative recovery of Chlamydia trachomatis from the human endocervix. Infect Immun 39:1491–1494
Brunham RC, Maclean IW, Binns B, Peeling RW (1985) Chlamydia trachomatis: its role in infertility J Infect Dis 152:1275–1282
Brunham RC, Peeling R, Maclean I, McDowell J, Persson K, Osser S (1987) Postabortal Chlamydia salpingitis: correlating risk with antigen-specific serological responses and neutralization. J Infect Dis 155:749–755

Brunham RC, Peeling R, Maclean I, Kosseim ML, Paraskevas M (1992) Chlamydia trachomatis-associated ectopic pregnancy: serologic and histologic correlates. J Infect Immun 165:1076–1081
Brunham R, Yang C, Maclean I, Kimani J, Maitha G, Plummer F (1994) Chlamydia trachomatis from individuals in an Sexually Transmitted Disease core group exhibit frequent esquence variation in the major outer membrane protein (omp1) gene. J Clin Invest 94:458–463
Burke RL, Goldbeck, Ng P, Stanberry L, Ott G, Van Nest G (1994) The influence of adjuvant on the therapeutic efficacy of a recombinant genital herpes vaccine. J Infect Dis 170:1110–1119
Butt NJ, Virji M, Vayreda F, Lambden PR, Heckels JE (1990) Gonococcal outer-membrane protein PIB: comparative sequence analysis and localization of epitopes which are recognized by type-specific and cross-reacting antibodies. J Gen Microbiol 136:2165–2172
Caldwell HD, Perry LJ (1982) Neutralization of Chlamydia trachomatis infectivity with antibodies to the major outer membrane protein. Infect Immun 38:745–754
Cameron DW, Simonsen JN, D'Coasta LJD, Ronald AR, Maitha GM, Gakinya MN, Cheang M, Brunham RC, Plummer FA (1989) Genital ulcer disese and lack of circumsion are cofactors in female to male sexual transmission of human immunideficiancy virus. Lancet ii:403–407
Carbonetti NH, Simnad VI, Seifert HS, So M, Sparling PF (1988) Genetics of protein I of Neisseria gonorrhoeae: construction of hybrid porins. Proc Natl Acad Sci USA 85:6841–6845
Carter JJ, Koutsky LA, Wipf GC, Christensen ND, Lee S-K, Kuypers J, Kiviat N, Galloway DA (1996) The natural history of human papillomavirus type 16 capsid antibodies among a cohort of university women. J Infect Dis 174:927–936
Champion CI, Miller JN, Borenstein LA, Lovett MA, Blanco DR (1990) Immunization with Treponema pallidum endoflagella alters the course of experimental rabbit syphilis. Infect Immun 58:3158–3161
Christensson B, Bottiger M, Svensson A, Jeansson SA (1992) A 15 year surveillance study of antibodies to herpes simplex types 1 and 2 in a cohort of young girls. J Infect Dis 25:147–154
Christensen ND, Reed CA, Cladel NM, Han R, Kreider JW (1996) Immunization with viruslike particles induces longterm protection of rabbits against challenge with cottontail rabbit papillomavirus infection. Virology 70:960–965
Cohen MS, Sparling PF (1992) Bacterial adaptation and mucosal defenses. J Clin Invest 89:1699–1705
Cohen MS, Cannon JG, Jerse AH, Charniga L, Isbey S, Whicker L (1994) Human experimentation with Neisseria gonorroeae: ratiolnale, methods, and implications for the biology of infection and vaccine development. I Infect Dis 169:532–537
Corey L, Adams HG, Brown ZA, Holmes KK (1983) Genital herpes simplex virus infections: clinical manifestations, course and complications. Ann Intern Med 98:958–972
Cornelissen CN, Biswas GD, Tsai J, Paruchuri D, Thompson SA, Sparling PF (1992) Gonococcal transferrin-binding protein 1 is required for transferrin utilitization and is homologous to TonB-dependent outer membrane receptors. J Bacteriol 174:5788–5797
Danilition SL, Maclean IW, Peeling R, Winston S, Brunham RC (1990) The 75 kDa protein of Chlamydia trachomatis: a member of the heat shock protein 70 family? Infect Immun 58:189–196
Desjardin M, Filion LG, Robertson S, Cameron DW (1995) Inducibel immunity with a pilus preparation booster vaccination in an animal model of Hemophilus ducreyi infection and disease. Infect Immun 63:2012–2020
Ducrey A (1889) Experimentelle Untersuchungen über den Ansteckensstoff des weichen Schankers und über die Bubonen. Monatsh Pract Dermatol 9:387–405

Eftahia MS, Amshel AL, Shonberg IL, Batshon B (1982) Giant and recurrent condylomata acuminata: apraisal of immunotherapy. Dis Colon Rectum 25:136–138
Elkins C, Carbonetti NH, Valera VA, Stirewalt D, Klapper DG, Sparling PF (1992) Antibodies to N-terminal peptides of gonococcal porin are bactericidal when lipopolysaccharide is not sialylated. Mol Microbiol 6:2617–2628
Elkins C, Carbonetti NH, Coimber AJ, Thomas CE, Sparling PF (1994) Cloning and consitutive expression of structural genes encoding gonococcal porin protein in Eschericha coli and attenuated Salmonella typhimurium vaccine strains. Gene 138:43–50
Farrell HE, McLean CS, Harley C, Efstathiou S, Inglis S, Minson AC (1994) Vaccine potential of a herpes simplex virus type 1 mutant with an essential glycoprotein deleted. J Virol 68:927–932
Fehniger TE, Radolf JD, Lovett MA (1986) Properties of an ordered ring structure formed by recombinant Treponema pallidum surface antigen 4D. J Bacteriol 165:732–739
Fitch WM, Peterson EM, de la Maza LM (1993) Phylogenetic analysis of the outer-membrane-protein genes of Chlamydiae, and its implication for vaccine development. Mol Biol Evol 10:892–913
Fitzgerald TJ (1991) Up-regulation of immunogenicity by cyclophosphamide, Ribi adjuvant, and an indomethacin confers significant protection against challenge in rabbits. Vaccine 9:266–272
Forsgren M, Malm G (1996) Herpes simplex virus and pregnancy. Scand J Infect Dis [Suppl] 100:14–19
Frazer IH (1996) The role of vaccines in the control of STDs: HPV vaccines. Genitourin Med 72:398–403
Gotschlich EC, Seiff ME, Blake MS, Koomey M (1987) Porin protein of Neisseria gonorrhoeae: cloning and gene structure. Proc Natl Acad Sci USA 84:8135–8139
Graystone JT, Wang S-P (1978) The potential for vaccine against infection of the genital tract with Chlamydia trachomatis. Sex Transm Dis 5:73–77
Greenstone HL, Nieland JD, de Visser KE, De Bruijn ML, Kirnbauer R, Roden RB, Lowy DR, Kast WM, Schiller JT (1998) Chimeric papillomavirus virus-like particles elicit antitumor immunity against E7 oncoprotein in HPB 16 tumor model. PNAS 95:1800–1805
Haas R, Meyer TF (1987) Molecular principles of antigenic variation in Neisseria gonorrhoeae. Antonie Van Leeuwenhoek 53:431–434
Hagblom P, Segal E, Billyard, So M (1985) Intragenic recombination leads to pilus antigenic variation in Neisseria gognorrhoeae. Nature 315:156–158
Hansen EJ, Lumbley SR, Richardson JA, Purcell BK, Stevens MK, Cope LD, Datte J, Radolf JD (1994) Induction of protective immunity to Haemephilus ducreyi in the temperature-dependent rabbit model of experimental chancroid. J Immunol 152:184–192
Hariharan MJ, Driver DA, Townsend K, Brumm D, Polo JM, Belli BA, Catton DJ, Hsu D, Mittelstaedt D, McCormack JE, Karavodin L, Dubensky TW, Chang SM, Banks TA (1998) DNA vaccination against herpes simplex virus: enhanced efficiency using a Sindbis virus-based vector. J Virol 72:950–958
Hazama M, Mayumi-Aono A, Miyazaki T, Hinuma S, Fujisawa Y (1993) Intranasal immunization against herpes simplex infection by using a recombinant glycoprotein D fused with immunomodulating proteins, the B subunit of Eschericha coli heat labile enterotoxin and interleukin-2. Immunology 78:643–649
Holmes KK, Mårdh P-A, Sparling PF, Wiesner PJ, Cates W, Lemon AM, Stamm WE (eds) (1990) Sexually transmitted diseases. McGraw-Hills, New York
Hook EW, Handsfield HH, Gonococcal infections in the adult (1990) Gonoccocal infections in the adult. In: Holmes KK, Mårdh P-A, Sparling PF, Wiesner PJ, Cates W, Lemon AM, Stamm WE (eds) Sexually transmitted diseases. McGraw-Hill, New York, pp 149–165

Jarrett WFH, O'Neil BW, Gaukroger JM, Laird HM, Smith KT, Campo MS (1990) Studies on vaccination against papillomaviruses: a comparision of purified virus, tumour extract and transformed cells in prophylactic vaccination. Vet Rec 126:449–452

Kellogg DS, Peacock WL, Deacon WE, Brown L, Prickle CI (1963) Neisseria gonorrhoeae. I. Virulence genetically linked to clonal variation. J Bacteriol 85:1274–1279

Kirnbauer R, Booy F, Cheng N, Lowy DR, Schiller JT (1992) Papillomavirus L1 major capsid protein self-assembles into virus-like particles that are highly immunogenic. Proc Natl Acad Sci USA 89:12180–12184

Knapp JS, Tam MR, Nowinsky RC, Holmes KK, Sandström EG (1984) Serological classification of Neisseria gonorrhoeae with use of the monoclonal antibodies to outer membrane protein I. J Infect Dis 150:44–48

Koomey M, Gotschlich EC, Robbins K, Bergström S, Swansson J (1987) Effect of recA mutations on pilus antigenic variation and phase transitions in Neisseria gonorrhoeae. Genetics 117:391–398

Kreisel JD, Spruance SL, Daynes RA, Aracano BA (1996) Nucleic acid vaccine encoding gD2 protects mice from herpes simplex virus type 2 disease. J Infect Dis 173:536–541

Krul MR, Tijhaar EJ, Kleijne JA, Van Loon AM, Nievers MG, Schipper H, Geerse L, Van der Kolk M, Steerenberg PA, Mooi FR, Den Otter (1996) Induction of an antibody response in mice against human papillomavirus (HPV) type 16 after immunization with HPV recombinant Salmonella strains. Cancer Immunol Immunotherapy 43:44–48

Kuklin N, Daheshia M, Karem K, Manickan E, Rouse BT (1997) Induction of mucosal immunity against herpes simplex virus by plasmid immunization. J Virol 71:3138–3145

Lagergård T, Purven M (1993) Neutralizing antibodies to Haemophilus ducreyi cytotoxin. Infect Immun 61:1589–1592

Langenberg AGM, Burke RL, Adair SF, Sekulovich R, Tigges M, Dekker CL, Corey L (1995) A recombinant glycoprotein vaccine for herpes simplex type 2: saftey and efficacy. Ann Intern Med 122:889–898

Lomholt H, Lind I, Kilian (1995) Neisseria gonorrhoeae IgA1 proteases share epitopes recognized by neutralizing antibodies. Proc Natl Acad Sci USA 89:2120–2124

Lukehart SA, Baker-Zander SA, Sell S (1980) Characterization of lymphocyte responses in early experimantal syphilis: in vitro response to mitogens and Treponema pallidum antigens. J Immunol 124:454–460

Lycke N, Bromander AK, Ekman L, Karlsson U, Holmgren J (1989) Cellular basis of immunomodulation by cholera toxin in vitro with possible association to the adjuvnat function in vivo. J Immunol 142:20–27

Lynch EC, Blake MS, Gotschlich EC, Mauro A (1984) Spontaneously transferred from whole cells and reconstituted from purified proteins of Neisseria gonorrhoeae and Neisseria meningitidis. Biophys J 45:104–107

Magnuson HJ, Thomas EW, Olansky S, Kaplan BI, de Mello L, Cutler JC (1956) Inoculation syphilis in human volonteers. Medicine 35:33–82

Makino S, van Putten JPM, Meyer TF (1991) Phase variation of the opacity membrane protein controls invasion by Neisseia gonorrhoeae into human epithelial cells. EMBO J 10:1307–1315

Mandrell RE, Griffis JM, Macher BA (1988) Lipooligosaccharides (LOS) of Neisseria gonorrhoeae and Neisseria meningitides have components that are immunochemically similar to precursors of human blood group antigens. Carbohydrate sequence specificity of mouse monoclonal antibodies that recognize crossreacting antigens on LOS and human erytrocytes. J Exp Med 168:107–126

Manickan E, Rouse RJD, Yu Z, Wire WS, Rouse BT (1995) Genetic immunization against herpes simplex virus. Protection is mediated by CD4+ lymphocytes. J Immunol 155:259–265

McLean CS, Erturk M, Jennings R, Ni Challanain D, Minson AC, Duncan I, Boursnell MEG, Inglis SC (1994) Protective vaccination against primary and recurrent disease caused by herpes simplex virus (HSV) type 2 using a genetically disabled HSV-1. J Infect Dis 170:1100–1109
Meignier B, Longnecker R, Roizman B (1988) In vivo behaviour of genetically engineered herpes simplex viruses R7017 and R7020. Construction and evaluation in rodents. J Infect Dis 158:602–614
Mertz GJ, Benedetti J, Ashley R, Selke SA, Corey L (1992) Risk factors for the sexual transmission of genital herpes. Ann Intern Med 116:197–202
Miller J (1973) Succesful vaccination of rabbits with Treponema pallidum, Nichols strain, attenuated by gamma-radiation. J Immunol 110:1206–1215
Morrison LA, Knipe DM (1994) Immunization with replication-defective mutants of herpes simplex type 1: sites of immune intervention in pathogenenesis of challenge virus infection. J Virol 68:689–696
Morrison RP, Belland RJ, Lyng K, Caldwell HD (1989) Chlamydial disease pathogenesis: the 57-kDa chlamydial hypersensitivity antigen is a stress response protein. J Exp Med 170:1271–1283
Morrison RP, Manning DS, Caldwell HD (1992) Immunology of Chlamydia trachomatis infections: immunoprotective and immunopathogenetic resonses. In: Quinn TC (ed) Sexually transmitted diseases. Raven, New York, pp 57–84 (Advances in host defence mechansisms, vol 8)
Morse SA (1989) Chancroid and Hamophilus ducreyi. Clin Microbiol Rev 2:137–157
Murdin AD, Su H, Klein MH, Caldwell HD (1995) Poliovirus hybrids expressing neutralizing epitopes from variable domains I and IV of the outer membrane protein from Chlamydia trachomatis elicit broadly cross-reactive C. trachomatis-neutralizing antibodies. Infect Immun 63:1116–1121
Naerdelli-Haefliger D, Roden RB, Benyacoub J, Sahli R, Kraehenbuhl JP, Shiller JT, Lachat P, Potts A, De Grandi P (1997) Human papillomavirus type 16 virus like particles expresed in Salmonella typhfimurium elicit mucosal and systemi neutralizing immunity. Infect Immun 65:3328–3336
Norris SJ, Treponema pallidum Polypeptide research Group (1993) Polypeptides of Treponema pallidum: progress toward understanding their structural, functional and immunologic roles. Microbiol Rev 57:750–779
Pal S, Fielder TJ, Peterson EM, de la Maza LM (1994) Protection against infertility in a BALB/c mouse salpingitis model by intranaasal immunization with the mouse pneumonitis biovar of Chlamydia trachomatis. Infect Immun 62:3354–3362
Palefsky J (1995) Human papillomavirus-associated malignancies in HIV-positive men and women. Curr Opin Oncol 7:437–441
Parkin DM (1994) Cancer in developing countries. Cancer Surv 19–20:519–516
Patton DL, Cosgrove Sweeney YT, Kuo C-C (1994) Demonstration of delayed hypersensitivity in Chlamydia trachomatis salpingitis in monkeys: a pathogenic mechanism of tubal dammage. J Infect Dis 169:680–683
Petry KU, Scheffel D, Bode U, Gabrysiak T, Kochel H, Kupsch E, Glaubitz M, Niesert S, Kuhnle H, Schedel I (1994) Cellular immunodeficieny enhances the progression of human papillomavirus-associated cervical lesions. Int J Cancer 57:836–840
Plourde PJ, Plummer PF, Pepin J, Agoki E, Moss G, Ombette J, Ronald AR, Cheang M, D'Costa L, Ndinya-Achola JO (1992) Human immunodeficiency virus type 1-infection in women attending a sexually transmitted disease clinic in Kenya. J Infect Dis 166:89–92
Plummer FA, Simonsen JN, Chubb H, Slaney L, Kimata J, Bosire M, Ndinya-Achola JO, Ngugi EN (1989) Epidemiologic evidence for the development of serovar-specific immunity after gonococcal infection. J Clin Invest 83:1472–1476
Plummer FA, Chubb H, Simonsen JN, Bosire M, Slaney L, Maclean I, Ndinya-Achola JO, Waiyaki P, Brunham RC (1993) Antibody to Rmp (outer membrane protein 3) increases susceptibility to gonococcal infection. J Clin Invest 91:339–343
Plummer F, Chubb H, Simonsen J, Bosire M, Slaney L, Nagelkerke NJ, Maclean, Ndinya-Achola JO, Waiyaki P, Brunham RC (1994) Antibodies to opacity

proteins (Opa) correlate with reduced risk for gonococcal salpingitis. J Clin invest 93:1748–1755
Qu Z, Cheng X, de la Maza LM, Petersen EM (1993) Characterization of a neutralizing monoclonal antibody directed at variable domain I of the major outer membrane protein of Chlamydia trachomatis C-complex serovars. Infect Immmun 61:1365–1370
Reeves WC, Corey L, Adams HG, Vontver LA, Holmes KK (1981) Risk of recurrence after first episodes of genital herpes: relation to HSV type and antibody response. N Engl J Med 305:315–319
Rice PA, Vayo HE, Tam MR, Blake MS (1986) Immunaglobulin G antibodies directed against protein III block killing of serum-resistant Neisseria gonorrhoeae by immune serum. J Exp Med 164:1735–1748
Robinson EJ, Clemens CM, Schoolnik GK, McGee ZA (1989) Probing the surface of Neisseria gonorrhoeae: immunoelectron microscopic studies to localize cyanogen bromide fragment 2 in gonococcal pili. Mol Microbiol 3:57–64
Roizman B, Sears A (1993) Herpes simplex viruses and their replication. In: Roizman B, Whitley RJ, Lopes C (eds) The human herpesviruses. Raven, New York, pp 11–68
Ronald AR, Plummer FA (1985) Chancroid and Haemophilus dycreyi. Ann Intern Med 102:705–708
Rose RC, Bonnez W, Reichman RC, Garcea RL (1993) Expression of human papillomavirus type 11 L1 protein in insect cells: in vivo and in vitro assembly of viruslike particles J Virol 67:1936–1944
Ross JDC, Smith IW, Elton RA (1993) The epidemiology of herpes simplex types 1 and 2 infection of the genital tract in Ediburgh 1978–1991. Genitourin Med 69:381–383
Sandström E, Danielsson D (1980) Serology of Neisseria gonorrhoeae. Classification with co-agglutination. Acta Pathol Microbiol Scand [B] 88:27–38
Sandström E, Chen KCS, Buchanan TM (1982) Serology of Neisseria gonorrrhoeae. Co-agglutination serogroups WI and WII/III correspond to different outer membrane Protein I molecules. Infect Immun 38:462–470
Sandström E, Venizelos N, Palmblad J (1983) Chemostatic activity of Nesseria gonorrhoeae. Br J Vener Dis 59:92–93
Scouler A, Leask BGS, Carrington D (1990) Changing trends in genital herpes due to herpes simplex virus type 1 in Glasgow. Genitourin Med 66:226–228
Segal E, Hagblom P, Seifert HS, So M (1986) Antigenic variation of gonococcal pilus involves assembly of separated silent gene segments. Proc Natl Acad Sci USA 83:2177–2181
Sell S, Norris (1983) The biology, pathology and immunology of syphilis. Int Rev Exp Pathol 24:203–276
Sköldenberg B (1996) Herpes simplex encephalits. Scand J Infect Dis [Suppl] 100:8–13
Sparling PF (1990) Biology of Neisseria gonorrhoeae. In: Holmes KK, Mårdh P-A, Sparling PF, Wiesner PJ, Cates W, Lemon AM, Stamm WE (eds) Sexually transmitted diseases. McGraw Hills, New York, pp 131–147
Sparling PF, Elkins C, Wyrick PB, Cohen MS (1994) Vaccines for bacterial sexually transmitted infections: A realistic goal? Proc Natl Acad Sci USA 61:2456–2463
Spinola S, Wild L, Apicella M, Gaspari, AA, Campagnari AA (1994) Experimental human infection with Hemophilus ducreyi. J Infect Dis 169:1146–1150
Stern A, Brown M, Nickel P, Meyer TF (1986) Opacity genes in Neisseria gonorrhoeae: control of phase and antigenic variation. Cell 47:61–71
Stevenson P, Williams P, Griffiths E (1992) Common antigenic domains in transferrin-binding protein 2 of Neisseria meningitides, Neisseria gonorrhoea and Haemophilus influenzae type b. Infect Immun 60:2391–3296
Strauss, SE, Savarese B, Tigges M, Freifeld AG, Krause PR, Margolis DM, Meier JL, Paar DP, Adair SF, Dina D, Dekker C, Burke RL (1993) Induction and enhancement of the immune responses to herpes simplex type 2 in humans by use of a recombinant glycoprotein D vaccine. J Infect Dis 167:1045–1052

Strauss SE, Corey l, Burke RL, Savarese B, Barnum G, Krause PR, Kost RG, Meier JL, Sekulovich R, Adair SF, Dekker CL (1994) Placebo-controlled trial of vaccination with recombinant glycopotein D of herpes simplex virus type 2 for immunotherapy of genital herpes. Lancet 343:1460–1463

Strauss SE, Wald A, Kost RG, McKenzie R, Langenberg AG, Hohman P, Lekstrom J, Cox E, Nakamura M, Selukovich R, Izu A, Dekker C, Corey L (1997) Immunotherapy of recurrent genital herpes with recombinant herpes simplex type 2 glycoproteins D and B: results of a placebo controlled trial. J Infect Dis 176:1129–1134

Su H, Caldwell HD (1993) Immunogenicity of a synthetic oligopeptide corresponding to antigenically common T-helper and B-cell neutrlizing epitopes of the outer membrane protein of Chlamydia trachomatis. Vaccine 11:1159–1166

Su H, Caldwell HD (1995) CD4+ T cell play a significant role in adoptive immunity to Chlamydia trachomatis infection of the mouse genital tract. Infect Immun 63:3302–3308

Su H, Zhang Y-X, Barrera O, Watkins NG, Caldwell HD (1988) Differential effect of trypsin on infectivity of Chlamydia trachomatis: loss of infectivity requires cleavage of major outer membrane protein variable domains II an IV. Infect Immun 56:2094–2100

Su H, Watkins NG, Zhang Y-X, Caldwell (1990) Chlamydia trachomatis-host cell interactions: role of the chalmydial major outer memebrane protein as an adhesin. Infec Immun 58:1017–1025

Su H, Parnell M, Caldwell HD (1995) Protective efficacy of a parenterally administred MOMP-derived vacine in a murine model, of Chlamydia trachomatis genital tract infection: serum neutralizing IgG antibodies do not protect against chlamydial genital tract infection. Vaccine 13:1023–1032

Swanson J, Barrera O, Sola J, Boslego J (1988) Expression of outer membrane protein II by gonococci in experimental gonorrhea. J Exp Med 168:2121–2129

Tam M, Buchanan T, Sandström E, Holmes KK, Knapp JS, Siadak AW, Nowinski R (1992) Serological classification of Neisseria gonorrhoeae with monoclonal antibodies. Infect Immun 36:1042–1053

Taylor HR, Maclean IW, Brunham RC, Pal S, Whittum-Hudson J (1990) Chlamydial heat shock proteins. Infect Immun 58:3061–3063

Toth M, Jeremias J, Ledger WJ, Witkin SS (1992) In vivo tumor necrosis factor production in women with salpingitis. Surg Gynecol Obstet 174:359–362

Toye B, Laferriee C, Claman P, Jessamine P, Peeling R (1993) Association between antibody to the chlamydial heat shock protein and tubal infertility. J Infect Dis 168:1236–1240

Tuffrey M, Alexander F, Conlan W, Woods C, Ward M (1992) Heterotypic protection of mice against chlamydial salpingitis and colonization of the lower genital tract with a human serovar F isolate of Chlamydia trachomatis by prior immunization with recombinant serovar L1 major outer membrane protein. J Gen Microbiol 138:1707–1715

Turner TB, Hollander DH (1957) Biology of the treponematoses. WHO, Geneva

Tyndall M, Malisa M, Plummer FA, Ombetti J, Ndinya-Achola JO, Ronald AR (1993) Ceftriaxone no longer predicably cures chancroid in Kenya. J Infect Dis 167:469–471

van der Ley P, Heckels JEnt killing by immune sera. Infect Immun 60:39–43

Wharton LR (1937) The criteria of cure of gonococal infections in women. Am J Syph Gonorrhea Vener Dis 21:593–608

Whitley RJ, Gnann JW (1993) The epidemiology and clinical manifestations of herpes simplex infections. In: Roizman B, Whitley RJ, Lopes C (eds) The human herpesviruses. Raven, New York, pp 69–105

Whitley RJ, Kern ER, Cha-628

Virji M, Zak K, Heckels JE (1986) Monoclonal antibodies to gonococcal outer membrane protein IB: use in investigation of the potential protective effect of

antibodies directed against conserved type-specific epitopes. J Gen Microbiol 132:1621–1629
Virji M, Fletcher JN, Zak K, Heckels JE (1987a) The potential protective effect of monoclonal antibodies to outer membrane IA. J Gen Microbiol 133:2639–2646
Virji M, Zak K, Heckels JE (1987b) Outer membrane protein III of Neisseria gonorrhoeae: variations in biological properiteis of antibodies directed against different epitopes. J Gen Microbiol 133:3393–3401
Wagar EA, Schachter J, Bavoil P, Stephens RS (1990) Differential human serologic response to two 60.000 molecular weight Chlamydia trachomatis antigens. J Infect Dis 162:992–927
Wald A, Zeh J, Selke S, Ashley RL, Corey L (1995) Virological chracteristics of subclinical and symptomatic genital herpes infections. N Engl J Med 333:770–775
Walker EM, Arnett JK, Heath JD, Norris SJ (1991) Treponema pallidum subsp. Pallidum has a singel, circular chromosome with a size ~900 kilobasepairs. Infect Immun 59:2476–2479
Weström L, Mårdh PA (1993) Chlamydial salpingitis. Br J Med Bull 39:145–150
Wetzler LM, Gotschlich, Blake MS, Koomey JM (1989) The construction and characterization of Neisseria gonorrhoeae lacking protein III in its outer membrane. J Exp Med 169:2199–2209
Wetzler LM, Blake MS, Barry K, Gotschlich EC (1992a) Gonococcal porin vaccine evaluation: Comparison of Por proteosomes, liposomes, and blebs isolated from Rmp deletion mutants. J Infect Dis 166:551–555
Wetzler LM, Barry K, Blake MS, Gotschlich EC (1992b) Gonococcal lipooligosaccharide sialylation prevents complement-dependent killing by immune sera. Infect Immun 60:39–43
Wharton LR (1937) The criteria of cure of gonococal infections in women. Am J Syph Gonorrhea Vener Dis 21:593–608
Whitley RJ, Gnann JW (1993) The epidemiology and clinical manifestations of herpes simplex infections. In: Roizman B, Whitley RJ, Lopes C (eds) The human herpesviruses. Raven, New York, pp 69–105
Whitley RJ, Kern ER, Chatterjee S, Chou J, Roizman B (1993) Replication, establishment of latency and induced reactivation of herpes simplex 34.5 deletion mutants in rodent models. J Clin Invest 91:2837–2843
Wicher V, Zabeck J, Wicher K (1991) Pathogen-specific humoral response in treponema pallidum-infected humans, rabbits, and guinea pigs. J Infect Dis 163:830–836
Zhang D, Yang X, Berry J, Shen C, McClarty G, Brunham RC (1997) DNA vaccination with the major outer-membrane protein gene induces acquired immunity to *Chlamydia trachomatis* (mouse pneumonitis) infection. J Infect Dis 176:1035–1040
Zhang Y-X, Stewart S, Joseph T, Taylor HR, Caldwell HD (1987) Protective monoclonal antibodies recognize epitopes located on the major outer membrane protein of Chlamydia trachomatis. J Immunol 138:575–581
Zhang Y-X, Stewart SJ, Caldwell HD (1989) Protective monoclonal antibodies to Chlamydia trachomatis serovar and serogroup-specific major outer membrane protein determinants. Infect Immun 57:636–638
Zhong G, Brunham RC (1992) Antibody responses to the chlamydial heat shock proteins hsp 60 and hsp 70 are H-2 linked. Infect Immun 60:3143–3149
Zhong G, Berry J, Brunham RC (1994) Antibody recognition of a neutralizing epitope on the major outer membrane protein of Chlamydia trachomatis. Infect Immun 62:1576–1583
Zhou J, Sun XY, Stenzel DJ, Frazer IH (1991) Expression of vaccinia recombinant HPV 16 L1 and L2 ORF proteins in epithelial cells is sufficient for assembly of HPV virion like particles. Virology 185:251–257

CHAPTER 14

Designing a Vaccine Against HIV

A.M. SCHULTZ

A. Perspective

I. The Need for a Vaccine

The viral cause of AIDS was not established until 1983 (BARRE-SINNOUSSI et al. 1983; GALLO et al. 1983; LEVY et al. 1984), although HIV could have existed in the human population for 50–200 years previously (LI et al. 1988; SMITH et al. 1988; YOKOYAMA et al. 1988) and may have found its way to Europe and North America as early the 1960s (CORBITT et al. 1990; GARRY et al. 1988; HUMINER et al. 1987). The epidemic continues to sweep across the world, making its most recent inroads in Asia (MANN and TARANTOLA 1996). Vaccines to prevent other viral diseases have been highly successful (PLOTKIN and MORTIMER 1994), and enthusiasm for obtaining one to contain AIDS, once HIV was identified as its cause, was high (RUSSELL 1984). Others pointed out apparently daunting obstacles (ADA 1988; COHN and STEINMAN 1988; FERDINAND et al. 1987; LEVY 1988; NATHANSON and GONZALEZ-SCARANO 1989), and the intervening years have seen swings of optimism and pessimism about a timetable for eventual success.

Recently HIV replication in some infected patients has been drastically suppressed through the use of complex antiretroviral drug therapy regimens (HAMMER et al. 1997; PALELLA et al. 1998; GULICK et al. 1997), raising hopes that HIV, until now an implacable opponent, could indeed be contained by therapeutic interventions. It is encouraging that this advance has finally come after a decade of painstakingly slow and step-by-step laboratory and clinical research. However, the greatest impact of the HIV epidemic is in the developing world, where the high cost and delivery complications of these drug regimens will prevent them from becoming available for widespread use in the foreseeable future. Attention is now renewed on the development of vaccines, which have always been highly cost-effective public health barriers against infectious disease.

II. The First Decade: 1985–1995

This 10-year-period of research in humans and primates underscored the enormous challenges of designing a successful HIV vaccine. It was recognized early that the immune system mounts easily measurable and durable immune responses, both humoral and cellular (Dalgleish et al. 1988; Ho et al. 1987; Lyerly et al. 1987; Robey et al. 1986; Walker et al. 1987), against HIV infection. Yet these responses appear to resolve only the acute phase of infection and are incapable of clearing or totally containing HIV infection. Thus, a gradual and inexorable progression to immunodeficiency and death occurred among virtually all HIV-seropositive individuals, with a median time of survival of about 10 years (Alcabes et al. 1993; Munoz et al. 1997). The extreme rarity of cases of long-term nonprogression who remain disease free (Buchbinder et al. 1994; Gange et al. 1997; Munoz et al. 1995; Sheppard et al. 1993; Schrager et al. 1994) only reinforces this observation.

This inexorable progression has had profound implications for vaccine development. Live-attenuated or whole-killed vaccines mimic natural exposure while avoiding disease, in the expectation that immunological memory and life-long immunity will be induced, just as in youngsters who recover from the usual childhood infections. However, if spontaneous recovery from HIV infection is extremely rare, it is possible that the natural immune response simply will not be effective against it. Furthermore, if HIV infection inevitably results in immunodeficiency and death, attenuated or killed virus vaccine empirical approaches must maintain zero tolerance for any infection that could be caused by an improperly prepared vaccine. The tolerance for rare cases of vaccine-induced disease, which until recently was deemed acceptable for live-attenuated polio (Centers for Disease Control 1986), is unlikely to be acceptable for HIV. Finally, the extreme difficulty in finding a population of patients successfully recovering from or coexisting with HIV infection hampers the rational approach to vaccine development as well. Without an opportunity to compare those who become infected and then recover to those who succumb to the disease, it will be a challenge to determine truly protective immune responses and the specific antigens against which they might be directed.

The first decade of HIV vaccine development side-stepped this dilemma by designing vaccines to prevent AIDS by completely preventing infection. For this paradigm, neutralizing antibody is the only relevant response, and therefore viral envelope is the only relevant antigen. Early attempts to induce neutralizing antibodies were made with synthetic peptides (Chanh et al. 1986), recombinant vaccinia (Zarling et al. 1986), and subunit proteins (Nara et al. 1988). The most effective antibodies raised in small animals and primates against the IIIB, MN, and SF2 strains of HIV-1 were induced by envelope subunit vaccines, which were subsequently developed and then evaluated in phase 1 human trials. Products based on yeast expression of an *env* fragment (Keefer et al. 1996; Wintsch et al. 1991), and mammalian-cell expressed gp160 (Belshe et al. 1993; Dolin et al. 1991; Graham et al. 1992; Kovacs et al. 1993; Salmon-Ceron et al. 1995) or gp120 (Belshe et al. 1994; Graham et al. 1996; Kahn et al. 1994) have been tested. Human volunteers indeed produced neutralizing antibodies, but they were narrow in specificity and short-lived (Graham 1994; Schwartz et al. 1993).

In the course of these studies it also became clear that HIV strains such as IIIB, MN, and SF2, on which vaccine candidates were based, were substantially different from wild-type, or "primary" HIV. Isolated after long passage in vitro, these laboratory viruses have a growth advantage in T-cell lines, exhibit prominent and immunodominant V3 regions (MOORE and HO 1995; ZHANG et al. 1997), and turn out to be rather susceptible to neutralization (SATTENTAU and MOORE 1995). In contrast, wild-type "primary" HIV strains that have had limited in vitro passage, and passage only in peripheral blood mononuclear cells (PBMC), do not have immunodominant V3 regions, and are resistant to neutralization (MOORE et al. 1995). The antibodies raised by the subunit IIIB, MN, or SF2 vaccines do not neutralize primary HIV well at all (DOLIN 1995; MASCOLA et al. 1996b; VAN COTT et al. 1995a). Therefore enthusiasm for subunit vaccines based on these strains is not widespread, and their protective potential is only now being given a test in a human efficacy trial (ALTMAN 1998; COHEN 1993).

III. New Paradigms

The first decade of vaccine development was fixed on completely preventing any trace of infection. Entering its second decade, AIDS vaccine design is accepting the more usual paradigm for vaccines, namely that an exposed person should be able to *recover* from infection if suitably immunized. Several independent lines of research have allowed this philosophical change to occur. First, a significant observation on HIV turnover in vivo has rehabilitated our thinking about the capacity of the immune system against this virus. From elegant kinetic analysis in clinical drug studies (BONHOEFFER et al. 1997; HO et al. 1995; PERELSON et al. 1997; WEI et al. 1995) it is now clear that billions of HIV virions are cleared from the body on a daily basis. Instead of being weak and ineffectual, the immune system copes for an average of 10 years (ALCABES et al. 1993; MUNOZ et al. 1997) in the face of this onslaught. Its eventual failure and collapse after such a titanic struggle now suggests that, with proper preparation of the immune system by vaccination, the balance might be tipped against HIV. Therefore, instead of focusing vaccine-induced immune responses only on antibody, vaccine design should utilize the full range of immune functions, with the goal of minimizing and then reducing virus burden to zero.

Second, primate experiments have demonstrated that, even against the most aggressive and pathogenic viruses, becoming infected after challenge is not necessarily a death sentence. In the simian immunodeficiency virus (SIV) model, depending on the vaccine approach, one can observe different degrees of clinical benefit against pathogenic challenge. Vaccinia containing only *env* and boosted with *env* protein significantly reduce the percentage of infected cells after challenge compared to control macaques (AHMAD et al. 1994); including *gag-pol* in the vector and boosting with inactivated SIV can leave some macaques virus isolation negative and remaining disease free for years, even though persistently polymerase chain reaction (PCR) positive for the challenge virus (HIRSCH et al. 1996). Recovery from initial infection now seems experimentally possible.

Third, it is possible that standard effector antiviral mechanisms [neutralizing antibody, cytotoxic T-cells (CTL)] are not the most effective ones. Impressive epidemiological data have been gradually accumulating that some individuals remain uninfected with HIV, as judged by serological testing, despite many years of high-risk behavior in an environment inundated with HIV (CLERICI et al. 1992; FOWKE et al. 1996). Whether such individuals truly have immune responses that are *protective* and thus hold the answer to the design of an effective HIV vaccine is not established. Nonetheless, HIV-specific cellular immune responses are being reported in some of these individuals (BERETTA et al. 1997; MAZZOLI et al. 1997; ROWLAND-JONES et al. 1995). Thus, it is conceivable that the individuals who should be studied for clues to immune protection against AIDS are those who have *not* seroconverted after exposure to HIV. If this hypothesis is true, protection from HIV might be mediated by immune responses that are hard to measure, and vaccine design should be modified accordingly. Clearly, drawing definitive conclusions about correlates of immune protection from studying such exposed but uninfected populations is controversial. Intriguingly, however, virus transmission studies in primates suggest that the vaginal route of exposure can lead to self-limiting infection even in naive, unvaccinated animals. Both vaginally in macaques (MILLER et al. 1994) and via the cervical *os* in chimpanzees (GIRARD et al. 1993), low-dose virus exposure has led to PCR-positive cells transiently observed in the circulation, without virus persistence or seroconversion. Although some cells have become infected, this infection was not sustained. If low-dose exposure without establishing infection also occurs in humans, it may act as a virtual immunization and create a protective state.

IV. What Should the Vaccine Do?

For those who still believe that the only way to achieve vaccine success is to prevent any cells from becoming infected at all (complete "sterile" immunity), ability to neutralize every virus strain will be required. Vaccines able to induce this quality and quantity of antibody appear far from realization, and this goal seems unattainable at present. However, an alternative and completely acceptable goal is to prevent the *establishment* of infection. One way to accomplish this is for cytotoxic cells to search out and destroy infected cells, in concert with neutralizing antibody to minimize reinfections and thus reduce the number of infected cells that are there to be destroyed. A rapid response of this type to initial infection could conceivably eliminate HIV. Even if some HIV persists in an integrated, latent form, neutralizing antibody and CTL will exert immune surveillance and keep the virus in check. Thus, the majority of scientists now focus vaccine design on the optimization of cytotoxic cellular responses in addition to neutralizing antibody, expecting that resolution of acute infection and then achieving virus clearance will be the key to success.

However, can other mechanisms also prevent the establishment of infection? There are subtler ways to interfere with HIV replication than cytotoxicity or neutralization. Some specific molecules that can mediate CD8 cell suppression of HIV in the absence of cytotoxicity (BARKER et al. 1998; BLACKBOURN et al. 1996; KANNAGI et al. 1988; YANG et al. 1997) are beginning to be investigated, as is chemokine blockade (COCCHI et al. 1995; ORAVECZ et al. 1996; PAL et al. 1997). Whether such responses could help tip the balance against HIV, especially during mucosal transmission which seems to pose inherent barriers to sustaining a chain of infectious events anyway, is an intriguing possibility. Those who believe that exposed but uninfected populations hold the key to truly protective responses against HIV look to cellular responses such as these that are hard to measure or may appear to be present at very low levels. This smaller cadre of scientists view containment of HIV as a problem for cellular responses, and not necessarily cytotoxic ones. An experimental example of such thinking is the protection of macaques against malaria by preexposure preparation of the immune system with interleukin 12, without requiring any malaria antigen (HOFFMAN et al. 1997). The signs of protective immunity existing in advance could be quite subtle for a vaccine operating on these principles.

Finally, a less than ideal but perhaps satisfactory outcome of vaccination, if it cannot prevent the establishment of infection, is to merely reduce the virus burden by several orders of magnitude. A vaccine that can accomplish only this may not prevent AIDS from developing, but it is likely to significantly extend the disease-free period and interfere with transmission *from* the infected vaccinee to others. Such a vaccine might even improve the effectiveness of antiretroviral therapy for that individual.

V. Design Considerations

The success of a vaccine will depend on two factors: adequate immunogenicity and appropriate antigen content. It must effectively induce and maintain protective responses, it must contain those antigens against which those protective responses are directed, and it must be able to provide the immune system with a breadth of recognition and of effect against the very heterogeneous virus that is HIV. Obviously, designing the vaccine is complicated by our lack of knowledge, and opinions differ according to the vaccine designer's expectations about mechanism. There are the traditionalists who favor attenuated-live or whole-killed HIV, because vaccines of this type have worked well against other diseases. Here the vaccines contain most if not all of HIV antigens, and the only design issue is the choice of strain(s) upon which to base the vaccine. The strictest empirical thinking proposes that these vaccines should be developed and tested for efficacy, almost independently of the measurable immune responses that they induce. In contrast, for vectored vaccines or subunit approaches, the choice of antigen and presentation system becomes crucial. It is possible to induce the "right" responses but waste them against the "wrong" antigens, and conversely to have the "right" constellation of antigens but fail to induce the "right" responses at all or not induce them at the effective level or location. Unfortunately, the pathways to determine what is "right" are less than clear, and the data prerequisite to initiating expensive and complicated efficacy trials are not established. These unknowns make for exciting science but contentious debate (WADMAN 1998).

This chapter considers in turn the potential roles of antibody, cytotoxic cells, other T-cell activities, and mucosal immunity. It attempts to give the reader some idea of the issues that occupy the field at present, reviews current data, and describes work in progress that may clarify and resolve them. It concludes with a brief overview of strengths and problems of major vaccine approaches. It does not include therapeutic use of vaccines, nor does it include the status and scope of human trials of currently available HIV vaccine candidates, which have recently been reviewed (CLEMENTS 1997; WALKER and FAST 1994).

B. The Role of Antibody

Effective neutralization of HIV was the first goal of vaccine design, but even now the concept of neutralization remains a thorny problem. Plainly stated, HIV is not especially susceptible to neutralization. Unlike poliomyelitis, where immune sera can reduce the titer of 10^5 infectious virions to zero (LEVINSON et al. 1954), HIV antisera are judged strong if they reduce the titer of 10^1–10^2 infectious virions by a factor of 10 (90% neutralization). HIV neutralization assays have been developed within these limits, and sera from infected as well as immunized individuals have been examined. Even by these standards, strong neutralizing antibody is not commonly found in patients. Infected individuals rarely can neutralize their contemporaneously isolated virus; sera taken months latter can sometimes neutralize the earlier virus, but in the intervening period variants have developed that evade the neutralization capacity of the contemporary serum (WRIN et al. 1994). These data show that HIV can be neutralized, but the breadth of this neutralization is narrow (BURTON and MONTEFIORI 1997).

I. Passive Transfer

It has been clear since the first experiments in chimpanzees (EMINI et al. 1992; PRINCE et al. 1990) that antibody in sufficient amounts, capable of neutralizing the challenge virus, can prevent infection. The antibody must be present at the time of infection, since passive antibody perfusion that starts 1 h after HIV exposure in chimpanzees fails to block infection (PRINCE et al. 1991). This experiment has also been successfully replicated with a primary, non-syncytium-inducing (NSI) virus which is harder to neutralize (CONLEY et al. 1996). Similarly, high levels of passive antibody prevent HIV infection in reconstituted SCID mice (ANDRUS et al. 1998; GAUDUIN et al. 1997). Neutralizing monoclonal antibody administered in advance, in concentrations predicted by in vitro neutralization assays, prevents infection.

Under certain conditions antibody alone can prevent infection even against highly pathogenic SIV. SIV challenge stocks tend to be very neutralization resistant, and in general antibody from long-term surviving macaques does not prevent infection with pathogenic SIV (ALMOND et al. 1997; KENT et al. 1994); two experiments have described success in some animals (CLEMENTS et al. 1995; PUTKONEN et al. 1991; BIBERFELD et al. 1992). However, in a unique case passive transfer from macaques immunized with whole-killed SIV was able to protect against pathogenic challenge (GARDNER et al. 1995). This experiment is complicated by the fact that the protective antigens turn out to be not SIV viral antigens but human cellular xenoantigens found in the vaccine preparation (LANGLOIS et al. 1992) and also present in the lipid envelope of the challenge virus (HENDERSON et al. 1987). Nonetheless the passively transferred antibody was successful; it must have prevented every infectious event (completely "sterile" immunity), since any cells that became infected in the challenged macaques would have produced virus devoid of the xenoantigens and not been susceptible to the passive antibody in a second round of reinfections.

More recently anti-HIV gp120 antibodies have prevented infection with chimeric SHIV virus, containing the envelope of the HIV strain DH12 (SHIBATA et al. 1997). High-titered neutralizing antibody came from chimpanzees infected with DH12 (SHIBATA et al. 1995); purified IgG was passively transferred to pigtailed macaques. Antibody levels achieved after the transfer could neutralize the SHIV/DH12, while IgG from a HIV_{IIIB}-infected chimp could not. The SHIV/DH12 causes CD4 decline and death in naive pigtailed macaques; when challenged, the passively immunized macaques receiving the highest levels of DH12-specific antibody were protected, but those receiving IIIB-induced IgG were not (SHIBATA et al. 1997). There is no question that neutralizing antibody works as well in vivo as it does in vitro; obtaining antibody of sufficient breadth and concentration remains the problem.

II. Does HIV Have Serotypes?

The extensive heterogeneity of the HIV *env* gene has long been recognized as an obstacle to obtaining broadly reactive neutralization. Classification of HIV into clades or subtypes (see KORBER et al. 1994) has been a valuable aid to virological epidemiology and tracing the evolution of HIV around the world. If each subtype represented a serotype of HIV, vaccine design could be centered around a polyvalent vaccine composed of five or so components. This hypothesis has now been fully investigated and disproven; extensive cross-neutralization tests covering subtypes (clades) A-E were performed with panels of sera from infected patients and the primary HIV isolates obtained from them. The results confirmed that most primary isolates are resistant to neutralization with most sera. However, when any person's serum could neutralize a primary strain, that strain was as likely to be from another clade as from the clade that person was infected with; a positive control serum, obtained from a clade B long-term nonprogressor termed FDA#2 (VUJCIC and QUINNAN 1995), was also able to neutralize some isolates from each of the clades (TRKOLA et al. 1995; WEBER et al. 1996). In this study there was a tendency for clade E isolates neither to be neutralized by sera from other clades nor to generate sera that neutralized viruses from other clades. In other work, pooled sera from clade B and from clade E infections used as reagents to neutralize viruses from those clades also indicate serological differentiation of E from B (MASCOLA et al. 1996a). Thus, subgroup E may be sufficiently distinct to constitute a separate serological group from all the other viruses, although it may not qualify as a serotype.

Serotyping is hampered by the lack of high-titer antibodies. It has been theorized that the fluidity of the HIV surface glycoprotein in a lipid-enveloped particle, in contrast to the repetitive surface structures of other viruses that induce good neutralization, may explain the difficulty in producing such reagents (BACHMANN and ZINKERNAGEL 1996). Immunizing animals with HIV antigens has not produced such reagents, and the few human monoclonal antibodies that can neutralize a variety of primary HIV strains do not define serotypes. The human monoclonal antibodies b12 (KESSLER et al. 1997), 2F5 (CONLEY et al. 1994; MUSTER et al. 1994; PURTSCHER et al. 1994), and 2G12 (TRKOLA et al. 1996) exhibit the broadest neutralization of wild-type HIV, but there is overlap of the isolates they neutralize (D'SOUZA et al. 1997); whatever epitopes they recognize do not define separate serotypes.

Attempts are being made to collect polyvalent sera from individuals who have been infected for many years and may have the strongest and broadest neutralizing activity. Using cluster analysis of neutralization data from panels of these sera and a variety of HIV strains, there is hope that serotypes may emerge (NYAMBI et al. 1996), much as transplantation antigens were originally described serologically using sera from multiparous women (PAYNE et al. 1964). However, the importance of serotyping for vaccine development is the expectation that any member of a given serotype will be able to induce neutralization of other members of that serotype. It remains to be seen whether common neutralization epitopes, if defined, can be rendered immunogenic enough to act as serotypes that are useful for vaccine design.

III. Do Coreceptor Families of HIV Represent Serotypes?

Interactions of HIV with chemokine coreceptors is a very complex subject and beyond the scope of this chapter (see BERGER [1997] and MOORE et al. [1997] for reviews), but these recent discoveries have given new insight into the relationship among laboratory HIV isolates (LAI, MN, SF2) on the one hand, and wild-type "primary" HIV isolates on the other (DENG et al. 1996; DRAGIC et al. 1996; JANSSON et al. 1996). The long-appreciated growth differences between HIV strains isolated early in infection and those isolated during clinical AIDS are now known to be reflected in gp120 structures related to coreceptor usage (BERGER et al. 1998). The "slow, low" viruses found early in infection (ASJO et al. 1986; FENYO et al. 1988) do not form syncytia in MT2 cells; their gp120s interact with the CCR5 chemokine receptor, which is expressed in PBMC but not in transformed T-cell lines such as MT2. Such strains are also resistant to neutralization and do not have immunodominant V3 structures (BOU-HABIB et al. 1994; VAN COTT et al. 1995b). In contrast, viruses isolated during clinical AIDS were described as "rapid, high" for replication and do form syncytia in MT2 cells. It is now known that these HIV strains can also use the chemokine coreceptor CXCR4, which is expressed on PBMC *and* on transformed T-cell lines as well. Their gp120 structures seem to have changed as a consequence of acquiring CXCR4-binding. Laboratory adaptation changes these HIV strains even further. The CXCR4-using isolates that have a selective growth advantage in culture evolve further, losing interaction with CCR5 and becoming dependent exclusively on CXCR4 (MOORE and HO 1995). In such laboratory strains the V3 variable domain is exposed (BEDDOWS et al. 1998) and has become an immunodominant neutralizing antibody epitope (MATSUSHITA et al. 1988; SATTENTAU and MOORE 1995).

All immunization studies to date have been carried out with laboratory-strain *env* immunogens; it is tempting to speculate that viruses using CCR5 could fall into a different serotype from those using CXCR4, and this could explain the failure of these vaccines to raise neutralizing antibodies against CCR5-using viruses. Of course, most infections occur with CCR5-using viruses, yet sera from infected persons are not particularly effective against those viruses either. On the other hand, it can be argued that a subtle immunodeficiency sets in immediately after infection, even while bulk CD4 counts are normal, and antibodies raised in response to infection may therefore not be the best that the human immune system can achieve. The problem is very complicated, as neutralization of HIV strains that can use either CCR5 or CXCR4 is independent of which coreceptor they are forced to use in an in vitro assay (LACASSE et al. 1998; MONTEFIORI et al. 1998). With respect to vaccine development this question may soon be resolved, as CCR5-using subunit proteins are now being produced, for human as well as macaque immunogenicity studies.

IV. Design Considerations

1. The Ideal Immunogen

Inducing broad and durable neutralization remains a hurdle for vaccine design. Since all HIV strains infect cells by binding to cell surface CD4, in theory an antibody directed against the CD4-binding face of gp120 would neutralize every HIV strain. This epitope would constitute in effect the ultimate single serotype. However, producing antibodies with sufficient avidity to compete with CD4 is quite a challenge, given the substantial binding constant that exists between gp120 and CD4 (IVEY-HOYLE et al. 1991). The fact that infected humans, HIV-infected chimpanzees (which experience no immunodeficiency), and macaques infected with attenuated SIV all fail to produce broadly neutralizing antibody suggests either that the avidity problem is not easily surmounted, or alternatively that the CD4-binding structure, even in a fully native envelope at the virion surface, is exposed so transiently during attachment that the immune system never has a chance to respond to it. The discovery of HIV coreceptors provides a possible new target for neutralizing antibodies, but transient appearance of their binding epitopes on gp120 may affect ability to generate those antibodies as well. Even if an immunogen could be created that permanently displays the transient binding structure(s) and thus could induce blocking antibodies, the very transience of exposure during binding and entry may render these antibodies ineffective at neutralization, except at extremely high concentrations. This theoretical discussion is moot, however, for information leading to the creation of such an ideal immunogen is totally lacking and must await elucidation of the dynamics of the three-dimensional structure for gp120 (KWONG et al. 1998; HENDRICKS and SODROSKI 1998), or especially of the native glycoprotein spike structure.

2. Obstacles to Inducing a Broad Response

Rare monoclonal antibodies can neutralize a significant fraction of primary HIV (D'SOUZA et al. 1997). Their existence proves that neutralization epitopes are exposed on primary viruses, and that individual IgG molecules can be made with sufficient avidity to prevent attachment and infection. Why have vaccines not produced antibodies with these properties? Possible answers include: (a) avidity of the induced antibodies to the epitopes is too weak; (b) the peptide sequence in the vaccine encodes the neutralization epitopes, but the conformation of the vaccine antigen is different from the one needed to display the epitopes as they exist on the virus; (c) antibodies are not made to all the neutralization epitopes that are present in the vaccine antigen; (d) no single genome contains all the epitopes required to achieve broad neutralization. Optimizing responses to epitopes present in the vaccine is a perennial problem. HIV gp120 produced in mammalian cells is able to bind to and remove antibodies from infected patient sera that can neutralize a wider range of HIV strains than can be neutralized with sera induced in vaccine volunteers by the same protein (STEIMER and HAIGWOOD 1991). This is an experimental demonstration of the classic nonreciprocity between antigenicity (content) and immunogenicity. While gp120 does contain the neutralization determinants, it simply does not induce adequate responses to them. A variety of adjuvants have been tried in humans to improve immunogenicity of envelope antigens. With laboratory-strain subunit envelope vaccines, novel adjuvants have accelerated the appearance of neutralizing antibodies and reduced the amount of antigen required for their induction, but only slightly raised the titer of antibody to laboratory viruses and did not improve the quality of the antibodies against primary HIV. Live-recombinant vectors expressing *env* proteins have been no more successful at inducing antibodies against primary HIV than does HIV infection (GRAHAM et al. 1992), but efforts are being directed to obtaining improved "boosting" with protein (CLEMENTS-MANN et al. 1998; GRAHAM et al. 1993). Encouraging results in mice for improved immunogenicity through direct addition of cytokines (KIM et al. 1998; OKADA et al. 1997) to the vaccine are being followed up in primates and in humans. Improving antigen presentation by targeting antigen to dendritic cells in vivo is also under active investigation in primates (O'DOHERTY et al. 1997).

What about conformation of the envelope component in the vaccine? Attention has been given to improving the form of envelope subunit vaccines. Mammalian cell expression systems have produced better immunogens for neutralizing antibody against laboratory strains than yeast or insect cell expression systems (GRAHAM 1994). However, even with an optimal expression system the gp120 protein as a soluble monomer may not adequately reflect the trimeric complex of gp120 and gp41 that is thought to comprise the glycoprotein spike (LU and KIM 1997). "Oligomeric" gp140 proteins have been proposed to better mimic the native structure and thus induce better antibodies (RICHARDSON et al. 1996). Results in primates are proving that current gp140 constructs do not induce significantly better neutralizing antibodies than do the corresponding gp120 proteins (VAN COTT 1997). Either these constructs do not express "native" structures, or alternatively the hypothesis that "native" conformations automatically induce broad neutralizing antibody is flawed. Live recombinant vectors or DNA, which should allows for synthesis and assembly of native structures in the cell, have not induced antibodies with particularly broad neutralizing properties. If "native" structure does not guarantee an adequate response to all the epitopes in the vaccine, perhaps the vaccine design must improve on nature. The high carbohydrate content of HIV gp120 led to the suggestion that shielding of the polypeptide backbone interferes with developing effective neutralizing antibodies (BENJOUAD et al. 1992). Recent results with SIV suggest that deletion of carbohydrate attachment sites greatly improves the titer of homologous neutralizing antibody (REITTER et al. 1998), although whether it increases the breadth of these sera is not yet known. Another improvement upon nature that has been suggested is to remove immunodominant variable regions so that antibodies are formed to conserved regions. Whether these remaining conserved structures could represent effective neutralization epitopes is under investigation.

Finally, it is possible that a multivalent approach must be taken if a single envelope component simply does not contain all of the neutralization epitopes found in HIV strains. Even though serotypes have not been defined, an obvious first attempt is to expand beyond the laboratory-adapted HIV strains that have dominated vaccine design since the beginning and include a primary HIV strain. Some vaccine manufacturers have already taken this step, in the expectation that the envelope proteins of primary viruses are more likely to have common neutralization epitopes with each other than with laboratory-selected isolates, and initial vaccine trials with bivalent gp120 vaccines are beginning soon in humans. A major drawback is that knowledge to guide the selection of one individual strain over another for inclusion in a multivalent mix is lacking. Some intelligent guesses can be made, but exhaustive analyses of a myriad of combinations may be necessary to optimize this approach.

3. Human Monoclonal Antibodies as Clues

The few human monoclonal antibodies that neutralize primary HIV were selected by screening hybridomas or libraries against monomeric gp120, and failure to find additional new ones is thought to be a shortcoming of the screening tool. Attempts are being made to use plasmavirus as a screen, in the hope that neutralization epitopes displayed on this native structure will allows for detection of antibodies with even better neutralization properties. It will be exciting if this works to find monoclonal antibodies that have new and complementary specificities to those that already exist, for the screening reagents themselves will provide information on new neutralization epitopes. Another approach is to expand upon the source of human antibodies. Preliminary data suggest that group "O" viruses from the Cameroon share broadly cross-reactive neutralizing epitopes with other HIV strains (NYAMBI et al. 1995). An attempt is being made to isolate monoclonal antibodies from this population, as well as from persons infected with other non-clade B viruses.

The remarkable character of these rare monoclonal antibodies is that they are efficient at neutralization when used singly. An encouraging observation found in studying a number of neutralizing human monoclonal antibodies is that substantial synergy occurs in mixtures (LI et al. 1998; MASCOLA et al. 1997; VIJH-WARRIER et al. 1996). That is to say, antibodies that by themselves are not extremely powerful and broad neutralizing agents, when taken even in pairs, demonstrate significantly greater neutralization potential. It is likely that efficient neutralization can be obtained in polyvalent sera without the individual monoclones comprising that sera needing to have the excellent properties of the most efficient monoclonal antibodies.

Effective neutralizing monoclonal antibodies can also be used as reagents to design a vaccine strategy. There is the possibility that mimotopes (BOOTS et al. 1997), or random peptide sequences from phage display libraries, can be found that bind to the combining sites of these efficient neutralizing antibodies. With considerable luck, a mimotope could be found that, as an immunogen, could efficiently induce antibodies with the neutralization properties of the original monoclonal.

4. Is Neutralization Required for Protection?

HIV heterogeneity remains a large obstacle for vaccine design, and the envelope component of a candidate HIV vaccine needs improvement in both content and immunogenicity. Live-attenuated vaccines are effective in macaques without apparently requiring antibody, as shown by passive transfer

(ALMOND et al. 1997; CLEMENTS et al. 1995). Heterologous challenge experiments in macaques are showing definite clinical benefit in the absence of neutralization against the challenge virus (VOGEL et al. 1998). There is also the possibility that neutralization is not the only property of antibody that may be important for protection. Apparently beneficial antibody-dependent cellular cytotoxicity against HIV has been measured in infected individuals (BAUM et al. 1996; CONNICK et al. 1996; TYLER et al. 1990) and induced by candidate vaccines (CLEMENTS-MANN et al. 1998; GORSE et al. 1998). Until strong neutralization responses can be raised by vaccines, other possible protective immune responses of antibody should not be neglected.

C. The Role of Cytotoxic T-Cells

Cytotoxicity assays are considerably more complicated and labor intensive than even neutralizing antibody tests, and only a relatively few laboratories are proficient. In the outbred populations of humans and macaques in which these assays are performed, transformed autologous B-cell lines must be created for each individual in order to have MHC-matched target cells. Distinctions must be made between effector cytotoxic cells that are already capable of lysing appropriate targets, and precursor cells, which require antigen-specific stimulation to increase their frequency and convert them into effector cells. In AIDS research, although infected humans and macaques can exhibit effector cells directly, immunized individuals always require the antigen-specific stimulation to reveal their cytotoxic capability (MCELRATH et al. 1994).

Thus, there is a paucity of CTL data in comparison to the extensive measurements of antibody, and definitive data on the role of CTL are often lacking. Also, the mechanistic meaning of "CTL activity" is imprecise. Ability to lyse *env*-expressing targets does not differentiate between recognizing only one epitope in the protein and several. As discussed below, this distinction may not be trivial for the predictive value of the CTL assay for protection. Quantitative measurements of precursor frequency, and not merely the ability to expand their numbers by stimulation, is becoming more common, and this precision may be a necessary refinement for accurate assessment of CTL. Specialized reagents that quantitate cells which present specific peptides in association with class I MHC molecules (ALTMAN et al. 1996) have the potential to revolutionize this field (MURALI-KRISHNA et al. 1998), but much research must be carried out before its impact becomes clear. Finally, CTL are routinely assayed from the peripheral circulation, while it is quite possible that effector cells must be in the lymphatics and local tissue to be effective for protection.

I. Evidence from Vaccination and Challenge

It is clear in many experiments that induction of CTL activity in primates, *as measured*, is not correlated with protection. Macaques with "CTL activity" are just as likely to be protected as they are to be unprotected (ANDERSSON et al. 1996; DANIEL et al. 1994; S. LU et al. 1996; YASUTOMI et al. 1995). However, this does not prove that cytotoxic activity is irrelevant to protection, for there can be many shortcomings of the assay results as just discussed. In the mouse, where powerful techniques for CTL assays exist, vaccination studies against the Friend murine leukemia virus show that relatively subtle changes in the dynamics of CTL responses to infection are the key to proving the role of CTL in protection (HASENKRUG and CHESEBRO 1997). Nonetheless, though the immunological power of the SIV/macaque model is not yet so sophisticated, there are a few suggestive experiments in the SIV model worth mentioning.

A vaccine based on the *nef* protein as the only antigen, expressed by recombinant vaccinia, was constructed in the hope that *nef*, as one of the first proteins synthesized early in infection, would mark the infected cell for destruction by CTL before any virus progeny had emerged to spread the infection. Seven macaques immunized with this vaccine, as well as four control macaques, were challenged with a lethal dose of SIV (GALLIMORE et al. 1995). All but one vaccinated macaque became infected, but the six that did exhibited a range of SIV burdens postchallenge. When the precursor frequencies of anti-*nef* CTL on the day of challenge were assessed, there was an inverse correlation of frequency with virus burden: the higher the precursor frequency, the lower the burden, and the macaque with the highest anti-*nef* precursor frequency was the one that was apparently completely protected.

In another vaccine study based on the *env*-protein alone, again expressed in vaccinia, macaques were immunized with the recombinant vaccinia and boosted with *env* protein, inducing both neutralizing antibody and anti-*env* CTL activity (KENT et al. 1996). When challenged, the animals were protected in apparently "sterile" fashion, with no challenge virus detected by isolation or by PCR. Nonetheless, one macaque developed anti-*gag* CTL after challenge, suggesting that subclinical infection had occurred and caused a measurable anti-*gag* CTL response to develop. It is tempting to speculate that de novo CTL arising after challenge contributed to protection.

II. Evidence from Infection

Passive transfer is the only way to prove unequivocally that immune cells are essential for protection. There is hope that passive cell transfers, on a limited scale, may become possible in rhesus macaques in the foreseeable future (MENG et al. 1997). Until such experiments can be performed, the best that can be done to elucidate the role of CTL is to make arguments based on only indirect correlations from the timing of CTL induction in infection with the resolution of acute viremia (JASSOY and WALKER 1997; PRICE et al. 1997; ROSENBERG et al. 1997; WALKER and PLATA 1990), and from long-term nonprogressing humans compared to symptomatic patients (GOULDER et al. 1997a; HAAS et al. 1997; LUBAKI et al. 1997; RINALDO et al. 1995; VAN DER BURG et al. 1997).

CTL can be shown to drive virus variation, with documented cases in both infected humans and in challenged macaques. Some infected patients had a strong early CTL response to an immunodominant gp160 epitope that caused the elimination of that epitope from the escape virus that survived (BORROW et al. 1997; GOULDER et al. 1997b; MCMICHAEL and PHILLIPS 1997). Similarly, macaques immunized with a *nef* peptide epitope vaccine responded to only a single epitope (MORTARA et al. 1998). The majority (84%) of viruses in the SIV challenge stock contained that epitope, but the vaccine did not prevent infection. Within 5 five months all of the viruses in the challenged macaques expressed the minor *nef* variant that the challenge stock contained, and the original *nef* epitope was gone. While it is encouraging that the CTL response eliminates the susceptible virus, the negative side of this observation is that the virus can escape CTL as easily as it can from neutralization responses, especially if this response is confined to a single epitope. A similar result occurred after passive transfer of autologous CTL in a human experiment (KOENIG et al. 1995). It appears that CTL responses to multiple epitopes will be essential for vaccine success.

III. Addressing the Heterogeneity of HIV

Whereas sufficient breadth in neutralization remains a problem, there is good news from CTL assays in infected and in immunized humans. Cytotoxic responses can be measured against not only *env* but also *gag*, *pol*, and accessory gene proteins. Both the *gag*- and *pol*-encoded proteins are considerably more conserved than the *env* gene, and there was thus the expectation that CTL, which are not limited to the *env* protein target, as neutralization is, would have a greater chance of demonstrating cross-reactivity. Recent results directly prove that this cross-reactivity can easily extend across clades, not only in infected persons (BETTS et al. 1997; DURALI et al. 1998), but encouragingly, in volunteers immunized with recombinant poxvirus vectors. CTL studies with vaccine volunteers in the United States have been performed with targets expressing HIV from clades A–F (FERRARI et al. 1997). While it is no surprise that viruses containing the same CTL epitope are equally susceptible to the same CTL effector, even if the viruses came from different clades, the encouraging observation is that multiple examples of such conservation are now demonstrable in functional, lytic assays against HIV-infected targets. Even more encouraging are results from other studies showing that infected individuals from West Africa, tested against clade B targets, can lyse not only on the basis of conserved epitopes but with epitopes mismatched in one or even two amino acids (CAO et al. 1997).

The data confirm that cytotoxic responses can be cross-reactive against a wide variety of HIV. Evidence, of course, that such responses are protective, or that candidate vaccines induce responses of sufficient intensities and persistence to be successful, is not yet available.

IV. Design Considerations

Suffice it to say that great interest has developed in the induction of CTL activity as a protective modality. Raising CTL responses to lipidated peptides has been successful in mice and macaques (Deprez et al. 1996; Martinon et al. 1992; Rouaix et al. 1994), but immunogenicity of such vaccines has been disappointing in humans. Although there is intellectual appeal to creating a vaccine made of only the specific CTL epitopes required, mapping these epitopes and assuring that the synthetic immunogen is complete enough to cover HLA types in the recipient population exceeds the knowledge base at the present time.

The simpler approach has been to provide whole proteins in the vaccine, especially including *gag* and *pol* in addition to *env*, and using recombinant vectors or DNA to achieve class I presentation and CTL induction. Although there is a reasonable database of HIV epitopes recognized by HLA types predominant in white populations, such information in incomplete for other populations. Until a more extensive catalog of specific CTL epitopes is compiled, putting full-length gene products into the vaccine is the best hope of maximizing epitope content. Small proteins encoded by accessory genes such as *nef* and *rev* have also been shown to contain CTL epitopes (Blazevic et al. 1995; Choppin et al. 1992; Hinkula et al. 1997; Koenig et al. 1990; Lieberman et al. 1997). Including such proteins in a vaccine candidate has the theoretical advantage that these early proteins may mark the infected cell for destruction by CTL before any virus progeny emerges to spread the infection. The database on sequence conservation of these proteins across many HIV strains is insufficient at present to predict whether HIV heterogeneity would limit the utility of focusing on these early gene products.

Candidate vaccines have not been very efficient at inducing CTL responses in humans and macaques. Proposals to improve those responses by adding specific cytokines such as interleukin 12 and interferon-γ to the vaccine (Kim et al. 1998), or by targeting antigen to dendritic cells to improve cellular immune responses, are in early stages of evaluation in primates.

D. Other T-Cell Activities

Noncytolytic suppression of HIV and SIV by CD8 cells has been known for a long time (Kannagi et al. 1988; Walker et al. 1986), but little progress has been made in exploiting this observation for vaccine design. Although this activity can be found in cytotoxic cells (Yang et al. 1997), it is different in that the target cells are not killed (Wiviott et al. 1990), and suppressor cells can be effective against HIV replication in MHC-mismatched cells (Walker 1993). Suppressor activity in infected patients is positively correlated with clinical status (Mackewicz et al. 1991), and HIV-exposed but uninfected human populations exhibit this activity (Furci et al. 1997; Zagury et al. 1998), suggesting that the phenomenon may in fact have protective potential. The discovery of a variety of chemokines (Cocchi et al. 1995; Pal et al. 1997) that can block the interaction of HIV with different coreceptors offers an additional array (Barker et al. 1998; Paliard et al. 1996; Rubbert et al. 1997) of secreted proteins that can block infection.

Direct assays of this suppressor activity are not as well developed as CTL or neutralization assays, and they currently do not provide sensitive tools for measuring the phenomenon. Quantitation of chemokine secretion in vitro, in response to antigen-specific stimulation of cells from immunized animals, may be a better way to link this type of activity to vaccination. There is a published report examining the correlation of protection against rectal challenge with SIV induced by a subunit vaccine targeted to lymph nodes with in vitro chemokine secretion after nonspecific stimulation with PHA (Lehner et al. 1996), and a similar observation has been made in a long-term (18 years) nonprogressing AIDS patient (Rosenberg et al. 1997).

T-cell help, as assessed by antigen-specific lymphoproliferation, is essential for the development of effector responses. Proliferation responses are so easily generated with existing HIV vaccines that they have been granted less importance than CTL and neutralization, but recent analysis of long-term nonprogressing patients and in patients who initiate triple-drug therapy during acute infection has implicated the central importance of maintaining T-cell help if HIV is to be contained (Kalams et al. 1998). In addition, thorough vaccine immunology studies of Friend murine erythroleukemia virus, which is another immunosuppressive retrovirus with an error-prone reverse transcriptase and ability to integrate into the mouse genome, has shown the central importance of T-cell help, even in the absence of other effector functions. Passive transfer of just the immune CD4 cells from successfully vaccinated mice could confer protection (for review see Hasenkrug and Chesebro 1997).

Refinement of proliferation assays, through quantitation of cytokines that are secreted in response to the antigen stimulation, allows for distinction between T-helper 1 and T-helper 2 classes of cells (Clerici and Shearer 1994; Romagnani et al. 1994). Assays such as these are now being used in primates and in humans to analyze the quality of T-cell help with greater subtlety than simple proliferation assays can provide. It remains to be seen whether this aspect of immunity proves to be more significant for protection than concentrating on the effector arms (neutralization, CTL) of the immune response.

E. Mucosal Immunity

Because HIV is spread primarily by sexual contact, and its major route of transmission is across mucosal surfaces, the need for specific mucosal immunity was felt to be required for a successful HIV vaccine. There was even some concern that mucosal transmission would be a particular problem for protection, since such responses were notoriously difficult to induce (Sabin 1992).

Subsequent experiments in primates have shown that mucosal transmission is not an especial problem for a vaccine (CRANAGE et al. 1997; LEHNER et al. 1996; MARTHAS et al. 1992; MILLER et al. 1997), and even that a vaccine which fails to protect against intravenous challenge is nonetheless partially successful against rectal challenge (BENSON et al. 1998).

Secretory IgA, the classic specific mucosal response, will have the same limitations that IgG recognition of HIV has. Although surface IgA could be effective against HIV by binding and opsonization without requiring neutralization, the time that it would have to interdict HIV during natural transmission might be quite brief. Secretory immune responses induced by candidate vaccines are beginning to be assessed in human volunteers (GORSE et al. 1995).

Is there a specific cellular mucosal response? If HIV simply crosses the mucosal barrier and enters into the circulation, immune responses that defend cells in the periphery and tissues could be as well induced by whole-body vaccination strategies, and many if not most scientists involved in HIV vaccine research are exploring this approach first. However, there may be a role for specific mucosal immunity against HIV in an intermediate arena, in the local tissues below the mucosa and in the draining lymph nodes, before infection is transmitted to the periphery. Infected primates have been shown to have CTL in the vaginal submucosa as well as in the circulation (LOHMAN et al. 1995), but a more interesting possibility is the existence of CTL specifically in those areas without being detectable in the periphery.

Rectal immunization of mice with a peptide epitope vaccine has induced HIV-specific CTL detectable in the rectal mucosa but not in the spleen (BELYAKOV et al. 1998). These mice are protected against rectal challenge with a lethal dose of recombinant vaccinia that expresses an HIV envelope gene including the CTL epitope, which sensitizes the killing of any vaccinia-infected cells because they also present this HIV epitope. In recent primate experiments low-dose infection with pathogenic SIV in the colon of macaques led to the induction of SIV-specific CTL in the lamina propria of the gut in some of the animals; no CTL were detected in the blood. When subsequently challenged rectally with much higher doses of the same SIV sufficient to establish persistent infection, the macaques with the local CTL were protected from infection, and the macaques that had failed to produce measurable CTL activity in the lamina propria became productively infected (MURPHEY-CORB et al. 1998).

These results show that protective immune responses can reside within the mucosa even if undetectable in the periphery and give some independent experimental credence to the possibility that protective immune responses reside in the exposed but uninfected human populations. The possibility exists that efficient immunization of this local cellular compartment can best be accomplished through a specific mucosal vaccination regimen. Vaccines that could induce such regionally restricted protective responses can be evaluated only through macaque challenge with pathogenic virus.

F. Vaccine Approaches

I. Live-Attenuated Vaccine

Attenuated vaccine strains make highly effective vaccines against other viruses (Plotkin and Mortimer 1994). In AIDS, primate experiments in the SIV model have confirmed that these are the most effective vaccines against highly virulent, homologous virus challenges (Warren and Levinson 1997). Successful protection against homologous virus by the rectal route also suggests that blocking mucosal transmission will not be a problem (Cranage et al. 1997). However, the strength of protection begins to wane if heterologous challenges are made. The first such experiment evaluated preventing infection as an endpoint (Putkonen et al. 1996) and showed that only 50% of macaques were protected against a divergent SIV strain. In the chimpanzee such an approach has been tested using infection with one HIV strain as a "vaccine" against subsequent exposure to another HIV strain, since almost all wild-type HIV strains in chimpanzees are already "attenuated" for disease. Within clade B one strain appears to be an effective vaccine against another, preventing either infection with or persistence of the second (Shibata et al. 1996). However, prior infection with a clade B virus failed to prevent either infection with or persistence of a clade E virus (Girard et al. 1996), suggesting in this model as well that the heterogeneity of HIV can be a problem even for this vaccine approach.

Production of such a vaccine for human use could be difficult. HIV requires CD4-positive cells for efficient replication (Weiss et al. 1989), and large-scale production in uncharacterized PBMC is unthinkable. There are no certified human T-cell lines available for vaccine production, and it is now known that growth in such cell substrates exerts selection pressure, favoring variants that use the chemokine receptor for CXCR4 as coreceptor, not CCR5 (Moore and Ho 1995). If it is true that an NSI, CCR5-using, attenuated HIV strain, similar to the majority of wild-type HIV, is needed for a successful vaccine, one possible solution would be to create well-behaved artificial CD4-positive cell lines that express CCR5 but not CXCR4 (see Trkola et al. 1998). Even with this approach, virus variants that might arise in each production lot would have to be characterized. Instead, utilizing an infectious DNA plasmid form of the attenuated strain might be the best approach for manufacturing.

Safety issues for an attenuated HIV vaccine are paramount. Attenuation may be less an intrinsic property of the virus than a reflection of its relationship with the host that it infects. For example, an attenuated SIV construct that initially appeared safe in juvenile and adult macaques grew to high titer and induced disease in neonatal macaques (Baba et al. 1995). Although there are rare cases of humans infected with spontaneously attenuated HIV that have been studied (Deacon et al. 1995; Kirchhoff et al. 1995; Learmont et al. 1992), there is really no way to assess long-term safety of such a vaccine candidate in advance of large human trials. The horrific impact of the epidemic in some areas of the developing world has led some to advocate taking calculated risks (Cohen 1997; Desrosiers 1994), yet daunting safety and ethical issues remain.

II. Whole, Inactivated Virus

The envelope protein is at the same time the strength and the weakness of this approach. The native conformation of the glycoprotein spikes in the virion, providing the proper three-dimensional structure for inducing antibodies to recognize wild-type HIV, is the strongest advantage of this vaccine concept. Unfortunately, retroviruses in general and HIV in particular easily shed this surface glycoprotein. Producing virions that retain a useful amount of glycoprotein, and establishing an inactivation procedure that does not destroy the native conformation, are two significant obstacles for this approach. Reflecting this problem, early tests of killed virus vaccines as preventatives in chimpanzees have not been encouraging from the standpoint of either envelope immunogenicity or protection from challenge (Gibbs et al. 1991; Niedrig et al. 1993). A series of successful protection experiments with whole-killed vaccines in the SIV model (Carlson et al. 1990; Johnson et al. 1992; Mills et al. 1992; Murphey-Corb et al. 1992; Putkonen et al. 1992) was, unfortunately, subsequently shown to depend not on the viral antigens but on cellular xenoantigens coming from the human cells that had been used to produce the vaccine (Langlois et al. 1992; Le Grand et al. 1992). Thus there is little evidence from primate studies that this approach can be easily accomplished. Developing a cell substrate for manufacturing this vaccine will also face the same hurdles as the live-attenuated approach. A related approach is to create a pseudovirion, or genetically inactivated virus that already is replication incompetent (Haynes et al. 1991; Rovinski et al. 1992). Such mutants could be manufactured in cell lines and would not require $CD4^+$ cells for production; they also would not require chemical inactivation. However, such a product still must solve the problem of retaining the envelope protein.

Until a fundamental understanding of *env* protein structure is obtained to allow for the design of an *env* sequence that maintains a stable and native association between $gp41^{env}$ and $gp120^{env}$ in the mature vaccine particle, an empirical search must be made for an HIV isolate that retains its envelope. The design question for this vaccine is whether the envelope gene which is best for retention is also the best for breadth of immune response. Although the structural proteins of this vaccine are by definition in native form, accessory proteins that might provide protective antigens, such as those encoded by *nef*, *tat*, or *rev*, are not found in the vaccine. Also, such vaccines have no special advantages for inducing cellular immune responses. Because such intact particles have been so hard to produce in sufficient quantities for immunization, at present there is no experimental data to confirm the expectation that these native glycoprotein spikes can induce better antibodies that will broadly neutralize a range of primary HIV isolates. Whether this vaccine approach can induce more subtle immune responses that result in protection, in the absence of broad neutralizing or cytotoxic immune responses, deserves an experimental test, but requires development of a product that retains envelope. An HIV vaccine cannot be tested in macaques and requires challenge in chimpanzees; the recent isolation of an HIV isolate pathogenic for chimpanzees (NOVEMBRE et al. 1997) may allow for an experimental test.

III. Recombinant Vectors

Viruses or bacteria modified to express HIV antigens as well as their own in the course of their infection hold the promise of striking a good balance between strong immunogenicity and safety. The biological properties of the vectors themselves may confer advantages. Vaccinia and vesicular stomatitis viruses infect many cell types and are generally cleared from the body by the immune system, while herpesviruses target the nerves and are able to persist there. Vectors that naturally infect the oro/nasal mucosa (e.g., adenovirus, influenza, rhinovirus) or gut (*Salmonella typhi*, *Shigella*, poliovirus) might deliver their antigens and induce good mucosal immune responses as a consequence, and an initial trial of a recombinant *S. typhi* vector (WU et al. 1997) to test this idea has begun in humans. While most vectors are intended as gene carriers to produce endogenous synthesis of antigen in cells of the vaccine recipient, there are a few examples of display vectors in which the HIV antigens are part of the vector structure (SMITH et al. 1998). If recombinant vectors can overcome the initial hurdles of inserted gene instability and prior immunity to the vector in the human population, they deserve to be investigated as vaccine candidates.

Although such recombinant vectors cannot cause AIDS, there are nonetheless safety concerns. Such a vaccine intended for places in the world where at-risk populations could have 10%–25% prevalence of HIV infection will need to have extremely limited replication capacity to avoid the possibility that some recipients could already be immunocompromised and unable to control infection by the vaccine vector. For this reason the best developed vector, vaccinia, is generally excluded from consideration as an HIV vaccine carrier despite having been safe enough, in a pre-HIV era, to inoculate hundreds of millions of persons as part of the world-wide campaign to eradicate smallpox 30 years ago.

Limited replication has been achieved in vectors under development for HIV vaccines in different ways. Despite not proceeding with extensive additional rounds of replication, these vectors can express the passenger antigens in the first cells that they infect and induce immune responses to them. Canarypox is a virus which is replication incompetent in mammalian cells (LIMBACH and PAOLETTI 1996), and multiple constructs containing various combinations of HIV genes are being evaluated in human trials (CLEMENTS-MANN et al. 1998; EGAN et al. 1995). One particular strain of vaccinia, through spontaneous deletion of 10%–15% of its genes, has lost almost all replication capacity in mammalian cells. This modified virus (modified vaccinia, Ankara strain (MVA); SUTTER and MOSS 1995), is increasingly being tested preclinically as a vector, even though its future as a commercial product is uncertain because it is in the public domain. Another approach is to alter the packaged nucleic acid of the vector virus so that it can replicate only one time in the host, and this "replicon" technology is being evaluated in primate experiments for poliovirus (ANDERSON et al. 1997), Semliki forest virus (ROLLS et al. 1996; ZHOU et al. 1994), and Venezuelan equine encephalitis virus (PUSHKO et al. 1997).

A fundamental uncertainty for all HIV vaccine research is the identity of the antigens or specific immune responses which are essential for protective immunity. Therefore exactly what HIV components need to be in the vector is a research emphasis as important as the particular advantages of a specific vector system to present those antigens.

IV. Plasmid Immunization

Immunization with DNA is a special case of vector presentation, wherein the vector organism is eliminated and only the encoding information is directly administered. The surprisingly strong immune responses to proteins encoded by small amounts of nucleic acid has become an area of intense research (see Chap. 3). Highly encouraging immunogenicity and protection results in rodents have not translated well into primates for HIV vaccine approaches (S. LU et al. 1996; YASUTOMI et al. 1996), although a DNA form of a known successful hepatitis B vaccine was effective in the chimpanzee (DAVIS et al. 1996). These vaccines have been good at inducing CTL, but good neutralizing antibodies do not appear until after protein boosting (LETVIN et al. 1997; SHIVER et al. 1997). Human trials of candidate HIV vaccines have begun only recently.

There is more to nucleic acid as a vaccine agent than attaching a strong promoter to the gene for the antigen of interest. Improvements in immunogenicity may depend on manipulating the antigen genes to optimize expression (ANDRE et al. 1998). Structures in the DNA itself, independent of the encoded information, appear to have self-adjuvanting properties (LIPFORD et al. 1997), and such features may differ between rodents and primates. Inclusion of cytokine DNA along with the antigen DNA has improved responses in mice (KIM et al. 1998) and is beginning to be evaluated in primates. If antibody responses to nucleic acid immunization can be improved, this approach would the fastest way to compare multivalent envelope combinations to address the heterogeneity problem. An intriguing concept is immunization with DNA libraries prepared from the pathogen's entire genome, bypassing the need to identify protective antigens in advance (BARRY et al. 1995). Although this approach saves time, it requires that protective epitopes be confined to contiguous peptide sequences. Formulations to facilitate direct mucosal administration of nucleic acid are being developed (Ishii et al. 1997; Jones et al. 1997; Klavinskis et al. 1997; Macklin et al. 1998; Sasaki et al. 1998). Plasmid immunization has a number of significant manufacturing advantages, including stability, low cost of production, and ease of purification.

V. Subunit Proteins and Peptides

Based on the model of the successful hepatitis B subunit vaccine, the first candidate HIV vaccines to be developed were envelope subunit proteins, since these proteins contain the neutralization epitopes. Manufacture and purification of these large, heavily glycosylated, internally disulfide-bridged proteins in market quantities is technologically demanding. Until recently these products were made exclusively from laboratory strains of HIV. Although better antibody responses were obtained with gp120 expressed in mammalian cells than from yeast or insect cell systems (Graham 1994), the value of these neutralization responses against wild-type primary strains of HIV has been questioned (Dolin 1995; Mascola et al. 1996b; van Cott et al. 1995a). CTL responses have been minimal (Stanhope et al. 1993), although good proliferative responses are seen (Gorse et al. 1998; Picard et al. 1992). Human evaluations are poised to begin with bivalent envelope vaccines that include a primary strain gp120 of HIV, from clade E virus for Thailand, and clade B for the United States (Altman 1998). However, there is wider interest in using these subunit vaccines in combination with other modalities than as standalone vaccines.

Peptide vaccines can focus even more precisely on B- and T-cell epitopes. The vaccines are rather simple to produce, and several human safety and immunogenicity trials have been performed (Kahn et al. 1992; Rubenstein et al. 1995). Antibodies induced by peptide vaccines in humans have been disappointing (Gorse et al. 1996; Kelleher et al. 1997; Phanuphak et al. 1997). Although strong neutralizing antibodies to the principal neutralizing domain, or V3-loop peptide, can be obtained (Haynes et al. 1993), such responses have little effect against primary HIV, whose neutralization epitopes appear to be only conformational and unable to be expressed by linear peptides (Moore et al. 1995). There have been some interesting recent developments in the synthesis of conformationally constrained synthetic peptides as immunogens (Frey et al. 1997; Robey et al. 1996) that may increase the utility of the peptide approach.

VI. Combinations

Live recombinant vaccines generate good cellular immune responses in humans (Perales et al. 1995), but the neutralizing antibodies that they induce against HIV are weak (Graham et al. 1992). Conversely, the best levels of neutralization are attained with multiple doses of adjuvanted preparations of envelope subunit or peptide vaccines, but CTL induction with these vaccines in human studies has been quite poor (STANHOPE et al. 1993). However, a single dose of envelope subunit administered after one or two priming doses of a live recombinant vector can yield the same levels of neutralization achieved after several doses of protein alone; CTL responses are not adversely affected by the protein boost (CLEMENTS-MANN et al. 1998; GRAHAM et al. 1993). Combining the two approaches appears to preserve the best immune properties of each. Priming with a live vector for T-cell responses followed by protein boosting to optimize the B-cell response has become a common strategy.

Nucleic acid immunization behaves as a replicating vector, priming for antibody responses which are brought out by the subsequent protein immunization (FULLER et al. 1997; LETVIN et al. 1997). When used in combination with live vector vaccines, DNA is more effective as a prime than as a boost (FULLER et al. 1997; SCHNEIDER et al. 1998). Because DNA and vectors are easier to produce and purify than proteins are, there may be some advantage to a vector/nucleic acid combination than a combination with protein antigens.

G. Prospectus

I. Design Considerations

Fifteen years of productive research have produced a remarkably detailed body of information about HIV, yet fundamental uncertainty still remains about the antigens or specific immune responses which are essential for protective immunity and thus would form the basis for rational vaccine design. The extensive heterogeneity of the envelope protein, and the virus' ability to persist in latent, integrated form to evade immune clearance from the body, pose the greatest problems for vaccine design. Table 1 summarizes a myriad of suggested approaches for improving HIV vaccines to address these problems. The range of approaches reflects the spectrum of thinking among developers.

Most work is focused on improving neutralizing and/or CTL responses in the expectation that the key to success lies in optimizing those effector responses. Active research is currently driven by a series of strategies to improve those responses, either by modifying the antigens which comprise the vaccine or by improving responses to those antigens. These strategies are not mutually exclusive, and candidate vaccines often incorporate several ideas. Strategies to specifically improve mucosal responses are much less developed.

Hypothesis IV in Table 1 covers a range of ideas which broaden the horizons of vaccine design. The first two strategies, as discussed above, propose that T-cell help is more important than either antibody or CTL. The third strategy reports a unique proposal to interfere with HIV replication by directing antibodies against the *tat* transactivating protein (GOLDSTEIN 1996). The final two strategies involve targeting cellular antigens to interfere with transmission, although there may be risks of developing autoimmunity.

Table 1. Rational vaccine design

I. Hypothesis: Induction of better neutralizing antibodies will improve the vaccine.
  A. Modify antigen form
    Strategy: employ macrophage-tropic (NSI), not laboratory strain, *env*
    Strategy: make *env* component more "native"
    Examples: oligomeric subunit proteins, whole-killed virus, gp120 complexed to CD4
    Strategy: alter glycosylation of the *env* protein
    Examples: delete specific glycosylation sites; insert glycosylation site into V3 loop
    Strategy: focus on specific epitopes
    Examples: V3 peptides; mimotopes; delete variable regions
    Strategy: employ multivalent *env* to increase breadth
  B. Improve "presentation"
    Strategy: test to find better adjuvants
    Strategy: coadminister cytokine (as protein or DNA)
    Strategy: optimize antigen expression by DNA
    Strategy: target dendritic antigen-presenting cells
    Examples: epithelial targeting of DNA, granulocyte-macrophage colony-stimulating factor as chemoattractant
    Strategy: combine vaccines in a "live" prime/protein boost regimen
II. Hypothesis: Making better CTL will improve the vaccine.
  A. Include multiple antigens
    Strategy: more gene products in the vaccine – more CTL targets
    Strategy: accessory gene products make early CTL targets
  B. Improve endogenous "presentation"
    Strategy: employ lipidated peptides for CTL epitopes
    Strategy: optimize antigen expression by DNA
    Strategy: employ granulocyte-macrophage colony-stimulating factor to improve cellular responses
    Strategy: target dendritic antigen-presenting cells
III. Hypothesis: Mucosal immunity will be helpful/essential to block sexual transmission.
    Strategy: employ adjuvants specific for mucosal responses
    Examples: Cholera toxin B subunit, *E. coli* labile toxin
    Strategy: target nonreplicating vaccines to draining nodes
    Strategy: administer live vaccines to mucosal sites, to stimulate mucosal responses
    Strategy: administer DNA to nasal mucosa
    Strategy: assess CTL in lamina propria/vaginal submucosa
IV. Hypothesis: Something other than neutralizing antibody or CTL to HIV is most important.
    Strategy: modulate immune response towards Th1 response
    Examples: coadminister γ-interferon or interleukin 12
    Strategy: immunize with T-helper peptides only
    Strategy: immunize with *tat* protein
    Strategy: immunize with alloantigens
    Strategy: immunize with coreceptor peptides

II. Evaluation Considerations

Everyone is concerned about "progress" in HIV vaccine development, but as this chapter shows, there is no single development pathway, and there are no established milestones by which to assess progress toward eventual success. We must develop and test a wide variety of vaccine candidates based on various assumptions. Comparative immunogenicity testing in animals and humans is essential, but interpretation is limited by the absence of a well-defined correlate of protective immunity for HIV. Because existing vaccine candidates are not inducing strong and broad neutralizing antibody at all, nor robust CTL in the great majority of volunteers, there has been reluctance to proceed with expanded human field trials. Therefore it is absolutely essential to perform challenge studies in primates, not only to compare vaccine concepts for potential efficacy but also to learn whether vaccines can provide clear clinical benefit despite unspectacular in vitro assay results. It is quite possible that no assay now in use can measure a specific response that is predictive of protective efficacy in a field trial.

Systematic comparisons are beginning to be made, especially in the SHIV model, where vaccines with HIV *env* components can be evaluated by challenge in macaques (Y. Lu et al. 1996). If vaccines are expected to prevent the establishment of infection, or especially to reduce virus burden after challenge, it is essential that challenge studies employ viruses which have an immunodeficiency disease endpoint and whose plasma loads in the test species are similar to those of HIV in humans (Mellors et al. 1997). While recent experiments in chimpanzees have generated much interest (Boyer et al. 1997; Lubeck et al. 1997), nonpathogenic viruses are all that have been available in that model.

Extremely active research is underway investigating a number of variables (Table 1) to improve vaccine designs. The best experimental approach is to proceed in animal and human studies of multiple products in a parallel path and to move into large-scale trials with all deliberate speed.

References

Ada G (1988) Prospects for HIV vaccines. J Acquir Immune Defic Syndr Hum Retrovirol 1:295–303

Ahmad S, Yilma T, Luciw PA et al (1994) Reduced virus load in rhesus macaques immunized with recombinant gp160 and challenged with simian immunodeficiency virus. AIDS Res Hum Retroviruses 10:195–204

Alcabes P, Munoz A, Vlahov D, Friedland GH (1993) Incubation period of human immunodeficiency virus. Epidemiol Rev 15:303–318

Almond N, Rose J, Sangster R et al (1997) Mechanisms of protection induced by attenuated simian immunodeficiency virus. I. Protection cannot be transferred with immune serum. J Gen Virol 78:1919–1922

Altman JD, Moss PAH, Goulder PJR et al (1996) Phenotypic analysis of antigen-specific T lymphocytes. Science 274:94–96

Altman LK (1998) FDA authorizes first full testing of HIV vaccine. The New York Times, 4 June 1998, p 1

Anderson MJ, Porter DC, Moldoveanu Z et al (1997) Characterization of the expression and immunogenicity of poliovirus replicons that encode simian immunodeficiency virus SIVmac239 Gag or envelope SU proteins. AIDS Res Hum Retroviruses 13:53–62

Andersson S, Makitalo B, Thorstensson R et al (1996) Immunogenicity and protective efficacy of a human immunodeficiency virus type 2 recombinant canarypox (ALVAC) vaccine candidate in cynomolgus monkeys. J Infect Dis 174:977–985

Andre S, Seed B, Eberle J et al (1998) Increased immune response elicited by DNA vaccination with a synthetic gp120 sequence with optimized codon usage. J Virol 72:1497–1503

Andrus L, Prince AM, Bernal I et al (1998) Passive immunization with a human immunodeficiency virus type 1-neutralizing monoclonal antibody in Hu-PBL-SCID mice: isolation of a neutralization escape variant. J Infect Dis 177:889–897

Asjo B, Morfeldt-Manson L, Albert J et al (1986) Replicative capacity of human immunodeficiency virus from patients with varying severity of HIV infection. Lancet 1986:660–662

Baba TW, Ruprecht RM, Greene MF et al (1995) Pathogenicity of live, attenuated SIV after mucosal infection of neonatal macaques. Science 267:1820–1825

Bachmann MF, Zinkernagel RM (1996) The influence of virus structure on antibody responses and virus serotype formation. Immunol Today 17:553–558

Barker E, Bossart KN, Levy JA (1998) Primary CD8+ cells from HIV-infected individuals can suppress productive infection of macrophages independent of beta-chemokines. Proc Natl Acad Sci USA 95:1725–1729

Barre-Sinnoussi F, Chermann JC, Rey F et al (1983) Isolation of a T-lymphotropic retrovirus from a patient at risk for acquired immune deficiency syndrome (AIDS). Science 220:868–871

Barry MA, Lai WC, Johnston SA (1995) Protection against mycoplasma infection using expression-library immunization. Nature 377:632–635

Baum LL, Cassutt KJ, Knigge K et al (1996) HIV-1 gp120-specific antibody-dependent cell-mediated cytotoxicity correlates with rate of disease progression. J Immunol 157:2168–2173

Beddows S, Louisirirotchanakul S, Cheingsong-Popov R et al (1998) Neutralization of primary and T-cell line adapted isolates of human immunodeficiency virus type 1: role of V3-specific antibodies. J Gen Virol 79:77–82

Belshe RB, Clements ML, Dolin R et al (1993) Safety and immunogenicity of a fully glycosylated recombinant gp160 human immunodeficiency virus type 1 vaccine in subjects at low risk of infection. National Institute of Allergy and Infectious Diseases AIDS Vaccine Evaluation Group Network. J Infect Dis 168:1387–1395

Belshe RB, Graham BS, Keefer MC et al (1994) Neutralizing antibodies to HIV-1 in seronegative volunteers immunized with recombinant gp120 from the MN strain of HIV-1. JAMA 272:475–480

Belyakov IM, Derby MA, Ahlers JD et al (1998) Mucosal immunization with HIV-1 peptide vaccine induces mucosal and systemic cytotoxic T lymphocytes and protective immunity in mice against intrarectal recombinant HIV-vaccinia challenge. Proc Natl Acad Sci USA 95:1709–1714

Benjouad A, Gluckman JC, Rochat H et al (1992) Influence of carbohydrate moieties on the immunogenicity of human immunodeficiency virus type 1 recombinant gp160. J Virol 66:2473–2283

Benson J, Franchini G, Tartaglia J et al (1998) Recombinant vaccine-induced protection against the highly pathogenic simian immunodeficiency virus SIV(mac251): dependence on route of challenge exposure. J Virol 72:4170–4182

Beretta A, Furci L, Burastero S et al (1997) HIV-1-specific immunity in persistently seronegative individuals at high risk for HIV infection. Immunol Lett 51:39–43

Berger EA (1997) HIV entry and tropism: the chemokine receptor connection. AIDS 11[Suppl A]:S3–16

Berger EA, Doms RW, Fenyo EM et al (1998) A new classification for HIV-1 (letter). Nature 391:240

Betts MR, Krowka J, Santamaria C et al (1997) Cross-clade human immunodeficiency virus (HIV)-specific cytotoxic T-lymphocyte responses in HIV-infected Zambians. J Virol 71:8908–8911
Biberfield G, Putkonen P, Thorstensson R, Norrby E (1992) Prevention of HIV-2 and SIVSM infection in cynomolgus monkeys by active or passive immunization. AIDS Res Hum Retroviruses 8:1511–1513
Blackbourn DJ, Mackewicz CE, Barker E et al (1996) Suppression of HIV replication by lymphoid tissue CD8+ cells correlates with the clinical state of HIV-infected individuals. Proc Natl Acad Sci USA 93:13125–13130
Blazevic V, Ranki A, Krohn KJ (1995) Helper and cytotoxic T cell responses of HIV type 1-infected individuals to synthetic peptides of HIV type 1 Rev. AIDS Res Hum Retroviruses 11:1335–1342
Bonhoeffer S, May RM, Shaw GM, Nowak MA (1997) Virus dynamics and drug therapy. Proc Natl Acad Sci USA 94:6971–6976
Boots LJ, McKenna PM, Arnold BA et al (1997) Anti-human immunodeficiency virus type 1 human monoclonal antibodies that bind discontinuous epitopes in the viral glycoproteins can identify mimotopes from recombinant phage peptide display libraries. AIDS Res Hum Retroviruses 13:1549–1559
Borrow P, Lewicki H, Wei X et al (1997) Antiviral pressure exerted by HIV-1-specific cytotoxic T lymphocytes (CTLs) during primary infection demonstrated by rapid selection of CTL escape virus. Nat Med 3:205–211
Bou-Habib DC, Roderiquez G, Oravecz T et al (1994) Cryptic nature of envelope V3 region epitopes protects primary monocytotropic human immunodeficiency virus type 1 from antibody neutralization. J Virol 68:6006–6013
Boyer JD, Ugen KE, Wang B et al (1997) Protection of chimpanzees from high-dose heterologous HIV-1 challenge by DNA vaccination. Nat Med 3:526–532
Buchbinder SP, Katz MH, Hessol NA et al (1994) Long-term HIV-1 infection without immunologic progression. AIDS 8:1123–1128
Burton DR, Montefiori DC (1997) The antibody response in HIV-1 infection. AIDS 11 [Suppl A]:S87–S98
Carlson JR, McGraw TP, Keddie E et al (1990) Vaccine protection of rhesus macaques against simian immunodeficiency virus infection. AIDS Res Hum Retroviruses 6:1239–1246
Cao H, Kanki P, Sankale JL et al (1997) Cytotoxic T-lymphocyte cross-reactivity among different human immunodeficiency virus type 1 clades: implications for vaccine development. J Virol 71:8615–8623
Centers for Disease Control (1986) Current trends poliomyelitis – United States, 1975–1984. MMWR 35:180–182
Chanh TC, Dreesman GR, Kanda P et al (1986) Induction of anti-HIV neutralizing antibodies by synthetic peptides. EMBO J 5:3065–3071
Choppin J, Guillet JG, Levy JP (1992) HLA class I binding regions of HIV-1 proteins. Crit Rev Immunol 12:1–16
Clements JE, Montelaro RC, Zink MC et al (1995) Cross-protective immune responses induced in rhesus macaques by immunization with attenuated macrophage-tropic simian immunodeficiency virus. J Virol 69:2737–2744
Clements ML (1997) Clinical trials of human immunodeficiency virus vaccines. In: DeVita VT Jr, Hellman S, Rosenberg SA (eds) AIDS: biology, diagnosis, treatment and prevention chap 32.2, 4th edn. Lippincott-Raven, Philadelphia, pp 617–626
Clements-Mann ML, Weinhold K, Matthews TJ et al (1998) Immune responses to human immunodeficiency virus (HIV) type 1 induced by canarypox expressing HIV-1MN gp120, HIV-1SF2 recombinant gp120, or both vaccines in seronegative adults. J Infect Dis 177:1230–1246
Clerici M, Shearer GM (1994) The Th1-Th2 hypothesis of HIV infection: new insights. Immunol Today 15:575–581

Clerici M, Giorgi JV, Chou CC et al (1992) Cell-mediated immune response to human immuno-deficiency virus (HIV) type 1 in seronegative homosexual men with recent sexual exposure to HIV-1. J Infect Dis 165:1012–1019
Cocchi F, DeVico AL, Garzino-Demo A et al (1995) Identification of RANTES, MIP-1 alpha, and MIP-1 beta as the major HIV-suppressive factors produced by CD8+ T cells. Science 270:1811–1815
Cohen J (1993) Jitters jeopardize AIDS vaccine trials. Science 262:980–981
Cohen J (1997) Novel campaign to test live HIV vaccine. Science 277:1035
Cohn ZA, Steinman RM (1988) The immunological and infectious sequelae of the acquired immune deficiency syndrome. J Exp Med 168:2415–2423
Conley AJ, Kessler JA Jr, Boots LJ et al (1994) Neutralization of divergent human immunodeficiency virus type 1 variants and primary isolates by IAM-41-2F5, an anti-gp41 human monoclonal antibody. Proc Natl Acad Sci USA 91:3348–3352
Conley AJ, Murthy KK, Rouse SR et al (1996) The consequence of passive administration of an anti-human immunodeficiency virus type 1 neutralizing monoclonal antibody before challenge of chimpanzees with a primary virus isolate. J Virol 70:6751–6758
Connick E, Marr DG, Zhang XQ et al (1996) HIV-specific cellular and humoral immune responses in primary HIV infection. AIDS Res Hum Retroviruses 12:1129–1140
Corbitt G, Williams G, Bailey AS (1990) HIV infection in Manchester, 1959. Lancet 336:51
Cranage MP, Hall GA, Dennis MJ et al (1997) Macaques infected with live attenuated SIVmac are protected against superinfection via the rectal mucosa. Virology 229:143–154
Dalgleish AG, Chanh TC, Kennedy RC et al (1988) Neutralization of diverse HIV-1 strains by monoclonal antibodies raised against a gp41 synthetic peptide. Virology 165:209–215
Daniel MD, Mazzara GP, Simon MA et al (1994) High-titer immune responses elicited by recombinant vaccinia virus priming and particle boosting are ineffective in preventing virulent SIV infection. AIDS Res Hum Retroviruses 10:839–851
Davis HL, McCluskie MJ, Gerin JL, Purcell RH (1996) DNA vaccine for hepatitis B: evidence for immunogenicity in chimpanzees and comparison with other vaccines. Proc Natl Acad Sci USA 93:7213–7218
Deacon NJ, Tsykin A, Solomon A et al (1995) Genomic structure of an attenuated quasi species of HIV-1 from a blood transfusion donor and recipients. Science 270:988–991
Deng H, Liu R, Ellmeier W et al (1996) Identification of a major co-receptor for primary isolates of HIV-1. Nature 381:661–666
Deprez B, Sauzet JP, Boutillon C et al (1996) Comparative efficiency of simple lipopeptide constructs for in vivo induction of virus-specific CTL. Vaccine 14:375–382
Desrosiers RC (1994) Controversies in science: yes, it is time to consider use of a live-attenuated virus vaccine against HIV-1. J NIH Res 6:54–60
Dolin R (1995) Human studies in the development of human immunodeficiency virus vaccines. J Infect Dis 172:1175–1183
Dolin R, Graham BS, Greenberg SB et al (1991) The safety and immunogenicity of a human immunodeficiency virus type 1 (HIV-1) recombinant gp160 candidate vaccine in humans. Ann Intern Med 114:119–127
Dragic T, Litwin V, Allaway GP et al (1996) HIV-1 entry into CD4+ cells is mediated by the chemokine receptor CC-CKR-5. Nature 381:667–673
D'Souza MP, Livnat D, Bradac JA, Bridges SH (1997) Evaluation of monoclonal antibodies to human immunodeficiency virus type 1 primary isolates by neutralization assays: performance criteria for selecting candidate antibodies for clinical trials. J Infect Dis 175:1056–1062

Durali D, Morvan J, Letourneur F et al (1998) Cross-reactions between the cytotoxic T-lymphocyte responses of human immunodeficiency virus-infected African and European patients. J Virol 72:3547–3553
Egan MA, Pavlat WA, Tartaglia J et al (1995) Induction of human immunodeficiency virus type 1 (HIV-1)-specific cytolytic T lymphocyte responses in seronegative adults by a nonreplicating, host-range-restricted canarypox vector (ALVAC) carrying the HIV-1MN env gene. J Infect Dis 171:1623–1627
Emini EA, Schlief WA, Nunberg JH et al (1992) Prevention of HIV-1 infection in chimpanzees by gp120 V3 domain-specific monoclonal antibody. Nature 355:728–730
Fenyo EM, Morfeldt-Manson L, Chiodi F et al (1988) Distinct replicative and cytopathic characteristics of human immunodeficiency virus isolates. J Virol 62:4414–4419
Ferdinand FJ, Dorner F, Kurth R (1987) Perspectives of HIV vaccine developments. J Virol Methods 17:63–67
Ferrari G, Humphrey W, McElrath MJ et al (1997) Clade B-based HIV-1 vaccines elicit cross-clade cytotoxic T lymphocyte reactivities in uninfected volunteers. Proc Natl Acad Sci USA 94:1396–1401
Fowke KR, Plummer FA, Ngugi EN et al (1996) Resistance to HIV-1 infection among persistently seronegative prostitutes in Nairobi, Kenya. Lancet 348:1347–1351
Frey A, Neutra MR, Robey (1997) Peptomer aluminum oxide nanoparticle conjugates as systemic and mucosal vaccine candidates: synthesis and characterization of a conjugate derived from the C4 domain of HIV-1MN gp120. Bioconjug Chem 8:424–433
Fuller DH, Corb MM, Barnett S et al (1997) Enhancement of immunodeficiency virus-specific immune responses in DNA-immunized rhesus macaques. Vaccine 15:924–926
Furci L, Scarlatti G, Burastero S et al (1997) Antigen-driven C-C chemokine-mediated HIV-1 suppression by CD4(+) T cells from exposed uninfected individuals expressing the wild-type CCR-5 allele. J Exp Med 186:455–460
Gallimore A, Stott J, Corcoran T et al (1995) Early suppression of SIV replication by $CD8^+$ nef-specific cytotoxic T cells in vaccinated macaques. Nat Med 1:1167–1173
Gallo RC, Sarin PC, Gelmann EP et al (1983) Isolation of human T-cell leukemia virus in acquired immunodeficiency syndrome (AIDS). Science 220:865–867
Gange SJ, Munoz A, Schrager LK et al (1997) Design of nested studies to identify factors related to late progression to HIV infection. J Acquir Immune Defic Syndr Hum Retrovirol 15 [Suppl 1]:S5–S9
Gardner M, Rosenthal A, Jennings M et al (1995) Passive immunization of rhesus macaques against SIV infection and disease. AIDS Res Hum Retroviruses 11:843–854
Garry RF, Drake WL Jr, Cole WR et al (1988) Documentation of an AIDS virus infection in the United States in 1968. JAMA 260:2085–2087
Gauduin MC, Parren PW, Weir R et al (1997) Passive immunization with a human monoclonal antibody protects hu-PBL-SCID mice against challenge by primary isolates of HIV-1. Nat Med 3:1389–1393
Gibbs CJ Jr, Peters R, Gravell M et al (1991) Observations after human immunodeficiency virus immunization and challenge of human immunodeficiency virus seropositive and seronegative chimpanzees. Proc Natl Acad Sci USA 88:3348–3352
Girard M, Fultz P, Barre-Sinnoussi F et al (1993) Further studies on HIV-1 vaccine protection in chimpanzees. In: Girard M, Valette L (eds) Retroviruses of human AIDS and related animal diseases: 8e Colloque des "Cent Gardes." Fondation Marcel Merieux, Lyon, France, pp 139–143
Girard M, Yue L, Barre-Sinoussi F et al (1996) Failure of a human immunodeficiency virus type 1 (HIV-1) subtype B-derived vaccine to prevent infection of chimpanzees by an HIV-1 subtype E strain. J Virol 70:8229–8233

Goldstein G (1996) HIV-1 Tat protein as a potential AIDS vaccine. Nat Med 2:960–964
Gorse GJ, Rogers JH, Perry JE et al (1995) HIV-1 recombinant gp160 vaccine induced antibodies in serum and saliva. Vaccine 13:209–214
Gorse GJ, Keefer MC, Belshe RB et al (1996) A dose-ranging study of a prototype synthetic HIV-1 MN V3 branched peptide vaccine. The National Institute of Allergy and Infectious Diseases AIDS Vaccine Evaluation Group. J Infect Dis 173:330–339
Gorse GJ, McElrath MJ, Matthews TJ et al (1998) Modulation of immunologic responses to HIV-1 MN recombinant gp160 vaccine by dose and schedule of administration. Vaccine 16:493–506
Goulder PJ, Phillips RE, Colbert RA et al (1997a) Late escape from an immunodominant cytotoxic T-lymphocyte response associated with progression to AIDS. Nat Med 3:212–217
Goulder P, Price D, Nowak M et al (1997b) Co-evolution of human immunodeficiency virus and cytotoxic T-lymphocyte responses. Immunol Rev 159:17–29
Graham BS (1994) Serological responses to candidate AIDS vaccines. AIDS Res Hum Retroviruses 10:S145–S148
Graham BS, Belshe RB, Clements ML et al (1992) Vaccination of vaccinia-naive adults with human immunodeficiency virus type 1 gp160 recombinant vaccinia virus in a blinded, controlled, randomized clinical trial. J Infect Dis 166:244–252
Graham BS, Matthews TJ, Belshe RB et al (1993) Augmentation of human immunodeficiency virus type 1 neutralizing antibody by priming with gp160 recombinant vaccinia and boosting with rgp160 in vaccinia-naive adults. J Infect Dis 167:533–537
Graham BS, Keefer MC, McElrath MJ et al (1996) Safety and immunogenicity of a candidate HIV-1 vaccine in healthy adults: recombinant glycoprotein (rgp)120. A randomized, double-blind trial. Ann Intern Med 125:270–279
Gulick RM, Mellors JW, Havlir D et al (1997) Treatment with indinavir, zidovudine, and lamivudine in adults with human immunodeficiency virus infection and prior antiretroviral therapy. N Engl J Med 337:734–739
Haas G, Hosmalin A, Hadida F et al (1997) Dynamics of HIV variants and specific cytotoxic T-cell recognition in nonprogressors and progressors. Immunol Lett 57:63–68
Hammer SM, Squires KE, Hughes MD et al (1997) A controlled trial of two nucleoside analogues plus indinavir in persons with human immunodeficiency virus infection and CD4 cell counts of 200 per cubic millimeter or less. AIDS Clinical Trials Group 320 Study Team. N Engl J Med 337:725–733
Hasenkrug KJ, Chesebro B (1997) Immunity to retroviral infection: the Friend virus model. Proc Natl Acad Sci USA 94:7811–7816
Haynes JR, Cao SX, Rovinski B et al (1991) Production of immunogenic HIV-1 viruslike particles in stably engineered monkey cell lines. AIDS Res Hum Retroviruses 7:17–27
Haynes BF, Torres JV, Langlois AJ et al (1993) Induction of HIVMN neutralizing antibodies in primates using a prime-boost regimen of hybrid synthetic gp120 envelope peptides. J Immunol 151:1646–1653
Henderson LE, Sowder R, Copeland TD et al (1987) Direct identification of class II histocompati-bility DR proteins in preparations of human T-cell lymphotropic virus type III. J Virol 61:629–632
Hinkula J, Svanholm C, Schwartz S et al (1997) Recognition of prominent viral epitopes induced by immunization with human immunodeficiency virus type 1 regulatory genes. J Virol 71:5528–5539
Hirsch VM, Lifson JD, Moss B et al (1996) Patterns of viral replication correlate with outcome in simian immunodeficiency virus (SIV)-infected macaques: effect of prior immunization with a trivalent SIV vaccine in modified vaccinia virus Ankara. J Virol 70:3741–3752

Ho DD, Sarngadharan MG, Hirsch MS et al (1987) Human immunodeficiency virus neutralizing antibodies recognize several conserved domains on the envelope glycoproteins. J Virol 61:2024–2028

Ho DD, Neumann AU, Perelson AS et al (1995) Rapid turnover of plasma virions and CD4 lymphocytes in HIV-1 infection. Nature 373:123–126

Hoffman SL, Crutcher JM, Puri SK et al (1997) Sterile protection of monkeys against malaria after administration of interleukin-12. Nat Med 3:80–83

Huminer D, Rosenfeld JB, Pitlik SD (1987) AIDS in the pre-AIDS era. Rev Infect Dis 9:1102–1108

Ishii N, Fukushima J, Kaneko T et al (1997) Cationic liposomes are a strong adjuvant for a DNA vaccine of human immunodeficiency virus type 1. AIDS Res Hum Retroviruses 13:1421–1428

Ivey-Hoyle M, Culp JS, Chaikin MA et al (1991) Envelope glycoproteins from biologically diverse isolates of immunodeficiency viruses have widely different affinities for CD4. Proc Natl Acad Sci USA 88:512–516

Jansson M, Popovic M, Karlsson A et al (1996) Sensitivity to inhibition by beta-chemokines correlates with biological phenotypes of primary HIV-1 isolates. Proc Natl Acad Sci USA 93:15382–15387

Jassoy C, Walker BD (1997) HIV-1-specific cytotoxic T lymphocytes and the control of HIV-1 replication. Springer Semin Immunopathol 18:341–354

Johnson PR, Montefiori DC, Goldstein S et al (1992) Inactivated whole-virus vaccine derived from a proviral DNA clone of simian immunodeficiency virus induces high levels of neutralizing antibodies and confers protection against heterologous challenge. Proc Natl Acad Sci USA 89:2175–2179

Jones DH, Corris S, McDonald S et al (1997) Poly(DL-lactide-co-glycolide)-encapsulated plasmid DNA elicits systemic and mucosal antibody responses to encoded protein after oral administration. Vaccine 15:814–817

Kahn JO, Stites DP, Scillian J et al (1992) A phase I study of HGP-30, a 30 amino acid subunit of the human immunodeficiency virus (HIV) p17 synthetic peptide analogue sub-unit vaccine in seronegative subjects. AIDS Res Hum Retroviruses 8:1321–1325

Kahn JO, Sinangil F, Baenziger J et al (1994) Clinical and immunologic responses to human immuno-deficiency virus (HIV) type 1SF2 gp120 subunit vaccine combined with MF59 adjuvant with or without muramyl tripeptide dipalmitoyl phosphatidylethanolamine in non-HIV-infected human volunteers. J Infect Dis 170:1288–1291

Kalams SA, Buchbinder SP, Billingsley JM et al (1998) Coordinate control of HIV replication mediated by CD4 and CD8 cells. (submitted)

Kannagi M, Chalifoux LV, Lord CI, Letvin NL (1988) Suppression of simian immunodeficiency virus replication in vitro by CD8+ lymphocytes. J Immunol 140:2237–2242

Keefer MC, Graham BS, McElrath MJ et al (1996) Safety and immunogenicity of Env 2–3, a human immunodeficiency virus type 1 candidate vaccine, in combination with a novel adjuvant, MTP-PE/MF59. AIDS Res Hum Retroviruses 12:683–693

Kelleher AD, Emery S, Cunningham P et al (1997) Safety and immunogenicity of UBI HIV-1MN octameric V3 peptide vaccine administered by subcutaneous injection. AIDS Res Hum Retroviruses 13:29–32

Kent KA, Kitchin P, Mills KH et al (1994) Passive immunization of cynomolgus macaques with immune sera or a pool of neutralizing monoclonal antibodies failed to protect against challenge with SIVmac251. AIDS Res Hum Retroviruses 10:189–194

Kent SJ, Greenberg PD, Morton WR et al (1996) Detection of simian immunodeficiency virus (SIV)-specific $CD8^+$ T-cells in macaques protected from SIV challenge by prior SIV subunit vaccination. J Virol 70:4941–4947

Kessler JA Jr, McKenna PM, Emini EA et al (1997) Recombinant human monoclonal antibody IgG1b12 neutralizes diverse human immunodeficiency virus type 1 primary isolates. AIDS Res Hum Retroviruses 13:575–582

Kim JJ, Trivedi NN, Nottingham LK et al (1998) Modulation of amplitude and direction of in vivo immune responses by co-administration of cytokine gene expression cassettes with DNA immunogens. Eur J Immunol 28:1089–1103
Kirchhoff F, Greenough TC, Brettler DB et al (1995) Brief report: absence of intact nef sequences in a long-term survivor with nonprogressive HIV-1 infection. N Engl J Med 332:228–232
Klavinskis LS, Gao L, Barnfield C et al (1997) Mucosal immunization with DNA-liposome complexes. Vaccine 15:818–820
Koenig S, Fuerst TR, Wood LV et al (1990) Mapping the fine specificity of a cytolytic T cell response to HIV-1 nef protein. J Immunol 145:127–135
Koenig S, Conley AJ, Brewah YA et al (1995) Transfer of HIV-1-specific cytotoxic T lymphocytes to an AIDS patient leads to selection for mutant HIV variants and subsequent disease progression. Nat Med 1:330–336
Korber BT, Osmanov S, Esparza J, Myers G (1994) The World Health Organization Global Programme on AIDS proposal for standardization of HIV sequence nomenclature. WHO Network for HIV Isolation and Characterization. AIDS Res Hum Retroviruses 10:1355–1358
Kovacs JA, Vasudevachari MB, Easter M et al (1993) Induction of humoral and cell-mediated anti-human immunodeficiency virus (HIV) responses in HIV sero-negative volunteers by immunization with recombinant gp160. J Clin Invest 92:919–928
Kwong PD, Wyatt R, Robinson J, et al (1998) Structure of an HIV gp120 envelope glycoprotein in complex with the CD4 receptor and a neutralizing human antibody. Nature 393:648–659
LaCasse RA, Follis KE, Moudgil T et al (1998) Coreceptor utilization by human immunodeficiency virus type 1 is not a primary determinant of neutralization sensitivity. J Virol 72:2491–2495
Langlois AJ, Weinhold KJ, Matthews TJ et al (1992) The ability of certain SIV vaccines to provoke reactions against normal cells. Science 255:292–293
Learmont J, Tindall B, Evans L et al (1992) Long-term symptomless HIV-1 infection in recipients of blood products from a single donor. Lancet 340:863–867
Le Grand R, Vogt G, Vaslin B et al (1992) Specific and non-specific immunity and protection of macaques against SIV infection. Vaccine 10:873–879
Lehner T, Wang Y, Cranage M et al (1996) Protective mucosal immunity elicited by targeted iliac lymph node immunization with a subunit SIV envelope and core vaccine in macaques. Nat Med 2:767–775
Letvin NL, Montefiori DC, Yasutomi Y et al (1997) Potent, protective anti-HIV immune responses generated by bimodal HIV envelope DNA plus protein vaccination. Proc Natl Acad Sci USA 94:9378–9383
Levinson SO, Milzer A, Shaughnessy HJ et al (1954) Studies on production of antipoliomyelitis serum in rabbits. Proc Soc Exp Biol Med 87:111–114
Levy JA (1988) Mysteries of HIV: challenges for therapy and prevention. Nature 333:519–522
Levy JA, Hoffman AD, Kramer S et al (1984) Isolation of lymphocytopathic retroviruses from San Franciso patients with AIDS. Science 225:840–842
Li A, Katinger H, Posner MR et al (1998) Synergistic neutralization of simian-human immuno-deficiency virus SHIV-vpu+ by triple and quadruple combinations of human monoclonal antibodies and high-titer anti-human immunodeficiency virus type 1 immunoglobulins. J Virol 72:3235–3240
Li WH, Tanimura M, Sharp PM (1988) Rates and dates of divergence between AIDS virus nucleotide sequences. Mol Biol Evol 5:313–330
Lieberman J, Fabry JA, Fong DM et al (1997) Recognition of a small number of diverse epitopes dominates the cytotoxic T lymphocytes response to HIV type 1 in an infected individual. AIDS Res Hum Retroviruses 13:383–392
Limbach KJ, Paoletti E (1996) Non-replicating expression vectors: applications in vaccine development and gene therapy. Epidemiol Infect 116:244–252

Lipford GB, Bauer M, Blank C et al (1997) CpG-containing synthetic oligonucleotides promote B and cytotoxic T cell responses to protein antigen: a new class of vaccine adjuvants. Eur J Immunol 27:2340–2344
Lohman BL, Miller CJ, McChesney MB (1995) Antiviral cytotoxic T lymphocytes in vaginal mucosa of simian immunodeficiency virus-infected rhesus macaques. J Immunol 155:5855–5860
Lu M, Kim PS (1997) A trimeric structural subdomain of the HIV-1 transmembrane glycoprotein. J Biomol Struct Dyn 15:465–471
Lu S, Arthos J, Montefiori DC et al (1996) Simian immunodeficiency virus DNA vaccine trial in macaques. J Virol 70:3978–3991
Lu Y, Salvato MS, Pauza CD et al (1996) Utility of SHIV for testing HIV-1 vaccine candidates in macaques. J Acquir Immune Defic Syndr Hum Retrovirol 12:99–106
Lubeck MD, Natuk R, Myagkikh M et al (1997) Long-term protection of chimpanzees against high-dose HIV-1 challenge induced by immunization. Nat Med 3:651–658
Lubaki NM, Ray SC, Dhruva B et al (1997) Characterization of a polyclonal cytolytic T lymphocyte response to human immunodeficiency virus in persons without clinical progression. J Infect Dis 175: 1360–1367
Lyerly HK, Reed DL, Matthews TJ et al (1987) Anti-gp120 antibodies from HIV seropositive individuals mediate broadly reactive anti-HIV ADCC. AIDS Res Hum Retroviruses 3:409–422
Mackewicz CE, Ortega HW, Levy JA (1991) CD8+ cell anti-HIV activity correlates with the clinical state of the infected individual. J Clin Invest 87:1462–1466
Macklin MD, McCabe D, McGregor MW et al (1998) Immunization of pigs with a particle-mediated DNA vaccine to influenza A virus protects against challenge with homologous virus. J Virol 72:1491–1466
Mann JM, Tarantola DJM (1995) AIDS in the World II. Oxford University Press, New York, pp 5–40
Marthas ML, Miller CJ, Sutjipto S et al (1992) Efficacy of live-attenuated and whole-inactivated simian immunodeficiency virus vaccines against vaginal challenge with virulent SIV. J Med Primatol 21:99–107
Martinon F, Gras-Masse H, Boutillon C et al (1992) Immunization of mice with lipopeptides bypasses the prerequisite for adjuvant. Immune response of BALB/c mice to human immunodeficiency virus envelope glycoprotein. J Immunol 149:3416–3422
Mascola JR, Louder MK, Surman SR et al (1996a) Human immunodeficiency virus type 1 neutralizing antibody serotyping using serum pools and an infectivity reduction assay. AIDS Res Hum Retroviruses 12:1319–1328
Mascola JR, Snyder SW, Weislow OS et al (1996b) Immunization with envelope subunit vaccine products elicits neutralizing antibodies against laboratory-adapted but not primary isolates of human immunodeficiency virus type 1. J Infect Dis 173:340–348
Mascola JR, Louder MK, van Cott TC et al (1997) Potent and synergistic neutralization of human immunodeficiency virus (HIV) type 1 primary isolates by hyperimmune anti-HIV immunoglobulin combined with monoclonal antibodies 2F5 and 2G12. J Virol 71:7198–7206
Matsushita S, Robert-Guroff M, Rusche J et al (1988) Characterization of a human immunodeficiency virus neutralizing monoclonal antibody and mapping of the neutralizing epitope. J Virol 62:2107–2114
Mazzoli S, Trabattoni D, Lo Caputo S et al (1997) HIV-specific mucosal and cellular immunity in HIV-seronegative partners of HIV-seropositive individuals. Nat Med 3:1250–1257
McElrath MJ, Corey L, Greenberg PD et al (1994) Evaluation of cytotoxic T cell responses to candidate HIV-1 vaccines in HIV-1-uninfected individuals. AIDS Res Hum Retroviruses 10 [Suppl 2]:S69–S72
McMichael AJ, Phillips RE (1997) Escape of human immunodeficiency virus from immune control. Annu Rev Immunol 15:271–296

Mellors JW, Munoz A, Giorgi JV et al (1997) Plasma viral load and CD4+ lymphocytes as prognostic markers of HIV-1 infection. Ann Intern Med 126:946–954
Meng L, Ely JJ, Stouffer RL, Wolf DP (1997) Rhesus monkeys produced by nuclear transfer. Biol Reprod 57:454–459
Miller CJ, Marthas M, Torten J et al (1994) Intravaginal inoculation of rhesus macaques with cell-free simian immunodeficiency virus results in persistent or transient viremia. J Virol 68:6391–6400
Miller CJ, Lu Y, Roberts B et al (1997) Rhesus macaques previously infected with simian/human immunodeficiency virus are protected from vaginal challenge with pathogenic SIVmac239. J Virol 71:1911–1921
Mills KH, Page M, Chan WL et al (1992) Protection against SIV infection in macaques by immunization with inactivated virus from the BK28 molecular clone, but not with BK28-derived recombinant env and gag proteins. J Med Primatol 21:50–58
Montefiori DC, Collman RG, Fouts TR et al (1998) Evidence that antibody-mediated neutralization of human immunodeficiency virus type 1 by sera from infected individuals is independent of coreceptor usage. J Virol 72:1886–1893
Moore JP, Ho DD (1995) HIV-1 neutralization: the consequences of viral adaptation to growth on transformed T cells. AIDS 9[Suppl A]:S117–S136
Moore JP, Cao Y, Qing L et al (1995) Primary isolates of human immunodeficiency virus type 1 are relatively resistant to neutralization by monoclonal antibodies to gp120, and their neutralization is not predicted by studies with monomeric gp120. J Virol 69:101–109
Moore JP, Trkola A, Dragic T (1997) Co-receptors for HIV-1 entry. Curr Opin Immunol 9:551–562
Mortara L, Letourneur F, Gras-Masse H et al (1998) Selection of virus variants and emergence of virus escape mutants after immunization with an epitope vaccine. J Virol 72:1403–1410
Munoz A, Kirby AJ, He YD et al (1995) Long-term survivors with HIV-1 infection: incubation period and longitudinal patterns of CD4+ lymphocytes. J Acquir Immune Defic Syndr Hum Retrovirol 8:496–505
Munoz A, Sabin CA, Phillips AN (1997) The incubation period of AIDS. AIDS 11 [Suppl A]:S69–S76
Murali-Krishna K, Altman JD, Suresh M et al (1998) Counting antigen-specific CD8 T cells: a reevaluation of bystander activation during viral infection. Immunity 8:177–187
Murphey-Corb M, Ohkawa S, Davison-Fairburn B et al (1992) A formalin-fixed whole SIV vaccine induces protective responses that are cross-protective and durable. AIDS Res Hum Retroviruses 8:1475–1478
Murphey-Corb M, Wilson LA, Trichel AM et al (1998) Selective induction of protective MHC class I restricted CTL in the intestinal lamina propria of rhesus monkeys by transient infections of the colonic mucosa. J Immunol in press
Muster T, Guinea R, Trkola A et al (1994) Cross-neutralizing activity against divergent human immunodeficiency virus type 1 isolates induced by the gp41 sequence ELDKWAS. J Virol 68:4031–4034
Nara PL, Robey WG, Pyle SW et al (1988) Purified envelope glycoproteins from human immunodefi-ciency virus type 1 variants induce individual, type-specific neutralizing antibodies. J Virol 62:2622–2628
Nathanson N, Gonzalez-Scarano F (1989) Human immunodeficiency virus: an agent that defies vaccination. Adv Vet Sci Comp Med 33:397–412
Niedrig M, Gregersen JP, Fultz PN et al (1993) Immune response of chimpanzees after immunization with the inactivated whole immunodeficiency virus (HIV-1), three different adjuvants and challenge. Vaccine 11:67–74
Novembre FJ, Saucier M, Anderson DC et al (1997) Development of AIDS in a chimpanzee infected with human immunodeficiency virus type. J Virol 71:4086–4091

Nyambi PN, Nkengasong J, Peeters M et al (1995) Reduced capacity of antibodies from patients infected with human immunodeficiency virus type 1 (HIV-1) group O to neutralize primary isolates of HIV-1 group M viruses. J Infect Dis 172:1228–1237
Nyambi PN, Nkengasong J, Lewi P et al (1996) Multivariate analysis of human immunodeficiency virus type 1 neutralization data. J Virol 70:6235–6243
O'Doherty U, Ignatius R, Bhardwaj N, Pope M (1997) Generation of monocyte-derived dendritic cells from precursors in rhesus macaque blood. J Immunol Methods 207:185–194
Okada E, Sasaki S, Ishii N et al (1997) Intranasal immunization of a DNA vaccine with IL-12- and granulocyte-macrophage colony-stimulating factor (GM-CSF)-expressing plasmids in liposomes induces strong mucosal and cell-mediated immune responses against HIV-1 antigens. J Immunol 159:3638–3647
Oravecz T, Pall M, Norcross MA (1996) Beta-chemokine inhibition of monocytotropic HIV-1 infection. Interference with a postbinding fusion step. J Immunol 157:1329–1332
Pal R, Garzino-Demo A, Markham PD et al (1997) Inhibition of HIV-1 infection by the beta-chemokine MDC. Science 278:695–698
Palella FJ Jr, Delaney KM, Moorman AC et al (1998) Declining morbidity and mortality among patients with advanced human immunodeficiency virus infection. HIV Outpatient Study Investigators. N Engl J Med 338:853–860
Paliard X, Lee AY, Walker CM (1996) RANTES, MIP-1 alpha and MIP-1 beta are not involved in the inhibition of HIV-1SF33 replication mediated by CD8+ T-cell clones. AIDS 10:1317–1321
Payne R, Tripp M, Weigle J et al (1964) A new leukocyte isoantigen system in man. Cold Spring Harbor Symp Quant Biol 29:285–295
Perales MA, Schwartz DH, Fabry JA, Lieberman J (1995) A vaccinia-gp160-based vaccine but not a gp160 protein vaccine elicits anti-gp160 cytotoxic T lymphocytes in some HIV-1 seronegative vaccinees. J Acquir Immune Defic Syndr Hum Retrovirol 10:27–35
Perelson AS, Essunger P, Ho DD (1997) Dynamics of HIV-1 and $CD4^+$ lymphocytes in vivo. AIDS 11 [Suppl]:S17–S24
Phanuphak P, Teeratakulpixarn S, Sarangbin S et al (1997) International clinical trials of HIV vaccines: I. Phase I trial of an HIV-1 synthetic peptide vaccine in Bangkok, Thailand. Asian Pac J Allergy Immunol 15:41–48
Picard O, Achour A, Bernard J et al (1992) A 2-year follow-up of an anti-HIV immune reaction in HIV-1 gp160-immunized healthy seronegative humans: evidence for persistent cell-mediated immunity. J Acquir Immune Defic Syndr Hum Retrovirol 5:539–546
Plotkin SA, Mortimer EA (1994) Vaccines, 2nd edn. Saunders, Philadelphia
Price DA, Goulder PJ, Klenerman P et al (1997) Positive selection of HIV-1 cytotoxic T lymphocyte escape variants during primary infection. Proc Natl Acad Sci USA 94:1890–1895
Prince AM, Horowitz B, Shulman RW et al (1990) Apparent protection of HIV infection by HIV immunoglobulin given prior to low-dose HIV challenge. In: Brown F, Chanock R, Ginsberg H, Lerner R (eds) Vaccines 90. Cold Spring Harbor Laboratory Press, Cold Spring Harbor, pp 347–351
Prince AM, Reesink H, Pascual D et al (1991) Prevention of HIV infection by passive immunization with HIV immunoglobulin. AIDS Res Hum Retroviruses 7:971–973
Purtscher M, Trkola A, Gruber G et al (1994) A broadly neutralizing human monoclonal antibody against gp41 of human immunodeficiency virus type 1. AIDS Res Hum Retroviruses 10:1651–1658
Pushko P, Parker M, Ludwig GV et al (1997) Replicon-helper systems from attenuated Venezuelan equine encephalitis virus: expression of heterologous genes in vitro and immunization against heterologous pathogens in vivo. Virology 239:389–401
Putkonen P, Thorstensson R, Ghavamzadeh L et al (1991) Prevention of HIV-2 and SIVsm infection by passive immunization in cynomolgus monkeys. Nature 352:436–438

Putkonen P, Thorstensson R, Cranage M et al (1992) A formalin inactivated whole SIVmac vaccine in Ribi adjuvant protects against homologous and heterologous SIV challenge. J Med Primatol 21:108–112
Putkonen P, Nilsson C, Makitalo B et al (1996) Immunization with live attenuated SIVmac can protect macaques against mucosal infection with SIVsm. In: Brown F, Chanock R, Ginsberg H, Lerner R (eds) Vaccines 96. Cold Spring Harbor Laboratory Press, Cold Spring Harbor, pp 1–5
Reitter JN, Means RE, Desrosiers RC (1998) A role for carbohydrates in immune evasion in AIDS. Nat Med 4:679–684
Richardson TM, Stryjewski BL, Broder CC et al (1996) Humoral response to oligomeric human immunodeficiency virus type 1 envelope protein. J Virol 70:753–762
Rinaldo CR Jr, Beltz LA, Huang XL et al (1995) Anti-HIV type 1 cytotoxic T lymphocyte effector activity and disease progression in the first 8 years of HIV type 1 infection of homosexual men. AIDS Res Hum Retroviruses 11:481–489
Robey FA, Harris-Kelson T, Robert-Guroff M et al (1996) A synthetic conformational epitope from the C4 domain of HIV Gp120 that binds CD4. J Biol Chem 271:17990–17995
Robey WG, Arthur LO, Matthews TJ et al (1986) Prospect for prevention of human immunodeficiency virus infection: purified 120 kDa envelope glycoprotein induces neutralizing antibody. Proc Natl Acad Sci USA 83:7023–7027
Rolls MM, Haglund K, Rose J (1996) Expression of additional genes in a vector derived from a minimal RNA virus. Virology 218:406–411
Romagnani S, Del Prete G, Manetti R et al (1994) Role of TH1/TH2 cytokines in HIV infection. Immunol Rev 140:73–92
Rosenberg ES, Billingsley JM, Caliendo AM et al (1997) Vigorous HIV-1-specific CD4+ T cell responses associated with control of viremia. Science 278:1447–1450
Rouaix F, Gras-Masse H, Mazingue C et al (1994) Effect of a lipopeptidic formulation on macrophage activation and peptide presentation to T cells. Vaccine 12:1209–1214
Rovinski B, Haynes JR, Cao SX et al (1992) Expression and characterization of genetically engineered human immunodeficiency virus-like particles containing modified envelope glycoproteins: implications for development of a cross-protective AIDS vaccine. J Virol 66:4003–4012
Rowland-Jones S, Corrah T, Gallimore A et al (1995) HIV-specific cytotoxic T-cells in HIV-exposed but uninfected Gambian women. Nat Med 1:59–64
Rubbert A, Weissman D, Combadiere C et al (1997) Multifactorial nature of noncytolytic CD8+ T cell-mediated suppression of HIV replication: beta-chemokine-dependent and -independent effects. AIDS Res Hum Retroviruses 13:63–69
Rubinstein A, Goldstein H, Pettoello-Mantovani M et al (1995) Safety and immunogenicity of a V3 loop synthetic peptide conjugated to purified protein derivative in HIV-seronegative volunteers. AIDS 9:243–251
Russell C (1984) Virus discovery could lead to test for AIDS. The Washington Post, 24 April 1984, section A, p 1
Sabin AB (1992) Improbability of effective vaccination against human immunodeficiency virus because of its intracellular transmission and rectal portal of entry. Proc Natl Acad Sci USA 89:8852–8855
Salmon-Ceron D, Excler JL, Sicard D et al (1995) Safety and immunogenicity of a recombinant HIV type 1 glycoprotein 160 boosted by a V3 synthetic peptide in HIV-negative volunteers. AIDS Res Hum Retroviruses 11:1479–1486
Sasaki S, Sumino K, Hamajima K et al (1998) Induction of systemic and mucosal immune responses to human immunodeficiency virus type 1 by a DNA vaccine formulated with QS-21 saponin adjuvant via intramuscular and intranasal routes. J Virol 72:4931–4939
Sattentau QJ, Moore JP (1995) Human immunodeficiency virus type 1 neutralization is determined by epitope exposure on the gp120 oligomer. J Exp Med 182:185–196

Schneider J, Gilbert SC, Blanchard TJ et al (1998) Enhanced immunogenicity for CD8+ T cell induction and complete protective efficacy of malaria DNA vaccination by boosting with modified vaccinia virus Ankara. Nat Med 4:397–402
Schrager LK, Young JM, Fowler MG et al (1994) Long-term survivors of HIV-1 infection: definitions and research challenges. AIDS 8 [Suppl 1]:S95–S108
Schwartz DH, Gorse G, Clements ML et al (1993) Induction of HIV-1-neutralising and syncytium-inhibiting antibodies in uninfected recipients of HIV-1 IIIB rgp120 subunit vaccine. Lancet 342:69–73
Sheppard HW, Lang W, Ascher MS et al (1993) The characterization of non-progressors: long-term HIV-1 infection with stable CD4+ T-cell levels. AIDS 7:1159–1166
Shibata R, Hoggan MD, Broscius C et al (1995) Isolation and characterization of a syncytium-inducing, macrophage/T-cell line-tropic human immunodeficiency virus type 1 isolate that readily infects chimpanzee cells in vitro and in vivo. J Virol 69:4453–4462
Shibata R, Seimon C, Cho MW et al (1996) Resistance of previously infected chimpanzees to successive challenges with a heterologous intraclade B strain of human immunodeficiency virus type 1. J Virol 70:4361–4369
Shibata R, Martin MA, Desrosiers RC et al (1997) Live, attenuated simian immunodeficiency virus vaccines elicit potent resistance against a challenge with a human immunodeficiency virus type 1 chimeric virus. J Virol 71:8141–8148
Shiver JW, Davies ME, Yasutomi Y et al (1997) Anti-HIV env immunities elicited by nucleic acid vaccines. Vaccine 15(8):884–887
Smith AD, Geisler SC, Chen AA et al (1998) Human rhinovirus type 14: human immuno-deficiency virus type 1 (HIV-1) V3 loop chimeras from a combinatorial library induce potent neutralizing antibody responses against HIV-1. J Virol 72:651–659
Smith TF, Srinivasan A, Schochetman G et al (1988) The phylogenetic history of immunodeficiency viruses. Nature 333:573–575
Stanhope PE, Clements ML, Siliciano RF (1993) Human CD4+ cytolytic T lymphocyte responses to a human immunodeficiency virus type 1 gp160 subunit vaccine. J Infect Dis 168:92–100
Steimer KS, Haigwood NL (1991) Importance of conformation on the neutralizing antibody response to HIV-1 gp120. Biotechnol Ther 2:63–89
Sutter G, Moss B (1995) Novel vaccinia vector derived from the host range restricted and highly attenuated MVA strain of vaccinia virus. Dev Biol Stand 84:195–200
Trkola A, Pomales AB, Yuan H et al (1995) Cross-clade neutralization of primary isolates of human immunodeficiency virus type 1 by human monoclonal antibodies and tetrameric CD4-IgG. J Virol 69:6609–6617
Trkola A, Purtscher M, Muster T et al (1996) Human monoclonal antibody 2G12 defines a distinctive neutralization epitope on the gp120 glycoprotein of human immunodeficiency virus type 1. J Virol 70:1100–1108
Trkola A, Ketas T, Kewalramani VN et al (1998) Neutralization sensitivity of human immunodeficiency virus type 1 primary isolates to antibodies and CD4-based reagents is independent of coreceptor usage. J Virol 72:1876–1885
Tyler DS, Stanley SD, Nastala CA et al (1990) Alterations in antibody-dependent cellular cytotoxicity during the course of HIV-1 infection. Humoral and cellular defects. J Immunol 144:3375–3384
van Cott TC (1997) Protection of rhesus macaques from homologous and heterologous SHIV challenge using oligomeric gp140. In: Girard M, Dodet B (eds) Retroviruses of human AIDS and related animal diseases: 11e Colloque des "Cent Gardes." Fondation Marcel Merieux, Lyon, France, pp 199–208
van Cott TC, Bethke FR, Burke DS et al (1995a) Lack of induction of antibodies specific for conserved, discontinuous epitopes of HIV-1 envelope glycoprotein by candidate AIDS vaccines. J Immunol 155:4100–4110

van Cott TC, Polonis VR, Loomis LD et al (1995b) Differential role of V3-specific antibodies in neutralization assays involving primary and laboratory-adapted isolates of HIV type 1. AIDS Res Hum Retroviruses 11:1379–1391
van der Burg SH, Klein MR, Pontesilli O et al (1997) HIV-1 reverse transcriptase-specific CTL against conserved epitopes do not protect against progression to AIDS. J Immunol 159:3648–3654
Vijh-Warrier S, Pinter A, Honnen WJ, Tilley SA (1996) Synergistic neutralization of human immunodeficiency virus type 1 by a chimpanzee monoclonal antibody against the V2 domain of gp120 in combination with monoclonal antibodies against the V3 loop and the CD4-binding site. J Virol 70:4466–4473
Vogel TU, Fournier T, Sherring A et al (1998) Presence of circulating CTL against SIV Env, RT or Nef induced by SIVmacJ5 or SIVmacC8 does not correlate with protection against SHIV89.6P challenge. J Med Primatol (in press)
Vujcic LK, Quinnan GV Jr (1995) Preparation and characterization of human HIV type 1 neutra-lizing reference sera. AIDS Res Hum Retroviruses 11:783–787
Wadman M (1998) Ex-UN AIDS chief is blasted for remarks on vaccine strategy. Nature 392:527–528
Walker BD, Plata F (1990) Cytotoxic T lymphocytes against HIV. AIDS 4:177–184
Walker BD, Chakrabarti S, Moss B et al (1987) HIV-specific cytotoxic T lymphocytes in seropositive individuals. Nature 328:345–348
Walker CM (1993) Non-cytolytic control of HIV replication by CD8+ T cells. Semin Immunol 5:195–201
Walker CM, Moody DJ, Stites DP, Levy JA (1986) CD8+ lymphocytes can control HIV infection in vitro by suppressing virus replication. Science 234:1563–1566
Walker MC, Fast PE (1994) Clinical trials of candidate AIDS vaccines. AIDS 8 [Suppl]:S213–S236
Warren JT, Levinson MA (1997) Preclinical AIDS vaccine development: formal survey of worldwide HIV, SIV and SHIV in vivo challenge studies in nonhuman primates. J Med Primatol 26:63–81
Weber J, Fenyo EM, Beddows S et al (1996) Neutralization serotypes of human immunodeficiency virus type 1 field isolates are not predicted by genetic subtype. The WHO Network for HIV Isolation and Characterization. J Virol 70:7827–7832
Wei X, Ghosh SK, Taylor ME et al (1995) Viral dynamics in human immunodeficiency virus type 1 infection. Nature 373:117–122
Weiss RA, Clapham PR, McClure M, Marsh M (1989) The CD4 receptor for the AIDS virus. Biochem Soc Trans 17:644–647
Wintsch J, Chaignat CL, Braun DG et al (1991) Safety and immunogenicity of a genetically engineered human immunodeficiency virus vaccine. J Infect Dis 163:219–225
Wiviott LD, Walker CM, Levy JA (1990) CD8+ lymphocytes suppress HIV production by autologous CD4+ cells without eliminating the infected cells from culture. Cell Immunol 128:628–634
Wrin T, Crawford L, Sawyer L et al (1994) Neutralizing antibody responses to autologous and heterologous isolates of human immunodeficiency virus. J Acquir Immune Defic Syndr 7:211–219
Wu S, Pascual DW, Lewis GK, Hone DM (1997) Induction of mucosal and systemic responses against human immunodeficiency virus type 1 glycoprotein 120 in mice after oral immunization with a single dose of a Salmonella-HIV vector. AIDS Res Hum Retroviruses 13:1187–1194
Yang OO, Walker BD, Johnson RP et al (1997) Suppression of human immunodeficiency virus type 1 replication by CD8+ cells: evidence for HLA class I-restricted triggering of cytolytic and noncytolytic mechanisms. J Virol 71:3120–3128
Yasutomi Y, Koenig S, Woods RM et al (1995) A vaccine-elicited, single viral epitope-specific cytotoxic T lymphocyte response does not protect against intravenous, cell-free simian immunodeficiency virus challenge. J Virol 69:2279–2284

Yasutomi Y, Robinson HL, Lu S et al (1996) Simian immunodeficiency virus-specific cytotoxic T-lymphocyte induction through DNA vaccination of rhesus monkeys. J Virol 70:678–681
Yokoyama S, Chung L, Gojobori T (1988) Molecular evolution of the human immunodeficiency and related viruses. Mol Biol Evol 5:237–251
Zagury D, Lachgar A, Chams V et al (1998) C-C chemokines, pivotal in protection against HIV type 1 infection. Proc Natl Acad Sci USA 95:3857–3861
Zarling JM, Morton W, Moran PA et al (1986) T-cell responses to human AIDS virus in macaques immunized with recombinant vaccinia viruses. Nature 323:344–346
Zhang YJ, Fredriksson R, McKeating JA et al (1997) Passage of HIV-1 molecular clones into different cell lines confers differential sensitivity to neutralization. Virology 238:254–264
Zhou X, Berglund P, Rhodes G et al (1994) Self-replicating Semliki Forest virus RNA as recombinant vaccine. Vaccine 12:1510–1514

CHAPTER 15

An Overview of Malaria Vaccine Development Efforts

S. KUMAR, D.C. KASLOW, and S.L. HOFFMAN

A. Epidemiology

Malaria is one of the most important diseases in the world. Currently about 2.4 billion people live in countries where malaria is transmitted. There are an estimated 300–500 million new cases of malaria with 1.5–2.7 million deaths annually (WHO FACT SHEET 1996). The majority of cases and deaths due to malaria occur in children under 5 years of age living in sub-Saharan Africa, but the disease exacts its toll in virtually all tropical and many subtropical areas of the world. A number of factors have led to the worsening of the situation in many parts of the world in recent years. Foremost among these factors has been the development of resistance of malaria parasites to drugs such as chloroquine that have been effectively used for treatment and prophylaxis for almost a half a century.

B. The Parasite

Human malaria is caused by infection with any of four protozoans of the genus *Plasmodium* (phylum Apicomplexia): *P. falciparum*, *P. vivax*, *P. ovale*, and *P. malariae*. Among these, *P. falciparum* is the deadliest and is responsible for more than 95% of deaths attributed to malaria.

The infection commences when a female *Anopheles* mosquito inoculates infectious sporozoites into the blood stream of the host while taking a blood meal. Within a brief period, which may be less than 5 min, the sporozoites invade hepatocytes. Inside the hepatocyte, in a period of 2–16 days (minimum of 5.5 days for parasites that infect humans), each sporozoite may develop into 10000–30000 uninucleated liver stage merozoites. After maturation the infected hepatocytes rupture and release merozoites that invade erythrocytes, initiating the erythrocytic stage of the parasite life cycle which is responsible for all malaria-associated pathology and clinical manifestations. The asexual erythrocytic stage cycle is 24–72 h (48–72 h in humans). Within an erythrocyte a merozoite matures into rings, trophozoites, and schizonts. The schizont stage is characterized by the initiation of nuclear division during which each merozoite divides into 12–36 individual merozoites. The merozoites are released following the rupture of the infected erythrocytes, and within seconds invade other erythrocytes. This cycle of invasion, development, rupture, and

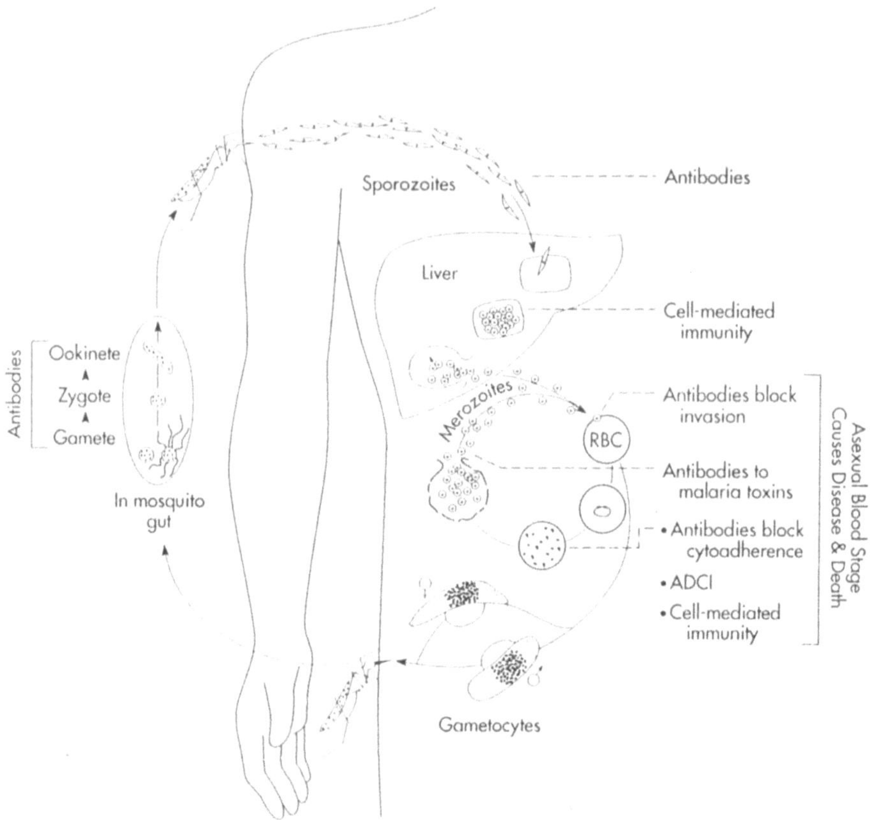

Fig. 1. The life cycle of *Plasmodium* species. (From GOOD and MILLER 1998)

reinvasion is associated with all clinical manifestations of the disease. Through unknown mechanism(s) some merozoites develop within erythrocytes into male or female gametocytes, the first cells in the sexual stage of parasite life cycle. While feeding on the host mosquitoes ingest erythrocytes containing gametocytes. In the midgut of the mosquito the gametocytes differentiate into extracellular gametes. Fertilization of the female by the male occurs, and a zygote is formed. The zygote form undergoes maturation into ookinete, oocyst and subsequently into infectious sporozoites. A diagrammatic representation of the life cycle of *Plasmodium* spp. is shown in Fig. 1.

C. Approaches to Malaria Vaccine Development: Preventing Erythrocytic Stage Infection or Reducing Morbidity and Mortality Without Preventing Infection

Several approaches are being pursued by investigators throughout the world to develop a malaria vaccine, and it would be a great simplification to imply that there are only two approaches to solving this problem. However, in the interests of clarity we find it useful to classify the approaches broadly into two areas. One approach is to prevent parasites from invading erythrocytes and thereby prevent any clinical manifestations of malaria. This is the type of vaccine being developed for use in nonimmune visitors to malarious areas. The aim of this approach is to prevent sporozoites from invading hepatocytes, and to inhibit development of liver stage parasites and the release of infectious liver stage merozoites. This is often termed a "preerythrocytic" stage malaria vaccine. A second approach focuses on reducing the millions of deaths caused by malaria each year without necessarily eliminating the erythrocytic stage infection. There are a number of ways this is being attempted and these will be described below.

D. Preerythrocytic Stage Vaccines

I. Preventing Sporozoite Invasion of Hepatocytes

It is estimated that during each infectious bite only 5–500 sporozoites are inoculated into the blood stream. These sporozoites are extracellular for a brief period before they reach the liver and invade hepatocytes. In hepatocytes there is an amplification process that results in fully mature liver stage schizonts with tens of thousands of uninucleate merozoites. The most practical way to prevent infection would be to neutralize sporozoite infectivity before the few sporozoites inoculated by mosquitoes enter hepatocytes and begin their enormous amplification process. Since sporozoites are extracellular, this requires antibodies against accessible proteins on the surface of the sporozoite. However, sporozoites are in the circulation for no more than 30–60 min (FAIRLEY 1947) before entering the liver, and therefore the appropriate antibodies must be at extremely high levels at the time of infection so as to neutralize the infectivity of inoculated sporozoites without boosting. In the case of a nonimmune individual who has only received a sporozoite vaccine, in whom the goal is to prevent erythrocytic stage infection, these antibodies must prevent all sporozoites from infecting hepatocytes. This is because a single sporozoite can cause infection (KHUSMITH et al. 1991, 1994) that can lead to all the clinical and pathological manifestations of malaria. This may not be the case among individuals who have naturally acquired or vaccine-induced partially effective immune responses against liver and/or erythrocytic stages of the parasite life cycle (HOFFMAN and MILLER 1996).

Based on the life cycle it is clear that to prevent sporozoite invasion of hepatocytes a vaccine must induce antibodies against surface proteins of the extracellular sporozoites. The most logical approach would be to identify parasite proteins that are involved in adhesion to and invasion of hepatocytes and to induce antibodies that interfere with these critical functions. Although there is now a body of data that has developed regarding the processes of adhesion and invasion, this has not yet resulted in an experimentally validated vaccine development strategy. However, it is hoped that this approach will eventually lead to a more effective vaccine (see below).

The primary strategy for inducing antibodies against sporozoites is based on the observations that passive transfer of mAbs against the central repeat domain of the *P. berghei* (Pb), and *P. yoelii* (Py) circumsporozoite proteins (CSP) protect mice against sporozoite challenge (Potocnjak et al. 1980; Charoenvit et al. 1991b), and passive transfer of a mAb against the *P. vivax* (Pv) CSP protects monkeys against *P. vivax* sporozoite challenge (Charoenvit et al. 1991a). In addition, it has been shown that incubation of Pv or *P. falciparum* (Pf) sporozoites with the appropriate mAbs prior to inoculation, neutralizes the infectivity of these sporozoites in chimpanzees (Nardin et al. 1982). There are good experimental data indicating that the antibodies function through direct neutralization of sporozoites and do not require complement or cellular-dependent cytotoxicity to mediate their anti-infective activity. An $F(ab)_2$ fragment of the anti-PbCSP mAb inhibits infection (Potocnjak et al. 1980), and the anti-PyCSP mAb is protective in mice depleted of complement and in mice deficient in macrophage function, and is not effective when transferred 5 min after sporozoite inoculation (Y. Charoenvit, personal communication).

The potential for the use of CSP as antisporozoite antibody inducing vaccine has been supported by a series of studies in mice in which immunization with purified PbCSP recombinant protein (Egan et al. 1987), and PbCSP or PyCSP linear synthetic or multiple antigenic peptide (MAP) vaccines in adjuvants have been shown to produce protective immunity (Tam et al. 1990; Zavala and Chai 1990; Wang et al. 1995). Based in part on these murine models there have been extensive studies of PfCSP purified recombinant protein and synthetic peptide vaccines in humans beginning more than a decade ago. In the first two studies a purified recombinant protein (Ballou et al. 1985) and a synthetic peptide (Herrington et al. 1987), both including CSP sequences only from the central repeat region, were shown to be modestly immunogenic, and to protect one of six (Ballou et al. 1985) and one of three (Herrington et al. 1987) individuals challenged by the bite of infected mosquitoes. Overall these early generation vaccines gave only 15%–20% protection.

There have been significant efforts over the last decade to improve on this protection by using new vaccines with new adjuvants based entirely on the central repeat region of the PfCSP (Rickman et al. 1991; Fries et al. 1992; Hoffman et al. 1994) and by incorporating the sequence from the C-terminus of the PfCSP into the immunogen. Although these C-terminal sequences may contain protective B cell epitopes, they have primarily been included to provide T cell epitopes that would allow boosting of vaccine-induced antibodies by natural exposure to sporozoites, and protective T cell responses against infected hepatocytes (see below). Recently a vaccine (designated RTS,S) in which the repeat region and entire C-terminus of the PfCSP are expressed on hepatitis B surface antigen particles and delivered in a complex adjuvant including QS-21, and monophosphoryl lipid A, induced extremely high levels of antibodies against sporozoites, and protected six of seven volunteers against experimental challenge (STOUTE et al. 1997). Although this vaccine induces high levels of antibodies, there was no clear-cut association between antibody level and protection, and no clear association between the capacity of these sera to inhibit sporozoite invasion of hepatocytes in vitro and protection. This has led to speculation that in some of the volunteers the protection was mediated all or in part by protective T cell responses against PfCSP peptides presented with host MHC molecules on the surface of infected hepatocytes. Work is in progress to determine how long this protection lasts, whether it is reproducible, and whether it is effective against natural infection in the field, and to further characterize T cell responses against the vaccine.

Scientists are pursuing two approaches to improving the protective antibody-mediated immunity induced by the CSP vaccines. The first is to identify other parasite proteins expressed on sporozoites and induce antibodies against these sporozoites. Using a mAb, CHAROENVIT and colleagues (1987) identified one such protein on the *P. yoelii* sporozoite surface and called it *P. yoelii* sporozoite surface protein 2 (PySSP2). The protein is presented in a patchy pattern on sporozoites (AIKAWA et al. 1990a). The partial (HEDSTROM et al. 1990) and the full-length PySSP2 gene was cloned and sequenced (ROGERS et al. 1992a). Subsequently ROGERS and colleagues (1992b) identified the *P. falciparum* homologue, PfSSP2, and found it to be the previously described protein, TRAP (ROBSON et al. 1988), which had been thought to be a blood-stage antigen. Polyclonal (ROGERS et al. 1992b) and monoclonal (CHAROENVIT et al. 1997) antibodies (mAbs) to SSP2 inhibit sporozoite invasion of hepatocytes in vitro. However, their inhibitory activity is significantly less than that of anti-PfCSP polyclonal and monoclonal antibodies. Nonetheless, combining PfSSP2 recombinant protein with PfCSP recombinant protein with the intention of inducing additive or synergistic antibodies is in progress.

Another approach to preventing sporozoite invasion of hepatocytes is to identify critical sporozoite-host receptor ligand interactions, and induce antibodies to these specific sporozoite ligands. Both CSP and SSP2 include domains called region II or region II$^+$ that bear similarity to adhesion molecules including thrombospondin and properdin, and these regions bind sulfated glyconjugates (CERAMI et al. 1992, 1994; PANCAKE et al. 1992; MULLER et al. 1993). It has subsequently been shown that these regions probably mediate attachment to hepatocytes through glycosaminoglycan chains of heparin sulfate proteoglycans on hepatocyte membranes (SINNIS et al. 1996). SSP2/TRAP also contains a region of approximately 200 amino acid residues with strong homology to the A-domain present on many proteins involved in cell-cell, cell-matrix, and matrix-matrix interactions (SULTAN et al. 1997). A rationale approach to vaccine development would be to induce antibodies against one or both of these regions. So far it has been extremely difficult to induce high levels of inhibitory antibodies against region II$^+$, and work is now in progress to target the A-domain-like region of SSP2/TRAP.

II. Attacking Infected Hepatocytes

After a sporozoite invades an hepatocyte it generally develops within 48 h (rodent malaria parasites, *P. yoelii*, and *P. berghei*) to 5.5–14 days (human malaria parasites, *P. falciparum*, *P. vivax*, *P. malariae*, *P. ovale*) to a mature liver stage schizont with thousands to tens of thousands uninucleate merozoites, each of which can invade an erythrocyte when the infected hepatocyte ruptures. The intrahepatic parasites which do not cause any clinical or pathological symptoms of malaria are considered excellent targets for vaccine-induced immunity. This is the only stage of the life cycle in which the parasite resides within a cell that consistently expresses major histocompatibility complex (MHC=HLA) molecules on its surface and therefore can directly interact with effector T cells. The elimination of all infected hepatocytes prevents all manifestations of the disease which arise from the cyclical invasion, development and rupture of erythrocytes.

The rationale for attacking infecting hepatocytes has been based on data from two different sets of experiments. The first is the observation that immunization of fowl (MULLIGAN et al. 1941), mice (NUSSENZWEIG et al. 1967), monkeys (GWADZ et al. 1979), and humans (CLYDE et al. 1973a, 1975, 1990; RIECKMANN et al. 1974, 1979; HERRINGTON et al. 1991; EGAN et al. 1993) with radiation-attenuated sporozoites induces sterile protective immunity. This protective immunity is stage specific; no protection was observed following challenges with infected erythrocytes. This means that the protective immunity is directed against the sporozoite or liver stage of the parasite life cycle. The immunity has generally been reported to be species-specific, suggesting that immune responses are antigen specific but not strain specific in the case of *P. falciparum* (CLYDE et al. 1973b; RIECKMANN et al. 1979). It was initially thought that antibodies against sporozoites are responsible for this protection since sera from immunized mice and humans precipitate the surface coat of sporozoites (COCHRANE et al. 1976). However, μ-suppressed mice that do not produce antibodies can be protected by immunization with irradiated sporozoites (CHEN et al. 1977), and adoptive transfer of purified T cells from mice immunized with radiation attenuated sporozoites into irradiated naive mice protects against challenge at a time when no antibodies are present (EGAN et al. 1987).

Most strikingly, mice immunized with irradiated sporozoites and depleted of $CD8^+$ T cell function by in vivo treatment with mAbs completely lose their protective immunity, indicating that $CD8^+$ T cells are required for the protection induced by the irradiated sporozoite vaccine (SCHOFIELD et al. 1987; WEISS et al. 1988). In some strains of mice interferon (IFN) γ and nitric oxide are also required for protection (SEGUIN et al. 1994). These findings suggest that $CD8^+$ T cells, either acting directly through cytotoxicity or perhaps, in a more likely scenario, by production of IFN-γ which induces the infected hepatocyte to produce nitric oxide which kills the parasite, are responsible for the protection induced by the irradiated sporozoite vaccine. In fact depending on the strain of mouse immunized, $CD8^+$ CTL, $CD8^+$ T cells producing IFN-γ with or without the involvement of nitric oxide, or a feedback mechanism in the liver for expansion of IFN-γ production, and $CD4^+$ T cells can all be required for protective immunity (Doolan and Hoffman 1997, unpublished).

A second set of experiments have produced findings entirely complementary to those in the irradiated sporozoite vaccine model systems. In vitro treatment of *P. berghei* (Ferreira et al. 1986) and *P. falciparum* (Mellouk et al. 1987) infected hepatocytes with IFN-γ eliminate infected hepatocytes from culture. It has also been shown that treatment of mice (Ferreira et al. 1986) and monkeys (Maheshwari et al. 1986) with IFN-γ reduces or prevents sporozoite infection. Subsequently it was shown that the activity of IFN-γ against infected hepatocytes requires the production of nitric oxide by the infected hepatocyte (Mellouk et al. 1991, 1994). Finally it has been shown that treatment of mice (Sedegah et al. 1994a) and monkeys (Hoffman et al. 1997a) with a single injection of recombinant interleukin (IL) 12 provides sterile protective immunity and that this protection is dependent on IFN-γ in mice and is associated with circulating IFN-γ levels in monkeys. These experiments demonstrate that IFN-γ induced production of nitric oxide by *Plasmodium* infected hepatocytes leads to elimination of the infected hepatocytes, and when combined with the observations from the irradiated sporozoite model, provide a clear-cut strategy for vaccine development.

Having established the immune effectors desired for induction of protective immunity, the next challenge is to identify the parasite proteins that are the targets of this T cell mediated protective immunity, and to develop vaccine delivery systems that induce protective T cell responses against infected hepatocytes. This work has been carried out primarily in the rodent malaria model systems.

There is ample evidence that both $CD8^+$ and $CD4^+$ T cell responses against infected hepatocytes can provide sterile protective immunity in rodents. The presence of a class I cytotoxic T cell epitope on a malaria protein was first demonstrated in studies by Kumar and colleagues (1988). $CD8^+$ T cells taken from the spleens of B10. BR ($H2^k$) mice immunized with recombinant vaccinia expressing PfCSP lysed fibroblast cells ($H2^k$) transfected with CSP gene. The CTL epitope was mapped in the polymorphic region which also contains T helper epitopes (Kumar et al. 1988). Since then a number of $CD8^+$ T cells epitopes have been identified on several preerythrocytic stage antigens. $CD8^+$ T cell clones against the PbCSP (Romero et al. 1989), PyCSP (Weiss et al. 1990; Rodrigues et al. 1991), and PySSP2 (Khusmith et al. 1994), and $CD4^+$ T cell clones against the PyCSP (Renia et al. 1991) adoptively transfer sterile protective immunity against sporozoite challenge.

Recombinant *Salmonella* (Aggarwal et al. 1991) and vaccinia expressing the PbCSP (Lanar et al. 1996), recombinant mastocytoma cells expressing the PyCSP and PySSP2 (Khusmith 1991), priming with recombinant influenza expressing a $CD8^+$ T cell epitope from the PyCSP following by boosting with recombinant vaccinia expressing the PyCSP (Li et al. 1993), and immunization with DNA plasmids expressing PyCSP (SEDEGAH et al. 1994b), PyHEP17 (DOOLAN et al. 1996), and PySSP2 (S. Hoffman and H. Wang, unpublished) all induce CD8$^+$ T cell dependent protective immunity against sporozoite challenge. Furthermore, immunization with a MAP vaccine in adjuvant based on a PyCSP CD8$^+$ T cell epitope also gives CD8$^+$ CTL and some protection (FRANKE et al. 1997). Immunization with a synthetic peptide based on a PbCSP CD4$^+$ T cell epitope has been shown to give protection (MIGLIORINI et al. 1993). Immunizations of mice with a linear synthetic peptide based on a CD4$^+$ T cell epitope on PySSP2 (WANG et al. 1996) and on PyHEP17 (Y Charoenvit, personal communication) induce sterile protective immunity that is dependent on CD4$^+$ T cells and IFN-γ, and is stage specific.

There have also been significant concurrent efforts to identify T cell epitopes on *P. falciparum* by studying T cells from individuals naturally exposed to malaria or those immunized with radiation-attenuated sporozoites. Because in all strains of mice studied thus far protective immunity has been dependent on CD8$^+$ T cells, work with *P. falciparum* has focused on identifying CD8$^+$ T cell epitopes; the first class I restricted CD8$^+$ CTL epitope on CSP from humans immunized with irradiated sporozoites was described in 1991 (MALIK et al. 1991). Since then there have been a number of reports documenting *P. falciparum* CTL epitopes in humans naturally exposed (DOOLAN et al. 1991, 1997; HILL et al. 1991; AIDOO et al. 1995) and immunized with irradiated sporozoites (WIZEL et al. 1995a,b). This work was tremendously expanded when 17 new CTL epitopes were identified on *P. falciparum* preerythrocytic stage proteins (DOOLAN et al. 1997).

Based on the rodent findings and the identification of multiple CD8$^+$ and CD4$^+$ T cell epitopes on *P. falciparum* proteins expressed in infected hepatocytes, there has been work to develop and test in humans *P. falciparum* vaccines designed to induce protective T cell responses against infected hepatocytes. A recombinant *S. typhi* expressing the PfCSP (GONZALEZ et al. 1994) and a recombinant attenuated vaccinia expressing seven *P. falciparum* proteins (at least five of which are expressed in hepatocytes) (OCKENHOUSE et al. 1998) have been evaluated for immunogenicity and in the case of the recombinant vaccinia for protective immunity. Both were only modestly immunogenic, and protection was minimal among those immunized with the recombinant vaccinia (OCKENHOUSE et al. 1998). The RTS,S vaccine described above (STOUTE et al. 1997) also induces excellent CD4$^+$ T cell responses against the C-terminus of the PfCSP, and it is possible that some of the protection found after immunization with this vaccine is due to protective CD4$^+$ T cell responses against infected hepatocytes. In rodents DNA vaccines induce extremely potent and protective CD8$^+$ T cell responses (SEDEGAH et al. 1994b; DOOLAN et al. 1996). A PfCSP DNA vaccine is now undergoing phase 1 clinical trial testing, and it is anticipated that a phase 1/2 trial of a DNA vaccine comprising five genes designed to induce protective CD8$^+$ T cell responses will be initiated in late 1999 (HOFFMAN et al. 1997). Synthetic peptide vaccines designed to induce $CD4^+$ T cell responses against infected hepatocytes will enter clinical trials in 1998 or 1999.

E. Erythrocytic Stage Vaccines: Reducing Parasite Burden and Blocking Pathogenesis

The asexual erythrocytic cycle of the malaria parasite begins with the release of merozoites from infected hepatocytes and the subsequent invasion of erythrocytes. The initial attachment of the merozoite to the erythrocyte is mediated by a specific receptor-ligand interaction. The parasite proteins involved in the process of invasion seem to be localized on the surface of merozoites and within the apical organelle complex. The molecular events involved in the process of invasion have been reviewed elsewhere (HADLEY et al. 1986; SIM 1995). Once in the erythrocyte, the parasite resides within a vacuole called the "parasitophorous vacuole," where it undergoes maturation from the ring stage to the trophozoite and finally to the schizont stage. In *P. falciparum* maturation takes about 48 h after which schizont-infected erythrocytes rupture and release extracellular merozoites which again invade fresh erythrocytes.

It may be possible to develop a vaccine that induces sustainable, sterile protective immune responses against the erythrocytic stage of the parasite similar to those proposed for preerythrocytic stage vaccines (see above). However, many malariologists view current erythrocytic stage vaccine development efforts to be aimed at reducing disease manifestations by reducing parasite burden or by blocking key pathogenic features of the infection but not by eliminating infection. There are a number of epidemiological and experimental observations that support this approach. One of the most important observation is that in areas of intense malaria transmission, if a child survives past the age of 5 years, probability of dying of malaria is dramatically reduced. One approach to vaccine development is therefore to try to delineate the immunological effector mechanisms and the target antigens responsible for this acquired protective immunity, and to develop vaccines that change the figure of 5 years old into one of over 10 years old from an immunological perspective.

A second observation is that passive transfer of purified IgG from adults with life-long exposure to malaria into children with acute *P. falciparum* infections leads to significant reduction in parasitemia (COHEN et al. 1961). If one can identify the antigenic targets of these antibodies and make a vaccine that induced such antibodies, a significant reduction in parasite burden might be achieved. With these observations in mind, and being cognizant of the life cycle of the parasite, investigators are developing vaccines designed to block merozoite invasion of erythrocytes, destroy infected erythrocytes, inhibit adherence of *P. falciparum*-infected erythrocytes to vascular endothelial cells, inhibit adherence of *P. falciparum*-infected erythrocytes to uninfected erythrocytes (rosettes), and to inhibit the effects of "toxins" released from parasite-infected erythrocytes.

I. Approaches to Reducing Parasite Burden

1. Preventing Merozoite Invasion of Erythrocytes

When merozoites rupture from infected hepatocytes or infected erythrocytes, they are extracellular for probably only a few seconds before attaching to and invading erythrocytes. The most obvious approach to inhibiting merozoite invasion of erythrocytes is to identify the parasite molecules (ligands) that bind to specific receptors on erythrocytes and to induce antibodies that prevent these receptor-ligand interactions. Another approach is to induce complement-fixing antibodies to merozoite surface proteins that would result in lysis of merozoites. In all likelihood any antibody against a merozoite surface protein would be capable of interfering with merozoite invasion by altering the merozoite surface protein structure in such a way that it can no longer invade effectively or by causing agglutination of merozoites so that they never reach erythrocytes.

2. Attacking Infected Erythrocytes

As the parasite undergoes maturation inside the erythrocyte, parasite proteins are released from the membrane-bound parasitophorous vacuole into the cytoplasm of the erythrocyte, and in some cases to the erythrocytic membrane. There is work underway to develop vaccines that induce antibodies that recognize parasite proteins on the surface of infected erythrocytes or actually enter infected erythrocytes. Antibodies are generally considered too large to traverse the erythrocyte membrane, but it has been reported that the parasitophorous vacuole is connected to the erythrocyte membrane through a "parasitophorous duct" (POUVELLE et al. 1994). Antibodies could pass through the duct and recognize parasite antigens within the vacuole itself. Complement fixing antibodies could lead to lysis of infected erythrocytes or antibodies could simply "neutralize" the parasites. Through another mechanism, within a certain microenvironment (for example in the spleen), biologically active molecules such as cytokines and free oxygen species enter the parasitophorous vacuole and neutralize or destroy the infected erythrocyte. These molecules could be released from T cells that are activated by presentation of released parasite proteins that are processed and presented by monocytes, macrophages and other antigen-presenting cells. They could also be released by monocytes or macrophages that are induced to release these bioactive molecules by direct interaction with a parasite product such as toxin(s), by binding the Fc domain of antibodies that bind parasite-derived peptides, or by products of activated T cells.

II. Antiparasite Immune Mechanisms that Contribute to Parasite Reduction

1. Reduction in Parasite Burden: The Direct Effects of Antibodies

The most persuasive demonstration of the effect of antibodies on parasite burden in humans comes from the passive transfer experiments described above. Similar studies have been carried out in rodent (DIGGS and OSLER 1969; JAYAWARDENA et al. 1978; WHITE et al. 1991) and nonhuman primate (DIGGS et al. 1995) malaria model systems. Sera from mice immune to *P. yoelii* transfer protection against parasite challenge to naive mice (JAYAWARDENA et al. 1978; WHITE et al. 1991). It has been shown that the protective effect of the immune sera resides in the IgG2a subclass; IgG1 and IgG2b antibodies have no protective effect (WHITE et al. 1991). While protection transferred through immune sera into naive recipients demonstrates the protective effect of antibody, its failure to transfer protection does not necessarily mean that antibody is not the mechanism of protective immunity. The quantity of antibody transferred may not be adequate for protection. Also, the parasite infection in a vaccinated or naturally exposed host may be essential to boost antibody titer or specificity of antibody. In the *P. berghei* model sera from immune mice taken 15 days after reinfection transfers protection into naive recipients. Immune sera taken at the time of reinfection does not transfer protection (P.B. Falanga, personal communication). Recently in two independent studies prechallenge sera from mice immunized with a subunit recombinant $MSP1_{19}$ vaccine transferred protection against lethal *P. yoelii* (DALY and LONG 1995; TIAN et al. 1996), suggesting that vaccination induced protective antibodies.

While antibodies can exert their protective effect by directly blocking the entry of merozoites into erythrocytes, results from several studies suggest that the antibody-dependent protection may not be simply due to neutralizing antibodies but could be mediated through binding to Fc receptors of monocytes. It has been demonstrated that antibodies or purified immunoglobulins from immune adults from endemic areas that transfer protection do not block invasion in vitro (BOUHAROUN-TAYOUN et al. 1990). These antibodies inhibit parasite development within erythrocytes only in the presence of monocytes, through antibody-dependent cellular inhibition (ADCI). It has been reported that the ADCI activity is dependent upon tumor necrosis factor (TNF) α and a yet undefined soluble factor produced by mononuclear cells (BOUHAROUN-TAYOUN et al. 1995).

2. Reduction in Parasite Burden: The Role of Cellular Mechanisms Through Direct Effects of Cytokines and Other Bioactive Molecules

a) The Role of $CD4^+$ T Cells

Immunity to erythrocytic stage parasites can be achieved in the absence of antibodies. μ-Suppressed mice (GRUN and WEIDANZ 1983; KUMAR et al. 1989) which lack B cells and antigen-specific antibodies develop immunity to certain species of *Plasmodium* suggesting that T cells alone can be sufficient for protection. In the murine malaria *P. chabaudi chabaudi*, μ-suppressed mice clear their primary infection at the same time as immunologically intact mice (GRUN and WEIDANZ 1983) showing that cellular mechanisms can control parasite multiplication. In contrast, μ-suppressed mice fail to clear infection with nonlethal *P. yoelii*, suggesting that immunity in this case is antibody dependent (GRUN and WEIDANZ 1983). μ-Suppressed mice have also been used to study the mechanism of reinfection immunity. Infection with another murine malaria, *P. vinckei*, is uniformly lethal in mice. Mice develop immunity to this parasite following repeated infections and drug cure. Both μ-suppressed and normal mice develop reinfection immunity following three cycles of infection followed by drug cure, suggesting the participation of antibody-independent mechanisms in immunity (KUMAR et al. 1989).

The factors responsible for antibody-independent mechanisms of immunity are not clearly defined. Studies performed about two decades ago demonstrated that athymic nude or T cell deficient mice are unable to control malaria infections (BROWN et al. 1968; JAYAWARDENA et al. 1978; ROBERTS et al. 1978), demonstrating a requirement for T cells in immunity. A clear requirement for $CD4^+$ T cells to control erythrocytic stage parasites was first demonstrated by studies in *P. vinckei* malaria (KUMAR et al. 1989). As mentioned above, both μ-suppressed mice and normal mice become immune to *P. vinckei* following reinfection and drug cure. Following challenge, μ-suppressed mice are able to suppress fulminant infection but fail to completely resolve infection and carry low-grade parasitemia. In vivo depletion of $CD4^+$ T cells abrogates immunity in both μ-suppressed and normal mice. Thus $CD4^+$ T cells in the absence of antibodies are able to control the infection to low grade parasitemia (KUMAR et al. 1989). However, both antibodies and $CD4^+$ T cells are required to completely clear malaria infection.

These results are reminiscent of earlier studies by GRUN and WEIDANZ (1983) who demonstrated that nonlethal *P. chabaudi adami* causes chronic infection in μ-suppressed mice. Further evidence for the role of $CD4^+$ T cells in protection against the erythrocytic stage of malaria comes from adoptive T cell transfer studies; a $CD4^+$ T cell clone of Th1 type can protect nude mice against *P. chabaudi adami* (BRAKE et al. 1986), suggesting that effector $CD4^+$ T cells can mediate immunity against this parasite. Analysis of cytokine production by parasite antigen-specific T cells from mice during the course of *P. chabaudi chabaudi* infection revealed that Th1 cells predominate during the early phase of infection whereas Th2 cells are dominant during the clearance phase (LANGHORNE et al. 1989). These results suggest that immunity to the early phase of parasite infection, at a time of rising parasitemia, is mediated by antibody-independent mechanisms whereas later clearance is antibody dependent. In this parasite model both $CD4^+$ T cells and B cells are required to adoptively transfer immunity in SCID mice; $CD4^+$ T cells alone have minimal effect on immunity. In another study *P. chabaudi chabaudi* specific $CD4^+$ T cell clones of both Th1 and Th2 type protected mice against the parasite challenge. Together these results suggest that immunity to the erythrocytic stage parasite is multifactorial and requires a combination of antibodies and $CD4^+$ T cells although in certain instances effector $CD4^+$ T cells alone are sufficient for protection.

How $CD4^+$ T cells perform their protective role is not completely clear. Since mouse erythrocytes do not express class II molecules, the effector arm of $CD4^+$ T cell mediated parasite killing seems to be nonspecific. The role of nonspecific, antigen-independent immunity against malaria has been well documented. Clark and colleagues have demonstrated that intravenous injection of mice with *Mycobacterium bovis* (bacille Calmette-Guérin strain) (CLARK et al. 1976) or killed *Corynebacterium parvum* (CLARK et al. 1977) protects against certain murine malarias. Intravenous or intraperitoneal injection is necessary for protection; subcutaneous injection fails to protect. Intravenous injection is known to prime immune cells in the spleen. In a later study immunization of mice with a combination of killed *P. vinckei* and live *Salmonella typhimurium* protected against lethal *P. vinckei*. (KUMAR et al. 1990). Immunization with *Salmonella* or antigen alone fails to protect.

Immunity is spleen dependent since splenectomy abrogated immunity (KUMAR et al. 1990). The role of the spleen in cellular immunity to malaria is well documented (GRUN and WEIDANZ 1985; KUMAR et al. 1989). These results suggest that a vaccine that can also exploit the benefits of nonspecific immunity may be better suited against erythrocytic stage malaria parasites. Whether antibody-independent mechanisms are sufficient to protect against human malarias is not clear. There are a few studies which suggest that cellular mechanisms participate in the clearance of parasites. A few individuals from Sudan who were clinically immune to malaria had a nonantibody serum factor that killed *P. falciparum* parasites in in vitro culture (JENSEN et al. 1983, 1984). Also, as discussed above, it appears that in some situations where immunity is antibody dependent, protection requires the participation of cellular mechanisms. Protective IgG from malaria immune individuals required cooperation with normal blood monocytes to inhibit the growth of *P. falciparum* parasites in vitro (BOUHAROUN-TAYOUN et al. 1990).

b) The Role of γ/δ T Cells

Although it is established that $CD4^+$ T cells play a major role in immunity to erythrocytic stage parasites, there is some evidence to indicate that γ/δ T cells also participate in immunity. Studies over the past several years have demonstrated a predominant increase in the number of γ/δ T cells during malaria infection, leading to the speculation that these cells contribute to immunity. During acute *P. falciparum* infection there is a 30%–40% increase in the number of circulating and splenic populations of γ/δ cells which persist up to 3–4 months after the infection is cleared (HO et al. 1990; ROUSSILHON et al. 1990; LANGHORNE et al. 1992). There is also an increase in the number of γ/δ cells during *P. vivax* infection which coincides with fever paroxysm (PERERA et al. 1994). Furthermore, γ/δ T cells have been reported to inhibit the growth of *P. falciparum* parasites in vitro (ELLOSO et al. 1994). Human γ/δ T cells from nonimmune individuals responded to *P. falciparum* parasites (BEHR and DUBOIS 1992; GOODIER et al. 1992). These cells produced high levels of IFN-γ, TNF-α, and TNF-β and no or small amounts of IL-4, IL-5, and IL-10 (HAAS et al. 1993; FERRICK et al. 1995) showing a Th1 type cytokine profile. However, these studies do not directly address the issue whether γ/δ T cells actually participate in parasite killing in vivo.

More direct information regarding the role of γ/δ T cells in immunity has been provided by studies performed in murine malarias. A proportional increase in the number of circulating and splenic γ/δ T cells has been reported during *P. chabaudi chabaudi*, *P. chabaudi adami* and *P. yoelii* infections (VAN DER HYDE 1993a; LANGHORNE et al. 1993). It has been shown that γ/δ T cells are alone not sufficient for protection since mice without $\alpha\beta$ TCR$^-$T cells fail to clear their infection (LANGHORNE et al. 1995). However, mice that are deficient in γ/δ T cells (due to deletion of TCR δ gene) have a prolonged period of acute-phase infection compared to immunologically intact mice (LANGHORNE et al. 1995). In situations where immunity to erythrocytic stage parasites can be accomplished in the absence of B cells, such as in B-cell deficient mice, and in SCID mice reconstituted with CD4$^+$ T cells, there is an increase in the population of γ/δ T cells (VAN DER HEYDE 1993b; VAN DER HEYDE et al. 1994). In one study B-cell deficient mice capable of resolving a primary *P. chabaudi adami* infection failed to control the infection if depleted of γ/δ T cells by antibody treatment (VAN DER HEYDE et al. 1995). Thus it appears that under special conditions γ/δ T cells are able to control erythrocytic stage infection but their contribution to immunity in immunologically intact host is not clearly known.

How γ/δ T cells contribute to malaria immunity remains unclear. γ/δ T cells have been shown to provide help to B cells for antibody production (MCMENAMIN et al. 1994; WEN et al. 1994) and perform other effector functions such as cytokine synthesis (FERRICK et al. 1995) and cytolytic activity (KOIZUMI et al. 1991). However, during *P. chabaudi adami* infection $\alpha\beta$ TCR deficient mice are unable to produce malaria-specific antibodies. Since the cytokine profile of γ/δ T cells is similar to that of Th1 type cells, it is possible that these cells exert their protective effect by using a similar pathway. γ/δ T cells are located in the red pulp of the spleen (BORDESSOULE et al. 1990; LANGHORNE et al. 1993), which is an excellent site to perform antiparasite function since the majority of the circulating parasites pass through the red pulp area (YADAVA et al. 1996) and thus become accessible for destruction by these cells. Because γ/δ T cells predominantly produce inflammatory cytokines such as IFN-γ, TNF-α, and TNF-β which are similar to the cytokines implicated in cerebral malaria, it has been postulated that these cells play a role in the pathogenesis of malaria.

c) The Role of Cytokines

Cytokines play a key role in malaria immunity and are produced by either malaria-specific T cells or by macrophages and other mononuclear cells. In murine malarias the role of a particular cytokine in host defense is dependent upon the species and strain of parasite and the genetic background of mouse strain studied. Of particular interest among various cytokines studied are TNF-α, IFN-γ, IL-4, IL-10, and IL-12. In the case of lethal *P. vinckei* malaria, reinfection immunity is primarily mediated by IFN-γ producing Th1 type cells (PERLMANN et al. 1995). The outcome of infection with *P. chabaudi chabaudi* is dependent upon the strain of mouse. In resistant strains of mice, the parasite causes a self-resolving infection which is associated with a Th1-type response during the acute phase and a Th2-type response in the chronic phase of infection (MEDING et al. 1990; STEVENSON et al. 1990). Thus in some malarias both the Th1 and Th2 type of responses may contribute to immunity during different phases of parasite clearance. In the case of *P. chabaudi* AS malaria high levels of TNF-α and IFN-γ expression in the spleen correlate with the ability of mice to self-clear their infections (JACOBS et al. 1996). Recently IL-12 has also been shown to protect against erythrocytic stage parasites (STEVENSON et al. 1995). In vivo injection of IL-12 protects susceptible strains of mice from *P. chabaudi AS* strain of parasite. IL-12 induced protection is dependent upon IFN-γ and TNF-α and utilizes an nitric oxide dependent pathway (STEVENSON et al. 1995). However, IL-12 is not a universal mediator of immunity against the erythrocytic stage parasites; in vivo injection of IL-12 has no effect on the course of infection with erythrocytic stage *P. yoelii* or *P. vinckei* malarias (S. Kumar, unpublished).

Cytokines also have a direct effect on the survival of gametocytes and render them noninfectious to mosquito. Crisis (the time of rapidly dropping parasitemia) and loss of infectivity during infection with *P. cynomolgi* in its natural host, the toque monkey, is mediated by TNF-α and IFN-γ. The killing activity of these cytokines also requires the presence of unidentified factor(s) in the crisis serum (NAOTUNNE et al. 1991).

All the effects of the cytokines are not protective in nature but are also known to have deleterious effects on the host defense. Clark and his colleagues were the first to suggest that the symptoms associated with the severe malaria have a striking resemblance to the toxicity caused by high levels of TNF-α (CLARK 1987; CLARK et al. 1989). The effect of TNF on malaria-associated pathology is discussed below.

III. Data Supporting Reduction in Parasite Burden and the Status of Experimental Erythrocytic Stage Vaccines Designed To Reduce Parasite Burden

The notion that it is possible to develop a malaria vaccine that will reduce or completely eliminate parasitemia comes from studies performed in rodent and monkey malaria models. Depending on the *Plasmodium* species and the mouse strain, mice develop immunity either during a primary infection or following repeated infections and drug cure. It is more difficult to develop natural immunity in some other hosts. In the rhesus *P. knowlesi* model, following reinfections and drug cure, the parasite undergoes antigenic variation and establishes a chronic infection (BROWN and BROWN 1965). However, vaccination of rhesus monkeys with merozoites in complete Freund's adjuvant (CFA) completely protects against challenge with homologous and heterologous strains of *P. knowlesi.* (MITCHELL et al. 1975). It is well documented that humans rarely develop sterile immunity to natural infection in endemic areas (HOFFMAN et al. 1987). However, a semi-immune child in Africa with high parasitemia often does not experience the complications of severe malaria. One of the remarkable features of malaria parasites are their capacity to survive over a long period of time in semi-immune hosts. It is not clear why humans fail to develop complete protection in endemic areas. Some possible explanations may include antigenic variation by targets of protective immunity, the existence of antigenically diverse clones within multiple parasite populations, the ability of the parasite to sequester in the deep tissues of host, and finally parasite induced modifications of the host immune responses.

A number of proteins from the erythrocytic stage of the parasite cycle protect monkeys from an otherwise lethal infection. Recombinant proteins also offer limited protection but immunity has required CFA as adjuvant. In the absence of an in vitro assay that predicts protective immunity the identification of potential vaccine candidates has been difficult. Currently selection as a vaccine candidate is largely based upon two criteria: (a) the ability to elicit antibodies that block parasite multiplication in vitro and (b) in vivo efficacy determined in experimental models. Since vaccine trials in monkeys are expensive and difficult to perform, one relies upon murine models for initial screening of vaccine candidates. There is an a priori sense that if an antigen from a murine malaria induces protective immunity in mice, the analogous molecule from human malarias is likely to protect humans. Several leading vaccine candidates from human malarias were first identified as protective antigens in murine and primate malaria models. Some of the promising candidates from the erythrocytic stage of the parasite life cycle are described below.

1. MSP1

Merozoite surface protein 1 (MSP1) is present on the merozoites of all *Plasmodium* and depending upon species varies in size between 190 and 230 kDa (HOLDER 1994). At the time of the release of the merozoite from the infected erythrocytes the precursor MSP1 is proteolytically cleaved into a series of distinct fragments (BLACKMAN et al. 1990). Of these fragments, the C-terminus 42-kDa fragment is of particular interest. During merozoites invasion of the erythrocyte, the 42-kDa fragment is further processed into a soluble 33-kDa and the membrane bound 19-kDa fragment. After invasion only the C-terminus $MSP1_{19}$ fragment, which contains two epidermal growth factor (EGF)-like domains, remains attached to the merozoite through a glycosylphosphatidyl inositol (GPI)-anchor and is expressed into freshly invaded rings (BLACKMAN et al. 1990, 1991). Several lines of evidence suggest that the C-terminus region of MSP1 is involved in protective immunity; anti-$MSP1_{19}$ mAbs (BLACKMAN et al. 1990) and anti-$MSP1_{42}$ polyclonal antibodies (CHANG et al. 1992) inhibit merozoite invasion of *P. falciparum* cultures. Also, seroepidemiological data from several field studies suggest an association between anti-MSP1 immune responses and resistance to clinical malaria (RILEY et al. 1992; EGAN et al. 1996). Furthermore, the sequencing of the MSP1 gene from several field isolates has revealed extensive polymorphism, suggesting that the molecule is under immune pressure (MILLER et al. 1993). Nonetheless, the most convincing evidence of its involvement in protective immunity is provided by data from in vivo challenge experiments conducted in rodent and monkey models described below.

Rodent Studies. The first evidence suggesting that MSP1 is capable of inducing protective immunity against erythrocytic stage parasites came from studies in the lethal *P. yoelii* challenge model. Immunization with native MSP1 confers protection against lethal *P. yoelii* in mice (HOLDER and FREEMAN 1981). Also, passive immunization with mAb 302, which recognizes an epitope in the first EGF-like domain on $MSP1_{19}$ protects mice against challenge with lethal and nonlethal strains of *P. yoelii* (BURNS et al. 1989). Subsequently it was shown by several groups that immunization with $MSP1_{19}$ of *P. yoelii* ($PyMSP1_{19}$) expressed as a fusion protein with glutathione-*S*-transferase in *E. coli* (DALY and LONG 1993; LING et al. 1994; TIAN et al. 1996) or as a nonfused protein in yeast (HIRUNPETCHARAT et al. 1997) induces protection in mice against lethal *P. yoelii*. In one study immunity was shown to be dependent upon the H-2 loci of mice (TIAN et al. 1996). Expression of $H\text{-}2^b$ loci was associated with the highest degree of protection whereas mice with $H\text{-}2^k$ loci had no immunity. The mechanism of $MSP1_{19}$-induced protective immunity appears to be antibody mediated since transfer of serum from immunized mice transfers partial protection (DALY and LONG 1995; TIAN et al. 1996). Furthermore, immunization with $MSP1_{19}$ fails to induce protection in B-cell knock out animals (mice with disruption in μ-chain of Ig) (HIRUNPETCHARAT et al. 1997).

Monkey Studies. Immunization with parasite derived MSP1 protects *Aotus* monkeys against *P. falciparum* parasites (SIDDIQUI et al. 1987). Earlier attempts to induce protection using the NH_2-terminus regions of MSP1 produced as synthetic peptides (CHEUNG et al. 1986) and recombinant proteins (HERRERA et al. 1990, 1992) were not very successful. However, in more recent years protection has been obtained with recombinant C-terminus regions of MSP1 which contain the two EGF-like domains. Immunization with yeast produced $MSP1_{19}$ ($yPfMSP1_{19}$) (KUMAR et al. 1995) or baculovirus produced $MSP1_{42}$ ($bPfMSP1_{42}$) (CHANG et al. 1996) induces protection against *P. falciparum* challenge when administered in CFA. The mechanism of PfMSP1-induced protective immunity remains unclear. In the $yPfMSP1_{19}$ study sera from the protected monkeys did not block invasion in vitro (KUMAR et al. 1995). The vaccinated monkeys experience two or more phases of recurring parasitemias before infection is completely cleared. This suggests that immunity is not due to neutralizing antibodies only and is boosted with each wave of parasitemia (KUMAR et al. 1995).

These in vivo challenge studies have clearly established that the recombinant $MSP1_{42}$ or $MSP1_{19}$ vaccines can protect against erythrocytic stage malaria parasites. However, the protective formulation requires CFA. In developing an MSP1-based vaccine the greatest challenge remains finding an effective adjuvant approved for human use. Also, whether an $MSP1_{19}$-based vaccine will induce protective immune responses in an HLA-diverse human population remains to be seen.

2. MSP2

The merozoite surface protein 2 (MSP2) of *P. falciparum* ranges in size between 45 and 55 kDa; based on its gene sequence the predicted molecular mass is approximately 28 kDa (SMYTHE et al. 1988). The protein is thought to be attached to the merozoite surface through a GPI-anchor (CLARK et al. 1989). MSP2 is comprised of domains of conserved and highly variable regions. Comparison of DNA sequences from different field isolates shows that variable regions of MSP2 exist as dimorphic forms (SMYTHE et al. 1990; MARSHALL et al. 1994) although the presence of some hybrid genes resulting from intragenic recombination has also been reported (MARSHALL et al. 1991). There is some evidence to suggest that MSP2 is involved in protection against malaria; mAbs which recognize epitopes from the variable regions of MSP2 inhibit parasite growth in vitro (RAMASAMY 1987; MIETTINEN-BAUMANN et al. 1988; CLARK et al. 1989; RAMASAMY et al. 1990). Homologues for the MSP2 gene from other *Plasmodium* species have not yet been identified. However, immunization with peptides from the conserved region of *P. falciparum* protects mice against challenge with *P. chabaudi* (SAUL et al. 1992), suggesting the presence of cross-reactive conserved protective epitopes on MSP2. The highly polymorphic nature of MSP2 also suggests that it may be involved in protective immunity. Further studies are needed to establish the vaccine potential of MSP2.

3. AMA1

The apical merozoite antigen 1 (AMA1) of *P. falciparum* is a 83-kDa protein that is processed into a 63-kDa polypeptide (CREWTHER et al. 1990; NARUM and THOMAS 1994). Initially AMA1 is localized in the neck of the rhoptries but after merozoite release the protein is also detected on the surface of the merozoites. The gene sequence of AMA1 encodes a polypeptide of 622 amino acids with the structural characteristics of an integral membrane protein (PETERSON et al. 1989; THOMAS et al. 1990). Based on the results obtained by sequencing laboratory and field isolates, it appears that AMA1 is relatively conserved among various parasite isolates (PETERSON et al. 1989). However, a more recent analysis of 11 field isolates suggests that the extent of diversity in AMA1 sequences is larger than previously thought (MARSHALL et al. 1996).

The homologue of AMA1 in *P. knowlesi* is a well-characterized molecule of 66 kDa (DEANS et al. 1984).

The vaccine potential of AMA1 was first identified in the studies performed with this 66-kDa molecule. mAbs which recognized conformation dependent epitopes on this 66-kDa protein block parasite invasion in vitro (DEANS et al. 1982). Deans and colleagues used affinity-purified protein to vaccinate rhesus monkeys. Following challenge infection four of six vaccinated monkeys were protected, and sera from vaccinated monkeys blocked merozoite invasion in vitro (DEANS et al. 1988). The gene encoding the homologue of AMA1 has been cloned and sequenced for other *Plasmodium* species including *P. vivax* (CHENG and SAUL 1994), *P. knowlesi* (WATERS et al. 1990, 1991), *P. chabaudi* (MARSHALL et al. 1989), *P. fragile* (PETERSON et al. 1990), and *P. yoelii*, and *P. berghei* (KAPPE and ADAMS 1996). The comparison of gene sequences reveals extensive homology among AMA1 from different *Plasmodium* species. The evidence for antigenic relatedness was apparent in a recent vaccine trial (COLLINS et al. 1994). *Saimiri* monkeys were immunized with baculovirus produced AMA1 of *P. fragile* in Montanide ISA 720 adjuvant. Following challenge with *P. fragile* vaccinated monkeys were able to self-resolve their infection. Upon rechallenge with heterologous *P. falciparum* vaccinated monkeys had no detectable parasitemia (COLLINS et al. 1994), suggesting the presence of heterologous protection.

4. EBA-175

The 175-kDa erythrocyte-binding protein (EBA-175) was identified in the supernatants of cultured *P. falciparum* (CAMUS and HADLEY 1985). EBA-175 is a merozoite ligand that binds to its receptor, glycophorin A, on erythrocytes (CAMUS and HADLEY 1985). The specificity of this receptor-ligand interaction involves both sialic acids and the protein backbone of glycophorin A (SIM et al. 1994). The gene encoding EBA-175 was cloned and sequenced in the late 1980s (SIM et al. 1990). Based on their gene structures EBA-175 and the Duffy binding proteins of *P. vivax* and *P. knowlesi* are classified as belonging to a family of EBPs (ADAMS et al. 1992). EBA-175 (SIM et al. 1992) and the *P. vivax* (FANG et al. 1991) and *P. knowlesi* (ADAMS et al. 1990) proteins are localized in the micronemes of the parasites. A characteristic feature of these three proteins is the presence of a cysteine-rich domain designated region II that has been identified as the erythrocyte binding domain of all three molecules (CHITNIS and MILLER 1994; SIM et al. 1994). This domain is highly conserved in *P. falciparum* (LIANG and SIM 1997). Antibodies against region II (B.K.L. Sim, personal communication), and against a C-terminal, conserved 42 amino acid peptide designated peptide 4 (SIM et al. 1990) inhibit merozoite invasion of erythrocytes in vitro. Based on these data EBA-175 is being developed as a human vaccine (SIM 1995). Studies are now in progress in *Aotus* monkeys with EBA-175 DNA and purified recombinant protein vaccines (B.K.L. Sim, personal communication).

5. SERA

The serine repeat antigen (SERA), a protein of approximately 130 kDa is expressed as a soluble protein in the parasitophorous vacuole surrounding trophozoites and early schizonts (DELPLACE et al. 1987). The deduced sequence consists of approximately 985 amino acids that contains a signal sequence but lacks a membrane anchor, and includes 37 consecutive serine residues (BZIK et al. 1988; KNAPP et al. 1989). SERA is structurally related to the cysteine proteases and may function as a protease during the rupture of the merozoite (HIGGINS et al. 1989). mAbs to SERA inhibit merozoite invasion in *P. falciparum* cultures (PERRIN et al. 1981). Vaccination with affinity-purified SERA from parasites induces partial protection from *P. falciparum* in *Saimiri* monkeys (PERRIN and DAYAL 1982). Various regions of SERA have been expressed with or without a fusion partner in yeast. Vaccination of *Aotus* monkeys with recombinant yeast expressed SERA1 (representing amino acids 24–285) and SERAN (representing amino acids 24–506) administered in CFA induced low levels of protection against *P. falciparum* challenge (INSELBURG et al. 1991). In another study SERA1, when delivered in CFA or MF75.2 adjuvants, gave partial protection (INSELBURG et al. 1993a,b). However, MF75.2 alone had some protective effect. More vaccine trials are needed to further assess the vaccine potential of SERA.

6. RESA

Ring-infected erythrocyte surface antigen (RESA or Pf155) was first described in 1984 (COPPEL et al. 1984; PERLMANN et al. 1984). The protein is first localized in the dense granule of the merozoite (AIKAWA et al. 1990b; CULVENOR et al. 1991) and immediately after the merozoite invasion is released into the lumen of parasitophorous vacuole and then eventually translocated to the cytoplasmic side of the erythrocyte membrane (FOLEY et al. 1991). The gene sequence of RESA predicts a polypeptide of 120 kDa which differs from its estimated size of 155 kDa (FAVALORO et al. 1986). RESA contains two tandem repeat sequences. These repeats contain immunodominant epitopes that are recognized by sera from humans living in malaria endemic areas (WAHLIN et al. 1984; PERLMANN et al. 1989). The vaccine potential of RESA is based on a number of findings. Polyclonal and monoclonal antibodies directed against the epitopes in the repeat regions inhibit merozoite invasion in vitro (BERZINS et al. 1986; RUANGJIRACHUPORN et al. 1988).

Passive immunization with affinity-purified human antibodies to RESA repeats lowers the parasite burden in *Aotus* monkeys (BERZINS et al. 1991). Active immunization of *Aotus* monkeys with *E. coli* produced β-galactosidase fusion proteins containing repeat regions of RESA and delivered in CFA gives only partial protection against *P. falciparum* challenge (COLLINS et al. 1986). In another trial immunization with synthetic RESA repeat peptides conjugated to diphtheria toxoid failed to induce clear-cut protection in *Aotus* monkeys (Collins et al. 1991). Recently a vaccine formulation containing recombinant MSP2, a part of MSP1 (190L), and a part of RESA in the Montanide ISA 720 adjuvant was administered to 12 human volunteers. The combination vaccine induced low antibody levels and a strong T cell response in most volunteers; none of the vaccinated volunteers had a reduction in parasite burden following challenge with erythrocytic stage parasites (Saul et al. 1997).

7. Synthetic SPf66 Vaccine

The most extensive malaria human trials conducted to date have been performed with a synthetic multivalent polymer vaccine containing *P. falciparum* sequences from three erythrocytic stage antigens and the repeats of the CSP (Patarroyo et al. 1988). The rationale for the design of this vaccine was based upon an earlier study showing that immunization with three synthetic peptides 83.1, 55.1 and 35.1 induce protective immunity in *Aotus* monkeys against *P. falciparum* challenge (Patarroyo et al. 1987). One of these peptides, 83.1, is derived from PfMSP1, but no *P. falciparum* proteins have been identified that contain the other two peptides. So far SPf66 vaccine has undergone five major double-blinded, placebo-controlled field trials conducted on three continents (Valero et al. 1993, 1996; Alonso et al. 1994; D'Alessandro et al. 1995; Nosten et al. 1996). These trials have produced varying results, making it difficult to assess the efficacy of this vaccine. In two trials in South America the vaccine gave more or less similar results. In the first trial the vaccine was estimated to be effective at 34% (95% CI 18.8%–45.7%) (Valero et al. 1993) whereas in the second trial the protective efficacy was 35% (95% CI 8.4%–54.2%) (Valero et al. 1996). In two vaccine trials in Africa the vaccine had varying results. In Tanzania, where malaria transmission is high, in children aged between 1–5 years the protective efficacy of the vaccine was reported to be 31% (95% CI 0%–52%) (Alonso et al. 1994). In a subsequent trial in the Gambia, an area of low parasite transmission, the vaccine offered only 8% (95% CI –18% to +29%) protection from malaria in children aged between 6–11 months (D'Alessandro et al. 1995). In the most recent trial conducted in Thailand, in children between the ages of 2 and 15 years the vaccine efficacy was estimated at –9% (95% CI –33% to +14%) (Nosten et al. 1996). There is no clear explanation for the variability in protective efficacy seen in different trials, and further testing of this vaccine is underway.

IV. Blocking Pathogenesis

One approach to reducing morbidity and mortality of malaria is to induce antibodies that inhibit key pathogenic events in the parasite life cycle without necessarily killing the parasite. Adherence of infected erythrocytes to endothelial cells (cytoadherence), adherence of infected erythrocytes to noninfected erythrocytes (rosetting), and the release of a malaria toxin are all thought to contribute to the pathogenesis of severe malaria. Groups throughout the world are working to develop vaccines that block these events.

1. Inhibiting Adherence of Infected Erythrocytes to Endothelial Cells

Among the four human malarias almost all cases of severe malaria especially cerebral malaria are caused by *P. falciparum.* A major difference between *P. falciparum* and other human malarias is that only erythrocytes containing young parasites (ring forms) generally are found in the peripheral circulation. Erythrocytes infected with the mature forms of the parasite adhere to endothelial cells lining venules in various organs, most importantly in the brain. It is thought that the infected erythrocytes adhere to endothelial cells in the microcirculation as they are developing to full maturity so that they will not have to pass through the spleen, where the more rigid infected erythrocytes can be removed from the circulation by nonspecific mechanisms. However, this phenomenon, known as cytoadherence, is thought to be associated with microcirculatory obstruction, decreased blood flow, and local anoxia, and with local release of cytokines and other bioactive materials by cells such as macrophages and endothelial cells. These events are attributed to be primary factors in the pathogenesis of severe and life-threatening malaria. A vaccine that induces antibodies that prevented cytoadherence would theoretically render parasites more susceptible to removal in the spleen and prevent the microcirculatory obstruction thought to be so critical to the pathogenesis of severe disease.

Scientists have worked for many years to identify the receptors for parasite molecules on endothelial cells, and the parasite molecules that act as ligands for these receptors. Several host molecules on endothelial cells including CD36, thrombospondin, ICAM-1, E-selectin, VCAM-1, and CSA serve as receptors (OCKENHOUSE 1993; FRIED and DUFFY 1996). To date only two proteins on the surface of parasitized erythrocytes, PfEMP1 and sequestrin have been identified as parasite ligands for these receptors.

P. falciparum erythrocyte membrane protein 1 (PfEMP1) is a clonally variant, size-diverse protein of 200–350 kDa present on the surface of parasitized erythrocytes that mediates binding to host endothelial cell receptors (BARUCH et al. 1995; SU et al. 1995). The cloning and sequencing of the PfEMP1 gene has revealed that this molecule exists as a extremely divergent family of *P. falciparum* genes (var). The deduced amino acid sequence of var genes has revealed the presence of extracellular domains with features resembling a receptor-binding motif, a transmembrane region, and a terminal domain (BARUCH et al. 1995; SU et al. 1995). It is estimated that there are 50–150 copies of var genes which are located on multiple *P. falciparum* chromosomes (SU et al. 1995). Proteins of the PfEMP1 (var) family contain multiple domains homologous to the cysteine-rich binding domains of a few malarial proteins such as EBA-175 of *P. falciparum* and the Duffy binding proteins of *P. vivax* and *P. knowlesi* which are involved in invasion (SU et al. 1995). The variability of var gene proteins has confounded vaccine development, but it is hoped that antibodies to conserved regions of PfEMP1 will block/reverse cytoadherence and prevent the clinical symptoms of cerebral malaria.

The other parasite ligand identified, sequestrin, is a 270-kDa protein found on the surface of parasitized erythrocytes and was identified based on its reactivity with an anti-idiotypic antibody that mimicked the binding region of CD36 (Ockenhouse et al. 1991). Based on its CD36 binding property and on the fact that sequestrin is expressed only in parasite isolates that bind to endothelial cells this molecule is now thought to be involved in sequestration/cytoadherence. One recent study suggests that parasites that bind to CD36 are associated with development of mild disease, not with severe malaria (Rogerson et al. 1997). Nonetheless, work is in progress to clone and sequence the gene encoding sequestrin so that this molecule can be adequately evaluated as an anti-cytoadherence immunogen.

2. Inhibiting Adherence of Infected Erythrocytes to Other Erythrocytes (Rosetting)

In addition to adhering to endothelial cells, *P. falciparum* infected erythrocytes adhere to other infected and noninfected erythrocytes. This phenomenon, which is called rosetting, is thought to contribute to microcirculatory obstruction and has been associated with increased severity of disease in field studies (Carlson et al. 1990a). One goal of vaccine developers is to identify the parasite molecules responsible for rosetting and to develop vaccines that induce antibodies that prevent it.

The parasite ligand(s) that are thought to be involved in the interaction with noninfected erythrocytes have been termed as "rosettins" (Wahlgren et al. 1994). These are a group of proteins (predominantly 28 and 22 kDa) expressed on the surface of parasitized erythrocytes and identified by immunoprecipitation with antisera that disrupt rosette formation (Carlson et al. 1990b; Helmby et al. 1993). The molecular mechanism of rosetting is not well understood but is suggested to be mediated by lectinlike interactions with blood group antigens on erythrocyte surface (Carlson et al. 1992). The fact that different parasites display different resetting characteristics in erythrocytes of different blood groups suggests that these blood groups are not the sole receptors for rosette formation.

The other molecule shown to mediate rosette formation is clonally variant PfEMP1. It has been reported that the first Duffy-binding-like domain (DBL-1) of var genes is a parasite ligand that interacts with complement-receptor 1 (CR1) on erythrocytes during rosette formation. COS-7 cells transfected with DBL-1 bind to normal erythrocytes but have reduced or no binding to CR1-deficient red blood cells (Rowe et al. 1997). CR1 polymorphism is common among Africans but is a rare trait in white populations. It has been postulated that this blood group polymorphism is a consequence of selection to protect against severe malaria (Rowe et al. 1997). Although there are considerable data indicating that rosetting is associated with severe disease, it is not clear whether the phenomenon of rosetting itself is responsible for the pathology of malaria. It remains to be seen whether a vaccine that prevents rosetting will have an impact on the occurrence of severe malaria.

3. Inhibiting Malaria Toxins

For nearly a century it has been suspected that the parasites that cause malaria produce a toxin responsible for the clinical manifestations and pathology of the disease. This view has been supported in recent years by the observation in a number of studies that there is an association between circulating levels of TNF-α and severity of malaria (GRAU et al. 1989; KERN et al. 1989; KWIATKOWSKI et al. 1990). Furthermore, parasite material which is considered by many to include a toxin released from rupturing schizonts has been shown to stimulate macrophages to produce TNF-α in vitro (TAVERNE et al. 1990). In addition to its potential direct effects, TNF-α upregulates the expression of the endothelial adhesion receptor ICAM (DUSTIN et al. 1986) and can cause increased cytoadherence and microcirculatory obstruction, as described above. This toxin has been shown to lead to increased production of nitric oxide (TACHADO et al. 1996), which is thought to be involved in the pathogenesis of severe malaria.

There are now considerable experimental data indicating that this toxin is a lipid molecule. It has been proposed that this toxin is a GPI moiety which is used as a membrane anchor by several malarial proteins (GEROLD et al. 1994). Schofield and colleagues have demonstrated that purified GPI isolated from *P. falciparum* parasites induces production of TNF-α and nitric oxide from macrophages and vasculo-endothelial cells and regulates expression of inducible nitric oxide synthase in macrophages (TACHADO et al. 1996). A vaccine which induces antibodies that neutralize such a toxin molecule might reduce or eliminate the clinical manifestations of malaria by blocking production of host molecules critical to pathogenesis. However, there has been concern that these same molecules are critical components of the host's response to infection, and blocking them could lead to an overwhelming increase in parasite burden.

F. Transmission Blocking Vaccines

Interventions directed at the mosquito vector or those that reduce the contact between the vector and the human host have been cornerstones in malaria control and eradication programs in the past. These previously highly successful measures are, for reasons of cost and insecticide resistance, now more difficult to use and sustain in developing countries (OAKS et al. 1991). Vaccines directed against the sexual stages of the parasite and the mosquito itself are some of the novel means of attacking the vector and blocking parasite transmission that are now being explored (KASLOW 1997). Because the immune response to these stages of the life cycle provide little if any direct protection to the vaccine recipient (despite offering a potentially immense indirect effect to the recipient and a direct effect to the community in which the vaccine is applied as a whole), they are often euphemistically called altruistic vaccines. The deployment of such vaccines may be more limited than that of the other malaria vaccines discussed above, in part because transmission-blocking vaccines are community-based interventions (KASLOW 1996).

Geographically isolated regions having seasonal or low level malaria transmission might benefit tremendously from transmission-blocking vaccines used as principal control or eradication interventions. It is doubtful, however, that the same vaccines would be effective on their own in areas of high transmission or with substantial flux of malaria-infected people, such as in much of sub-Saharan Africa. The contribution of transmission-blocking vaccines in these regions would more likely be as a component of a cocktail vaccine or in combination with other modalities, such as bed nets and chemotherapy. In cocktail vaccines the transmission-blocking component would be included to prevent the spread of mutant parasites resistant to the protective malaria vaccines and to enhance an otherwise partially effective protective component by reducing the level of transmission. Transmission-blocking vaccines might also prove useful in areas that surround unstable malaria and uncontrolled chemotherapy, so-called hot spots of drug-resistant malaria parasite selection, by controlling the spread of resistant parasites in the region.

All malaria vaccines that block parasite development ultimately block parasite transmission; however, only those that specifically reduce parasite infectivity to mosquitoes are referred to as transmission-blocking vaccines. The reason that antisexual and antisporogonic stage vaccines are so termed is because the sole purpose of developing such vaccines is to focus the human immune system on attacking the parasite as it leaves the human host and begins its complex transition to the poikilothermic primary host, the mosquito vector. To prepare for the transition from the human host to the mosquito vector the parasite undergoes sexual differentiation into male and female gametocytes (CARTER and GRAVES 1988). Similar to trophozoites and schizonts, but perhaps by a different or modified mechanism, the immature gametocytes remain sequestered in the deep vasculature of the human host and thus do not circulate. Undoubtedly there are potential target antigens on the surface of the sequestered immature gametocytes; however, none has been identified.

So that it is available in the circulation for ingestion by a female mosquito taking a blood meal, the mature gametocytes must be released from the deep vasculature. The process by which immature gametocytes maintain sequestration and release themselves once mature are not yet understood; whatever the mechanism(s), the cell surfaces of erythrocytes harboring mature male and female gametocytes, unlike asexual blood stage parasites and presumably the immature gametocytes, appear to be devoid of antigens (CARTER and GRAVES 1988). Despite the lack of detectable antigens on the surface of mature gametocytes, T cells and humoral responses can diminish infectivity to mosquitoes (VERMEULEN et al. 1985; TARGETT et al. 1990). Cytokines, such as TNF and a "crisis factor," contribute to the reduction in infectivity of circulating gametocytes (NAOTUNNE et al. 1991); however, eliciting and sustaining such a cytokine response by vaccination remains nonobvious, and thus for the most part efforts to develop transmission-blocking vaccines have focused on the later sexual stages and sporogonic stages.

I. Gamete and Early Zygote Surface Target Antigens

Antibodies that recognize the surface of male and/or female gametes, the initial forms of the parasite that emerge in the mosquito midgut after ingestion by the mosquito of gametocyte-infected erythrocytes, can block infectivity to mosquitoes (RENER et al. 1983; VERMEULEN et al. 1986). Antibodies to these surface proteins can be detected in populations residing in malaria endemic regions and may in part explain the poor infectivity to mosquitoes of infected adults (GRAVES et al. 1988a,b). The mechanisms by which antibodies to gametes block infectivity appear to be diverse and obviously reflect the biology of the parasite at this point in its life cycle. Antibodies to some of the surface proteins, for example, Pfs48/45, appear to block fertilization (RENER et al. 1983); others, such as Pfs230, appear to make the naturally complement-resistant gametes sensitive to antibody-dependent complement-mediated lysis (QUAKYI et al. 1987; WILLIAMSON et al. 1995). In the case of the later mechanism convincing evidence suggests that the antibody response after vaccination must be isotype specific to be effective (READ et al. 1994).

The major advantage in development of vaccine against the early extracellular sexual forms, as opposed to the later sporogonic stages, is that some of the surface antigens of gametes and the subsequent zygotes that result from fertilization in the mosquito midgut are present in the circulating gametocytes (e.g., antibodies to Pfs230 and Pfs48/45, see Table 1) and boosting after natural infection may occur. The disadvantage is of course that these proteins may have already been under prolonged, significant immune selection to limit their potency as transmission-blocking target antigens (GOOD et al. 1988). Evidence for this is suggested by the limited immune response to these lead target antigens in populations known to have been exposed to gametocytes as reflected in a ubiquitous immune response to an immunodominant intracellular protein, Pfg27/25 (CARTER et al. 1989a; RILEY et al. 1994).

II. Late Zygote and Ookinete Surface Target Antigens

The surface proteins as well as the morphology of the parasite change as the round, sedentary zygote transforms into the oblong, motile ookinete. The major surface target antigens of transmission-blocking antibodies of the gamete and early zygotes are replaced first by Pfs25, a cysteine-rich GPI-anchored protein (KASLOW et al. 1988; FRIES et al. 1989), and subsequently by Pfs28, a structurally similar cysteine-rich GPI-anchored protein (DUFFY and KASLOW 1997). Because neither of these proteins is abundantly expressed by

Table 1. Transmission-blocking vaccine candidate antigens

Antigen	Epitope			Epitope Conservation		Sequence status		Recombinant Protein Expression		Reference(s)
	No.	Confrom.[a]	T-B[b]	Ab[c]	PCR[d]	AA[e]	DNA[f]	Expressed	T-B[b]	
Pf11.1/Pfs2400	?	?	+	?	?	?	+	+	?	(Scherf et al. 1992; Feng et al. 1993)
Pfs230	I	C	+	Yes	?	+	+	+	+	(Quakyi et al. 1987; Williamson et al. 1993)
	II	C	+	Yes						
Pfs48/45	I	C	+	Yes						(Rener et al. 1983; Kocken et al. 1993)
	II	C	+	No						
	III	C	+	Yes	Yes	+	+	+	?	
	IV	C	+	Yes						
	V	L	–	Yes						
Pfs40/10	?	?	?	?	?	?	?	?	?	(Rawlings and Kaslow 1992; Templeton et al. 1998)
Pfg27/25	I	L	+	?	?	+	+	+	?	(Carter et al. 1989a; Wizel and Kumar 1991)
	II	L	–	?	?					
Pfs28	?	?	+	?	Yes	+	+	+	+	(Duffy and Kaslow 1997)
Pfs25	I	C	+	?	Yes	+	+	+	+	(Kaslow et al. 1988; Kaslow et al. 1989)
	II	C	–	?						
Pfs16	I	L	–	?	Yes	+	+	+	?	(Moelans et al. 1991a; Moelans et al. 1991b)
	II	L	–	?						
Chitinase	?	?	?	?	?	?	?	?	?	(Shahabuddin et al. 1993)

[a] Epitope conformation: L, linear; C, conformational.
[b] Transmission-blocking activity (T-B): +, transmission-blocking activity; –, no transmission-blocking activity; ?, unclear or not yet determined.
[c] Epitope conservation according to immunological data.
[d] Epitope conservation according to nucleotide sequence from PCR-amplified DNA.
[e] Amino acid sequence (AA) directly from peptide sequencing.
[f] Nucleotide sequence (DNA) available.

gametocytes, it is not surprising that antibodies to these proteins have not been detected in populations known to have been exposed to gametocytes (CARTER et al. 1989). Whether enough of either protein is expressed in gametocytes to boost a vaccine-induced immune response after natural infection remains to be seen, but may be unlikely. In the absence of a substantial naturally acquired immune response to either of these proteins it is not surprising that the genes encoding both of these proteins have limited polymorphism (KASLOW et al. 1989; DUFFY and KASLOW 1997).

Because the function of these two proteins has not been determined, it has been difficult to elucidate the mechanisms by which anti-Pfs25 and anti-Pfs28 sera disrupt the development of the zygote to oocysts. Preliminary evidence suggest that polyclonal antisera to Pfs25 interferes with zygote development, whereas mAbs to the same protein appear to inhibit the ookinete from further developing, perhaps by blocking its ability to interact with or traverse the two potential barriers of the mosquito midgut the peritrophic matrix (a chitinous saclike structure that forms around the blood meal) and the midgut epithelium (KASLOW et al. 1994). Mixing antisera directed against Pfs25 and Pfs28 appears to be at least additive if not synergistic in its ability to block mosquito infectivity (DUFFY and KASLOW 1997). Interestingly, mixing the two antigens together for vaccination does not enhance the potency of either alone; whereas, fusing the two into a single immunogen does (GOZAR and KASLOW 1998). Measurements of the antibody titers to each component and the ability of each of the fusion partners to stimulate T cell proliferation suggest that a dominant T cell epitope resides in the Pfs28 portion and drives a higher antibody titer to the Pfs25 portion (GOZAR and KASLOW 1998). The important concept here is that combining antigens do in practice enhance immunogenicity, but not necessarily for reasons that one might predict.

III. Ookinete Secreted and Mosquito Derived Target Antigens

Recent studies of the basic biology of the interactions of the parasite and the mosquito midgut have revealed two other transmission-blocking target antigens, a parasite-produced chitinase and a mosquito midgut produced late trypsin (SHAHABUDDIN and KASLOW 1994). During its egress from the blood meal the highly invasive ookinete must traverse the peritrophic matrix (PM) and the midgut epithelium. As the ookinete penetrates the PM, a dense material appears at its apical end and is associated with disruption of the normally laminated chitin fibrils that comprise the bulk of the PM (SIEBER et al. 1991). In fact, the parasite secretes a chitinase, inhibition of which by allosamidin, a potent inhibitor of chitinases, completely blocks parasite infectivity (SHAHABUDDIN et al. 1993). In the absence of highly purified parasite-produced chitinase or the cloning of the gene encoding it remains to be determined whether antibodies to chitinase will block malaria parasite transmission. If the gene sequences encoding this apparently critical enzyme from the major human malaria parasites are conserved, there is a chance that a single subunit vaccine will prove effective against more than one species of malaria parasites.

The parasite-derived chitinase is secreted as an inactive or partially active proenzyme and appears to rely on the mosquito midgut trypsin for activation (SHAHABUDDIN et al. 1996). Trypsin inhibitors when added to infectious blood meals substantially diminish the parasite's infectivity to mosquitoes. The mechanism of this blocking effect was initially thought to be a result of a failure of midgut proteases to inactivate complement leading to complement-mediated lysis of the complement-sensitive ookinete midgut proteases (SHAHABUDDIN and KASLOW 1994); however, recently this hypothesis has been challenged by the fact that chemical removal of the PM (by addition of fungal chitinase to the infectious blood meal) completely reverses the blocking-activity of the trypsin inhibitors (SHAHABUDDIN and KASLOW 1994). Perhaps the best current explanation is that the trypsin inhibitors block the activation of parasite-produced chitinase by late midgut trypsin. Indeed, antibodies that inhibit trypsin activity block transmission, but only in the presence of an intact PM (SHAHABUDDIN et al. 1996). Similar to trypsin inhibitors, disruption of the PM by exogenous chitinase completely reverses the blocking effect of antitrypsin sera.

Another potential transmission-blocking vaccine that has yet to be completely explored is an antimosquito vaccine. Clearly any vaccine that kills the mosquito, any time between taking a blood meal and salivary gland infection with sporozoites, will block transmission. The goal of such a vaccine would not necessarily be to reduce the numbers of mosquitoes in a population but rather eliminate only infected mosquitoes or mosquitoes that have taken a human blood meal. Deployment of such a vaccine would immediately begin a selection either for mosquitoes that are resistant to the antibody response or perhaps for mosquitoes that have an altered feeding behavior (preferring nonhuman hosts for blood meals). The former would limit the usefulness of such an approach, but the latter could have a beneficial effect with regard to the nuisance of being bitten by mosquitoes. In some areas of the tropics the bite rates on humans can reach a 100 or more per person per night; to individuals living in these areas mosquito bites are more than just a slight annoyance, and selection of mosquitoes with altered feeding behavior would be a welcomed relief from bites as well as malaria.

IV. Progress Towards a Transmission Blocking Vaccine

Seven lead candidate antigens have been identified (Pfs230, Pfs48/45, Pfg27/25, Pfs28, Pfs25, PfChitinase and mosquito midgut late trypsin; see Table 1). Each represents a unique opportunity for the development of transmission-blocking vaccines. Because there is little commercial interest in development of these so-called altruistic vaccines, and the resources necessary to optimally develop these vaccines has not been available, only two, Pfs25 and a fusion protein Pfs25-Pfs28, have been manufactured as clinical-grade recombinant protein in a form that elicits transmission-blocking activity in experimental laboratory animals. Although two human phase 1 safety, immunogenicity, and in vitro efficacy trials have been completed with Pfs25-based vaccines, much more work on this and the other candidate antigens will be necessary in order to develop this important component of malaria control interventions and, eventually, eradication.

G. Conclusions

We conclude, as have many of our colleagues, that development and deployment of highly effective malaria vaccines that provide sustainable protection in multiple epidemiological settings, against a parasite with numerous allelic variants, will be difficult and complex. At the core of any process that results in the development of such vaccines will be increased understanding of:

- The biology of *P. falciparum* and the other parasites that cause human malaria
- The epidemiology, pathogenesis, and clinical manifestations of malaria
- The immune mechanisms responsible for protective immunity
- The parasite antigens that are the targets of these protective immune responses
- Vaccine delivery systems that induce the required immune responses against the identified targets.

There are a number of different strategies proposed for how to produce an effective multivalent, multistage vaccine that induces protective immunity against all stages of the life cycle. Remarkable progress toward achievement of that goal has been made during the past 20 years, and we look forward to continued progress in the future.

Acknowledgements. We thank Drs. Trevor Jones, David Haynes, and Joao Aguiar for critically reviewing this manuscript. The work reported here was supported by the Naval Medical Research and Development Command work units 61152NMR0004001, 61102AA0101BFX, 62787A00101EFX. The views and assertions herein are those of the authors and are not to be construed as official or as reflecting the views of the United States Navy or naval services at large.

References

Adams JH, Hudson D, Torii M, Ward GE, Wellems TE, Aikawa M, Miller LH (1990) The Duffy receptor family of *Plasmodium knowlesi* is located within the micronemes of invasive malaria merozoites. Cell 63:141–153

Adams JH, Sim BKL, Dolan SA, Fang X, Kaslow D, Miller LH (1992) A family of erythrocyte binding proteins of malaria parasites. Proc Natl Acad Sci USA 89:7085–7089

Aggarwal A, Kumar S, Jaffe R, Hone D, Gross M, Sadoff J (1991) Oral *Salmonella*: malaria circumsporozoite recombinants induce specific $CD8^+$ cytotoxic T cells. J Exp Med 172:1083–1090

Aidoo M, Lalvani A, Allsopp CE, Plebanski M, Meisner SJ, Krausa P, Browning M, Morris-Jones S, Gotch F, Fidock DA, Takiguchi M, Robson KJH, greenwood BM, Druilhe P, Wittle HC, Hill AVS (1995) Identification of conserved antigenic components for a cytotoxic T lymphocyte-inducing vaccine against malaria. Lancet 345:1003–1007

Aikawa M, Adkinson CT, Beaudoin LM, Sedegah M, Charoenvit Y, Beaudoin R (1990a) Localization of CS and non-CS antigens in the sporogonic stages of *Plasmodium yoelii*. Bull World Health Organ 68 [Suppl]:165–171

Aikawa M, Torii M, Sjolander A, Berzins K, Perlmann P, Miller LH (1990b) Pf155/RESA antigen is localized in dense granules of *Plasmodium falciparum* merozoites. Exp Parasitol 71:326–329

Alonso PL, Tanner M, Smith T, Hayes RJ, Schellenberg JA, Lopez MC, Bastos de Azevedo I, Menendez C, Lyimo E, Weiss N, Kilama WL, Teuscher T (1994) A trial of the synthetic malaria vaccine SPf66 in Tanzania: rationale and design. Vaccine 12:181–186

Ballou WR, Rothbard J, Wirtz RA, Gordon DM, Williams JS, Gore RW, Schneider I, Hollingdale MR, Beaudoin RL, Maloy WL, Miller LH, Hockmeyer WT (1985) Immunogenicity of synthetic peptides from circumsporozoite protein of *Plasmodium falciparum*. Science 228:996–999

Baruch DI, Pasloske BL, Singh HB, Bi X, Ma XC, Feldman M, Taraschi TF, Howard RJ (1995) Cloning the *Plasmodium falciparum* gene encoding PfEMP1, a malarial variant antigen and adherence receptor on the surface of parasitized human erythrocytes. Cell 82:77–87

Behr C, Dubois P (1992) Preferential expansion of $V\gamma 9V\delta 2$ T cells following stimulation of peripheral blood lymphocytes with extracts of *Plasmodium falciparum*. Int Immunol 4:361–366

Berzins K, Perlmann H, Wahlin B, Carlsson J, Wahlgren M, Udomsangpetch I, Bjorkman A, Patarroyo ME, Perlman P (1986) Rabbit and human antibodies to a repeated amino acid sequence of a *Plasmodium falciparum* antigen, Pf155, react with the native protein and inhibit merozoite invasion. Proc Natl Acad Sci USA 83:1065–1069

Berzins K, Perlmann H, Wahlin B, Ekre HP, Hogh B, Petersen E, Wellde B, Schoenbechler M, Williams J, Chulay J, Perlmann P (1991) Passive immunization of *Aotus* monkeys with human antibodies to the *Plasmodium falciparum* antigen Pf155/RESA. Infect Immun 59:1500–1506

Blackman MJ, Heidrich HG, Donachie S, McBride JS, Holder AA (1990) A single fragment of a malaria merozoite surface protein remains on the parasite during red cell invasion and is the target of invasion-inhibiting antibodies. J Exp Med 172:379–382

Blackman MJ, Ling IT, Nicholls SC, Holder AA (1991) Proteolytic processing of *Plasmodium falciparum* merozoite surface protein-1 produces a membrane-bound fragment containing two epidermal growth factor-like domains. Mol Biochem Parasitol 49:29–33

Bordessoule D, Gaulard P, Mason DY (1990) Preferential localisation of human lymphocytes bearing gamma delta T cell receptors to the red pulp of the spleen. J Clin Pathol 43:461–464

Bouharoun-Tayoun H, Attanath P, Sabchareon A, Chongsuphajaisiddhi T, Druilhe P (1990) Antibodies that protect humans against *Plasmodium falciparum* blood stages do not on their own inhibit parasite growth and invasion in vitro, but act in cooperation with monocytes. J Exp Med 172:1633–1641

Bouharoun-Tayoun H OC, Lunel F, Druilhe P (1995) Mechanisms underlying the monocyte-mediated antibody-dependent killing of *Plasmodium falciparum* asexual blood parasites. J Exp Med 182:409

Brake DA, Weidanz WP, Long CA (1986) Antigen-specific, interleukin 2-propagated T lymphocytes confer resistance to a murine malaria parasite, *Plasmodiumchabaudiadami*. J Immunol 137:347–352

Brown IN, Allison A, Taylor RB (1968) *Plasmodium berghei* infection in thymectomized rats. Nature 219:292–293
Brown KN, Brown IN (1965) Immunity to malaria: antigenic variation in chronic infections of *Plasmodium knowlesi*. Nature (London) 208:1286–1288
Burns JM, Majarian WR, Young JF, Daly TM, Long CA (1989) A protective monoclonal antibody recognizes an epitope in the carboxyl-terminal cysteine-rich domain in the precursor of the major merozoite surface antigen of the rodent malarial parasite, *Plasmodium yoelii*. J Immunol 69:351–356
Bzik DJ, Li WB, Horii T, Inselburg J (1988) Amino acid sequence of the serine-repeat antigen (SERA) of *Plasmodium falciparum* determined from cloned cDNA. Mol Biochem Parasitol 30:279–288
Camus D, Hadley TJ (1985) A *Plasmodium falciparum* antigen that binds to host erythrocytes and merozoites. Science 230:553–556
Carlson J, Wahlgren M (1992) *Plasmodium falciparum* erythrocyte rosetting is mediated by promiscuous lectin-like interactions. J Exp Med 176:1311–1317
Carlson J, Helmby H, Hill AVS, Brewster D, Greenwood BM, Walhgren M (1990a) Human cerebral malaria: association with erythrocyte rosetting and lack of anti-rosetting antibodies. Lancet 336:1457–1460
Carlson J, Holmquist G, Taylor DW, Perlmann P, Wahlgren M (1990b) Antibodies to histidine-rich protein (PfHRP1) disrupt spontaneously formed *Plasmodium falciparum* erythrocyte rosettes. Proc Natl Acad Sci USA 87:2511–2515
Carter R, Graves PM (1988) Gametocytes. In: Wernsdorfer WH, McGregor I (eds) Malaria: Principles and practice of maloriology. Churchill Livingstone, Edinburgh, pp 253–305
Carter R, Graves PM, Creasey A, Byrne K, Read D, Alano P, Fenton B (1989a) *Plasmodium falciparum*: an abundant stage-specific protein expressed during early gametocyte development. Exp Parasitol 69:140–149
Carter R, Graves PM, Quakyi IA, Good MF (1989b) Restricted or absent immune responses in human populations to *Plasmodium falciparum* gamete antigens that are targets of malaria transmission-blocking antibodies. J Exp Med 169:135–147
Cerami C, Frevert U, Sinnis P, Takacs B, Clavijo P, Santos MJ, Nussenzweig V (1992) The basolateral domain of the hepatocyte plasma membrane bears receptors for the circumsporozoite protein of *Plasmodium falciparum* sporozoites. Cell 70:1021–1033
Cerami C, Frevert U, Sinnis P, Takacs B, Nussenzweig V (1994) Rapid clearance of malaria circumsporozoite protein (CS) by hepatocytes. J Exp Med 179:695–701
Chang SP, Gibson HL, Lee-NG CT, Barr PJ, Hui GSN (1992) A carboxyl-terminal fragment of *Plasmodium falciparum* gp195 expressed by a recombinant baculovirus induces antibodies that completely inhibit parasite growth. J Immunol 149:548–555
Chang SP, Case SE, Gosnell WL, Hashimoto A, Kramer KJ, Tam LQ, Hashiro CQ, Nikaido CM, Gibson HL, Lee-Ng CT, Barr PJ, Yokota BT, Hui GSN (1996) A recombinant baculovirus 42-kilodalton C- terminal fragment of *Plasmodium falciparum* merozoite surface protein-1 protects *Aotus* monkeys against malaria. Infect Immun 64:253–261
Charoenvit Y, Leef ML, Yuan LF, Sedegah M, Beaudoin RL (1987) Characterization of *Plasmodium yoelii* monoclonal antibodies directed against stage-specific sporozoite antigens. Infect Immun 55:604–608
Charoenvit Y, Collins WE, Jones TR, Millet P, Yuan L, Campbell GH, Beaudoin RL, Broderson JR, Hoffman SL (1991a) Inability of malaria vaccine to induce antibodies to a protective epitope within its sequence. Science 251:668–671
Charoenvit Y, Mellouk S, Cole C, Bechara R, Leef MF, Sedegah M, Yuan LF, Robey FA, Beaudoin RL, Hoffman SL (1991b) Monoclonal, but not polyclonal antibodies protect against *Plasmodium yoelii* sporozoites. J Immunol 146:1020–1025
Charoenvit Y, Fallamare V, Rogers WO, Sacci JB, Kaur M, Aguiar JC, Yuan LF, Corradin G, Anderson E, Wizel B, Houghten RA, Oloo A, Delavega P, Hoffman SL (1997) Development of two monoclonal antibodies against *Plasmodium*

falciparum sporozoite surface protein 2 and mapping of B-cell epitopes. Infect Immun 65:3430–3437

Chen DH, Tigelaar RE, Weinbaum FI (1977) Immunity to sporozoite-induced malaria infection in mice. I. The effect of immunization of T and B cell-deficient mice. J Immunol 118:1322–1327

Cheng Q, Saul A (1994) Sequence analysis of the apical membrane antigen 1 (AMA-1) of *Plasmodium vivax*. Mol Biochem Parasitol 65:183–187

Cheung A, Leban J, Shaw AR, Merkil B, Stocker J, Chizzolini C, Sander C, Perrin LH (1986) Immunization with synthetic peptides of a *Plasmodium falciparum* surface antigen induces antimerozoite antibodies. Proc Natl Acad Sci USA 83:8328–8332

Chitnis CE, Miller LH (1994) Identification of the erythrocyte binding domains of *Plasmodium vivax* and *Plasmodium knowlesi* proteins involved in erythrocyte invasion. J Exp Med 180:497–506

Clark IA (1987) Cell mediated immunity in protection and pathology of malaria. Parasitol Today 3:300–305

Clark IA, Allison A, Cox FE (1976) Protection of mice against *Babesia* and *Plasmodium* with BCG. Nature 259:309

Clark IA, Cox FE, Allison A (1977) Protection of mice against *Babesia* spp. and *Plasmodiums* pp with killed *Corynebacterium parvum*. Parasitology 74:9

Clark IA, Chaudri G, Cowden WB (1989) Roles of tumor necrosis factor in the illness and pathology of malaria. Trans R Soc Trop Med Hyg 83:436–440

Clark JT, Donachie S, Anand R, Wilson CF, Heidrich HG, McBride JS (1989) 46–53 Kilodalton glycoprotein from the surface of *Plasmodium falciparum* merozoites. Mol Biochem Parasitol 32:15–24

Clyde DF (1990) Immunity to *falciparum*and *vivax* malaria induced by irradiated sporozoites: a review of the University of Maryland studies, 1971–75. Bull World Health Organ 68 [Suppl]:9–12

Clyde DF, McCarthy VC, Miller RM, Hornick RB (1973b) Specificity of protection of man immunized against sporozoite-induced *falciparum* malaria. Am J Med Sci 266:398–401

Clyde DF, Most H, McCarthy VC, Vanderberg JP (1973a) Immunization of man against sporozoite-induced *falciparum* malaria. Am J Med Sci 266:169–177

Clyde DF, McCarthy VC, Miller RM, Woodward WE (1975) Immunization of man against *falciparum*and *vivax* malaria by use of attenuated sporozoites. Am J Trop Med Hyg 24:397–401

Cochrane AH, Masamichi A, Jeng M, Nussenzweig RS (1976) Antibody-induced ultra-structural changes of malarial sporozoites. J Immunol 116:859–867

Cohen S, McGregor IA, Carrington S (1961) Gamma-globulin and acquired immunity to human malaria. Nature 192:733

Collins WE, Anders RF, Pappaioanou M, Campbell GH, Brown GV, Kemp DJ, Coppel RL, Skinner JC, Andrysiak PM, Favaloro JM, Corcoran LM, Broderson JR, Mitchell GF, Campbell CC (1986) Immunization of *Aotus* monkeys with recombinant proteins of an erythrocyte surface antigen of *Plasmodium falciparum*. Nature 323:259–262

Collins WE, Anders R, Ruebush TI, Kemp DJ, Woodrow GC, Campbell GH, Brown GV, Irving DO, Goss N, Filipski VK, Coppel RL, Broderson JR, Thomas LM, Pye D, Skinner JC, Wilson C, Stanfill PS, Procell PM (1991) Immunization of owl monkeys with the ring-infected erythrocyte surface of *Plasmodium falciparum*. Am J Trop Med Hyg 44:34–41

Collins WE, Pye D, Crewther PE, Vanderberg KL, Galland GG, Sulzer AJ, Kemp DJ, Edwards SJ, Coppel RL, Sullivan JS, Morris CL, Anders RE (1994) Protective immunity induced in sqirrel monkeys with recombinant apical membrane antigen-1 of *Plasmodium fragile*. Am J Trop Med Hyg 51:711–719

Coppel R, Cowman AF, Anders RE, Bianco AE, Saint RB, Lingelbach KR, Kemp DJ, Brown GV (1984) Immune sera recognize on erythrocytes a *Plasmodium falciparum* antigen composed of repeated amino acid sequences. Nature 310:789–792

Crewther PE, Culvenor JG, Silva A, Cooper JA, Anders RE (1990) *Plasmodium falciparum*-2 antigens of similar size are located in different compartments of the rhoptry. Exp Parasitol 70:193–206

Culvenor JG, Day KP, Anders RF (1991) *Plasmodium falciparum* ring-infected-erythrocyte surface antigen is released from merozoite dense granules after erythrocyte invasion. Infect Immun 59:1183–1187

D'Alessandro U, Leach A, Drakeley CJ, Bennett S, Olaleye BO, Fegan GW, Jawara M, Langerock P, George MO, Targett GA, Greenwood BM (1995) Efficacy trial of malaria vaccine SPf66 in Gambian infants. Lancet 346:462–467

Daly TM, Long CA (1993) A recombinant 15-kilodalton carboxyl-terminal fragment of *Plasmodium yoelii yoellii* 17XL merozoite surface protein-1 induces a protective immune response in mice. Infect Immun 61:2462–2467

Daly TM, Long C (1995) Humoral immune response to a carboxyl-terminal region of the merozoite surface protein-1 plays a predominant role in controlling blood stage infection in rodent malaria. J Immunol 155:236–243

Deans JA, Alderson T, Thomas AW, Mitchell GH, Lennox ES, Cohen S (1982) Rat monoclonal antibodies which inhibit the in vitromultiplication of *Plasmodium knowsi*. Clin Exp Immunol 49:297–309

Deans JA, Thomas AW, Alderson T, Cohen S (1984) Biosynthesis of a putative protective *Plasmodium knowlesi* merozoite antigen. Mol Biochem Parasitol 11:189

Deans JA, Knight AM, Jean WC, Waters AP, Cohen S, Mitchell GH (1988) Vaccination trials in rhesus monkeys with a minor, invariant, *Plasmodium knowlesi* 66 kD merozoite antigen. Parasite Immunol 10:535–552

Delplace P, Fortier B, Tronchin G, Dubermetz JF, Vernes A (1987) Localization, biosynthesis, processing and isolation of a major 126 Da antigen of the parasitophorous vacuole of *Plasmodium falciparum*. Mol Biochem Parasitol 23:193–201

Diggs CL, Osler AG (1969) Humoral immunity in rodent malaria. II. Inhibition of parasitemia by serum antibody. J Immunol 102:298

Diggs CL, Hines F, Wellde BT (1995) *Plasmodium falciparum*: passive immunization of *Aotus lemurinus griseimembra* with immune serum. Exp Parasitol 80:291–296

Doolan DL, Hoffman SL (1997) Pre-erythrocytic-stage immune effector mechanisms in *Plasmodium* spp. infections. Phil Trans R Soc Lond B 352:1361–1367

Doolan DL, Houghten RA, Good MF (1991) Location of human cytotoxic T cell epitopes within a polymorphic domain of the *Plasmodium falciparum* circumsporozoite protein. Int Immunol 3:511–516

Doolan DL, Sedegah M, Hedstrom RC, Hobart P, Charoenvit Y, Hoffman SL (1996) Circumventing genetic restriction of protection against malaria with multigene DNA immunization: $CD8^+$ T cell-, interferon gamma-, and nitric oxide-dependent immunity. J Exp Med 183:1739–1746

Doolan DL, Hedstrom SL, Southwood S, Wenworth PA, Sidney J, Chesnut RW, Keogh E, Appella E, Nutman TB, Lal AA, Gordon DM, Oloo A, Sette A (1997) Degenerate cytotoxic T cells epitopes from *P. falciparum* restricted by multiple HLA-A and HLA-B supertype alleles. Immunity 7:97–112

Duffy PE, Kaslow DC (1997) A novel malaria protein, Pfs28, and Pfs25 are genetically linked and synergistic as *falciparum* malaria transmission-blocking vaccines. Infect Immun 65:1109–1113

Dustin ML, Rothlein R, Bhan AK, Dinarello CA, Springer TA (1986) Induction by IL 1 and interferon- gamma: tissue distribution, biochemistry, and function of natural adherence molecule (ICAM- 1). J Immunol 137:245–254

Egan JE, Weber JL, Ballou WR, Hollingdale MR, Majarian WR, Gordon DM, Maloy WL, Hoffman SL, Wirtz RA, Schneider I, Woollett GR, Young JF, Hockmeyer WT (1987) Efficacy of murine malaria sporozoite vaccines: implications for human vaccine development. Science 236:453–456

Egan JE, Hoffman SL, Haynes JD, Sadoff JC, Schneider I, Grau GE, Hollingdale MR, Ballou WR, Gordon DM (1993) Humoral immune response in volunteers immunized with irradiated *Plasmodium falciparum* sporozoites. Am J Trop Med Hyg 49:166–173

Egan AF, Morris J, Barnish G, Allen S, Greenwood BM, Kaslow DC, Holder AA, Riley EM (1996) Clinical immunity to *Plasmodium falciparum* malaria is associated with serum antibodies to the 19 Da C- terminal fragment of the merozoite surface antigen, PfMSP1. J Infect Dis 173:765–759

Elloso MM, van der Heyde HC, Vande Waa JA, Manning DD, Weidanz WP (1994) Inhibition of *Plasmodiumfalciparum in vitro* by human gamma delta T cells. J Immunol 153:1187

Fairley NH (1947) Sidelights on malaria in man obtained by subinoculation experiments. Trans R Soc Trop Med Hyg 40:621–676

Fang XD, Kaslow DC, Adams JH, Miller LH (1991) Cloning of *Plasmodium vivax* Duffy receptor. Mol Biochem Parasitol 44:125

Favaloro JM, Coppel RL, Corcoran LM, Foote SJ, Brown GV, Anders RF, Kemp DJ (1986) Structure of the RESA gene of *Plasmodium falciparum*. Nucleic Acids Res 14:8265–8277

Feng Z, Hoffmann RN, Nussenzweig RS, Tsuji M, Fujioka H, Aikawa M, Lensen TH, Ponnudurai T, Pologe LG (1993) Pfs2400 can mediate antibody-dependent malaria transmission inhibition and may be the *Plasmodium falciparum* 11.1 gene product. J Exp Med 177:273–281

Ferreira A, Schofield L, Enea V, Schellekens H, van der Meide P, Collins WE, Nussenzweig RS, Nussenzweig V (1986) Inhibition of development of exoerythrocytic forms of malaria parasites by gamma-interferon. Science 232:881–884

Ferrick DA, Schrenzel MD, Mulvania T, Hsieh B, Ferlin WG, Lepper H (1995) Differential production of interferon-gamma and interleukin-4 in response to Th1 and Th2 stimulating pathogens by gamma delta T cells in vivo. Nature 373:255–257

Foley M, Tilley L, Sawyer WH, Anders RF (1991) The ring-infected erythrocyte surface antigen of *Plasmodium falciparum* associates with spectrin in the erythrocyte membrane. Mol Biochem Parasitol 46:137–148

Franke ED, Corradin G, Hoffman SL (1997) Induction of protective CTL responses against the *Plasmodium yoelii* circumsporozoite protein by immunization with peptides. J Immunol 1997:3424–3433

Fried M, Duffy P (1996) Adherence of *Plasmodium falciparum* chondroitin sulphate A in the human placenta. Science 272:1502

Fries HC, Lamers MB, Smits MA, Ponnudurai T, Meuwissen JH (1989) Characterization of epitopes on the 25 kD protein of the macrogametes/zygotes of *Plasmodium falciparum*. Parasite Immunol 11:31–45

Fries LF, Gordon DM, Richards RL, Egan JE, Hollingdale MR, Gross M, Silverman C, Alving CR (1992) Liposomal malaria vaccine in humans: a safe and potent adjuvant strategy. Proc Natl Acad Sci USA 89:358–362

Gerold P, Dieckmann Schuppert A, Schwarz RT (1994) Glycosylphosphatidylinositols synthesized by asexual erythrocytic stages of the malarial parasite, *Plasmodium falciparum*. Candidates for *plasmodial* glycosylphosphatidylinositol membrane anchor precursors and pathogenicity factors. J Biol Chem 269:2597–2606

Gonzalez C, Hone D, Noriega FR, Tacket CO, Davis JR, Losonsky G, Nataro JP, Hoffman S, Malik A, Nardin E, Sztein MB, Heppner DG, Fouts TR, Isibasi A, Levine MM (1994) *Salmonella typhi* vaccine strain CVD 908 expressing the circumsporozoite protein of *Plasmodium falciparum*: strain construction and safety and immunogenicity in humans. J Infect Dis 169:927–931

Good MF, Miller LH, Kumar S, Quakyi IA, Keister D, Adams JH, Moss B, Berzofsky JA, Carter R (1988) Limited immunological recognition of critical malaria vaccine candidate antigens. Science 242:574–577

Good MF, Kaslow DC, Miller LH (1998) Pathways and strategies for developing a malaria blood-stage vaccine. Ann Rev Immunol 16:57–87

Goodier M, Fey P, Eichmann K, Langhorne J (1992) Human peripheral blood gamma delta T cells respond to antigens of *Plasmodium falciparum*. Int Immunol 4:33–41

Gozar MMG, Kaslow DC (1998) *Saccharomyces cerevisiae*-secreted fusion proteins Pfs25 and Pfs28 elicit potent *Plasmodium falciparum* transmission-blocking antibodies. Infect Immun 66:59–64

Grau GE, Taylor TE, Molyneux ME, Wirima JJ, Vassalli P, Hommel M, Lambert PH (1989) Tumor necrosis factor and disease severity in children with *falciparum* malaria. N Engl J Med 320:1586–1591

Graves PM, Burkot TR, Carter R, Cattani JA, Lagog, Parker J, Brabin BJ, Gibson FD, Bradley DJ, Alpers MP (1988a) Measurement of malarial infectivity of human populations to mosquitoes in the Madang area, Papua, New Guinea. Parasitology 96:251–263

Graves PM, Carter R, Burkot TR, Quakyi IA, Kumar N (1988b) Antibodies to *Plasmodium falciparum* gamete surface antigens in Papua New Guinea sera. Parasite Immunol 10:209–218

Grun JL, Weidanz WP (1983) Antibody independent immunity to reinfection malaria in B-cell deficient mice. Infect Immun 41:1197–1204

Grun JL, Long CA, Weidanz WP (1985) Effects of splenectomy on antibody-independent immunity to *Plasmodium chabaudi adami* malaria. Infect Immun 48:853–858

Gwadz RW, Cochrane AH, Nussenzweig V, Nussenzweig RS (1979) Preliminary studies on vaccination of rhesus monkeys with irradiated sporozoites of *Plasmodium knowlesi* and characterization of surface antigens of these parasites. Bull World Health Organ 57 [Suppl 1]:165–173

Haas W, Perera P, Tonegawa S (1993) Gamma /delta T cells. Annu Rev Immunol 11:637–685

Hadley TJ, Klotz FW, Miller LH (1986) Invasion of erythrocytes by malaria parasites: a cellular and molecular overview. Annu Rev Microbiol 40:451–477

Hedstrom RC, Campbell JR, Leef ML, Charoenvit Y, Carter M, Sedegah M, Beaudoin RL, Hoffman SL (1990) A malaria sporozoite surface antigen distinct from the circumsporozoite protein. Bull World Health Organ 68 [Suppl]:152–157

Helmby H, Cavelier L, Pettersson U, Wahlgren M (1993) Rosetting *Plasmodium falciparum*-infected erythrocytes express unique strain-specific antigens on their surface. Infect Immun 61:284–288

Herrera MARF, Herera S, Caspers P, Rotmann D, Sinigaglia F, Certa U (1992) Protection against malaria in *Aotus* monkeys immunized with a recombinant blood-stage antigen fused to a universal T cell epitope: correlation of serum gamma interferon levels with protection. Infect Immun 60:154–158

Herrera S, Perlaza BL, Burki Y, Caspers P, Dobeli H, Rotmann D, Certa U (1990) Immunization of *Aotus* monkeys *Plasmodium falciparum* blood stage recombinant proteins. Proc Natl Acad Sci USA 87:4017–4021

Herrington DA, Clyde DF, Losonsky G, Cortesia M, Murphy JR, Davis J, Baqar S, Felix AM, Heimer EP, Gillessen D, Nardin E, Nussenzweig RS, Nussenzweig V, Hollingdale MR, Levine MM (1987) Safety and immunogenicity in man of a synthetic peptide malaria vaccine against *Plasmodium falciparum* sporozoites. Nature (London) 328:257–259

Herrington DA, Davis J, Nardin E, Beier M, Cortese J, Eddy H, Losonsky G, Hollingdale M, Sztein M, Levine M, Nussenzweig RS, Clyde D, Edelman R (1991) Successful immunization of humans with irradiated sporozoites: humoral and cellular responses of the protected individuals. Am J Trop Med Hyg 45:539

Higgins DG, Mcconnell DJ, Sharp PM (1989) Malarial proteases. Nature (London) 340:604

Hill AVS, Allsopp CEM, Kwiatkowki D, Anstey NM, Twumasi P, Rowe PA, Bennett S, Brewster D, McMichael A, Greenwood BM (1991) Common West African HLA antigens are associated with protection from severe malaria. Nature (London) 352:595–600

Hirunpetcharat C, Tian JH, Kaslow DC, Rooijen NV, Kumar S, Berzofsky JA, Miller LH, Good MF (1997) Complete protective immunity induced in mice by immunization with the 19-kilodalton carboxyl-terminal fragment of the merozoite surface protein-1 ($MSP1_{19}$) of *Plasmodium yoelii* expressed in *Saccharomyces cerevisiae*. J Immunol 159:3400–3411

Ho M, Webster K, Tongtowe P, Pattanpanyasat K, Weidanz WP (1990) Increased gamma delta T cells in acute *Plasmodium falciparum* malaria. Immunol Lett 25:139–141

Hoffman SL, Oster CN, Plowe CV, Woollett GR, Beier JC, Chulay JD, Wirtz RA, Hollingdale MR, Mugambi M (1987) Naturally acquired antibodies to sporozoites do not prevent malaria: vaccine development implications. Science 237:639

Hoffman SL, Edelman R, Bryan J, Schneider I, Davis J, Sedegah M, Gordon D, Church P, Gross M, Silverman C, Hollingdale M, Clyde D, Sztein M, Losonsky G, Paparello S, Jones TR (1994) Safety, immunogenicity, and efficacy of a malaria sporozoite vaccine administered with monophosphoryl lipid A, cell wall skeleton of mycobacteria and squalene as adjuvant. Am J Trop Med Hyg 51:603–612

Hoffman SL, Miller LH (1996) Perspectives on malaria vaccine development. In: Hoffman SL (ed) Malaria vaccine development. ASM, Washington DC pp 1–13

Hoffman SL, Crutcher JM, Puri SK, Ansari AA, Villinger F, Franke ED, Singh PP, Finkelman F, Gatley MK, Dutta GP, Sedegah M (1997a) Sterile protection of monkeys against malaria after administration of interleukin-12. Nat Med 3:80–83

Hoffman SL, Doolan D, Sedegah M, Wang R, Scheller LF, Kumar A, Weiss WR, Le TP, Klinman DM, Hobart P, Norman JA, Hedstrom RC (1997b) Towards clinical trials of DNA vaccines against malaria. Immunol Cell Biol 75:376–381

Holder AA (1994) Proteins on the surface of the malaria parasite and cell invasion. Parasitology 108 [Suppl]:85

Holder AA, Freeman RR (1981) Immunization against blood-stage rodent malaria using purified parasite antigens. Nature 294:361–364

Inselburg J, Bzik DJ, Li WB, Green KM, Kansopon J, Hahm BK, Bathurst IC, Barr PJ, Rossan RN (1991) Protective immunity induced in *Aotus* monkeys by recombinant SERA proteins of *Plasmodium falciparum*. Infect Immun 59:1247–1250

Inselburg J, Bathurst IC, Kansopon J, Barchfeld GL, Barr PJ, Rossan RN (1993a) Protective immunity induced in *Aotus* monkeys by a recombinant SERA protein of *Plasmodium falciparum*: adjuvant effects on induction of protective immunity. Infect Immun 61:2041–2047

Inselburg J, Bathurst IC, Kansopon J, Barr PJ, Rossan R (1993b) Protective immunity induced in *Aotus* monkeys by a recombinant SERA protein of *Plasmodium falciparum*: further studies using SERA 1 and MF75.2 adjuvant. Infect Immun 61:2048–2052

Jacobs P, Radzioch D, Stevenson MM (1996) A Th1-associated increase in tumor necrosis factor alpha expression in the spleen correlates with resistance to blood-stage malaria in mice. Infect Immun 64:535–541

Jayawardena AN, Targett GAT, Leuchars E, Davies AJS (1978) The immunological response of CBA mice to *P. yoelii*. II. The passive transfer of immunity with serum and cells. Immunology 33:157

Jensen JB, Boland MT, Allan JS, Carlin JM, Van de Waa JA, Divo AA, Akood MAS (1983) Association between human serum induced crisis forms in cultured *Plasmodium falciparum* and clinical immunity to malaria in Sudan. Infect Immun 41:1302–1311

Jensen JB, Hoffman SL, Boland MT, Akood MA, Laughlin LW, Kurniawan L, Marwoto MA (1984) Comparison of immunity to malaria in Sudan and Indonesia: crisis-form versus merozoite invasion inhibition. Proc Natl Acad Sci USA 81:922–925

Kappe SH, Adams JH (1996) Sequence analysis of the apical membrane antigen-1 genes (ama-1) of *Plasmodium yoelii yoelii* and *Plasmodium berghei*. Mol Biochem Parasitol 78:279–283

Kaslow DC (1996) Transmission-blocking vaccine. In: Hoffman SL (ed) Malaria vaccine development. ASM, Washington, pp 181–227
Kaslow DC (1997) Transmission-blocking vaccines: Uses and current status of development. Int J Parisitol 27:183–189
Kaslow DC, Quakyi IA, Syin C, Raum MG, Keister DB, Coligan JE, McCutchan TF, Miller LH (1988) A vaccine candidate from the sexual stage of human malaria that contains EGF-like domains. Nature 333:74–76
Kaslow DC, Quakyi IA, Keister DB (1989) Minmal variation in a vaccine candidate from the sexual stage of *Plasmodium falciparum*. Mol Biochem Parasitol 32:101–103
Kaslow DC, Bathurst IA, Lensen T, Ponnudurai T, Barr J, Keister DB (1994) *Saccharomyces cerevisiae* recombinant Pfs25 adsorbed to alum elicits antibodies that block transmission of *Plasmodium falciparum*. Infect Immun 62:5576–5548
Kern PC, Hemmer J, Van Damme J, Gruss H-J, Dietrich M (1989) Elevated tumor necrosis factor alpha and interleukin 6 serum levels as markers for complicated *Plasmodium falciparum* malaria. Am J Med 57:139–143
Khusmith S, Charoenvit Y, Kumar S, Sedegah M, Beaudoin RL, Hoffman SL (1991) Protection against malaria by vaccination with sporozoite surface protein 2 plus CS protein. Science 252:715–718
Khusmith S, Sedegah M, Hoffman SL (1994) Complete protection against *Plasmodium yoelii* by adoptive transfer of a $CD8^+$ cytotoxic T cell clone recognizing sporozoite surface protein 2. Infect Immun 62:2979–2983
Knapp B, Hundt E, Kupper HA (1989) A new blood stage antigen of *Plasmodium falciparum* transported to the erythrocyte surface. Mol Biochem Parasitol 37:47–58
Kocken CH, Jansen J, Kaan AM, Beckers PJ, Ponnudurai T, Kaslow DC, Konings RN, Schoenmakers JG (1993) Cloning and expression of the gene coding for the transmission blocking target antigen Pfs48/45 of *Plasmodium falciparum*. Mol Biochem Parasitol 61:59–68
Koizumi H, Liu CC, Zheng LM, Joag SV, Bayne NK, Holoshitz J, Young JD (1991) Expression of perforin and serine estrase by human gamma/delta T cells. J Exp Med 173:499–502
Kumar S, Miller LH, Quakyi IA, Keister DB, Houghten RA, Maloy WL, Moss B, Berzofsky JA, Good MF (1988) Cytotoxic T cells specific for the circumsporozoite protein of *Plasmodium falciparum*. Nature (London) 334:258–260
Kumar S, Good MF, Dontfraid F, Vinetz JM, Miller LH (1989) Interdependence of $CD4^+$ T cells and malarial spleen in immunity to *Plasmodium vinckei vinckei*. Relevance to vaccine development. J Immunol 143:2017–2023
Kumar S, Gorden J, Flynn JL, Berzofsky JA, Miller LH (1990) Immunization of mice against *Plasmodium vinckei* with a combination of attenuated *Salmonella typhimurium* and malarial antigen. Infect Immun 58:3425–3429
Kumar S, Yadava A, Keister DB, Tian JH, Ohl M, Perdue-Greenfield KA, Miller LH, Kaslow DC (1995) Immunogenicity and in vivo efficacy of recombinant *Plasmodium falciparum* merozoite surface protein-1 in *Aotus* monkeys. Mol Med 1:325–332
Kwiatkowski D, Hills AVS, Sambou I, Twumasi P, Castracane J, Manogue KR, Cerami A, Brewster DR, Greenwood BM (1990) TNF concentration in fatal cerebral, non-fatal cerebral, and uncomplicated *Plasmodium falciparum* malaria. Lancet 336:1201–1204
Lanar DE, Tine JA, de Taine C, Seguin MC, Cox WI, Winslow JP, Ware LA, Kauffman EB, Gordon D, Ballou WR, Paoletti E, Sadoff JC (1996) Attenuated vaccinia virus-circumsporozoite protein recombinants confer protection against rodent malaria. Infect Immun 64:1666–1671
Langhorne J, Gillard S, Simon B, Slade S, Eichmann K (1989) Frequencies of $CD4^+$ T cells reactive with *Plasmodium chabaudi chabaudi*: Distinct response kinetics for cells with Th1 and TH2 characteristics during infection. Int Immunol 1:416–424

Langhorne J, Goodier M, Behr C, Dubois P (1992) Is there a role for gamma delta T cells in malaria? Immunol Today 13:298–300

Langhorne J, Pells S, Eichmann K (1993) Phenotypic characterisation of splenic T cells from mice infected with *Plasmodium chabaudi chabaudi*. Scand J Immunol 38:521–528

Langhorne J, Mombaerts P, Tonegawa S (1995) Alpha beta and gamma delta T cells in the immune response to the erythrocytic stages of malaria in mice. Int Immunol 7:1005–1011

Li S, Rodrigues M, Rodriguez D, Rodriguez JR, Esteban M, Palese P, Nussenzweig RS, Zavala F (1993) Priming with recombinant influenza virus followed by administration of recombinant vaccinia virus induces $CD8^+$ T-cell- mediated protective immunity against malaria. Proc Natl Acad Sci USA 90:5214–5218

Liang H, Sim BKL (1997) Conservation of structure and function of the erythrocyte-binding domain of *Plasmodium falciparum*. Mol Biochem Parasitol 84:241–245

Ling IT, Ogun SA, Holder AA (1994) Immunization against malaria with a recombinant protein. Parasite Immunol 16:63–67

Maheshwari RK, Czarniecki CW, Dutta GP, Puri SK, Dhawan BN, Friedman RM (1986) Recombinant human gamma interferon inhibits simian malaria. Infect Immun 53:628–630

Malik A, Egan JE, Houghton RA, Sadoff JC, Hoffman SL (1991) Human cytotoxic T lymphocytes against *Plasmodium falciparum* circumsporozoite protein. Proc Natl Acad Sci USA 88:3300–3304

Marshall VM, Peterson MG, Lew AM, Kemp DJ (1989) Structure of the apical membrane antigen-1 (AMA-1) of *Plasmodium chabaudi*. Mol Biochem Parasitol 37:281–283

Marshall VM, Coppel RL, Martin RK, Oduola AMJ, Anders RF, Kemp DJ (1991) A *Plasmodium falciparum* MSA-2 gene apparently generated by intragenic recombination between the two allelic families. Mol Biochem Parasitol 45:349–351

Marshall VM, Anthony RL, Bangs Purnomo MJ, Anders RF, Coppel RL (1994) Allelic variants of the *Plasmodium falciparum* merozoite surface antigen 2 (MSA-2) in a geographically restricted area of Irian Jaya. Mol Biochem Parasitol 63:13–21

Marshall VM, Zhang LX, Anders RF, Coppel RL (1996) Diversity of the vaccine candidate AMA-1 of *Plasmodium falciparum*. Mol Biochem Parasitol 77:109–113

McMenamin C, Pimm C, McKersey M, Holt PG (1994) Regulation of IgE responses to inhaled antigen in mice by antigen-specific gamma delta T cells. Science 265:1869–1871

Meding SJ, Cheng SC, Simon-Haarhaus B, Langhorne J (1990) Role of gamma interferon during infection with *Plasmodium chabaudi chabaudi*. Infect Immun 58:3671–3976

Mellouk S, Maheshwari RK, Rhodes-Feuillette A, Beaudoin RL, Berbiguier N, Matile H, Miltgen F, Landau I, Pied S, Chigot JP, Friedman RM, Mazier D (1987) Inhibitory activity of interferons and interleukin 1 on the development of *Plasmodium falciparum* in human hepatocyte cultures. J Immunol 139:4192–4195

Mellouk S, Green SJ, Nacy CA, Hoffman SL (1991) IFN-gamma inhibits development of *Plasmodium berghei* exoerythrocytic stages in hepatocytes by an L-arginine-dependent effector mechanism. J Immunol 146:3971–3976

Mellouk S, Hoffman SL, Liu ZZ, De la Vega P, Billiar TR, Nussler AK (1994) Nitric oxide-mediated antiplasmodial activity in human and murine hepatocytes induced by gamma interferon and the parasite itself: enhancement by exogenous tetrahydrobiopterin. Infect Immun 62:4043–4046

Miettinen-Baumann A, Strych W, McBride J, Heidrich H-G (1988) A 46,000 dalton *Plasmodium falciparum* merozoite surface glycoprotein not related to the 185,000–195,000 dalton schizont precursor molecule: isolation and characterization. Parasitol Res 74:317–323

Migliorini P, Betschart B, Corradin G (1993) Malaria vaccine: immunization of mice with a synthetic T cell helper epitope alone leads to protective immunity. Eur J Immunol 23:582–585

Miller LH, Roberts T, Shahabuddin M, Mccutchan TF (1993) Analysis of sequence diversity in the *Plasmodium falciparum* merozoite surface protein-1 (MSP-1). Mol Biol Parasitol 59:1–14

Mitchell GH, Butcher GA, Cohen S (1975) Merozoite vaccination against *Plasmodium knowlesi* malaria. Immunology 29:397–401

Moelans II, Klaassen CH, Kaslow DC, Konings RN, Shoenmakers JG (1991a) Minimal variation in Pfs16, a novel protein located in the membrane of gametes and sporozoites of *Plasmodium falciparum*. Mol Biochem Parasitol 46:311–313

Moelans II, Meis JF, Kocken C, Konings RN, Shoenmakers JG (1991b) A novel protein antigen of the malaria prasite *Plasmodium falciparum*, located on the surface of gametes and sporozoites. Mol Biochem Parasitol 45:193–204

Muller HM, Reckmann I, Hollingdale MR, Bujard H, Robson KJ, Crisanti A (1993) Thrombospondin related anonymous protein (TRAP) of *Plasmodium falciparum* binds specifically to sulfated glycoconjugates and to HepG2 hepatoma cells suggesting a role for this molecule in sporozoite invasion of hepatocytes. EMBO J 12:2881–2889

Mulligan HW, Russell P, Mohan BN (1941) Active immunization of fowls against *Plasmodium gallinaceum* by injections of killed homologous sporozoites. J Mal Inst India 4:25–34

Naotunne TS, Karunaweera ND, Del Giudice G, Kularatne MU, Grau GE, Carter R, Mendis KN (1991) Cytokines kill malaria parasites during infection crisis: extracellular complementary factors are essential. J Exp Med 173:523–529

Nardin EH, Nussenzweig V, Nussenzweig RS, Collins WE, Harinasuta KT, Tapchaisri P, Chomcharn Y (1982) Circumsporozoite proteins of human malaria parasites *Plasmodium falciparum* and *Plasmodium vivax*. J Exp Med 156:20–30

Narum DL, Thomas AW (1994) Differential localization of full length and processed forms of PF83/AMA-1 an apical membrane antigen of *Plasmodium falciparum* merozoites. Mol Biochem Parasitol 67:59–68

Nosten F, Luxemburger C, Kyle DE, Ballou WR, Wittes J, Wah E, Chongsuphajaisiddhi T, Gordon DM, White NJ, Sadoff JC, Heppner DG (1996) Randomised double-blind placebo-controlled trial of SPf66 malaria vaccine in children in northwestern Thailand. Lancet 348:701–707

Nussenzweig RS, Vanderberg J, Most H, Orton C (1967) Protective immunity produced by the injection of X-irradiated sporozoites of *Plasmodium berghei*. Nature (London) 216:160–162

Oaks SC, Mitchell VS, Pearson GW (1991) Malaria: Obstacles and opportunities. In: Carpenter CCI (ed) A report of the Committee for the Study on Malaria Prevention and Control. National Academy Press, Washington

Ockenhouse CF (1993) The molecular basis for the cytoadherence of *Plasmodium falciparum*-infected erythrocytes to endothelium. Semin Cell Biol 4:297–303

Ockenhouse CF, Klotz FW, Tandon NN, Jamieson GA (1991) Sequestrin, a CD36 recognition protein on *Plasmodium falciparum* malaria-infected erythrocytes identified by anti-idiotype antibodies. Proc Natl Acad Sci USA 88:3175–3179

Ockenhouse CF, Sun PF, Lanar DE, Wellde BT, Hall T, Kester K, Stoute JA, Magill A, Kryzch U, Farley L, Wirtz RA, Sadoff JC, Kaslow DC, Kumar S, Church LWP, Crutcher JM, Wizel B, Hoffman S, Lalvani A, Hill AVS, Tine JA, Guito KP, de Taisne C, Anders A, Horii T, Paoletti EB (1998) Phase I/IIa safety, immunogenicity, and efficacy of NYVAC Pf7, a Pox-vectored, multiantigen, multistage vaccine candidate for *Plasmodium falciparum* malaria. J Infect Dis 177:1664–1673

Pancake SJ, Holt GD, Mellouk S, Hoffman SL (1992) Malaria sporozoites and circumsporozoite proteins bind specifically to sulfated glycoconjugates. J Cell Biol 117:1351–1357

Patarroyo ME, Romero P, Torres ML, Clavijo P, Moreno A, Martinez A, Rodrigues R, Guzman F, Cabezas E (1987) Induction of protective immunity against experimental infetion with malaria using synthetic peptides. Nature (London) 328:629–632

Patarroyo ME, Romero P, Amador R, Clavijo P, Moreno A, Guzman F, Romero P, Tascor, Franco A, Murillo LA, Ponton G, Trujillo G (1988) A synthetic vaccine protects humans against challenge with asexual blood stages of *Plasmodium falciparum* malaria. Nature (London) 158–161

Perera MK, Carter R, Goonewardene R, Mendis KN (1994) Transient increase in circulating gamma/delta T cells during *Plasmodium vivax* malarial paroxysms. J Exp Med 179:311–315

Perlmann H, Berzins K, Wahlgren M, Carlsson J, Bjorkman A, Patarroyo ME, Perlmann P (1984) Antibodies in malarial sera to parasite antigens in the membrane of erythrocytes infected with early asexual stages of *Plasmodium falciparum*. J Exp Med 159:1686–1704

Perlmann H, Perlmann P, Berzins K, Wahlin B, Troye Blomberg M, Hagstedt M, Andersson I, Hogh B, Petersen E, Bjorkman A (1989) Dissection of the human antibody response to the malaria antigen Pf155/RESA into epitope specific components. Immunol Rev 112:115–135

Perlmann H, Kumar S, Vinetz JM, Kullberg M, Miller LH, Perlmann P (1995) Cellular mechanisms in the immune response to malaria in *Plasmodium vinckei*-infected mice. Infect Immun 63:3987–3993

Perrin LH, Dayal R (1982) Immunity to asexual erythrocytic stages of *Plasmodium falciparum*: role of defined antigens in the humoral response. Immunol Rev 61:245–269

Perrin LH, Ramirez E, Lambert PH, Miescher PA (1981) Inhibition of *P. falciparum* growth in human erythrocytes by monoclonal antibodies. Nature (London) 289:301–303

Peterson MG, Marshall VM, Smythe JA, Crewther PE, Lew A, Silva A, Anders RF, Kemp DJ (1989) Integral membrane protein located in the apical complex of *Plasmodium falciparum*. Mol Biochem Parasitol 39:3151–3154

Peterson MG, Nguyendinh P, Marshall VM, Elliott JF, Collins WE, Anders RF, Kemp DJ (1990) Apical membrane antigen of *Plasmodium fragile*. Mol Biochem Parasitol 39:279–284

Potocnjak P, Yoshida N, Nussenzweig RS, Nussenzweig V (1980) Monovalent fragments (Fab) of monoclonal antibodies to a sporozoite surface antigen (Pb44) protect mice against malaria infection. J Exp Med 151:1504–1513

Pouvelle B, Gormley JA, Taraschi TF (1994) Characterization of trafficking pathways and membrane genesis in malaria-infected erythrocytes. Mol Biochem Parasitol 66:83–96

Quakyi IA, Carter R, Rener J, Kumar N, Good M, Miller LH (1987) The 230-kDa gamete surface protein of *Plasmodium falciparum* is also a target for transmission-blocking antibodies. J Immunol 139:4213–4217

Ramasamy R (1987) Studies on glycoproteins in the human malaria parasite *Plasmodium falciparum*-lectin binding properties and the possible carbohydrate-protein linkage. Immunol Cell Biol 65:147–152

Ramasamy RJ, Jones G, Lord R (1990) Characterization of an inhibitory monoclonal antibody defined epitope on a malaria vaccine candidate antigen. Immunol Lett 23:305–309

Rawlings DJ, Kaslow DC (1992) A novel 40-kDa membrane-associated EF-hand calcium-binding protein in *Plasmodium falciparum*. J Biol Chem 267:3976–3982

Read D, Lensen AHW, Bergarnie S, Haley S, Raza A, Carter R (1994) Transmission-blocking antibodies against multiple non-variant target epitopes of the *Plasmodium falciparum* gamete surface antigen Pfs230 are all complement-fixing. Parasite Immunol 16:511–520

Rener J, Graves PM, Carter R, Williams JL, Burkot TR (1983) Target antigens of transmission- blocking immunity on gametes of *Plasmodium falciparum*. J Exp Med 158:976–981

Renia L, Marussig MS, Grillot D, Pied S, Corradin G, Miltgen F, Del Giudice G, Mazier D (1991) In vitro activity of $CD4^+$ and $CD8^+$ T lymphocytes from mice immunized with a synthetic malaria peptide. Proc Natl Acad Sci USA 88:7963–7967

Rickman LS, Gordon DM, Wistar R, Krzych U, Gross M, Hollingdale MR, Egan JE, Chulay JD, Hoffman SL (1991) Use of adjuvant containing mycobacterial cell-wall skeleton, monophosphoryl lipid A, and squalane in malaria circumsporozoite protein vaccine. Lancet 337:998–1001

Rieckmann KH, Carson PE, Beaudoin RL, Cassells JS, Sell KW (1974) Sporozoite induced immunity in man against an Ethiopian strain of *Plasmodium falciparum*. Trans R Soc Trop Med Hyg 68:258–259

Rieckmann KH, Beaudoin RL, Cassells JS, Sell DW (1979) Use of attenuated sporozoites in the immunization of human volunteers against *falciparum* malaria. Bull World Health Organ 57 [Suppl 1]:261–265

Riley EM, Allen SJ, Wheeler J, Blackman MJ, Bennett S, Takacs B, Schonfeld HJ, Holder AA, Greenwood BM (1992) Naturally aquired cellular and humoral immune responses to the major merozoite surface antigen (PfMSP1) of *Plasmodium falciparum* are associated with reduced malaria morbidity. Parasite Immunol 14:321–337

Riley EM, Bennett S, Jepson A, Hassan-King M, Whittle H, Olerup O, Carter R (1994) Human antibody responses to Pfs230, a sexual stage-specific surface antigen of *Plasmodium falciparum*: non- responsiveness is a stable phenotype but does not appear to be genetically regulated. Parasite Immunol 16:55–62

Roberts DW, Rank RG, Weidanz WP, Finnerty JF (1978) Prevention of recrudescent malaria in nude mice by thymic grafting or by treatment with hyperimmune serum. Infect Immun 16:821

Robson KJ, Hall JR, Jennings MW, Harris TJ, Marsh K, Newbold CI, Tate VE, Weatherall DJ (1988) A highly conserved amino-acid sequence in thrombospondin, properdin and in proteins from sporozoites and blood stages of a human malaria parasite. Nature (London) 335:79–82

Rodrigues MM, Cordey A-S, Arreaza G, Corradin G, Romero P, Maryanski JL, Nussenzweig RS, Zavala F (1991) $CD8^+$ cytolytic T cell clones derived against the *Plasmodium yoelii* circumsporozoite protein protect against malaria. Int Immunol 3:579–588

Rogers WO, Malik A, Mellouk S, Nakamura K, Rogers MD, Szarfman A, Gordon DM, Nussler AK, Aikawa M, Hoffman SL (1992b) Characterization of *Plasmodium falciparum* sporozoite surface protein 2. Proc Natl Acad Sci USA 89:9176–9180

Rogers WO, Rogers MD, Hedstrom RC, Hoffman SL (1992a) Characterization of the gene encoding sporozoite surface protein 2, a protective *Plasmodium yoelii* sporozoite antigen. Mol Biochem Parasitol 53:45–52

Rogerson SJ, Tenbenu T, Dobano C, Molyneux ME, Taylor TE (1997) Cytoadherence characterization of P-falciparum infected erythrocytes from Malawian children. Abstract number S29.2, Second Global Meet on Parasitic Diseases, Hyderabad, India

Romero P, Maryanski JL, Corradin G, Nussenzweig RS, Nussenzweig V, Zavala F (1989) Cloned cytotoxic T cells recognize an epitope in the circumsporozoite protein and protect against malaria. Nature (London) 341:323–325

Roussilhon C, Agrapart M, Ballet JJ, Benussan A (1990) T lymphocytes bearing gamma delta T cell receptor in patients with acute *Plasmodium falciparum* malaria. J Infect Dis 162:283–285

Rowe JA, Moulds JM, Newbold CI, Miller LH (1997) *P. falciparum* rosetting mediated by a parasite- variant erythrocyte membrane protein and complement-receptor 1. Nature (London) 388:292–295

Ruangjirachuporn W, Wahlin B, Perlmann H, Carlsson J, Berzins K, Wahlgren M, Udomsangpetch R, Wigzell H, Perlmann P (1988) Monoclonal antibodies to a synthetic peptide corresponding to a repeated sequence in the *Plasmodium falciparum* antigen Pf155. Mol Biochem Parasitol 29:19–28

Saul A, Lord R, Jones GL, Spencer L (1992) Protective immunization with invariant peptides of the *Plasmodium falciparum* antigen MSA2. J Immunol 148:208–211

Saul A, Lawrence G, Cheng Q, Reed C, Cloonan N, Taylor, Stowers AW, Rzepczyk CA, Smillie A, Anderson, Pombo D, Eisen D (1997) Efficacy trial of a Plasmodium falciparum asexual stage vaccine in human volunteers. Abstract number 299, 46th annual meeting of American Society of Tropical Medicine and Hygiene, Lake Buena Vista, Florida USA

Scherf A, Carter R, Petersen C, Alano P, Nelson R, Aikawa M, Mattei D, Pereira da Silva L, Leech J (1992) Gene activation of Pf11-1 of *Plasmodium falciparum* by chromosome breakage and healing: indentification of gametocyte-specific protein role in gametogenesis. EMBO J 11:2293–2301

Schofield L, Villaquiran J, Ferreira A, Schellekens H, Nussenzweig RS, Nussenzweig V (1987) Gamma-interferon, $CD8^+$ T cells and antibodies required for immunity to malaria sporozoites. Nature (London) 330:664–666

Sedegah M, Finkelman F, Hoffman SL (1994a) Interleukin-12 induction of interferon gamma-dependent protection against malaria. Proc Natl Acad Sci USA 91:10700–10702

Sedegah M, Hedstrom R, Hobart P, Hoffman SL (1994b) Protection against malaria by immunization with plasmid DNA encoding circumsporozoite protein. Proc Natl Acad Sci USA 91:9866–9870

Seguin MC, Klotz FW, Schneider I, Weir JP, Goodbary M, Slayter M, Raney JJ, Aniagolu JU, Green SJ (1994) Induction of nitric oxide synthase protects against malaria in mice exposed to irradiated *Plasmodium berghei* infected mosquitoes: involvement of interferon gamma and $CD8^+$ T cells. J Exp Med 180:353–358

Shahabuddin M, Kaslow DC (1994) Biology of the development of *Plasmodium* in the mosquito midgut: a molecular and cellular view. Bull Inst Pasteur (92):119–132

Shahabuddin M, Toyoshima T, Aikawa M, Kaslow DC (1993) Transmission-blocking activity of a chitinase inhibitor and activation of malarial parasite chitinase by mosquito protease. Proc Natl Acad Sci USA 90:4266–4270

Shahabuddin M, Lemons FJ, Kaslow D, Jacobson-Lorena M (1996) Antibody mediated inhibition of *Aedes egypti* midgut trypsins blocks sporogonic development of *Plasmodium gallinaceum*. Infect Immun 64:739–743

Siddiqui WA, Tam LQ, Kramer KJ, Hui GS, Case SE, Yamaga KM, Chang SP, Chan EB, Kan SC (1987) Merozoite surface coat precursor protein completely protects *Aotus* monkeys against *Plasmodium falciparum*malaria. Proc Natl Acad Sci USA 84:3014–3018

Sieber KP, Huber M, Kaslow D, Banks SM, Torii M, Aikawa M, Miller LH (1991) The peritrophic membrane as a barrier: its penetration by *Plasmodium gallinaceum* and the effect of a monoclonal antibody to ookinetes. Exp Parasitol 72:145–156

Sim BKL (1995) EBA-175: an erythrocyte-binding ligand of *Plasmodium falciparum*. Parasitol Today 11:213

Sim BKL, Orlandi PA, Haynes JD, Klotz FW, Carter JM, Camus D, Zegans ME, Chulay JD (1990) Primary structure of the 175K *Plasmodium falciparum* erythrocyte binding antigen and identification of a peptide which elicits antibodies that inhibit malaria merozoite invasion. J Cell Biol 111:1877–1884

Sim BKL, Toyoshima T, Hayes J D, Aikawa M (1992) Localization of the 175-kilodalton erythrocyte binding antigen in micronemes of *Plasmodium falciparum* merozoites. Mol Biochem Parasitol 51:157–160

Sim BKL, Chitnis CE, Wasniowska K, Hadley TJ, Miller LH (1994) Receptor and ligand domains for invasion of erythrocytes by *Plasmodium falciparum*. Science 264:1941–1944

Sinnis P, Nussenzweig V (1996) Preventing sporozoite invasion of hepatocytes. In: Hoffman SL (ed) Malaria vaccine development. ASM, Washington, pp 15–33

Smythe JA, Coppel RL, Brown GV, Ramasamy R, Kemp DJ, Anders RF (1988) Identification of two integral membrane proteins of *Plasmodium falciparum*. Proc Natl Acad Sci USA 85:5195–5199

Smythe JA, Peterson MG, Coppel RL, Saul AJ, Kemp DJ, Anders RF (1990) Structural diversity in the 45-kilodalton merozoite surface antigen of *Plasmodium falciparum*. Mol Biochem Parasitol 39:227–234

Stevenson MM, Tam MF, Belosevic M, van der Meide PH, Podoba JE (1990) Role of endogenous gamma interferon in host response to infection with blood-stage *Plasmodium chabaudi AS*. Infect Immun 58:3225–3232

Stevenson MM, Tam MF, Wolf SF, Sher A (1995) IL-12-induced protection against blood-stage *Plasmodium chabaudi AS* requires IFN-gamma and TNF-alpha and occurs via a nitric oxide- dependent mechanism. J Immunol 155:2545–2556

Stoute JA, Slaoui M, Hepner G, Momin P, Kester KE, Desmons P, Wellde BT, Nathalie G, Krzych U, Marchand M, Ballou WR, Cohen JE, For the malaria evalutation group (1997) A preliminary evaluation of a recombinant circumsporozoite protein vaccine against *Plasmodium falciparum* malaria. N Engl J Med 336:86–91

Su XZ, Heatwole VM, Wertheimer SP, Guinet F, Herrfeldt JA, Peterson DS, Ravetch JA, Wellems TE (1995) The large diverse gene family var encodes proteins involved in cytoadherence and antigenic variation of *Plasmodium falciparum*-infected erythrocytes. Cell 82:89–100

Sultan AA, Thathy V, Frevert U, Robson KJ, Crisanti A, Nussenzweig V, Nussenzweig RS, Menard R (1997) TRAP is necessary for gliding motility and infectivity of *Plasmodium* sporozoites. Cell 1997:511–522

Tachado SD, Gerold P, McConville MJ, Baldwin T, Quilici D, Schwartz RT, Schofield L (1996) Glycosylphosphatidylinositol toxin of *Plasmodium* induces nitric oxide synthase expression in macrophages and vascular endothelial cells by a protein tyrosine kinase-dependent and protein kinase C-dependent signaling pathway. J Immunol 156:1897–1901

Tam JP, Clavijo P, Lu Y, Nussenzweig V, Nussenzweig R, Zavala F (1990) Incorporation of T and B epitopes of the circumsporozoite protein in a chemically defined synthetic vaccine against malaria. J Exp Med 171:299–306

Targett GA, Harte PG, Eida S, Rogers NC, Ong CS (1990) *Plasmodium falciparum* sexual stage antigens: immunogenicity and cell-mediated responses. Immunol Lett 25:77–81

Taverne J, Bate AW, Sarkar DA, Meager A, Rook GAW, Playfair JH (1990) Human and murine macrophages produce TNF in response to soluble antigens of *Plasmodium falciparum*. Parasite Immunol 12:33–43

Templeton TJ, Aikawa M, Parker KC, Kaslow DC (1998) *Plasmodium falciparum* Pfs40, renamed Pfs39, is located to an intracellular membrane-bound compartment and is not sexual stage- specific. Mol Biochem Parasitol 90:359–365

Thomas AW, Waters AP, Carr D (1990) Analysis of variation in Pf83, an erythrocytic merozoite vaccine candidate antigen of *Plasmodium falciparum*. Mol Biochem Parasitol 42:285–287

Tian JH, Miller LH, Kaslow DC, Ahlers J, Good MF, Alling DW, Berzofsky JA, Kumar S (1996) Genetic regulation of protective immune responses in congenic strains of mice vaccinated with a subunit malaria vaccine. J Immunol 157:1176–1181

Valero MV, Amador LR, Galindo C, Figueroa J, Bello MS, Murillo LA, Mora AL, Patarroyo G, Rocha CL, Rojas M, Aponte JJ, Sarmiento LE, Lozada DM, Coronell CG, Ortega NM, Rosas JE, Alonso PL, Patarroyo ME (1993) Vaccination with SPf66, a chemically synthesised vaccine, against *Plasmodium falciparum* malaria in Columbia. Lancet 341:705–710

Valero MV, Amador R, Aponte JJ, Narvaez A, Galindo C, Silva Y, Rosas J, Guzman F, Patarroyo ME (1996) Evaluation of SPf66 malaria vaccine during a 22-month follow-up field trial in the Pacific coast of Colombia. Vaccine 14:1466–1470

van der Heyde HC, Huszar D, Woodhouse C, Manning DD, Weidanz WP (1994) The resolution of acute malaria in a definitive model of B-cell deficiency, the JHD mouse. J Immunol 152:4557–4562

van der Heyde HC, Elloso MM, Chang WL, Kaplan M, Manning DD, Weidanz WP (1995) Gamma delta T cells function in cell-mediated immunity to acute blood-stage *Plasmodium chabaudi adami* malaria. J Immunol 154:3985–3990

van der Heyde HC, Elloso MM, Roopenian DC, Manning DD, Weidanz WP (1993a) Expansion of the $CD4^-$, $CD8^-$ gamma/delta T cell subset in the spleens of mice during blood stage malaria. Eur J Immunol 23:1846–1850

van der Heyde HC, Manning DD, Weidanz WP (1993b) Role of $CD4^+$ T cells in the expansion of the $CD4^-$, $CD8^-$, gamma/delta T cell subset in the spleens of mice during blood stage malaria. J Immunol 151:6311–6317

Vermeulen AN, Ponnudurai T, Beckers PJA, Verhave J, Smits MA, Meuwissen HE (1985) Sequential expression of antigens on sexual stage of Plasmodium falciparum accessible to tansmission-blocking antibodies in the mosquito. J Exp Med 162:1460–1476

Vermeulen AN, van Deursen J, Brakenhoff RH, Lensen TH, Ponnudurai T, Meuwissen JH (1986) Characterization of *Plasmodium falciparum* sexual stage antigens and their biosynthesis in synchronised gametocyte cultures. Mol Biochem Parasitol 20:155–163

Wahlgren M, Fernandez V, Scholander C, Carlson J (1994) Rosetting. Parasitol Today 10:73–79

Wahlin B, Wahlgren M, Perlmann H, Berzins K, Bjorkman A, Patarroyo ME, Perlmann P (1984) Human antibodies to a Mr 155,000 *Plasmodium falciparum*antigen efficiently inhibit merozoite invasion. Proc Natl Acad Sci USA 81:7912–7916

Wang R, Charoenvit Y, Corradin G, Porrozzi R, Hunter RL, Glenn G, Alving CR, Church P, Hoffman SL (1995) Induction of protective polyclonal antibodies by immunization with a *Plasmodium yoelii* circumsporozoite protein multiple antigen peptide vaccine. J Immunol 154:2784–2793

Wang R, Charoenvit Y, Corradin G, De La Vega P, Franke ED, Hoffman SL (1996) Protection against malaria by *Plasmodium yoelii* sporozoite surface protein 2 linear peptide induction of $CD4^+$ T cell-and IFN-gamma-dependent elimination of infected heparocytes. J Immunol 157:4061–4067

Waters AP, Thomas AW, Deans JA, Mitchell GH, Hudson DE, Miller LH, Mccutchan TF, Cohen S (1990) A merozoite receptor protein from *Plasmodium knowlesi* is highly conserved and distributed throughout *Plasmodium*. J Biol Chem 265:17974–17979

Waters AP, Thomas AW, Mitchell GH, McCutchan TF (1991) Intra-generic conservation and limited inter-strain variation in a protective minor surface antigen of *Plasmodium knowlesi* merozoites. Mol Biochem Parasitol 44:141–144

Weiss WR, Sedegah M, Beaudoin RL, Miller LH, Good MF (1988) $CD8^+$ T cells (cytotoxic/suppressors) are required for protection in mice immunized with malaria sporozoites. Proc Natl Acad Sci USA 85:573–576

Weiss WR, Houghten RA, Good MF, Berzofsky JA, Miller LH, Hoffman SL (1990) A CTL epitope on the circumsporozoite protein of *P. yoelii*. Bull World Health Organ 68 [Suppl]:99–103

Wen L, Roberts SJ, Viney RJ, Wong FS, Mallick C, Findly RC, Peng Q, Craft JE, Owen MJ, Hayday AC (1994) Immunoglobulin synthesis and generalised autoimmunity in mice congenitally deficient in alpha beta (+) T cells. Nature (London) 369:654–658

White W, Evans C, Taylor DW (1991) Antimalarial antibodies of the immunoglobulin G2a isotype modulate parasitemia in mice infected with *Plasmodium yoelii.* Infect Immun 59:3547–3554

Williamson KC, Criscio MD, Kaslow DC (1993) Cloning and expression of the gene for *Plasmodium falciparum*transmission-blocking target antigen Pfs230. Mol Biochem Parasitol 58:355–358

Williamson KC, Keister D, Morutova O, Kaslow DC (1995) Recombinant Pfs230, a *Plasmodium falciparum* gametocyte protein, induces antisera that reduce infectivity of *Plasmodium falciparum*in mosquitos. Mol Biochem Parasitol 75:33–42

Wizel B, Kumar N (1991) Identification of a continuous and cross-reacting epitope for *Plasmodium flaciparum* transmission-blocking immunity. Proc Natl Acad Sci USA 88:9533–9537

Wizel B, Houghten R, Church P, Tine JA, Lanar DE, Gordon DM, Ballou WR, Sette A, Hoffman SL (1995a) HLA-A2-restricted cytotoxic T lymphocyte responses to multiple *Plasmodium falciparum*sporozoite surface protein 2 epitopes in sporozoite-immunized volunteers. J Immunol 155:766–775

Wizel B, Houghten RA, Parker K, Coligan JE, Church P, Gordon DM, Ballou WR, Hoffman SL (1995b) Irradiated sporozoite vaccine induces HLA-B8-restricted cytotoxic T lymphocyte responses against two overlapping epitopes of the *Plasmodium falciparum* surface sporozoite protein 2. J Exp Med 182:1435–1445

Yadava A, Kumar S, Dvorak JA, Milon G, Miller LH (1996) Trafficking of *Plasmodium chabaudi adami*-infected erythrocytes within the mouse spleen. Proc Natl Acad Sci USA 93:4595–4599

Zavala F, Chai S (1990) Protective anti-sporozoite antibodies induced by a chemically defined synthetic vaccine. Immunol Lett 25:27–274

CHAPTER 16

Antifertility Vaccines

V.C. STEVENS

A. Introduction

The notion that fertility might be disrupted by immunological mechanisms was conceived nearly a century ago (LANDSTEINER 1899; METCHINIKOFF 1899). During the next three or four decades numerous studies in animals and humans supported this contention, but there was insufficient understanding of the reproductive and immune systems to seriously pursue the development of a product or procedure for use as a practical antifertility method. Towards the middle of the twentieth century immunologists were beginning to obtain an understanding of the concept of tolerance to self antigens and means to alter antigens to render them immunogenic (CINADER and DUBERT 1955; BURNETT 1961; MEDAWAR 1961). Simultaneously, reproductive scientists began to define the precise function and chemical structure of certain antigens vital for successful reproduction (CLOSSET et al. 1973; MORGAN et al. 1973). These pioneering research contributions led to serious consideration of the potential for developing a safe and effective antifertility method based upon immunological principles during the past three decades.

The process for developing a vaccine for the regulation of fertility has many features in common with those used for developing vaccines against infectious diseases but also has some distinct differences. A principal difference is that most of the reproductive antigens which are the targets for immunological attack are self components (isoantigens) exhibiting tolerance and must be chemically modified to render them immunogenic. Also, vaccines against pathogens are designed, if possible, to convey permanent immunity whereas it is desirable for an antifertility vaccine to be predictably reversible. An efficacy rate of 80%–90% is often satisfactory for a vaccine against infectious micro-organisms, but an immunocontraceptive method must be effective in nearly 100% of recipients. Further, a moderate degree of local reactivity, fever, and/or pain following the administration of an anti-infection vaccine might be acceptable for general use. An antifertility product must be essentially free from such side effects if it is to be attractive for birth control purposes since its users are primarily healthy adults with alternative contraceptive methods available. One may logically ask, "With all of these difficulties and problems, and with the availability of other birth control methods, why is an antifertility vaccine needed?"

Immunological contraception has several potential collective advantages over currently available methods of family planning (JONES 1995). These include:

- Choice of several sites of action in the reproductive process, including possibly, in the male
- Freedom from menstrual, systemic, metabolic, or endocrine sequelae, or undesirable local physical effects
- No interference with sexual response or activity
- Ease of administration (systemic or oral), making it suitable for distribution, if necessary, by a variety of health care providers
- Sustained and defined duration of action (e.g., 6, 12, or 18 months).
- High intrinsic efficacy and low user failure
- Economical manufacture and distribution

The rationale for antifertility vaccines in practical terms dictates that the induction of antibodies against reproductive antigens, rather than cell-mediated immunity, is preferable for an acceptable and effective method. In addition, the ideal characteristics of an immunological approach to contraception favors a method involving the production of antibodies endogenously (active immunization) rather than administering antibodies produced exogenously (passive immunization). The current view of most workers in this field is that the immunological disruption of fertility is more feasible and safer in the female than the male. There are some exceptions to these views and examples of approaches to male immunocontraception are presented here.

The immunological abrogation of fertility, theoretically, might involve a number of molecules functioning at different stages of the human reproductive process. Tissue antigens and hormones derived from the hypothalamus, pituitary, gonads, gametes, blastocyst, and placenta have been proposed and studied as potential vaccine antigens (DICZFALUSY 1974; EDWARDS and JOHNSON 1976; JONES 1982; ADA and GRIFFIN 1991). However, in practice, logistic and safety considerations dictate caution in the selection of target antigens for vaccine development. Interference with the function of the selected antigen must not only be highly specific to the reproductive process but must also result in a physiological state acceptable for contraception. For example, the method should not disrupt ovulation, menstrual cycle events, or libido or alter any metabolic process. Where possible, the antigen should be present in the body intermittently in low concentrations and should be allogenetically foreign to the recipient (i.e., sperm antigens administered to women). When these considerations are taken into account, the list of potential vaccine antigens is limited considerably. The following discussion limits itself to those vaccine candidate antigens that may meet the requirements described above and are currently being actively assessed in vaccine development programs.

B. Current Status of Vaccine Development

I. Antisperm Vaccines

The first recorded efforts to disrupt fertility by antibodies involved the injection of animals and humans with sperm or sperm antigens from a variety of species (BASKIN 1932; HENLE et al. 1940). Antigens of likely relevance to antisperm vaccine development appear mostly to be acquired within the testis. Some are species specific, but others are highly conserved. The majority of antigens on the surface of the sperm are membrane-bound glycoproteins, and some contain carbohydrate epitopes recognized by sperm antibodies. Antigens involved in "naturally occurring" immunity to sperm are usually tissue specific, but some degree of allogenic restriction may occur. They may be altered, revealed, or released by sperm capacitation and the acrosome reaction. The contribution of potentially relevant antigens from the epididymis and seminal plasma to the immunogenicity of ejaculated sperm is uncertain.

Sperm enzymes have been studied extensively over many years but only a testes-specific lactic dehydrogenase (LDH-C_4) has yet shown promise for vaccine development (O'HERN et al. 1995). This acrosomal enzyme was identified as a candidate antigen over two decades ago. Fertility reduction accompanies immunization in a variety of experimental animals. A human testis expression library has been screened with polyclonal and monoclonal antibodies, and the nucleotide sequence coding for human LDH-C_4 has been deduced and engineered into an expression vector system. Vaccines involving recombinant and synthetic peptide antigens have shown promise of efficacy in preliminary studies in female baboons (Table 1).

Glycoprotein antigens intrinsic to, or expressed on, the sperm membrane have attracted considerable, though relatively unsophisticated, attention until recent years when more refined techniques in immunochemistry and molecular biology have allowed a number of defined immunogens to be evaluated as vaccine candidates. Several such molecules have shown encouraging preliminary antifertility effects in immunized experimental animals or in in vitro systems. They include FA-1 (NAZ and MENGE 1990; NAZ et al. 1991; COONROD et al. 1994), RSA (O'RAND et al. 1990), HSA 11 and JS 63 (LEE et al. 1990; ARCHIBONG et al. 1995), PH 20 and PH 30 (PRIMAKOFF et al. 1988; PRIMAKOFF and MYLES 1990; RAMARAO et al. 1994), SP-10 (HERR et al. 1990; WRIGHT et al. 1993), and epididymal protein DE (PEREZ-MARTINEZ et al. 1995). Two other unnamed molecules of possible interest are a 71-kDa antigen described by SURI et al. (1996) with homologues in several species and a 95-kDa antigen which may have a role in sperm activation during fertilization in the mouse (LEYTON and SALING 1989) and human (NAZ et al. 1991). Some of these antigens have been found wanting in immunogenicity and/or efficacy on more advanced animal testing. Others await convincing evidence of efficacy in immunized subhuman primates before they can be considered for use in antifertility vaccines.

Table 1. Serum antibody titers versus fertility[a] in female baboons immunized with LDH-C (adapted from O'Hern et al. 1995 with permission)

Animal no.	Titer		Fertility trial outcome	Reversal trial outcome
	Week 10	Week 15		
875	100	74	NP	P
1268	90	30	NP	NP
1271	60	1200	NP	P
1293	2000	600	P	P
1296	20	22	NP	P
1309	0	0	NP	NP
1345	120	2800	NP	NP
1408	0	0	NP	P
1410	0	0	NP	NP
1430	27	710	NP	–b
1437	0	0	NP	NP
1457	0	0	NP	NP
1474	180	410	P	P
1476	180	780	NP	NP

[a] Males were introduced to the colony during week 11.
[b] Animal no. 1430 died before the reversal trial.
NP, not pregnant; P, pregnant.

Researchers in this area have also addressed the prospect of optimizing the efficacy of an antisperm vaccine by enhancing the local secretory immune response in the female genital tract. Immunization at mucosal sites have been investigated. Current evidence indicates that intra vaginal or intrauterine immunization with sperm or model antigens is ineffective unless combined with concomitant systemic antigen administration (Drew et al. 1993; Parr and Parr 1993). There is other evidence, however, that intragastric or oral immunization of mice or rats with whole sperm, defined sperm antigens, or a recombinant antigen in a live viral vector provokes a substantial local genital immune response, in some instances with antifertility effects (Allardyce 1984; Suri et al. 1993; Srinivasan et al. 1995). These responses are always accompanied by systemic as well as local immunity, and the utility and rationale of this approach need to be studied further before an oral antisperm vaccine can be considered seriously.

While the prospect for an antisperm immunocontraceptive remains viable, none has yet reached the stage of clinical testing.

II. Ovum Antigens

The ovum is an attractive target for contraceptive vaccination considering that a specific antibody response might inhibit fertilization by blocking receptors for sperm on or in the zona pellucida or oolemma. An antiovum vaccine combines the advantages of a prefertilization method (as an antisperm vaccine) with a one-cell target present only following ovulation (unlike an antisperm method). The major caveat is the potential for autoimmune damage should target antigens be present and accessible on ovarian oocytes. The earliest studies of the effects of antibodies against ova were conducted using marine invertebrates (PEARLMAN 1954) and later extended to mammals (SHIVERS 1974).

The zona pellucida (ZP) is highly immunogenic, and there is clear and long-standing evidence, from in vitro studies and from a variety of animal models that immunity to ZP antigens induces antifertility effects (SHIVERS and DUNBAR 1977; SACCO 1987; HENDERSON et al. 1988; DEAN 1992; AITKEN et al. 1993). ZP genes seem to be conserved across mammalian species, so that the results of studies in lower animal models where the genes have been identified can potentially be extrapolated to primates, including man (DEAN 1992). Several relevant ZP antigens have been isolated and characterized, including glycoproteins ZP2 and ZP3 which, respectively, are the secondary and primary receptors for sperm (DEAN 1992). ZP3 has attracted particular interest; its genetic and molecular features and its immunogenic characteristics have been described in several species (DEAN and MILLAR 1990; SACCO et al. 1991; DEAN 1992). Continuing efforts are directed at "dissecting out" specific peptide antigens from ZP3 and other molecules in order to induce a safe and effective immune response to the ovum in the absence of ovarian pathology or alteration of ovarian function (DEAN and MILLAR 1990; JONES et al. 1992; RHIM et al. 1992; TUNG et al. 1994). Early experiments in primates employing porcine zona antigens were unsuccessful in avoiding these problems (Fig. 1).

Other approaches to ZP vaccine development have involved the use, as immunogens, of recombinant ZP proteins (SCHWOEBEL et al. 1992; AITKEN et al. 1993; PRASAD et al. 1995). This strategy has brought into focus the importance of the carbohydrate side chains of ZP glycoprotein antigens in both immunogenicity and contraceptive efficacy. With the use of appropriate expression vectors this problem should be resolved; the imperative remain, however, to scrupulously exclude the induction of autoimmune oophoritis at every stage of vaccine development.

In contrast to the ZP, little attention has been paid to the identification of other candidate antigens in the ovum. Swedish workers (JIN and NIILSSON 1992; JIN 1994) have identified the presence of phase-specific antigens in the oolemma using monoclonal antibodies which exhibit antifertility activity in vitro and in vivo. Such antigens would be particularly attractive targets in the event that they are essential to sperm-ovum interaction, and that they have some degree of temporal restriction. Even so, if they are expressed in maturing oocytes as well as ova, their relatively secluded location beneath the ZP would make untoward intraovarian autoimmunity an unlikely complication of antiovum vaccination.

Whether an antiovum immunocontraceptive that will provide women safe and effective protection from pregnancy is feasible has yet to be established,

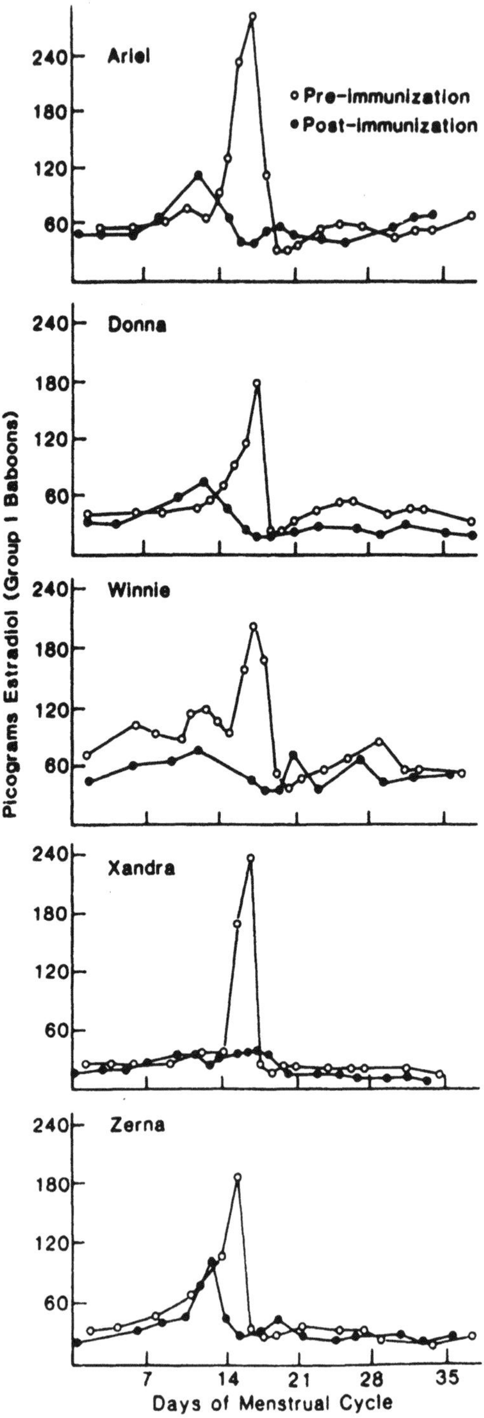

Fig. 1. Serum estradiol levels in female baboons during a menstrual cycle prior to immunization with a major porcine zona pellucida antigen and during a postimmunization cycle 9–10 weeks after the last immunization. (Reproduced from DUNBAR 1989, with permission of the publisher, the American Society for Reproductive Medicine [formerly the American Fertility Society])

but research now underway should clarify this question within the next decade.

III. Antihormone Vaccines

The various protein and steroid hormones of the reproductive system have been studied widely and are chemically well characterized. Their potential use as immunogens in antifertility vaccines has been extensively reviewed (DICZFALUSY 1974; EDWARDS 1976; JONES 1982). While gonadal steroid hormones have been assessed as possible vaccine antigens, these have been generally abandoned for a variety of reasons (see reviews), and current efforts are focused on protein hormones produced by the hypothalamus, pituitary and blastocyst.

1. Gonadotropin-Releasing Hormone

Gonadotropin-releasing hormone (GnRH) has been selected as a target for the immunological inhibition of reproduction in many nonhuman animal species as this hypothalamic peptide hormone regulates the secretion of the pituitary gonadotropins follicle-stimulating hormone (FSH) and luteinizing hormone (LH), which in turn control the development and function of the gonads in both males and females. While GnRH has other physiological functions, its inhibition by antibodies has been shown to result in gonadal atrophy and sterility (FRASER et al. 1974). These effects from successful active immunization against GnRH or the passive administration of GnRH antibodies provide attractive methods to control estrus and fertility in domestic pets, livestock, and certain wildlife species. The total arrest of gonadal function is useful for these applications where the loss of gonadal steroid secretion (estrogen and progesterone in females and testosterone in males) is either desirable, temporary, or of no concern. This same application of anti-GnRH therapy may be applicable to human use in persons suffering from prostate or breast cancer. In such cases gonadal steroid inhibition is desirable inasmuch as these hormones cause growth and proliferation of tumors in a high percentage of these patients.

The application of immunological inhibition of GnRH to humans to achieve acceptable contraception poses significant problems. Inhibition of gonadal function in women denies them estrogen and progesterone secretion needed to regulate menstrual cycles and to provide necessary support of physiological functions such as bone metabolism, libido, and psychological well-being. Men denied testosterone experience loss of libido and changes in secondary sex characteristics such as hair growth and breast development. Such a state of nonsurgical castration would result in a menopausal condition for women and a feminizing effect on men. To overcome these problems investigators have proposed to apply a GnRH immunocontraceptive to men together with androgen replacement therapy to compensate for the loss of natural testosterone. The concept for this approach is to supply sufficient androgen to restore libido and secondary sex characteristics but insufficient to restore spermatogenesis and fertility.

Currently, successful development of GnRH immunocontraceptive methods have been achieved for applications to domestic pets and farm animals (LADD et al. 1994; MELOEN et al. 1994). Products for application as a human male contraceptive are still in the development stage. The basic premise of GnRH inhibition together with androgen replacement has, however, been demonstrated in laboratory animals. LADD et al. (1989) and AWONIYI et al. (1992) have used these combined treatments successfully in male rats and have shown that fertility can be suppressed while maintaining normal mating behavior. Human clinical trials using GnRH immunogens have been quite limited to date. There is little doubt that GnRH immunizations will affect infertility in men and thus the likely efficacy of this method is not questioned. However, the overall safety and acceptability of this approach as a contraceptive for men has not yet been fully established. Although no health hazards have been reported by those developing animal immunocontraceptives, one report has been published suggesting that hypothalamic lesions can be induced from immunizing pigs against GnRH (MOLENAAR et al. 1993). While this study was limited in scope and technology, it raises concern that autoimmune pathology might result from the widespread use of GnRH immunizations. Perhaps more likely to cause problems to the application of the proposed product, is the androgen replacement portion of the contraceptive. The titration of the appropriate dose of androgen, applicable to all normal men, that will maintain male sex characteristics but not restore spermatogenesis, may be difficult to achieve. Whether these problems will be of significant severity or frequency to prevent the application of such a product from being used can only be answered after adequate clinical testing has been completed.

2. Follicle-Stimulating Hormone

The traditional concept of the regulation of testicular function by the pituitary gland in male mammals is that LH controls the secretion of testosterone and its subsequent maturation of spermatozoa while FSH initiates spermatogenesis. It was thus assumed that these functions of the two gonadotropins are separate entities, and that the selective inhibition of FSH would arrest spermatogenesis and lead to azoospermia and thus infertility while permitting normal testosterone secretion. Such a state in men would render them infertile while not affecting libido or secondary sex characteristics regulated by testosterone. The possibility of achieving this condition by specifically immunizing men against FSH has stimulated research during the past two decades to develop an immunocontraceptive for men based on FSH neutralization by antibodies.

Despite the attractiveness of this concept for developing a male immunocontraceptive, basic research conducted during recent years has raised some doubts of whether FSH inhibition would cause the arrest of spermatogenesis. Although studies in male rats demonstrated that FSH is required for the initiation of spermatogenesis, testosterone alone could maintain spermatogenesis in adult animals (CUNNINGHAM and HUCKINS 1979). Clinical studies in men have failed to present a clear answer to this controversial question as some evidence is available that FSH is required for the maintenance of sperm production while others have shown that androgens alone can maintain spermatogenesis once it has been initiated by FSH (JOHNSEN 1966; SHERINS 1974). Inasmuch as there are differences in the findings in rats and men, and that ethical considerations limit human experimentation, several studies in nonhuman primates have been conducted in an attempt to elucidate this question (WICKINGS et al. 1980; RAJ et al. 1982).

Studies in male rhesus monkeys using passive immunization with antibodies against ovine FSH showed a reduction in testicular volume and sperm counts although azoospermia was not achieved (WICKINGS et al. 1980). Two groups of investigators actively immunized male monkeys with ovine FSH and observed the effect of immunizations on sperm counts and testosterone production (MOUDGAL 1981; RAJ et al. 1982). In both studies sperm counts were reduced, but not to zero; testosterone levels remained normal. One of the groups of monkeys was used in mating studies. These male monkeys failed to cause a pregnancy in any female although the number tested was small. Thus after many years of investigation the question of whether FSH inhibition is a viable approach to fertility suppression in men has not been conclusively answered since some scientists believe that azoospermia is an absolute requirement for complete fertility blockage in men (BARFIELD et al. 1979).

Although the feasibility of an immunocontraceptive for men based on FSH immunization has not been established, investigators in this field continue to pursue this goal. A clinical trial using ovine FSH in men has been conducted in India, but the results of this trial have not been published or otherwise reported. Still other scientists have perused the development of immunogens using fragments of human, rat, and ovine FSH. These studies have been designed to select a portion of the FSH molecule that will elicit FSH specific antibodies and to define a product that can be produced synthetically. In one study (BUTTERSTEIN et al. 1993) human FSH β-subunit was used to immunize female rats, and the effects on endogenous FSH and fertility were tested. Immunoneutralizing antibodies were induced, but only those animals where complete FSH neutralization occurred were infertile. It was found that all rats immunized with human FSH-β produced antibodies that bound a peptide representing amino acid residues 33–53 of the subunit. This latter finding was confirmed by another group of investigators who found that hormone specific antisera raised to ovine and human FSH also bound a peptide with the β-hFSH sequence of 37–55 (WESTHOFF et al. 1996). This conserved region of all of the glycoprotein protein hormones was previously described to contain a hormone-specific epitope within the human chorionic gonadotropin (hCG) molecule (STEVENS et al. 1986). It seems likely that this rational approach to FSH immunogen development will be fruitful should the concept of FSH immunocontraception for men prove to be a feasible proposition.

In summary, it still remains to be demonstrated whether FSH inhibition by immunological methods will yield a useful male contraceptive. Much more research and testing of potential product prototypes is required before any suggestion is made that this prospect will become reality. Clearly such a development will not occur in the near future.

3. Human Chorionic Gonadotropin

By far the most progress towards developing a successful immunocontraceptive has been the research and development of a product(s) for immunization with antigens derived from hCG. Active immunization of women against hCG is potentially a promising new option for women to choose for family planning purposes. From a biological perspective the specific inhibition of the function of this molecule by immunological means during early gestation would provide a highly safe and effective procedure for preventing pregnancy. Expression of the hormone by the conceptus occurs within 2–3 days following sperm penetration of the egg and secretion from preimplantation blastocysts is believed to have important function(s) in the establishment of pregnancy (FUCHS et al. 1982; BONDUELLE et al. 1988; DIMITRIADOU et al. 1992). In addition to the well-known function of stimulating progesterone secretion by the ovarian corpus luteum following implantation, other physiological activities of hCG earlier in gestation have been reported (MILWIDSKY et al. 1993; STEWART et al. 1993). Evidence is available to indicate that interference with hCG function by antibodies can disrupt pregnancy prior to implantation. The quantity of hCG present in biological fluids during the critical time period for pregnancy establishment is sufficiently small to be inhibited by levels of antibodies that can be induced by active immunization. Also, hCG is not present in these fluids in physiological amounts except during pregnancy, and antibodies specific for hCG would not effect the function of other hormones in nonpregnant women.

Three main types of hCG immunocontraceptives have been subjected to clinical trials. While more than three formulations have been tested, the principal immunogen components evaluated to date are the following:

- hCG β-subunit conjugated to tetanus toxoid (β-hCG-TT) (SINGH et al. 1989).
- hCG β-subunit–ovine LH α-subunit heterodimer conjugated to tetanus toxoid and diphtheria toxoid (HSD-TT-DT) (SINGH et al. 1989).
- hCG β-subunit C-terminal 37 residue synthetic peptide conjugated to diphtheria toxoid (CTP-DT) (JONES et al. 1988).

All three immunocontraceptive types have undergone phase 1 (safety) trials. While antibody levels elicited were quite variable, all of these products generated sufficient levels to suggest antifertility effects could be expected following immunization. No serious side effects were experienced in the study subjects, and despite cross-reaction of antibodies raised by the β-hCG-TT and HSD-TT-DT immunocontraceptives with human luteinizing hormone (hLH) no effects of immunizations on ovulation or menstrual cycle patterns were reported (Singh et al. 1989). Serological and clinical assessment of the women in these trials revealed no health hazards from the use of any of these products.

Only the HSD-TT-DT immunocontraceptive has completed a phase 2 (efficacy) trial to date. The details of the trial and its results are presented in detail elsewhere (Talwar et al. 1994). In brief, approximately 150 fertile women were immunized. Approximately 80% of the women elicited antibody levels of sufficient magnitude to prevent pregnancy, and approximately 60% sustained protective levels for 6 months or more. However, in those women with sufficient and sustained antibody levels, only one pregnancy was observed in more than 1200 menstrual cycles of otherwise unprotected exposure to pregnancy. This vaccine, while not representing a product acceptable for general use, demonstrated that immunizations against hCG can be effective in preventing pregnancy. This milestone was very important for justifying further research and development of hCG immunocontraceptives.

The development and testing of the prototype vaccines discussed above have served a useful purpose in (a) demonstrating the successful immunization of humans against an antigen to which the immune system does not usually respond, (b) indicating that these responses can be elicited with no apparent adverse side effects, and (c) showing that hCG vaccines can be effective in preventing pregnancy when sufficient levels of neutralizing antibodies are generated in fertile women. However, none of these formulations is suitable in its current form for manufacture and widespread application to family planning. First of all, none of these vaccines can establish effective immunity with a single inoculation which will provide protection from pregnancy for at least 6 months in greater than 90% of women before another immunization is required. Currently, available products require 2–4 immunizations to establish effective antibody levels and booster injections are needed at 3- to 6-month intervals to maintain titers (Talwar et al. 1994). Such frequent applications would clearly not be attractive to women as a birth control method, and therefore the need for a formulation to provide sustained antibody levels from a single immunization is paramount to further vaccine development.

Vaccines that are manufactured for distribution and use by the general population must be produced under strict quality control and at an affordable cost, particularly in developing countries. Recognizing that improved vaccines must become available before women can be offered this method of contraception, scientists working in this area are continuing the development of advanced prototypes of the earlier versions of their hCG immunocontraceptive formulations in order to provide suitable products. In India a government-appointed task force has asked researchers to refine production and delivery methods to ensure that the approach is both practical and cost effective. Principal among the improvements sought are the use of a more effective adjuvant than was used in the HSD immunogen already tested and a reduction in the number of immunizations needed for protection from pregnancy. Another goal is to develop a live recombinant anti-hCG immunogen using a fowlpox virus vector (JAYARAMAN 1995).

Researchers associated with a WHO-sponsored program for hCG immunocontraceptive development have also made improvements in their prototype formulation. The original product (the CTP/DT immunogen described above) looked promising following a phase 1 clinical trial but a reformulation of the product, and the use of a higher dose in a phase 2 trial resulted in unacceptable local reactions at the injection site (unpublished data). While dose-reduction studies in animals indicated that the formulation could be used without producing these undesirable side effects, the availability of an improved product led the WHO to abandon the CTP/DT prototype and proceed with the testing of a more advanced version of an hCG immunocontraceptive.

The new WHO product utilizes DT conjugates of two peptide segments of b-hCG. The CTP portion (residues 109–145) is retained with the addition of a second peptide representing the amino acid 38–57 region. An analogue of this latter peptide sequence will be used rather than the native sequence as a low level of antibodies reactive with hLH is produced in some animals administered conjugates of DT/β-hCG 38–57, but this level is reduced when the analogue instead is conjugated to DT. The resultant combined immunogen is more potent in inducing hCG-reactive antibodies, but, more importantly, these antibodies are much more effective in the bioneutralization of hCG in vitro and in vivo (Fig. 2). Further, the weak reaction of antisera to hLH in vitro may not be of significance in vivo (Fig. 3).

Advances have also been made in the delivery of hCG immunogens for sustained immunity from a single application of the product. A method for delivering the CTP/DT immunogen in biodegradable microspheres (STEVENS et al. 1992) has demonstrated the capability of eliciting high levels of hCG antibodies for more than 1 year from a single injection (Fig. 4). In addition, improvements in emulsion technology have offered another option for this mode of product delivery without unacceptable local reactions in animal studies. While emulsion delivery is unlikely to elicit protective antibody levels for a 1-year period as with microspheres, a "once every 6-month" product appears feasible with the new emulsion formulations. These delivery system options may permit the development of a series of products with varying periods of efficacy.

Further clinical testing of hCG immunocontraceptives is anticipated in the near future, and no doubt new versions will be subsequently developed. The main goal of future research will be to develop products with higher efficacy and no side effects (STEVENS 1992). Therefore it seems probable that an hCG immunocontraceptive will be the first product based on immune mechanisms available for the purpose of family planning.

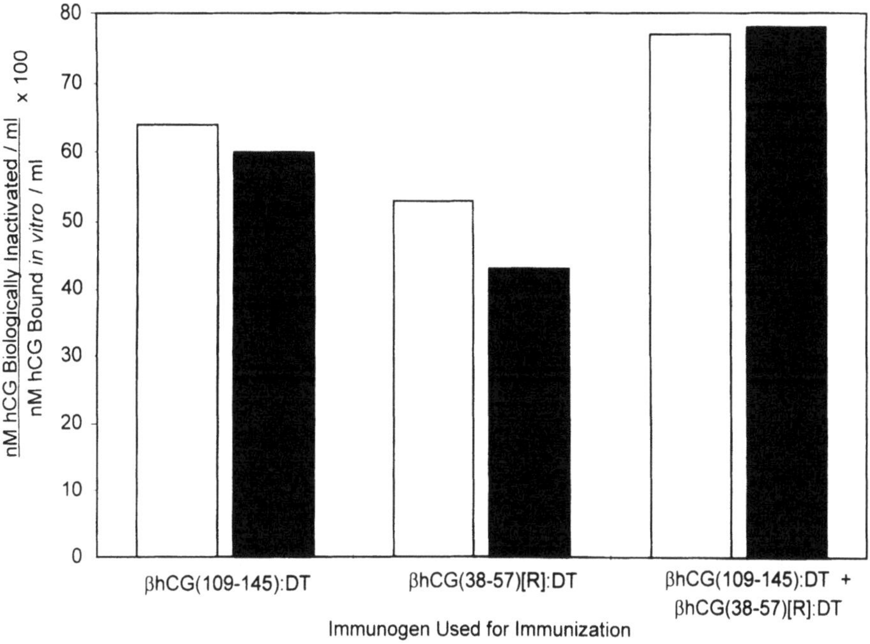

Fig. 2. A comparison of the biological neutralization capacities of antisera from rabbits immunized with β-hCG (109–145):DT, β-hCG (38–57)[R]:DT or immunized with a combination of the two conjugates as measured in vitro by a rat testes receptor binding assay (*open bars*) or in vivo by the mouse uterine weight assay (*solid bars*). Neutralizing capacity is expressed as a percentage of total binding capacity

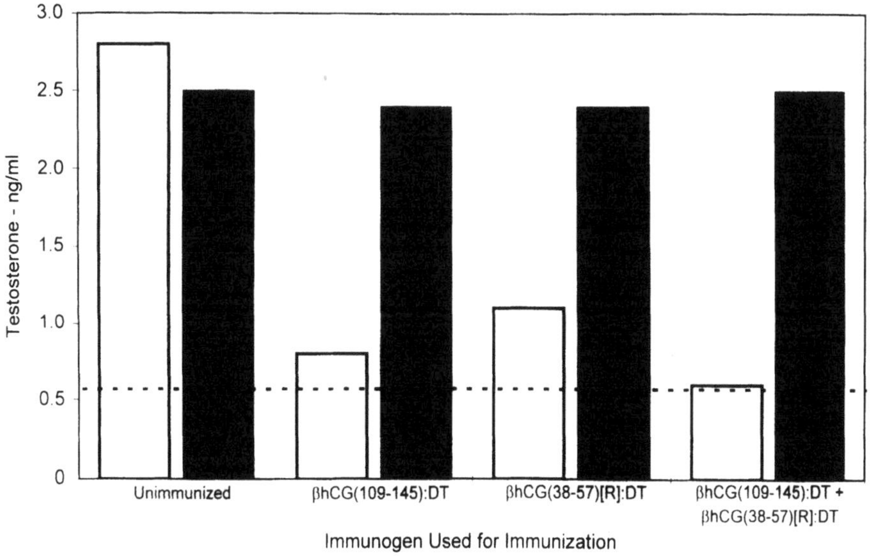

Fig. 3. Mean testosterone levels in male rabbits 6 h after the injection of hCG (*open bars*) or hLH (*solid bars*) immunized with β-hCG (109–145):DT, β-hCG (38–57)[R]:DT, a combination of the two conjugates or nonimmunized controls. *Broken lines*, mean testosterone level in untreated animals

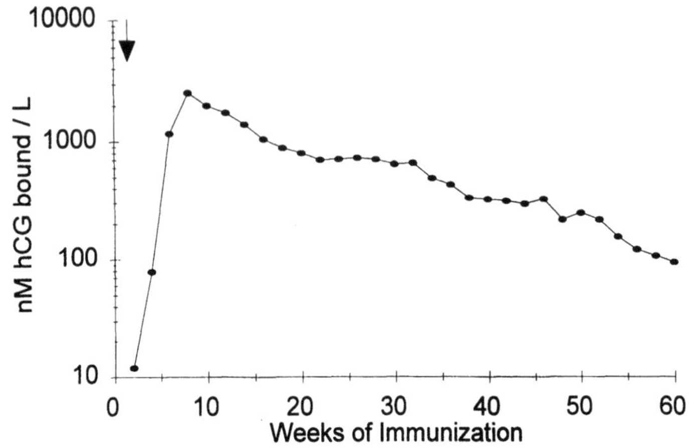

Fig. 4. Mean antibody levels in rabbits following a single immunization with a β-hCG peptide 109–145 diphtheria toxoid conjugate and an MDP adjuvant, both encapsulated in a blend of polylactide/glycolide microspheres

C. Problems and Prospects

The immune response to any vaccine is under genetic control and will therefore vary among members of an outbred population such as humans. Inasmuch as antifertility vaccines require a high level of efficacy to be acceptable products, few nonresponders to immunization can be tolerated. In quantitative terms the level of antibodies required to block fertility using a particular immunogen will dictate the percentage of the human population protected from fertility since the number completely devoid of an immune response will likely be small. For vaccines eliciting weak antibody responses in a significant proportion of recipients efforts will need to be made to reduce MHC restrictions to T-cell activation by the use of "promiscuous" T-cell epitopes (KAUMAYA et al. 1993) or to bypass the need for T-cells in eliciting elevated antibody levels (STEVENS 1996). These approaches are probably feasible but will make vaccine development more difficult if required.

Once a safe, effective, and acceptable vaccine has been formulated that meets the criteria discussed in this report, still other problems must be overcome before practical application to family planning is feasible. First of all, administration of the vaccine will not provide "instant" protection against fertility. A lag period of 3–6 weeks will be required for antibody levels to become sufficiently high to provide antifertility effects. During this interval other contraceptive methods will need to be used to avoid pregnancy. There are a number of currently available methods that would be suitable, and some of these could be applied at the same time as the first vaccine immunization.

Another problem for successful application of these vaccines is the determination of when vaccine effects have waned, and protection from fertility no longer exists. For those who wish to employ the method continuously for the remainder of their reproductive lives, a booster injection at regular intervals may be sufficient to assure protection. However, should a couple desire to have a baby after vaccine use, the timing of regained fertility will be important for child spacing. In certain settings it will be feasible for the user to have serum antibody level determinations performed to provide information on his or her fertility status. In other settings, where such laboratory services are not readily available or are unaffordable, a simple test for antibody levels may be developed. The technology for such tests from a finger prick drop of blood exists and the application of this technology to the detection of antibodies against reproductive antigens is clearly feasible.

Finally, efforts must be made to devise means to neutralize vaccine effects upon demand in order to allow vaccine recipients to "change their mind" about having a baby after he or she has received an antifertility vaccine. These events may be few provided the duration of vaccine-induced immunity can be made highly predictable and the time interval between immunization and the return to fertility is no longer than, say, 6 months. However, should safe methods for vaccine reversal become available, a higher level of method acceptability would likely result.

Not all the problems associated with the introduction of immunocontraceptives are technical. There are several cultural, religious, legal, and gender issues surrounding the development and distribution of any contraceptive or fertility regulating drug. These issues and concerns are significant and must be dealt with in a rational and sensitive manner (Griffin et al. 1994). Notwithstanding the technical and social obstacles yet to be overcome for the provision of one or more antifertility vaccines, the development of such technology and its eventual application will offer a wider array of options available for family planning.

References

Ada GL, Griffin PD (1991) Vaccines for fertility regulation: the assessment of their safety and efficacy. Cambridge University Press, Cambridge

Aitken RJ, Paterson M, Koothan PT (1993) Contraceptive vaccines. Br Med Bull 49:88–99

Allardyce RA (1984) Effect of ingested sperm on fecundity in the rat. J Exp Med 159:1548–1553

Archibong AE, Lee CY, Wolf DP (1995) Functional characterization of the primate sperm acrosomal antigen (PSA-63). J Androl 16:318–236

Awoniyi CA, Kim WK, Hurst BS, Schlaff WD (1992) Immunoneutralization of gonadotropin-releasing hormone and subsequent treatment with testosterone Silastic implants in rats: an approach toward developing a male contraceptive. Fertil Steril 58(2):403–408

Barfield A, Melo J, Coutinho E, Alvarez-Sanchez F, Faundes A, Brache V, Leon P, Frick J, Bartsch G, Weiske WH, Berner P, Mishell D Jr, Bernstein G, Oritiz A (1979) Pregnancies associated with sperm concentrations below 10 million/ml in clinical studies of a potential male contraceptive method, monthly depot medroxyprogesterone acetate and testosterone esters. Contraception 20:121–127

Baskin MJ (1932) Temporary sterilization by the injection of human spermatozoa (a preliminary report). Am J Obstet Gynecol 24:892
Bonduelle ML, Dodd R, Liebaers I, VanSteirteghem A, Williamson R, Akhurst R (1988) Chorionic gonadotropin-β mRNA, a trophoblast marker, is expressed in human 8-cell embryos derived from tripronucleate zygotes. Hum Reprod 3:909–914
Burnett FM (1961) Immunological recognition of self. Science 133:307
Butterstein GM, Sachar D, Dias JA (1993) Immunoneutralization of heterodimeric FSH using FSH beta subunit as the immunogen. J Reprod Immunol 29:48–55
Cinader B, Dubert JM (1955) Acquired immune tolerance to human albumin and the response to subsequent injections of diazo human albumin. J Exp Pathol (Oxford) 36:515–529
Closset J, Hennen G, Lequin RM (1973) Human luteinizing hormone: the amino acid sequence of the beta subunit FEBS Lett 29:97–100
Coonrod SA, Westhusin ME, Naz RK (1994) Monoclonal antibody to human fertilization antigen (FA-1) inhibits bovine fertilization in vivo: application in immunocontraception. Biol Reprod 51:14–23
Cunningham GR, Huckins C (1979) Persistence of complete spermatogenesis in the presence of low intratesticular concentrations of testosterone. Endocrinology 105:177–186
Dean J (1992) Biology of mammalian fertilization: role of the zona pellucida. J Clin Invest 89:1055–1059
Dean J, Millar SE (1990) Zona pellucida: target for a contraceptive vaccine. In: Alexander NJ, Griffin D, Spieler JM, Waites GMH (eds) Gamete interaction: prospects for immunocontraception. Wiley-Liss, New York, pp 313–226
Diczfalusy E (1974) Steps in the human reproductive process susceptible to immunological interference. In: Diczfalusy E (ed) Immunological approaches to fertility control. Karolinska Institutet, Stockholm, pp 13–36
Dimitriadou F, Phocas I, Mantazavinos T, Sarandakou A, Rizos D, Zourlas P (1992) Discordant secretion of pregnancy specific β_1-glycoprotein and human chorionic gonadotropin by human pre-embryos cultured in vitro. Fertil Steril 57:631–636
Dunbar BS, Lo C, Powell J, Stevens VC (1989) Use of a synthetic peptide adjuvant for the immunization of baboons with denatured and deglycosylated pig zona pellucida glycoproteins. Fertil Steril 52:311–318
Drew MD, Rosenthal KL, McDermott MR (1993) The design of vaccines for inducing secretory immunity in the reproductive tract. In: Griffin PD, Johnson PM (eds) Local immunity in reproductive tract tissues. Oxford University Press, Oxford, pp 484–514
Edwards RG (1976) Immunity and the control of human fertility. In: Scott JS, Jones WR (eds) Immunology of human reproduction. Grune & Stratton, New York, pp 415–470
Edwards RG, Johnson MH (1976) Physiological effects of immunity against reproductive hormones. Cambridge University Press, London
Fraser HM, Gunn A, Jeffcoate SL, Holldan, DT (1974) Effect of active immunization to luteinizing hormone releasing hormone on serum and pituitary gonadotrophins, testes and accessory sex organs in the male rat. J Endocrinol 63:399–406
Fuchs T, Hammarstrom L, Smith CI, Brundin J (1982) Sex dependent induction of human suppressor T cells by chorionic gonadotropin. J Reprod Immunol 4:185–190
Griffin PD, Jones WR, Stevens VC (1994) Anti-fertility vaccines: current status and implications for family planning programmes. Reprod Health Matters 3:108–113
Henderson CJ, Hulme MJ, Aitken RJ (1988) Contraceptive potential of antibodies to zona pellucida. J Reprod Fertil 83:325–243
Henle W, Henle G, Church CF, Foster I (1940) Spermatozoal antibodies and fertility: attempts to induce temporary sterility in female white mice by passive immunization with spermatoxoal antisera. J Immunol 38:97

Herr JC, Wright RM, John E, Foster J, Kays T, Flickinger CJ (1990) Identification of human acrosomal antigen SP-10 in primate and pigs. Biol Reprod 42:377–382
Jayaraman KS (1995) India forges ahead with contraceptive vaccine. Nat Med 1 (7):609–610
Jin M (1994) Anti-oocyte-antibodies as candidates for immuno-contraception. Thesis, Uppsala University
Jin M, Nilsson OB (1992) Enhanced production of monoclonal antibodies against zona-free, unfertilized mouse oocytes by intra-splenic immunization. Hybridoma 11:689–699
Johnsen SG (1966) A study of human testicular function by the use of human menopausal gonadotrophin and of human chorionic gonadotrophin in male hypogonadotrophic eunuchoidism and infantilism. Acta Endocr 53:489
Jones GR, Sacco AG, Subramanian MG Kruger M, Zhang S, Yurewicz EC, Moghissi KS (1992) Histology of ovaries of female rabbits immunized with deglycosylated zona pellucida manromolecules of pigs. J Reprod Fertil 95:513–525
Jones WR (1982) Immunological fertility regulation. Blackwell Scientific, Oxford
Jones WR (1995) Anti fertility vaccines. In: Nilsson O, Mattsson R (eds) Frontiers in endocrinology: immunocontraception. Ares-Serono Symposia, Rome, pp 5–21
Jones WR, Judd SJ, Ing RMY, Powell JE, Bradley J, Denholm EH, Mueller UW, Griffin PD, Stevens VC (1988) Phase I clinical trial of a World Health Organization birth control vaccine. Lancet 1:1295–1298
Kaumaya PTP, Kobs-Conrad S, Seo YH, Lee H, VanBuskirk AM, Feng N, Sheridan JF, Stevens VC (1993) Peptide vaccines incorporating a 'promiscuous' T-cell epitope bypass certain haplotype restricted immune responses and provide broad spectrum immunogenicity. J Mol Recognit 6:81–94
Ladd A, Tsong YY, Prabhu G, Thau R (1989) Effects of long-term immunization against LHRH and androgen treatment on gonadal function. J Reprod Immunol 5:85–101
Ladd A, Tsong YY, Walfield, Thau, R (1994) Development of an antifertility vaccine for pets based on active immunization against luteinizing hormone-releasing hormone. Biol Reprod 51:1076–1083
Landsteiner K (1899) Zur Kenntnis der spezifisch Auf Blukorperchen wirkenden Sera. Zentralbl Baktiol 25:546–558
Lee C-YG, Liu M-S, Su M-W, Zhu JB (1990) Studies of sperm antigens reactive to HS-11 & JS-63 monoclonal antibodies. In: Alexander NJ, Griffin D, Speiler JM, Waites GMH (eds) Gamete interaction: prospects for immunocontraception, Wiley-Liss, New York, pp 37–51
Leyton L, Sailing P (1989) 95kD sperm proteins bind ZP3 and serve as tyrosine kinase substrates in response to zona binding. Cell 57:1123–1130
Medawar PB (1961) Immunological tolerance. Science 133:303
Meloen RH, Turkstra JA, Lankhof WC, Puijk WC, Schaaper WMM, Dijkstra G, Wensing CJG, Oonk RB (1994) Efficient immunocastration of male piglets by immunoneutralization of GnRH using a new GnRH-like peptide. Vaccine 12:741–746.
Metchinikoff E (1899) Etudes sur la resortion des cellules. Ann Inst Pasteur 13:737–746
Milwidsky A, Finci-Yeheskel Z, Yagel S, Mayer, M (1993) Gonadotropin-mediated inhibition of proteolytic enzymes produced by human trophoblast in culture. J Clin Endocrinol Metab 76:1101–1105
Molenaar GJ, Lugard-Kok C, Meloen RH, Oonk RB, de Koning, J, Wensing CJG (1993) Lesions in the hypothalamus after active immunisation against GnRH in the pig. J Neuroimmunol 48:1–11
Morgan FJ, Birken S, Canfield RE (1973) Human chorionic gonadotropin: a proposal for the animo acid sequence. Mol Cell Biochem 2:97–99
Moudgal NR (1981) A need for FSH in maintaining fertility of adult male subhuman primates. Arch Androl 7:117–125

Naz R, Menge A (1990) Development of antisperm contraceptive vaccines for humans: why and how? Hum Reprod 5:511–518
Naz RK, Ahmad K, Kumar R (1991) Human spermatozoa FA-I binds with ZP3 of porcine zona pellucida. J Reprod Immunol 20:43–58
O'Hern PA, Bambra CS, Ishakia M, Goldberg E (1995) Reversible contraception in female baboons immunized with a synthetic epitope of sperm-specific lactate dehydrogenase. Biol Reprod 52:331–339
O'Rand MG, Widgren EE, Nikolajezk BS, Richardson RT, Shabanowitz RB (1990) Receptors for zona pellucida on sperm. In: Alexander NJ, Griffin D, Spieler JM, Waites GMH (eds) Gamete interaction: prospects for immunocontraception. Wiley-Liss, New York, pp 213–224
Parr EL, Parr, MB (1993) Local immunization for antifertility immunity. In: Griffin PD, Johnson PM (eds) Local immunity in reproductive tract tissues. Oxford University Press, Oxford, pp 441–458
Pearlman P (1954) Study on the effect of antisera on unfertilized sea urchin eggs. Exp Cell Res 6:485
Perez-Martinez S, Conesa D, Cuasrucu PD (1995) Potential contraceptive use of epididymal proteins: evidence for the participation of specific antibodies against rat epididymal protein DE in male and female fertility inhibition. J Reprod Immunol 29:31–45
Prasad SV, Mujtake S, Lee VH, Dunbar BS (1995) Immunogenicity enhancement of recombinant rabbit 55-kilodalton zona pellucida protein expressed using the baculovirus expression system. Biol Reprod 52:467–478
Primakoff P, Myles DG (1990) Progress towards a birth control vaccine that blocks sperm function. In: Alexander NJ, Griffin D, Speiler JM, Waites GMH (eds) Gamete interaction: prospects for immunocontraception, Wiley-Liss, New York, pp 89–102
Primakoff P, Lathrop W, Woolman L, Cowen L, Myles D (1988) Fully effective contraception in male and female guinea pigs immunized with the sperm protein PH-20. Nature 335:543–546
Raj HGM, Murty GSRC, Sairam MR, Talbert LM (1982) Control of spermatogenesis in primates: effects of active immunization against FSH in the monkey. Int J Androl [Suppl] 5:27–33
Ramarao CS, Myles DG, Primakoff P (1994) Multiple roles for Ph-20 and fertilin in sperm-egg interaction. Dev Biol 5:265–271
Rhim SH, Millar SE, Robey F, Lau AM, Lou YH, Yule T, Allen P, Dang J, Tung KSK (1992) Autoimmune disease of the ovary induced by a ZP3 peptide from the mouse zona pellucida. J Clin Invest 89:28–35
Sacco A (1987) The zona pellucida: current status as a candidate antigen for contraceptive vaccine development. J Reprod Immunol 15:122–130
Sacco AG, Yurewicz EC, Subramaniaan MG, deMayo FJ, Dukelow (1991) Immunologic response and ovarian histology of squirrel monkeys (*Saimeri sciureus*) immunized with porcine zona pellucida ZP3 macromolecules. Am J Primatol 24:15–28
Schwoebel ED, Vandervoort CA, Lee VH, Lo YK, Dunbar BS (1992) Molecular analysis of the antigenicity and immunogenicity of recombinant zona pellucida antigens in a primate model. Biol Reprod 47:857–865
Sherins RJ (1974) Clinical aspects of treatment of male infertility with gonadotrophins: testicular response of some men given hCG with and without Pergonal. In: Mancini RE, Martini L (eds) Male fertility and sterility. Raven, New York, pp 545–556
Shivers CA (1974) Immunological interference with fertilization. In: Diczfalusy E (ed) Immunological approaches to fertility control. Karolinska Institutet, Stockholm, pp 223–244
Shivers CA, Dunbar BS (1977) Auto antibodies to zona pellucida: a possible cause for infertility in women. Science 192:1082–1084

Singh O, Rao LV, Gaur A, Sharma NC, Alam A, Talwar GP (1989) Antibody response and characteristics of antibodies in women immunized with three contraceptive vaccines inducing antibodies against human chorionic gonadotropin. Fertil Steril 52:739–744

Srinivasan J, Tinge S, Wright R, Herr JC, Curtiss III R (1995) Oral immunization with attenuated Salmonella expressing human sperm antigen induces antibodies in serum and the reproductive tract. Biol Reprod 53:462–471

Stevens VC (1992) Future perspectives for vaccine development. Scand J Immunol 36 [Suppl 11]:137–143

Stevens VC (1996) Progress in the development of human chorionic gonadotropin antifertility vaccines. J Reprod Immunol 35:148–155

Stevens VC, Chou WS, Powell JE, Lee AC, Smoot J (1986) The identification of peptide sequences of human chorionic gonadotropin containing a confirmational epitope. Immunol Lett 12:11–18

Stevens VC, Powell JE, Lee AE, Kaumaya PTP, Lewis DH, Rickey M, Atkins TJ (1992) Development of a delivery system for a birth control vaccine using biodegradable microspheres. In: Kopecek (ed) Proceedings of the 19th International Symposium on Controlled Release of Bioactive Materials. Controlled Release Society, Orlando, pp 112–113

Stewart DR, Overstreet JW, Nakajima ST, Lasley BL (1993) Enhanced ovarian steroid secretion before implantation in early human pregnancy. J Clin Endocrinol Metab 76:1470–1476

Suri A, Talwar GP, Shaba C (1993) Oral immunization with sperm antigens. In: Griffin PD, Johnson PM (eds) Local immunity in reproductive tract tissues. Oxford University Press, Oxford, pp 427–440

Suri A, Chhabra S, Upadhay S (1996) Identification of human sperm antigen recognized by serum of an immunoinfertile woman: a candidate for immunocontraception. J Reprod Immunol 36:317–326

Talwar GP, Singh O, Pal R, Chatterjee N, Sahai P, Dhall K, Kaur J, Das SK, Suri S, Buckshee K, Saraya L, Saxena BN (1994) A vaccine that prevents pregnancy in women. Proc Natl Acad Sci USA 91:8532–8536

Tung KSK, Lou Y-H, Luo A-M, Ang J (1994) Contraceptive vaccine assessment based on murine ZP3 mini-autoantigen. Reprod Fertil Dev 6:340–355

Westhoff WE, Slootstra JW, Puijk WC, Kuperus D, Flinterman JF, Schaaper WMM, Oonk HB, Meloen RH (1996) Detection of epitopes on follicle-stimulating hormone and FSH-antiserum-induced suppression of bioactivity of follicle-stimulating hormone and luteinizing hormone. J Reprod Immunol 30:133–149

Wickings EJ, Usadel KH, Dathe G, Nieschlag E (1980) The role of follicle-stimulating hormone in testicular function of the mature rhesus monkey. Acta Endocr 95–117

Wright RM, Suri AK, Komreich B, Flinkenger CJ, Herr JC (1993) Cloning and characterization of the gene coding for the human acrosomal protein SP-10. Biol Reprod 49:316–325

CHAPTER 17

Cancer Vaccines

K.E. HELLSTRÖM and I. HELLSTRÖM

A. Introduction

The idea of immunizing (vaccinating) against cancer has been around since the days of Paul Ehrlich (HIMMELWEIT 1957). For most of this century, however, it was not taken seriously. Although targets against which to immunize were identified long ago in inbred mice (GROSS 1945; FOLEY 1953; PREHN and MAIN 1957; KLEIN et al. 1960b; SJÖGREN et al. 1961; HABEL 1962; OLD and BOYSE 1964), they were detected among tumors that were induced either with much larger doses of chemical carcinogens than those playing a role normally or by viruses which do not cause cancer under natural conditions (KLEIN and KLEIN 1977). In contrast, none of 27 different mouse neoplasms of spontaneous origin induced a tumor destructive immune response when tested by HEWITT et al. (1976) using methods similar to those successfully employed with the experimentally induced cancers. Although spontaneous mammary carcinomas in mice were known to be immunogenic under certain experimental conditions (MORTON 1964; WEISS et al. 1964), and in vitro studies had demonstrated lymphocyte- and/or antibody-mediated immunological reactions to many of the common human cancers (HELLSTRÖM and HELLSTRÖM 1969; BUBENIK et al. 1970; HERBERMAN 1974; MCCOY et al. 1974; COCHRAN et al. 1976; SHIKU et al. 1976), the findings of HEWITT et al. cast doubts about immunological approaches to cancer therapy and prevention.

These views have changed over the past few years towards a guarded enthusiasm. Human tumor-associated, and even tumor-specific, antigens have been identified molecularly, and effective immunogens have been prepared against several of the mouse tumors originally considered to be non-immunogenic (BOON et al. 1994). Some promising clinical results have also been obtained (LIVINGSTON et al. 1994; MORTON and BARTH 1996). We have recently reviewed this area (HELLSTRÖM and HELLSTRÖM 1993, 1997; HELLSTRÖM et al. 1997a,b) and focus here on some of the major conceptual issues. We make no attempt to cover the vast literature in this field and base most of the discussion on work carried out either in mouse models or with human material in vitro.

B. Tumor Antigens

The first question is the extent to which cancer cells have molecules which can serve as antigenic targets by being either qualitatively unique or detectable at much higher levels in neoplastic than in normal cells. The second question is the frequency with which any such antigens are expressed at the cell surface (for recognition by antibodies) or presented by MHC molecules (for recognition by T lymphocytes).

I. Recognition by Antibodies

While the immune response to tumor antigens is mediated primarily by T lymphocytes (HELLSTRÖM and HELLSTRÖM 1969; PERLMANN and CEROTTINI 1979; GREENBERG 1991; BOON 1992; MELIEF and KAST 1994), the role of antibodies acting alone or in consort with complement, NK cells, or macrophages must not be ignored. The value of anti-tumor antibodies has been shown most clearly for lymphomas, both in animal models (BERNSTEIN et al. 1980) and in man (MILLER et al. 1982). However, RIETHMÜLLER et al. (1994) reported prolonged tumor-free survival also in colon carcinoma patients following treatment with monoclonal antibody (Mab) 17.1 A.

One of the more convincing clinical demonstrations of a beneficial effect of tumor vaccination utilized a vaccine which induces antibodies to the melanoma associated GM2 antigen (LIVINGSTON et al. 1994). Vaccinated melanoma patients had a longer tumor-free survival than did parallel controls. This was shown in a randomized study in which all patients had minimal disease following surgery. Most likely the beneficial effects were due to complement activation and/or antibody-dependent cellular cytotoxicity. Tumor vaccines directed against other carbohydrates, such as the Tn antigen (LONGENECKER et al. 1994; SPRINGER et al. 1994), probably also work by inducing antibodies, although a possible involvement of lymphocyte-mediated immune reactions should not be dismissed.

Antibodies to certain cell surface molecules can induce apoptosis, acting in synergy with chemotherapeutic drugs. This has been best demonstrated for antibodies to various members of the epidermal growth factor receptor family (BASELGA et al. 1993; FAN et al. 1993; PIETRAS et al. 1994). By binding to growth stimulatory molecules, such as hormones and growth factors, antibodies can interfere with growth controlling mechanisms and therefore be inhibitory to tumor cells and/or induce apoptosis. Antibodies induced by cancer vaccines can thus be clinically beneficial in several ways.

II. Recognition by T Lymphocytes

T lymphocytes have immunological "memory" making the effects of tumor vaccination long-lasting (GREENBERG 1991; CHEEVER et al. 1995). Their ability to recognize processed intracellular proteins, including those which are modified as a result of the neoplastic transformation, provides opportunities for high tumor-selectivity.

T lymphocytes can act both as cytolytic T cells (CTL) and by producing lymphokines which expand CTL populations and/or activate tumor-destructive macrophages and NK cells. The CTL activity is mediated primarily by $CD8^+$ cells while the lymphokines are primarily made by $CD4^+$ helper cells (KERN et al. 1986), with those of the T helper 1 (Thl) type likely to play a larger role than Th2 cells in tumor destruction.

The CTL response has exquisite specificity for antigen-positive cells, while activated macrophages and NK cells also kill tumor cells which have lost the target antigen or the ability to present it via MHC. It is noteworthy that neoplastic cells are often more sensitive than their normal counterparts to such killing, a finding that can be taken advantage of therapeutically by inducing an inflammatory response at the tumor site (KLEIN et al. 1976).

In most mouse models tumor rejection is caused by $CD8^+$ T lymphocytes with in vitro CTL activity (PERLMANN and CEROTTINI 1979; MELIEF 1992). However, in some models $CD4^+$ cells play the major role (GREENBERG 1991), and destruction of tumor-metastases in the lungs of mice has been demonstrated subsequent to systemic administration of tumor-reactive $CD4^+$ Th clones (KAHN et al. 1991).

III. Tumor Peptides as T Cell Targets

CTL recognize peptides as presented by MHC class I molecules. These peptides generally consist of nine amino acids and result from degradation of intracellular protein antigens. Various enzymes, particularly TAP-1, are involved in the processing of protein and transport of peptides to MHC (ELLIOTT 1996). A repetitive mucin peptide which is expressed by many carcinomas forms an exception to this general rule, since it can serve as a target for CTL without need for presentation by MHC class I (FINN et al. 1995). Its potential advantage as a target is that the same vaccine can be used in patients irrespective of their MHC haplotype and that class I losses from tumor cells will not eliminate their sensitivity to CTL.

There are many tumor proteins which after processing into peptides can serve as epitopes for T cells both in vivo and in vitro. The most ideal among them have absolute tumor specificity and are directly involved in the neoplastic transformation (HELLSTRÖM and HELLSTRÖM 1989). The polyoma tumor rejection (T) antigen, which is encoded by the polyoma virus, is such a protein, as are antigens encoded by several other oncogenic viruses and studied in animals (SJÖGREN 1965).

CHEN et al. (1991, 1992a,b) showed in a mouse model that the E6 and E7 open reading frames of HPV16 encode tumor rejection antigens which serve as CTL targets in vitro. This was the first demonstration that a virus causing a common form of human cancer, cervical carcinoma (ZUR HAUSEN 1987), can encode such antigens. Subsequent work showed that HPVl6-associated cervical carcinomas in humans can be killed by CTL specific for peptides encoded by E6 or E7 (RESSING et al. 1995). Some other cancer-specific T cell epitopes are peptides encoded by a mutated ras p21 (PEACE et al. 1991), a mutated p53 (WALKER and LEVINE 1996), a mutated cyclin CDK4 (WOLFEL et al. 1995), and a translocated bcr-abl (CHEEVER et al. 1995). Epitopes encoded by amplified cellular oncogenes, such as HER-2, are likewise of great interest (INANNIDES et al. 1993) since they are closely associated with the neoplastic transformation, although their tumor selectivity is quantitative rather than qualitative.

Tumors also express differentiation (oncofetal) antigens, which can serve as T cell targets. A cancer cell often has more than one such specificity, and most of the human tumor antigens characterized by Mabs belong to this category (HELLSTRÖM and HELLSTRÖM 1993). While the first evidence of lymphocyte-mediated reactivity to human oncofetal antigens was published more than 25 years ago (HELLSTRÖM et al. 1970), the discovery by Boon's group of the MAGE series of CTL epitopes (VAN DER BRUGGEN et al. 1991; BOON et al. 1994) played a key role by starting the identification of oncofetal antigens at the molecular level, many of which have now been characterized (KAWAKAMI et al. 1994b; BOON and VAN DER BRUGGEN 1996). While their tumor specificity is not absolute, several oncofetal antigens are likely to have sufficient selectivity to provide clinically useful therapeutic targets.

Most T cell epitopes in human cancers are shared by many tumors, often independently of their histological type. This is in contrast to those antigens which were initially discovered by transplantation techniques among chemically induced sarcomas of mice and rats. These were individually unique for each individual neoplasm, even when they originated in the same mouse strain and did have the same histology (PREHN and MAIN 1957; OLD and BOYSE 1964); it is not clear whether the antigens were secondary to the use of a very large dose of a (mutagenic) carcinogen for tumor induction. With the exception of idiotypic determinants on B cell lymphomas (KWAK et al. 1992) and some studies on melanoma (SHIKU et al. 1976) there is limited evidence for individually unique antigens among human cancers. Although such antigens have the advantage of absolute specificity for a particular tumor, they are sometimes lost during its progression. Furthermore, their uniqueness makes it difficult to prepare standardized vaccines for use in large patient populations.

We conclude that lack of molecules that can serve as potential tumor antigens is not a factor limiting cancer vaccination.

IV. Antigen Presentation

To *induce* an immune response, tumor peptides can be displayed either by the neoplastic cells or by professional antigen-presenting cells (APC) after these have taken up and processed antigen released from the tumor cells, an event that commonly occurs in tumor-draining lymph nodes. To serve as a CTL *target*, a processed antigen must be presented by MHC class I molecules as a peptide. However, much less peptide is needed. Thus a tumor which is MHC class I negative according to immunohistology, and which cannot induce a tumor-directed CTL response, may still serve as a CTL target in vitro and be rejected by CD8 positive T cells in vivo (PORGADOR et al. 1995).

Since most T cell epitopes are presented only by those MHC molecules to which they can be anchored, a tumor epitope is recognized as a CTL target only by T cells from patients with the matching class I haplotype. There is thus a need to MHC type patients to be given cancer vaccines and to verify that their tumors can present at least some of the peptides vaccinated against.

Carcinomas in situ normally express MHC class I and II to the extent needed for detection by immunohistology. However, this expression is commonly downregulated during their progression, and metastatic tumors often contain large fractions of cells which lack detectable MHC, and sometimes they lack detectable MHC altogether (MARINCOLA et al. 1994; STERN and DUGGAN-KEEN 1994; FERRONE and MARINCOLA 1995; GARRIDO et al. 1997).

Expression of MHC class I is more pertinent than that of class II. This is in view of the key role of class I for presenting peptides as CTL targets and also because peptides recognized by $CD4^+$ Th cells can be presented not only by class II positive tumor cells but also by tumor-infiltrating APC (which are class II positive) to induce the production of various lymphokines, some of which have antitumor activity. Since we have reviewed this area elsewhere (HELLSTRÖM et al. 1997a), we confine the discussion here to three points.

First, if *complete* MHC class I losses occur in vivo, they are rare, at least in mouse systems where extensive studies have been performed by KLEIN et al. (1960a) with a large number of different mouse sarcomas, lymphomas, and carcinomas. Although stringent immunoselection procedures were applied to isolate tumor variants which entirely lacked MHC class I (then referred to as H-2), not a single such variant was detected. It is possible that class I negative cells were eliminated by NK cells in vivo as can be expected on the basis of studies by Kärre's group (KÄRRE 1995).

Second, the studies by Klein et al. (1960a) demonstrated relatively frequent *partial* class I losses. For example, tumors originating in Fl hybrids between two congenic mouse strains, which differed only at MHC class I, gave rise to variant lines which could grow in one parental strain because of a loss of class I from the other parent. Losses of part of one haplotype were also observed. The pattern and frequency of these losses suggested that they were caused by genetic events, such as a chromosome deletion, somatic crossing over, or mutation.

Third, one should realize that if indeed class I losses are as frequent as suggested by immunohistology, most antigenic mouse tumors would lose their immunogenicity following serial transplantation to immunocompetent hosts, which is the way by which they are normally propagated. However, that is very rarely the case. Furthermore, studies on transplanted polyoma virus induced mouse tumors have failed to select any variants that could not be rejected by syngeneic mice immunized to the polyoma tumor antigen in spite of immunoselection over many passages (SJÖGREN 1965). Such variants would have occurred had those class I sequences been lost that were needed to present the polyoma tumor T antigen as a CTL target, since rejection of polyoma tumors is primarily mediated by $CD8^+$ T cell with CTL activity (BERKE et al. 1996).

We conclude that most tumors retain at least some MHC class I expression in vivo. However, since partial losses of class I sequences do frequently occur, the sequence needed to present a particular peptide is sometimes lost from a cancer cell. Therefore tumor immunogens aiming to induce a CTL response should contain whole protein antigens, or at least several peptides that are anchored by different class I sequences.

C. Types of Tumor Vaccines

Tumor immunity was first discovered with neoplasms that had been induced experimentally. The immunizing cells were lethally irradiated or allowed to grow for a few days after which tumor nodules were removed. As stated in Sect. A, the same procedure was ineffective against spontaneous tumors.

MORTON (1964) and WEISS et al. (1964) were the first to demonstrate an immune response to spontaneous neoplasms, working with mammary tumor virus (MTV) associated mammary carcinomas in mice. The carcinomas grew well in mice harboring MTV after immunization similar to that which is effective against many experimentally induced cancers. They were, however, rejected if the mice to be immunized had been removed from their mothers prior to birth so as not to be exposed to MTV. It was thus evident that the spontaneous mammary carcinomas had MTV-related antigens, and that these antigens were immunogenic unless the mice had become "tolerant" by exposure to MTV.

Many approaches are now known to make tumors sufficiently immunogenic that immunization with them induces the rejection of transplanted wild type cells, even when the tumors are nonimmunogenic, according to the criteria of HEWITT et al. (1976). They include mixtures of tumor cells and a highly immunogenic agent such as BCG (OETTGEN and OLD 1991) as well as tumor cells infected with certain viruses (AUSTIN and BOONE 1979; WALLACK and BASH 1992), hybridized with an allogeneic cell (MITCHELL et al. 1988), mutagenized (BOON 1992), or transfected with a gene encoding a proper lymphokine, such as granulocyte-macrophage colony-stimulating factor (PARDOLL 1993), or a costimulatory ligand, such as B7-1 (CHEN et al. 1992a). The common denominator of this approach is the presentation of "weak" tumor antigen(s) under conditions where an immune response can be expanded by lymphokine(s), formed, for example, in response to a "strong" antigen.

Construction of tumor immunogens should take into account what has been learned about antigen presentation by MHC (commented on above) but also the needs for costimulation. This is because two signals are necessary to initiate an effective immune response. First, the T lymphocyte must recognize an antigenic peptide, presented to its T cell receptor (TCR) by MHC class I (to

$CD8^+$ cells) or class II (to $CD4^+$ cells). Second, there must be a second, costimulatory signal (BRETSCHER and COHN 1970; LAFFERTY et al. 1983; SCHWARTZ 1992). The most important second signal is one provided by interaction between molecules of the B7 family on APC and their receptor, CD28, on T lymphocytes (LINSLEY and LEDBETTER 1993); B7-1 (CD80) and B7-2 (CD86) are expressed as well as some other costimulatory molecules, on professional APC such as dendritic cells and activated macrophages. Triggering of the TCR in the absence of this signal does not induce an effective immune response (immunological ignorance). It can even lead to anergy, unless one provides some key lymphokines, including IL-2, whose production is being enhanced by the costimulation. Interaction between CD2 and CD48 (in the mouse; the human counterparts are CD58 and CD59) provides another costimulatory signal (LI et al. 1996).

Cancer cells do not express B7-1 or B7-2 with the exception of many B cell lymphomas and a few solid tumors, particularly some melanomas. They thus present antigens in a way which maintains immunological ignorance and can even cause anergy. Failure of tumor cells to express B7 may explain why even highly immunogenic tumors can develop in an immunocompetent host (HELLSTRÖM et al. 1995). A tumor-destructive immune response is initiated first after tumor antigens have been released and is taken up, processed, and presented by professional APC in the regional lymph nodes. Inhibition of the B7-CD28 interaction prevents the development of antitumor immunity and facilitates the outgrowth of small doses of highly immunogenic tumor cells (YANG et al. 1997).

Improved costimulation can be achieved by transfecting B7-1 or B7-2 into tumor cells. Such transfected cells have vastly greater immunogenicity than their nontransfected counterparts and are regularly rejected by an immunocompetent syngeneic host. Most importantly, they can serve as effective immunogens against wild-type (i.e., $B7^-$) cells from the respective tumors inducing a strong immune response which is mediated primarily by $CD8^+$ T cells with CTL activity (CHEN et al. 1992a, 1994a; TOWNSEND and ALLISON 1993).

Costimulation by the B7-CD28 interaction also expands the repertoire of CTL epitopes which can be recognized by the immune system from the small number of immunodominant epitopes revealed by immunizing with B7- tumor cells (JOHNSTON et al. 1996). This is an important advantage, since the fact that tumors sometimes loose MHC class I sequences favors vaccines that generate an immune response to many peptides with different anchoring sequences.

The beneficial effects of transfecting, into tumor cells, B7-1, B7-2, or a combination of the two are seen only with tumors that are immunogenic according to the classical criteria of being rejected by syngeneic mice immunized with wild type tumor cells (CHEN et al. 1994a,b). However, both of two tumors of very low to nonexistent immunogenicity, sarcoma Ag104 and melanoma K1735, became immunogenic after B7-1 was transfected into them together with the second costimulatory ligand CD48 (LI et al. 1996).

Recent data indicate that the immune response to tumors can be further increased by triggering the T cell activation molecule 4-1BB (Melero et al. 1997a). The 4-1BB glycoprotein is a member of the tumor necrosis factor receptor superfamily (Smith et al. 1994) and binds to a high-affinity ligand (4-1BBL) which is expressed on APC such as macrophages (Pollok et al. 1994). Expression of 4-1BB is restricted to primed (but not naive) $CD4^+$ and $CD8^+$ T cells (Pollok et al. 1993), and 4-1BB signaling either by binding to 4-1BBL or by antibody ligation delivers a comitogenic signal for T cell activation and proliferation (Goodwin et al. 1993). This suggests an important role for 4-1BB in the amplification of a T cell mediated immune response. Mabs to mouse 4-1BB enhance T cell proliferation in vitro when added together with a suboptimal dose of an anti-CD3 Mab, and they hasten the rejection of allogeneic skin grafts and increase the severity of graft versus host disease (Shuford et al. 1997).

Experiments were performed to investigate whether treatment of mice with an anti 4-1BB Mab had in vivo antitumor activity (Melero et al. 1997). Regressions were observed of disseminated P815 mastocytoma and Ag104 sarcoma as well as of relatively large (>10 mm diameter) P815 or Ag104 tumors growing subcutaneously. The findings with well-established Ag104 sarcomas are particularly noteworthy in view of their very low immunogenicity. Mice given a Mab to 4-1BB had a vastly increased proliferation of $CD8^+$ T cells with strong tumor-directed CTL activity.

It was subsequently shown that transfection of immunogenic mouse tumors with the 4-1BB ligand had an effect similar to that of the anti 4-1BB Mab (Melero et al. 1998). In contrast, transfection of this ligand alone did not increase the immunogenicity of sarcoma Agl04, and neither did transfection of B7-1 alone. However, tumor cells transfected with both the 4-1BB ligand and B7-1 were rejected by immunocompetent syngeneic mice, and when used as immunogens, they induced the rejection of transplanted Ag104 wild-type cells.

The findings which we have reviewed indicate a beneficial effect of cell-based tumor vaccines which combine cells expressing tumor antigens with either the costimulatory ligand B7-1 (or B7-2) and the T cell activating 4-1BB ligand, or the two costimulator ligands B7-1 and CD48. However, cell-based vaccines are hard to standardize, and they may induce an immune response also against normal tissue antigens expressed on the immunizing cells. A recombinant virus, which contains a gene encoding a tumor antigen, gene(s) encoding costimulatory molecule(s), including B7-1 and/or B7-2, and perhaps a gene encoding the 4-1BB ligand, is much more attractive. The target antigen is then defined at the molecular level, and a standardized preparation can be made. The virus of choice may be vaccinia or one of its relatives among the fowl-pox viruses (Hu et al. 1991), but adenoviruses (Sparer et al. 1997) or deficient HIV viruses (Naldini et al. 1996) may prove even better. Recombinant viruses combining a tumor antigen (CEA) with B7 have been tested preclinically and found to induce effective immune responses (Kantor et al. 1992). DNA vaccines combining genes encoding tumor antigens, costimulatory molecules and, maybe, T cell "triggering" molecules such as 4-1BB ligand (or certain lymphokines) seem attractive.

Another approach is to deliver the tumor antigen, either as a protein or a combination of several different peptides, to professional APC. It can be given with an adjuvant or enclosed in liposomes (ALVING 1995). In the latter case the liposomes may also contain a lymphokine, such as granulocyte-macrophage colony-stimulating factor, to upregulate the expression of various costimulatory molecules in macrophages. Immunization with dendritic cells which have been pulsed with tumor peptides in vitro has attracted much attention lately. It has worked well in preclinical models (HSU et al. 1996) and is now being tried clinically with some promising results reported (NESTLE et al. 1998). However, this approach is not easy to standardize for large-scale use.

Anti-idiotypic antibodies (HELLSTRÖM et al. 1997b) have been considered as alternative cancer vaccines for more than a decade. There are supportive preclinical data (LEE et al. 1986), and some cancer patients who have received Mabs and subsequently made an anti-idiotypic response do clinically better than patients who do not (KOPROWSKI et al. 1984; FAGERBERG et al. 1995). There are also data showing that some patients receiving anti-idiotypic antibodies have responded favorably (HERLYN et al. 1989; BHATTACHARYA-CHATTERJEE et al. 1990; IRVINE and SCHLOM 1993). However, the clinical progress with anti-idiotypic antibodies has been slow.

D. Downregulatory Mechanisms

While many cancer vaccines are effective in preclinical models with a low tumor load, they are generally ineffective against bulky tumors. This constitutes a problem, since these are normally the tumors with which trials of new therapeutic entities are performed clinically, because studies on patients with minimal disease are time consuming and costly.

Tumor-bearers often display concomitant tumor immunity, i.e., animals (GERSHON et al. 1969) and human cancer patients (SOUTHAM 1967) with established tumors often reject a small dose of subcutaneously transplanted cells from their respective neoplasms, and their lymphocytes can destroy tumor cells in vitro (HELLSTRÖM and HELLSTRÖM 1967, 1969). The antitumor immune response is thus inhibited (blocked) at the tumor site.

Some of the T cell signaling molecules are downregulated in lymphocytes from animals with large tumors, particularly when the lymphocytes are derived from the tumor site (MIZOGUCHI et al. 1992; FINKE et al. 1993; NAKAGOMI et al. 1993). Their expression increases upon in vitro culturing of the lymphocytes, and in vitro expanded, tumor-infiltrating lymphocytes can have anticancer activity, even in man (ROSENBERG 1992).

A simple explanation of these data is that tumor-bearers have some soluble factors which downregulate T cell responses, and various blocking factors have indeed been detected in tumor-bearer serum (HELLSTRÖM et al. 1969; BALDWIN 1973) and at much higher levels within the growing tumor mass (HELLSTRÖM and HELLSTRÖM 1979).

Both antigen-specific and -nonspecific blocking factors have been demonstrated. Among the nonspecific factors, transforming growth factor-β (SULITZEANU 1993) has attracted particular interest, together with Fas ligand (HAHNE et al. 1996) and pl5e (SNYDERMAN and PIKE 1979). The relative importance of these (and other) factors, how they may relate to each other and to antigen-specific factors has not been clarified. It is not known, for example, whether release of tumor antigen and its uptake by certain host cells initiates the production/release of nonspecifically acting inhibitors, some of which are also known to be made by the tumor cells.

Drugs capable of abrogating the effects of the blocking phenomenon need to be sought once the most important blockers have been identified. Alternatively, the immune response needs to be strengthened to the extent that it is less sensitive to inhibition.

E. Conclusions and Outlook

Cancer cells express various molecules with sufficient selectivity to be targets for therapeutic vaccination. Some of these molecules, particularly those belonging to the epidermal growth factor receptor family, are present at the cell surface and accessible to antibodies which can synergize with chemotherapeutic drugs to induce apoptosis. Antibodies can be therapeutic also by fixing complement and/or mediating antibody-dependent cellular cytotoxicity. Therapeutic vaccines which induce antibody formation can thus be clinically beneficial.

Most tumor vaccines are, however, considered because of their ability to induce T lymphocyte-mediated immune responses. Since T cells can "see" intracellular proteins, following their processing and presentation as peptides by MHC molecules, they can recognize (and destroy) tumor cells with a high level of selectivity.

While there is an abundance of potential T cell targets in most (all?) tumors, the extent to which such targets can be presented, via MHC, remains an issue. Complete loss of MHC class I from tumor cells is probably much more rare than indicated by immunohistology. However, partial class I losses have been amply documented, particularly in studies performed on mouse tumors. Therefore vaccines which induce an immune response towards many tumor peptides (with different class I anchoring sequences) are the ones of choice. Furthermore, the epitopes need to be presented so there is optimal costimulation. Attention should also be given to immunogens which induce a strong T helper cell response to initiate the production, at the tumor site, of lymphokines which expand CTL population and activate macrophages and NK cells.

Downregulation of antitumor immune reactions by various blocking factors remains a problem, and there is a need for drugs able to circumvent it. At this time the major therapeutic application for cancer vaccines is for "mopping up" tumor cells remaining after surgery, chemotherapy, etc. rather than for treatment of bulky tumors.

Finally, we like to add that it may soon be justified to consider also prophylactic cancer vaccines, not merely for tumors associated with a virus, such as HPV-16, hepatitis B, Epstein-Barr virus, as more is being learned about T cell targets appearing as a result of cancer-initiating early genetic events.

References

Alving CR (1995) Liposomal vaccines: clinical status and immunological presentation for humoral and cellular immunity. Ann NY Acad Sci 754:143–152

Austin FC, Boone CW (1979) Virus augmentation of the antigenicity of tumor cell extracts. Adv Cancer Res 30:301–345

Baldwin RW (1973) Immunological aspects of chemical carcinogenesis. Adv Cancer Res 18:1–75

Baselga J, Norton L, Masui H, Pandiella A, Coplan K, Miller WH, Mendelsohn J (1993) Antitumor effects of doxorubicin in combination with anti-epidermal growth factor receptor monoclonal antibodies. J Natl Cancer Inst 85:1327–1333

Berke Z, Palmer S, Bergman T, Wester D, Svedmyr J, Linder S, Jornvall H, Dalianis T (1996) A short peptide eluted from the H-2 b molecule of a polyomavirus-positive tumor corresponds to polyomavirus large T antigen peptide at amino acids 578 to 585 and induces polyomavirus-specific immunity. J Virol 70:3093–3097

Bernstein JP, Tam MR, Nowinski RC (1980) Mouse leukemia: therapy with monoclonal antibody against a thymus differentiation antigen. Science 207:68–71

Bhattacharya-Chatterjee M, Mukerjee S, Biddle W, Foon KA, Kohler H (1990) Murine monoclonal anti-idiotype antibody as a potential network antigen for human carcinoembryonic antigen. J Immunol 145:2758–2765

Boon T (1992) Toward a genetic analysis of tumor rejection antigens. Ann Inst Pasteur Immunol 58:177–210

Boon T, van der Bruggen P (1996) Human tumor antigens recognized by T lymphocytes. J Exp Med 183:1173–1183

Boon T, Certottini J, van den Eynde B, van der Bruggen P, Van Pel A (1994) Tumor antigens recognized by T lymphocytes. Annu Rev Immunol 12:337–365

Bretscher P, Cohn M (1970) A theory of self-nonself discrimination. Science 169:1042–1049

Bubenik J, Perlmann P, Helmstein K et al (1970) Cellular and humoral innume responses to human urinary bladder carcinomas. Int J Cancer 5:310–319

Cheever M, Disis ML, Bernhard H, Gralow JR, Hand SL, Huseby ES, Qin HL, Takahashi M, Chen W (1995) Immunity to oncogenic proteins. Immunol Rev 145:33–59

Chen L, Kinney-Thomas E, Hu S-L, Hellström I, Hellström KE (1991) Human papillomavirus type 16 nucleoprotein is a tumor rejection antigen. Proc Natl Acad Sci USA 88:110–114

Chen L, Ashe S, Brady WA, Hellström I, Hellström KE, Ledbetter JA, McGowan P, Linsley PS (1992a) Costimulation of antitumor immunity by the B7 counter-receptor for the T lymphocyte molecules CD28 and CTLA-4. Cell 71:1093–1102

Chen L, Mizuno MT, Singhal MC, Hu S-L, Galloway DA, Hellström I, Hellström KE (1992b) Induction of cytotoxic T lymphocytes specific for a syngeneic tumor expressing the E6 oncoprotein of human papillomavirus type 16. J Immunol 146:2617–2621

Chen L, McGowan P, Ashe S, Johnston J, Li Y, Hellström I, Hellström KE (1994a) Tumor immunogenicity determines the effect of B7 co-stimulation on T cell-mediated tumor immunity. J Exp Med 179:523–532

Chen L, McGowan P, Ashe S, Johnston JV, Hellström I, Hellström KE (1994b) B7-1/CD80-transduced tumor cells elicit better systemic immunity than wild type tumor cells admixed with Corynebacterium parvum. Cancer Res 54:5420–5423
Cochran AJ, Grant RM, Spilg WJ (1976) Sensitization to tumor-associated antigens in human breast carcinoma. Int J Cancer 14:19
Elliott T (1996) Tapping into tumours. Nat Genet 13:139–140
Fagerberg J, Hjelm AL, Ragnhammer P, Frodin JE, Wigzell H, Mellstedt H (1995) Tumor regression in monoclonal antibody-treated patients correlates with the presence of anti-idiotype-reactive T lymphocytes. Cancer Res 55:1824–1827
Fan Z, Baselga J, Masui H, Mendelsohn J (1993) Antitumor effect of anti-epidermal growth factor receptor monoclonal antibodies plau ces-diamminedichloroplatinum on well established A431 cell xenografts. Cancer Res 53:4637–4642
Ferrone S, Marincola FM (1995) Loss of HLA class I antigens by melanoma cells: molecular mechanisms, functional significance and clinical relevance. Immunol Today 16:487–494
Finke JH, Zea AH, Stanley J, Longo DL, Mizoguchi H, Tubbs RR, Wiltrout RH, O'Shea JJ, Kudoh S, Klein E, Bukowski RM, Ochoa AC (1993) Loss of T-cell receptor ν chain and $p56^{lck}$ in T-cells infiltrating human renal cell carcinoma. Cancer Res 53:5613–5616
Finn OJ, Jerome KR, Henderson RA, Pecher G, Domenech N, Magrian-Blander J, Barratt-Boyes SM (1995) MUC-1 epithelial tumor mucin-based immunity and cancer vaccines. Immunol Rev 145:61–89
Foley EJ (1953) Antigenic properties of methylcholanthrene-induced tumor in mice of the strain of origin. Cancer Res 13:835–837
Garrido F, Ruiz-Cabello F, Cabrera T, Perez-Villar JJ, Lopex-Botet M, Duggan-Keen M, Stern PL (1997) Implications for immunosurveillance of altered HLA class I phenotypes in human tumors. Immunol Today 18:89–95
Gershon RK, Carter RL, Kondo K (1969) Nature 224:277
Goodwin RG, Din WS, Davis-Smith T, Anderson DM, Gimpel SD, Sato TA, Maliszewski CR, Brannan CE, Copeland BG, Jenkins NA et al (1993) Molecular cloning of a lignad for the inducible T cell gene 4-1BB: a member of an emerging family of cytokines with homology to tumor necrosis factor. Eur J Immunol 23:2631–2641
Greenberg PD (1991) Adoptive T cell therapy of tumors: mechanisms operative in the recognition and elimination of tumor cells. Adv Immunol 49:281–355
Gross L (1945) The specificity of acquired tumor immunity. J Immunol 50:91
Habel K (1962) Immunological determinants of polyoma virus oncogenesis J Exp Med 115:181
Hahne M, Rimoldi D, Schroter M, Romero P, Schreier M, French LE, Schneider P, Bornand T, Fontana A, Lienard D, Cerottini J-C, Tschopp J (1996) Melanoma cell expression of Fas (Apo-1/CD95) ligand: implications for tumor immune escape. Science 274:1363–1366
Hellström I, Hellström KE (1967) In vitro demonstration of cell-bound immunity against autologous MCA and plastic disc-induced mouse tumors. Science 156:981–983
Hellström K, Hellström I (1969) Cellular immunity against tumor specific antigens. Adv Cancer Res 12:167–223
Hellström KE, Hellström I (1979) Enhancement of tumor outgrowth by tumor-associated blocking factors. Int J Cancer 23:366–373
Hellström KE, Hellström I (1989) Oncogene-associated tumor antigens as targets for immunotherapy. FASEB J 3:1715–1720
Hellström I, Hellström KE (1993) Tumor immunology: an overview. Specific immunotherapy of cancer with vaccines. Ann NY Acad Sci 690:24–33
Hellström I, Hellström KE (1998) Tumor vaccines- a reality at last? J Immunother 21:119–126

Hellström I, Hellström KE, Evans CA, Heppner G, Pierce GE, Yang JPS (1969) Serum mediated protection of neoplastic cells from inhibition by lymphocytes immune to their tumor specific antigens. Proc Natl Acad Sci USA 62:362–369
Hellström K, Hellström I, Shepard TH (1970) Cell-mediated immunity against antigens common to human colonic carcinomas and fetal gut epithelium. Int J Cancer 6:346–351
Hellström KE, Hellström I, Chen L (1995) Can co-stimulated tumor immunity be therapeutically efficacious? Immunol Rev 145:123–145
Hellström KE, Gladstone P, Hellström I (1997a) Cancer vaccines: challenges and potential solutions. Mol Med Today 3:286–290
Hellström KE, Hellström I, Chen L (1997b) Antitumor vaccines. In: Levine MM, Woodrow GC, Kaper JB, Cobon GS (eds) New generation vaccines. Marcel Dekker, New York, pp 1095–1115
Herberman RB (1974) Cell-mediated immunity to tumor cells Adv Cancer Res 19:207
Herlyn D, Wettendorff M, Iliopoulos D, Schmoll E, Schedel I, Koprowski H (1989) Modulation of cancer patients' immune responses by administration of anti-idiotypic antibodies. Virol Immunol 2:271–276
Hewitt HB, Blake ER, Walder ES (1976) A critique of the evidence for active host defense against cancer based on personal studies of 27 murine tumors of spontaneous origin. Br J Cancer 33:241–259
Himmelweit B (1957) The collected papers of Paul Ehrlich. Pergamon, Oxford, England
Hsu FJ, Benike C, Fagnoni F, Liles TM, Czerwinski D, Taidi B, Engleman EG, Levy R (1996) Vaccination of patients with B-cell lymphoma using autologous antigen-pulsed dendritic cells. Nat Med 2:52–58
Hu S-L, Hellström I, Hellström K (1991) Recent advances in antitumor vaccines. In: Ellis RW (ed) Vaccines: new approaches to immunological problems. Butterworths, Boston, pp 327–343
Inannides CG, Fisk B, Fan D, Biddison WE, Wharton JT, O'Brian CA (1993) Cytotoxic T cells isolated from ovarian malignant ascites recognize a peptide derived from the HER-2/neu proto-oncogene. Cell Immunol 151:225–233
Irvine K, Schlom J (1993) Induction of delayed-type hypersensitivity responses by monoclonal anti-idiotypic antibodies to tumor cells expressing carcinoembronic antigen and tumor-associated glycoprotein-72. Cancer Immunol Immunother 36:281–292
Johnston J, Malaco AR, Mizuno MT, McGowan P, Hellström I, Hellström KE (1996) B7-CD28 costimulation unveils the hierarchy of tumor epitopes recognized by MHC class I-restricted CD8+ cytolytic T lymphocytes. J Exp Med 183:791–800
Kahn M, Sugawara H, McGowan P, Okuno K, Nagoya S, Hellström KE, Hellström I, Greenberg P (1991) CD4+ T cell clones specific for the human p97 melanoma associated antigen can eradicate pulmonary metastases from a murine tumor expressing the p97 antigen. J Immunol 146:3235–3241
Kantor J, Irvine K, Abrams S, Snoy P, Olsen R, Greiner J, Kaufman H, Eggensperger D, Schlom J (1992) Immunogenicity and safety of a recombinant vaccinia virus vaccine expressing the carcinoembryonic antigen gene in a nonhuman primate. Cancer Res 52:6917–6925
Kärre K (1995) Express yourself or die: peptides, MHC molecules, and NK cells. Science 267:978–979
Kawakami Y, Eliyahu S, Sakaguchi K, Robbins PF, Rivoltini L, Yanelli JR, Appella E, Rosenberg SA (1994b) Identification of the immunodominant peptides of the MART-1 human melanoma antigen recognized by the majority of HLA-A2 restricted tumor infiltrating lymphocytes. J Exp Med 180:347–352
Kern DE, Klarnet JP, Jensen MC, Greenberg PD (1986) Requirement for recognition of class II molecules and processed tumor antigen for optimal generation of syngeneic tumor-specific class I-restricted CTL. J Immunol 136:4303–4310

Klein G, Klein E (1977) Rejectability of virus-induced tumors and non-rejectability of spontaneous tumors. Transplant Proc 9:1095–1104
Klein E, Klein G, Hellström KE (1960a) Further studies on isoantigenic variation in mouse carcinomas and sarcomas. J Natl Cancer Inst 25:271–274
Klein G, Sjögren HO, Klein E, Hellström KE (1960b) Demonstration of resistance against methylcholanthrene-induced sarcomas in the primary autochthonous host. Cancer Res 20:1561–1572
Klein E, Rosner D, Holtermann OA, Stoll HL, Song SY (1976) Nonspecific antigen reactions. In: Lewison EF (ed) Conference on spontaneous regression of cancer. NIH, Bethesda, pp 87–97
Koprowski H, Herlyn D, Lubeck M, DeFreitas E, Sears HF (1984) Human anti-idiotype antibodies in cancer patients: is the modulation of the immune response beneficial for the patient? Proc Natl Acad Sci USA 81:216–219
Kwak L, Campbell MJ, Czerwinski DK, Hart S, Miller RA, Levy R (1992) Induction of immune responses in patients with B-cell lymphoma against the surface-immunoglobulin idiotype expressed by their tumors. N Engl J Med 327:1209–1215
Lafferty KJ, Prowse SJ, Simeonovic CJ, Warren HS (1983) Immunobiology of tissue transplantation: a return to the passenger leukocyte concept. Annu Rev Immunol 1:143–173
Lee VK, Hellstrom KE, Nepom GT (1986) Idiotypic interactions in immune responses to tumor-associated antigens. Biochim Biophys Acta 865:127–139
Li Y, Hellström KE, Ashe Newby S, Chen L (1996) Costimulation by CD48 and B7-1 induces immunity against poorly immunogenic tumors. J Exp Med 183:639–644
Linsley PS, Ledbetter JA (1993) The role of the CD28 receptor during T cell responses to antigen. Annu Rev Immunol 11:191–212
Livingston PO, Wong GYC, Adluri S, Tao Y, Padavan M, Parente R, Hanlon C, Calves MJ, Helling F, Ritter G, Oettgen HF, Old LJ (1994) Improved survival in stage III melanoma Patients with GM2 antibodies: a rancomized trial of adjuvant vaccination with GM2 ganglioside. J Clin Oncol 12:1036–1044
Longenecker BM, Reddish M, Koganty R, MacLean GD (1994) Specificity of the IgG response in mice and human breast cancer patients following immunization against synthetic sialyl-Tn, an epitope with possible functional significance in metastasis. Adv Exp Med Biol 353:105–124
Marincola FM, Shamamian P, Alexander RB, Gnarra JR, Turetskaya RL, Nedospasov SA, Simonis TB, Taubenberger JK, Yannelli J, Mixon A, Retifo NP, Herlyn M, Rosenberg SA (1994) Loss of HLA haplotype and B locus down-regulation in melanoma cell lines. J Immunol 153:1225–1237
McCoy JL, Jerome LF, Dean JH, Cannon GB, Aldord TC, Doering T, Herberman RB (1974) Inhibition of leukocyte migration by tumor-associated antigens in soluble extracts of human malignant melanoma. J Natl Cancer Inst 53:11–17
Melero I, Shuford WW, Ashe-Newby S, Aruffo A, Ledbetter JA, Hellström KE, Mittler RS, Chen L (1997) Anti 4–1BB monoclonal antibodies activate T cells and eradicate established tumors. Nat Med 3:682–685
Melero I, Bach M, Hellstrom KE, Aruffo A, Mittler RS, Chen L (1998) Amplification of cytolytic T lymphocyte-mediated tumor immunity by expression of the costimulatory 4-1BB ligand on tumors: requirement for and synergy with the CD28 costimulatory pathway. Eur J Immunol 28:1116–1121
Melief CJM (1992) Tumor eradication by adoptive transfer of cytotoxic T lymphocytes. Adv Cancer Res 58:143–175
Melief CJM, Kast WM (1994) Prospects for T cell immunotherapy of tumors by vaccination with immunodominant and subdominant peptides. Ciba Found Symp 187:97–104
Miller RA, Maloney OG, Waruke R, Levy R (1982) New Engl J Med 306:517–522
Mitchell MS, Kan-Mitchell J, Kempf RA et al (1988) Active specific immunotherapy for melanoma: phase I trial of allogeneic lysates and a novel adjuvant. Cancer Res 48:5883–5893

Mizoguchi H, O'Shea JJ, Longo DL, Loeffler CM, McVicar DW, Ochoa AC (1992) Alterations in signal transduction molecules in T-lymphocytes from tumor-bearing mice. Science 258:1795–1798
Morton DL (1964) Acquired immunological tolerance to spontaneous mammary adenocarcinomas following neonatal infection with mammary tumor agent (MTA) Proc Am Assoc Cancer Res 5:A46
Morton DL, Barth A (1996) Vaccine therapy for malignant melanoma. CA Cancer J Clin 46:225–244
Nakagomi H, Petersson M, Magnusson I, Juhlin C, Matsuda M, Mellstedt H, Taupin J, Vivier E, Anderson P, Kiessling R (1993) Decreased expression of the signal-transducing z chains in tumor-infiltrating T-cells and NK cells of patients with clorectal carcinoma. Cancer Res 53:5610–5612
Naldini L, Blomer U, Gallay P, Ory D, Mulligan R, Gage FH, Verma IM, Trono D (1996) In vivo gene delivery and stable transduction of nondiving cells by a lentiviral vector. Science 272:263–267
Nestle FO, Alijigic S, Gilliet M, Sun Y, Grabbe S, Dummer R, Burg G, Schadendorf D (1998) Vaccination of melanoma patients with peptide or tumor lysate-pulsed dendritic cells. Nat Med 4:328–332
Oettgen HF, Old LJ (1991) The history of cancer immunotherapy. In: DeVita VT, Hellman S, Rosenberg SA, (eds) Biologic therapy of cancer. Lippincott, Philadelphia pp 87–119
Old LJ, Boyse EA (1964) Immunology of experimental tumors Annu Rev Med 15:167–186
Pardoll DM (1993) Genetically engineered tumor vaccines. Ann NY Acad Sci 609:301
Peace DJ, Chen W, Nelson H, Cheever MA (1991) T cell recognition of transforming proteins encoded by mutated ras proto-oncogenes. J Immunol 146:2059–2065
Perlmann P, Cerottini JC (1979) Cytotoxic lymphocytes. In: Sela M (ed) The Antigens. Academic, New York, 5:173–281
Pietras RJ, Fendly BM, Chazin VR, Pegram MD, Howell SB, Slamon DJ (1994) Antibody to HER-2/neu receptor blocks DNA repair after cisplatin in human breast and ovarian cancer cells. Oncogene 9:1829–1838
Pollok KE, Kim YJ, Zhou Z, Hurtado J, Kim KK, Pickard RT, Kwan BS (1993) Inducible T cell antigen 4-1BB. Analysis of expression and function. J Immunol 150:771–781
Pollok KE, Kim YJ, Hurtada J, Zhou Z, Kim KK, Kwon BS (1994) 4-1BB T-cell antigen binds to mature B cells and macrophages, and costimulates anti-mu-primed splenic B cells. Eur J Immunol 24:367–374
Porgador A, Tzehoval E, Vadai E, Feldman M, Eisenbach L (1995) Combined vaccination with major histocompatibiility class I and interleukin 2 gene-transduced melanoma cells synergizes the cure of potsurgical established lung metastases. Cancer Res 55(21):4941–4949
Prehn R, Main D (1957) Immunity to methylcholanthrene-induced sarcomas. J Natl Cancer Inst 18:769–778
Ressing ME, Sette A, Brandt RM, Ruppert J, Wentworth PA, Hartman M, Oseroff C, Grey HM, Melief CJ, Kast WM (1995) Human CTL epitopes encoded by human papillomavirus type 16 E6 and E7 identified through in vivo and in vitro immunogenicity studies of HLA-A*0201-binding peptides. J Immunol 154:5934–5943
Riethmüller G, Schneider-Gadicke E, Schlimok GWS, Raab R, Hoffken K, Gruber R, Pichlmaier H, Hirche H, Pichlmayr R et al (1994) Randomized trial of monoclonal antibody for adjuvant therapy of resected Dukes' C colorectal carcinoma. Lancet 343:1177–1183
Rosenberg SA (1992) The immunotherapy and gene therapy of cancer. J Clin Oncol 10:180–199
Schwartz RH (1992) Costimulation of T lymphocytes: the role of CD28, CTLA-4, and B7/BB1 in interleukin-2 production and immunotherapy. Cell 71:1065–1068

Shiku H, Takahashi T, Oettgen HF et al (1976) Cell surface antigens of human malignant melanoma. II. Serological typing with immune adherence test. J Exp Med 144:873
Shuford W, Klussman K, Tritchler D, Loo D, Chalupny J, Siadak A, Brown TJ, Emswiler J, Raecho H, Larsen CP, Ledbetter JA, Aruffo A, Mittler R (1997) 4-1BB costimulatory signals preferentially induce CD8+ T cell proliferation and lead to the amplification in vivo of cytotoxic T cell responses. J Exp Med 186:47–55
Sjögren HO (1965) Transplantation methods as a tool for detection of tumor specific antigens. Prog Exp Tumor Res 6:289–296
Sjögren HO, Hellström I, Klein G (1961) Transplantation of polyoma virus-induced tumors in mice. Cancer Res 21:329–337
Smith CA, Farrah T, Goodwin RG (1994) The TNF receptor superfamily of cellular and viral proteins: activation, costimulation, and death. Cell 76:959–962
Snyderman R, Pike M (1979) A macrophage chemotaxis inhibitors produced by neoplasms: characterization of its biological activity. Adv Exp Med Biol 121B:471–483
Southam CM (1967) Evidence for cancer-specific antigens in man. Prog Exp Tumor Res 9:1–39
Sparer RE, Wynn SG, Clark DJ, Kaplan JM, Cardoza LM, Wadsworth SC, Smith AE, Gooding LR (1997) Generation of cytotoxic T lymphocytes against immunorecessive epitopes after multiple immunizations with adenovirus vectors is dependant on haplotype. J Virol 71:2277–2284
Springer GF, Desai PR, Tegtmeyer H, Carlstedt SC, Scanlon EF (1994) T/Tn antigen vaccine is effective and safe in preventing recurrence of advanced human breast carcinoma. Cancer Biother 9:7–15
Stern PL, Duggan-Keen M (1994) MHC expression in the natural history of cervical cancer. In: Stern P, Stanley M (eds) Human papillomaviruses and cervical cancer: biology and immunology. Oxford University Press, New York, pp 162
Sulitzeanu D (1993) Immunosuppressive factors in human cancer. Adv Cancer Res 60:247–267
Townsend SE, Allison JP (1993) Tumor rejection after direct costimulation of CD8+ T cells by B7-transfected melanoma cells. Science 259:368–370
Van der Bruggen P, Traversari C, Chomez P, Lurquin C, De Plaen E, Van den Eynde, Knuth A, Boon T (1991) A gene encoding an antigen recognized by cytolytic T lymphocytes on a human melanoma. Science 254:1643–1647
Walker KK, Levine AJ (1996) Identification of a novel p53 functional domain that is necessary for efficient growth suppression. Proc Natl Acad Sci USA 93:15335–15340
Wallack MK, Bash J (1992) Vaccinia oncolysates as melanoma vaccines. In: Balch CM (ed) Cutnaeous melanoma. V. Distant metastases. Lippincott, Philadelphia, pp 548–550
Weiss DW, Faulkin LJ, DeOme KB (1964) Acquisition of heightened resistance and susceptibility to spontaneous mouse mammary carcinomas in the original host. Cancer Res 24:732–741
Wolfel T, Hauer M, Schneider J, Serrano M, Wolfel C, Klehmann-Hieb E, De Plaen E, Hankeln T, Meyer zum Buschenfelde KH, Beach D (1995) A p16INK4a-insensitive CDK4 mutant targeted by cytolytic T lymphocytes in a human melanoma. Science 269:1281–1284
Yang G, Mizuno MT, Hellström KE, Chen L (1997) B7-negative vs B7 Positive P815 Tumor: differential requirements for priming of an anti-tumor immune response in lymph nodes J Immunol 158:851–858
zur Hausen H (1987) Papillomaviruses in human cancer. Cancer 59:1692–1696

CHAPTER 18

Prevention of Autoimmunity

A.J. SLAVIN and H.L. WEINER

A. Overview of Autoimmune Diseases

There are several hypotheses for the pathogenesis of autoimmune diseases. The organ may become infected with a virus or other infectious agent resulting in the infiltration of reactive cells targeted to the infectious agent initiating the events which begin the cascade leading to organ specific autoimmunity. Alternatively, the initial infiltrating cells may be autoimmune in nature and recognize the organ specific proteins presented by local antigen presenting cells in the context of major histocompatibility complex (MHC). Either hypothesis is possible; however, once the autoimmune cascade begins, it is likely that T cells recognizing other organ specific proteins will be recruited and epitope spreading will occur. Perhaps viruses or infectious agents trigger or drive the autoimmune process as opposed to being the primary target of infiltrating cells. Thus understanding the mechanisms of initiating autoimmune disorders and the antigens recognized by self reactive T cells is critical for developing antigen specific therapies for use as vaccines for these disorders.

B. Autoreactive T Cells

Although it is presumed that organ-specific autoimmune disease is mediated by autoantigen reactive T cells, these cells are found at about the same frequency in normal individuals. Extensive studies using animal models indicate that it is the activation state and cytokine secretion pattern by autoreactive T cells that leads to tissue damage and disease. This concept is supported by investigations which reveal that in subjects with organ specific autoimmune disease, autoreactive cells are found in an activated state compared to those T cells found in normal individuals (HAFLER and WEINER 1995; BIEGANOWSKA et al. 1997). An important question in the treatment of autoimmune disease is the role of these autoreactive T cells in mediating the immune response. Autoreactive T cells can be both destructive in the induction of autoimmune disease or protective in downregulating immune responses, thus providing a potential mechanism whereby vaccination may be a viable way to manipulate the autoimmune response for effective antigen specific treatment of autoimmunity.

C. Antigen-Specific Therapy

Two general approaches have been used to address antigen specific immunotherapy of autoimmune diseases (HAFLER and WEINER 1995). Firstly, to block the initial activation and subsequent recognition of autoantigen by the autoreactive T cells. This strategy requires that the immune response to the self-antigen is either restricted in the number of epitopes recognized, or in the T-cell receptor (TCR) repertoire used. Examples of these therapies include MHC blocking peptides, TCR antagonists and TCR peptides which inhibit specific T cell function (COHEN 1988; KUMAR et al. 1996). The second approach uses antigen-specific targeting of T cells that regulate immune responses at the inflammatory sites. The benefit of this approach is that neither the specificity of the antigen that elicits the immune response nor restriction of the TCR in terms of antigen recognition or usage is required. Recently myelin basic protein (MBP) specific T cells transduced with a retroviral vector expressing interleukin (IL) 4 have been used to deliver regulatory cytokines in a site specific manner and were able to delay the onset and reduce the severity of experimental autoimmune encephalomyelitis (EAE) (SHAW et al. 1997). Other examples of this method are oral tolerization which generates regulatory cells secreting transforming growth factor (TGF) β, IL-4, and IL-10 which migrate to the sites of inflammation and suppress the immune response or the use of altered peptide ligands that switch the cytokine profile secreted by T cells.

I. Altered Peptide Ligands

The ability of altered peptide ligands to induce different signaling events (ELVAVOLD et al. 1993; SLOAN-LANCASTER et al. 1993) has prompted examination of their effects upon cytokine secretion in a number of experimental models of autoimmunity and autoreactive human T cell clones. WINDHAGEN et al. (1995) demonstrated that stimulation of human autoreactive T cell clones with altered peptide ligand induced a switch in the cytokine pattern secreted from a Th0 to a TGF-β secreting phenotype. These findings have potential therapeutic implications for antigen specific modification of immune responses. Indeed altered peptide ligands can effectively inhibit the development of experimental EAE as demonstrated by the use of peptides derived from proteolipid protein peptide 139–151 and substituted at the primary TCR contact points (NICHOLSON et al. 1995). Thus T cell differentiation may be significantly altered by a single amino acid substitution in an antigenic peptide and this immune deviation may be a mechanism by which altered peptide ligands could be used as a vaccine to inhibit autoimmune disease.

II. Oral Tolerance

Oral tolerance describes the phenomenon that a state of systemic hyporesponsiveness follows immunization with a protein that has previously been fed. This observation was first described in 1911 by Wells who fed hen egg proteins to guinea pigs and found them resistant to anaphylaxis when challenged (WELLS 1911). Subsequently, numerous investigators have found that when fed antigens such as contact sensitizing agents, ovalbumin (OVA), sheep red blood cells, viruses, or bacteria animals do not respond as well to these antigens when subsequently immunized, but respond normally to other antigens (MOWAT 1987). The phenomenon of oral tolerance has also been observed in humans fed and immunized with keyhole limpet hemocyanin (KLH) (HUSBY et al. 1994).

Historically, immunological tolerance was a fundamental property of the immune system providing a mechanism for self/nonself discrimination. Thus the immune system could protect the host from external pathogens (i.e., against nonself) without eliciting autoimmune disease (i.e., against self) (reviewed by LANGMAN and COHN 1996; MATZINGER and FUCHS 1996). Initially it was thought that self/nonself discrimination was a simple matter of deleting autoreactive cells in the thymus, but it is now clear that the maintenance of immunological tolerance is a much more dynamic process involving regulatory T cells (HAMMERLING et al. 1991; GOODNOW 1996). Indeed, autoreactive cells, such as those reacting with brain, are not deleted and can be found in normal individuals (KERLERO DE ROSBO et al. 1993; ZHANG et al. 1993) Why these cells become activated and cause disease in some individuals while remaining harmless in others is a major question in basic immunology. How to control the autoimmune process once it has been initiated is major a problem in clinical medicine.

Recently, oral tolerance has been used successfully to treat autoimmune diseases in several experimental animal models of autoimmunity (reviewed by WEINER 1997) and is currently being applied to the treatment of human diseases (TRENTHAM et al. 1993; WEINER et al. 1993; BARNETT et al. 1996; SIEPER et al. 1996; NUSSENBLATT et al. 1997). The mechanisms by which orally administered antigens induce immune tolerance are beginning to be elucidated (WEINER and MAYER 1996). Oral tolerance, as with immunological tolerance induced by other methods, has been found to involve multiple mechanisms and is not a single immunological event (WHITACRE et al. 1991; GREGERSON et al. 1993; FRIEDMAN and WEINER 1994; CHEN et al. 1995a)

1. Mechanisms of Oral Tolerance

The primary mechanisms by which oral tolerance is mediated include deletion, anergy and active cellular suppression (Fig. 1). The determining factor in this process is the dose of antigen fed (WHITACRE et al. 1991; GREGERSON et al. 1993; FRIEDMAN and WEINER 1994). Low doses favor the induction of active suppression, mediated by the induction of regulatory T cells in the gut-associated lymphoid tissue (GALT), for example, in the Peyer's patches, a unique site since it favors the induction of T helper 2 (Th2) cells and T cells that secrete TGF-β (SANTOS et al. 1994). In contrast, the administration of high doses of antigen favors the utilization of deletion and anergy (CHEN et al. 1995a).

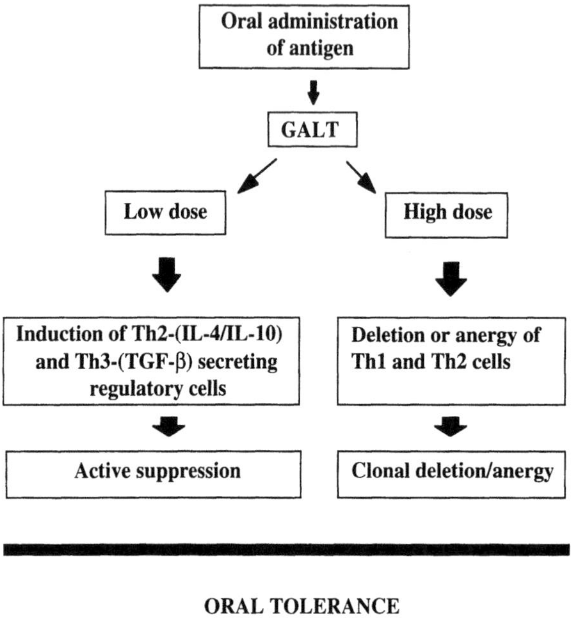

Fig. 1. Mechanisms of oral tolerance

The characterization of active cellular suppression of immune responses has remained ill-defined due to difficulties in cloning suppressor cells and defining their mechanism of action despite extensive studies. However, recent evidence indicates that one of the primary mechanisms of active cellular suppression is via the secretion of suppressive cytokines such as transforming growth factor β (TGF-β), IL-4, and IL-10 following antigen-specific triggering (CHEN et al. 1994). The potent immunosuppressive and anti-inflammatory cytokine TGF-β plays an important role in local function of the gut as it serves as a switch factor for IgA production in the mucosa (KIM and KAGNOFF 1990) and may also be involved in the homing mechanism of cells to high endothelial venules (CHIN et al. 1992). Produced by both $CD4^+$ and $CD8^+$ GALT-derived T cells (CHEN et al. 1994; SANTOS et al. 1994) TGF-β is an important mediator of the active suppression component of oral tolerance. TGF-β deficient mice have inflammation that affects multiple organ systems (SHULL et al. 1992), demonstrating the importance of TGF-β in maintaining immune homeostasis. TGF-β is also associated with natural recovery from EAE (KHOURY et al. 1992) and regulatory cells that suppress certain forms of inflammatory colitis act via secretion of TGF-β (NEURATH et al. 1996; POWRIE et al. 1996).

TGF-β secreting $CD4^+$ cells specific for MBP have been cloned from the mesenteric lymph nodes of SJL mice (CHEN et al. 1994). These clones suppressed disease although they were found to be structurally identical to Th1 disease-inducing clones with respect to TCR usage, MHC restriction and epitope recognition. TGF-β secreting $CD4^+$ cells were also cloned from mice

Table 1. Characteristics of the T-cell subsets

	Th1	Th2	Th3
Cytokine profile			
IFN-γ	++++	–	+
IL-4	–	++++	+
TGF-β	+	+	++++
Growth/differentiation factors	IL-2	IL2/IL4	IL-4
Functions			
Help	DTH/IgG2a	IgG1/IgE	IgA
Suppression	Th2	Th1	Th1/Th2

DTH, delayed-type hypersensitivity.

transgenic for an MBP-specific TCR by culturing in the presence of IL-4, but not IL-2; these clones did not secrete IL-2, interferon γ (IFN-γ), IL-4, or IL-10, and used IL-4 as a growth/differentiation factor (Weiner et al. 1996; Inobe et al. 1988). Seder et al. have also recently found that IL-4, and in some instances IL-10, serves to help differentiation of TGF-β secreting cells (Seder et al. 1998).

Thus $CD4^+$ cells that primarily produce TGF-β appear to be a unique T-cell subset that includes mucosal T helper function and downregulatory properties for Th1 and other immune cells. In contrast to Th1 and Th2 cells, these cells provide help for IgA production and have been termed Th3 cells (Chen et al. 1994; Mosmann and Sad 1996) (Table 1). MBP- and proteolipid protein (PLP) specific TGF-β secreting Th3 type cells have been observed in multiple sclerosis (MS) patients orally administered myelin proteins (Fukaura et al. 1996). These clones grow very poorly, and the fact that Th3 cells do not proliferate well may also account for their relative resistance to deletion, as compared with Th1 and Th2 cells, following feeding of high-dose antigen to mice transgenic for an OVA-specific TCR (Chen et al. 1995a). Thus it appears that Th3 type cells are cells which primarily secrete TGF-β, provide help for IgA and have suppressive properties. Th3 type cells appear distinct from the Th2 cells as $CD4^+$ TGF-β secreting cells with suppressive properties in the gut have been generated from IL-4 deficient animals (Powrie et al. 1996). In addition, administration of anti-IL-12 to OVA TCR transgenic animals is associated with inhibition of Th1 responses and high TGF-β production without increased IL-4 production (Marth et al. 1996).

Orally administered antigen is known to suppress IgE antibody responses, which is rather paradoxical if Th2 responses are generated preferentially in the gut. The suppression of IgE responses following oral antigen may relate to the generation of TGF-β-secreting cells. In addition, $\gamma\delta$ T cells are reported to be involved in the suppression of IgE responses following aerosol administration of antigen in suppression of diabetes following aerosol insulin (Harrison et al. 1996) and appear to be involved in oral tolerance although their mechanism of action is poorly understood (McMenamin et al. 1994; Mengel et al. 1995).

Animals deficient in $\gamma\delta$ T cells have impaired mucosal IgA responses (FUJIHASHI et al. 1996), and it has been reported to be more difficult to induce oral tolerance in animals depleted of such cells (KE and KAPP 1996; Weiner and Spahn, unpublished).

2. Bystander Suppression

Bystander suppression was originally described during an investigation of the regulatory cells induced by oral administration of low doses of MBP (MILLER et al. 1991). It solves a major conceptual problem in the design of antigen or T-cell specific therapy for inflammatory autoimmune diseases such as MS, type 1 diabetes and rheumatoid arthritis in which the autoantigen is unknown, or where there are reactivities to a burgeoning number of autoantigens in the target tissue. During the course of chronic inflammatory autoimmune processes in animals there is intra- and interantigenic spread of autoreactivity at the target organ (LEHMANN et al. 1992; CROSS et al. 1993; KAUFMAN et al. 1993; TISCH et al. 1993). Similarly, in human autoimmune diseases there are reactivities to multiple autoantigens in the target tissue. For example, in MS, there is immune reactivity to multiple myelin antigens including: MBP, PLP, and myelin oligodendrocyte glycoprotein (KERLERO DE ROSBO et al. 1993; ZHANG et al. 1993). In type 1 diabetes there are multiple islet cell antigens that could be the target of autoreactivity including glutamic acid decarboxylase, insulin, and heat shock proteins (HARRISON 1992). Because regulatory cells induced by oral antigen secrete antigen-nonspecific cytokines after being triggered by the fed antigen, they suppress inflammation in the microenvironment where the fed antigen is localized. Thus, for a human organ-specific inflammatory disease, it is not necessary to know the specific antigen that is the target of an autoimmune response but only to orally administer an antigen capable of inducing regulatory cells, which then migrate to the target tissue and suppress inflammation.

Bystander suppression was demonstrated in vitro when it was shown that cells from animals fed MBP suppressed proliferation of an ovalbumin line across a transwell, but only when triggered by the fed antigen (MILLER et al. 1991). The soluble factor was identified as TGF-β. Bystander suppression has since been demonstrated in a number of autoimmune disease models. For instance, PLP-induced EAE can be suppressed by feeding MBP (AL-SABBAGH et al. 1994) or by administering TGF-β-secreting MBP-specific T-cell clones from orally tolerized animals (CHEN et al. 1994). In the Lewis rat model of EAE disease induced by MBP peptide 71–90 can be suppressed by feeding peptide 21–40 (MILLER et al. 1993a). In arthritis models adjuvant- and antigen-induced arthritis can be suppressed by feeding type II collagen (ZHANG et al. 1990; YOSHINO et al. 1995). In a virus-induced model of diabetes, whereby a lymphocytic choriomeningitis virus (LCMV) protein is expressed on the insulin promoter and the animal is then infected with the LCMV, diabetes can be suppressed by feeding insulin (VON HERRATH et al. 1996). Bystander suppression in vitro has recently been shown in humans using common dietary proteins (OVA, bovine gammaglobulin) to stimulate cells in vitro and then using these cells to suppress in vitro tetanus toxoid responses (ZIVNY et al. 1996). However, bystander suppression may be more difficult to induce in uveitis models (WILDNER and THURAU 1995), and feeding OVA has been reported to induce bystander suppression in adult animals but anergy in young animals (LUNDIN et al. 1996).

In theory, bystander suppression could be applied for the treatment of organ-specific inflammatory conditions that are not classic autoimmune diseases, such as psoriasis, or could be used to target anti-inflammatory cytokines to an organ where inflammation may play a role in disease pathogenesis even if the disease is not primarily inflammatory in nature, such as stroke (BECKER et al. 1997). Although bystander suppression was initially described in association with regulatory cells induced by oral antigen, the process could in principle be induced by any immune manipulation that induces Th2- or Th3-type regulatory cells.

3. Modulation of Oral Tolerance

Current studies have addressed the use of antigen alone to induce oral tolerance and factors that may enhance or modulate oral tolerance are not well understood. It is well known that immune responses can be enhanced with appropriate adjuvants. This has been most clearly shown for the generation of Th1 responses in which the antigen is injected in an emulsion of oil and mycobacteria (complete Freund's adjuvant). One of the major challenges facing the use of oral antigen to induce tolerance is to find ways to enhance this natural form of immune tolerance. Investigating these factors has both theoretical and practical importance for the generation of tolerance to treat autoimmune diseases. If oral tolerance is broadly defined as the inhibition of Th1 responses in the periphery, either by large doses of antigen that induce anergy and/or deletion or by gut-induced Th2- or Th3-type regulatory cells, anything that favors Th1 versus Th2 or Th3 responses would abrogate oral tolerance and the converse (Table 2).

Recently a study in which anti-IL-12 antibodies were administered to OVA TCR transgenic mice were shown to enhance oral tolerance with the inhibition of Th1 type responses (MARTH et al. 1996), indeed investigators have found that anti-IL-12 enhances oral tolerance in animals transgenic for an OVA-specific TCR and is associated both with increased TGF-β secretion and T cell apoptosis (MARTH et al. 1996). Large doses of IFN-γ abrogate oral tolerance (ZHANG and MICHAEL 1990) and in the uveitis model, IL-2 potentiates oral tolerance and is associated with increased production of TGF-β, IL-4, and IL-10 in Peyer's patches (RIZZO et al. 1994). Furthermore, IL-4 administration enhances low-dose oral tolerance to MBP in the EAE model and is associated with increased fecal IgA anti-MBP antibodies (INOBE et al.

Table 2. Modulation of oral tolerance

Modulation	Reference
Augments	
IL-4/IL-10	INOBE et al. (1998)
Anti-IL-12	MARTH et al. (1996)
CTB antigen	SUN et al. (1994)
LPS	KHOURY et al. (1990)
IFN-β	NELSON et al. (1996)
IL-2	RIZZO et al. (1994)
Multiple emulsions	ELSON et al. (1996)
Decreases	
IFN-γ	ZHANG and MICHAEL (1990)
IL-12	MARINARO et al. (1997)
CTB	ELSON and EALDING (1984)
Anti-MCP-1	KARPUS and LUKACS (1996)
Anti-$\gamma\delta$-A6	MENGEL et al. (1995)
GVH	STROBEL et al. (1985)

1998). Orally administered cytokines retain their biological activity (ROLLWAGON and BAQAR 1996), and oral IL-4 and IL-10 can enhance oral tolerance, suppressing EAE (Fig. 2). Lipopolysaccaride (LPS) enhances oral tolerance to MBP (KHOURY et al. 1990) and is associated with increased expression of IL-4 in the brain (KHOURY et al. 1992), and IFN-β synergizes with the induction of oral tolerance in SJL/J mice fed low doses of MBP (NELSON et al. 1996).

Cholera toxin (CT) is one of the most potent mucosal adjuvants and feeding CT abrogates oral tolerance when fed with an unrelated protein antigen (ELSON and EALDING 1984). However, when a protein is coupled to recombinant cholera toxin B subunit (CTB) and given orally, there is enhancement of peripheral immune tolerance (SUN et al. 1994). Although it is not known how CTB enhances oral tolerance, it may do so by the more efficient induction of regulatory cells due to the special binding properties of the CTB subunit. The abrogation of oral tolerance by CT may also be the result of an active immune process in the gut, as it has been shown that CT given orally with KLH primes both Th1 and Th2 responses (HORNQUIST and LYCKE 1993), even though Th2 responses may be preferentially generated (XU-AMANO et al. 1990; WILSON et al. 1991). The priming of Th1 responses via the gut may make it impossible to suppress systemic Th1 responses following peripheral immunization. It has also been reported that antibodies to the chemokine monocyte chemotactic protein 1 abrogate oral tolerance, perhaps by interfering with chemotaxis of the Th2- or Th3-type regulatory cells generated in the gut or by acting as a factor required for generation of such cells (KARPUS and LUKACS 1996). Oral antigen delivery using a multiple emulsion system also enhances oral tolerance (ELSON et al. 1996).

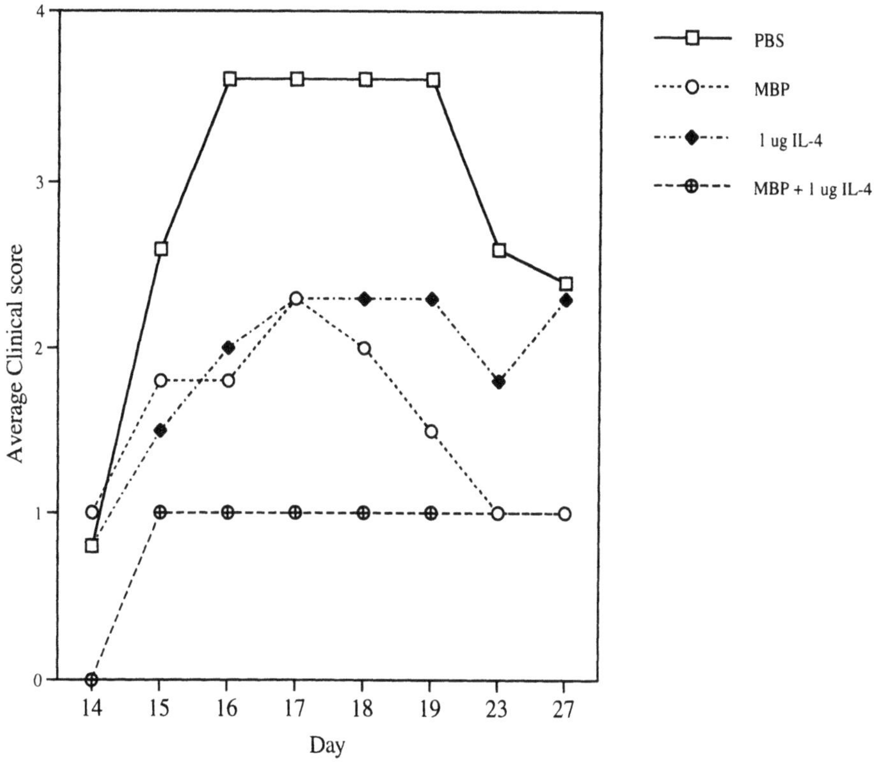

Fig. 2. Oral IL-4 and/or MBP suppresses EAE induced with MBP. Mice were fed murine MBP (0.5 mg) and/or recombinant murine IL-4 (1 μg) on 3 consecutive days. Twenty four hours after the last feeding mice were immunized subcutaneously with 200 μg MBP emulsified in complete Freund's adjuvant containing 2 mg/ml *Mycobacterium tuberculosis* (H37RA). Immediately thereafter and again 48 h later 150 ng pertussis toxin was administered intravenously. Mice were examined daily for clinical signs of disease, which were scored as follows: *1*, limp tail; *2*, limp tail and hindlimb weakness; *3*, hindlimb paralysis; *4*, hindlimb paralysis with forelimb involvement; *5*, moribund

Commonly used drugs for autoimmune diseases may also affect oral tolerance. For example, methotrexate, a drug used to treat rheumatoid arthritis enhances oral tolerance (A. Al-Sabbagh and H.L. Weiner, unpublished).

III. Nasal and Aerosol Mucosal Tolerance

Recently other forms of mucosal tolerance have been investigated, specifically the administration of antigen via the nasal or aerosol route. These routes appear equally efficient and in some instances may be more effective in suppressing autoimmune diseases in animal models (Metzler and Wraith 1993; Ma et al. 1995; Al-Sabbagh et al. 1996a; Harrison et al. 1996; Staines et al. 1996). Smaller doses of antigen can be administered than orally, perhaps because there is less degradation of the proteins before they reach the intestinal mucosal surface. The degree to which the mechanisms of oral versus nasal tolerance are the same remains to be elucidated.

IV. Treatment of Autoimmune Diseases in Animals

1. Experimental Autoimmune Encephalomyelitis

Several studies have demonstrated the effectiveness of orally administered myelin antigens (e.g., MBP, PLP) in both the rat and mouse models of EAE. In the Lewis rat high doses of MBP result in clonal anergy (BITAR and WHITACRE 1988; WHITACRE et al. 1991) whereas lower doses induce transferable cellular suppression (HIGGINS and WEINER 1988; LIDER et al. 1989; MILLER et al. 1993a,b). In the nervous system of low-dose fed animals, inflammatory cytokines such as tumor-necrosis factor and IFN-γ are downregulated and TGF-β is upregulated (KHOURY et al. 1992). Administration of myelin to sensitized animals in the chronic guinea pig model or larger doses of MBP in the murine EAE model is protective and does not exacerbate disease (BROD et al. 1991), and long-term (6-month) administration of myelin in the chronic EAE model was beneficial (AL-SABBAGH et al. 1996b) EAE can also be suppressed in animals transgenic for an MBP specific TCR following feeding with MBP (CHEN et al. 1996). Nasally administered MBP peptides have been reported to suppress EAE whereas orally administered peptides do not, and this may relate to binding affinity (METZLER and WRAITH 1993) or breakdown of peptide in the gut.

2. Arthritis

Oral administration of type II collagen suppresses several models of arthritis including collagen-induced arthritis (CIA) (THOMPSON et al. 1993a), adjuvant arthritis (ZHANG et al. 1990), pristine arthritis (THOMPSON et al. 1993b), and antigen-induced arthritis (YOSHINO et al. 1995). One of the first studies to demonstrate that an orally administered autoantigen can suppress an autoimmune disease was the use of oral type II collagen in CIA (THOMPSON et al. 1993a). Of note is that oral type I collagen has also been shown to suppress adjuvant arthritis (ZHANG et al. 1990). Oral administration of an immunodominant human collagen peptide modulates CIA in mice (KHARE et al. 1995) and type II collagen peptides given nasally also suppresses CIA in mice (STAINES et al. 1996). Oral mycobacterial 65-kDa heat shock protein suppresses adjuvant arthritis (HAQUE et al. 1996).

3. Uveitis

Oral administration of S antigen (S-Ag), a retinal autoantigen that induces experimental autoimmune uveitis, or S-Ag peptides prevents or markedly diminishes the clinical appearance of S-Ag induced disease as measured by ocular inflammation (NUSSENBLATT et al. 1990; GREGERSON et al. 1993). S-Ag induced experimental autoimmune uveitis can also be suppressed by feeding an HLA peptide (WILDNER and THURAU 1994). Feeding interphotoreceptor binding protein (IRBP) suppresses IRBP-induced disease and is potentiated by IL-2 (RIZZO et al. 1994; WILDNER and THURAU 1995).

4. Myasthenia Gravis

Although myasthenia gravis is an antibody-mediated disease, oral and nasal (MA et al. 1995) administration of the *Torpedo* acetylcholine receptor to Lewis rats prevented or delayed the onset of myasthenia gravis (WANG et al. 1993a; OKUMURA et al. 1994) although large doses of antigen were required.

5. Diabetes

Oral insulin has been shown to delay and, in some instances, prevent diabetes in the nonobese diabetic mouse model. Such suppression is transferable (ZHANG et al. 1995), primarily with $CD4^+$ cells (BERGEROT et al. 1994). Immunohistochemistry of pancreatic islets of Langerhans isolated from insulin fed animals demonstrates decreased insulitis associated with decreased IFN-γ, as well as increased expression of tumor necrosis factor, IL-4, IL-10, TGF-β, and prostaglandin (HANCOCK et al. 1995). Recently it was also reported that nasal administration of the insulin B chain or glutamic acid decarboxylase and aerosol insulin suppresses diabetes in the nonobese diabetic mouse (DANIEL and WEGMAN 1996; HARRISON et al. 1996; TIAN et al. 1996). Under special experimental conditions large doses of OVA given to OVA double transgenic mice resulted in diabetes mediated by OVA-specific CTL (BLANAS et al. 1996). These animals expressed OVA on the islets under the rat insulin promoter and were made chimeric to enrich for OVA-specific transgenic TCR CTL. Oral insulin suppressed diabetes in a viral-induced model of diabetes in which LCMV was expressed under the insulin promoter and animals infected with LCMV to induce diabetes (VON HERRATH et al. 1996) In this model protection was associated with protective cytokine shifts (IL-4 / IL-10, TGF-β) in the islets.

6. Transplantation and Other Models

Oral administration of allogeneic cells prevents sensitization by skin grafts and changes accelerated rejection of vascularized cardiac allografts to an acute form typical of nonsensitized recipients (SAYEGH et al. 1992a). DTH responses to the peptide can be reduced in the Lewis rat by orally administered allopeptides, accompanied by elevated intragraft IL-4. This was not observed following intravenous administration of the alloantigen (SAYEGH et al. 1992b). Similarly, oral alloantigen enhanced corneal allograft survival even in preimmune hosts (HE et al. 1996). Oral thyroglobulin has been shown to suppress autoimmune thyroiditis (GUIMARAES et al. 1995; PETERSON and BRAYLEY-MULLEN 1995), and feeding peptides of the Der p I allergen suppressed responses to the whole allergen (HOYNE et al. 1994). Recently it has been shown that experimental granulomatous colitis in mice is abrogated by TGF-β mediated oral tolerance after administration of haptenized colonic proteins (NEURATH et al. 1996).

V. Treatment of Autoimmune Diseases in Humans

Investigators have shown that exposure of a contact-sensitizing agent via the mucosa prior to subsequent skin challenge led to unresponsiveness in a portion of patients studied (LOWNEY 1968). KLH administered orally to human subjects has been reported to decrease subsequent cell-mediated immune responses although antibody responses were not affected (HUSBY et al. 1994). Nasal KLH has also been reported to induce tolerance in humans (WALDO et al. 1994).

On the basis of the long history of oral tolerance and the safety of the approach, human trials have been initiated in MS, rheumatoid arthritis (RA), uveitis, and diabetes. These initial trials suggest that there has been no systemic toxicity or exacerbation of disease, although reproducible clinical efficacy has yet to be demonstrated. Results in humans, however, have paralleled several aspects of what has been observed in animals.

In MS patients MBP- and PLP-specific TGF-β secreting Th3-type cells have been observed in the peripheral blood of patients treated with an oral bovine myelin preparation and not in patients who were untreated (FUKAURA et al. 1996). There was no increase in MBP- or PLP-specific IFN-γ secreting cells in treated patients. These results demonstrate that it is possible to immunize via the gut for autoantigen-specific TGF-β-secreting cells in a human autoimmune disease by oral administration of the autoantigen. A pilot study of oral myelin in MS suggested positive effects related to DR type or sex; however, a recently completed 515 patient, placebo-controlled, double-blind phase 3 trial of single-dose bovine myelin in relapsing-remitting MS showed no differences between placebo and treated groups in the number relapses, a large placebo effect being observed (AutoImmune, Inc., Lexington, MA, USA). Preliminary analysis of magnetic resonance imaging data from this trial, however, showed significant changes favoring oral myelin in one of the four prospectively defined and randomized patient subgroups (DR$^+$ males).

In RA a single center double-blind study of oral collagen showed statistically significant differences favoring oral collagen over placebo. A recently completed 280 patient double-blind phase II dosing trial of chicken type II collagen in doses ranging from 20 to 2500 μg demonstrated statistically significant positive effects in the group treated with the lowest dose (BARNETT et al. 1998). Oral administration of larger doses of bovine type II collagen (1–10 mg) did not show a significant difference between tested and placebo groups, although a higher prevalence of responders was reported for the groups treated with type II collagen (SIEPER et al. 1996). These results are consistent with animal studies of orally administered type II collagen in which protection against adjuvant- and antigen-induced arthritis, and bystander suppression was observed only at the lower doses (Brod et al. 1991; Al-Sabbagh et al. 1996a,b) An open-label pilot study of oral collagen in juvenile RA gave positive results with no toxicity (Barnett et al. 1996). This lack of systemic toxicity is an important feature for the clinical use of oral tolerance, especially in children for whom the long-term effects of immunosuppressive drugs is unknown. Five phase II randomized studies of oral type II collagen have been performed and based on the results to date, a phase III trial of oral collagen using 60 μg in patients with RA is underway.

In uveitis, a pilot trial of S-Ag and an S-Ag mixture has recently been completed at the National Eye Institute (Bethesda, MD, USA) and showed positive trends with oral bovine S-Ag but not the retinal mixture (Nussenblatt et al. 1997). Trials have been initiated in new-onset diabetes in which recombinant human insulin is administered orally, and trials are underway in subjects at risk for diabetes as part of the diabetes prevention trial (DPT-1). Preliminary analysis of a randomized double blind placebo controlled study of oral insulin in newly diagnosed immune-mediated (type I) diabetes demonstrated preserved beta cell function as measured by endogenous C-peptide insulin responses in adult new onset diabetics fed 10 mg of recombinant human insulin as compared to those fed placebo (Coutant et al. 1998). Oral desensitization to nickel-allergy in humans induces a decrease in nickel-specific T cells and affects cutaneous eczema (Guimaraes et al. 1995). Finally, a pilot immunological study of oral MHC peptides has been initiated in transplantation patients. Based on results to date in humans, it appears that the clinical application of oral antigen for the treatment of human conditions will depend on the specific disease, the nature, and doses of proteins administered and may require the use of synergists or mucosal adjuvants to enhance biological effects. Also, recombinant human proteins may be more efficacious than animals proteins (Hoyne et al. 1994).

D. Future Directions

Although it is clear that oral antigen can suppress autoimmunity in animals, and results to date suggest that this approach may have utility in humans, much remains to be learned. Under certain experimental conditions worsening of autoimmune diseases in animals by oral antigen has been seen (Miller et al. 1994; Blanas et al. 1996; Meyer et al. 1996; Terato et al. 1996) although this has not been observed in human trials.

We are yet to understand the role of cell surface molecules and cytokines associated with inductive events in the gut that generate and that modulate oral tolerance. The cytokine milieu, antigen presentation and costimulation requirements, routes of antigen processing, form of the antigen, role of the liver, the effect or oral antigens on antibody and IgE responses and on CTLs, and the role of $\gamma\delta$ T-cells also represent important areas of investigation. As these molecular events associated with generation and modulation of oral tolerance are better understood, the ability to apply mucosal tolerance successfully for the treatment of human autoimmune and other diseases will be further enhanced.

References

Al-Sabbagh A, Miller A, Santos LMB, Weiner HL (1994) Antigen-driven tissue-specific suppression following oral tolerance: orally administered myelin basic protein suppresses proteolipid induced experimental autoimmune encephalomyelitis in the SJL mouse. Eur J Immunol 24:2104–2109

Al-Sabbagh A, Nelson P, Sobel RA, Weiner HL (1996a) Antigen-driven peripheral immune tolerance: suppression of experimental autoimmune encephalomyelitis and collagen induced arthritis by aerosol administration of myelin basic protein or type II collagen. Cell Immunol 171:111–119

Al-Sabbagh AM, Goad MEP, Weiner HL, Nelson PA (1996b) Decreased CNS inflammation and absence of clinical exacerbation of disease after six months oral administration of bovine myelin in diseased SJL/J mice with chronic relapsing experimental autoimmune encephalomyelitis. J Neurol Res 45:424–429

Barnett ML, Combitchi D, Trentham DE (1996) A pilot trial of oral type II collagen in the treatment of juvenile rheumatoid arthritis. Arthritis Rheum 39:623–628

Barnett ML, Kremer JM, St. Clair EW, Clegg DO, Furst D, Weisman M, Fletcher MJF, Lavin PT, Finger E, Morales A, Le CH, Trentham DE (1998) Treatment of rheumatoid arthritis with oral type II collagen: results of a multicenter, double-blind, placebo-controlled trial. Arthritis Rheum 41:290

Becker KJ, McCarron RM, Ruetzler C, Laban O, Sternberg E, Flanders KC, Hallenbeck JM (1997) Immunologic tolerance to myelin basic protein decreases stroke size after transient focal cerebral ischemia. Proc Natl Acad Sci 94:10873–10878

Bergerot J, Fabien N, Maguer V, Thivolet C (1994) Oral administration of human insulin to NOD mice generates $CD4^+$ T cells that suppress adoptive transfer of diabetes. J Autoimmun 7:655–663

Bieganowska KD, Ausubel LJ, Moddaber Y, Slovik E, Messersmith W, Hafler DA (1997) Direct ex vivo analysis of activated, Fas-sensitive autoreactive T cells in human autoimmune disease. J Exp Med 185:1585–1594

Bitar D, Whitacre CC (1988) Suppression of experimental autoimmune encephalomyelitis by the oral administration of myelin basic protein. Cell Immunol 112:364–370

Blanas E, Carbone FR, Allison J, Miller JFAP, Heath WR (1996) Induction of autoimmune diabetes by oral administration of autoantigen. Science 274:1707–1709

Brod SA, Al-Sabbagh A, Sobel RA, Hafler DA, Weiner HL (1991) Suppression of experimental autoimmune encephalomyelitis by oral administration of myelin antigens. IV. Suppression of chronic relapsing disease in the Lewis rat and strain 13 guinea pig. Ann Neurol 29:615–622

Chen Y, Kuchroo VK, Inobe JI, Hafler DA, Weiner HL (1994) Regulatory T cell clones induced by oral tolerance: suppression of autoimmune encephalomyelitis. Science 265:1237–1240

Chen Y, Inobe JI, Marks R, Gonnella P, Kuchroo VK, Weiner HL (1995a) Peripheral deletion of antigen-reactive T cells in oral tolerance. Nature 376:177–180

Chen Y, Inobe JI, Weiner HL (1995b) Induction of oral tolerance to myelin basic protein in CD-8 depleted mice: both CD4+ and CD8+ T cells mediate active suppression. J Immunol 156:910–916

Chen Y, Inobe J-I, Kuchroo VK, Baron JL, Janeway CA, Weiner HL (1996) Oral tolerance in myelin basic protein T-cell receptor transgenic mice: suppression of autoimmune encephalomyelitis and dose-dependent induction of regulatory cells. Proc Natl Acad Sci USA 93:388–391
Chin YH, Cai JP, Xu XM (1992) Transforming growth factor-b1 and IL-4 regulate the adhesiveness of Peyer's patch high endothelial venule cells for lymphocytes. J Immunol 148:1106–1112
Cohen I (1988) The self, the world and autoimmunity. Sci Am 258:52–60
Coutant R, Zeidler A, Rappaport R, Schatz D, Schwartz S, Raskin P, Rogers D, Bode B, Crockett S, Marks J, Deeb L, Chalew S, MacLaren N (1998) Oral insulin therapy in newly-diagnosed immune mediated (type I) diabetes. Preliminary analysis of a randomized double blind placebo controlled study. Diabetes 47 [Suppl 1]:A97
Cross AH, Tuohy VK, Raine CS (1993) Development of reactivity to new myelin antigens during chronic relapsing autoimmune demyelination. Cell Immunol 146:261–270
Daniel D, Wegmann DR (1996) Protection of nonobese diabetic mice from diabetics by intranasal or subcutaneous administration of insulin peptide B-(9-23). Proc Natl Acad Sci USA 93:956–960
Elson CO, Ealding W (1984) Cholera toxin feeding did not induce oral tolerance in mice and abrogated oral tolerance to an unrelated protein antigen. J Immunol 133:2892–2897
Elson CO, Tomasi M, Dertzbaugh MT, Thaggard G, Hunter R, Weaver C (1996) Oral antigen delivery by way of a multiple emulsion system enhances oral tolerance. Ann NY Acad Sci 778:156–162
Elvavold BD, Sloan-Lancaster J, Hsu BL, Allen PM (1993) Separation of T helper 1 clone cytolysis from proliferation and lymphokine production using analog peptides. J Immunol 150:3131
Friedman A, Weiner HL (1994) Induction of anergy or active suppression following oral tolerance is determined by antigen dosage. Proc Natl Acad Sci USA 91:6688–6692
Fujihashi K, McGhee JR, Kweon MN, Cooper MD, Tonegawa S, Takahashi I, Hiroi T, Mestecky J, Kiyono H (1996) g/d T Cell-deficient mice have impaired mucosal immunoglobulin A responses. J Exp Med 183:1929–1935
Fukaura H, Kent SC, Pietrusewicz MJ, Khoury SJ, Weiner HL, Hafler DA (1996) Induction of circulating myelin basic protein and proteolipid protein specific TGF-b1 secreting Th3 T cells by oral administration of myelin in multiple sclerosis patients. J Clin Invest 98:70–77
Goodnow CC (1996) Balancing immunity and tolerance: deleting and tuning lymphocyte repertoires. Proc Natl Acad Sci USA 93:2264–2271
Gregerson DS, Obritsch WF, Donoso LA (1993) Oral tolerance in experimental autoimmune uveoretinitis: Distinct mechanisms of resistance are induced by low versus high dose feeding protocols. J Immunol 151:5751–5761
Guimaraes VC, Quintans J, Fisfalen M-E, Straus FH, Wilhelm K, Medeiros-Neto GA, DeGroot LJ (1995) Suppression of experimental autoimmune thyroiditis by oral administration of thyroglobulin. Endocrinology 136:3353–3359
Hafler DA, Weiner HL (1995) Immunologic mechanisms and therapy in multiple sclerosis. Immunol Rev 144:75–107
Hammerling GJ, Schonrich G, Momberg F, Auphan N, Malissen B, Schmitt-Verhlust AM, Arnold B (1991) Non-deletional mechanisms of peripheral and central tolerance: studies with transgenic mice with tissue-specific expression of a foreign MHC class-I antigen. Immunol Rev 122:47–67
Hancock WW, Polanski M, Zhang ZJ, Blogg N, Weiner HL (1995) Suppression of insulitis in NOD mice by oral insulin administration is associated with selective expression of IL-4, IL-10, TGF-b and prostaglandin-E. Am J Pathol 147:1193–1199

Haque MA, Yoshino S, Inada S, Nomaguchi H, Tokunaga O, Kohashi O (1996) Suppression of adjuvant arthritis in rats by induction of oral tolerance to mycobacterial 65-kDa heat shock protein. Eur J Immunol 26:2650–2656

Harrison LC (1992) Islet cell antigens in insulin-dependent diabetes: Pandora's box revisited. Immunol Today 13:348–352

Harrison LC, Collier-Dempsey M, Kramer DR, Takahashi K (1996) Aerosol insulin induces regulatory CD8 gδ T cells that prevent murine insulin-dependent diabetes. J Exp Med 184:2167–2174

He YG, Mellon J, Niederkorn JY (1996) The effect of oral immunization on corneal allograft survival. Transplantation 61:920–926

Higgins P, Weiner HL (1988) Suppression of experimental autoimmune encephalomyelitis by oral administration of myelin basic protein and its fragments. J Immunol 140:440–445

Hohol MJ, Khoury SJ, Cook SL, Orav J, Hafler DA, Weiner HL (1996) Three-year open protocol continuation study of oral tolerization with myelin antigens in multiple sclerosis and design of a phase III pivotal trial. Ann NY Acad Sci 778:243–250

Hörnquist E, Lycke N (1993) Cholera toxin adjuvant greatly promotes antigen priming of T cells. Eur J Immunol 23:2136–2143

Hoyne GF, Callow MG, Kuo MC, Thomas WR (1994) Inhibition of T-cell responses by feeding peptides containing major and cryptic epitopes: studies with the Der p I allergen. Immunology 83:190–195

Husby S, Mestecky J, Moldoveanu Z, Holland S, Elson CO (1994). Oral tolerance in humans: T cell but not B cell tolerance after antigen feeding. J Immunol 152:4663–4670

Ilan Y, Prakash R, Davidson A, Jona V, Droguett G, Horwitz M, Chowdhury NR, Chowdhury JR (1997) Oral tolerization to adenoviral antigens permit long-term gene expression using recombinant adenoviral vectors. J Clin Invest 99:1098–1106

Inobe JI, Slavin AJ, Komagata Y, Chen Y, Liu L, Weiner HL (1998). IL-4 is a differentiation factor for TGF-β secreting Th3 cells and oral administration of IL-4 enhances oral tolerance in experimental allergic encephalomyelitis. Eur J Immunol, in press

Karpus WJ, Lukacs NW (1996) The role of chemokines in oral tolerance: abrogation of nonresponsiveness by treatment with antimonocyte chemotactic protein-1. Ann NY Acad Sci 778:133–144

Kaufman DI, Clare-Salzler M, Tian J, Forsthuber T, Ting GSP, Robinson P, Atkinson MA, Sercarz EE, Tobin AJ, Lehmann PV (1993) Spontaneous loss of T-cell tolerance to glutamic acid decarboxylase in murine insulin-dependent diabetes. Nature 366:69–72

Ke Y, Kapp JA (1996) Oral antigen inhibits priming of CD8+ CTL, CD4+ T cells, and antibody responses while activating CD8+ suppressor T cells. J Immunol 156:916–921

Kerlero de Rosbo N, Milo R, Lees MB, Burger D, Bernard CCA, Ben-Nun A (1993) Reactivity to myelin antigens in multiple sclerosis: peripheral blood lymphocytes respond predominantly to myelin oligodendrocyte glycoprotein. J Clin Invest 92:2602–2608

Khare SD, Krco CJ, Griffiths MM, Luthra HS, David CS (1995) Oral administration of an immunodominant human collagen peptide modulates collagen-induced arthritis. J Immunol 155:3653–3659

Khoury SJ, Lider O, Al-Sabbagh A, Weiner HL (1990) Suppression of experimental autoimmune encephalomyelitis by oral administration of myelin basic protein. III. Synergistic effect of lipopolysaccharide. Cell Immunol 131:302–310

Khoury SJ, Hancock WW, Weiner HL (1992) Oral tolerance to myelin basic protein and natural recovery from experimental autoimmune encephalomyelitis as associated with downregulation of inflammatory cytokines and differential upregulation

of transforming growth factor β, interleukin 4, and prostaglandin E expression in the brain. J Exp Med 176:1355–1364
Kim P-H, Kagnoff MF (1990) Transforming growth factor-β1 is a costimulator for IgA production. J Immunol 144:3411–3416
Kumar V, Stellrecht K, Sercarz E (1996) Inactivation of T cell receptor peptide specific CD4 regulatory T cells induces chronic experimental autoimmune encephalomyelitis (EAE). J Exp Med 184:1609–1617
Langman RE, Cohn M (1996) Terra firma a retreat from danger. J Immunol 157:4273–4276
Lehmann P, Forsthuber T, Miller A, Sercarz E (1992) Spreading of T-cell autoimmunity to cryptic determinants of an autoantigen. Nature 358:155–157
Lider O, Santos LMB, Lee CSY, Higgins PJ, Weiner HL (1989) Suppression of experimental allergic encephalomyelitis by oral administration of myelin basic protein. II. Suppression of disease and in vitro immune responses is mediated by antigen-specific CD8+T lymphocytes. J Immunol 142:748–752
Lowney ED (1968) Immunologic unresponsiveness to a contact sensitizer in man. J Invest Dermatol 51:411–417
Lundin BS, Dahlgren UIH, Hanson LA, Telemo E (1996) Oral tolerization leads to active suppression and bystander tolerance in adult rats while anergy dominates in young rats. Scand J Immunol 43:56–63
Ma C-G, Zhang G-X, Xiao B-G, Link J, Olsson T, Link H (1995) Suppression of experimental autoimmune myasthenia gravis by nasal administration of acetylcholine receptor. J Neuroimmunol 58:51–60
Marth T, Strober W, Kelsall BL (1996) High dose oral tolerance in ovalbumin TCR-transgenic mice: systemic neutralization of IL-12 augments TGF-beta secretion and T cell apoptosis. J Immunol 157:2348–2357
Matzinger P, Fuchs EJ (1996) Beyond self and non-self: immunity is a conversation not a war. J NIH Res 8:35–39
McMenamin C, Pimm C, McKersey M, Holt PG (1994) Regulation of IgE responses to inhaled antigen in mice by antigen-specific gd T cells. Science 265:1869–1871
Mengel J, Cardillo F, Aroeira LS, Williams O, Russo M, Vaz NM (1995) Anti-gd T cell antibody blocks the induction and maintenance of oral tolerance to ovalbumin in mice. Immunology 48:97–102
Metzler B, Wraith DC (1993) Inhibition of experimental autoimmune encephalomyelitis by inhalation but not oral administration of the encephalitogenic peptide: influence of MHC binding affinity. Int Immunol 5:1159–1165
Meyer AL, Benson JM, Gienapp IE, Cox KL, Whitacre CC (1996) Suppression of murine chronic relapsing experimental autoimmune encephalomyelitis by the oral administration of myelin basic protein. J Immunol 157:4230–4238
Miller A, Lider O, Weiner HL (1991) Antigen-driven bystander suppression following oral administration of antigens. J Exp Med 174:791–798
Miller A, Al-Sabbagh A, Santos L, Das MP, Weiner HL (1993a) Epitopes of myelin basic protein that trigger TGF-β release following oral tolerization are distinct from encephalitogenic epitopes and mediate epitope driven bystander suppression. J Immunol 151:7307–7315
Miller A, Zhang ZJ, Sobel RA, Al-Sabbagh A, Weiner HL (1993b) Suppression of experimental autoimmune encephalomyelitis by oral administration of myelin basic protein. VI. Suppression of adoptively transferred disease and differential effects of oral versus intravenous tolerization. J Neuroimmunol 46:73–82
Miller A, Lider O, Abramsky O, Weiner HL (1994) Orally administered myelin basic protein in neonates primes for immune responses and enhances experimental autoimmune encephalomyelitis in adult animals. Eur J Immunol 24:1026–1032
Mosmann TR, Sad S (1996) The expanding universe of T-cell subsets: Th1, Th2, and more. Immunol Today 17:138–146

Mowat AM (1987) The regulation of immune responses to dietary protein antigens. Immunol Today 8:93–98
Nelson PA, Akselband Y, Dearborn S, Al-Sabbagh A, Tian ZJ, Gonnella P, Zamvil S, Chen Y, Weiner HL (1996) Effect of oral beta interferon on subsequent immune responsiveness. Ann NY Acad Sci 778:145–155
Neurath MF, Fuss I, Kelsall BL, Presky DH, Waegell W, Strober W (1996) Experimental granulomatous colitis in mice Is abrogated by induction of TGF-b-mediated oral tolerance. J Exp Med 183:2605–2616
Nicholson LB, Greer JM, Sobel RA, Lees MB, Kuchroo VK (1995) An altered peptide ligand mediates immune deviation and prevents autoimmune encephalomyelitis. Immunity 3:397–405
Nussenblatt RB, Caspi RR, Mahdi R, Chan CC, Roberge F, Lider O, Weiner HL (1990) Inhibition of S-antigen induced experimental autoimmune uveoretinitis by oral induction of tolerance with S-antigen. J Immunol 144:1689–1695
Nussenblatt RB, Gery I, Weiner HL, Ferris F, Shiloach J, Ramaley N, Perry C, Caspi R, Hafler DA, Foster S, Whitcup SM (1997) Treatment of uveitis by oral administration of retinal antigens: results of a phase I/II randomized masked trial. Am J Ophthalmol 123:583–592
Okumura S, McIntosh K, Drachman DB (1994) Oral administration of acetylcholine receptor: effects on experimental myasthenia gravis. Ann Neurol 36:704–713
Peterson KE, Braley-Mullen H (1995) Suppression of murine experimental autoimmune thyroiditis by oral administration of porcine thyroglobulin. Cell Immunol 166:123–130
Powrie F, Carlino J, Leach MW, Mauze S, Coffman RL (1996) A critical role for transforming growth factor-beta but not interleukin 4 in the suppression of T helper type 1-mediated colitis by CD45RB (low) CD4+ T cells. J Exp Med 183:2669–2674
Rizzo LV, Miller-Rivero NE, Chan C-C, Wiggert B, Nussenblatt RB, Caspi RR (1994) Interleukin-2 treatment potentiates induction of oral tolerance in a murine model of autoimmunity. J Clin Invest 94:1668–1672
Rollwagon FM, Baqar S (1996) Oral cytokine administration. Immunol Today 17:548–550
Santos LMB, Al-Sabbagh A, Londono A, Weiner HL (1994) Oral tolerance to myelin basic protein induces regulatory TGF-b-secreting T cells in Peyer's patches of SJL mice. Cell Immunol 157:439–447
Sayegh MH, Khoury SJ, Hancock WH, Weiner HL, Carpenter CB (1992a) Induction of immunity and oral tolerance with polymorphic class II major histocompatability complex allopeptides in the rat. Proc Natl Acad Sci USA 89:7762–7766
Sayegh MH, Zhang ZJ, Hancock WW, Kwok CA, Carpenter CB, Weiner HL (1992b) Down-regulation of the immune response to histocompatibility antigen and prevention of sensitization by skin allografts by orally administered alloantigen. Transplantation 53:163–166
Seder RA, Marth T, Sieve MC, Strober W, Letterio JJ, Roberts AB, Kelsall B (1998) Factors involved in the differentiation of TGF-β-producing cells from naive CD4$^+$ T cells: IL-4 and IFN-γ have opposing effects, while TGF-β positively regulates its own production. J Immunol 160:5719–5728
Shaw MK, Lorens JB, Dalcanto R, Tse HY, Tran AB, Bonpane C, Eswaran SL, Brocke S, Sarvetnick N, Steinmann L, Nolan GP, Fathman CG (1997) local delivery of interleukin-4 by retrovirus transduced T lymphocytes ameliorates experimental autoimmune encephalomyelitis. J Exp Med 185:1711–1714
Shull MM, Ormsby I, Kier AB, Pawlowski S, Diebold RJ, Yin M, Allen R, Sidman C, Proetzel G, Calvin D, Annunziata N, Doetschman T (1992) Targeted disruption of the mouse transforming growth factor-β1 gene results in multifocal inflammatory disease. Nature 359:693–699
Sieper J, Kary S, Sörensen H, Alten R, Eggens U, Hüge W, Hiepe F, Kühne A, Listing J, Ulbrich N, Braun J, Zink A, Mitchison NA (1996) Oral type II collagen treatment in early rheumatoid arthritis. Arthritis Rheum 39:41–51

Sloan-Lancaster J, Evavold BD, Allen PM (1993) Induction of T cell anergy by altered T cell receptor ligand on live antigen presenting cells. Nature 363:156–159

Staines NA. Harper N, Ward FJ, Malmström V, Holmdahl R, Bansal S (1996) Mucosal tolerance and suppression of collagen-induced arthritis (CIA) induced by nasal inhalation of synthetic peptide 184–198 of bovine type II collagen (CII) expressing a dominant T cell epitope. Clin Exp Immunol 103:368–375

Sun JB, Holmgren C, Czerkinsky C (1994) Cholera toxin B subunit: an efficient transmucosal carrier-delivery system for induction of peripheral immunological tolerance. Proc Natl Acad Sci USA 91:10795–10799

Terato K, Xiu JY, Miyahara H, Cremer MA, Griffiths MM (1996) Induction of chronic autoimmune arthritis in DBA/1 mice by oral administration of type II collagen and escherichia coli lipopolysaccharide. Br J Rheumatol 35:1–11

Thompson HSG, Harper N, Bevan DJ, Staines NA (1993a) Suppression of collagen induced arthritis by oral administration of type II collagen: changes in immune and arthritic responses mediated by active peripheral suppression. Autoimmunity 16:189–199

Thompson SJ, Thompson HSG, Harper N, Day MJ, Coad AJ, Elson CJ, Staines NA (1993b) Prevention of pristane-induced arthritis by the oral administration of type II collagen. Immunology 79:152–157

Tian J, Atkinson MA, Clare-Salzler M, Herschenfeld A, Forsthuber T, Lehmann PV, Kaufman DL (1996) Nasal administration of glutamate decarboxylase (GAD65) peptides induces Th2 responses and prevents murine insulin-dependent diabetes. J Exp Med 183:1561–1567

Tisch R, Yang X-D, Singer SM, Liblau RS, Fugger L, McDevitt HO (1993) Immune response to glutamic acid decarboxylase correlates with insulitis in non-obese diabetic mice. Nature 366:72–75

Trentham DE, Dynesius-Trentham RA, Orav EJ, Combitchi D, Lorenzo C, Sewell KL, Hafler D, Weiner HL (1993) Effects of oral administration of type II collagen on rheumatoid arthritis. Science 261:1727–1730

von Herrath MG, Dyrberg T, Oldstone MBA (1996) Oral insulin treatment suppresses virus-induced antigen-specific destruction of beta cells and prevents autoimmune diabetes in transgenic mice. J Clin Invest 98:1324–1331

Waldo FB, Van Den Wall Bake AWL, Mestecky J, Husby S (1994) Suppression of the immune response by nasal immunization. Clin Immunol Immunopathol 72:30–34

Wang ZY, Qiao J, Link H (1993a) Suppression of experimental autoimmune myasthenia gravis by oral administration of acetylcholine receptor. J Neuroimmunol 44:209–214

Weiner HL (1997) Oral tolerance: immune mechanisms and treatment of autoimmune diseases. Immunol Today 18:335–343

Weiner HL, Mayer LF (1996) Oral tolerance: mechanisms and applications. Ann NY Acad Sci 778:1–453

Weiner HL, Mackin GA, Matsui M, Orav EJ, Khoury SJ, Dawson DM, Hafler DA (1993) Double-blind pilot trial of oral tolerization with myelin antigens in multiple sclerosis. Science 259:1321–1324

Weiner HL, Inobe J-I, Kuchroo V, Chen Y (1996) Induction and characterization of TGF-beta secreting Th3 cells. FASEB 10:A1444

Wells HG (1911) Studies on the chemistry of anaphylaxis (III). Experiments with isolated proteins, especially those of the hen's egg. J Infect Dis 8:147–171

Whitacre CC, Gienapp IE, Orosz CG, Bitar D (1991) Oral tolerance in experimental autoimmune encephalomyelitis. III. Evidence for clonal anergy. J Immunol:2155–2163

Wildner G, Thurau SR (1994) Cross reactivity between an HLA-B27-derived peptide and a retinal autoantigen peptide: a clue to major histocompatibility complex association with autoimmune disease. Eur J Immunol 24:2579–2585

Wildner G, Thurau SR (1995) Orally induced bystander suppression in experimental autoimmune uveoretinitis occurs only in the periphery and not in the eye. Eur J Immunol 25:1292–1297

Wilson AD, Bailey M, Williams NA, Stokes CR (1991) The in vitro production of cytokines by mucosal lymphocytes immunized by oral administration of keyhole limpet hemocyanin using cholera toxin as an adjuvant. Eur J Immunol 23:2333–2339
Windhagen A, Scholz C, Hollsberg P, Fukaura H, Sette A, Hafler DA (1995) Modulation of cytokine patterns of human autoreactive T cell clones by a single amino acid substitution of their peptide ligand. Immunity 2:373–380
Xu-Amano J, Kiyono H, Jackson RJ, Staats HF, Fujihashi K, Burrows PD, Elson CO, Pillai S, McGhee JR (1990) Helper T cell subsets for immunoglobulin A responses: oral immunization with tetanus toxoid and cholera toxin as adjuvant selectively induces Th2 cells in mucosa associated tissues. J Exp Med 178:1309–1320
Yoshino S, Quattrocchi E, Weiner HL (1995) Oral administration of type II collagen suppresses antigen-induced arthritis in lewis rats. Arthritis Rheum 38:1092–1096
Zhang J, Markovic S, Raus J, Lacet B, Weiner HL, Hafler DA (1993) Increased frequency of IL-2 responsive T cells specific for myelin basic protein and proteolipid protein in peripheral blood and cerebrospinal fluid of patients with multiple sclerosis. J Exp Med 179:973–984
Zhang JZ, Lee CSY, Lider O, Weiner HL (1990) Suppression of adjuvant arthritis in Lewis rats by oral administration of type II collagen. J Immunol 145:2489–2493
Zhang Z, Michael J (1990) Orally inducible immune unresponsiveness is abrogated by IFN-g treatment. J Immunol 144:4163–4165
Zivny JH, Russell MW, Vu HL, Moldoveanu J, Mestecky J, Elson CO (1996) Multiple mechanisms of oral tolerance to food antigens in humans. FASEB J 10:A1116

CHAPTER 19

Vaccines Against Allergies

L. HELLMAN

A. Introduction

A dramatic increase in the incidence of atopic (IgE mediated) allergies has occurred during the past 20–30 years. Estimates show that the allergic symptoms in 20%–30% of the population in some western countries are so severe that they constitute a medical problem. This situation has, together with better management of many of the major infectious diseases, placed atopic allergies among the major medical issues of the industrialized world. At present the traditional treatment of allergies centers mainly around limiting the effect of substances already released from the effector cells of atopic allergies, or treatment with relatively crude allergen extracts by hyposensitization therapy. However, the increased incidence and doubts about the efficacy of traditional immunotherapy have boosted the development of several novel strategies for treatment of atopic disease. Here I discuss recent advances in these novel strategies. Only some of these strategies are based on principles comparable to classical vaccines. However, since many of them represent interesting strategies for diverting or redirecting the immune system, they may be of general interest in the context of vaccine development. The major issues which are addressed are: (a) modifications of allergen extracts or purified recombinant allergens by monomethoxy-polyethylene glycol (mPEG) coupling, allergen cross-linking or immune complex formation; (b) oral administration of allergens or allergen extracts; (c) treatment with immunodominant peptides of major allergens to induce unresponsiveness (anergy) in allergen specific T cells; (d) administration of cytokines or cytokine antagonists to limit allergic sensitization or to block late phase reactions; (e) treatment with substances that block either IgE or the IgE receptor: (f, g) immune intervention aimed at depletion of circulating and mast cell bound IgE, by treatment with monoclonal antibodies or by vaccination against IgE, using selected parts of the IgE molecule coupled to a foreign carrier protein.

B. The Allergic Immune Response

The progress in our understanding of the basic mechanisms involved in regulating the immune system has led to a relatively detailed picture of the major regulatory events during both the early and late phase of an allergic reaction.

To facilitate the understanding of the underlying immunological mechanisms of the various novel treatment strategies I give here a relatively detailed description of the present view of some of these central events. A schematic drawing of these events is shown in Fig. 1.

The regulation of IgE-mediated allergies is highly T cell dependent and with only a few exceptions, such as penicillin allergies, almost all allergens are proteins. There is a fundamental difference in how T cells and B cells recognize antigen. B cells recognize surface epitopes of soluble native molecules and there is no direct restriction to the type of antigen. B cells can recognize proteins as well as carbohydrates, lipids, and other types of molecules or even combinations of various structures. In contrast, antigen recognition by T cells require that the antigen, almost exclusively proteins, is internalized digested and presented as short linear peptides in association with MHC class I or class II antigens.

One of the first events in the immune response against an allergen is the capture and processing of the allergen by antigen-presenting cells (APCs). This probably occurs immediately after the allergen has entered the tissues, for example, within the skin or the bronchial or intestinal mucosa (Fig. 1). Dendritic cells (Langerhans cells and other dendritic cells) and B cells are the most efficient APCs, while monocytes and macrophages functions as APCs to a lesser extent (KNIGHT and STAGG 1993).

T cells can be divided into two major populations depending on the way in which they recognize antigen and on their expression of certain surface molecules: cells expressing CD4 ($CD4^+$) compose the T helper population (T_H cells), while cells expressing CD8 ($CD8^+$) compose the T killer (T_K) or cytotoxic T cell (T_C) population. T helper cells recognize peptides when presented on MHC class II molecules whereas T killer cells recognize antigen presented on MHC class I molecules. Due to the structure of the peptide binding pocket, the size of the MHC class I associated peptides are relatively restricted and most likely limited to 8–10 amino acids in length (APPELLA et al. 1995; KOOPMANN et al. 1997). A larger flexibility in size is apparently allowed for peptides presented on MHC class II molecules (10–34 amino acids) (APPELLA et al. 1995; KROPSHOFER et al. 1996). Peptides with a size of up to 74 amino acids have actually been reported to bind to MHC class II molecules (CHICZ et al. 1994). With some exceptions, the general rule is that extracellular antigens are internalized and presented on MHC class II molecules, while intracellular antigens are digested by a protease complex within the cytoplasm, the proteosome, transported into the endoplasmic reticulum and presented on MHC class I molecules.

Interleukins and cell-to-cell contacts play key roles in the early events of an allergic immune response by determining whether the immune system will react via a humoral or cell-mediated immunity. This process is regulated by the preferential activation of either of two specific subsets of T helper cells, type 1 (T_H1) or type 2 (T_H2) subsets (Fig. 1) (ABBAS et al. 1996). Interleukin-12 (IL-12) and interferon-α (IFN-α) seem to be the most important cytokines in

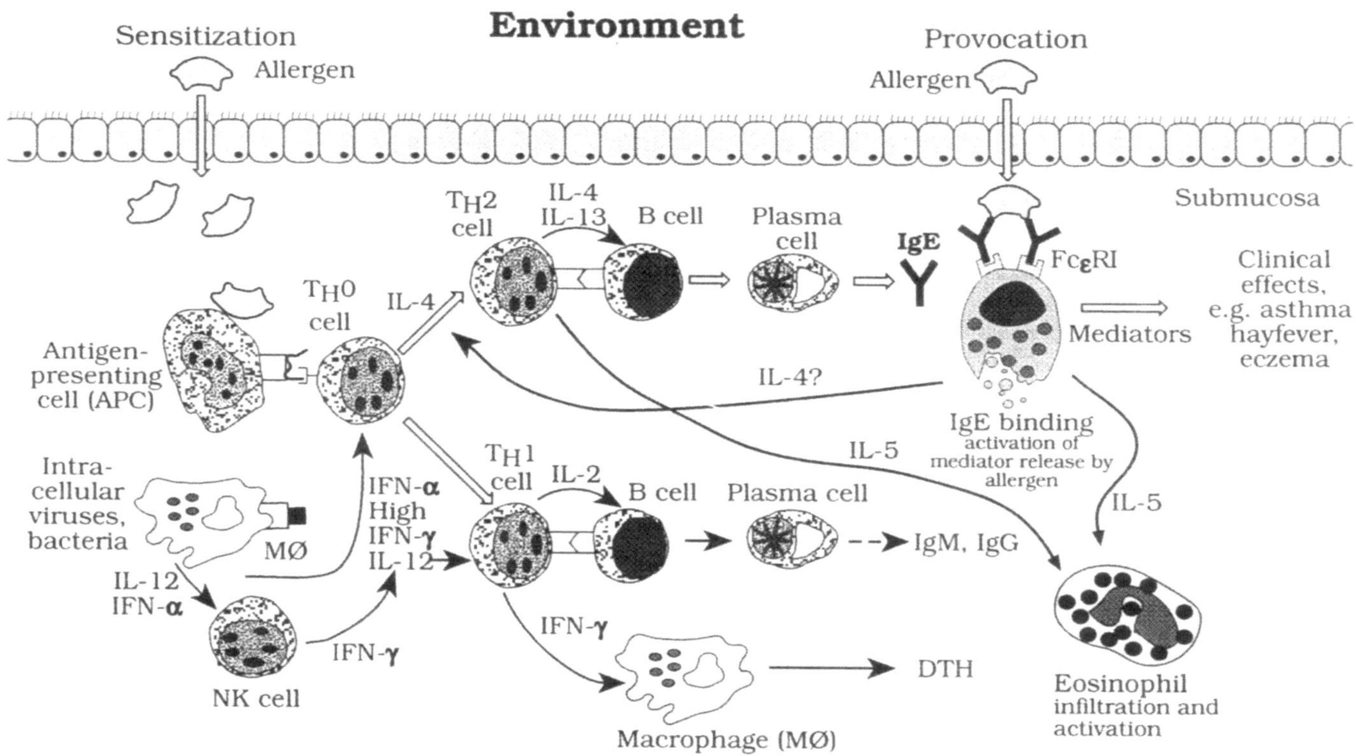

Fig. 1. The allergic immune response. *FcεRI*, High-affinity receptor for the Fc portion of IgE; *IFN*, interferon; *IL*, interleukin; T_H, T helper

activating T_H1 cells. Activated T_H1 cells produce a restricted number of cytokines, mainly IL-2 and IFN-γ but also TNF-β (Fig. 1) (Abbas et al. 1996). Macrophages infected with intracellular parasites are the major producers of IL-12, and most virus infected cells produce IFN-β and IFN-α. A highly specialized IFN-α producing cell type is also present in the blood, the natural interferon producing (NIP) cells (Fitzgerald-Bocarsly 1993). The production of IL-2 and IFN-γ by activated T_H1 cells leads to preferential activation of cell-mediated immunity through activation of T killer cells (by IL-2), and activation of macrophages (by IFN-γ; Fig. 1). However, the T_H1 cytokines are also involved in regulation of B cell isotype switch to certain isotypes such as IgM and IgG2a in the mouse. In addition, IFN-γ is of importance for down regulating a T_H2 type response (Fig. 1).

Activation of T_H2 type helper cells induces production of a different set of cytokines namely IL-4, IL-5, and IL-13 (Fig. 1) (Abbas et al. 1996). Many of these interleukins are important for the development of a humoral response, for example, by strongly influencing the isotype switch in B lymphocytes. Both IL-4 and IL-13 are important switch factors for IgE in human B cells (Fig. 1). In contrast, IL-13 does not seem to have this particular effect in rodents. IL-4 also has an important role in inducing differentiation of the early uncommitted T_H0 cell to a T_H2 phenotype. The presence of small amounts of IL-4 during the initial encounter with antigen preferentially induces a T_H2 response (Fig. 1) (Abbas et al. 1996; Breit et al. 1996). During an immune response small amounts of IL-4 and several other cytokines have been detected within a few hours after antigen contact. T cell depletion does not seem to influence this cytokine production, indicating that other cell types may be involved. Both mast cells and basophils have been shown to be potent producers of IL-4, suggesting that these cell types may contribute to the induction of a T_H2 response and to the switch to IgE (Bradding et al. 1992; Arock et al. 1993; Brunner et al. 1993). Interestingly, basophils also express the CD40 ligand and can induce IgE switch in vitro without addition of exogenous IL-4 (Gauchat et al. 1993). The actual role for both mast cells and basophils in the regulation of T_H1 versus T_H2 responses in vivo has, however, not yet been fully elucidated (Fig. 1).

IL-10 is an anti inflammatory cytokine which has been shown to be produced by both T_H1 and T_H2 cells and also by macrophages. IL-10 has an important role in down regulating T_H1 responses (Abbas et al. 1996), and has been shown to induce long term antigen specific anergic states in human $CD4^+$ T cells (Groux et al. 1996). IL-10 most likely also regulates the T cell population by negatively affecting the IL-12 production from monocytes and macrophages.

The strict division of helper cells into three subtypes have been very important for our understanding of the regulatory processes involved in the immune response. However, this picture is probably an oversimplification of the actual process in vivo. A more accurate description would probably involve several intermediate types of T helper cells expressing other cytokine combinations. However, the overall pattern of cytokines produced during a certain immune reaction often has a clear T_H1 or T_H2 profile.

The preferential regulation of a T_H1 or T_H2 response does not seem to be dependent exclusively on the cytokine environment, but also on the type and the concentration of the antigen. Generally low concentrations of soluble monomeric antigens as well as low-affinity antigens seem to favor T_H2 responses, while peptides with high-affinity interactions with MHC antigens and multimeric antigens preferentially induce T_H1 responses. Thus it is most likely that both the cytokine environment and the type of antigen are of major importance for the initial sensitization phase during an allergic immune response. Genetic factors such as the tendency to produce IL-4 and IL-12 during the initial antigen encounter are probably of importance when determining whether an individual is at high risk of developing an atopic allergy or not. In addition, environmental factors such as stress level (cortisol level), antigen load, the presence of substances which irritate skin and bronchi (e.g., formaldehyde) and the extent of T_H1 driving infections (e.g., mycobacterial infections) are most likely of central importance in determining whether an individual actually will develop allergic symptoms (Shirakawa et al. 1997; Cookson and Moffatt 1997).

The cytokine environment in local tissues in contact with antigen and in the draining lymph nodes affects the switching pattern of B cells that recognize surface epitopes of the allergen. If IL-4 and IL-13 are present, a substantial fraction of the B cells switch to IgE production and mature into IgE producing plasma cells (Fig. 1). The secreted IgE then diffuses into the tissue and binds to the high-affinity receptors for IgE expressed on the surface of mast cells and basophils. However, a fraction of the IgE remain as free circulating plasma IgE. IgE production, and increased reactivity in the local tissue has been observed in the nasal mucosa of patients with hayfever, indicating that there may be differences in the mast cell reactivity to different allergens in various tissues depending on the site of allergen encounter (Durham et al. 1997). Plasma IgE has the relatively short half-life of 48 h or less, whereas receptor bound IgE is relatively stable, a half life of 14 days or more (Fig. 1). Depending on the cell type mast cells and basophils have 30000–300000 IgE receptors on their surface. The concentration of IgE in the circulation has been shown to influence the number of receptors expressed by mast cells and basophils (MacGlashan et al. 1997a,b; Shaikh et al. 1997). Approximately 220000 receptors have been detected on basophils in normal subjects. The receptor number decrease to approximately 8500 as a result of depletion of circulating IgE (MacGlashan et al. 1997a,b). A large fraction of the IgE produced in response to a particular allergen binds to the IgE receptors present in the local tissue and probably remain in the tissue as cell bound IgE. The individual has now become sensitized to this particular allergen. Upon a second encounter with the same allergen, cross-linking of two or several IgE receptors occurs. This cross-linking triggers release of preformed material stored in cytoplasmic secretory granules of mast cells and basophils, including histamine, heparin, and several serine proteases and cytokines. In addition, specific enzyme systems become activated which lead to the production of leukotrienes and prostaglandins (Fig. 1). Cross-linking of only a small number (two to ten) of the possibly 300 000 IgE high-affinity receptors is sufficient to trigger mast cells to release their entire granule content. This system is thus an extremely potent amplifier of inflammation, triggered by antigens present in very small amounts, for example, certain pollen allergens.

The competition between antigen specific IgG and IgE in the involved tissue is likely to determine the magnitude and type of the inflammatory response. Large amounts of IgG result in a milder immune complex mediated inflammation with a limited mast cell contribution. A limited mast cell degranulation takes place due to activation of mast cells by IgE-nondependent mechanisms, for example, through anaphylotoxins and neurogenic activation by the action of substance P. In contrast, high IgE concentrations lead to massive mast cell activation, which gives rise to an anaphylactic shock if the antigen is distributed systemically.

The released mast cell material initiates an inflammatory reaction and attracts other immune cells to the area: primarily eosinophils, sometimes neutrophils, and if the inflammation continues into a more chronic phase, T cells and macrophages (Fig. 1). T cells are probably very important in the more chronic phase where IL-5 produced by T cells (produced also by mast cells and eosinophils), activates and attracts eosinophils to the inflamed tissue (YAMAGUCHI et al. 1988a,b). This pattern of reactions occurs during most types of IgE mediated allergies, for example, reactions to pollen, cats, dogs, dust mites, and many food allergens. In addition, extrinsic asthma is an IgE mediated reaction. However, in the case of intrinsic asthma, no clear evidence for the involvement of IgE has been found; therefore T cells are most likely the major immune cells involved in this type of inflammatory disorder.

Based on the present knowledge of the various regulatory events during the early and the late phase of an allergic reaction, several potential levels of immune intervention have emerged. Some strategies are presently used in allergy therapy while others are in the phase of pre clinical or at the initial levels of clinical studies. A schematic drawing of these potential levels of intervention is shown in Fig. 2.

Allergen avoidance is one very important, although often neglected factor in reducing symptoms, which is very important in every day practice (Fig. 2). However, in many situations it is not possible to completely avoid contact with the allergen and even very small concentrations can have fatal consequences if proper medication is not obtained.

Most of the present methods for management of atopic allergies are based on attempts to reduce the effects of histamine, leukotrienes, and prostaglandins already released from mast cells or basophils, for example, the use of antihistamines, antileukotrienes and β_2-adrenergic agonists or more general anti-inflammatory drugs such as corticosteroids (Fig. 2). However, several different approaches with the aim of interfering with allergen induced activa-

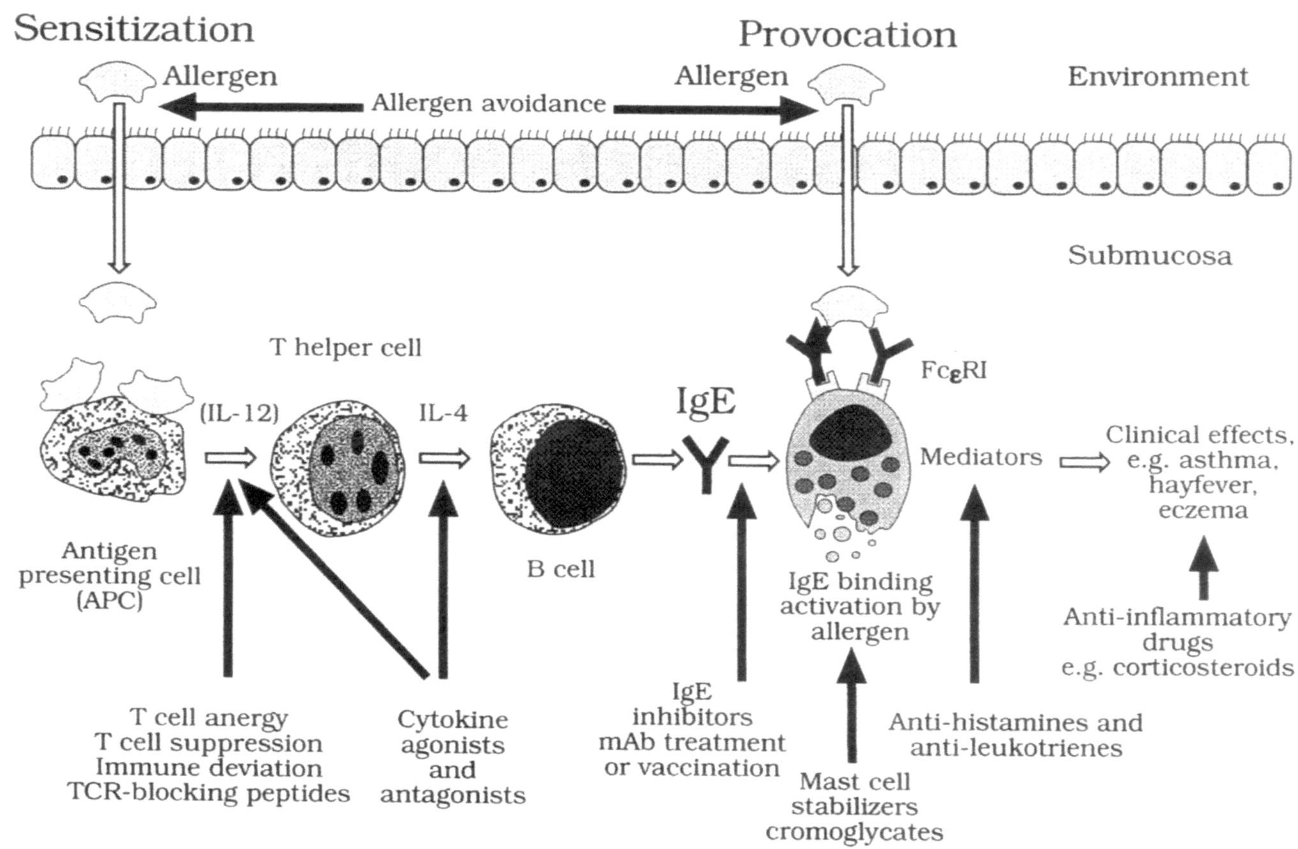

Fig. 2. Important levels of intervention in the allergic immune response. *FcεRI*, high-affinity receptor for the Fc portion of IgE; *IL*, interleukin; *TCR* T cell receptor

tion of mast cells at a stage prior to the release of granule stored material are emerging.

Traditional immunotherapy by hyposensitization as well as several of the novel strategies, described in more detail in this review, are centered around very early events in the allergic sensitization. Novel approaches to traditional hyposensitization involves modifications of allergens or allergen extracts by cross-linking, immune complex formation or by mPEG coupling, as well as the use of recombinant allergens and the possibility of using bacteria expressing the specific allergen as vaccine carriers for oral vaccines (Fig. 2).

Furthermore, several novel strategies in progress are aiming at interfering with the activation of specific T cell populations or with the isotype switching event. For example, the peptide based strategies are aimed at interfering with the activation of the T_H0 and T_H2 populations. Non responsiveness (anergy) is induced in the target T cell population by administration of large amounts of immunodominant T cell epitopes from dominant allergens (Fig. 2). In addition, strategies involving the use of cytokines or cytokine antagonists interfere at the T cell or B cell levels of sensitization and thereby divert or block the early events of the sensitization phase (Fig. 2).

Other novel strategies of immune intervention is aimed at the interaction of IgE with its high-affinity receptor, by using either low molecular weight compounds that interfere directly with the interaction or by directly targeting the IgE molecule (mast cell bound and circulating IgE) for depletion (Fig. 2).

C. Traditional Immunotherapy

Subcutaneous injections of progressively increasing doses of crude allergen extracts has been used in the treatment of atopic allergies since the beginning of this century (FREEMAN 1911; NOON 1911). Although there are many doubts regarding the efficacy of conventional immunotherapy (hyposensitization), several well-controlled double-blind studies have demonstrated reduced severity of the clinical symptoms in atopic allergy. The most convincing results originate from studies with insect venom hypersensitivities where selected studies show an efficacy as high as 95% (NAKAGAWA and GERSHWIN 1993). Positive results have also been documented for cat and house dust mite allergens and several seasonal inhaled allergens (OHMAN 1992). The results are, however, less convincing for many other IgE mediated allergies (NELSON et al. 1993; CRETICOS et al. 1996). The varied efficacy, together with the long duration of the treatment that is needed to reduce symptoms (6–12 months), limits the applicability.

In addition, positive results have only been obtained with high-dose immunotherapy, a situation which further limits its applicability due to the high sensitivity to the allergen found in most patients with severe atopic allergies. Subcutaneous injections with even small amounts of unmodified allergens can lead to severe adverse reactions, ranging from local erythrema and pain to more systemic reactions such as urticaria, asthma and anaphylactic shock. According to reports from the United States Food and Drug administration an average of two persons die in the United States per year from anaphylactic shock caused by allergen immunotherapy or skin testing (FDA NOTES 1988) and the total number of patients severely affected is probably much higher. Due to the severe adverse reactions, a very strict procedure is recommended for immunotherapy in the United Kingdom (NAKAGAWA and GERSHWIN 1993). The Committee on Safety of Medicines in the United Kingdom recommends that immunotherapy be administered only in facilities with full cardiopulmonary resuscitation capabilities, and that the patient be observed for at least 20 min following the injection of the allergen extract (NAKAGAWA and GERSHWIN 1993). Depending on the difficulties connected with traditional immunotherapy there are marked differences between different countries with respect to the number of patients being treated with this type of therapy. However, immunotherapy has clearly been demonstrated to be effective for seasonal pollenosis. It is widely used in the United States by specialists and is, with careful administration, quite safe.

The mechanisms underlying the positive effects observed after treatment by traditional immunotherapy are not fully understood. However, an increase in IgG is considered as one of the most important factors. This conclusion is based on the finding that an increase in the levels of allergen-specific IgG is often correlated with reduced symptom scores in the patient (GOLDEN et al. 1992). Following intravenous injections of IgG antivenom in honey-bee venom sensitive patients, protection against systemic reactions have also been observed (MÜLLER et al. 1986). However, improvement in symptoms is often correlated neither with increases in the concentration of antigen specific IgG nor with the decrease in the concentration of antigen specific IgE (OHMAN 1992). It is therefore possible that other factors than an increase in antigen specific IgG are more important for the positive effects observed in many patients. Several recent studies have addressed this question and the results indicate that a shift in cytokine profile, with decreasing IL-4 and IL-5 and increasing or unaffected levels of IFN-γ and IL-10, is important for the observed clinical effects (DURHAM 1995; JUTEL et al. 1995; BELLINGHAUSEN et al. 1997; SÖDERLUND et al. 1997).

I. Modified Allergens or Allergen Extracts

Since successful traditional immunotherapy usually requires high doses of antigen and therefore has potential severe side effects, the use of this type of therapy is often prevented in patients with severe allergies. With the aim of improving traditional immunotherapy, several modifications of allergens and allergen extracts have been made and their effect on the outcome of therapy have been carefully studied.

The desired effect of the modifications is to reduce the mast cell activating properties but still maintaining the epitopes important for T cell recognition, and the aim of such modifications is mainly to redirect the immune response to a T_H1 IgG response and to inactivate sensitized mast cells. However, the possibility of inducing allergen specific tolerance is also of great interest.

Modification of allergens or allergen extracts involve either aggregation or shielding of the allergen. One of the most intensively studied modifications is shielding of the allergen by mPEG, work pioneered by SEHON and coworkers (1982). After mPEG addition the mast cell bound IgE may not recognize the allergen, or alternatively the allergen may not have the capacity to bind more than one receptor bound IgE molecule. This type of modification makes it possible to increase the amount of allergen that can be injected without causing severe side effects. Clinical trials with mPEG conjugated allergen extracts have verified that an increase in the amount of modified allergen compared with native antigen is tolerated by the patient during immunotherapy without significant adverse effects (DREBORG and ÅKERBLOM 1990). However, the clinical benefit of this approach has not yet been proven superior to that of conventional immunotherapy (DREBORG and ÅKERBLOM 1990). However, experiments in rats have shown that passive cutaneous anaphylaxis caused by intravenous challenge with allergen injection can be effectively inhibited by preceding injections in the skin of mPEG conjugates of the same allergen (LANG et al. 1997). Similar experiments in dogs have shown that the blocking effect of mPEG conjugated allergen can be achieved in this species as well (LANG et al. 1997).

The immunological effects of aggregation of an allergen or allergen extract have been studied by HayGlass and coworkers. A study of glutaraldehyde polymerized (aggregated) allergen has shown that high molecular weight polymers can modify interleukin production, leading to preferential induction of IFN-γ production (HAYGLASS and STEFURA 1991a,b; GIENI et al. 1993; YANG et al. 1993). The increase in IFN-γ favors cell mediated T_H1 immunity and resulted in a relatively long-lasting (more than 260 days) down regulation of the IgE response in some strains of mice (GIENI and HAYGLASS 1991; HAYGLASS and STEFURA 1991). However, no effect was seen in the IgE high responder strain BALB/c (GIENI and HAYGLASS 1991). Recent in vitro studies of antigen specific T cells isolated from mice injected with polymerized or unpolymerized antigen, showed that the polymerized antigen gave rise to 10 fold higher IFN-γ levels and five fold lower IL-10 levels, while the IL-4 production was unaffected. However, two forms of antigen elicited T cell activation of comparable intensity (GIENI et al. 1996). Although somewhat limited to certain strains of animals, this type of modification thus seems to influence how the immune system recognizes the allergen and thereby the outcome of the vaccination.

Saint-Remy and associates have during a number of years studied an approach involving immune complexes of allergen and allergen specific IgG to redirect the immune response. Treatment over several years, of more than 250 patients with chronic asthma or atopic dermatitis has shown significant clinical improvement for these patients (MACHIELS et al. 1990; SAINT-REMY 1994). The beneficial effect was also shown to be maintained for a long time after the discontinuation of the therapy. The immunological basis for the positive effect of immune complexes most likely lies in altered antigen presentation by changing the target cell for antigen presentation. Immune complexes probably preferentially target antigens to B cells and macrophages and to a lesser extent to dendritic cells, possibly resulting in a T_H1 cytokine profile during the initial contact between antigen and the naive T cell (THEPEN et al. 1991). Differences in antigen processing may also be of importance for the observed effects (MANCA et al. 1988, 1991).

The various types of allergen modifications described above have given very interesting primary results. However, further experimentation is needed to determine their role in future treatment programs.

II. Oral Administration of Recombinant Allergens or Allergen Extracts

The administration route of the allergen has been shown to be of importance for the type of immune response elicited in the patient during immunotherapy. By changing the site of allergen entry of the allergen, for example, substantial reduction the adverse effects in high-dose immunotherapy has been obtained by oral administration of allergens or allergen extracts. In addition, several studies indicate that oral administration may induce tolerance to that particular allergen (HANSON 1981; HOYNE and LAMB 1997; MOWAT 1987). However, the different studies performed to date have led to quite different views concerning the clinical benefit of this kind of therapy (TAUDORF et al. 1985, 1987, 1994; OPPENHEIMER et al. 1994).

Compared to traditional subcutaneous administration of allergen, about 100-fold higher amounts of the allergen are needed to obtain significant clinical effect by oral immunization, which in most cases probably reduces its usefulness for economical reasons (SAINT-REMY 1994). However, the recent cloning of a large number of gene sequences encoding major allergen proteins may solve this potential problem (VALENTA et al. 1998). Expression of the allergens as recombinant proteins in apathogenic bacteria is a very economical way of producing large amounts of antigen. Such bacterial strains may also have the potential of being used as live vaccines administered to the patient or to young children as a food additive. A recent study by VRTALA et al. (1995) has shown that feeding mice with Bet vI (the major birch pollen allergen) expressed in apathogenic *Salmonella* strains, induces a predominant T_H1 type response against the allergen, accompanied by specific IgG2a/b, without detectable IgG1 or IgE.

Different strategies for oral immunization, possibly involving live vaccines are thus of potential importance and deserve to be studied in more detail for clinical usefulness (VALENTA et al. 1993).

III. Peptide Vaccines

Due to the high risk of severe side effects connected with traditional immunotherapy the approach of digesting allergens or allergen extracts into peptides with pepsin has been tested as a conceptually new strategy for treatment of atopic allergies. Such peptide mixtures have the advantage of displaying up to 1000-fold reduced cross-linking activity on mast cells, although maintaining the capacity of reducing the levels of antigen specific IgE and IgG in the patient (LITWIN et al. 1988). Due to major difficulties in obtaining substantial amounts of the immunodominant peptides and problems with reproducibly being able to produce identical peptide mixtures derived from native antigens, most studies are currently performed with synthetic peptides. Synthetic peptides are more easily obtained in relatively large quantities and in a pure form, which is a prerequisite for a well controlled treatment program. The rationale behind this type of treatment is the induction of allergen specific T cell unresponsiveness, a phenomenon called "anergy" (O'HEHIR et al. 1991a,b; NAKAGAWA and GERSHWIN 1993; SCHWARTZ 1993). During induction of T cell responses the interaction between the peptide-MHC complexes on the surface of APCs and the TCRs on the surface of the T cells (signal 1) is insufficient for activation of naive T cells. It has been shown that the CD28 molecules on T cells must interact with B7-1 or B7-2 molecules on APCs to deliver a second signal required for the activation of T cells (LANZAVECCHIA 1993; ABBAS et al. 1996). Activation of T cells by APCs which lack B7, or interaction of CTLA-4 on the T cells with B7-1 or B7-2 leads to a functional inactivation of the T cell (PEREZ et al. 1997). Administration of large amounts of peptide has been shown to induce tolerance (LAMB et al. 1983). The high peptide concentration probably result in that many T cells encounter the antigenic peptide on the surface of nonprofessional APCs lacking the expression of B7-1/-2. The T cells thereby become anergic, a state which has been shown to be relatively stable even if professional APCs are added to the cultures after the initial exposure to the peptide (LAMB et al. 1983). The state of anergy can, however, be relieved by exposure to IL-2. In addition, down regulation of CD28 has been reported to occur as a consequence of induction of the anergic state (LAKE et al. 1993).

The experimental models of peptide-induced anergy were initiated by in vitro studies of human peripheral blood lymphocytes using peptides from *Staphylococcus aureus* enterotoxins, influenza hamaglutinin and a house dust mite allergen (*Der pI*) (O'HEHIR and LAMB 1990; O'HEHIR et al. 1991a,b). Supra optimal concentrations of an immunodominant epitope from the *Der pI* allergen was found to decrease T cell proliferation even in the presence of APCs. During restimulation of the unresponsive allergen specific T cells by antigen challenge these cells failed to induce secretion of IL-2 and IL-4 while the secretion of IFN-γ was maintained (O'HEHIR et al. 1991).

A research group at the Immulogic Corporation (United States) has cloned, sequenced and mapped immunodominant epitopes of the major allergen from cats, *Fel dI* (MORGENSTERN et al. 1991; ROGERS et al. 1993; COUNSELL et al. 1996). In an in vivo model using the mice strains B6CBAF1 ($H2^{bxd}$) and B6D2F1 ($H2^{bxa}$) it was shown that peptides comprising the *Fel dI* immunodominant epitopes could induce tolerance against the native protein (BRINER et al. 1993). It was also demonstrated that antigen challenge by intravenous administration of the peptide during an ongoing immune response in *Fel dI*-primed animals could induce tolerance.

Clinical trials with two peptides of 27 amino acids (Allergvax Cat), corresponding to immunodominant epitopes of the major cat antigen has showed that intradermal injections of doses up to 1500 μg were tolerated without any severe adverse effects (NORMAN et al. 1994). After establishing the safety of the peptides, subsequent clinical studies by double-blind placebo-controlled human trials were performed. The initial results indicated significantly reduced symptom scores and that the improvement was correlated with the dose of peptide given (NORMAN et al. 1995, 1996, 1997). However, for unknown reasons the placebo group also showed a considerable improvement. In addition, recent results from clinical trials performed with the same peptides have not been able to confirm these results. Subjects receiving peptide immunotherapy did not improve either on the dose of cat extract tolerated in skin tests or in the magnitude of the late phase responses elicited by the extract (SIMONS et al. 1996).

The difference in results obtained from different clinical trials may lie in the difficulty connected with this type of strategy, namely that the treatment is limited to one or a small number of immunodominant epitopes. Due to large variation in the peptide binding regions of MHC molecules of different MHC haplotypes, the immunodominant epitope(s) from a particular allergen usually differs markedly from one individual to another. Thus a peptide may be effective in the treatment of one individual while the same peptide, due to inability to interact with the MHC is without therapeutic effect in another individual. Hence a single immunodominant peptide is therefore useful only in the treatment of single haplotypes, out of many hundred possible haplotypes, within the human (outbred) population. At least theoretically this is a major limitation of this strategy. The fact that individuals that are allergic, for example, to cats normally also are allergic to several other allergens makes the treatment of most atopic allergies even more complicated. A number of immunodominant peptides must be identified for each of the major allergens, and the peptides that are recognized by each individual MHC haplotype must be identified.

The strong dependence on MHC haplotype for allergen recognition has been well documented in many independent studies. One particularly interesting example is the use of class II MHC binding hapten conjugates in downregulation of poison ivy/oak induced contact sensitivity (GELBER et al. 1997). Covalent linkage of 3-pentadecyl-cathecol to the 7th position in a $I\text{-}A^k$ binding peptide resulted in a strong T cell response after in vitro stimulation and an induction of anergy after intravenous administration of the same peptide in the I-A^k expressing strains of mice, but not in strains with other MHC haplotypes (GELBER et al. 1997).

At present peptide based strategies aimed at redirecting or anergizing T cell responses seem to have their major potential in treatment of T cell mediated diseases where the number of antigenic epitopes are highly limited. Contact hypersensitivity to single low molecular weight antigens, and possibly also autoimmune disorders in response to a limited number of antigenic epitopes constitute the most promising examples for a future role of peptide based strategies in clinical treatment. However, in the majority of atopic allergies the immunodominant epitopes of an allergen differs markedly between individuals. Thus the task to cover even a minor fraction of the population is very limited. However, progress in solving technical problems involved in determining the immunodominant epitopes of different allergens may in the future quite dramatically change the possibility of using immunodominant peptides in the treatment of atopic allergies.

A more detailed review on epitope based strategies has recently been published (VAN NEERVEN et al. 1996).

IV. Cytokine Agonists and Antagonists

The use of cytokines and cytokine antagonists compose an interesting potential strategy in future treatment of atopic allergies. As mentioned above, IL-4 and IL-13 are key factors in the induction of IgE synthesis. In addition, IL-4 is directly involved in stimulating naive T cells into the T_H2 phenotype. Further, recent in vitro experiments have shown that IL-4 is almost essential for the development of T cells secreting high amounts of IL-4 (BREIT et al. 1996). IL-4 may thereby be an ideal target for immunotherapeutic intervention, provided that no adverse effects on hematopoiesis arise. A high-affinity IL-4 antagonist has been developed by a single amino acid substitution in the activation domain of the IL-4 molecule (KRUSE et al. 1993). This IL-4 antagonist has been shown to inhibit both IL-4 and IL-13 induced IgG4 and IgE secretion from human B cells in vitro (AVERSA et al. 1993). In order to investigate the in vivo effects of treatment with this IL-4 antagonist, a mouse analogue to the antagonist was recently developed (GRUNEWALD et al. 1997). This molecule, which completely inhibits IL-4 induced cell proliferation, differentiation and signal transduction can serve as an attractive tool in studying the therapeutic potential of the IL-4 antagonist using a mouse system. Recent in vivo experiments in mice has shown that this IL-4 antagonist efficiently inhibits IL-4 dependent sensitisation to an allergen in vivo (GRUENEWALD et al. 1998). These very promising results indicate that IL-4 antagonist may also function to inhibit ongoing sensitisations in man. No results has, however, been presented on the efficacy of this antagonist in reducing levels of already pre-existing allergen specific IgE. Most patient contact a physician when their levels of antigen specific IgE already are high, why this question needs to be addressed in detail before more definitive conclusions can be drawn concerning the use of IL-4 antagonists in clinical treatment.

IL-12 is, due to its T_H1 potentiating activity, another molecule of potential interest in treatment of atopic allergies. Recent immunization experiments have shown that IL-12 administered together with the antigen can redirect an ongoing T_H2 response to a T_H1 dominated response in C57Bl/6 mice (REMPEL et al. 1997). Administration of IL-12 together with ovalbumin in bulk cultures resulted in elevated serum IFN-γ levels, enhanced IFN-γ synthesis, and inhibited IL-4 synthesis. In addition, IL-12 inhibited antigen specific IgE and IgG1 responses by 80%–99% and enhanced IgG2a synthesis 15- to 20-fold. However, each of these effects were highly transient and therefore failed to give a long-lasting impact on OVA-specific T and B cell responses (REMPEL et al. 1997). A similar shift in T cell cytokine profile and in antibody isotype pattern was seen in a recent study of the immune response against tetanus toxin after peritoneal administration of IL-12 in mice (MARINARO et al. 1997). In addition, oral IL-12 administration resulted in an almost identical effect on the Ig isotype pattern and cytokine expression as in the case of peritoneal administration except for the level of secretory IgA, which was not affected. This in contrast to intraperitoneal injections which resulted in reduced secretory IgA levels. No detectable increase in serum IL-12 was observed after oral administration, indicating that oral delivery only affected lymphocytes in the Payer's patch region (MARINARO et al. 1997).

IL-5 is a central cytokine in the recruitment and activation of eosinophils during the allergic late phase reaction (Fig. 1). Consequently, IL-5 is an additional potential target for therapeutic intervention (YAMAGUCHI et al. 1988a,b). Studies of the effect on helminth induced eosinophilia has shown that the infiltration of eosinophils can be blocked by injection of an anti IL-5 monoclonal antibody (COFFMAN et al. 1989; PARSONS et al. 1993). The pharmaceutical companies, Schering Plough and SmithKline Beecham are currently developing an anti-IL-5 antibody for human trials. Results from animal studies, including primates, indicate that treatment with anti-IL-5 monoclonal antibodies can reduce both the infiltration of eosinophils and prevent airway constriction (VOGEL 1997).

These strategies described above using cytokines or cytokine antagonists are presently being assessed for their potential use in clinical treatment of atopic allergies. Although the primary results are promising, it is too early to predict their role in future treatment programs.

V. Low Molecular Weight Compounds Interfering with the Interaction Between IgE and its High-Affinity Receptor

In several laboratories attempts are presently being made to use various low molecular weight compounds which interfere with the interaction between IgE and its high-affinity receptor on mast cells and basophils. However, only a few reports on such strategies have yet appeared in the literature. Initial studies of the receptor binding region of IgE, indicated that a 76 amino acid peptide derived from the N terminal region of the CH3 domain of IgE blocked the interaction between native IgE and its high-affinity receptor (HELM et al.

1988, 1989). However, no results describing attempts to reduce the size of this peptide or attempts to stabilize it have yet been presented.

At least two additional studies has been published on low molecular weight substances which interfere with or block the interaction between IgE and its high-affinity receptor. By using the available information on the primary structure of the receptor binding region of IgE, and by modeling this structure on the three dimensional structure of the second Ig-like domain of CD2, which show high structural similarity to the IgE binding region of the IgE receptor α chain, McDonnell et al. have designed two dodeca peptides which bind with relatively high affinity to human IgE (MCDONNELL et al. 1996). In order to stabilize these peptides they were cyclized by disulfide bonds between terminal cysteine residues. Further, D-amino acids were used in one of these peptides to increase its half life in plasma. These two peptides were shown to produce a dose-dependent inhibition of hexosamidase (a granule associated mast cell enzyme) release from RBL cells. At the highest peptide concentrations tested (200 μM), the cellular degranulation process was virtually ablated (a small proportion compared to almost 100% using an unrelated control peptide) (MCDONNELL et al. 1996). Using an in vitro system for systematic evolution by exponential enrichment WIEGAND et al. have developed several oligonucleotide ligands which bind with high specificity and relatively high affinity to human IgE. These oligonucleotides were shown to specifically block IgE mediated seretonin release from RBL cells in vitro (WIEGAND et al. 1996). However, both the oligonucleotides and the previously described peptide ligand bind IgE with an affinity several orders of magnitude lower than the interaction between native IgE and its high affinity receptor, which may cause significant problems in inhibiting IgE from binding to the receptor in vivo (HELM et al. 1997). Due to practical and economical reasons connected with regular injections in the patient of the relatively high concentrations of such compounds to efficiently compete with IgE for binding to the receptor this may also limit the applicability of this type of treatment.

A more detailed review on peptide ligands blocking the interaction between IgE and its high affinity receptor has recently been published (HELM et al. 1997).

VI. Depletion of Plasma and Mast Cell Bound IgE by Treatment with Monoclonal Anti-IgE Antibodies

In the following sections two strategies that aim at reducing the entire pool of plasma IgE are presented. This can be achieved either by the use of monoclonal antibodies (mAbs) directed against the receptor binding region of IgE, or by induction of an anti-IgE response by vaccination with specific regions of the constant domain of IgE coupled to a foreign carrier protein. The advantage of these approaches is that they are not limited to a specific allergen and can thereby, at least in theory, be used to treat practically all IgE mediated diseases. In addition, these strategies can be used in the treatment of patients sensitive to multiple allergens and in the treatment of allergies caused by allergens originating from different allergen sources.

The former strategy is based on injection of purified monoclonal anti-IgE antibodies in the patient. To avoid anaphylactic reactions the mAb should interact with the receptor binding region of IgE, without interacting with, and cross-linking, IgE already bound to the surface of the mast cell. The receptor binding region of IgE has been mapped to the N terminal region of the CH3 domain (HELM et al. 1988, 1996, 1997; PRESTA et al. 1994).

The mAbs used in this type of strategy are produced in large scale and are carefully analyzed for their chemical and biological activities. To avoid immune reactions against these antibodies, the mAb, usually originating from mouse can be "humanized." The humanization of a mouse mAb is usually achieved by isolation of cDNAs encoding the desired mouse antibody, and subsequent transfer of the whole V region, or only fragments encoding the complementarity-determining regions, to the framework of a human IgG antibody by genetic engineering (KOLBINGER et al. 1993; PRESTA et al. 1993). This "humanization" procedure reduces the risk of an immune response against the antibody upon repeated injections in a patient. Such a response can otherwise lead to immune complex disease, which in its most severe form can result in kidney failure and death.

The strategy of using monoclonal anti-IgE antibodies for treatment of human allergy patients is primarily being exploited by a collaborative effort between three different companies, Tanox Biosystems (United States), Novartis (Switzerland), and Genentech (United States). Several high-affinity mAbs have been isolated and humanized by investigators from these companies (KOLBINGER et al. 1993; PRESTA et al. 1993). One of the isolated anti-IgE antibodies has been reported to have a k_d of 7.4×10^{-10} M, which is close to or even greater than the affinity of IgE for its high-affinity receptor (DAVIS et al. 1993). A high affinity is essential in order to obtain an efficient depletion of IgE, since the anti-IgE-antibodies must compete with the high-affinity IgE receptor for binding to circulating IgE. At least one of the humanized mAbs has been analyzed for toxicity. This mAb was shown to have little or no antigenicity and no toxicity, even after injection of very large amounts of the antibody (FROEHLICH et al. 1995). Intravenous injections of this mAb in cynomolgus monkeys reduced circulating IgE to undetectable levels (BATES et al. 1995), without any reduction in allergen specific skin reactivity during a short term study (FROEHLICH et al. 1995). A natural explanation for this apparent discrepancy may be that long-term treatment is required to reduce the amount of mast cell bound IgE to sufficiently low levels to achieve measurable clinical effects. However, studies in mice with a nonanaphylogenic anti-mouse IgE antibody showed that antigen-induced skin reactions were blocked within 3–4 days of treatment, accompanied by a 90% reduction in the antigen induced infiltration of eosinophils into the lung (HEUSSER et al. 1997). Studies in mice have also shown that the inhibitory effect on the IgE sensitization was maintained for more than 2 months after discontinuation of the mAb treatment, even though the half-life of the anti-IgE mAb was only 7 days (HAAK-FRENDSCHO et al. 1994). In addition, promising results have been obtained from phase I and phase II of clinical trials (SABAN et al. 1994, 1997; SHIELDS et al. 1995; VOGEL 1997; RACINE-POON et al. 1997). A large phase III clinical trial is presently also being performed. The results of treatment with mAb must thereby be considered as promising, suggesting that it may, within a few years, become a part in the clinical treatment of allergy.

However, there are several potential risks that may be encountered following treatment with mAbs for extended periods. Regular injections of purified mAb may lead to induction of an anti-idiotypic response against the variable region of the mAb, which may cause life-threatening immune complex formation. The risk of this complication is always present, requiring careful monitoring of the patients condition after each injection. Injection of very large amounts of mAb may overcome such complications by inducing B cell tolerance against the idiotype. However, this approach may be too expensive for large-scale treatment. If each injection must be monitored by a physician, due to the potential risk of immune complex formation, this may also add additional costs to the treatment program. At present it is unclear it and in which form the mAb therapy will reach the clinic. This must await the results from phase III clinical trials. In addition, since patients treated with this kind of therapy probably will continue to receive mAb injections for the rest of their lives, the cost of this treatment must be carefully evaluated.

VII. Induction of a Strong Anti-IgE Response by Vaccination

The strategy described in this section is based on vaccination against IgE by using specific regions of the constant region of the IgE molecule, coupled to a foreign carrier protein to induce a strong autoimmune anti-IgE response in the patient. This approach will probably solve some of the problems connected with the strategy described in the previous section, but it may also involve some new difficulties. Vaccination induces a polyclonal response against self IgE, why the risk of immune complex formation is minimal. However, a potential problem with the vaccination procedure is that it may be difficult to overcome the strong tolerance exhibited towards self molecules, leading to problems in reducing IgE to such an extent that it results in a significant clinical effect.

Studies aimed at assessing the possibility of inducing an auto-anti-IgE response have been performed in mouse and rat by several investigators. However, most of these studies have been performed with unmodified IgE, with the disadvantage that an autoimmune response against IgE was obtained in animals only between days 1 and 10 after birth (MARSHALL and BELL 1985; HABA and NISONOFF 1988, 1990, 1994) (HABA and NISONOFF 1994). Injections of IgE after the second or third week after birth did not induce an autoantibody response. This was probably due to the rising plasma levels of IgE in the animal with subsequent tolerization (HABA and NISONOFF 1992). However,

Haba and Nisonoff (1987) have extended these studies by covalent coupling of intact monoclonal IgE to a foreign carrier protein, keyhole limpet hemocyanin, and were able to elicit a strong autoimmune response in adult mice using this antigen. A potential risk with this type of antigen is that it contains the complete IgE molecule and thereby large regions outside of the receptor binding region. Antibodies directed against such regions may have the capacity to cross-link IgE bound to the surface receptors on mast cells and basophils and thereby induce massive cellular degranulation, which may lead to an anaphylactic shock in the patient.

To improve the vaccination procedure we have used domains 2 and 3 of the constant region of IgE (CH2 and CH3), covalently coupled to a foreign carrier protein, the glutathione-*S*-transferase (GST) from the parasitic trematode *Schistosoma japonicum* as the immunogen. By using this smaller region of IgE, mainly comprising the receptor-binding region (Helm et al. 1988, 1996) (Presta et al. 1994), the risk of generating anaphylactic antibodies should be substantially reduced. This protein immunogen was produced as a fusion protein in a bacterial host. The covalent coupling of the CH2-CH3 region of IgE to a foreign carrier molecule may circumvent the tolerance against the animal's self IgE molecules by facilitating the recruitment of a nontolerized T-cell population which normally would give help only to B cells directed against the carrier. These T cells may, due to the coupling of the IgE (or region of IgE) to the carrier, also support B cells with specificities against the IgE molecules of the vaccinated subject.

To study the strain dependence of the anti IgE response and the influence of the plasma levels of IgE at the time of vaccination, the GST-CH2-CH3 fusion protein was administered together with a potent adjuvant in four different strains of rats by intraperitoneal injections. A strong autoimmune anti-IgE response was detected in low and medium responder strains (referring to the baseline IgE levels in respective strains and their tendency to produce IgE upon antigen stimulation) but not in the high responder strain Brown Norway (Hellman 1994). When extending the analysis to several high responder strains of rats (Hooded Lister, Louvain, Dark Agouti, and Brown Norway) we were able to induce a strong autoimmune response in two of the four strains (Hooded Lister and Dark Agouti; Carlsson et al. unpublished observations). The anti-IgE response in this study was correlated with the baseline levels of circulating plasma IgE. High baseline IgE concentrations probably result in a strong B-cell tolerance against the CH2-CH3 region of the injected fusion protein and thereby a weak anti-IgE response.

This tolerance, or anergy, in the B cell compartment is probably caused by the interaction of membrane IgG or IgM on virgin auto reactive B cells with soluble IgE (antigen) in the absence of a secondary signal such as CD40 ligand interaction and interleukins. To overcome this problem in animals with high IgE titers at the initiation of the vaccination program we are presently analyzing the possibility of using fusion protein produced in eucaryotic hosts in order to obtain a higher fraction of the immunogen in its native conformation. By increasing the number of native surface epitopes in this antigen we hope to obtain a stronger anti-IgE response, that generates antibodies of sufficient affinity to compete with the high-affinity receptor for the soluble free IgE. However, in individuals with initially very high amounts of circulating plasma IgE it may be necessary to use a combination of the two methods described above: treatment with nonanaphylogenic anti-IgE monoclonal antibodies for a few weeks to deplete the patient of free plasma IgE, combined with vaccination a few days after the initiation of the monoclonal treatment, when IgE has dropped below detectable levels. Hopefully, increased levels of anti IgE antibodies generated in the immune response of the host will, within a few weeks of combination treatment, make the monoclonal treatment obsolete. The vaccination effect may be maintained by administration of booster doses of the vaccine every 2–3 months. An interesting possibility of simplifying the vaccination procedure may be the use of a DNA based vaccine as an alternative strategy for inducing a strong autoimmune response against IgE (work in progress) (DONNELLY et al. 1997). However, the use of DNA vaccines needs extensive evaluation. The potential dangers associated with tumorogenic effects, due to possible random integrations of vaccine DNA in host cells must be investigated further.

To study the clinical effect of reduced levels of IgE, Wistar rats (a medium responder strain) previously sensitized to ovalbumin, were vaccinated with the GST-CH2-CH3 fusion protein, and used as a model system to study allergen- and anti-IgE provoked skin reactions. After 10 weeks of treatment a profound reduction in total IgE levels and in both allergen and anti-IgE elicited inflammatory responses was observed (HELLMAN 1994). In comparison with blank vaccinated animals no reduction in mast cell numbers was detected, and no difference in the health condition was observed as a result of the vaccination (HELLMAN 1994). This suggests that it is possible to reduce IgE levels in an animal by vaccination, to such an extent that it gives a clear clinical effect, and this probably without affecting the health status of the animal.

Evaluation of ongoing studies of the effects of IgE vaccination in high responder strains with fusion proteins produced in eucaryotic hosts, and of vaccination in combination with initial monoclonal treatment will reveal the potential of this novel strategy for immunotherapy.

An obvious question connected with both of these latter strategies is whether a complete depletion of IgE will be connected with severe adverse effects. Based on the conservation of IgE and the binding of IgE to its high-affinity receptors on mast cells in all placental mammals studied so far, IgE most likely plays an important role in our immune system. The general view is that IgE is of importance in the immune defence against, at least, certain parasitic infections (HAGAN et al. 1991; DUNNE et al. 1992; SUTTON and GOULD 1993). It has also been suggested that IgE has a function in the early events during antigen contact. This is based on experiments in which IgE in combination with CD23, the low-affinity receptor for IgE, was shown to provide a potent enhancing system for antigen presentation at low antigen concentrations (Heyman et al. 1993; Gustavsson et al. 1994). The IgE mast cell system may thereby function as a doorkeeper, possibly needed in a pathogen rich environment, and as a powerful inflammatory potentiator in the initial contact and in the defence against certain parasites.

However, due to the redundancy of the mammalian immune system, other molecules and immune cells compensate for the loss of IgE in most experimental systems. A minor delay of the initiation of the immune response may not be easily detected, which may be the reason for difficulties in observing any adverse effects in rodents and humans lacking IgE (Levy and Chen 1970; Kühn et al. 1991). Possibly it is only in an environment rich in parasites and other pathogens, and under strong selective pressure, where only a small fraction of the population survive to reproductive age, that we are able to detect an advantage of having IgE. In modern society and in animals kept under hygienic laboratory conditions this is not the case. A possible negative correlation between reduced IgE levels and cancer susceptibility was detected in a clinical study of potential correlation between atopy and cancer (Fisherman 1960; Johnson 1966; Alderson 1974; Ford 1978; Cockcroft et al. 1979; Vena et al. 1985), indicating that it may be hazardous to completely eliminate IgE. However, several later studies addressing this question have not been able to establish any statistically significant correlation between low amounts of IgE and an increase in tumor frequency (McKee et al. 1967; Shapiro et al. 1971; Cavigioli et al. 1991; McDuffie 1991; Michils et al. 1992). Furthermore, some reports have actually indicated a positive correlation (Logan and Saker 1953; Robinette and Fraumeni 1978; McWhorter 1988). Thus the collected information from various studies of the physiological role of IgE, indicates that a complete deletion of IgE most likely will not lead to any detectable adverse effects in humans living in an environment almost completely devoid of severe parasitic worm infections. In addition, efficient drugs for the treatment of most parasitic worm infections are presently available.

D. Conclusions

Although a number of very interesting novel strategies for the treatment of atopic allergies are emerging, neither preclinical nor clinical studies have reached definitive answers regarding the final form and their potential reaching the clinic. However, for some of these strategies this information will most likely be available within a few years.

Acknowledgements. I thank Dr. Gunnar Pejler, Molly Vernersson, and Claudia Baule for critically reading the manuscript.

References

Abbas, AK, Murphy KM, Sher A (1996) Functional diversity of helper T lymphocytes. Nature 383:787–793

Alderson M (1974) Mortality from malignant disease in patients with asthma. Lancet 2:1475–1477

Appella E, Padlan EA, Hunt DF (1995) Analysis of the structure of naturally processed peptides bound by class I and class II major histocompatibility complex molecules. EXS 73:105–119

Arock M, Merle-Beral H, Dugas B, Ouaaz F, Legoff L, Vouldoukis I, Menciahuerta JM, Schmitt C, Leblondmissenard V, Debre P, Mossalayi MD (1993) IL-4 release by human leukemic and activated normal basophils. J Immunol 151:1441–1447

Aversa G, Punnonen J, Cocks BG, De Waal Malefyt R, Vega F, Zurawski SM, Zurawski G, de Vries JE (1993) An interleukin 4 (IL-4) mutant protein inhibits both IL-4 or IL-13-induced human immunoglobulin G4 (IgG4) and IgE synthesis and B cell proliferation: support for a common component shared by IL-4 and IL-13. J Exp Med 178:2213–2218

Bates D, Ruppel J, Fei BAD, Fox JA, Thomas D, Jardieu P (1995) Pharmacokinetics/dynamics following administration of recombinant humanized monoclonal anti IgE antibody in cynomolgus monkey. J Allergy Clin Immunol 95:365. Abstr no 864

Bellinghausen I, Metz G, Enk AH, Christmann S, Knop J, Saloga J (1997) Insect venom immunotherapy induces interleukin-10 production and a Th2- to-Th1 shift, and changes surface marker expression in venom-allergic subjects. Eur J Immunol 27:1131–1139

Bradding P, Feather IH, Howarth PH, Mueller R, Roberts JA, Britten K, Bews JPA, Hunt TC, Okayama Y, Heusser CH, Bullock GR, Church MK, Holgate ST (1992) Interleukin 4 is localized to and released by human mast cells. J Exp Med 176:1381–1386

Breit S, Steinhoff M, Blaser K, Heusser CH, Sebald W, Levine AD, Rocken M (1996) A strict requirement of interleukin-4 for interleukin-4 induction in antigen-stimulated human memory T cells. Eur J Immunol 26:1860–1865

Briner TJ, Kuo MC, Keating KM, Rogers BL, Greenstein JL (1993) Peripheral T-cell tolerance induced in naive and primed mice by subcutaneous injection of peptides from the major cat allergen *Fel dI*. Proc Natl Acad Sci USA 90:7608–7612

Brunner T, Heusser CH, Dahinden CA (1993) Human peripheral blood basophils primed by interleukin 3 (IL-3) produce IL-4 in response to immunoglobulin E receptor stimulation. J Exp Med 177:605–611

Cavigioli G, Mastropasqua B, Pelucchi A, Mazza P, Marazzini L (1991) Cutaneous response to intradermal histamine in lung cancer. Agents Actions 34:309–315

Chicz RM, Lane WS, Robinson RA, Trucco M, Strominger JL, Gorga JC (1994) Self-peptides bound to the type I diabetes associated class II MHC molecules HLA-DQ1 and HLA-DQ8. Int Immunol 6:1639–1649

Cockcroft DW, Klein GJ, Donevan RE, Copland GM (1979) Is there a negative correlation between malignancy and respiratory atopy? Ann Allergy 43:345–347

Coffman RL, Seymour BW, Hudak S, Jackson J, Rennick D (1989) Antibody to interleukin-5 inhibits helminth-induced eosinophilia in mice. Science 245:308–310

Cookson WO, Moffatt MF (1997) Asthma: an epidemic in the absence of infection? Science 275:41–42

Counsell CM, Bond JF, Ohman JL Jr, Greenstein JL, Garman RD (1996) Definition of the human T-cell epitopes of Fel d 1, the major allergen of the domestic cat. J Allergy Clin Immunol 98:884–894

Creticos PS, Reed CE, Norman PS, Khoury J, Adkinson N F Jr, Buncher CR, Busse WW, Bush RK, Gadde J, Lz JT et al (1996) Ragweed immunotherapy in adult asthma. N Engl J Med 334:501–506

Davis FM, Gossett LA, Pinkston KL, Liou R, Sun LK, Kim YW, Chang NT, Chang TW, Wagner K, Bews J, Brinkmann V, Towbin H, Subramanian N, Heusser C (1993) Can anti-IgE be used to treat allergy? Semin Immunopathol 15:51–73

Donnelly JJ, Ulmer JB, Shiver JW, Liu MA (1997) DNA vaccines. Annu Rev Immunol 15:617–648

Dreborg S, Åkerblom E (1990) Immunotherapy with monomethoxypolyethylene glycol modified allergens. Crit Rev Ther Drug Carrier Syst 6:315–365
Dunne DW, Butterworth AE, Fulford AJC, Kariuki HC, Langley JG, Ouma JH, Capron A, Pierce RJ, Sturrock RF (1992) Immunity after treatment of human schistosomiasis: association between IgE antibodies to adult worm antigens and resistance to infection. Eur J Immunol 22:1483–1494
Durham SR (1995) New insights into the mechanisms of immunotherapy. Eur Arch Otorhinolaryngol Suppl 1:64–67
Durham SR, Gould HJ, Hamid QA (1997) Local IgE production in nasal allergy. Int Arch Allergy Immunol 113:128–130
FDA notes (1988) Food and Drug Administration. Washington, November
Fisherman EW (1960) Does the allergic diathesis influence malignancy? J Allergy 31:74–78
Fitzgerald-Bocarsly P (1993) Human natural interferon-alpha producing cells. Pharmacol Ther 60:39–62
Ford RM (1978) Primary lung cancer and asthma. Ann Allergy 40:240–242
Freeman J (1911) Further observations on the treatment of hay fever by hypodermic inoculations of pollen vaccine. Lancet II:814–817
Froehlich J, Schoenhoff M, Jardieu P, Ruppel J, Fei BAD, Buckley P, Bush R (1995) Multiple doses of a recombinant humanized monoclonal anti IgE antibody are safely tolerated and decrease free serum IgE to undetectable levels. J Allergy Clin Immunol. 95:356. Abstr no 863
Gauchat JF, Henchoz S, Mazzei G, Aubry JP, Brunner T, Blasey H, Life P, Talabot D, Floresromo L, Thompson J, Kishi K, Butterfield J, Dahinden C, Bonnefoy JY (1993) Induction of human IgE synthesis in B-cells by mast cells and basophils. Nature 365:340–343
Gelber C, Gemmell L, McAteer D, Homola M, Swain P, Liu A, Wilson KJ, Gefter M (1997) Down-regulation of poison ivy/oak-induced contact sensitivity by treatment with a class II MHC binding peptide:hapten conjugate. J Immunol 158:2425–2434
Gieni RS, HayGlass KT (1991) Regulation of murine IgE responses: induction of long-lived inhibition of allergen-specific responses is genetically restricted. Cell Immunol 138:64–78
Gieni RS, Yang X, HayGlass KT (1993) Allergen-specific modulation of cytokine synthesis patterns and IgE responses in vivo with chemically modified allergen. J Immunol 150:302–310
Gieni RS, Yang X, Kelso A, Hayglass KT (1996) Limiting dilution analysis of CD4 T-cell cytokine production in mice administered native versus polymerized ovalbumin: directed induction of T-helper type-1-like activation. Immunology 87:119–126
Golden DBK, Lawrence ID, Hamilton RH, Kagey-Sobotka A, Valentine MD, Lichtenstein LM (1992) Clinical correlation of the venom-specific IgG antibody level during maintenance venom immunotherapy. J Allergy Clin Immunol 90:386–393
Groux H, Bigler M, de Vries JE, Roncarolo MG (1996) Interleukin-10 induces a long-term antigen-specific anergic state in human CD4+ T cells. J Exp Med 184:19–29
Grunewald SM, Kunzmann S, Schnarr B, Ezernieks J, Sebald W, Duschl A (1997) A murine interleukin-4 antagonistic mutant protein completely inhibits interleukin-4-induced cell proliferation, differentiation, and signal transduction. J Biol Chem 272:1480–1483
Grunewald SM, Werthmann A, Schnarr B, Klein CE, Bröcker EB, Mohrs M, Brombacher F, Sebald W, Duschl A (1998) An antagonistic IL-4 mutant prevents type I allergy in the mouse: inhibition of the IL-4/IL-13 receptor system completely abrogates humoral immune response to allergen and development of allergic symptoms in vivo. J Immunol 160:4004–4009

Gustavsson S, Hjulstrom S, Liu T, Heyman B (1994) CD23/IgE-mediated regulation of the specific antibody response in vivo. J Immunol 152:4793–4800
Haak-Frendscho M, Robbins K, Lyon R, Shields R, Hooley J, Schoenhoff M, Jardieu P (1994) Administration of an anti-IgE antibody inhibits CD23 expression and IgE production in vivo. Immunology 82:306–813
Haba S, Nisonoff A (1987) Induction of high titers of anti-IgE by immunization of inbred mice with syngeneic IgE. Proc Natl Acad Sci USA 84:5009–5013
Haba S, Nisonoff A (1988) Immunological responsiveness of neonatal A/J mice to isotypic determinants of syngeneic IgE. J Exp Med 168:713–724
Haba S, Nisonoff A (1990) Inhibition of IgE synthesis by anti-IgE: Role of long-term inhibition of IgE synthesis by neonatally administered soluble IgE. Proc Natl Acad Sci USA 87:3363–3367
Haba S, Nisonoff A (1992) IgE-secreting cells in the thymus: correlation with induction of tolerance to IgE. Proc Natl Acad Sci USA 89:5185–5187
Haba S, Nisonoff A (1994a) Effect of syngeneic anti-IgE antibodies on the development of IgE memory and on the secondary IgE response. J Immunol 152:51–57
Haba S, Nisonoff A (1994b) Role of antibody and T cells in the long-term inhibition of IgE synthesis. Proc Natl Acad Sci USA 91:604–608
Hagan P, Blumenthal UJ, Dunn D, Simpson AJG, Wilkins HA (1991) Human IgE, IgG4 and resistance to infection with *Schistosoma haematobium*. Nature 349:243–245
Hanson DG (1981) Ontogeny of orally induced tolerance to soluble proteins in mice I. Priming and tolerance in newborns. J Immunol 127:1518–1524
HayGlass KT, Stefura BP (1991a) Antigen-specific inhibition of ongoing murine IgE responses II. Inhibition of IgE responses induced by treatment with glutaraldehyde-modified allergens is paralleled by reciprocal increases in IgG2a synthesis. J Immunol 147:2455–2460
HayGlass KT, Stefura BP (1991b) Anti-interferon g treatment blocks the ability of glutaraldehyde-polymerized allergens to inhibit specific IgE responses. J Exp Med 173:279–285
Hellman L (1994) Profound reduction in allergen sensitivity following treatment with a novel allergy vaccine. Eur J Immunol 24:415–420
Helm B, Marsh P, Vercelli D, Padlan E, Gould H, Geha R (1988) The mast cell binding site on human immunoglobulin E. Nature 331:180–183
Helm B, Kebo D, Vercelli D, Glovsky M, Gould H, Ishizaka K, Geha R, Ishizaka T (1989) Blocking of passive sensitization of human mast cells and basophil granulocytes with IgE antibodies by a recombinant human ε-chain fragment of 76 amino acids. Proc Natl Acad Sci USA 86:9465–9469
Helm BA, Sayers I, Higginbottom A, Machado DC, Ling Y, Ahmad K, Padlan EA, Wilson AP (1996) Identification of the high affinity receptor binding region in human immunoglobulin E. J Biol Chem 271:7494–7500
Helm BA, Spivey AC, Padlan EA (1997) Peptide blocking of IgE/receptor interaction: possibilities and pitfalls. Allergy 52:1155–1169
Heusser CH, Wagner K, Bews JP, Coyle A, Bertrand C, Einsle K, Kips J, Eum SY, Lefort J, Vargaftig BB (1997) Demonstration of the therapeutic potential of non-anaphylactogenic anti-IgE antibodies in murine models of skin reaction, lung function and inflammation. Int Arch Allergy Immunol 113:231–235
Heyman B, Tianmin L, Gustavsson S (1993) In vivo enhancement of the specific antibody response via the low-affinity receptor for IgE. Eur J Immunol 23:1739–1742
Hoyne GF, Lamb JR (1997) Regulation of T cell function in mucosal tolerance. Immunol Cell Biol 75:197–201
Johnson KJ (1966) The relation of cancer to allergy. J Lancet 86:5–11
Jutel M, Pichler WJ, Skrbic D, Urwyler A, Dahinden C, Muller UR (1995) Bee venom immunotherapy results in decrease of IL-4 and IL-5 and increase of IFN-gamma secretion in specific allergen-stimulated T cell cultures. J Immunol 154:4187–4194

Knight SC, Stagg AJ (1993) Antigen-presenting cell types. Curr Opin Immunol 5:374–382
Kolbinger F, Saldanha J, Hardman N, Bendig MM (1993) Humanization of a mouse anti-human IgE antibody: a potential therapeutic for IgE-mediated allergies. Protein Eng 6:971–980
Koopmann JO, Hammerling GJ, Momburg F (1997) Generation, intracellular transport and loading of peptides associated with MHC class I molecules. Curr Opin Immunol 9:80–88
Kropshofer H, Vogt AB, Moldenhauer G, Hammer J, Blum Hammerling GJ (1996) Editing of the HLA-DR-peptide repertoire by HLA-DM. EMBO J 15:6144–6154
Kruse N, Tony HP, Sebald W (1993) Conversion of human interleukin-4 into high affinity antagonist by a single amino acid replacement. EMBO J 11:3237–3244
Kühn R, Rajevsky K, Müller W (1991) Generation and analysis of IL-4 deficient mice. Science 254:707–710
Lake RA, O'Hehir RE, Verhoef A, Lamb JR (1993) CD28 messenger RNA rapidly decays when activated T-Cells are functionally anergized with specific peptide. Int Immunol 5:461–466
Lamb JR, Skidmore BJ, Green N, Chiller JM, Feldmann M (1983) Induction of tolerance in influenza virus-immune T lymphocyte clones with synthetic peptides of influenza hemagglutinin. J Exp Med 157:1434–1447
Lang GM, Bitoh S, Becker AB, Sehon AH (1997) Potential therapeutic efficacy of allergen-monomethoxypolyethylene glycol conjugates for in vivo inactivation of sensitized mast cells responsible for common allergies and asthma. Int Arch Allergy Immunol 113:58–60
Lanzavecchia A (1993) Identifying strategies for immune intervention. Science 14:937–944
Levy DA, Chen J (1970) Healthy IgE-deficient person. N Engl J Med 283:541
Litwin A, Pesce AJ, Michael JG (1988) Regulation of the immune response to allergens by imunosuppressive allergenic fragments. I. Peptic fragment of honey bee venom phospholipase A2. Int Arch Allergy Appl Immunol 87:361–366
Logan J, Saker D (1953) The incidence of allergic disorders in cancer. N Z Med J 52:210–212
MacGlashan DW Jr, Bochner BS, Adelman DC, Jardieu PM, Togias A, Lichtenstein LM (1997a) Serum IgE level drives basophil and mast cell IgE receptor display. Int Arch Allergy Immunol 113:45–57
MacGlashan DW Jr, Bochner BS, Adelman DC, Jardieu PM, Togias A, McKenzie-White J, Sterbinsky SA. Hamilton RG, Lichtenstein LM (1997b) Down-regulation of Fc(epsilon)RI expression on human basophils during in vivo treatment of atopic patients with anti-IgE antibody. J Immunol 158:1438–1445
Machiels JJ, Somville MA, Lebrun PM, Lebecque SJ, Jacquemin MG, Saint-Remy J-MR (1990) Allergic bronchial asthma due to Drematophagoides pteronyssinus hypersensitivity can be efficiently treated by inoculation of allergen-antibody complexes. J Clin Invest 85:1024–1035
Manca F, Fenoglio D, Kunkl A, Cambiaggi C, Sasso M, Celada F (1988) Differential activation of T cell clones stimulated by macrophages exposed to antigen complexed with monoclonal antibodies. J Immunol 140:2893–2898
Manca F, Fenoglio D, Li Pira G, Kunkl A, Celada F (1991) Effect of antigen/antibody ratio on macrophage uptake, processing and presentation to T cells of antigen complexed with polyclonal antibodies. J Exp Med 173:37–48
Marinaro M, Boyaka PN, Finkelman FD, Kiyono H, Jackson, RJ, Jirillo E, McGhee JR (1997) Oral but not parenteral interleukin (IL)-12 redirects T helper 2 (Th2)-type responses to an oral vaccine without altering mucosal IgA responses. J Exp Med 185:415–427
Marshall JS, Bell EB (1985) Induction of an auto-anti-IgE response in rats I. Effects on serum IgE concentrations. Eur J Immunol 15:272–277

McDonnell JM, Beavil AJ, Mackay GA, Jameson BA, Korngold R, Gould HJ, Sutton BJ (1996) Structure based design and characterization of peptides that inhibit IgE binding to its high-affinity receptor. Nat Struct Biol 3:419–426
McDuffie HH (1991) Atopy and primary lung cancer. Histology and sex distribution. Chest 99:404–407
McKee WD, Arnold CA, Perlman MD (1967) A double-blind study of the comparative incidence of malignancy and allergy. J Allergy 39:294–301
McWhorter WP (1988) Allergy and risk of cancer. A prospective study using NHANESI followup data. Cancer 62:451–455
Michils A, Yernault JC, Noel E, Gossart B, Servais G, Duchateau J (1992) Abnormal humoral immune response to mucosal antigenic stimulation in patients with lung cancer. Cancer 69:2252–2257
Morgenstern JP, Griffith IJ, Brauer AW, Rogers BL, Bond JF, Chapman MD, Kuo MC (1991) Amino acid sequence of Fel dI, the major allergen of the domestic cat: protein sequence and cDNA cloning. Proc Natl Acad Sci USA 88:9690–9694
Mowat AM (1987) The regulation of immune responses to dietary protein antigens. Immunol Today 8:93–98
Müller U, Morris T, Bischof M, Friedli H, Skarvil F (1986) Combined active and passive immunotherapy in honey-bee-sting allergy. J Allergy Clin Immunol 78:115–122
Nakagawa T, Gershwin ME (1993) Immunotherapy of allergic diseases. Int Arch Allergy Immunol 102:117–120
Nelson HS, Oppenheimer J, Vatsia GA, Buchmeier A (1993) A double-blind, placebo-controlled evaluation of sublingual immunotherapy with standardized cat extract [see comments]. J Allergy Clin Immunol 92:229–236
Noon L (1911) Prophylactic inoculation against hay fever. Lancet II:1572–1573
Norman PS, Ohman JL, Long AA, Creticos PS, Shaked Z, Wood RA, Eggleston PA, Jones NH, Nicodemus CF (1994) Early clinical experiance with T cell reactive peptides from cat allergen *Fel dI*. J Allergy Clin Immunol 93:231 (Abstr)
Norman PS, Ohman JL, Long AA, Creticos PS, Gefter ML, Shaked Z, Wood RA, Eggleston PA, Lichtenstein LM, Jones NH, Nicodemus CF (1995) Follow on study of the first clinical trial with T cell defined peptides from cat allergen *Fel dI* (abstract). J Allergy Clin Immunol 95:259
Norman PS, Ohman JL Jr, Long AA, Creticos PS, Gefter M. A, Shaked Z, Wood RA, Eggleston PA, Hafner KB, Rao P, Lichtenstein LM, Jones NH, Nicodemus CF (1996) Treatment of cat allergy with T-cell reactive peptides. Am J Respir Crit Care Med 154:1623–1628
Norman PS, Nicodemus CF, Creticos PS, Wood RA, Eggleston PA, Lichtenstein LM, Kagey-Sobotka A, Proud D (1997) Clinical and immunologic effects of component peptides in Allervax Cat. Int Arch Allergy Immunol 113:224–236
O'Hehir RE, Lamb JR (1990) Induction of specific clonal anergy in human T lymphocytes by *Staphylococcus aureus* enterotoxins. Proc Natl Acad Sci USA 87:8884–8888
O'Hehir RE, Garman RD, Greenstein JL, Lamb JR (1991a) The specificity and regulation of T-cell responsiveness to allergens. Annu Rev Immunol 9:67–95
O'Hehir RE, Yssel H, Verma S, de Vries JE, Spits H, Lamb JR (1991b) Clonal analysis of differenetial lymphokine production in peptide and superantigen induced T cell anergy. Int Immunol 3:819–826
Ohman JL (1992) Allergen immunotherapy. Clin Allergy 76:977–991
Oppenheimer J, Areson JG, Nelson HS (1994) Safety and efficacy of oral immunotherapy with standardized cat extract. J Allergy Clin Immunol 93:61–67
Parsons JC, Coffman RL, Grieve RB (1993) Antibody to interleukin 5 prevents blood and tissue eosinophilia but not liver trapping in murine larval toxocariasis. Parasite Immunol 15:501–508
Perez VL, Van Parijs L, Biuckians A, Zheng XX, Strom TB, Abbas AK (1997) Induction of peripheral T cell tolerance in vivo requires CTLA-4 engagement. Immunity 6:411–417

Presta LG, Lahr SJ, Shields RL, Porter JP, Gorman CM, Fendly BM, Jardieu PM (1993) Humanization of an antibody directed against IgE. J Immunol 151:2623–2632

Presta L, Shields R, O'Connell L, Lahr S, Porter J, Gorman C, Jardieu P (1994) The binding site on human immunoglobulin E for its high affinity receptor. J Biol Chem 269:26368–26373

Racine-Poon A, Botta L, Chang TW, Davis FM, Gayax D, Liou RS, Rohane P, Staehelin T, van Steijn AMP, Frank W (1997) Efficacy, pharmacodynamics, and pharmacokinetics of CGP 51901, an anti-immunoglobulin E chimeric monoclonal antibody, in patients with seasonal allergic rhinitis. Clin Pharmacol Ther 62:675–690

Rempel JD, Wang M, HayGlass KT. (1997) In vivo IL-12 administration induces profound but transient commitment to T helper cell type 1-associated patterns of cytokine and antibody production. J Immunol 159:1490–1496

Robinette CD, Fraumeni JF Jr (1978) Asthma and subsequent mortality in World War II veterans. J Chronic Dis 31:619–624

Rogers BL, Morgenstern JP, Garman RD, Bond JF, Kuo MC (1993) Recombinant Fel dI: Expression, purification, IgE binding and reaction with cat-allergic human T cells. Mol Immunol 30:559–568

Saban R, Haak-Frendscho M, Zine M, Ridgway J, Gorman C, Presta LG, Bjorling D, Saban M, Jardieu P (1994) Human FcεRI-IgG and humanized anti-IgE monoclonal antibody MaE11 block passive sensitization of human and rhesus monkey lung. J Allergy Clin Immunol 94:836–843

Saban R, Haak-Frendscho M, Zine M, Presta LG, Bjorling DE, Jardieu P (1997) Human anti-IgE monoclonal antibody blocks passive sensitization of human and rhesus monkey bladder. J Urol 157:689–693

Saint-Remy J-M (1994) Novel approaches in immunotherapy. Clin Rev Allergy 12:23–42

Schwartz RH (1993) T cell anergy. When cells of the immune system "see" antigens in the absence of the right cosignals, they shut themselves down instead of attacking. Further therapies might capitalize on that reaction. Sci Am (Aug):48–54

Sehon AH (1982) Suppression of IgE antibody responses with tolerogenic conjugates of allergens and haptens In: Ishizaka K (ed) Regulation of the IgE antibody response. Progress in allergy, vol 32. Karger, Baltimore, pp 161–202

Shaikh N, Rivera J, Hewlett BR, Stead RH, Zhu FG, Marshall JS (1997) Mast cell Fc epsilonRI expression in the rat intestinal mucosa and tongue is enhanced during Nippostrongylus brasiliensis infection and can be up-regulated by in vivo administration of IgE. J Immunol 158:3805–3812

Shapiro S, Heinonen OP, Siskind V (1971) Cancer and allergy. Cancer 28:396–400

Shields RL, Whether WR, Zioncheck K, O'Connell L, Fendly B. Presta LG, Thomas D, Saban R, Jardieu P (1995) Inhibition of allergic reactions with antibodies to IgE. Int Arch Allergy Immunol 107:308–312

Shirakawa T, Enomoto T, Shimazu S, Hopkin JM (1997) The inverse association between tuberculin responses and atopic disorder. Science 275:77–79

Simons FE, Imada M, Li Y, Watson WT, HayGlass KT (1996) Fel d 1 peptides: effect on skin tests and cytokine synthesis in cat-allergic human subjects. Int Immunol 8:1937–1945

Söderlund A, Gabrielsson S, Paulie S, Hammarström ML, Rak S, Troye-Blomberg M (1997) Allergen induced cytokine profiles in type I allergic individuals before and after immunotherapy. Immunol Lett 57:177–181

Sutton BJ, Gould HJ (1993) The human IgE network. Nature 366:421–428

Taudorf E, Laursen LC, Djurup R, Kappelgaard E, Pedersen CT, Söborg M, Wilkinson P, Weeke B (1985) Oral administration of grass pollen to hay fever patients. An efficacy study in oral hyposensitization. Allergy 40:321–335

Taudorf E, Laursen LC, Lanner Å, Björkstén B, Dreborg S, Söborg M, Weeke B (1987) Oral immunotherapy in birch pollen hay fever. J Allergy Clin Immunol 80:153–161

Taudorf E, Möller C, Russell MW (1994) Secretory IgA response in oral immunotherapy. Investigation in birch pollinosis. Allergy 49:760–765
Thepen T, McMenamin C, Oliver J, Kraal G, Holt PG (1991) Regulation of immune response to inhaled antigen by alveolar macrophages: differential effects of in vivo alveolar macrophage elimination on the induction of tolerance vs. immunity. Eur J Immunol 21:2845–2850
Valenta R, Duchêne M, Vrtala S, Ball T, Ferreira F, Laffer S, Hirschwehr R, Ebner C, Sperr W, Valent P, Kraft D, Scheiner O (1993) Recombinant allergens as candidates for immunotherapy of type I allergic diseases. Vaccines 93:37–42
Valenta R, Vrtala S, Laffer S, Spitzauer S, Kraft D (1998) Recombinant Allergens. Allergy 53:552–561
van Neerven RJ, Ebner C, Yssel H, Kapsenberg ML, Lamb JR (1996) T-cell responses to allergens: epitope-specificity and clinical relevance. Immunol Today 17:526–532
Vena JE, Bona JR, Byers TE, Middleton E Jr, Swanson MK, Graham S (1985) Allergy-related diseases and cancer: an inverse association. Am J Epidemiol 122:66–74
Vogel G (1997) New clues to asthma therapies. Science 276:1643–1646
Vrtala S, Grote M, Ferreira F, Susani M, Stocker B, Kraft D, Valenta R (1995) Humoral immune responses to recombinant tree pollen allergens (Bet v I and Bet v II) in mice: construction of a live oral allergy vaccine. Int Arch Allergy Immunol 107:290–294
Wiegand TW, Williams BP, Dreskin SC, Jouvin MH, Kinet JP, Tasset D (1996) High-affinity oligonucleotide ligands to human IgE inhibit binding to Fcε receptor I. J Immunol 157:221–230
Yamaguchi Y, Hayashi Y, Sugama Y, Miura Y, Kasahara T, Kitamura S, Torisu M, Mita S, Tominaga A, Takatsu K (1988a) Highly purified murine interleukin 5 (IL-5) stimulates eosinophil function and prolongs in vitro survival. IL-5 as an eosinophil chemotactic factor. J Exp Med 167:1737–1742
Yamaguchi Y, Suda T, Suda J, Eguchi M, Miura Y, Harada N, Tominaga A, Takatsu K (1988b) Purified interleukin 5 supports the terminal differentiation and proliferation of murine eosinophilic precursors. J Exp Med 167:43–56
Yang X, Gieni RS, Mosmann TR, Hay Glass KT (1993) Chemically modified antigen preferentially elicits induction of Th1-like cytokine synthesis patterns in vivo. J Exp Med 178:349–353

Subject Index

MIX
Papier aus verantwortungsvollen Quellen
Paper from responsible sources
FSC® C105338
FSC
www.fsc.org

If you have any concerns about our products,
you can contact us on
ProductSafety@springernature.com

In case Publisher is established outside the EU,
the EU authorized representative is:
Springer Nature Customer Service Center GmbH
Europaplatz 3, 69115 Heidelberg, Germany

Printed by Libri Plureos GmbH
in Hamburg, Germany